Adaptive Estimation and Control

Partitioning Approach

Prentice Hall International
Series in Systems and Control Engineering

M. J. Grimble, Series Editor

BANKS, S. P., *Control Systems Engineering: modelling and simulation, control theory and microprocessor implementation*

BANKS, S. P., *Mathematical Theories of Nonlinear Systems*

BENNETT, S., *Real-time Computer Control: an introduction*

BITMEAD, R. R., GEVERS, M. and WERTZ, V., *Adaptive Optimal Control*

CEGRELL, T., *Power Systems Control*

COOK, P. A., *Nonlinear Dynamical Systems*

ISERMANN, R., LACHMANN, K. H. and MATKO, D., *Adaptive Control Systems*

LUNZE, J., *Feedback Control of Large-Scale Systems*

LUNZE, J., *Robust Multivariable Feedback Control*

McLEAN, D., *Automatic Flight Control Systems*

PATTON, R., CLARK, R. N. and FRANK, P. M. (editors), *Fault Diagnosis in Dynamic Systems*

PETKOV, P. H., CHRISTOV, N. D. and KONSTANTINOV, M. M., *Computational Methods for Linear Control Systems*

SÖDERSTROM, T. and STOICA, P., *System Identification*

WARWICK, K., *Control Systems: an introduction*

ADAPTIVE ESTIMATION AND CONTROL

Partitioning Approach

Keigo Watanabe

*Department of Mechanical Engineering,
Saga University, Japan*

Prentice Hall
New York • London • Toronto • Sydney • Tokyo • Singapore

First published 1991 by
Prentice Hall International (UK) Ltd
66 Wood Lane End, Hemel Hempstead
Hertfordshire HP2 4RG
A division of
Simon & Schuster International Group

Typeset in 10/12pt Times
by Keytec Typesetting Ltd, Bridport, Dorset

Printed and bound in Great Britain by
BPCC Wheatons Ltd, Exeter

Library of Congress Cataloging-in-Publication Data

Watanabe, Keigo, 1952–
Adaptive estimation and control: partitioning approach/Keigo Watanabe.
p. cm.
Includes index.
ISBN 0-13-005422-4
1. Adaptive control systems. I. Title.
TJ217.W16 1991
629.8′36–dc20 91-8819
CIP

British Library Cataloguing in Publication Data

Watanabe, Keigo, *1952–*
Adaptive estimation and control.
1. Stochastic systems
I. Title
519.2

ISBN 0-13-005422-4

---

1 2 3 4 5 95 94 93 92 91

To my wife, Kimie, and my daughters, Anna and Yuri

To my wife, Kittie, and my daughters, Anna and Yan

Contents

Preface xiii

Acknowledgements xv

Glossary xvii

1 Introduction 1

1.1 Adaptive Estimation and Control 1
1.2 Why Partitioned Adaptive Filtering and Control? 2
1.3 Synopsis of the Book 8
References 11

2 Review of Kalman Filtering Theory 15

2.1 Introduction 15
2.2 The Discrete-Time Kalman Filter 15
2.3 Stability of the Discrete-Time Kalman Filter 26
2.4 Computational Approaches to the Discrete-Time Riccati Equation 34
2.5 The Continuous-Time Kalman Filter 39
2.6 Stability of the Continuous-Time Kalman Filter 45
2.7 Computational Approaches to the Continuous-Time Riccati Equation 51
2.8 Summmary 61
Exercises 62
References 70

3 Structure and Parameter Adaptive Estimation 73

3.1 Introduction 73
3.2 Structure and Parameter Adaptive Estimation in Continuous Time 73
3.3 Structure and Parameter Adaptive Estimation in Discrete Time 85
3.4 Controlling the Number of Elemental Kalman Filters 90

3.5 Application to Inertial Navigation Systems 96
3.6 Summary 108
Appendix 3A Some Properties of Innovation Processes 108
Exercises 113
References 121

4 Asymptotic and Convergence Properties of Partitioned Adaptive Filters 125

4.1 Introduction 125
4.2 Discrete Time and Discrete Parameter Sets 125
4.3 Discrete Time and Continuous Parameter Sets 129
4.4 Continuous Time and Discrete Parameter Sets 133
4.5 Summary 143
Exercises 144
References 145

5 The Partitioning Filter: a Probabilistic Approach 147

5.1 Introduction 147
5.2 The Partitioning Filter (Continuous-Time Systems) 147
5.3 The Multipartitioning Filter (Continuous-Time Systems) 155
5.4 The Partitioning Filter (Discrete-Time Systems) 160
5.5 The Multipartitioning Filter (Discrete-Time Systems) 169
5.6 Partitioned Numerical Algorithms for Riccati Equations 175
5.7 Summary 184
Appendix 5A Matrix Inversion Lemma 184
Appendix 5B Proof of the Conditional Partition Theorem 185
Appendix 5C Proof of the Continuous-Time Dependent Multipartitioning Filter 186
Exercises 193
References 198

6 Partitioning Estimators: the Scattering Approach 201

6.1 Introduction 201
6.2 Redheffer's Scattering Theory 201
6.3 The Forward Partitioning Filter 205
6.4 The Forward Partitioning Smoother 212
6.5 The Backward Partitioning Filter 220
6.6 Summary 246
Exercises 246
References 253

7 The Pseudolinear Partitioning Filter and Tracking Motion Analysis 257

7.1 Introduction 257
7.2 Problem Formulation 257
7.3 Derivation of the Pseudolinear Partitioning Filter 259
7.4 Application to Tracking Motion Analysis 264
7.5 Summary 268
Appendix 7A Equation of Motion for Bearing-Only Tracking Motion Analysis 273
Exercises 274
References 276

8 Two-Stage Bias Correction Estimators Based on the Generalized Partitioning Filter 278

8.1 Introduction 278
8.2 A Structurally Partitioning Filter for Any System 278
8.3 Bias Correction Filter and Predictor 284
8.4 A Bias Correction Smoother 293
8.5 Stability of the Bias Correction Filter 302
8.6 A Bias Correction Filter and Predictor in Discrete-Time Systems 309
8.7 Summary 325
Exercises 326
References 327

9 Forward-Pass Fixed-Interval Smoothers in Discrete-Time Systems 330

9.1 Introduction 330
9.2 The Lagrange Multiplier Method and the Two-Filter Smoother 330
9.3 A Forward-Pass Fixed-Interval Smoother 334
9.4 U–D Information Matrix Factorization 348
9.5 The Steady-State Solution 365
9.6 Summary 377
Appendix 9A Triangular UDU^{T} Factorization 377
Appendix 9B U–D Filter Algorithms 378
Appendix 9C Rank-One Factorization Update Algorithm 380
Appendix 9D The Bierman Rank-Two U–D Smoother 381
Appendix 9E Arithmetic Operation Counts for Rank-Two U–D Smoothers 385

Exercises 392
References 399

10 Decentralized Smoothers Based on the Two-Filter Smoother Formula 402

10.1 Introduction 402
10.2 Decentralized Smoothing in Continuous-Time Systems 402
10.3 Decentralized Smoothing in Discrete-Time Systems 418
10.4 Summary 430
Exercises 431
References 433

11 Multiple Model Adaptive Control in Discrete-Time Systems 436

11.1 Introduction 436
11.2 Linear-Quadratic-Gaussian Control 437
11.3 Some Passive-Type MMAC Algorithms 448
11.4 The Joint Detection–Control Problem 457
11.5 Application to a Fault-Tolerant Control System 460
11.6 Summary 466
Exercises 468
References 474

12 Multiple Model Adaptive Filtering and Control for Markovian Jump Systems 478

12.1 Introduction 478
12.2 Statement of the Estimation Problem 479
12.3 Optimal Solution of the Estimation Problem 480
12.4 Generalized Pseudo-Bayes Algorithms 481
12.5 Structure Detection 502
12.6 The Problem of Control of Systems with Faulty Sensors 503
12.7 Summary 507
Exercises 508
References 511

Appendices 515

A Brief Review of Linear Algebra 515

A.1 Matrices and Vectors 515
A.2 Equality, Addition and Multiplication 516
A.3 Transposition 517
A.4 Determinants 518

A.5 Matrix Inversion and Singularity 519
A.6 Linear Independence and Rank 520
A.7 The Range Space and Null Space of a Matrix 521
A.8 Eigenvalues and Eigenvectors 521
A.9 Quadratic Forms and Positive (Semi)definiteness 522
A.10 Exponential of a Square Matrix 523
A.11 Trace 524
A.12 Similarity 524
A.13 Jordan Canonical Form 525
A.14 Norms and Inner Products 525
A.15 Differentiation and Integration 527
References 528

B Brief Review of Probability Theory 530

B.1 Probability Theory 530
B.2 Stochastic Processes 540
B.3 Gaussian Random Variables, Vectors and Processes 544
B.4 Estimates and Estimators 547
References 552

C Brief Review of Linear System Theory 554

C.1 Continuous-Time Linear System Models 554
C.2 Sampling of a Continuous-Time Linear System 555
C.3 Discrete-Time Linear System Models 556
C.4 Reachability and Observability for Continuous-Time Systems 557
C.5 Reachability and Observability for Discrete-Time Systems 561
C.6 Canonical Decomposition of Linear Systems 563
References 563

D Brief Review of Stability Theory 565

D.1 Local Equilibrium Stability Conditions 565
D.2 Stability in the Large 568
D.3 Some Results Due to the Second Method of Lyapunov 569
D.4 Some Stability Results on Discrete-Time Systems 571
References 572

Index 575

Preface

This book is about partitioned adaptive estimators for stochastic systems. Its aims are to unify those basic theories in continuous-time and/or discrete-time linear dynamic systems and to apply them to other estimation or control problems. Because the system models are so broad, we confine ourselves to lumped-parameter systems with unknown constant parameters which allow the parameters to vary slowly with time. However, the techniques used are by no means restricted to these systems: that is, most of the material will serve as a starting point for more complicated problems, e.g., partitioned adaptive filtering and control problems for lumped-parameter systems with abruptly changing parameters and for distributed-parameter systems.

Partitioned adaptive filtering and control (or multiple model adaptive filtering and control) has been quoted in a number of books in English in the control literature during the last decade. It has been felt, however, that several problems such as the partitioning filter (which is a special class of the partitioned adaptive filter with continuous parameter sets), the partitioning smoother, and applications to synthesizing other estimators have not been given systematic treatment in book form.

This book summarizes recent work in such fields, but also includes much other material in order to give a self-contained treatment; it is aimed at professional researchers and engineers in the fields of control, operations research, system theory and applied mathematics, as well as the related application areas; in addition, it is intended to serve graduate students as a self-contained theoretical reference work. The reader is assumed to have some basic knowledge of state-space, linear systems and stochastic processes, and also to know a little about Kalman filtering theory (which may include elementary stochastic control theory). However, the book contains appendices which summarize the relevant background material.

Keigo Watanabe
Saga, Japan
December, 1990

Acknowledgements

Of those who have been most closely associated with my investigations, I feel especially grateful to the former president of the University of Tokushima, Professor Takashi Soeda, who introduced me to the fascinating study of estimators. I also thank Professors Haruo Kimura (Nihon Bunri University), Setsuo Sagara, Hiroshi Kunita, Akira Mohri (Kyushu University), Toshio Yoshimura, Shigeru Omatu, Katsunobu Konishi (University of Tokushima), Masatoshi Nakamura (Saga University), Masanori Sugisaka (Oita University), Katsuji Uosaki (Osaka University) and Toshio Fukuda (Nagoya University) for their encouragement and interest.

I thank Professor T. Kailath (Stanford University) and Professor Toshimitsu Nishimura (Institute of Space and Astronautical Science) for providing several references. I also thank Professor J. K. Tugnait (Auburn University), Dr R. C. Montgomery (Langley Research Center, NASA) and Professor D. Andrisani II (Purdue University) for permission to include in this book considerable material from their publications.

I wish to acknowledge Professor D. G. Lainiotis (University of Patras) for discussions on the contents of the book. I extend my thanks to Professor Spyros G. Tzafestas (National Technical University of Athens) for his encouragement and support in preparing the manuscript, since partial modification of the manuscript was carried out while I was an invited Research Fellow at his university. I would finally like to thank two reviewers for their valuable comments, and Professor Mike Grimble (University of Strathclyde), Andrew Binnie (formerly Acquisitions Editor, Prentice Hall, UK) and Allison King (Acquisitions Editor, Prentice Hall, UK) for their kind and patient editing of the book.

Glossary

SUBSCRIPTS

b: (formal) backward time; bias
B: (strict) backward time
c: control
d: discrete time
f: final time; forward time
k: discrete time
K: Kalman filter
n: nominal; negative
r: remainder
rs: real-time smoothing
s: stabilizing; strong; sensor; smoothing; steady-state
S: sensor
t: continuous time
T: target
x: x-coordinate; state x
y: y-coordinate; state y
z: z-coordinate

SUPERSCRIPTS

T: transpose
$\bar{}$: steady-state value; mean value; complement
$\tilde{}$: estimation error
*: true; complex conjugate; optimal
$\hat{}$: estimate
-1: inverse

NOTATION

a acceleration

A	system matrix in continuous time
$\tilde{A}$	closed-loop system matrix in continuous time
$\mathcal{A}$	augmented system matrix in continuous time
b	bias state vector
B	distribution matrix of system noise; bias distribution matrix
B_{ij}	Bhattacharyya distance
$\mathcal{B}$	Bayes risk; Borel field; augmented distribution matrix of system noise
C	distribution matrix of system noise (Section 8.3 only); cost function associated with Bayes risk; $H^{\mathrm{T}} R^{-1} H = C^{\mathrm{T}} C$
$\mathbb{C}^n$	complex space of dimension n
d	information filtered estimate
D	$BQB^{\mathrm{T}} = DD^{\mathrm{T}}$; system matrix for bias b; diagonal matrix in U–D factorization; distribution matrix of measurement noise
f_{b}	backward-time compensative smoothing estimate
F	system matrix for state y (Chapter 8 only); LQG control gain
$\mathcal{F}$	σ-algebra
G	distribution matrix of control; $BQB^{\mathrm{T}} = GG^{\mathrm{T}}$ (Chapter 9 only); distribution matrix of system noise (Chapter 8 only)
h_i	ith hypothesis
H	observation matrix
i	integer
I	identity matrix; Kullback information measure
$I(k)$	Markov chain state sequence
$I_j(k)$	specific Markov chain state sequence
I^i	indicator variable
$\tilde{I}_k$	information data $\{Z_k, U_{k-1}\}$
$\mathcal{I}$	Fisher information matrix
j	integer
J	cost function; divergence
k	integer; discrete-time variable
K	filter gain; smoother gain
$\mathcal{K}$	filter gain set
l	discrete-time variable
L	unconditional likelihood ratio function
m	dimension of the observation z; mean value; interval of data compression; integer
$\hat{m}$	predicted estimate; initial source vector

M	symplectic matrix; Hamiltonian matrix; observation Gramian matrix; deterministic input; infinitesimal generator of scattering matrix S; integer
$\mathcal{M}$	infinitesimal generator of scattering matrix $\mathcal{S}$
n	dimension of the state x; positive integer
N	final time in discrete time
p	probability; probability density function; dimension of the system noise w
$p(a\|b)$	probability density function of a given b
$p(a, b)$	joint probability density function of a and b
P	probability; estimation error covariance matrix; right reflection operator of scattering matrix S
$\mathcal{P}$	right reflection operator of scattering matrix $\mathcal{S}$; partition set
q	backward-time information filtered estimate (Chapter 9 only); compensative information filtered estimate
Q	covariance matrix of system noise w
Q_c	weighting matrix in control cost function
r	(relative) position; compensative filtered estimate; dimension of the control vector u
R	covariance matrix of the measurement noise v
R_c	weighting matrix in control cost function
$\mathbb{R}^n$	Euclidean space of dimension n
s	time variable in continuous time; number of Markov chain state; variable of the Laplace transform
S	backward-time information matrix; scattering matrix; covariance matrix of noises w and v; solution of control Riccati equation
$\mathfrak{S}$	positive number set
$\mathcal{S}$	scattering matrix; system set
t	time variable in continuous time
t_0	initial time in continuous time
T	final time in continuous time; eigenvector matrix associated with symplectic (or Hamiltonian) matrix M
u	control (or input) vector
U	upper triangular factor with unit diagonal elements; eigenvector matrix
U_k	control data $\{u(0), \ldots, u(k)\}$
v	measurement noise vector
V	(relative) velocity; initial covariance; Lyapunov function; optimal cost-to-go for the case of exactly known state; eigenvector matrix
w	system noise vector

W	information matrix; left reflection operator of scattering matrix S; optimal cost-to-go for the case of inexactly known state; control Gramian matrix; array matrix for MWGS time update algorithm
$\mathcal{W}$	left reflection operator of scattering matrix $\mathcal{S}$
x	state variable (vector); coordinate x
X	eigenvector matrix
X_k	state data $\{x(0), \ldots, x(k)\}$
$\mathcal{X}$	augmented state vector
y	state vector (Chapter 8 only); coordinate y; observation
Y	eigenvector matrix
Y_i	observation space (or decision region) in which hypothesis h_i is accepted; observation data up to time i
$\mathcal{Y}_t$	σ-algebra generated by Y_t
z	observation vector; coordinate z
z_i	observation at ith local station
Z	observation data
Z_i	observation data up to time i
Z^i	observation data at ith local station
Z^i_j	observation data up to time j at ith local station
β	indicator variable; bearing measurement
δ	small value
δ_{kj}	Kronecker delta function ($\delta_{kj} = 1$ if $k = j$, else $\delta_{kj} = 0$)
$\delta(t - s)$	Dirac delta function ($\delta(t - s) = 1$ if $t = s$, else $\delta(t - s) = 0$)
δx	perturbed value of vector x
δP	perturbed value of matrix P
Δ	constant step size; fixed constant
Δt	(small) sampling interval
ε	small value
ζ	initial source vector in scattering matrix $\mathcal{S}$; damping ratio
η	compensative smoothing estimate in smoothing update; Brownian motion
θ	unknown parameter
Θ	parameter space (or set)
λ	Lagrange multiplier (or adjoint) vector; eigenvalue
λ_i	ith eigenvalue
$\lambda_i(M)$	ith eigenvalue of a matrix M
Λ	diagonal matrix with eigenvalues as elements; second-order moment of Lagrange multiplier vector λ; Lagrange multiplier vector associated with state vector $\mathcal{X}$

μ	measure; mean value
ν	innovation process
ξ	compensative smoothing estimate in decentralized smoothing; Brownian motion
π	transition probability matrix $[p_{ij}]$
Π	initial scattering matrix; predicted error covariance matrix
ρ	Bhattacharyya coefficient; a priori likelihood ratio
σ	standard deviation; time variable in continuous time
τ	time variable in continuous time; correlation time
Φ	state transition matrix; system matrix in discrete time; forward transmission operator of scattering matrix S
Φ^{T}	backward transmission operator of scattering matrix S
$\tilde{\Phi}$	closed-loop matrix in discrete time; forward transmission operator of scattering matrix $\mathcal{S}$; blending matrix
$\tilde{\Phi}^{\mathrm{T}}$	backward transmission operator of scattering matrix $\mathcal{S}$
ω	element of sample space Ω; natural frequency (or angular frequency)
Ω	sample space

NOTATIONAL CONVENTIONS

$a \ll b$	a is much less than b; measure a is absolutely continuous with respect to measure b
$a \bullet b$	dot product of vectors a and b
$\mathrm{adj}\,[A]$	adjoint of matrix A
$A^{-\mathrm{T}}$	$[A^{-1}]^{\mathrm{T}}$ (or $[A^{\mathrm{T}}]^{-1}$)
$A \geqslant B$	the difference matrix $(A - B)$ is nonnegative definite (or positive semidefinite)
$A > B$	the difference matrix $(A - B)$ is positive definite
block $\mathrm{diag}\,(A, B)$	block diagonal matrix with diagonal blocks A and B; block $\mathrm{diag}\,(A, B) \equiv \mathrm{diag}\,(A, B)$ if both A and B matrices are diagonal
$c \sim N(a, b)$	c is distributed as $N(a, b)$
$\dim a$	dimension of vector a
$\mathrm{diag}\,(x_1, \ldots, x_n)$	diagonal matrix with diagonal elements $x_1, \ldots, x_n$
$\exp\,(\cdot)$	exponential
$E[a]$	mean value (or expectation) of a
$E[a \mid b]$	mean value (or expectation) of a conditioned on b

$\text{Im}\{A\}$	image of A
j	$\sqrt{-1}$
$\text{Ker}\{A\}$	kernel of A
lim	limit
$\ln(\cdot)$	natural logarithm
$\max_a$	the maximum with respect to a
$\min_a$	the minimum with respect to a
$\min(a, b)$	the smallest of the numbers a and b
$N(a, b)$	Gaussian (or normal) distribution with mean a and covariance (or variance) b
$\mathcal{N}(A)$	null space of A
$o(x)$	small ordo x: function tending to zero faster than x
$O(x)$	ordo x: function tending to zero at the same rate as x
$Q^{1/2}$	matrix square root of a nonnegative definite matrix Q: $Q^{1/2}(Q^{1/2})^{\mathrm{T}} = Q$
$Q^{\mathrm{T}/2}$	$[Q^{1/2}]^{\mathrm{T}}$
$Q^{-1/2}$	$[Q^{1/2}]^{-1}$ (or $[Q^{-1}]^{1/2}$)
rank A	rank of A
$\text{Re}(a)$	real part of a complex value a
$\mathcal{R}(A)$	range space of A
$\text{sign}(\cdot)$	signum (or sign of)
sup	supremum
$S_1 * S_2$	star product of scattering matrices S_1 and S_2
$\text{tr}(\cdot)$	trace
$\dot{x}(t)$	time derivative of $x(t)$
$\|a\|$	absolute value of scalar a
$\|A\|$	determinant of a square matrix A
$\Vert \cdot \Vert$	Euclidean norm of a matrix or vector
$\Vert a \Vert_X^2$	$a^{\mathrm{T}} X a$ with X a symmetric positive definite (or nonnegative definite) matrix
$\sqrt{\cdot}$	square root of
$\forall$	for all
$\in$	element of
$\notin$	not element of
$\subset$	subset of
$\subseteq$	subset of or equal to
$\triangleq$	defined as
$:=$	assignment operator
$\cap$	cap (or intersection)
$\cup$	cup (or union)
$\{\cdot\}$	set of; sequence of

ABBREVIATIONS

a.e.	almost everywhere
AF	Ackerson–Fu
ALE	algebraic Lyapunov equation
ARE	algebraic Riccati equation
Cov(·)	covariance
DEA	detection–estimation algorithm
DP	dynamic programming
DUL	Deshpande–Upadhyay–Lainiotis
FDI	failure (or fault) detection and identification
GPBA	generalized pseudo-Bayes algorithm
IMMA	interacting multiple model algorithm
INS	inertial navigation system
JG	Jaffer–Gupta
JLQ	jump-linear-quadratic
JLQG	jump-linear-quadratic-Gaussian
l.m.t.	limit in the mean
LQ	linear-quadratic
LQG	linear-quadratic-Gaussian
LS	least-squares
MAP	maximum a posteriori
ML	maximum likelihood
MMAC	multiple model adaptive control
MMAF	multiple model adaptive filter
MMSE	minimum mean-square error
MSE	mean-square error
MV	minimum variance
MWGS	modified weighted Gram–Schmidt
RDE	Riccati difference (or differential) equation
RSA	random sampling algorithm
RTS	Rauch–Tung–Striebel
TMA	tracking motion analysis
TPBVP	two-point boundary-value problem
U–D	covariance (or information) matrix factors: U upper triangular and D diagonal
VSTOL	vertical short takeoff and landing
WLS	weighted least-squares
w.p.1	with probability one

1

Introduction

1.1 ADAPTIVE ESTIMATION AND CONTROL

The area of estimation, in particular state estimation and parameter estimation in dynamic systems, is important in many areas of science and engineering such as econometrics, medicine and electronic engineering. In particular, optimal filtering, also known as *Kalman* filtering [1–3], continues to be a fruitful area of research and has been substantially developed in many different directions. For a general survey, we refer the reader to Sorenson [4], Bucy and Joseph [5], Bryson and Ho [6], Meditch [7], Jazwinski [8], Sage and Melsa [9], Gelb [10], Kailath [11, 12], Lainiotis [13], Bierman [14], Maybeck [15], Anderson and Moore [16], Arimoto [17], Katayama [18], Brown [19] and Balakrishnan [20]. Recent progress in the application of Kalman filtering is described in the special issue of *IEEE Trans. Automatic Control* [21] and in three volumes in the series *Control and Dynamic Systems, Advances in Theory and Applications* edited by Leondes [22–24].

When system models and Gaussian white noise statistics are completely known, we know that minimum variance (MV) estimation is easily obtainable as the Kalman filter for linear stochastic systems. In some engineering applications, however, we often encounter the estimation problem for systems with completely or partially unknown dynamics operating in a deterministic or a stochastic environment. In order to estimate such systems optimally or suboptimally, we must use an *adaptive* estimator [25] or a *self-organizing* estimator [26], which is able to identify the dynamics or statistics during a given interval of time and to make appropriate adjustments in the estimates to compensate for uncertainties in system models. Self-organization, with its active accumulation of information intended to improve the performance of the system on-line, can be regarded as the first step towards an advanced learning and intelligent control system.

The theory of adaptive estimation and control has been an active area of research over the past few decades, and it has found application in several fields. Thus, it has been of considerable interest to academic research communities, systems and control researchers and high-technology industries. In particular, the *partitioning* approach to adaptive estimation and control has recently been gaining popularity. However, its theory has been documented in only a small number of articles scattered in international journals, though there are a few books on the self-tuning and model-reference approaches.

1.2 WHY PARTITIONED ADAPTIVE FILTERING AND CONTROL?

Many techniques are available for the adaptation of systems. Most identification approaches (see, for example, Sorenson [27], Sinha and Kusta [28], Ljung and Söderström [29], Graupe [30], Mendel [31], and Ljung [32]) can probably be applied to construct an adaptive mechanism. Among existing methods, we are particularly interested in the partitioned adaptive technique that is mathematically based on Bayesian estimation theory [33], since it is useful not only for identifying noise statistics but also for estimating unknown system parameters, as well as for estimating the order of systems, which is sometimes called *structure adaptation*. Pioneering work in this area can be found in Magill [34] and Lainiotis [35–39]. It is also interesting to note that Bayesian estimation can be regarded as a candidate for learning methods [40, 41].

It is known that the linear filtering problem with unknown time-invariant or time-varying parameters, i.e. the adaptive filtering problem, reduces to a nonlinear filtering problem, which has major difficulties in its realization. In particular, it is extremely difficult to assess the effect of approximations made in the suboptimal realization of nonlinear filters. However, partitioned adaptive filtering constitutes a partitioning of the original nonlinear filter into a bank or set of much simpler linear elemental Kalman or Kalman–Bucy filters. In other words, the minimum mean-square error estimate (MMSE) of the state is given by a weighted sum of the model- or parameter-conditional estimates; this realization is very simple to implement physically. Along the same lines, partitioned adaptive control is given as the weighted sum of the model-conditional linear-quadratic-Gaussian (LQG) controllers, each of which is matched to a particular admissible model or parameter [38, 39].

To illustrate the need for partitioned adaptive filtering and control in practical applications, we shall consider some examples from different areas. This will also bring out what partitioned adaptive filtering and control are and why they are of interest.

Example 1.1 Detection and classification of cardiac arrhythmias

We consider the problem of automated rhythm analysis of electrocardiograms (ECGs) and vector-cardiograms (VCGs). In order to apply the partitioned adaptive filtering technique, we must develop dynamic models describing various arrhythmic events. In this example, we use simple models that accurately describe the macroscopic behaviour of certain observations—the time intervals between certain cardiac events, specifically the R waves.

For example, we consider four persistent rhythm classes. The categorization of rhythms we consider is based solely on intervals between R waves (R–R intervals). The models are of the form

$$x(k) = \Phi x(k-1) \tag{1.1}$$

$$z(k) = Hx(k) + v(k) \tag{1.2}$$

where $z(k)$ denotes the kth R–R interval and $x(k)$ is the n-dimensional pattern state vector. The additive noise $v(k)$ represents the deviations from the ideal R–R pattern, but has zero-mean and variance R_d. Note that the $n \times n$ matrix Φ models the periodicity of the particular persistent rhythm.

Small variation: This is the category for R–R intervals which exhibit small but random deviations from their mean value. This ideal pattern is identically constant, and therefore leads to the model

$$x(k) = x(k-1) \tag{1.3}$$

$$z(k) = x(k) + v(k) \tag{1.4}$$

in which $\Phi \equiv H \equiv 1$, and $R_d = R_s$.

Large variation: This class is characterized by a large but random variation in the R–R interval sequence from the mean value. The model for large variation is identical in form to that for small variation. The only difference is that the variance, R_l, of $v(k)$ in the large variation case is set to be substantially larger than R_s.

Period-two oscillator (P_2): This class is characterized by R–R intervals which are alternately long and short. A second-order model which represents this oscillating rhythm pattern is decribed by

$$x(k) = \begin{bmatrix} 0 & 1 \\ 1 & 0 \end{bmatrix} x(k-1) \tag{1.5}$$

$$z(k) = [1 \quad 0]x(k) + v(k) \tag{1.6}$$

where the initial state $x(0)$ is a two-dimensional random vector with a mean $m(0)$ and covariance $P(0)$, and the noise $v(k)$ is a zero-mean Gaussian white noise process with variance R_2.

Period-three oscillator (P_3): This class is characterized by an R–R interval sequence which repeats over a period of three beats. This model is described by

$$x(k) = \begin{bmatrix} 0 & 0 & 1 \\ 1 & 0 & 0 \\ 0 & 1 & 0 \end{bmatrix} x(k-1) \tag{1.7}$$

$$z(k) = [1 \quad 0 \quad 0]x(k) + v(k) \tag{1.8}$$

where it is assumed that $x(0)$ has mean $m(0)$ and covariance $P(0)$, and $v(k)$ has variance R_3.

We can now solve the persistent rhythm classification problem by applying the partitioned adaptive filtering approach, in which each elemental filter generates the innovation sequence $\nu(k)$ and consequently the a posteriori probabilities for all of the hypothesized models are calculated. Then, we can identify the true rhythm by comparing the probabilities with a prespecified threshold. The resulting persistent rhythm classifier is shown in Figure 1.1. See Gustafson *et al.* [42] for a more detailed discussion.

Example 1.2 Failure detection and identification

In many dynamic systems such as aircraft, space vehicles, and nuclear power plants, it is often necessary to detect possible failures (or faults) which are modelled, for example, by abrupt changes in system parameters, or to decide when a repair or replacement must be made. For this purpose, a failure detection and identification (FDI) system generally consists of two basic stages: residual generation and decision-making. In the first stage, sensor outputs are processed to form residuals that typically have distinct characteristics under normal modes and some possible failure modes. The task of the second stage is to monitor the residuals (or innovations) and make decisions concerning the occurrence and identification of failure modes. The partitioned adaptive filtering technique is naturally applicable to the design of FDI systems which have mechanisms as mentioned above. Such FDI systems are discussed, for example, by Willsky *et al.* [43, 44] and Watanabe [45].

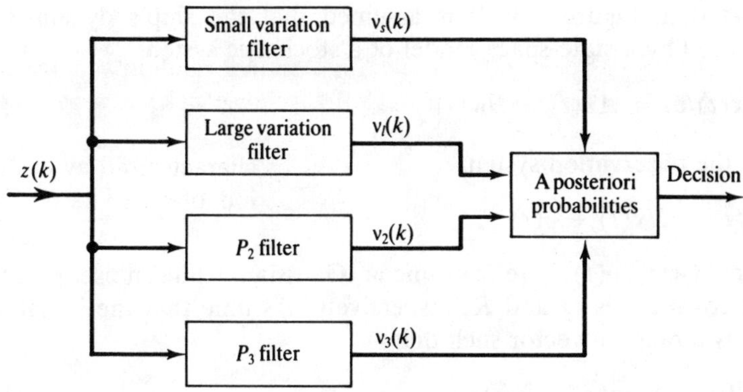

Figure 1.1 Rhythm identification system.

Many other problems exhibit features similar to the foregoing examples. The problems of tracking an aircraft making a sudden manoeuvre (Ricker and Williams [46], Moore *et al.* [47]), as depicted in Figure 1.2, and of tracking a ballistic re-entry vehicle making a sudden aerodynamic manoeuvre (Chang *et al.* [48]) belong to the wide class of algorithms commonly known as *target tracking*, an example of which will now be given.

Example 1.3 Tracking of ship position and identification of destination

A ship-surveillance system has two primary objectives: the first is to keep track of all ships in a given area, and the second is to maintain a record of all the information related to these ships—status, nationality, possible destination, etc. In this example, we consider the problem of tracking a ship with multiple possible destinations, as

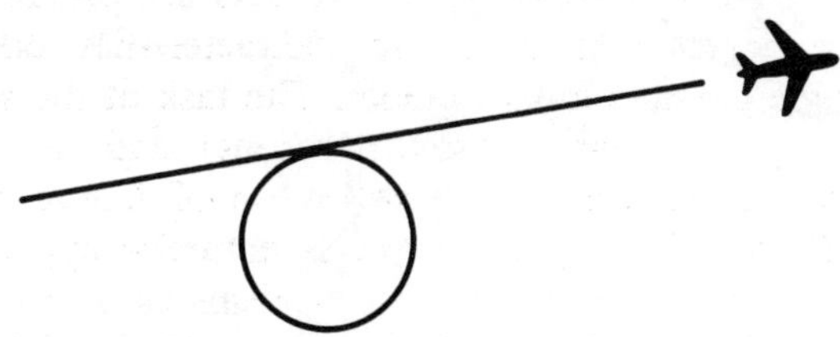

Figure 1.2 The motion of an aircraft making a sudden manoeuvre.

depicted in Figure 1.3. It is assumed that the ship's dynamics are governed by a state-space model of a stochastic system

$$\mathrm{d}x(t)/\mathrm{d}t = Ax(t) + Bw(t) \tag{1.9}$$

with the observation system

$$z(t) = Hx(t) + v(t) \tag{1.10}$$

where $\{w(t), v(t)\}$ are zero-mean Gaussian white noise processes with covariances Q and R, respectively. Assume that the initial state $x(0)$ is a random vector such that

$$x(0) \sim N(m(t_0), P(t_0)) \tag{1.11}$$

where $N(a, b)$ denotes the Gaussian distribution with mean a and covariance b, and $x(0)$ is independent of $w(t)$ and $v(t)$ for all t in $[0, t_f]$. Furthermore, assume that we have the following additional predictive information, at time t_f, concerning the ship's possible destinations:

$$x(t_f) \sim N(m^{(i)}(t_f), P^{(i)}(t_f)),\ i = 1, \ldots, M \tag{1.12}$$

We can derive a partitioned adaptive filter for this problem, in which the filter can estimate the state of the ship and simultaneously identify the true destination, by remodelling expressions (1.9)–(1.12) as forward Markovian models, which already incorporate the predictive information (1.12), or by using the concept of two-filter smoothing. Such a tracking filter has been presented by Castanon *et al.* [49].

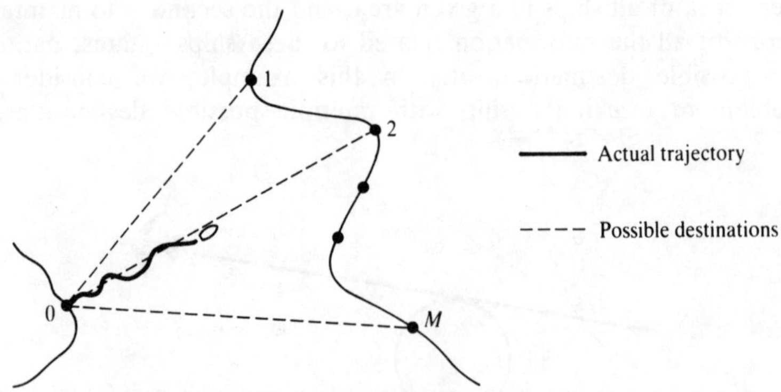

Figure 1.3 Tracking a ship's position and the identification of its destination.

One can also treat with a reverse problem which is schematically shown in Figure 1.4. The problem of tracking a ship with multiple possible departures, but with a single piece of predictive information concerning its destination, is solved in Watanabe [50].

To illustrate a situation where the partitioned adaptive control approach is appropriate, we consider the adaptive control of an aircraft.

Example 1.4 Adaptive flight control

In longitudinal or lateral aircraft dynamics, the aerodynamic derivatives (i.e. stability derivatives in the linearized version) change in a natural way with each flight condition where the number of possible flight conditions is assumed to be M. For example, in linearized longitudinal dynamics, the time-invariant characterization for each flight condition is of the form

$$\mathrm{d}x(t)/\mathrm{d}t = A_i x(t) + Gu(t) + B_i w_\mathrm{d}(t),\ i = 1, \ldots, M \tag{1.13}$$

where

$$x^\mathrm{T}(t) \triangleq [q(t), v(t), \alpha(t), \theta(t), \delta_\mathrm{e}(t), \delta_\mathrm{ec}(t), w(t)] \tag{1.14}$$

in which $q(t)[\mathrm{rad\,s^{-1}}]$ is the pitch rate, $v(t)[\mathrm{ft\,s^{-1}}]$ is the velocity error, $\alpha(t)$[rad] is the perturbed angle of attack from its trimmed value, $\theta(t)$[rad] is the pitch attitude deviation from its trimmed value, $\delta_\mathrm{e}(t)$[rad] is the deviation of the elevator from its trimmed value, $\delta_\mathrm{ec}(t)$[rad] is the actual commanded elevator position, and $w(t)$ is a

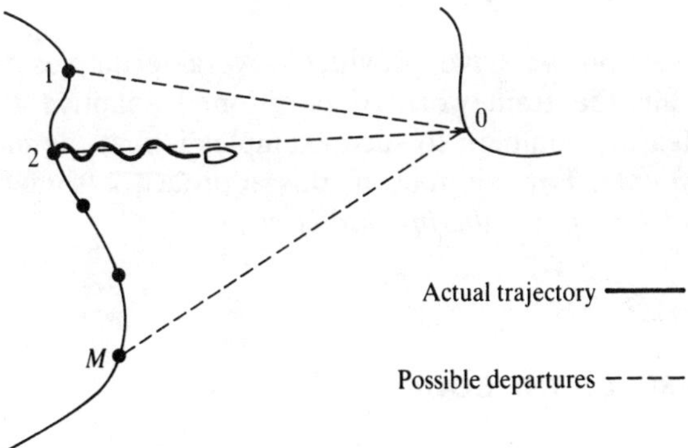

Figure 1.4 Tracking a ship's position and the identification of its departure.

wind disturbance state. The control variable is the commanded elevator rate

$$u(t) \triangleq \dot{\delta}_{ec}(t) \tag{1.15}$$

and $w_d(t)$ is zero-mean Gaussian white noise with unity covariance. Note that the elements of A_i and B_i change with each flight condition while

$$G^T = [0 \ \ 0 \ \ 0 \ \ 0 \ \ 0 \ \ 1 \ \ 0] \tag{1.16}$$

The associated observation equation is

$$z(t) = H_i x(t) + v_d(t), \ i = 1, \ldots, M \tag{1.17}$$

where the measurement data $z(t)$ consist of pitch rate and normal acceleration, and $v_d(t)$ is zero-mean Gaussian white noise with appropriate covariance.

A quadratic cost function may be specified by the following equation:

$$J = \lim_{t_f \to \infty} E\left\{\frac{1}{t_f}\int_0^{t_f} [x^T(t) Q_c^i x(t) + u^T(t) R_c^i u(t)] \mathrm{d}t\right\} \tag{1.18}$$

where the weighting matrices, Q_c^i and R_c^i, have to be different from flight condition to flight condition reflecting in a natural way the fact that the pilot wants different handling qualities as the speed changes. Thus, adaptive flight control such that the controller $u(t)$ minimizes the cost given by equation (1.18) can be realized by applying the partitioned adaptive control approach. Such adaptive control problems are discussed, for example, in Athans and Willner [51] and Athans *et al.* [52]. Figure 1.5 shows the basic structure of an adaptive control system.

In this section we have provided several examples that can be handled within the framework of partitioned adaptive filtering and control. A feature common to such examples is that the mathematical model is *multiple*. For this reason, this approach is sometimes called *multiple model adaptive filtering and control*.

1.3 SYNOPSIS OF THE BOOK

The first ten chapters are mainly concerned with filtering; the last two chapters are devoted to the control of two specific types of system. The twelve chapters are organized into four parts. The first part

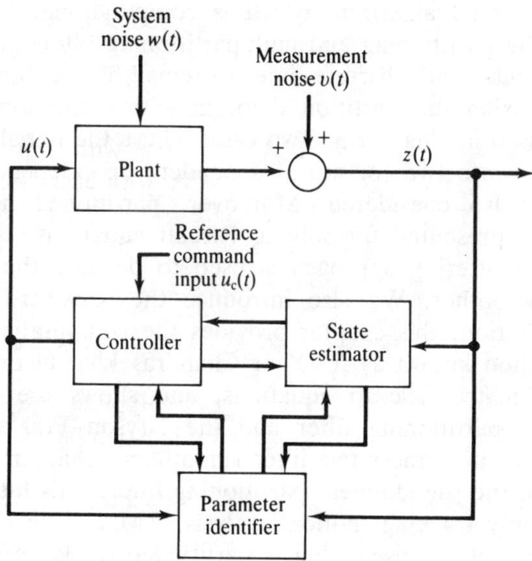

Figure 1.5 The basic structure of an adaptive control system.

provides an introduction to the theory of adaptive estimation and control using the partitioning approach and lays the foundations for a consideration of Kalman filtering. Chapter 1 is introductory. Chapter 2 reviews some basic results and definitions of Kalman filtering theory. This review is essential, since it is to serve as the background for subsequent derivations of partitioned adaptive filtering, control and other estimator algorithms.

The second part of the book presents the core topic of the theory, Lainiotis' partition theorem in continuous-time and discrete-time systems, and discusses the properties of partitioned adaptive filters. Chapter 3 deals with the problem of structure and parameter adaptive estimation. Partitioned adaptive filters are introduced for linear continuous- and discrete-time systems with unknown constant parameters. In particular, the innovation approach is applied to prove some theorems for continuous-time systems. The chapter concludes with an application of the discrete-time partitioned adaptive filter to the error estimation of inertial navigation systems. Chapter 4 is devoted to the asymptotic and convergence analysis of partitioned adaptive filters.

The third part of the book gives derivations of partitioned filters and estimators from probability theory and scattering theory; then, it presents the pseudolinear partitioning filter for a tracking problem and an associated algorithm which is computationally efficient. In Chapter 5 the partitioning and multipartitioning filters are developed for continuous- and discrete-time systems. These algorithms are derived by using the partition theorem with continuous parameter space discussed in Chapter 3. Two cases where the initial state vector is partitioned into two (or more) dependent or independent random vectors are also considered. Moreover, partitioned numerical algorithms are presented for solving Riccati equations. In Chapter 6 Redheffer's scattering approach is used to develop the partitioning filter and smoother. We also introduce the backward partitioning filter. In addition, this chapter provides the computationally advantageous method known as $X-Y$ or Chandrasekhar algorithms to the solution of matrix Riccati equations, and shows the relationships between the partitioning filter and the Bryson–Frazier, Wienert–Desai and Mayne–Fraser two-filter smoothers. Chapter 7 is devoted to a study of the pseudolinear partitioning filter. This filter is applied to bearing-only tracking motion analysis (TMA). We also illustrate the properties of the pseudolinear partitioned tracking filter through numerical experimentations.

The final part of the book presents applications of the theory so far developed: three applications to specific types of filter; and two applications to specific types of control system. Two-stage bias correction estimators, based on the initial dependent-type partitioning filter, are developed in Chapter 8 for linear continuous-time systems. After showing a condition for the possibility of synthesizing a two-stage bias correction filter, we derive generalized two-stage bias correction filter, predictor and smoother algorithms. Moreover, we analyze the stability of the two-stage bias correction filter by applying a generalized controllability matrix. Finally, some of the results are extended to the case of discrete-time systems. In Chapter 9, a forward-pass fixed-interval smoother is developed for linear discrete-time systems. Using the orthogonal projection lemmas, the fixed-interval smoother is derived algebraically. A $U-D$ factorization-based algorithm and the analysis of the steady-state error covariance are highlighted. Chapter 10 considers the development of some decentralized smoothing algorithms for continuous- and discrete-time systems, particularly where the two-filter smoother form is adopted as a global smoother. The smoothing update and real smoothing problems are also solved.

Chapter 11 applies the partition theorem of Chapter 3 to the passive-type multiple model adaptive control in discrete-time systems. After reviewing the linear-quadratic-Gaussian (LQG) control strategy, some multiple model adaptive control (MMAC) algorithms are derived using Bellman's dynamic programming approach. As a typical example of the use of MMAC algorithms, the joint detection–control problem is investigated. A fault-tolerant control system is also discussed in the framework of the MMAC approach. In Chapter 12, a multiple model adaptive filtering method is extended to jump parameter systems, in which parameters are modelled by Markov chains with completely known transition probabilities. Generalized pseudo-Bayes algorithms (GPBAs) are introduced as an approximate approach to reducing the complexity of the optimal multiple model adaptive filter in the state-estimation and structure-detection problems. The estimation results are also applied to control problems for systems with Markovian faulty sensors.

The book concludes with four appendices which present basic results from linear algebra, probability theory, linear system theory, and stability theory.

REFERENCES

[1] KALMAN, R. E., A New Approach to Linear Filtering and Prediction Theory, *Trans. ASME, J. Bas. Engng., Series D*, vol. 82, no. 1, 1960, pp. 34–45.

[2] KALMAN, R. E. and BUCY, R. S., New Results in Linear Filtering and Prediction Theory, *Trans. ASME, J. Bas. Engng., Series D*, vol. 83, no. 1, 1961, pp. 95–108.

[3] KALMAN, R. E., New Methods in Wiener Filtering Theory. In J. L. Bogdanoff and F. Kozin (eds), *Proc. 1st Symp. on Engineering Appl. of Random Function Theory and Probability*, Wiley, New York, 1963.

[4] SORENSON, H. W., Kalman Filtering Techniques. In C. T. Leondes (ed.), *Advances in Control Systems, Theory and Applications*, Vol. 3, Academic Press, New York, 1966, pp. 219–92.

[5] BUCY, R. S. and JOSEPH, P. D., *Filtering for Stochastic Processes with Applications to Guidance*, Wiley, New York, 1968.

[6] BRYSON, A. E. and HO, Y. C., *Applied Optimal Control*, Hemisphere, New York, 1975.

[7] MEDITCH, J. S., *Stochastic Optimal Linear Estimation and Control*, McGraw-Hill, New York, 1969.

[8] JAZWINSKI, A. H., *Stochastic Processes and Filtering Theory*, Academic Press, New York, 1970.

[9] SAGE, A. P. and MELSA, J. L., *Estimation Theory with Applications to Communications and Control*, Robert E. Krieger, New York, 1979.

[10] GELB, A., *Applied Optimal Estimation*, MIT Press, Cambridge, Massachusetts, 1974.

[11] KAILATH, T., A View of Three Decades of Linear Filtering Theory, *IEEE Trans. Inf. Theory*, vol. IT-20, no. 2, 1974, pp. 146–81.

[12] KAILATH, T., *Lectures on Linear Least-Squares Estimation*, CISM Courses and Lectures, no. 140, Springer-Verlag, New York, 1976.

[13] LAINIOTIS, D. G., Estimation: A Brief Survey, *Inf. Sci.*, vol. 7, 1974, pp. 191–201.

[14] BIERMAN, G. J., *Factorization Methods for Discrete Sequential Estimation*, Academic Press, New York, 1977.

[15] MAYBECK, P. S., *Stochastic Models, Estimation and Control*, Vols 1, 2 and 3, Academic Press, New York, 1979, 1982.

[16] ANDERSON, B. D. O. and MOORE, J. B., *Optimal Filtering*, Prentice Hall, Englewood Cliffs, New Jersey, 1979.

[17] ARIMOTO, S., *Kalman Filters*, Sangyo Tosho, Tokyo, 1977 (in Japanese).

[18] KATAYAMA, T., *Applied Kalman Filters*, Asakura Shoten, Tokyo, 1983 (in Japanese).

[19] BROWN, R. G., *Introduction to Random Signal Analysis and Kalman Filtering*, Wiley, New York, 1983.

[20] BALAKRISHNAN, A. V., *Kalman Filtering Theory*, Optimization Software, Inc., New York, 1984.

[21] SORENSON, H. W. (ed.), A Special Issue, On Applications of Kalman Filtering, *IEEE Trans. Aut. Control*, vol. AC-28, no. 3, 1983.

[22] LEONDES, C. T. (ed.), *Control and Dynamic Systems, Advances in Theory and Applications*, Vol. 19, Academic Press, New York, 1983.

[23] LEONDES, C. T. (ed.), *Control and Dynamic Systems, Advances in Theory and Applications*, Vol. 20, Academic Press, New York, 1983.

[24] LEONDES, C. T. (ed.), *Control and Dynamic Systems, Advances in Theory and Applications*, Vol. 21, Academic Press, New York, 1984.

[25] CHIN, L., Advances in Adaptive Filtering. In C. T. Leondes (ed.) *Control and Dynamic Systems, Advances in Theory and Applications*, Vol. 15, Academic Press, New York, 1979, pp. 277–356.

[26] SARIDIS, G. N., *Self-organizing Control of Stochastic Systems*, Marcel Dekker, New York, 1977.

[27] SORENSON, H. W., *Parameter Estimation*, Marcel Dekker, New York, 1980.

[28] SINHA, N. K. and KUSTA, B., *Modeling and Identification of Dynamic Systems*, Van Nostrand Reinhold, New York, 1983.

[29] LJUNG, L. and SÖDERSTRÖM, T., *Theory and Practice of Recursive Identification*, MIT Press, Cambridge, Massachusetts, 1984.

[30] GRAUPE, D., *Time Series Analysis, Identification and Adaptive Filtering*, Robert E. Krieger, Malabar, Florida, 1984.

[31] MENDEL, J. M., *Lessons in Digital Estimation Theory*, Prentice Hall, Englewood Cliffs, New Jersey, 1987.

[32] LJUNG, L., *System Identification: Theory for the User*, Prentice Hall, Englewood Cliffs, New Jersey, 1987.

[33] SPALL, J. C. (ed.), *Bayesian Analysis of Time Series and Dynamic Models*, Marcel Dekker, New York, 1988.

[34] MAGILL, D. T., Optimal Adaptive Estimation of Sampled Stochastic

Processes, *IEEE Trans. Aut. Control*, vol. AC-10, October 1965, pp. 434–9.

[35] LAINIOTIS, D. G., Sequential Structure and Parameter-Adaptive Pattern Recognition—Part I: Supervised Learning, *IEEE Trans. Inf. Theory*, vol. IT-16, no. 5, Sept. 1970, pp. 548–56.

[36] LAINIOTIS, D. G., Optimal Non-linear Estimation, *Int. J. Control*, vol. 14, 1971, pp. 273–95.

[37] LAINIOTIS, D. G., Optimal Adaptive Estimation: Structure and Parameter Adaptation, *IEEE Trans. Aut. Control*, vol. AC-16, April 1971, pp. 160–70.

[38] LAINIOTIS, D. G., Partitioning: A Unifying Framework for Adaptive Systems, I: Estimation, *Proc. IEEE*, vol. 64, no. 8, 1976, pp. 1126–43.

[39] LAINIOTIS, D. G., Partitioning: A Unifying Framework for Adaptive Systems, II: Control, *Proc. IEEE*, vol. 64, no. 8, 1976, pp. 1182–98.

[40] FU, K., Learning Control Systems—Review and Outlook, *IEEE Trans. Aut. Control*, vol. AC-15, no. 2, 1970, pp. 210–21.

[41] SARIDIS, G. N., Toward the Realization of Intelligent Controls, *Proc. IEEE*, vol. 67, no. 8, 1979, pp. 1115–33.

[42] GUSTAFSON, D. E., WILLSKY, A. S., WANG., J. Y., LANCASTER, M. C. and TRIEBWASSER, J. H., ECG/VCG Rhythm Diagnosis Using Statistical Signal Analysis—I. Identification of Persistent Rhythms, *IEEE Trans. Biomedical Engng.*, vol. BME-25, no. 4, 1978, pp. 344–53.

[43] WILLSKY, A. S., DEYST, J. J. and CRAWFORD, B. S., Adaptive Filtering and Self-test Methods for Failure Detection and Compensation, *Proceedings of the 1974 JACC*, Austin, Texas, 1974, pp. 637–45.

[44] WILLSKY, A. S., DEYST, J. J. and CRAWFORD, B. S., Two Self-test Methods Applied to an Inertial Systems Problem, *J. Spacecraft and Rockets*, vol. 12, no. 7, 1975, pp. 434–7.

[45] WATANABE, K., A Multiple Model Adaptive Filtering Approach to Fault Diagnosis in Stochastic Systems. In R. J. Patton *et al.* (eds), *Fault Diagnosis in Dynamic Systems: Theory and Applications*, Prentice Hall, London, 1989.

[46] RICKER, G. G. and WILLIAMS, J. R., Adaptive Tracking Filter for Maneuvering Targets, *IEEE Trans. Aero. and Electron. Syst.*, vol. AES-14, no. 1, 1978, pp. 185–93.

[47] MOOSE, R. L., VAN LANDINGHAM, H. F. and MCCABE, D. H., Modeling and Estimation for Tracking Maneuvering Targets, *IEEE Trans. Aero. and Electron. Syst.*, vol. AES-15, no. 3, 1979, pp. 448–56.

[48] CHANG, C. B., WHITING, R. H. and ATHANS, M., On the State and Parameter Estimation for Maneuvering Reentry Vehicles, *IEEE Trans. Aut. Control*, vol. AC-22, February 1977, pp. 99–105.

[49] CASTANON, D. A., LEVY, B. C. and WILLSKY, A. S., Algorithms for the Incorporation of Predictive Information in Surveillance Theory, *Int. J. Syst. Sci.*, vol. 16, no. 3, 1985, pp. 367–82.

[50] WATANABE, K., A Method of Backward-Pass Adaptive Filtering, *Syst. and Control*, vol. 30, no. 6, 1986, pp. 367–74 (in Japenese).

[51] ATHANS, M. and WILLNER, D., A Practical Scheme for Adaptive Aircraft

Flight Control Systems, *Proc. Symp. on Parameter Estimation Techniques and Appl. in Aircraft Flight Testing*, NASA Flight Research Center, Edwards, California, TN-D-7647, April 1973, pp. 315–36.

[52] ATHANS, M., CASTANON, D., DUNN, K. P., GREENE, C. S., LEE, W. H., SANDELL JR, N. R. and WILLSKY, A. S., The Stochastic Control of the F-8C Aircraft Using a Multiple Model Adaptive Control (MMAC) Method—Part I: Equilibrium Flight, *IEEE Trans. Aut. Control*, vol. AC-22, no. 5, October 1977, pp.768–80.

2

Review of Kalman Filtering Theory

2.1 INTRODUCTION

The Kalman filter is known as an optimal estimator which can estimate the state of a system using noisy measurement data (and known input data), for the case where the model is linear, the system and measurement noises are Gaussian white processes, and the initial state is a Gaussian random vector. Since the Kalman filtering algorithm is frequently used in all the following chapters, it is important to review the basic results. Thus, this chapter aims to provide a background summary for the reader who has not studied Kalman filtering theory, in order quickly to build a basis for the derivations which follow.

Section 2.2 begins with a review of the discrete-time Kalman filter. Section 2.3 discusses the stability of the time-invariant Kalman filter in discrete-time systems, from the viewpoint of *stabilizability* and *detectability*. In Section 2.4, the solution of the discrete-time algebraic Riccati equation (ARE) is discussed. The continuous-time versions of the results presented in Sections 2.2, 2.3 and 2.4 are summarized in Sections 2.5, 2.6 and 2.7, respectively.

2.2 THE DISCRETE-TIME KALMAN FILTER

We consider a linear stochastic system described by the following state-space model:

$$x(k+1) = \Phi(k+1, k)x(k) + G(k)u(k) + B(k)w(k) \tag{2.1}$$

$$z(k) = H(k)x(k) + v(k) \tag{2.2}$$

where $x(k) \in \mathbb{R}^n$, $u(k) \in \mathbb{R}^r$, $w(k) \in \mathbb{R}^p$, $v(k)$, $z(k) \in \mathbb{R}^m$, $x(0) \sim N(\hat{x}(0), P(0))$ and the zero-mean Gaussian white noise processes $\{w(k), v(k)\}$ have covariances given by

$$E\left\{\begin{bmatrix} w(k) \\ v(k) \end{bmatrix} [w^{\mathrm{T}}(j)\ v^{\mathrm{T}}(j)]\right\} = \begin{bmatrix} Q(k) & S(k) \\ S^{\mathrm{T}}(k) & R(k) \end{bmatrix} \delta_{kj} \tag{2.3}$$

in which E denotes the expectation operators, superscript T represents the transpose of a matrix or vector and δ_{kj} denotes the Kronecker delta function. Further, it is assumed that the matrices $\Phi(\cdot)$, $G(\cdot)$, $B(\cdot)$, $H(\cdot)$, $Q(\cdot)$, $R(\cdot)$ and $S(\cdot)$ are completely known, $u(k)$ is also a known input (or control) vector, and $\{w(k), v(k)\}$ are uncorrelated with the initial state $x(0)$.

2.2.1 The Kalman Filter

The Kalman filter [1, 2] provides a way of estimating the state $x(k)$ of the model given by equations (2.1)–(2.3). This filter gives the minimum mean-square error (MMSE) or minimum variance (MV) estimate† of the state, i.e. it evaluates the conditional mean of $x(t)$ given $Z_k \triangleq \{z(0), \ldots, z(k)\}$.

Theorem 2.1 The discrete-time Kalman filter: predictive type

Consider the model given by equations (2.1)–(2.3) and assume that the initial state and noise processes are jointly Gaussian. Let

$$\hat{x}(k+1|k) = E[x(k+1)|\,Z_k]$$

be the one-step predicted estimate, and

$$\hat{x}(k|k) = E[x(k)|\,Z_k]$$

be the filtered estimate. Then $\hat{x}(k+1|k)$ satisfies the following recursion (see also Figure 2.1):

$$\begin{aligned} \hat{x}(k+1|k) = {} & \Phi(k+1, k)\hat{x}(k|k-1) \\ & + K^*(k)[z(k) - H(k)\hat{x}(k|k-1)] \\ & + G(k)u(k), \qquad \hat{x}(0|-1) = \hat{x}(0) \end{aligned} \tag{2.4}$$

where $K^*(k)$ is the *Kalman filter gain* given by

$$\begin{aligned} K^*(k) = {} & [\Phi(k+1, k)P(k|k-1)H^{\mathrm{T}}(k) + R(k)S(k)] \\ & \times [H(k)P(k|k-1)H^{\mathrm{T}}(k) + R(k)]^{-1} \end{aligned} \tag{2.5}$$

† If the Gaussian assumption is removed, the filter gives the *linear* minimum variance estimate of the state, but this will not generally be the conditional mean.

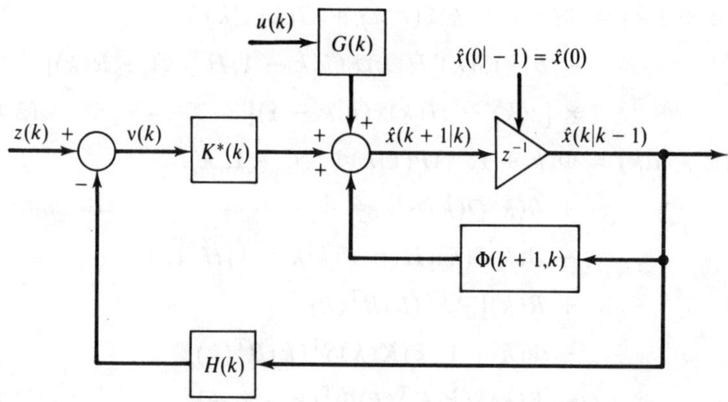

Figure 2.1 Block diagram of the discrete-time Kalman one-step predictor.

and $P(k|k-1) \triangleq E\{[x(k) - \hat{x}(k|k-1)][x(k) - \hat{x}(k|k-1)]^{\mathrm{T}}|Z_{k-1}\}$, the one-step predicted error covariance, is subject to the following *Riccati difference equation* (RDE):

$$\begin{aligned} P(k+1|k) = \Phi(k+1, k)P(k|k-1)\Phi^{\mathrm{T}}(k+1, k) \\ + B(k)Q(k)B^{\mathrm{T}}(k) \\ - K^*(k)[H(k)P(k|k-1)H^{\mathrm{T}}(k) \\ + R(k)]K^{*\mathrm{T}}(k), \qquad P(0|-1) = P(0) \end{aligned} \tag{2.6}$$

We can also go from predicted estimates to filtered estimates by means of the *measurement-update* equations:

$$\hat{x}(k|k) = \hat{x}(k|k-1) + K(k)[z(k) - H(k)\hat{x}(k|k-1)], \qquad \hat{x}(0|-1) = \hat{x}(0) \tag{2.7}$$

$$P(k|k) = [I - K(k)H(k)]P(k|k-1), \qquad P(0|-1) = P(0) \tag{2.8}$$

where $P(k|k) \triangleq E\{[x(k) - \hat{x}(k|k)][x(k) - \hat{x}(k|k)]^{\mathrm{T}}|Z_k\}$ is the *filtered error covariance,* and

$$K(k) = P(k|k-1)H^{\mathrm{T}}(k)[H(k)P(k|k-1)H^{\mathrm{T}}(k) + R(k)]^{-1} \tag{2.9}$$

and then from the filtered estimates to predicted estimates by the *time-update* equation:

$$\begin{aligned}
\hat{x}(k+1|k) = {} & \Phi(k+1,k)\hat{x}(k|k) + G(k)u(k) \\
& + B(k)S(k)[H(k)P(k|k-1)H^{\mathrm{T}}(k) + R(k)]^{-1} \\
& \times [z(k) - H(k)\hat{x}(k|k-1)] \qquad (2.10)
\end{aligned}$$

$$\begin{aligned}
P(k+1|k) = {} & \Phi(k+1,k)P(k|k)\Phi^{\mathrm{T}}(k+1,k) \\
& + B(k)Q(k)B^{\mathrm{T}}(k) \\
& - B(k)S(k)[H(k)P(k|k-1)H^{\mathrm{T}}(k) \\
& + R(k)]^{-1}S^{\mathrm{T}}(k)B^{\mathrm{T}}(k) \\
& - \Phi(k+1,k)K(k)S^{\mathrm{T}}(k)B^{\mathrm{T}}(k) \\
& - B(k)S(k)K^{\mathrm{T}}(k)\Phi^{\mathrm{T}}(k+1,k) \qquad (2.11)
\end{aligned}$$

Proof

We first note that the initial conditions are true, because they are given. Then, assuming that the conditional distribution for $x(k)$ given Z_{k-1} is Gaussian with mean $\hat{x}(k|k-1)$ and covariance $P(k|k-1)$, we can determine the conditional distribution of $x(k+1)$ given Z_k. From the properties of Gaussian random variables (or vectors) (see Appendix B.3.3), the joint distribution of

$$\begin{bmatrix} x(k+1) \\ z(k) \end{bmatrix}$$

given Z_{k-1} is Gaussian with mean

$$\begin{bmatrix} \Phi(k+1,k)\hat{x}(k|k-1) + G(k)u(k) \\ H(k)\hat{x}(k|k-1) \end{bmatrix}$$

and covariance

$$\begin{bmatrix} \Phi(k+1,k)P(k|k-1)\Phi^{\mathrm{T}}(k+1,k) + B(k)Q(k)B^{\mathrm{T}}(k) & \Phi(k+1,k)P(k|k-1)H^{\mathrm{T}}(k) + B(k)S(k) \\ H(k)P(k|k-1)\Phi^{\mathrm{T}}(k+1,k) + S^{\mathrm{T}}(k)B^{\mathrm{T}}(k) & H(k)P(k|k-1)H^{\mathrm{T}}(k) + R(k) \end{bmatrix}$$

Hence the desired equations (2.4)–(2.6) follow by the application of Appendix B.3.6. By setting $\Phi(k+1,k) = I$, $G(k) = 0$ and $B(k) = 0$ in equations (2.4)–(2.6) (so that $x(k+1) = x(k)$) we have equations (2.7)–(2.9). Finally, substituting equations (2.7) and (2.8) into (2.4) and (2.6) gives the desired results (2.10) and (2.11). □

Note that we can take $S = 0$ in (2.3) without loss of generality, because the case $S \neq 0$ can be handled by rewriting the original model as follows:

$$\begin{aligned} x(k+1) &= \Phi(k+1)x(k) + G(k)u(k) + B(k)w(k) \\ &\quad + B(k)S(k)R^{-1}(k) \\ &\quad \times [z(k) - H(k)x(k) - v(k)] \\ &= \Phi^*(k+1, k)x(k) + G(k)u(k) \\ &\quad + B(k)S(k)R^{-1}(k)z(k) + w^*(k) \end{aligned} \tag{2.12}$$

where

$$\Phi^*(k+1, k) = \Phi(k+1, k) - B(k)S(k)R^{-1}(k)H(k) \tag{2.13}$$

$$w^*(k) = B(k)w(k) - B(k)S(k)R^{-1}(k)v(k) \tag{2.14}$$

$$E\left\{\begin{bmatrix} w^*(k) \\ v(k) \end{bmatrix} [w(k)^{*\mathrm{T}}(j) \; v^{\mathrm{T}}(j)]\right\} = \begin{bmatrix} Q^*(k) & 0 \\ 0 & R(k) \end{bmatrix} \delta_{kj} \tag{2.15}$$

$$Q^*(k) = B(k)Q(k)B^{\mathrm{T}}(k) - B(k)S(k)R^{-1}(k)B^{\mathrm{T}}(k). \tag{2.16}$$

Since the terms $G(k)u(k)$ and $B(k)S(k)R^{-1}(k)z(k)$ are known, we can implement the standard Kalman filter algorithm with $S = 0$ using the system model given by (2.12) and (2.2). The following algorithms are easily obtained.

Theorem 2.2 The discrete-time Kalman filter: filtering type

The MMSE filtered estimate $\hat{x}(k|k)$ for the model (2.1)–(2.3) is obtained by the following recursion:

$$\begin{aligned} \hat{x}(k+1|k) &= \Phi(k+1, k)\hat{x}(k|k) + G(k)u(k) \\ &\quad + K_{\mathrm{p}}(k)[z(k) - H(k)\hat{x}(k|k)], \qquad \hat{x}(0|0) = \hat{x}(0) \end{aligned} \tag{2.17}$$

$$K_{\mathrm{p}}(k) \triangleq B(k)S(k)R^{-1}(k) \tag{2.18}$$

$$\begin{aligned} \hat{x}(k+1|k+1) &= \hat{x}(k+1|k) \\ &\quad + K(k+1)[z(k+1) - H(k+1)\hat{x}(k+1|k)] \end{aligned} \tag{2.19}$$

where the filter gain $K(k+1)$ is specified by the set of relations

$$K(k+1) = P(k+1|k)H^{\mathrm{T}}(k+1) \times [H(k+1)P(k+1|k)H^{\mathrm{T}}(k+1) + R(k+1)]^{-1} \tag{2.20}$$

$$P(k+1|k) = [\Phi(k+1, k) - K_{\mathrm{p}}(k)H(k)]P(k|k) \times [\Phi(k+1, k) - K_{\mathrm{p}}(k)H(k)]^{\mathrm{T}} + B(k)Q(k)B^{\mathrm{T}}(k) - B(k)S(k)R^{-1}(k)S^{\mathrm{T}}(k)\mathrm{B}^{\mathrm{T}}(k), \quad P(0|0) = P(0) \tag{2.21}$$

$$P(k+1|k+1) = [I - K(k+1)H(k+1)]P(k+1|k) \tag{2.22}$$

Proof

The proof of the theorem can proceed as for Theorem 2.1. Consider the joint distribution of

$$\begin{bmatrix} x(k+1) \\ z(k+1) \end{bmatrix}$$

given Z_k, which is Gaussian with mean

$$\begin{bmatrix} \Phi^*(k+1, k)\hat{x}(k|k) + G(k)u(k) + B(k)S(k)R^{-1}(k)z(k) \\ H(k)\hat{x}(k+1|k) \end{bmatrix}$$

and covariance

$$\begin{bmatrix} \Phi^*(k+1, k)P(k|k)\Phi^{*\mathrm{T}}(k+1, k) + Q^*(k) & P(k+1|k)H^{\mathrm{T}}(k+1) \\ H(k+1)P(k+1|k) & H(k+1)P(k+1|k)H^{\mathrm{T}}(k+1) + R(k+1) \end{bmatrix}$$

The theorem then follows immediately by the application of Appendix B.3.6. □

Figure 2.2 illustrates the above Kalman filter with $S(k) \equiv 0$.

The following are some properties of the Kalman filter (see also Appendix B.4):

1. Since $x(k)$ and $\hat{x}(k|k)$ (or $\hat{x}(k|k-1)$) are jointly Gaussian, conditioned on Z_k (or Z_{k-1}), $\tilde{x}(k|k) \triangleq x(k) - \hat{x}(k|k)$ (or $\tilde{x}(k|k-1) \triangleq x(k) - \hat{x}(k|k-1)$) is a Gaussian random variable, and its density function is then completely specified by its mean and covariance. The mean is

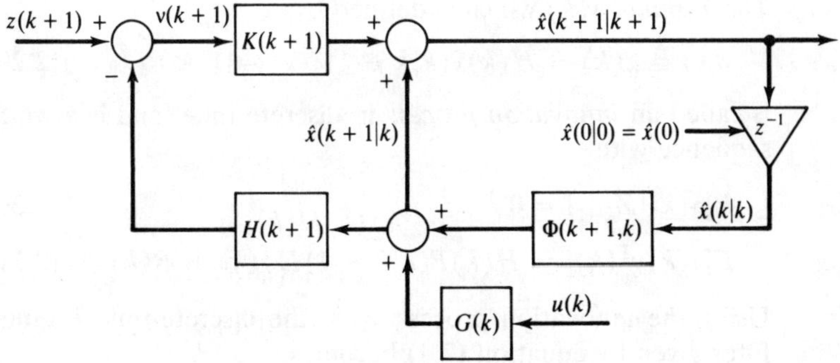

Figure 2.2 Block diagram of the discrete-time Kalman filter, where $S(k) = 0$.

$$
\begin{aligned}
E[\tilde{x}(k|k)|Z_k] &= E[x(k)|Z_k] - E[\hat{x}(k|k)|Z_k] \\
&= \hat{x}(k|k) - \hat{x}(k|k) = 0
\end{aligned}
\tag{2.23}
$$

that is, the estimator is *unbiased* since the error has zero mean. The variance is given by

$$
P(k|k) \leqslant P_{\mathrm{F}}(k|k) \tag{2.24}
$$

i.e. the estimator is the MV estimator, where $P_{\mathrm{F}}(k|k)$ is the error covariance given by any other filter.

2. The input of the filter is the process $\{z(k)\}$, the output is $\{\hat{x}(k|k)\}$ (or $\{\hat{x}(k|k-1)\}$). The estimator error covariance matrix $P(k|k)$ (or $P(k|k-1)$) is actually independent of Z_k (or Z_{k-1}). The gain $K(k)$ (or $K^*(k)$) is also independent of Z_{k-1}. Because of this, $K(k)$ (or $K^*(k)$), $P(k|k)$ and $P(k|k-1)$ are precomputable.
3. If the Gaussian assumption is removed, the estimation errors $\tilde{x}(k|k)$ and $\tilde{x}(k|k-1)$ must satisfy the orthogonal projection lemmas:

$$
E[\tilde{x}(k|k)\hat{x}^{\mathrm{T}}(k|k)] = 0 \tag{2.25}
$$

$$
E[\tilde{x}(k|k-1)\hat{x}(k|k-1)] = 0 \tag{2.26}
$$

or

$$
E[x(k)\hat{x}^{\mathrm{T}}(k|k)] = E[\hat{x}(k|k)\hat{x}^{\mathrm{T}}(k|k)] \tag{2.27}
$$

$$
E[x(k)\hat{x}^{\mathrm{T}}(k|k-1)] = E[\hat{x}(k|k-1)\hat{x}^{\mathrm{T}}(k|k-1)] \tag{2.28}
$$

so that the estimator gives the *linear* MV estimator.

4. The *residual* $\nu(k)$ which is defined by

$$\nu(k) \triangleq z(k) - H(k)\hat{x}(k|k-1) \tag{2.29}$$

is called an *innovation process* in discrete time, and is a white sequence with

$$E[\nu(k)|Z_{k-1}] = 0 \tag{2.30}$$

$$E[\nu(k)\nu^{\mathrm{T}}(k)] = H(k)P(k|k-1)H^{\mathrm{T}}(k) + R(k) \tag{2.31}$$

Using the innovation process $\nu(k)$, the discrete-time Kalman filter given by equation (2.4) becomes

$$\hat{x}(k+1|k) = \Phi(k+1, k)\hat{x}(k|k-1) + G(k)u(k) + K^*(k)\nu(k) \tag{2.32}$$

$$z(k) = H(k)\hat{x}(k|k-1) + \nu(k) \tag{2.33}$$

which can be thought of as an alternative model for $\{z(k)\}$ driven by the innovation process $\{\nu(k)\}$. This model is termed the *innovation model*. Thus the filter equations with input $z(k)$ and output $\nu(k)$ are as follows:

$$\hat{x}(k+1|k) = [\Phi(k+1, k) - K^*(k)H(k)]\hat{x}(k|k-1) + G(k)u(k) + K^*(k)z(k) \tag{2.34}$$

$$\nu(k) = z(k) - H(k)\hat{x}(k|k-1) \tag{2.35}$$

This is called a *whitening filter*.

Note that the discrete-time Kalman filter algorithm with $S = 0$ presented in Theorem 2.2 is used to develop the most theories in this book. However, it is convenient in analyzing the stability of the discrete-time Kalman filter to use the algorithm of Theorem 2.1.

Note also that the use of the innovation model has been recently shown by some researchers to be problematic in forming MV controllers [3–5]. To overcome the limitation of the innovation model, the use of a *noise-free measurement model* [3] is recommended:

$$\hat{x}(k+1|k+1) = \Phi(k+1,k)\hat{x}(k|k)+G(k)u(k)+K(k+1)\nu(k+1) \tag{2.32$'$}$$

$$z(k) = H(k)\hat{x}(k|k) \tag{2.33$'$}$$

The so-called *smoothing* problem can be treated as an extension of the Kalman filtering problem. There are three types of smoothing:

1. *Fixed-point* smoothing: $\hat{x}(k_*|k) \triangleq E[x(k_*)|Z_k]$, where k_* is a fixed integer such that $k_* \leq k$.
2. *Fixed-interval* smoothing: $\hat{x}(k|N) \triangleq E[x(k)|Z_N]$, where $0 \leq k \leq N$ and N is a fixed positive integer.
3. *Fixed-lag* smoothing: $\hat{x}(k-N|k) \triangleq E[x(k-N)|Z_k]$, where N is a fixed positive integer.

The fixed-point and fixed-lag smoothers are derived in Exercises 2.10 and 2.11, respectively. The fixed-interval smoother will be discussed in detail in Chapter 9.

Example 2.1 From Meditch [6]

Consider the scalar system

$$x(k+1) = ax(k) + w(k)$$

$$z(k) = x(k) + v(k), \qquad k = 1, 2, \ldots$$

where $\{w(k),\ k = 0, 1, \ldots\}$ is a zero-mean Gaussian white sequence with constant variance Q, $\{v(k),\ k = 1, 2, \ldots\}$ is a zero-mean Gaussian white sequence with constant variance R. $x(0)$ is a zero-mean Gaussian random variance with variance $P(0)$, and a is a constant. It is assumed that the two Gaussian white sequences and $x(0)$ are independent.

The discrete-time Kalman filter (filtering type) is

$$\hat{x}(k+1|k+1) = a\hat{x}(k|k) + K(k+1)[z(k+1) - a\hat{x}(k|k)]$$

From (2.20) and (2.21), we obtain

$$K(k+1) = [a^2P(k|k) + Q][a^2P(k|k) + Q + R]^{-1}$$

$$= \frac{a^2P(k|k) + Q}{a^2P(k|k) + Q + R}$$

and

$$P(k+1|k) = a^2P(k|k) + Q$$

The filtering error variance equation is then found to be

$$P(k+1|k+1) = \left[1 - \frac{a^2P(k|k) + Q}{a^2P(k|k) + Q + R}\right][a^2P(k|k) + Q]$$

$$= \frac{R[a^2P(k|k) + Q]}{a^2P(k|k) + Q + R}$$

subject to the initial condition $P(0|0) = P(0)$.

From the above results, we see that $P(k+1|k) \geqslant Q$, since $P(k|k) \geqslant 0$. This implies that the variance of the system disturbance sets the performance limit on prediction accuracy.

The range on the filter gain is found to be $0 \leqslant K(k+1) \leqslant 1$ except for the trivial case $P(k|k) = Q = R \equiv 0$. It is also seen that

$$P(k+1|k+1) = RK(k+1)$$

and it is obvious that $0 \leqslant P(k+1|k+1) \leqslant R$ for $k = 0, 1, \ldots$. If $P(0) \gg R$, the first measurement $z(1)$ will lead to a large reduction in the filtering error variance from $P(0)$ to $P(1|1) \leqslant R \ll P(0)$.

Example 2.2

Consider a four-dimensional system with state

$$\begin{bmatrix} r_x \\ r_y \\ V_x \\ V_y \end{bmatrix} = \begin{bmatrix} x\text{-position} \\ y\text{-position} \\ x\text{-velocity} \\ y\text{-velocity} \end{bmatrix}$$

which represents a vehicle motion constrained to a plane and evolves according to the equation

$$x(k+1) = \begin{bmatrix} 1 & 0 & \delta t & 0 \\ 0 & 1 & 0 & \delta t \\ 0 & 0 & 1 & 0 \\ 0 & 0 & 0 & 1 \end{bmatrix} x(k) + \begin{bmatrix} 0 & 0 \\ 0 & 0 \\ 1 & 0 \\ 0 & 1 \end{bmatrix} w(k)$$

where $\delta t = 50$ s, $w(k)$ is random velocities consisting of zero-mean white Gaussian sequences with covariances $Q = \text{diag}(1.0 \times 10^{-6}\ \text{km}^2, 1.0 \times 10^{-6}\ \text{km}^2)$ and

$$x^{\mathrm{T}}(0) = [-14\ \text{km} \quad 6\ \text{km} \quad 0.014\ \text{km}\,\text{s}^{-1} \quad -0.006\ \text{km}\,\text{s}^{-1}]$$

The observation equation is

$$z(k) = \begin{bmatrix} 1 & 0 & 0 & 0 \\ 0 & 1 & 0 & 0 \end{bmatrix} x(k) + v(k), \; k = 0, 1, \ldots, 5$$

where measurement error is zero-mean white Gaussian sequences with covariance $R = \text{diag}(5 \times 10^{-3}\ \text{km}^2, 5 \times 10^{-3}\ \text{km}^2)$, and $\{w(k)\}$ and $\{v(k)\}$ are mutually independent.

The Kalman filter given by (2.7)–(2.11) was initialized as follows:

$$\hat{x}^{\mathrm{T}}(0) = [-20\ \text{km} \quad 10\ \text{km} \quad 0.1\ \text{km}\,\text{s}^{-1} \quad 0.1\ \text{km}\,\text{s}^{-1}]$$

$$P(0) = \text{diag}(1\ \text{km}^2, 1\ \text{km}^2, 1 \times 10^{-3}\ \text{km}^2\,\text{s}^{-2}, 1\ \text{km}^2\ \text{s}^{-2})$$

and simulated using a digital computer program.

Table 2.1 summarizes the true trajectories (at $k = 0, 1, \ldots, 5$) and filtered estimates. Note that the true states are estimated quite accurately after sampling twice.

2.2.2 The Extended Kalman Filter

For nonlinear problems the Kalman filter is not strictly applicable since linearity plays an important role in its derivation and performance as an optimal filter. The *extended* Kalman filter attempts to overcome this difficulty by applying a linearized technique where the linearization is performed about the current state estimate.

Consider a nonlinear model of the form

$$x(k+1) = f(x(k), u(k), w(k), k) \tag{2.36}$$

$$z(k) = h(x(k), u(k), v(k), k) \tag{2.37}$$

with

$$E\left\{\begin{bmatrix} w(k) \\ v(k) \end{bmatrix} [w^{\mathrm{T}}(j) \quad v^{\mathrm{T}}(j)]\right\} = \begin{bmatrix} Q & 0 \\ 0 & R \end{bmatrix} \delta_{kj} \tag{2.38}$$

Taking the Taylor expansion of equation (2.36) around $x(k) = \hat{x}(k|k)$, $w(k) = 0$, and of equation (2.37) around $x(k) = \hat{x}(k|k-1)$, $v(k) = 0$ yields the following:

$$\begin{aligned} x(k+1) = {} & f(\hat{x}(k|k), u(k), 0, k) + F(k)[x(k) - \hat{x}(k|k)] \\ & + G(k)w(k) + (\text{higher-order terms}) \end{aligned} \tag{2.39}$$

Table 2.1 Vehicle motion in a plane

k	*0*	*1*	*2*	*3*	*4*	*5*
r_x	−14.000	−13.300	−12.704	−12.033	−11.414	−10.794
r_y	6.000	5.700	5.434	5.079	4.680	4.302
V_x	0.014	0.012	0.013	0.012	0.012	0.013
V_y	−0.006	−0.005	−0.007	−0.008	−0.008	−0.007
$\hat{r}_x$	−13.904	−13.271	−12.658	−11.994	−11.381	−10.800
$\hat{r}_y$	6.065	5.807	5.555	5.035	4.581	4.262
$\hat{V}_x$	0.100	0.013	0.013	0.013	0.013	0.012
$\hat{V}_y$	−0.100	−0.005	−0.005	−0.008	−0.009	−0.007

$$z(k) = h(\hat{x}(k|k-1), u(k), 0, k) + H(k)[x(k) - \hat{x}(k|k-1)] + J(k)v(k) + (\text{higher-order terms}) \tag{2.40}$$

where

$$F(k) = \left.\frac{\partial f(x(k), u(k), w(k), k)}{\partial x(k)}\right|_{x(k)=\hat{x}(k|k), w(k)=0} \tag{2.41}$$

$$G(k) = \left.\frac{\partial f(x(k), u(k), w(k), k)}{\partial w(k)}\right|_{x(k)=\hat{x}(k|k), w(k)=0} \tag{2.42}$$

$$H(k) = \left.\frac{\partial h(x(k), u(k), v(k), k)}{\partial x(k)}\right|_{x(k)=\hat{x}(k|k-1), v(k)=0} \tag{2.43}$$

$$J(k) = \left.\frac{\partial h(x(k), u(k), v(k), k)}{\partial v(k)}\right|_{x(k)=\hat{x}(k|k-1), v(k)=0} \tag{2.44}$$

After neglecting the higher-order terms, applying the usual Kalman filter algorithm given in Theorem 2.2 to equations (2.39) and (2.40), we have extended the Kalman filter (filtering type):

$$\hat{x}(k+1|k) = f(\hat{x}(k|k), u(k), 0, k), \qquad \hat{x}(0|0) = \hat{x}(0) \tag{2.45}$$

$$P(k+1|k) = F(k)P(k|k)F^{\mathrm{T}}(k) + G(k)Q(k)G^{\mathrm{T}}(k), \qquad P(0|0) = P(0) \tag{2.46}$$

$$K(k+1) = P(k+1|k)H^{\mathrm{T}}(k+1)[H(k+1)P(k+1|k)H^{\mathrm{T}}(k+1) + J(k+1)R(k+1)J^{\mathrm{T}}(k+1)]^{-1} \tag{2.47}$$

$$\hat{x}(k+1|k+1) = \hat{x}(k+1|k) + K(k+1)[z(k+1) - h(\hat{x}(k+1|k), u(k+1), 0, k+1)] \tag{2.48}$$

$$P(k+1|k+1) = [I - K(k+1)H(k+1)]P(k+1|k) \tag{2.49}$$

2.3 STABILITY OF THE DISCRETE-TIME KALMAN FILTER

Let us consider the discrete-time Kalman filter with $S = 0$, presented by Theorem 2.1, for the time-invariant system. The Kalman filter is summarized as follows:

$$\hat{x}(k+1|k) = \Phi\hat{x}(k|k-1) + Gu(k) + K^*(k)[z(k) - H\hat{x}(k|k-1)], \qquad \hat{x}(0|-1) = \hat{x}(0) \tag{2.50}$$

or, equivalently, as follows:

$$\hat{x}(k+1|k) = \tilde{\Phi}(k)\hat{x}(k|k-1) + Gu(k) + K^*(k)z(k) \tag{2.51}$$

where $\tilde{\Phi}(k)$ and $K^*(k)$ are the closed-loop transition and filter gain matrices, respectively, given by

$$\tilde{\Phi}(k) = \Phi - K^*(k)H \tag{2.52}$$

$$K^*(k) = \Phi P(k|k-1)H^{\mathrm{T}}[HP(k|k-1)H^{\mathrm{T}} + R]^{-1} \tag{2.53}$$

$$\begin{aligned} P(k+1|k) = {} & \Phi P(k|k-1)\Phi^{\mathrm{T}} \\ & - \Phi P(k|k-1)H^{\mathrm{T}}[HP(k|k-1)H^{\mathrm{T}} + R]^{-1} \\ & \times HP(k|k-1)\Phi^{\mathrm{T}} + BQB^{\mathrm{T}}, \\ & P(0|-1) = P(0) \end{aligned} \tag{2.54}$$

If $P(k+1|k)$ converges as $k \to \infty$, then the limiting solution P will satisfy the following discrete-time algebraic Riccati equation (ARE), obtained from equation (2.54) by putting $P(k+1|k) = P(k|k-1) \triangleq P$:

$$P - \Phi P \Phi^{\mathrm{T}} + \Phi P H^{\mathrm{T}}[HPH^{\mathrm{T}} + R]^{-1} HP\Phi^{\mathrm{T}} - BQB^{\mathrm{T}} = 0 \tag{2.55}$$

Using the solution of this equation, we obtain the *steady-state* Kalman filter in discrete time:

$$\hat{x}(k+1|k) = \tilde{\Phi}\hat{x}(k|k-1) + Gu(k) + K^*z(k) \tag{2.56}$$

where

$$\tilde{\Phi} \triangleq \Phi - K^*H \tag{2.57}$$

$$K^* \triangleq \Phi P H^{\mathrm{T}}[HPH^{\mathrm{T}} + R]^{-1} \tag{2.58}$$

To proceed further, it is convenient to set

$$H^{\mathrm{T}}R^{-1}H = C^{\mathrm{T}}C$$

$$BQB^{\mathrm{T}} = DD^{\mathrm{T}}$$

where C and D are matrices of full rank such that

$$\text{rank } C = \text{rank } H^{\mathrm{T}}R^{-1}H,$$

$$\text{rank } D = \text{rank } BQB^{\mathrm{T}}$$

Then the discrete-time ARE becomes

$$P - \Phi P \Phi^{\mathrm{T}} + \Phi P C^{\mathrm{T}}[CPC^{\mathrm{T}} + I]^{-1}CP\Phi^{\mathrm{T}} - DD^{\mathrm{T}} = 0 \tag{2.59}$$

We shall be particularly interested in those solutions of the discrete-time ARE which are real, symmetric and positive semidefinite, and which give a steady-state filter having roots on or inside the unit circle. We therefore introduce the following two definitions.

Definition 2.1

A real symmetric positive semidefinite solution of the discrete-time ARE is said to be a *stabilizing solution* if the eigenvalues of the corresponding filter closed-loop transition matrix $\tilde{\Phi}$ all lie *inside* the unit circle in the complex plane (i.e. $|\lambda_i(\tilde{\Phi})| < 1,\ i = 1, \ldots, n$).

Definition 2.2

A real symmetric positive semidefinite solution of the discrete-time ARE is said to be a *strong solution* if the eigenvalues of the corresponding filter closed-loop transition matrix $\tilde{\Phi}$ all lie *inside or on* the unit circle in the complex plane (i.e. $|\lambda_i(\tilde{\Phi})| \leqslant 1,\ i = 1, \ldots, n$).

Equation (2.59) is closely related to the $2n \times 2n$ *symplectic* matrix:

$$M = \begin{bmatrix} \Phi^{-T} & \Phi^{-T}C^{T}C \\ DD^{T}\Phi^{-T} & \Phi + DD^{T}\Phi^{-T}C^{T}C \end{bmatrix} \tag{2.60}$$

which is the *forward* Hamiltonian system matrix for the discrete-time filtering problem (see also Chapter 9), i.e.

$$\begin{bmatrix} \lambda(k) \\ x(k+1) \end{bmatrix} = M \begin{bmatrix} \lambda(k-1) \\ x(k) \end{bmatrix}$$

where $\lambda(k)$ is a Lagrange multiplier vector associated with the state $x(k+1)$. The following definition describes an important property of M.

Definition 2.3

$M \in \mathbb{R}^{2n \times 2n}$ is symplectic if $J^{-1}M^{T}J = M^{-1}$, where

$$J = \begin{bmatrix} 0 & I \\ -I & 0 \end{bmatrix} \in \mathbb{R}^{2n \times 2n}$$

and $J^{T} = J^{-1} = -J$.

We note from this definition that the associated *backward* symplectic matrix M^{-1} is just

$$M^{-1} = \begin{bmatrix} \Phi^{\mathrm{T}} + C^{\mathrm{T}}C\Phi^{-1}DD^{\mathrm{T}} & -C^{\mathrm{T}}C\Phi^{-1} \\ -\Phi^{-1}DD^{\mathrm{T}} & \Phi^{-1} \end{bmatrix} \tag{2.60'}$$

Moreover, M^{-1} is similar to M^{T}, and both M^{-1} and M^{T} must have the same eigenvalues; equivalently, the eigenvalues of M occur in reciprocal pairs. This leads to the following lemma for M.

Lemma 2.1

Let $M \in \mathbb{R}^{2n\times 2n}$ be symplectic. Then $\lambda \in \sigma(M)$† implies $1/\lambda \in \sigma(M)$ with the same multiplicity.

Let

$$M\xi_i = \lambda_i \xi_i,\ i = 1, \ldots, 2n$$

and write

$$\xi_i = \begin{bmatrix} x_i \\ y_i \end{bmatrix}$$

where x_i and y_i are elements of $\mathbb{C}^n$.

The following theorem summarizes some conditions for the existence and uniqueness of the stabilizing solution of discrete-time ARE.

Theorem 2.3

(i) Each solution P of equation (2.59) takes the form

$$P = YX^{-1},$$

where

$$X = [x_1, x_2, \ldots, x_n],\ Y = [y_1, y_2, \ldots, y_n]$$

correspond to a choice of eigenvalues $\lambda_1, \lambda_2, \ldots, \lambda_n$ of M such that X^{-1} exists.

(ii) The matrix $\tilde{\Phi}^{-\mathrm{T}}$, where $\tilde{\Phi}$ is yielded by $P = YX^{-1}$, has λ_i as its eigenvalues and x_i, $i = 1, \ldots, n$, as the corresponding eigenvectors‡.

† $\sigma(M)$ denotes the spectrum of M (the set of $2n$ eigenvalues).

‡ If the backward symplectic matrix M^{-1} is used, then $\tilde{\Phi}^{-\mathrm{T}}$ is replaced by $\tilde{\Phi}^{\mathrm{T}}$ in this condition.

(iii) There exists a stabilizing solution $P \geqslant 0$ to equation (2.59) if and only if (C, Φ) is *detectable* and $|\lambda| \neq 1$ for all eigenvalues λ of M.
(iv) The stabilizing solution is the only positive semidefinite solution of equation (2.59) if and only if (Φ, D) is *stabilizable*.†
(v) Stabilizability of (Φ, D) and detectability of (C, Φ) are necessary and sufficient conditions for equation (2.59) to have a unique stabilizing solution.

Proof

See Kučera [7]. □

Some of the key properties of the strong solution to equation (2.59) are discussed in the following theorem.

Theorem 2.4

Provided that (C, Φ) is detectable:

(i) The strong solution of the discrete-time ARE exists and is unique.
(ii) If (Φ, D) is stabilizable, the strong solution is the only positive semidefinite solution of the discrete-time ARE.
(iii) If (Φ, D) has no *controllable* modes on the unit circle, the strong solution coincides with the stabilizing solution.
(iv) If (Φ, D) has an *uncontrollable* mode on the unit circle, then, although the strong solution exists, there is no stabilizing solution.
(v) If (Φ, D) has an uncontrollable mode inside or on the unit circle, the strong solution is not positive definite.
(vi) If (Φ, D) has an uncontrollable mode outside the unit circle, then, as well as the strong solution, there is at least one other positive semidefinite solution of the discrete-time ARE.

† When requiring the only positive definite solution of equation (2.59), (Φ, D) must be reachable.

Proof

See Chan *et al.* [8]. □

The following two theorems state the convergence of the solution of the RDE to the stabilizing solution of the discrete-time ARE.

Theorem 2.5

Subject to the following conditions:

(i) (Φ, D) is stabilizable;
(ii) (C, Φ) is detectable;
(iii) $P(0) \geqslant 0$;

then the limit

$$\lim_{k \to \infty} P(k) = P_s$$

is reached exponentially fast; where $P(k)$ is the solution of the RDE with initial condition $P(0)$ and P_s is the unique stabilizing solution of the discrete-time ARE.

Proof

See Cains and Mayne [9] (see also Anderson and Moore [10, p.76]). □

Theorem 2.6

Subject to the following conditions:

(i) there are no uncontrollable modes of (Φ, D) on the unit circle;
(ii) (C, Φ) is detectable;
(iii) $P(0) > 0$;

then the limit

$$\lim_{k \to \infty} P(k) = P_s$$

is reached exponentially fast; where $P(k)$ and P_s are as in Theorem 2.5.

Proof

See Chan *et al.* [8] (or Goodwin and Sin [11, p. 513]). □

We also have the following theorem on the convergence of the solution of the RDE to the strong solution of the discrete-time ARE.

Theorem 2.7

Subject to the following conditions:

(i) (C, Φ) is observable,
(ii) $(P(0) - P_s) > 0$ or $P(0) = P_s$,

then

$$\lim_{k\to\infty} P(k) = P_s$$

where $P(k)$ is as in Theorem 2.5, but now P_s is the (unique) strong solution of the discrete-time ARE.

Proof

See Chan *et al.* [8] (or Goodwin and Sin [11, p. 514]). □

Example 2.3 Chan *et al.* [8]; Goodwin and Sin [11]

Consider the following simple scalar system:

$$x(k+1) = ax(k) + w(k); \qquad \Phi = a$$

$$z(k) = x(k) + v(k); \qquad H = 1$$

where

$$E\left\{\begin{bmatrix} w(k) \\ v(k) \end{bmatrix} [w(j)\ v(j)]\right\} = \begin{bmatrix} Q & 0 \\ 0 & R \end{bmatrix} \delta_{kj}; \qquad Q = D^2, R = 1$$

Note that

1. If $Q \neq 0$, then (Φ, D) is stabilizable for all Φ.
2. If $|a| < 1$, then (Φ, D) is stabilizable for all Φ.
3. If $|a| = 1$ and $Q = 0$, then the system has an uncontrollable root on the unit circle, and hence is not stabilizable.

For the case $Q \neq 0$ and/or $|a| < 1$, we can apply Theorem 2.5 to find that the solution of the discrete-time ARE converges exponentially fast to the stabilizing solution for all $P(0) \geqslant 0$.

For the case $Q = 0$, $|a| > 1$, the system is unstable and thus from Theorem 2.4(vi) there are at least two positive semidefinite solutions to the ARE (in fact there are exactly two in this case, $P_s = 0$ and $P_s = a^2 - 1$). The strong (in fact stabilizing) solution is $P_s = a^2 - 1$ since it gives an asymptotic stable filter with $\tilde{\Phi} = 1/a$ where $|1/a| < 1$. With $P_s = 0$ the filter is unstable since $\tilde{\Phi} = a$ with $|a| > 1$. To ensure that $\lim_{k\to\infty} P(k|k-1) = P_s$, Theorem 2.6 requires $P(0) > 0$, otherwise $P(k|k-1) = 0$ for all k giving rise to an unstable filter.

For the case $a = 1$, $Q = 0$, the system becomes

$$x(k+1) = x(k), \qquad x(k=1) = x(1)$$

$$z(k) = x(k) + v(k)$$

where $\{v(k)\}$ is an independent and identically distributed sequence having unit variance. This system is not stabilizable since it has an uncontrollable root on the unit circle. The initial state estimate is assumed to have mean $\hat{x}(1)$ and variance $P(1) = 1$ at time $k = 1$.

The Kalman filter (predictive type) for this system is

$$\hat{x}(k+1|k) = \hat{x}(k|k-1) + K^*(k)[z(k) - \hat{x}(k|k-1)], \qquad \hat{x}(1|0) = \hat{x}(1)$$

where

$$K^*(k) = \frac{P(k|k-1)}{P(k|k-1) + 1}$$

and the associated RDE is

$$P(k+1|k) = \frac{P(k|k-1)}{P(k|k-1) + 1}, \qquad P(1|0) = 1$$

The analytic solution of the RDE is

$$P(k|k-1) = \frac{1}{k}$$

It is clear that the only possible solution to the ARE is $P_s = 0$. It is also observed that $P(k|k-1)$ converges to P_s as $1/k$ (not exponentially fast) and the resulting steady-state Kalman filter is

$$\hat{x}(k+1|k) = \hat{x}(k|k-1)$$

Of course, the estimate generated by the time-varying Kalman filter converges (in mean squares) to $x(0)$ as $k \to \infty$, and thus asymptotically the correct initial conditions are generated for the steady-state filter $\hat{x}(k+1|k) = \hat{x}(k|k-1)$.

2.4 COMPUTATIONAL APPROACHES TO THE DISCRETE-TIME RICCATI EQUATION

In this section, we shall review some computational methods for solving the discrete-time ARE (2.59).

2.4.1 Eigenvalue–eigenvector method

Since good procedures now exist for computing the eigenvalues and eigenvectors of a matrix, this method has come into increasing prominence in recent years. The essence of the procedure is described in the following theorem.

Theorem 2.8

It is assumed for simplicity that the eigenvalues of M are distinct. The stabilizing solution, P_s, to equation (2.59) is obtained as follows:

Step 1: Formulate the symplectic matrix using equation (2.60).

Step 2: Compute the $2n$ eigenvalues of M, $\lambda_i(M)$, $i = 1, \ldots, 2n$.

Step 3: Compute the n $2n$-dimensional eigenvectors† ξ_i such that

$$|\lambda_i(M)| > 1, \qquad i = 1, \ldots, n$$

Step 4: Partition the eigenvectors matrix (or modal matrix) as follows:

$$[\xi_1, \ldots, \xi_n] = \begin{matrix} & n \\ n & \\ n & \end{matrix}\!\!\begin{pmatrix} X \\ Y \end{pmatrix}$$

Step 5: Solve for P_s:

$$P_s = YX^{-1}$$

† For the eigenvalues of M^{-1}, one must take the eigenvectors $\lambda_i(M^{-1})$ such that $|\lambda_i(M^{-1})| < 1$.

Proof

See Vaughan [12] and Kučera [7]. □

For the case when an eigenvalue of M has a multiplicity, we refer the reader to Pappas *et al.* [13] and Chan *et al.* [8] who use the generalized eigenvector or Shur vector method.

2.4.2 Kleinman's Iterative Method

This approach is a direct modification of the discrete-time analogue of Kleinman's algorithm which will be discussed in Section 2.8 and is due to Hewer [14]. The basis of the method is a variant of the Newton approximation method for finding the root of a system of nonlinear equations by successive linear approximations. We assume that equation (2.59) possesses a unique stabilizing solution P_s, i.e. (Φ, D) is stabilizable and (C, Φ) is detectable.

Theorem 2.9

The algorithm due to Kleinman consists of the iterative solution of a linear matrix equation

$$P_i = \hat{\Phi}_i P_i \hat{\Phi}_i^T + L_i R L_i^T + BQB^T, \qquad i = 0, 1, \ldots \tag{2.61}$$

where

$$\hat{\Phi}_i = \Phi - L_i H, \qquad i = 0, 1, \ldots \tag{2.62}$$

$$L_i = \Phi P_{i-1} H^T [H P_{i-1} H^T + R]^{-1}, \qquad i = 1, 2, \ldots \tag{2.63}$$

to obtain P_s from

$$\lim_{i \to \infty} P_i = P_s$$

The matrix L_0 must be chosen so that the eigenvalues of $\hat{\Phi}_0$ are inside the unit circle.

Proof

We briefly derive an equation for the algorithm. Equation (2.6) with $S = 0$ reduces to

$$P(k+1|k) = \Phi P(k|k-1)\Phi^T + BQB^T - K^*(k)\tilde{R}(k)K^{*T}(k) \tag{2.64}$$

$$\widetilde{R}(k) \triangleq HP(k|k-1)H^{\mathrm{T}} + R$$

for the time-invariant system, where

$$K^*(k) \triangleq \Phi P(k|k-1)H^{\mathrm{T}}\widetilde{R}^{-1}(k)$$

Equation (2.64) can also be written as

$$\begin{aligned}
P(k+1|k) &= [\Phi - K^*(k)H]P(k|k-1)[\Phi - K^*(k)H]^{\mathrm{T}} \\
&\quad + \Phi P(k|k-1)H^{\mathrm{T}}K^{*\mathrm{T}}(k) \\
&\quad + K^*(k)HP(k|k-1)\Phi^{\mathrm{T}} \\
&\quad - K^*(k)HP(k|k-1)H^{\mathrm{T}}K^{*\mathrm{T}}(k) \\
&\quad + BQB^{\mathrm{T}} - K^*(k)\widetilde{R}(k)K^{*\mathrm{T}}(k) \\
&= [\Phi - K^*(k)H]P(k|k-1)[\Phi - K^*(k)H]^{\mathrm{T}} \\
&\quad + K^*(k)\widetilde{R}(k)K^{*\mathrm{T}}(k) \\
&\quad + K^*(k)\widetilde{R}(k)K^{*\mathrm{T}}(k) \\
&\quad - K^*(k)HP(k|k-1)H^{\mathrm{T}}K^{*\mathrm{T}}(k) \\
&\quad + BQB^{\mathrm{T}} - K^*(k)\widetilde{R}(k)K^{*\mathrm{T}}(k) \\
&= [\Phi - K^*(k)H]P(k|k-1)[\Phi - K^*(k)H]^{\mathrm{T}} \\
&\quad + BQB^{\mathrm{T}} + K^*(k)RK^{*\mathrm{T}}(k). \qquad (2.65)
\end{aligned}$$

Hence, regarding $P(k+1|k)$ (or $P(k|k-1)$) and $K^*(k)$ appearing explicitly in equation (2.65) as the steady-state values P_i and L_i at the ith iteration, we obtain equations (2.61) and (2.62), where $P(k|k-1)$ involved in $K^*(k)$ is regarded as the steady-state value P_{i-1} at the $(i-1)$th iteration. □

This method gives monotonic and quadratic convergence to P_s and is believed to be one of the best methods for finding P_s, but the total number of iterations depends on the initial choice L_0. Moreover, it requires the solution of a Lyapunov equation, for which we refer to Pace and Barnett [15] and Štecha *et al.* [16] (see also Exercise 5.14 and Appendix C).

Example 2.4

We consider the scalar discrete-time system

$$x(k+1) = x(k) + w(k)$$

$$z(k) = x(k) + v(k), \qquad k = 0, 1, \ldots$$

with variances $Q = R = 1$ and $P(0) = 10$.

(i) *Direct computation method*
The discrete-time RDE can be solved recursively in the forward direction. That is, from equation (2.54) we have

$$P(k + 1|k) = [1 + 2P(k|k - 1)][1 + P(k|k - 1)]^{-1},$$
$$P(0|-1) = 10$$

and the sequence $P(k|k - 1)$ for $k = 0, 1, 2, \ldots$ is 10, 1.909, 1.656, 1.624, 1.619, 1.618, 1.618, ...

(ii) *Eigenvalue–eigenvector method*
The symplectic matrix M is given by

$$M = \begin{bmatrix} 1 & 1 \\ 1 & 2 \end{bmatrix}$$

with eigenvalues $\frac{1}{2}(3 \pm \sqrt{5})$. The eigenvector corresponding to the eigenvalue outside the unit circle is

$$\begin{bmatrix} X \\ Y \end{bmatrix} = \begin{bmatrix} 1 \\ \frac{1}{2}(1 + \sqrt{5}) \end{bmatrix}$$

so that

$$P_s = YX^{-1} = \tfrac{1}{2}(1 + \sqrt{5})$$

(iii) *Kleinman's iterative method*
In this case, a choice of $L_0 = 0.5$ yields the steady-state solution P_s in two iterations:

$$P_0 = 1.667, \quad P_1 = 1.618$$

while a choice of $L_0 = 0.1$ yields

$$P_0 = 5.316, \; P_1 = 1.752, \; P_2 = 1.627, \; P_3 = 1.618$$

This demonstrates that while near the solution the convergence is fast, different choices of L_0 may require substantially different computational effort.

2.4.3 Chandrasekhar-type Algorithm

In this algorithm, the nonlinear Riccati equation is replaced by certain nonlinear difference equations which are called *Chandrasekhar-type equations* because they are analogous to certain differential equations encountered in the corresponding continuous-time problem (Kailath [17]). The algorithm is summarized in the following theorem.

Theorem 2.10

For the time-invariant model given by equations (2.1)–(2.3), the Kalman gain $K_i^* \triangleq K^*(k)$ is provided by the following recursions:

$$\Omega_{i+1} = \Omega_i + HY_iM_iY_i^{\mathrm{T}}H^{\mathrm{T}} \tag{2.66}$$

$$K_{i+1}^* = (K_i^*\Omega_i + \Phi Y_iM_iY_i^{\mathrm{T}}H^{\mathrm{T}})\Omega_{i+1}^{-1} \tag{2.67}$$

$$Y_{i+1} = (\Phi - K_{i+1}^*H)Y_i \tag{2.68}$$

$$M_{i+1} = M_i + M_iY_i^{\mathrm{T}}H^{\mathrm{T}}\Omega_i^{-1}HY_iM_i \tag{2.69}$$

where Ω_{i+1}, K_{i+1}^*, Y_{i+1}, M_{i+1} are matrices of size $m \times m$, $n \times m$, $n \times \alpha$, $\alpha \times \alpha$, respectively, and α will be specified below. The initial values for equations (2.66) and (2.67) are obtained from the expressions

$$\begin{aligned} \Omega_0 &= HP(0)H^{\mathrm{T}} + R \\ K_0^* &= [\Phi P(0)H^{\mathrm{T}} + BS]\Omega_0^{-1} \end{aligned} \tag{2.70}$$

while $Y_0 = \bar{Y}$ and $M_0 = \bar{M}$ are found by factoring

$$\left.\begin{aligned} \Delta P(0) &\triangleq \Phi P(0)\Phi^{\mathrm{T}} + BQB^{\mathrm{T}} - K_0^*\Omega_0^{-1}K_0^{*\mathrm{T}} - P(0) \\ \operatorname{rank}[\Delta P(0)] &= \alpha \end{aligned}\right\} \tag{2.71}$$

as

$$\Delta P(0) = \bar{Y}\bar{M}\bar{Y}^{\mathrm{T}}, \; \bar{M} = \begin{bmatrix} M_+ & 0 \\ 0 & M_- \end{bmatrix}, \; M_+ > 0, \; M_- < 0 \tag{2.72}$$

In the case where $P(0) \equiv 0$, $S \equiv 0$, and Q has full rank, we can set $Y_0 = B$ and $M_0 = Q$.

Proof

See Morf *et al.* [18]. □

In the discrete-time Kalman filter algorithm, the gain $K^*(k)$ is computed via the solution of the difference equation (2.6) for the

$n(n+1)/2$ elements of the auxiliary symmetric matrix P. In the above algorithm we have to update equations (2.66)–(2.69) for $n(m+\alpha)+\frac{1}{2}(m^2+m+\alpha^2+\alpha)$ different entries of K^*, Y, Ω and M. It often happens that $P(0)$ is such or can be chosen to be such that $n \gg \alpha$. If in addition $n \gg m$, the algorithm has to update far fewer variables. Furthermore, in discrete-time systems it is the total number of operators rather than the number of variables that is important, and on this score the algorithm seems to have a definite edge. We also observe that although the matrix $P(k+1|k)$ does not enter into the algorithm, it may be computed from it as follows:

$$P(i+1|i) = P(i|i-1) + Y_i M_i Y_i^{\mathrm{T}}, \quad P(0|-1) = P(0) \tag{2.73a}$$

or

$$P(i+1|i) = P(0) + \sum_{j=0}^{i} Y_j M_j Y_j^{\mathrm{T}} \tag{2.73b}$$

2.4.4 Doubling Algorithm

This algorithm will be discussed in detail in Chapter 5; in addition, we refer the reader to Anderson [19] or Anderson and Moore [10] for another derivation of the algorithm (see also Exercise 5.12).

2.5 THE CONTINUOUS-TIME KALMAN FILTER

We consider a formal linear stochastic system described by the following continuous-time model:

$$\mathrm{d}x(t)/\mathrm{d}t = A(t)x(t) + G(t)u(t) + B(t)w(t) \tag{2.74}$$

$$z(t) = H(t)x(t) + v(t) \tag{2.75}$$

where $x(t) \in \mathbb{R}^n$, $u(t) \in \mathbb{R}^r$, $w(t) \in \mathbb{R}^p$, $v(t)$, $z(t) \in \mathbb{R}^m$, $x(t_0) \sim N(\hat{x}(0), P(0))$ and the zero-mean Gaussian white noise processes $\{w(t), v(t)\}$ have covariance given by

$$E\left\{\begin{bmatrix} w(t) \\ v(t) \end{bmatrix} [w^{\mathrm{T}}(s) \quad v^{\mathrm{T}}(s)]\right\} = \begin{bmatrix} Q(t) & S(t) \\ S^{\mathrm{T}}(t) & R(t) \end{bmatrix} \delta(t-s) \tag{2.76}$$

where $\delta(t-s)$ denotes the Dirac delta function, $Q(t) \geqslant 0$ and $R(t) > 0$. In addition, it is assumed that $u(t)$ is a known input vector and $\{w(t), v(t)\}$ are independent of $x(t_0)$.

2.5.1 The Kalman Filter

Given the observation data $Z_t \triangleq \{z(s), t_0 \leqslant s \leqslant t\}$, the MMSE estimate of $x(t)$, i.e. $\hat{x}(t|t) \triangleq E[x(t)|Z_t]$, is provided by the continuous-time Kalman filter† equation.

Theorem 2.11 The continuous-time Kalman filter

The solution of the MMSE (or optimal) filtering problem given by equations (2.74)–(2.76) is given by the following relations:

$$\begin{aligned} \mathrm{d}\hat{x}(t|t)/\mathrm{d}t = {} & A(t)\hat{x}(t|t) + G(t)u(t) \\ & + K^*(t)[z(t) - H(t)\hat{x}(t|t)], \qquad \hat{x}(t_0|t_0) = \hat{x}(0) \end{aligned} \tag{2.77}$$

where the filter gain $K^*(t)$ is defined as

$$K^*(t) \triangleq [P(t|t)H^{\mathrm{T}}(t) + B(t)S(t)]R^{-1}(t) \tag{2.78}$$

and the filter error covariance, $P(t|t) \triangleq E\{[x(t) - \hat{x}(t|t)]$ $[x(t) - \hat{x}(t|t)]^{\mathrm{T}}|Z_t\}$, satisfies the Riccati differential equation (RDE):

$$\begin{aligned} \mathrm{d}P(t|t)/\mathrm{d}t = {} & A(t)P(t|t) + P(t|t)A^{\mathrm{T}}(t) + B(t)Q(t)B^{\mathrm{T}}(t) \\ & - K^*(t)R(t)K^{*\mathrm{T}}(t), \qquad P(t_0|t_0) = P(0) \end{aligned} \tag{2.79}$$

Proof

There are many methods of deriving the continuous-time Kalman filter. Since we already have a solution in discrete time, it is natural to try to use a *limiting procedure* [6, 20] to obtain the continuous-time formulas. To this end, we consider a discrete-time model described by the following equations:

$$x(t + \Delta t) = (I + A(t)\Delta t)x(t) + G(t)\Delta t u(t) + B(t)\Delta t w(t) \tag{2.80}$$

$$z(t) = H(t)z(t) + v(t) \tag{2.81}$$

where t denotes the discrete-time instants $\{t = t_0 + j\Delta t, j = 0, 1, 2, \ldots\}$ and

$$E\left\{\begin{bmatrix} w(t) \\ v(t) \end{bmatrix} [w^{\mathrm{T}}(s) \quad v^{\mathrm{T}}(s)]\right\} = \begin{bmatrix} Q(t)/\Delta t & S(t)/\Delta t \\ S^{\mathrm{T}}(t)/\Delta t & R(t)/\Delta t \end{bmatrix} \delta_{jk} \tag{2.82}$$

† Often called the Kalman–Bucy filter.

where s is the discrete-time index $\{s = t_0 + k\Delta t,\ k = 0, 1, \ldots\}$. From Theorem 2.1, the filtered estimate of this model is

$$\hat{x}(t + \Delta t|t) = (I + A(t)\Delta t)\hat{x}(t|t - \Delta t) + K^*(t)\Delta t[z(t) - H(t)\hat{x}(t|t - \Delta t)] + G(t)\Delta t u(t) \tag{2.83}$$

and

$$K^*(t) = [(I + A(t)\Delta t)P(t|t - \Delta t)H^T(t) + B(t)S(t)]\tilde{R}^{-1}(t)/\Delta t \tag{2.84}$$

$$\tilde{R}(t) = [H(t)P(t|t - \Delta t)H^T(t) + R(t)/\Delta t] \tag{2.85}$$

Now in the limit as $\Delta t \to 0$ we see that

$$\begin{aligned}\lim_{\Delta t \to 0} K^*(t) &= \lim_{\Delta t \to 0} [(I + A(t)\Delta t)P(t|t - \Delta t)H^T(t) + B(t)S(t)] \\ &\quad \times [H(t)P(t|t - \Delta t)H^T(t)\Delta t + R(t)]^{-1} \\ &= [P(t|t)H^T(t) + B(t)S(t)]R^{-1}(t)\end{aligned} \tag{2.86}$$

The estimator equation (2.83) can be written as

$$\frac{\hat{x}(t + \Delta t|t) - \hat{x}(t|t - \Delta t)}{\Delta t} = A(t)\hat{x}(t|t - \Delta t) + K^*(t)[z(t) - H(t)\hat{x}(t|t - \Delta t)] + G(t)u(t) \tag{2.87}$$

which in the limit becomes the differential equation (2.77). Similarly, from Theorem 2.1, the filtered error covariance associated with equation (2.83) is

$$\begin{aligned}P(t + \Delta t|t) &= (I + A(t)\Delta t)P(t|t - \Delta t)(I + A(t)\Delta t)^T \\ &\quad + B(t)Q(t)B^T(t)\Delta t \\ &\quad - K^*(t)\Delta t[H(t)P(t|t - \Delta t)H^T(t) \\ &\quad + R(t)/\Delta t]\Delta t K^{*T}(t) \\ &= P(t|t - \Delta t) + [A(t)P(t|t - \Delta t) \\ &\quad + P(t|t - \Delta t)A^T(t) + B(t)Q(t)B^T(t) \\ &\quad - K^*(t)R(t)K^{*T}(t)]\Delta t + O(\Delta t^2)\end{aligned} \tag{2.88}$$

which can be rewritten as

$$\frac{P(t+\Delta t|t) - P(t|t-\Delta t)}{\Delta t}$$
$$= A(t)P(t|t-\Delta t) + P(t|t-\Delta t)A^{\mathrm{T}}(t) + B(t)Q(t)B^{\mathrm{T}}(t)$$
$$- K^*(t)R(t)K^{*\mathrm{T}}(t) + O(\Delta t^2)/\Delta t \tag{2.89}$$

Taking the limit as $\Delta t \to 0$ in equation (2.89) gives the desired result (2.79). □

The above filter is shown diagrammatically in Figure 2.3. Note that equations (2.74) and (2.75) may be manipulated by mathematically rigorous rules, known as *Ito stochastic calculus* [21]. That is, models (2.74) and (2.75) can also be described by Ito stochastic differential equations:

$$\mathrm{d}x(t) = A(t)x(t)\,\mathrm{d}t + G(t)u(t)\,\mathrm{d}t + B(t)\,\mathrm{d}\xi(t), \qquad t \geq t_0 \tag{2.90}$$

$$\mathrm{d}y(t) = H(t)x(t)\,\mathrm{d}t + \mathrm{d}\eta(t) \tag{2.91}$$

where $\{\xi(t), \eta(t)\}$ are Wiener (or Brownian motion) processes† with

$$E[\xi(t)] = 0$$
$$E[\eta(t)] = 0 \tag{2.92}$$

$$E\left\{\begin{bmatrix} \mathrm{d}\xi(t) \\ \mathrm{d}\eta(t) \end{bmatrix} [\mathrm{d}\xi^{\mathrm{T}}(t) \quad \mathrm{d}\eta^{\mathrm{T}}(t)]\right\} = \begin{bmatrix} Q(t)\,\mathrm{d}t & S(t)\,\mathrm{d}t \\ S^{\mathrm{T}}(t)\,\mathrm{d}t & R(t)\,\mathrm{d}t \end{bmatrix} \tag{2.93}$$

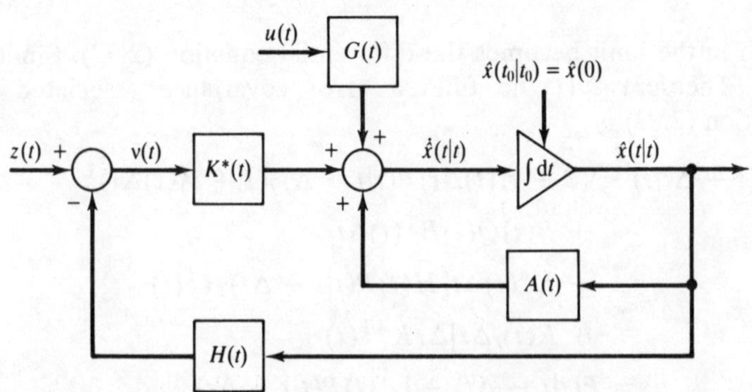

Figure 2.3 Block diagram of the continuous-time Kalman filter.

† Formally letting $\mathrm{d}\xi/\mathrm{d}t = w$, $\mathrm{d}\eta/\mathrm{d}t = v$ and $\mathrm{d}y/\mathrm{d}t = z$ gives the less rigorous forms (2.74) and (2.75).

In this case, the continuous-time Kalman filter equation (2.77) is replaced by

$$\mathrm{d}\hat{x}(t|t) = A(t)\hat{x}(t|t)\,\mathrm{d}t + G(t)u(t)\,\mathrm{d}t + K^*(t)[\mathrm{d}y(t) - H(t)\hat{x}(t|t)\,\mathrm{d}t], \qquad \hat{x}(t_0|t_0) = \hat{x}(0) \tag{2.94}$$

or, in the stochastic integral form,

$$\hat{x}(t|t) = \hat{x}(0) + \int_{t_0}^{t} [A(\tau)\hat{x}(\tau|\tau) + G(\tau)u(\tau)]\,\mathrm{d}\tau + \int_{t_0}^{t} K^*(\tau)\,\mathrm{d}\nu(\tau) \tag{2.95}$$

where $\nu(\tau)$ is the innovation process in continuous-time (see Appendix 3A for details) defined by

$$\mathrm{d}\nu(\tau) \triangleq \mathrm{d}y(\tau) - H(\tau)\hat{x}(\tau|\tau)\,\mathrm{d}\tau \tag{2.96}$$

Example 2.5 Meditch [6]

Consider a scalar stochastic system which is defined by the differential equation

$$\dot{x}(t) = -ax(t) + w(t)$$

for $t \geqslant 0$ where $a > 0$, $X(0) \sim N(0, P(0))$, and $\{w(t), t \geqslant 0\}$ is a zero-mean Gaussian white noise process which is independent of $x(0)$ and has variance $Q > 0$. Suppose further that the observation equation is of the form:

$$z(t) = x(t) + v(t)$$

where $\{v(t), t \geqslant 0\}$ is a zero-mean Gaussian white noise process independent of $x(0)$ and $\{w(t), t \geqslant 0\}$, and with variance $R > 0$.

For this problem, the filtering error covariance (or variance) equation is

$$\dot{P} = -2aP - \frac{1}{R}P^2 + Q, \qquad t \geqslant 0$$

with $P(0)$ the initial condition. This equation can be solved quite readily by first separating the variables to obtain

$$\frac{\mathrm{d}P}{P^2 + 2aRP - RQ} = -\frac{1}{R}\,\mathrm{d}t$$

Letting ρ_1 and ρ_2 denote the roots of $P^2 + 2aRP - RQ = 0$, we have

$$\frac{\mathrm{d}P}{[P-\rho_1][P-\rho_2]} = \frac{1}{(\rho_1-\rho_2)}\left[\frac{\mathrm{d}P}{P-\rho_1} - \frac{\mathrm{d}P}{P-\rho_2}\right] = -\frac{1}{R}\,\mathrm{d}t$$

where

$$\rho_{1,2} = \left[-a \pm \sqrt{a^2 + \frac{Q}{R}}\right]R$$

Since

$$\rho_1 - \rho_2 = 2R\sqrt{a^2 - \frac{Q}{R}}$$

we have

$$\frac{\mathrm{d}P}{P-\rho_1} - \frac{\mathrm{d}P}{P-\rho_2} = -2\mathrm{d}t\sqrt{a^2 + \frac{Q}{R}}$$

Defining $\mu = \sqrt{a^2 + Q/R}$, we obtain

$$\frac{P(t|t)-\rho_1}{P(t|t)-\rho_2} = c\exp(-2\mu t)$$

where c is the constant of integration.

Utilizing the fact that $P(0|0) = P(0)$, we have

$$c = \frac{P(0)-\rho_1}{P(0)-\rho_2}$$

The filtering error covariance is then found to be

$$P(t|t) = \frac{\rho_1 - \rho_2 c\exp(-2\mu t)}{1 - c\exp(-2\mu t)}$$

for $t \geqslant 0$. We also have

$$K(t) = \frac{1}{R}P(t|t)$$

and

$$\dot{\hat{x}} = -a\hat{x} + K(t)[z(t) - \hat{x}]$$

2.5.2 The Extended Kalman Filter

The continuous-time version of the extended Kalman filter can be based on the following model:

$$\dot{x}(t) = f(x(t), u(t), w(t), t) \tag{2.97}$$

$$z(t) = h(x(t), u(t), v(t), t) \tag{2.98}$$

where $w(t)$ and $v(t)$ are zero-mean white Gaussian noise processes with covariances

$$E\left\{\begin{bmatrix} w(t) \\ v(t) \end{bmatrix} [w^{\mathrm{T}}(s) \quad v^{\mathrm{T}}(s)]\right\} = \begin{bmatrix} Q(t) & 0 \\ 0 & R(t) \end{bmatrix} \delta(t - s) \tag{2.99}$$

The estimator equation becomes

$$\dot{\hat{x}}(t|t) = f(\hat{x}(t|t), u(t), 0, t) + K(t)[z(t) - h(\hat{x}(t|t), u(t), 0, t)], \quad \hat{x}(0|0) = \hat{x}(0) \tag{2.100}$$

where the gain $K(t)$ is given by

$$K(t) = P(t|t)H^{\mathrm{T}}(t)[J(t)R(t)J^{\mathrm{T}}(t)]^{-1} \tag{2.101}$$

and $P(t|t)$ satisfies the following time-varying RDE:

$$\begin{aligned}\dot{P}(t|t) = {} & F(t)P(t|t) + P(t|t)F^{\mathrm{T}}(t) + G(t)Q(t)G^{\mathrm{T}}(t) \\ & - P(t|t)H^{\mathrm{T}}(t)[J(t)R(t)J^{\mathrm{T}}(t)]^{-1}H(t)P(t|t), \\ & P(0|0) = P(0)\end{aligned} \tag{2.102}$$

where $F(t)$ and $G(t)$ are the partials of $f(\cdot)$ with respect to $x(t)$ and $w(t)$, respectively, evaluated at $x(t) = \hat{x}(t|t)$ and $w(t) = 0$; and $H(t)$ and $J(t)$ are the partials of $h(\cdot)$ with respect to $x(t)$ and $v(t)$, respectively, evaluated at $x(t) = \hat{x}(t|t)$ and $v(t) = 0$.

Note that a filter linearized about a nominal value $x_n(t)$ is called a *linearized* Kalman filter, or *perturbation* Kalman filter. The extended Kalman filter discussed above is clearly a special case of the linearized Kalman filter. Other aspects on the extended Kalman filter are described by Gelb [22] and Maybeck [23]. Some estimators for general nonlinear models (e.g. conditional moment estimators, conditional mode estimators, and statistically linearized filter) can also be found in Maybeck [23].

2.6 STABILITY OF THE CONTINUOUS-TIME KALMAN FILTER

Now consider the continuous-time Kalman filter with $S = 0$, given by Theorem 2.11, for the time-invariant system. The Kalman filter becomes

$$\mathrm{d}\hat{x}(t|t)/\mathrm{d}t = A\hat{x}(t|t)\, G(t)u(t) + K(t)[z(t) - H\hat{x}(t|t)], \quad \hat{x}(t_0|t_0) = \hat{x}(0) \tag{2.103}$$

or equivalently

$$d\hat{x}(t|t)/dt = \tilde{A}(t)\hat{x}(t|t) + G(t)u(t) + K(t)z(t) \tag{2.104}$$

where $\tilde{A}(t)$ and $K(t)$ are the closed-loop (dynamic) system and filter gain matrices, respectively, which are subject to

$$\tilde{A}(t) = A - K(t)H \tag{2.105}$$

$$K(t) = P(t|t)H^{T}R^{-1} \tag{2.106}$$

$$\begin{aligned} dP(t|t)/dt = {} & AP(t|t) + P(t|t)A^{T} + BQB^{T} \\ & - P(t|t)H^{T}R^{-1}HP(t|t), \qquad P(t_0|t_0) = P(0) \end{aligned} \tag{2.107}$$

If $P(t|t)$ converges as $t \to \infty$, then the limiting solution P will satisfy the following continuous-time ARE, obtained from equation (2.107) by setting $dP(t|t)/dt = 0$ and $P(t|t) \triangleq P$:

$$AP + PA^{T} + BQB^{T} - PH^{T}R^{-1}HP = 0 \tag{2.108}$$

Applying the solution of this equation, we have the steady-state Kalman filter in continuous-time:

$$d\hat{x}(t|t)/dt = \tilde{A}\hat{x}(t|t) + G(t)u(t) + Kz(t) \tag{2.109}$$

where

$$\tilde{A} \triangleq A - KH \tag{2.110}$$

$$K \triangleq PH^{T}R^{-1} \tag{2.111}$$

As for the discrete-time problem, defining

$$H^{T}R^{-1}H = C^{T}C$$

$$BQB^{T} = DD^{T}$$

we have the continuous-time ARE:

$$AP + PA^{T} + DD^{T} - PC^{T}CP = 0 \tag{2.112}$$

Definition 2.4

A real symmetric positive semidefinite solution of the continuous-time ARE is said to be a *stabilizing* solution if all the eigenvalues of the corresponding filter closed-loop (dynamic) system matrix, $\tilde{A}$, lie inside the left half of the complex plane ($\mathrm{Re}\,\lambda_i(\tilde{A}) < 0$, $i = 1, \ldots, n$).

Definition 2.5

A real symmetric positive semidefinite solution of the continuous-time ARE is said to be a *strong* solution if all the eigenvalues of the corresponding filter closed-loop (dynamic) system matrix, $\widetilde{A}$, lie inside the left half of the complex plane or on the imaginary axis ($\operatorname{Re}\lambda_i(\widetilde{A}) \leq 0,\ i = 1, \ldots, n$).

To develop the stability theory in continuous-time systems, it is convenient to introduce the following $2n \times 2n$ Hamiltonian matrix:

$$M = \begin{bmatrix} -A^{\mathrm{T}} & C^{\mathrm{T}}C \\ DD^{\mathrm{T}} & A \end{bmatrix} \tag{2.113}$$

This matrix represents the *forward* Hamiltonian system matrix for the continuous-time filtering problem (see also Chapter 6),

$$\frac{\mathrm{d}}{\mathrm{d}t}\begin{bmatrix} \lambda(t) \\ x(t) \end{bmatrix} = M \begin{bmatrix} \lambda(t) \\ x(t) \end{bmatrix} \tag{2.114}$$

Corresponding to Definition 2.3, we get the following definition.

Definition 2.6

$M \in \mathbb{R}^{2n\times 2n}$ is Hamiltonian if $J^{-1}M^{\mathrm{T}}J = -M$.

From this definition, we can see that the associated *backward* matrix $-M$ is easily obtained by

$$-M = \begin{bmatrix} A^{\mathrm{T}} & -C^{\mathrm{T}}C \\ -DD^{\mathrm{T}} & -A \end{bmatrix} \tag{2.113$'$}$$

In addition, $-M$ is similar to M^{T}, and both $-M$ and M^{T} must have the same eigenvalues. Thus, we obtain the following lemma.

Lemma 2.2

Let $M \in \mathbb{R}^{2n\times 2n}$ be Hamiltonian. Then $\lambda \in \sigma(M)$ implies $-\lambda \in \sigma(M)$ with the same multiplicity.

Write λ_i for the eigenvalues of M and let ξ_i denote the eigenvector of M associated with λ_i, i.e.

$$M\xi_i = \lambda_i \xi_i, \qquad i = 1, \ldots, 2n$$

and let us decompose each eigenvector (or generalized eigenvector) ξ_i of M into two n-vectors

$$\xi_i = \begin{bmatrix} x_i \\ y_i \end{bmatrix}, \qquad i = 1, \ldots, 2n$$

Then we have the following theorem giving conditions for the stabilizing solution of the continuous-time ARE:

Theorem 2.12

(i) Each solution of equation (2.112) can be expressed as

$$P = YX^{-1}$$

where

$$X = [x_1, \ldots, x_n]$$

$$Y = [y_1, \ldots, y_n]$$

and where the inverse is assumed to exist for certain combinations of eigenvectors. Conversely, if X is nonsingular, then $P = YX^{-1}$ satisfies equation (2.112).

(ii) The matrix $-\widetilde{A}^{\mathrm{T}}$, where $\widetilde{A}$ is yielded by $P = YX^{-1}$, has eigenvalues λ_i associated with the eigenvectors $x_1, \ldots, x_n$.†

(iii) $\mathrm{Re}\,\lambda \neq 0$, for all eigenvalues λ of M, and detectability of (C, A) are necessary as well as sufficient for the existence of a stabilizing solution $P \geqslant 0$ to equation (2.112).

(iv) The stabilizing solution P is the only positive semidefinite solution of equation (2.112) if and only if (A, D) is stabilizable.

(v) Equation (2.112) possesses a unique stabilizing solution $P \geqslant 0$ if and only if (A, D) is stabilizable and (C, A) is detectable.

Proof

See Mårtensson [24] and Kučera [25, 26]. □

† If the backward Hamiltonian matrix $-M$ is used, then $-\widetilde{A}^{\mathrm{T}}$ is replaced by $\widetilde{A}^{\mathrm{T}}$ in this condition.

We summarize many properties of the strong solution to equation (2.112) in the following theorem:

Theorem 2.13

Provided that (C, A) is detectable:

(i) The strong solution of the continuous-time ARE exists and is unique.
(ii) If (A, D) is stabilizable, the strong solution is the only positive semidefinite solution of the continuous-time ARE.
(iii) If (A, D) has no uncontrollable modes on the imaginary axis, the strong solution coincides with the stabilizing solution.
(iv) If (A, D) has an uncontrollable mode on the imaginary axis, then, although the strong solution exists, there is no stabilizing solution.
(v) If (A, D) has an uncontrollable mode lying in the left half of the complex plane, or on the imaginary axis, the strong solution is not positive definite.
(vi) If (A, D) has an uncontrollable mode lying in the right half of the complex plane, as well as the strong solution, there is at least one other positive semidefinite solution of the continuous-time ARE.

Proof

The theorem can be proved using a continuous-time version of the results of Chan *et al.* [8, Theorem 3.1]. □

Our next results deal with the convergence of the solution of the RDE to the stabilizing solution of the continuous-time ARE.

Theorem 2.14

Subject to the following conditions:

(i) (A, D) is stabilizable,
(ii) (C, A) is detectable,
(iii) $P(0) \geqslant 0$,

then the limit

$$\lim_{t\to\infty} P(t) = P_s$$

is reached exponentially fast; where $P(t)$ is the solution of the RDE with initial condition $P(0)$ and P_s is the unique stabilizing solution of the continuous-time ARE.

Proof

See Kučera [26]. □

Theorem 2.15

Subject to the following conditions:

(i) there are no uncontrollable modes of (A, D) on the imaginary axis,
(ii) (C, A) is detectable,
(iii) $P(0) > 0$,

then the limit

$$\lim_{t\to\infty} P(t) = P_s$$

is reached exponentially fast; where $P(t)$ and P_s are as in Theorem 2.14.

Proof

The theorem may be proved using a continuous-time version of the results of Chan *et al.* [8, Theorem 4.2]. □

Moreover, under certain circumstances the solution of the RDE converges to the strong solution of the continuous-time ARE.

Theorem 2.16

Subject to the following conditions:

(i) (C, A) is observable,
(ii) $(P(0) - P_s) > 0$ or $P(0) = P_s$

then

$$\lim_{t \to \infty} P(t) = P_s$$

where $P(t)$ is as in Theorem 2.14, but P_s is the (unique) strong solution of the continuous-time ARE.

Proof

This theorem can also be proved using a continuous-time version of the results of Chan *et al.* [8, Theorem 4.3]. □

Theorems 2.14–2.16 (or Theorems 2.5–2.7) give sufficient conditions for the RDE to converge to the stabilizing or strong solution. Hence, they also provide sufficient conditions for the stability and time-invariance of the Kalman filter.

2.7 COMPUTATIONAL APPROACHES TO THE CONTINUOUS-TIME RICCATI EQUATION

Practical implementation of the continuous-time Kalman filter requires that we devote substantial attention to the question of efficient computation of the RDE, or of the positive semidefinite, stabilizing or strong solution of the continuous-time ARE. In this section, we briefly review various techniques for finding the solution to equations (2.107) and (2.108). It is very difficult to say that one method is superior to another, since a given method may prove better in one application but fail in another. Our aim is to bring the most important methods to the reader's attention and indicate their ranges of applicability, strengths, and weaknesses.

In Sections 2.7.1–2.7.4 we consider four approaches used to obtain the nonsteady-state solution of the RDE (equation (2.107)). Then, in Sections 2.7.5–2.7.8, we consider four approaches used to obtain the numerical solution of the continuous-time ARE (equation (2.108) or (2.112)).

2.7.1 Numerical Integration

The most natural approach to the solution of (2.107) is by direct numerical integration using, for example, the fourth-order Runge–Kutta or an Adams–Bashforth scheme. Clearly, this method can also be applied to compute a desired steady-state solution. However, for this purpose the scheme is usually time-consuming and not sufficiently accurate.

2.7.2 Hamiltonian Matrix Method (or Transition Matrix Method)

This procedure is important from the theoretical point of view since it enables us to obtain an explicit representation of the solution of equation (1.107) even in the time-varying case. In some exceptional cases, the solution even admits an explicit closed-form (or analytic) solution.

Theorem 2.17

The solution of equation (2.107) can be obtained by the following steps:

Step 1: Set up the Hamiltonian matrix

$$M = \begin{bmatrix} -A^{\mathrm{T}} & H^{\mathrm{T}}R^{-1}H \\ BQB^{\mathrm{T}} & A \end{bmatrix}$$

Step 2: Compute the $2n \times 2n$ transition matrix $\Phi(t, t_0)$:

$$\mathrm{d}\Phi(t, t_0)/\mathrm{d}t = M\Phi(t, t_0), \qquad \Phi(t, t_0) = I$$

Step 3: Decompose $\Phi(t, t_0)$ into four submatrices $\Phi_{ij}(t, t_0)$, $i, j = 1, 2$:

$$\Phi(t, t_0) = \begin{bmatrix} \Phi_{11}(t, t_0) & \Phi_{12}(t, t_0) \\ \Phi_{21}(t, t_0) & \Phi_{22}(t, t_0) \end{bmatrix}$$

Step 4: Compute $P(t|t)$ as follows:

$$P(t|t) = [\Phi_{21}(t, t_0) + \Phi_{22}(t, t_0)P(0)] \times [\Phi_{11}(t, t_0) + \Phi_{12}(t, t_0)P(0)]^{-1}$$

Proof

The proof of the theorem is mainly attributed to Kalman and Bucy [2]. Let $X(t)$, $Y(t)$ be the solution matrix pair for equation (2.114), i.e.

$$\frac{d}{dt}\begin{bmatrix} X(t) \\ Y(t) \end{bmatrix} = M\begin{bmatrix} X(t) \\ Y(t) \end{bmatrix}, \qquad \begin{bmatrix} X(t_0) \\ Y(t_0) \end{bmatrix} = \begin{bmatrix} I \\ P(0) \end{bmatrix} \tag{2.115}$$

Then we have the following identity

$$Y(t) = P(t|t)X(t), \qquad t \geqslant t_0 \tag{2.116}$$

which is readily verified by differentiating equation (2.116) and using (2.107) and (2.115). On the other hand, from the fact that

$$\begin{bmatrix} X(t) \\ Y(t) \end{bmatrix} = \begin{bmatrix} \Phi_{11}(t, t_0) & \Phi_{12}(t, t_0) \\ \Phi_{21}(t, t_0) & \Phi_{22}(t, t_0) \end{bmatrix}\begin{bmatrix} X(t_0) \\ Y(t_0) \end{bmatrix} \tag{2.117}$$

we can see that

$$X(t) \equiv \Phi_{11}(t, t_0) + \Phi_{12}(t, t_0)P(0) \tag{2.118}$$

$$Y(t) \equiv \Phi_{21}(t, t_0) + \Phi_{22}(t, t_0)P(0) \tag{2.119}$$

because of the initial conditions of equation (2.115). □

Example 2.6 [27]

Consider the following scalar system:

$$\dot{x}(t) = -x(t) + \sqrt{2}w(t)$$

$$z(t) = x(t) + v(t)$$

where $w(t)$ and $v(t)$ are zero-mean Gaussian white sequences with variances $Q = 1$ and $R = 1$, and $w(t)$ is independent of $v(t)$. The differential equations for $X(t)$ and $Y(t)$ are then

$$\dot{X}(t) = X(t) + Y(t), \qquad X(0) = 1$$

$$\dot{Y}(t) = -Y(t) + 2X(t), \qquad Y(0) = P(0)$$

These equations can be solved readily using Laplace transform techniques. The result is

$$X(t) = \cosh\sqrt{3}t + \frac{(P(0) + 1)}{\sqrt{3}}\sinh\sqrt{3}t$$

$$Y(t) = P(0)\cosh\sqrt{3}t + \frac{(2 - P(0))}{\sqrt{3}}\sinh\sqrt{3}t$$

The solution for $P(t|t)$ is now given by $P(t|t) = Y(t)X^{-1}(t)$:

$$P(t|t) = \frac{P(0)\cosh\sqrt{3}t + \dfrac{(2 - P(0))}{\sqrt{3}}\sinh\sqrt{3}t}{\cosh\sqrt{3}t + \dfrac{(P(0)+1)}{\sqrt{3}}\sinh\sqrt{3}t}$$

Letting $t \to \infty$ and setting $P(0) = 0$, since the steady-state solution is independent of $P(0)$, we have the steady-state solution

$$P(t|t) \to \frac{P(0)\exp(\sqrt{3}t) + \dfrac{(2 - P(0))}{\sqrt{3}}\exp(\sqrt{3}t)}{\exp(\sqrt{3}t) + \dfrac{(P(0)+1)}{\sqrt{3}}\exp(\sqrt{3}t)} = \sqrt{3} - 1$$

This method has the problem that $\Phi(t, t_0)$ contains stable as well as unstable modes. As time proceeds, the unstable modes tend to dominate and solution accuracy suffers. The next method bypasses this difficulty.

2.7.3 Negative Exponential Method

The basic trick of this technique is to recast the equations so that only negative exponentials occur in the computations. The algorithm is summarized in the following theorem.

Theorem 2.18

It is assumed for simplicity that the eigenvalues of M are distinct. Then, we obtain $P(t|t)$ through the following steps:

Step 1: Formulate the Hamiltonian matrix by equation (2.113).
Step 2: Compute the $2n$ eigenvalues of M, $\lambda_i(M)$, $i = 1, \ldots, 2n$.
Step 3: Compute the n $2n$-dimensional eigenvectors ξ_i corresponding to $\operatorname{Re}\lambda_i(M) > 0$, $i = 1, \ldots, n$ and further compute the n eigenvectors ξ_i corresponding to $\operatorname{Re}\lambda_i(M) < 0$, $i = n+1, \ldots, 2n$.
Step 4: Partition the eigenvectors as follows:

$$[\xi_1, \ldots, \xi_n, \xi_{n+1}, \ldots, \xi_{2n}] = \begin{bmatrix} T_{11} & T_{12} \\ T_{21} & T_{22} \end{bmatrix} \triangleq T$$

Step 5: Compute the following:

$$L = -[T_{22} - P(0)T_{12}]^{-1}[T_{21} - P(0)T_{11}]$$

$$\Lambda_1 = \begin{bmatrix} \lambda_1 & & \\ & \ddots & \\ & & \lambda_n \end{bmatrix}, \qquad \mathrm{Re}\,\lambda_i > 0,\ i = 1, \ldots, n$$

$$G(t) = \exp(-\Lambda_1 t) L \exp(-\Lambda_1 t)$$

Step 6: Compute $P(t|t)$ via:

$$P(t|t) = [T_{21} + T_{22}G(t)][T_{11} + T_{12}G(t)]^{-1}$$

Proof

This proof is based on Vaughan [28]. Let Λ be a $2n \times 2n$ diagonal matrix of the eigenvalues of M arranged so that Λ_1 is an $n \times n$ diagonal matrix of the eigenvalues lying in the right half of the complex plane and

$$\Lambda = \begin{bmatrix} \Lambda_1 & 0 \\ 0 & -\Lambda_1 \end{bmatrix} \tag{2.120}$$

Then there is a nonsingular eigenvector matrix T, such that

$$M = T\Lambda T^{-1} \tag{2.121}$$

Therefore, equation (2.115) can be rewritten as follows:

$$\frac{\mathrm{d}}{\mathrm{d}t}\begin{bmatrix} y(t) \\ m(t) \end{bmatrix} = \Lambda \begin{bmatrix} y(t) \\ m(t) \end{bmatrix} \tag{2.122}$$

where $y(t)$ and $m(t)$ are a new $n \times n$ matrix solution satisfying

$$\begin{bmatrix} X(t) \\ Y(t) \end{bmatrix} = \begin{bmatrix} T_{11} & T_{12} \\ T_{21} & T_{22} \end{bmatrix} \begin{bmatrix} y(t) \\ m(t) \end{bmatrix} \tag{2.123}$$

Then the solution of equation (2.122) becomes

$$\begin{bmatrix} y(t) \\ m(t) \end{bmatrix} = \begin{bmatrix} \exp(\Lambda_1 t) & 0 \\ 0 & \exp(-\Lambda_1 t) \end{bmatrix} \begin{bmatrix} y(t_0) \\ m(t_0) \end{bmatrix} \tag{2.124}$$

or

$$\begin{bmatrix} y(t_0) \\ m(t) \end{bmatrix} = \begin{bmatrix} \exp(-\Lambda_1 t) & 0 \\ 0 & \exp(-\Lambda_1 t) \end{bmatrix} \begin{bmatrix} y(t) \\ m(t_0) \end{bmatrix} \tag{2.125}$$

Using equation (2.123), the initial condition at $t = t_0$ of equation (2.115) can be written in terms of $m(t_0)$ and $y(t_0)$ as follows:

$$m(t_0) = Ly(t_0) \tag{2.126}$$

where L is defined by

$$L = -[T_{22} - P(0)T_{12}]^{-1}[T_{21} - P(0)T_{11}] \tag{2.127}$$

Further, using equations (2.125) and (2.126), the relationship between $m(t)$ and $y(t)$ can be written in the form

$$m(t) = G(t)y(t) \tag{2.128}$$

where

$$G(t) = \exp(-\Lambda_1 t) L \exp(-\Lambda_1 t) \tag{2.129}$$

Finally, using equations (2.123) and (2.128), the desired relationship between $Y(t)$ and $X(t)$ can be written in the form

$$Y(t) = P(t|t)X(t), \qquad t \geqslant t_0 \tag{2.130}$$

where

$$Y(t) = [T_{21} + T_{22}G(t)] \tag{2.131}$$

$$X(t) = [T_{11} + T_{12}G(t)] \tag{2.132}$$

This completes the proof of the theorem. □

2.7.4 Chandrasekhar-type Algorithm

This approach will be discussed in detail in Chapter 6. Hence, its presentation is omitted here.

2.7.5 Eigenvalue–Eigenvector Method

As in the discrete-time case, we summarize the approach in the form of a theorem.

Theorem 2.19

It is assumed for simplicity that the eigenvalues of M are distinct. Then we proceed as follows:

Step 1: Formulate the Hamiltonian matrix as in equation (2.113).
Step 2: Compute the $2n$ eigenvalues of M, $\lambda_i(M)$, $i = 1, \ldots, 2n$.
Step 3: Compute the n $2n$-dimensional eigenvectors ξ_i which satisfy

$\operatorname{Re}\lambda_i(M) > 0,\ i = 1, \ldots, n.$†

Step 4: Partition the eigenvectors matrix as follows:

$$[\xi_1, \ldots, \xi_n] = \begin{bmatrix} X \\ Y \end{bmatrix}$$

Step 5: Solve for P:

$$P = YX^{-1}$$

Proof

In equations (2.131) and (2.132), $\lim_{t\to\infty} G(t) = 0$ because Λ_1 is a diagonal matrix consisting of unstable eigenvalues. Hence, the algorithm of Theorem 2.18 in steady-state is identical to the present algorithm. □

An interesting side-aspect of this method is that it is the only procedure available for computing *all* the solutions of equation (2.108), regardless of whether it is positive semidefinite or not.

2.7.6 Kleinman's Iterative Method

This approach is due to Kleinman [29]. We assume that equation (2.108) possesses a unique stabilizing solution P_s.

Theorem 2.20

Let P_i, $i = 0, 1, \ldots$ be the unique positive semidefinite solution of the linear algebraic equation

$$\hat{A}_i P_i + P_i \hat{A}_i^{\mathrm{T}} = -(L_i R L_i^{\mathrm{T}} + BQB^{\mathrm{T}}) \tag{2.133}$$

where

$$\hat{A}_i = A - L_i H, \qquad i = 0, 1, \ldots \tag{2.134}$$

$$L_i = P_{i-1} H^{\mathrm{T}} R^{-1}, \qquad i = 1, 2, \ldots \tag{2.135}$$

and L_0 is chosen such that the matrix $\hat{A}_0 = A - L_0 H$ is a stable

† For the eigenvalues of $-M$, we must take the eigenvectors which satisfy $\operatorname{Re}\lambda_i(M) < 0$.

matrix, i.e. all the eigenvalues of $\hat{A}_0$ have negative real part. Then

$$P_s \leqslant P_{i+1} \leqslant P_i < \ldots, \qquad i = 0, 1, \ldots$$

and

$$\lim_{i\to\infty} P_i = P_s \tag{2.136}$$

Proof

We first derive an equation for the algorithm. Equation (2.108) can be rewritten as follows:

$$(A - KH)P + P(A - KH)^T + KHP + PH^TK^T + BQB^T - KRK^T = 0 \tag{2.137}$$

where K is defined as in equation (2.111). This equation may also be written as

$$(A - KH)P + P(A - KH)^T = -(KRK^T + BQB^T) \tag{2.138}$$

Regarding P and K in equation (2.138) as the steady-state values P_i and L_i at the ith iteration we obtain equations (2.133) and (2.134). Further, we obtain (2.135) by regarding P in equation (2.111) as the steady-state value P_{i-1} at the $(i-1)$th iteration. Indeed, L_0 being stable, equation (2.135) has a unique positive semidefinite solution P_0. This yields an L_1 and, in turn, a P_1. A little manipulation with the associated minimum variance costs reveals that $P_s \leqslant P_1 \leqslant P_0$, P_1 being bounded, $\hat{A}_1$ is a stable matrix, etc. A theorem on monotonic convergence of positive semidefinite matrices generates the existence of a limit. Then equation (2.135) is identical to equation (2.108) when $i \to \infty$ and equation (2.136) holds by uniqueness of P_s. □

2.7.7 Matrix Sign Function Method

The matrix sign function method is described by Roberts [30] and may be viewed as a simplification of the eigenvalue–eigenvector method (Section 2.7.5), designed to find the stabilizing solution P_s.

Definition 2.7

Let T be the nonsingular transformation matrix of a matrix $A \in \mathbb{R}^{n\times n}$ and

$$D = T^{-1}AT = \text{block diag}\,[D_+,\ D_-]$$

where $D_+ \in \mathbb{R}^{n_1 \times n_1}$ and $D_- \in \mathbb{R}^{n_2 \times n_2}$, with $n_1 + n_2 = n$, are square matrices which have eigenvalues $\operatorname{Re}\lambda_i(D_+) > 0$, $i = 1, \ldots, n_1$ and $\operatorname{Re}\lambda_j(D_-) < 0$, $j = 1, \ldots, n_2$, respectively.† Then

$$\begin{aligned} \text{sign}\,(A) &= T\begin{bmatrix} \text{sign}\,(D_+) & 0 \\ 0 & \text{sign}\,(D_-) \end{bmatrix} T^{-1} \\ &= T\begin{bmatrix} I & 0 \\ 0 & -I \end{bmatrix} T^{-1} \end{aligned}$$

The matrix sign function sign (A) can also be defined as

$$\begin{aligned} \text{sign}\,(A) &= 2\,\text{sign}^+\,(A) - I \\ &= I - 2\,\text{sign}^-\,(A) \end{aligned}$$

where the sign-plus and sign-minus functions can be written as

$$\begin{aligned} \text{sign}^+\,(A) &= \tfrac{1}{2}[\text{sign}\,(A) + I] \\ &= T\begin{bmatrix} I & 0 \\ 0 & 0 \end{bmatrix} T^{-1} \\ \text{sign}^-\,(A) &= \tfrac{1}{2}[I - \text{sign}\,(A)] \\ &= \text{sign}^+\,(A) - \text{sign}\,(A) \\ &= I - \text{sign}^+\,(A) \\ &= T\begin{bmatrix} 0 & 0 \\ 0 & I \end{bmatrix} T^{-1} \end{aligned}$$

Theorem 2.21

The sign function method algorithm is summarized as follows:

Step 1: Find the matrix sign (M) such that

$$\text{sign}\,(M) = \lim_{i \to \infty} M_{i+1}$$

where

$$\begin{aligned} M_{i+1} &= \tfrac{1}{2}(M_i + M_i^{-1}), \qquad i = 0, 1, \ldots \\ M_0 &= M \end{aligned}$$

† If T can be taken as the modal matrix of A, then D_+ and D_- become the collections of Jordan blocks associated with the eigenvalues $\operatorname{Re}\lambda_i(A) > 0$, $i = 1, \ldots, n_1$ and $\operatorname{Re}\lambda_j(A) < 0$, $j = 1, \ldots, n_2$.

Step 2: Obtain $\text{sign}^+(M)$ as

$$\text{sign}^+ M = \tfrac{1}{2}(I + \text{sign}\, M)$$

Step 3: From $\text{sign}^+(M)$, extract the right two blocks V and PV (or the left two blocks $I - VP$ and $P(I - VP)$) where

$$\text{sign}^+(M) = \begin{bmatrix} I - VP & V \\ P(I - VP) & PV \end{bmatrix}$$

Step 4: Invert V to solve P_s from PV (or invert $(I - VP)$ to solve P_s from $P(I - VP)$), if V (or $(I - VP)$) is invertible.

Proof

The Hamiltonian matrix given by equation (2.113) can be block-triangularized as

$$\begin{bmatrix} -A^{\mathrm{T}} & C^{\mathrm{T}}C \\ DD^{\mathrm{T}} & A \end{bmatrix} = \begin{bmatrix} I & 0 \\ P & I \end{bmatrix} \begin{bmatrix} -\tilde{A}^{\mathrm{T}} & C^{\mathrm{T}}C \\ 0 & \tilde{A} \end{bmatrix} \begin{bmatrix} I & 0 \\ P & I \end{bmatrix}^{-1} \tag{2.139}$$

where $\tilde{A}$ has already been defined in equation (2.110), and $\tilde{A}$ is assumed to be asymptotically stable. If we now define a matrix V by

$$\tilde{A}^{\mathrm{T}} V + V\tilde{A} + C^{\mathrm{T}}C = 0 \tag{2.140}$$

then the central factor of equation (2.139) can be block-diagonalized further as follows:

$$\begin{bmatrix} -\tilde{A}^{\mathrm{T}} & C^{\mathrm{T}}C \\ 0 & \tilde{A} \end{bmatrix} = \begin{bmatrix} I & -V \\ 0 & I \end{bmatrix} \begin{bmatrix} -\tilde{A}^{\mathrm{T}} & 0 \\ 0 & \tilde{A} \end{bmatrix} \begin{bmatrix} I & -V \\ 0 & I \end{bmatrix}^{-1} \tag{2.141}$$

Combining equations (2.139) and (2.141) yields†

$$\begin{bmatrix} -\tilde{A}^{\mathrm{T}} & C^{\mathrm{T}}C \\ DD^{\mathrm{T}} & A \end{bmatrix} = \begin{bmatrix} I & -V \\ P & I - PV \end{bmatrix} \begin{bmatrix} -\tilde{A}^{\mathrm{T}} & 0 \\ 0 & \tilde{A} \end{bmatrix} \begin{bmatrix} I - VP & V \\ -P & I \end{bmatrix} \tag{2.142}$$

† See Appendix A for the inversion of block matrices.

Again noting that $\widetilde{A}$ is stable, we obtain

$$\text{sign}^+(M) = \begin{bmatrix} I & -V \\ P & I - PV \end{bmatrix} \begin{bmatrix} I & 0 \\ 0 & 0 \end{bmatrix} \begin{bmatrix} I - VP & V \\ -P & I \end{bmatrix}$$

$$= \begin{bmatrix} I - VP & V \\ P(I - VP) & PV \end{bmatrix} = \begin{bmatrix} I \\ P \end{bmatrix} [(I - VP) \quad V] \tag{2.143}$$

This immediately gives a solution for $P = P_s$ as the quotient of two submatrices if V or $(I - VP)$ are nonsingular. □

If we do not assume the nonsingularity of V (or $(I - VP)$), we can still show that $W^T \triangleq [(I - VP) \quad V]$ is of full rank since

$$W^T = \begin{bmatrix} I \\ P \end{bmatrix} = I$$

It follows that $W^T W$ is also of full rank, and therefore the solution can always be obtained by defining

$$\text{sign}^+(M) \triangleq \begin{bmatrix} U_1 & U_2 \\ U_3 & U_4 \end{bmatrix} \equiv \begin{bmatrix} W^T \\ Z^T \end{bmatrix} \begin{bmatrix} I \\ P \end{bmatrix} W^T$$

and using

$$P = Z^T W[W^T W]^{-1}$$
$$= [U_3 U_1 + U_4 U_2][U_1^2 + U_2^2]^{-1}$$

A fast algorithm to compute sign(A) can also be found in Balzer [31] and Shieh *et al.* [32].

2.7.8 Doubling Algorithm

This algorithm will be presented, together with the result of the discrete-time case, in Chapter 5.

2.8 SUMMARY

This chapter has reviewed the results of Kalman filtering for both discrete-time and continuous-time system models. The associated stability properties were also discussed. Several computational methods, useful in solving the discrete-time and continuous-time algebraic Riccati equations, were then presented.

It was assumed, throughout this chapter, that all parameter matrices are completely known. However, when they are not known, adaptive features are called for. In the following chapter such cases will be discussed in detail, using the partitioned adaptive technique.

EXERCISES

2.1 Show that the filtering error covariance matrix $P(k+1|k+1)$ can be computed from the following equation:

$$\begin{aligned} P(k+1|k+1) = {} & [I - K(k+1)H(k+1)]P(k+1|k) \\ & \times [I - K(k+1)H(k+1)]^{\mathrm{T}} \\ & + K(k+1)R(k+1)K^{\mathrm{T}}(k+1) \end{aligned}$$

This is called the *Joseph form* (or *stabilized form*) of covariance update, which is much less sensitive to numerical errors from the prior calculation of the gain matrix $K(k+1)$ than is the standard form given by equation (2.22).

2.2 Verify equations (2.25)–(2.28).

2.3 Show that the gain matrix $K(k+1)$ given in equation (2.20) can also be represented by the following equation:

$$K(k+1) = P(k+1|k+1)H^{\mathrm{T}}(k+1)R^{-1}(k+1)$$

2.4 (Information Filter) Let us consider the discrete-time Kalman filter (predictive type), with $S = 0$, given by Theorem 2.1. Show that the definitions

$$d(k|k-1) \triangleq P^{-1}(k|k-1)\hat{x}(k|k-1)$$

$$d(k|k) \triangleq P^{-1}(k|k)\hat{x}(k|k)$$

can also be written as follows:

$$d(k|k) = d(k|k-1) + H(k)R^{-1}(k)z(k), \qquad d(0|-1) = P^{-1}(0)\hat{x}(0)$$

$$\begin{aligned} d(k+1|k) = {} & [I - L(k)B^{\mathrm{T}}(k)]\Phi^{-\mathrm{T}}(k+1, k)[d(k|k) \\ & + P^{-1}(k|k)\Phi^{-1}(k+1, k)G(k)u(k)] \end{aligned}$$

where

$$L(k) \triangleq F(k)B(k)[B^{\mathrm{T}}(k)F(k)B(k) + Q^{-1}(k)]^{-1}$$

$$F(k) \triangleq \Phi^{-\mathrm{T}}(k+1, k)P^{-1}(k|k)\Phi^{-1}(k+1, k)$$

Here, the information matrices $P^{-1}(k|k-1)$ and $P^{-1}(k|k)$ satisfy the following recursions:

$$P^{-1}(k|k) = P^{-1}(k|k-1) + H^{\mathrm{T}}(k)R^{-1}(k)H(k),$$

$$P^{-1}(0|-1) = P^{-1}(0)$$

$$P^{-1}(k+1|k) = [I - L(k)B^{\mathrm{T}}(k)]F(k)$$

These constitute the information filter in discrete time.

2.5 Extend the results of Exercise 2.4 to the case where system and measurement noise processes are correlated, i.e. $S \neq 0$.

2.6 Show that in Exercise 2.4 the Joseph form of $P^{-1}(k+1|k)$ is represented by

$$P^{-1}(k+1|k) = [I - L(k)B^{\mathrm{T}}(k)]F(k)[I - L(k)B^{\mathrm{T}}(k)]^{\mathrm{T}} + L(k)Q^{-1}(k)L^{\mathrm{T}}(k)$$

2.7 (Fisher Information Matrix) The so-called Fisher information matrix in discrete time is defined by

$$\mathcal{J}(k, 1) = \sum_{i=1}^{k} \Phi^{\mathrm{T}}(i, k)H^{\mathrm{T}}(i)R^{-1}(i)H(i)\Phi(i, k)$$

where $\Phi(i, k)$, for $i < k$, is the transition matrix for propagating the system state backwards in time: $\Phi(i, k) = \Phi^{-1}(k, i)$.

(a) Ignoring the system noise: supposing that there is no $w(k)$ sequence, or equivalently, that $Q(k) \equiv 0$ for all k, show that

$$\mathcal{J}(k, 0) = P^{-1}(k|k) - \Phi^{\mathrm{T}}(0, k)P^{-1}(0)\Phi(0, k).$$

This means that if there is no a priori information about the state, or formally if $P^{-1}(0) = 0$, then the information matrix is the inverse of the corresponding filter error covariance.

(b) Show that $\mathcal{J}(k, 0)$ satisfies the following recursive relation:

$$\mathcal{J}(k+1, 0) = \Phi^{\mathrm{T}}(k, k+1)\mathcal{J}(k, 0)\Phi(k, k+1) + H^{\mathrm{T}}(k+1)R^{-1}(k+1)H(k+1)$$

2.8 Consider the formal backward-time model of equations (2.1) and (2.2) with $S \equiv 0$, which is given by

$$x(k) = \Phi^{-1}(k+1, k)x(k+1) - \Phi^{-1}(k+1, k)G(k)u(k) - \Phi^{-1}(k+1, k)B(k)w(k)$$

$$z(k) = H(k)x(k) + v(k)$$

where Φ is assumed to be nonsingular. It is also assumed that the information on the final state $x(N)$ is given by

$$x(N) \sim N(\hat{x}(N), P(N))$$

(a) Derive a *formal* backward-time Kalman filter for the model above:

$$\hat{x}_b(k|k) = \hat{x}_b(k|k+1) + K_b(k) \times [z(k) - H(k)\hat{x}_b(k|k+1)],$$

$$\hat{x}_b(N|N+1) = \hat{x}(N)$$

$$K_b(k) = P_b(k|k+1)H^T(k) \times [H(k)P_b(k|k+1)H^T(k) + R(k)]^{-1}$$

$$P_b(N|N+1) = P(N)$$

$$P_b(k|k) = [I - K_b(k)H(k)]P_b(k|k+1)$$

$$\hat{x}_b(k-1|k) = \Phi^{-1}(k, k-1)\hat{x}_b(k|k) - \Phi^{-1}(k, k-1)G(k-1)u(k-1)$$

$$P_b(k-1|k) = \Phi^{-1}(k, k-1)P_b(k|k)\Phi^{-T}(k, k-1) + \Phi^{-1}(k, k-1)B(k-1)Q(k-1) \times B^T(k-1)\Phi^{-T}(k, k-1)$$

(b) Show that, if we define

$$d_b(k|k+1) \triangleq P_b^{-1}(k|k+1)\hat{x}_b(k|k+1)$$

$$d_b(k|k) \triangleq P_b(k|k)\hat{x}_b(k|k)$$

it follows that

$$d_b(k|k) = d_b(k|k-1) + H(k)R^{-1}(k)z(k),$$

$$d_b(N|N+1) = P^{-1}(N)\hat{x}(N)$$

$$d_b(k-1|k) = \Phi^T(k, k-1) \times [I - L_b(k)B^T(k-1)][d_b(k|k) - P_b^{-1}(k|k)G(k-1)u(k-1)]$$

where

$$L_b(k) \triangleq P_b^{-1}(k|k)B(k-1)[B^T(k-1)P_b^{-1}(k|k)B(k-1) + Q^{-1}(k-1)]^{-1}$$

$$P_b^{-1}(k|k) = P_b^{-1}(k|k+1) + H^T(k)R^{-1}(k)H(k),$$

$$P_b^{-1}(N|N+1) = P^{-1}(N)$$

$$P_b^{-1}(k-1|k) = \Phi^T(k, k-1)[I - L_b(k)B^T(k-1)] \times P_b^{-1}(k|k)\Phi(k, k-1)$$

where it is assumed that $Q(k) > 0$. These constitute the formal backward-time information filter in discrete time.

(c) For the case when $Q(k) \geqslant 0$, show that $L_b(k)$ in part (b) can be rewritten as follows:

$$L_b(k) = P_b^{-1}(k|k)D(k-1)[D^T(k-1) \times P_b^{-1}(k|k)D(k-1) + I]^{-1}Q^{T/2}(k-1)$$

where $D(k)D^T(k) \triangleq B(k)Q(k)B^T(k)$, $Q^{T/2}(k) = (Q^{1/2}(k))^T$, $Q^{1/2}(k)$ is the square root of $Q(k)$, and hence $D(k) = B(k)Q^{1/2}(k)$.

2.9 (Linear Least-Squares (LS) Estimation) Let the unknown constant vector x be obtained from the measurements

$$z(i) = H(i)x + v(i), \qquad i = 1, \ldots, k$$

Define the following quadratic error

$$J(x, k) = \sum_{i=1}^{k} [z(i) - H(i)x]^T R^{-1}(i)[z(i) - H(i)x]$$
$$= [Z_k - H_k x]^T R_k^{-1} [Z_k - H_k x]$$

where

$$Z_k \triangleq \begin{bmatrix} z(1) \\ \vdots \\ z(k) \end{bmatrix} = H_k x + v_k$$

$$H_k \triangleq \begin{bmatrix} H(1) \\ \vdots \\ H(k) \end{bmatrix}, \qquad v_k \triangleq \begin{bmatrix} v(1) \\ \vdots \\ v(k) \end{bmatrix}$$

$$R_k \triangleq \begin{bmatrix} R(1) & \cdots & 0 \\ \vdots & \ddots & \vdots \\ 0 & \cdots & R(k) \end{bmatrix} > 0$$

Note that the least-squares (LS) criterion above implicitly assumes that the $v(i)$ are independent zero-mean processes with covariance $R(i)$. Thus, the problem is regarded as *weighted* least-squares (WLS) estimation with a positive definite weighting matrix R_k^{-1}.

(a) Show that performing the following minimization

$$\left. \frac{\partial J(x, k)}{\partial x} \right|_{x = \hat{x}(k)} = 0$$

yields

$$\hat{x}(k) = [H_k^T R_k^{-1} H_k]^{-1} H_k^T R_k^{-1} Z_k$$

which is the batch form of the LS estimator.

(b) Show that

$$P(k) \triangleq E[\tilde{x}(k)\tilde{x}^T(k)]$$
$$= [H_k^T R_k^{-1} H_k]^{-1}$$

where $\tilde{x}(k) = x - \hat{x}(k)$.

(c) Show that

$$\hat{x}(k+1) = \hat{x}(k) + K(k+1)[z(k+1) - H(k+1)\hat{x}(k)]$$
$$P(k+1) = [I - K(k+1)H(k+1)]P(k)$$

which is the recursive form of the LS estimator, where

$$K(k+1) \triangleq P(k)H^{\mathrm{T}}(k+1)V^{-1}(k+1)$$
$$\equiv P(k+1)H^{\mathrm{T}}(k+1)R^{-1}(k+1)$$
$$V(k+1) \triangleq H(k+1)P(k)H^{\mathrm{T}}(k+1) + R(k+1)$$

Hint: Use the following relationships:

$$Z_{k+1} = \begin{bmatrix} Z_k \\ z(k+1) \end{bmatrix}, \; H_{k+1} = \begin{bmatrix} H_k \\ H(k+1) \end{bmatrix}$$

$$v_{k+1} = \begin{bmatrix} v_k \\ v(k+1) \end{bmatrix}, \; R_{k+1} = \begin{bmatrix} R_k & 0 \\ 0 & R(k+1) \end{bmatrix}$$

(d) Verify that when defining

$$\nu(k+1) \triangleq z(k+1) - H(k+1)\hat{x}(k)$$

it follows that

$$K(k+1) = E[x\nu^{\mathrm{T}}(k+1)]\{E[\nu(k+1)\nu^{\mathrm{T}}(k+1)]\}^{-1}$$

2.10 Let us consider the *fixed-point smoothing* problem, defining

$$\hat{x}(k_*|k) \triangleq E[x(k_*)|Z_k]$$
$$P(k_*|k) \triangleq E\{[x(k_*) - \hat{x}(k_*|k)][x(k_*) - \hat{x}(k_*|k)]^{\mathrm{T}}|Z_k\}$$

where $k_* \leqslant k$ is fixed. The resulting algorithm is summarized as follows:

$$\hat{x}(k_*|k) = \hat{x}(k_*|k-1) + K(k_*|k)[z(k) - H(k)\hat{x}(k|k-1)],$$
$$\hat{x}(k_*|k-1)|_{k=k_*} = \hat{x}(k_*|k_* - 1)$$
$$K(k_*|k) = \Sigma(k|k-1)H^{\mathrm{T}}(k)[H(k)P(k|k-1)H^{\mathrm{T}}(k) + R(k)]^{-1}$$
$$\Sigma(k+1|k) = \Sigma(k|k-1)[\Phi(k+1, k) - K^*(k)H(k)]^{\mathrm{T}},$$
$$\Sigma(k_*|k_* - 1) = P(k_*|k_* - 1)$$
$$P(k_*|k) = P(k_*|k-1) - \Sigma(k|k-1)H^{\mathrm{T}}(k)K^{\mathrm{T}}(k_*|k),$$
$$P(k_*|k-1)|_{k=k_*} = P(k_*|k_* - 1)$$

where the initial conditions $\hat{x}(k_*|k_* - 1)$ and $P(k_*|k_* - 1)$ are given by the Kalman filter. Here, $\hat{x}(k|k-1)$, $P(k|k-1)$, and $K^*(k)$ are the state estimate, state estimation error covariance, and Kalman gain appearing in the usual predictive-type Kalman filter (see Theorem 2.1). Using the idea of an augmented model:

$$\begin{bmatrix} x(k+1) \\ x^{\mathrm{a}}(k+1) \end{bmatrix} = \begin{bmatrix} \Phi(k+1, k) & 0 \\ 0 & I \end{bmatrix} \begin{bmatrix} x(k) \\ x^{\mathrm{a}}(k) \end{bmatrix} + \begin{bmatrix} B(k) \\ 0 \end{bmatrix} w(k)$$

$$z(k) = [H(k) \quad 0] \begin{bmatrix} x(k) \\ x^{\mathrm{a}}(k) \end{bmatrix} + v(k)$$

with the state vector at $k = k_*$ satisfying

$$\begin{bmatrix} x(k_*) \\ x^{\mathrm{a}}(k_*) \end{bmatrix} = \begin{bmatrix} x(k_*) \\ x(k_*) \end{bmatrix}$$

and then applying the Kalman filter to this, and further identifying $\hat{x}^{\mathrm{a}}(k+1|k) = \hat{x}(k_*|k)$, derive the above fixed-point smoother.

2.11 Consider the following augmented model

$$\begin{bmatrix} x(k+1) \\ x^{(1)}(k+1) \\ x^{(2)}(k+1) \\ \vdots \\ x^{(N+1)}(k+1) \end{bmatrix} = \begin{bmatrix} \Phi(k+1, k) & 0 & \dots & 0 & 0 \\ I & 0 & \dots & 0 & 0 \\ 0 & I & \dots & 0 & 0 \\ \vdots & \vdots & \ddots & \vdots & \vdots \\ 0 & 0 & \dots & I & 0 \end{bmatrix}$$

$$\times \begin{bmatrix} x(k) \\ x^{(1)}(k) \\ x^{(2)}(k) \\ \vdots \\ x^{(N+1)}(k) \end{bmatrix} + \begin{bmatrix} B(k) \\ 0 \\ 0 \\ \vdots \\ 0 \end{bmatrix} w(k)$$

$$z(k) = [H(k) \quad 0 \dots 0 \quad 0] \begin{bmatrix} x(k) \\ x^{(1)}(k) \\ x^{(2)}(k) \\ \vdots \\ x^{(N+1)}(k) \end{bmatrix} + v(k)$$

Determine, for all k and some fixed-lag N, recursive equations for the estimate

$$\hat{x}(k-N|k) = E[x(k-N)|Z_k]$$

and the associated estimation error covariance

$$P(k-N|k) = E\{[x(k-N) - \hat{x}(k-N|k)] \times [x(k-N) - \hat{x}(k-N|k)]^{\mathrm{T}}|Z_k\}$$

The resulting algorithm for $0 \leqslant i \leqslant N$ becomes

$$\hat{x}(k-i|k) = \hat{x}(k-i|k-1) + K^{(i+1)}(k)[z(k) - H(k)\hat{x}(k|k-1)],$$

$$\hat{x}(0|-1) = \hat{x}(0) \quad \text{and} \quad \hat{x}(-i|-1) = 0 \quad \text{for} \quad i \neq 0$$

$$K^{(i+1)}(k) = \Sigma^{(i)}(k|k-1)H^{\mathrm{T}}(k)[H(k)P(k|k-1)H^{\mathrm{T}}(k) + R(k)]^{-1}$$

$$\Sigma^{(i+1)}(k+1|k) = \Sigma^{(i)}(k|k-1)[\Phi(k+1, k) - K^*(k)H(k)]^{\mathrm{T}},$$

$$\Sigma^{(0)}(k|k-1) = P(k|k-1)$$

$$P(k-i|k) = P(k-i|k-1) - \Sigma^{(i)}(k|k-1)H^{\mathrm{T}}(k)K^{(i+1)^{\mathrm{T}}}(k),$$

$$P(0|-1) = P(0) \quad \text{and} \quad P(-i|-1) = 0 \text{ for } i \neq 0$$

where $P(k|k-1)$ satisfies the standard matrix RDE and

$$K^*(k) = \Phi(k+1, k)P(k|k-1)H^{\mathrm{T}}(k) \times [H(k)P(k|k-1)H^{\mathrm{T}}(k) + R(k)]^{-1}$$

To derive this *fixed-lag smoother*, apply the usual predictive-type Kalman filter and identify $\hat{x}^{(i)}(k+1|k) = \hat{x}(k+1-i|k)$.

2.12 Consider the following discrete-time model:

$$x(k+1) = \Phi x(k) + Bw(k)$$

$$z(k) = Hx(k) + w(k)$$

where $\{w(k)\}$ is a zero-mean Gaussian white process with covariance Q.

(a) Transform this model to the form of equations (2.12)–(2.16).
(b) With $DD^{\mathrm{T}} \triangleq Q^*$ for the transformed model, show that (Φ^*, D) is always uncontrollable.
(c) Show that $P = 0$ is always one possible solution of the ARE.
(d) If Φ is stable, show that the RDE will converge to $P = 0$ for all initial conditions $P(0) > 0$.
(e) If Φ has unstable roots, will $P = 0$ be the strong solution of the ARE?
(f) If Φ has no eigenvalues with modulus equal to 1, show that the solutions of the RDE will converge to the stabilizing solution for all $P(0) > 0$.

These problems were considered in Goodwin and Sin [11].

2.13 Use the eigenvalue–eigenvector method to find the ARE solution for the following discrete-time system:

$$\Phi = \begin{bmatrix} 0.9 & 0.1 \\ 0 & 0.4 \end{bmatrix}, \quad B = \begin{bmatrix} 1 \\ 1 \end{bmatrix}, \quad Q = 1$$

$$H = [1, \ 0], \qquad R = 10, \qquad S = 0$$

2.14 Use Kleinman's iterative method to solve the ARE for the system given in Exercise 2.13.

2.15 Obtain the Kalman gains $K^*(k)$ for the system given in Exercise 2.13 by using the Chandrasekhar-type algorithm.

2.16 (Fisher Information Matrix) The Fisher information matrix in continuous time is defined by

$$\mathcal{J}(t, t_0) = \int_{t_0}^{t} \Phi^{\mathrm{T}}(\tau, t)H^{\mathrm{T}}(\tau)R^{-1}(\tau)H(\tau)\Phi(\tau, t)\,\mathrm{d}\tau$$

(a) Show that

$$\frac{\mathrm{d}\mathcal{J}(t, t_0)}{\mathrm{d}t} = -A^{\mathrm{T}}(t)\mathcal{J}(t, t_0) - \mathcal{J}(t, t_0)A(t) + H^{\mathrm{T}}(t)R^{-1}(t)H(t)$$

(Hint: After differentiating $\mathcal{J}(t, t_0)$ with respect to t, use the derivative of $\Phi(t, \tau)\Phi(\tau, t) = I$ with respect to t.)

(b) Show that $\mathcal{J}^{-1}(t, t_0)$ satisfies the covariance equation in the case of no system noise. (Hint: Apply the derivative of $\mathcal{J}(t, t_0)\mathcal{J}^{-1}(t, t_0) = I$ with respect to t.)

2.17 (Information Filter) Consider the continuous-time Kalman filter given in Theorem 2.11 with $S \equiv 0$. Defining $d(t) \triangleq P^{-1}(t|t)\hat{x}(t|t)$, show that

$$\dot{d}(t) = -[A(t) + B(t)Q(t)B^{\mathrm{T}}(t)P^{-1}(t|t)]^{\mathrm{T}}d(t) + P^{-1}(t|t)G(t)u(t) + H^{\mathrm{T}}(t)R^{-1}(t)z(t), \qquad d(t_0|t_0) = P^{-1}(0)\hat{x}(0)$$

$$\dot{P}^{-1}(t|t) = -P^{-1}(t|t)A(t) - A^{\mathrm{T}}(t)P^{-1}(t|t) - P^{-1}(t|t)B(t)Q(t)B^{\mathrm{T}}(t)P^{-1}(t|t) + H^{\mathrm{T}}(t)R^{-1}(t)H(t), \qquad P^{-1}(t_0|t_0) = P^{-1}(0)$$

which is the information filter in continuous time.

2.18 Consider the formal backward-time model of equations (2.74), (2.75) with $S \equiv 0$:

$$-\mathrm{d}x(t)/\mathrm{d}t = -A(t)x(t) - G(t)u(t) - B(t)w(t)$$

$$z(t) = H(t)x(t) + v(t)$$

It is assumed that the information on the final state $x(T)$ is given by

$$x(T) \sim N(\hat{x}(T), P(T))$$

(a) Construct a discrete-time model for the above model.

(b) By applying the procedure used in the proof of Theorem 2.11 to the formal backward-time Kalman filter for the model of part (a), show that

$$-\mathrm{d}\hat{x}_{\mathrm{b}}(t|t)/\mathrm{d}t = -A(t)\hat{x}_{\mathrm{b}}(t|t) - G(t)u(t) + P_{\mathrm{b}}(t|t)H^{\mathrm{T}}(t)R^{-1}(t) \times [z(t) - H(t)\hat{x}_{\mathrm{b}}(t|t)], \qquad \hat{x}_{\mathrm{b}}(T|T) = \hat{x}(T)$$

$$-\mathrm{d}P_{\mathrm{b}}(t|t)/\mathrm{d}t = -A(t)P_{\mathrm{b}}(t|t) - P_{\mathrm{b}}(t|t)A^{\mathrm{T}}(t) + B(t)Q(t)B^{\mathrm{T}}(t) - P_{\mathrm{b}}(t|t)H^{\mathrm{T}}(t)R^{-1}(t)H(t)P_{\mathrm{b}}(t|t), \qquad P_{\mathrm{b}}(T|T) = P(T)$$

(c) Show that the formal backward-time information filter in continuous time becomes

$$-\dot{d}_b(t) = [A(t) - B(t)Q(t)B^T(t)P_b^{-1}(t|t)]^T d_b(t) - P_b^{-1}(t|t)G(t)u(t) + H^T(t)R^{-1}(t)z(t),$$

$$d_b(T|T) = P^{-1}(T)\hat{x}(T)$$

$$-\dot{P}_b^{-1}(t|t) = P_b^{-1}(t|t)A(t) + A^T(t)P_b^{-1}(t|t) - P_b^{-1}(t|t)B(t)Q(t)B^T(t)P_b^{-1}(t|t) + H^T(t)R^{-1}(t)H(t),$$

$$P_b^{-1}(T|T) = P^{-1}(T)$$

where $d_b(t) \triangleq P_b^{-1}(t|t)\hat{x}_b(t|t)$.

2.19 Consider a robot manipulator with a single degree of freedom which is controlled by feeding back a position measurement only. The manipulator is subject to zero-mean disturbance forces, and the position sensor has zero-mean encoder error, leading to the following system model [33]:

$$\begin{bmatrix} \dot{x}_1(t) \\ \dot{x}_2(t) \end{bmatrix} = \begin{bmatrix} 0 & 1 \\ -100 & -7 \end{bmatrix} \begin{bmatrix} x_1(t) \\ x_2(t) \end{bmatrix} + \begin{bmatrix} 0 \\ 100 \end{bmatrix} u(t) + \begin{bmatrix} 0 \\ 100 \end{bmatrix} w(t)$$

$$z(t) = x_1(t) + v(t)$$

where $Q = 1$, $R = 0.1$, $S = 0$ and $P(0) = I$. Obtain the solution to the RDE for this estimation problem by using (i) the Hamiltonian matrix method and (ii) the negative exponential method.

2.20 Apply (i) the eigenvalue–eigenvector method, (ii) Kleinman's iterative method, and (iii) the matrix sign function method to obtain the solution of the ARE for the system in Exercise 2.19.

REFERENCES

[1] KALMAN, R. E., A New Approach to Linear Filtering and Prediction Theory, *Trans. ASME, J. Bas. Engng., Series D,* vol. 82, no. 1, 1960, pp. 34–45.

[2] KALMAN, R.E. and BUCY, R.S., New Results in Linear Filtering and Prediction Theory, *Trans. ASME, J. Bas. Engng., Series D,* vol. 83, no. 1, 1961, pp. 95–108.

[3] WARWICK, K., Optimal Observers for ARMA Models, *Int. J. Control,* vol. 46, no. 5, 1987, pp. 1493–503.

[4] LAM, K. P., State-Space Frameworks for the Filtered CARMA Model, *Int. J. Control*, vol. 50, no. 3, 1989, pp. 871–81.

[5] PERTERKA, V., Predictive and LQG Optimal Control: Equivalences, Differences, Improvements, *Proc. Workshop on Control of Uncertain Systems*, Univ. of Bremen, Germany, 12–15 June 1989.

[6] MEDITCH, J. S., *Stochastic Optimal Linear Estimation and Control*, McGraw-Hill, New York, 1969.

[7] KUČERA, V., The Discrete Riccati Equation of Optimal Control, *Kybernetika,* vol. 8, no. 5, 1972, pp. 430–47.

[8] CHAN, S. W., GOODWIN, G. C. and SIN, K. S., Convergence Properties of the Riccati Difference Equation in Optimal Filtering of Nonstabilizable Systems, *IEEE Trans. Aut. Control,* vol. AC-29, no. 2, 1984, pp. 110–18.

[9] CAINES, P. E. and MAYNE, D. Q., On the Discrete Time Matrix Riccati Equation of Optimal Control, *Int. J. Control,* vol. 12, no. 5, 1970, pp. 785–94.

[10] ANDERSON, B. D. O. and MOORE, J. B., *Optimal Filtering,* Prentice Hall, Englewood Cliffs, New Jersey, 1979.

[11] GOODWIN, G. C. and SIN, K. S., *Adaptive Filtering Prediction and Control*, Prentice Hall, Englewood Cliffs, New Jersey, 1984.

[12] VAUGHAN, D. R., A Nonrecursive Algebraic Solution for the Discrete Riccati Equation, *IEEE Trans. Aut. Control,* vol. AC-15, no. 5, 1970, pp. 597–9.

[13] PAPPAS, T., LAUB, A. J. and SANDELL, N. R., JR, On the Numerical Solution of the Discrete-Time Algebraic Riccati Equation, *IEEE Trans. Aut. Control,* vol. AC-25, no. 4, pp. 631–41.

[14] HEWER, G. A., An Iterative Technique for the Computation of the Steady-State Gains for the Discrete Optimal Regulator, *IEEE Trans. Aut. Control,* vol. AC-16, no. 4, 1971, pp. 382–3.

[15] PACE, I. S. and BARNETT, S., Comparison of Numerical Methods for Solving Liapunov Matrix Equations, *Int. J. Control,* vol. 15, no. 5, 1972, pp. 907–15.

[16] ŠTECHA, J., KOZÁČIKOVÁ, A. and KOZÁČIK, J., Algorithms for Solution of Equations $PA + A^{T}P = -Q$ and $M^{T}PM - P = -Q$ Resulting in Lyapunov Stability Analysis of Linear Systems, *Kybernetika,* vol. 9, no. 1, 1973, pp. 62–71.

[17] KAILATH, T., Some New Algorithms for Recursive Estimation in Constant Linear Systems, *IEEE Trans. Inf. Theory*, vol. IT-19, no. 6, 1973, pp. 750–60.

[18] MORF, M., SIDHU, G. S. and KAILATH, T., Some New Algorithms for Recursive Estimation in Constant, Linear, Discrete-Time Systems, *IEEE Trans. Aut. Control,* vol. AC-19, no. 4, 1974, pp. 315–23.

[19] ANDERSON, B. D. O., Second-Order Convergent Algorithms for the Steady-State Riccati Equation, *Int. J. Control,* vol. 28, no. 2, 1978, pp. 295–306.

[20] KAILATH, T., *Lectures on Wiener and Kalman Filtering*, CISM Courses and Lectures, no. 140, Springer-Verlag, New York, 1981.

[21] JAZWINSKI, A. H., *Stochastic Processes and Filtering Theory*, Academic Press, New York, 1970.

[22] GELB, A., *Applied Optimal Estimation,* MIT Press, Cambridge, Massachusetts, 1974.

[23] MAYBECK, P. S., *Stochastic Models, Estimation, and Control*, vol. 2, Academic Press, New York, 1982.

[24] MÅRTENSSON, K., On the Matrix Riccati Equation, *Inf. Sciences,* vol. 3, 1971, pp. 17–49.

[25] KUČERA, V., A Contribution to Matrix Quadratic Equations, *IEEE Trans. Aut. Control*, vol. AC-17, no. 3, 1972, pp. 344–7.

[26] KUČERA, V., A Review of the Matrix Riccati Equation, *Kybernetika,* vol. 9, no. 1, 1973, pp. 42–61.

[27] BROWN, R. G., *Introduction to Random Signal Analysis and Kalman Filtering*, Wiley, New York, 1983.

[28] VAUGHAN, D. R., A Negative Exponential Solution for the Matrix Riccati Equation, *IEEE Trans. Aut. Control,* vol. AC-14, no. 1, 1969, pp. 72–5.

[29] KLEINMAN, D., On an Iterative Technique for Riccati Equation Computations, *IEEE Trans. Aut. Control*, vol. AC-13, no. 1, 1968, pp. 114–15.

[30] ROBERTS, J. D., Linear Model Reduction and Solution of the Algebraic Riccati Equation by Use of the Sign Function, *Int. J. Control,* vol. 32, no. 4, 1980, pp. 677–87.

[31] BALZER, L. A., Accelerated Convergence of the Matrix Sign Function Method of Solving Lyapunov, Riccati and Other Matrix Equations, *Int. J. Control,* vol. 32, no. 6, 1980, pp. 1057–78.

[32] SHIEH, L. S., TSAY, Y. T., LIN, S. W. and COLEMAN, N. P., Block-Diagonalization and Block-Triangularization of a Matrix via the Matrix Sign Function, *Int. J. Syst. Sci.,* vol. 15, no. 11, 1984, pp. 1203–20.

[33] STENGEL, R. F., *Stochastic Optimal Control*, Wiley, New York, 1986.

3

Structure and Parameter Adaptive Estimation

3.1 INTRODUCTION

The purpose of this chapter is mainly to provide *Lainiotis' partition theorem* [1, 2] on partitioned adaptive filtering, and its proof in continuous-time and discrete-time stochastic systems.

We start in Section 3.2 with a presentation of structure and parameter adaptive estimation for continuous-time linear stochastic systems. Although the original proofs make use of Bucy's representation theorem [3], the development here relies on the *innovation theorem* from the work of Kailath [4, 5].

In Section 3.3, we shall present a discrete-time version of Lainiotis' partition theorem, i.e. an algorithm of structure and parameter adaptive estimation for discrete-time linear stochastic systems. In Section 3.4 we discuss approximate schemes for use in controlling the number of parallel Kalman filters. Finally, we apply partitioned adaptive filtering to the problem of estimation of inertial navigation system (INS) error states in Section 3.5.

3.2 STRUCTURE AND PARAMETER ADAPTIVE ESTIMATION IN CONTINUOUS TIME

Our dynamical system is described by the following linear vector stochastic differential equation

$$dx(t) = A(t, \theta)x(t)\,dt + B(t, \theta)\,d\xi(t), \qquad t \geqslant t_0 \tag{3.1}$$

$$dy(t) = H(t, \theta)x(t)\,dt + d\eta(t) \tag{3.2}$$

where $x(t)$ is the n-vector state; A, B, and H are, respectively, $n \times n$, $n \times p$, and $m \times n$ continuous matrix time-functions, but including the unknown parameter vector θ; and $\{\xi(t), t \geqslant t_0\}$ is a

p-vector Wiener process with the statistics

$$E[\xi(t)] = 0 \tag{3.3}$$

$$E[\xi(t_1)\xi^{\mathrm{T}}(t_2)] = \int_{t_0}^{\min(t_1,t_2)} Q(\tau)\,\mathrm{d}\tau \tag{3.4}$$

Furthermore, $y(t)$ is an m-vector observation, and $\{\eta(t), t \geqslant t_0\}$ is an m-vector Wiener process with the statistics

$$E[\eta(t)] = 0 \tag{3.5}$$

$$E[\eta(t_1)\eta^{\mathrm{T}}(t_2)] = \int_{t_0}^{\min(t_1,t_2)} R(\tau)\,\mathrm{d}\tau \tag{3.6}$$

The distribution of $x(t_0)$ is Gaussian, $x(t_0) \sim N(\hat{x}(0, \theta), P(0, \theta))$, and $x(t_0)$, $\{\xi(t)\}$ and $\{\eta(t)\}$ are assumed to be mutually independent. In addition, it is assumed that an a priori probability for θ, $p(\theta)$, is available.

Given the observation data $\{y(\tau), t_0 \leqslant \tau \leqslant t\}$, the minimum mean-square error (MMSE) estimate $\hat{x}(t|t)$ of $x(t)$ is given by the following theorem on what is usually called *partitioned adaptive filtering*.

Theorem 3.1 Lainiotis' partition theorem

The optimal mean-square estimate of $x(t)$ is given by

$$\hat{x}(t|t) = \int \hat{x}(t|t, \theta)p(\theta|t)\mathrm{d}\theta \tag{3.7}$$

and the corresponding estimation error covariance is given by

$$P(t|t) = \int \{P(t|t, \theta) + [\hat{x}(t|t, \theta) - \hat{x}(t|t)] \times [\hat{x}(t|t, \theta) - \hat{x}(t|t)]^{\mathrm{T}}\}p(\theta|t)\mathrm{d}\theta \tag{3.8}$$

where $\hat{x}(t|t) = E[x(t)|\mathcal{Y}_t]$, $\hat{x}(t|t, \theta) \triangleq E[x(t)|\mathcal{B}(t, \theta) = \mathcal{Y}_t \times \mathcal{B}(\theta)]$, $P(t|t) = E\{[x(t) - \hat{x}(t|t)][x(t) - \hat{x}(t|t)]^{\mathrm{T}}|\mathcal{Y}_t\}$, and $P(t|t, \theta) \triangleq E\{[x(t) - \hat{x}(t|t, \theta)][x(t) - \hat{x}(t|t, \theta)]^{\mathrm{T}}|\mathcal{B}(t, \theta)\}$, in which $\mathcal{Y}_t$ is the σ-field† generated by $\{y(\tau), t_0 \leqslant \tau \leqslant t\}$, $\mathcal{B}(\theta)$ is the Borel field induced by θ, and $p(\theta|t) \triangleq p(\theta|\mathcal{Y}_t)$ denotes the a posteriori probability for θ.

† For a brief review of probability theory, see Appendix B. See also, for example, Wong and Hajek [6] or Maybeck [7] for the mathematical details.

Proof

Noting that $\mathcal{Y}_t \subset \mathcal{B}(t, \theta)$, and utilizing the smoothing property of the conditional expectation operator, we see that

$$\begin{aligned}\hat{x}(t|t) &= E[x(t)|\mathcal{Y}_t] \\ &= E\{E[x(t)|\mathcal{B}(t, \theta)]|\mathcal{Y}_t\} \\ &= \int \hat{x}(t|t, \theta)p(\theta|t)\mathrm{d}\theta \end{aligned} \tag{3.9}$$

The estimation error covariance matrix is given, in the same manner, as

$$\begin{aligned}P(t|t) &= E[\tilde{x}(t|t)\tilde{x}^{\mathrm{T}}(t|t)|\mathcal{Y}_t] \\ &= \int P_{\mathrm{a}}(t|t, \theta)p(\theta|t)\mathrm{d}\theta \end{aligned} \tag{3.10}$$

where $\tilde{x}(t|t) \triangleq x(t) - \hat{x}(t|t)$ and $P_{\mathrm{a}}(t|t, \theta)$ is obtained as follows:

$$\begin{aligned}P_{\mathrm{a}}(t|t, \theta) &\triangleq E\{[x(t) - \hat{x}(t|t)][x(t) - \hat{x}(t|t)]^{\mathrm{T}}|\mathcal{B}(t, \theta)\} \\ &= E[x(t)x^{\mathrm{T}}(t)|\mathcal{B}(t, \theta)] + \hat{x}(t|t)\hat{x}^{\mathrm{T}}(t|t) \\ &\quad - \hat{x}(t|t)\hat{x}^{\mathrm{T}}(t|t, \theta) - \hat{x}(t|t, \theta)\hat{x}^{\mathrm{T}}(t|t) \\ &= E\{[x(t) - \hat{x}(t|t, \theta)][x(t) - \hat{x}(t|t, \theta)]^{\mathrm{T}}|\mathcal{B}(t, \theta)\} \\ &\quad + [\hat{x}(t|t, \theta) - \hat{x}(t|t)][\hat{x}(t|t, \theta) - \hat{x}(t|t)]^{\mathrm{T}} \\ &= P(t|t, \theta) + [\hat{x}(t|t, \theta) - \hat{x}(t|t)][\hat{x}(t|t, \theta) - \hat{x}(t|t)]^{\mathrm{T}} \end{aligned} \tag{3.11}$$

in which the fact that $\hat{x}(t|t)$ and $\hat{x}(t|t, \theta)$ are $\mathcal{B}(t, \theta)$-measurable has been employed. The expression for $p(\theta|t)$ will be given in Theorem 3.2. □

Before proceeding to the derivation of $p(\theta|t)$, hypothesis testing problems for the observation system (3.2) are discussed from the viewpoint of innovation processes [4, 5].

Consider the following binary testing (or detection) problem:

$$h_1: \mathrm{d}y(t) = X(t)\,\mathrm{d}t + \mathrm{d}\eta(t), \qquad t_0 \leqslant t \leqslant t_{\mathrm{f}} \tag{3.12a}$$

$$h_2: \mathrm{d}y(t) = \mathrm{d}\eta(t) \tag{3.12b}$$

in which the decision as to which hypothesis to accept is made based on a set of observations $\mathcal{Y}_t$. Here for convenience $X(t) \triangleq H(t)x(t)$ is assumed to be a measurable zero-mean process. Further, it is assumed that $X(t)$ is statistically independent of $\eta(t)$. Then, the following lemma is obtained.

Lemma 3.1

The binary hypothesis testing problem given by equations (3.12) can be replaced, without changing the probability of error of Bayes' risk (see also Appendix B.4.3), by the problem

$$\tilde{h}_1\text{:}\mathrm{d}y(t) = \hat{X}_1(t|t)\,\mathrm{d}t + \mathrm{d}\nu(t),\ t_0 \leqslant t \leqslant t_\mathrm{f} \tag{3.13a}$$

$$\tilde{h}_0\text{:}\mathrm{d}y(t) = \mathrm{d}\nu(t) \tag{3.13b}$$

where $\hat{X}_1(t|t) \triangleq E[X(t)|\mathscr{Y}_t, h_1]$ and $\nu(t)$ is an innovation process (see Appendix 3A) such that

$$E[\nu(t)\nu^\mathrm{T}(s)|\mathscr{Y}_s] = \int_{t_0}^{\min(t,s)} R(\tau)\,\mathrm{d}\tau, \qquad s \leqslant t \tag{3.14}$$

Proof

It is noted from Appendix 3A that the hypothesis h_1 can be replaced by the hypothesis $\tilde{h}_1$. Also, the hypotheses h_0 and $\tilde{h}_0$ are statistically identical because the Wiener processes $\eta(t)$ and $\nu(t)$ have the same statistics. Hence, the problem $[h_1, h_0]$ is the same as the problem $[\tilde{h}_1, \tilde{h}_0]$. □

In addition to the above hypothesis testing problem, consider the following θ-conditioned binary hypothesis testing problem

$$\tilde{h}_\theta\text{:}\mathrm{d}y(t) = \hat{X}_1(t|t, \theta)\,\mathrm{d}t + \mathrm{d}\nu(t), \qquad t_0 \leqslant t \leqslant t_\mathrm{f} \tag{3.15a}$$

$$\tilde{h}_1\text{:}\mathrm{d}y(t) = \hat{X}_1(t|t)\,\mathrm{d}t + \mathrm{d}\nu(t) \tag{3.15b}$$

for $\theta = \theta^*$ (which is the true parameter), where $\hat{X}_1(t|t, \theta) \triangleq E[X(t)|\mathscr{B}(t, \theta), h_1]$. Then, let μ_1 and μ_0 denote the probability measures induced in the space of continuous functions by the observations $\{y(\cdot)\}$ under the hypotheses h_1 and h_0, and let μ_θ indicate the measure under the hypothesis $\tilde{h}_\theta$. Additionally, it is assumed that $\tilde{\mu}_\theta \ll \tilde{\mu}_1$, i.e. $\tilde{\mu}_\theta$ is absolutely continuous with respect to $\tilde{\mu}_1$, where $\tilde{\mu}_1$ represents the measure corresponding to the hypothesis $\tilde{h}_1$. With these assumptions, the following theorem for the a posteriori probability $p(\theta|t) \equiv p(\theta|\mathscr{Y}_t, h_1)$ is obtained.

Theorem 3.2

If in the binary hypothesis testing problems given by equations (3.13) and (3.15),

(i) $\hat{X}_1(t|t, \omega)$ and $\hat{X}_1(t|t, \omega, \theta)$, where ω is used to indicate explicitly the probability space variable, are measurable in t and ω;
(ii) $\hat{X}_1(t|t, \omega)$ and $\hat{X}_1(t|t, \omega, \theta)$, for every t, are measurable with respect to the past $\{y(s), s \leq t\}$ and $\{y(s), \theta; s \leq t\}$, respectively;
(iii) $\int_{t_0}^{t_f} \| \hat{X}_1(\tau|\tau, \omega)\|^2 \mathrm{d}\tau < \infty$, $\int_{t_0}^{t_f} \| \hat{X}_1(\tau|\tau, \omega, \theta)\|^2 \, \mathrm{d}\tau < \infty$, for almost all ω;
(iv) $\tilde{\mu}_1 \ll \tilde{\mu}_0$ and $\tilde{\mu}_\theta \ll \tilde{\mu}_0$, where $\tilde{\mu}_0$ denotes the measure corresponding to the hypothesis $\tilde{h}_0$;

then the a posteriori probability $p(\theta|t)$ is given by

$$p(\theta|t) = \frac{\Lambda(t|\theta)p(\theta)}{\Lambda(t)} \tag{3.16a}$$

or

$$p(\theta|t) = \frac{\Lambda(t|\theta)p(\theta)}{\int \Lambda(t|\theta)p(\theta)\mathrm{d}\theta} \tag{3.16b}$$

where $\Lambda(t|\theta)$ is the θ-conditional likelihood ratio function (i.e. the Radon–Nikodym derivative of $\tilde{\mu}_\theta$ with respect to $\tilde{\mu}_0$) such that

$$\begin{aligned} \Lambda(t|\theta) &\triangleq \mathrm{d}\tilde{\mu}_\theta / \mathrm{d}\tilde{\mu}_0 \\ &= \exp\left\{ \int_{t_0}^{t} \langle H(\tau, \theta)\hat{x}(\tau|\tau, \theta), R^{-1}(\tau)\,\mathrm{d}y(\tau)\rangle \right. \\ &\qquad \left. - \frac{1}{2}\int_{t_0}^{t} \|H(\tau, \theta)\hat{x}(\tau|\tau, \theta)\|^2_{R^{-1}(\tau)} \, \mathrm{d}\tau \right\} \end{aligned} \tag{3.17}$$

and $\Lambda(t)$ is the usual likelihood ratio function

$$\begin{aligned} \Lambda(t) &\triangleq \mathrm{d}\tilde{\mu}_1 / \mathrm{d}\tilde{\mu}_0 \\ &= \exp\left\{ \int_{t_0}^{t} \langle H(\tau)\hat{x}(\tau|\tau), R^{-1}(\tau)\mathrm{d}y(\tau)\rangle \right. \\ &\qquad \left. - \frac{1}{2}\int_{t_0}^{t} \|H(\tau)\hat{x}(\tau|\tau)\|^2_{R^{-1}(\tau)} \, \mathrm{d}\tau \right\} \end{aligned} \tag{3.18}$$

in which $\langle \cdot \rangle$ and $\|\cdot\|$ denote the inner product and norm, respectively (see Appendix A.14), in the Euclidean space $\mathbb{R}^m$.

Proof

The proofs of conditions (i)–(iv) are omitted here, because their results can easily be deduced from the Radon–Nikodym theorem for the usual unconditional likelihood ratio function [5] (see Appendix 3A). Therefore, only the result of equations (3.16) is proved. In the binary hypothesis testing problem given by equation (3.15), the Radon–Nikodym derivative with respect to measures $\tilde{\mu}_\theta$ and $\tilde{\mu}_1$, that is, the likelihood ratio function for this problem, can be represented by

$$\begin{aligned} \mathrm{d}\tilde{\mu}_\theta/\mathrm{d}\tilde{\mu}_1 &= p(\mathcal{Y}_t|\tilde{h}_\theta)/p(\mathcal{Y}_t|\tilde{h}_1) \\ &\equiv p(\mathcal{Y}_t|\tilde{h}_1, \theta)/p(\mathcal{Y}_t|\tilde{h}_1) \end{aligned} \tag{3.19}$$

where $p(\mathcal{Y}_t|\cdot)$ is the likelihood function in continuous time. Noting Bayes' rule, given by

$$p(\mathcal{Y}_t|\tilde{h}_1, \theta) = \frac{p(\theta|\mathcal{Y}_t, \tilde{h}_1)p(\mathcal{Y}_t|\tilde{h}_1)p(\tilde{h}_1)}{p(\tilde{h}_1, \theta)} \tag{3.20}$$

and substituting equation (3.20) into (3.19), gives

$$\mathrm{d}\tilde{\mu}_\theta/\mathrm{d}\tilde{\mu}_1 = \frac{p(\theta|\mathcal{Y}_t, \tilde{h}_1)p(\tilde{h}_1)}{p(\tilde{h}_1, \theta)} \tag{3.21}$$

or

$$\begin{aligned} p(\theta|\mathcal{Y}_t, \tilde{h}_1) &= \frac{\mathrm{d}\tilde{\mu}_\theta}{\mathrm{d}\tilde{\mu}_1}\frac{p(\tilde{h}_1, \theta)}{p(\tilde{h}_1)} \\ &= \frac{\mathrm{d}\tilde{\mu}_\theta/\mathrm{d}\tilde{\mu}_0}{\mathrm{d}\tilde{\mu}_1/\mathrm{d}\tilde{\mu}_0}\, p(\theta|\tilde{h}_1) \end{aligned} \tag{3.22}$$

where the measure transformation technique has been used. Here note that a lower-case p is used interchangeably to denote a probability or probability density. Furthermore, the denominator of equation (3.22), i.e. the usual likelihood ratio function, can be rewritten as

$$\begin{aligned} \mathrm{d}\tilde{\mu}_1/\mathrm{d}\tilde{\mu}_0 &= p(\mathcal{Y}_t|\tilde{h}_1)/p(\mathcal{Y}_t|\tilde{h}_0) \\ &= \frac{\int p(\mathcal{Y}_t|\tilde{h}_1, \theta)p(\theta|\tilde{h}_1)\,\mathrm{d}\theta}{p(\mathcal{Y}_t|\tilde{h}_0)} \\ &= \int \frac{\mathrm{d}\tilde{\mu}_\theta}{\mathrm{d}\tilde{\mu}_0}\, p(\theta|\tilde{h}_1)\,\mathrm{d}\theta \end{aligned} \tag{3.23}$$

Hence, redefining $p(\theta|\mathcal{Y}_t, \tilde{h}_1) \triangleq p(\theta|t)$ and $p(\theta|\tilde{h}_1) \triangleq p(\theta)$ gives the desired results. □

Note that the expression $\hat{x}(t|t, \theta)$ in Theorem 3.1 can be produced by a Kalman filter in continuous time as discussed in Chapter 2 for a particular parameter vector value. Unfortunately, the integrations involved in Theorems 3.1 and 3.2 make the estimate computationally unfeasible for online usage.

To overcome this problem, let the parameter space consist of only a discrete finite space. This finite space might be the result of discretizing a continuous parameter space, selecting a set of values $\{\theta_1, \ldots, \theta_M\}$. Or a problem of interest might naturally be represented by a discrete parameter space, as for the detection, estimation and control of systems discussed in Chapter 1.

Where the unknown parameter θ is known to belong to the discrete space, the following corollary of Theorems 3.1 and 3.2 is obtained:

Corollary 3.1

If the unknown parameter belongs to the discrete parameter space, then the optimal MMSE estimate of $x(t)$ and the associated error covariance are

$$\hat{x}(t|t) = \sum_{i=1}^{M} \hat{x}(t|t, \theta_i)p(\theta_i|t) \tag{3.24}$$

$$p(\theta_i|t) = \frac{\Lambda(t|\theta_i)p(\theta_i)}{\sum_{j=1}^{M}\Lambda(t|\theta_j)p(\theta_j)} \tag{3.25}$$

and

$$\begin{aligned} P(t|t) = \sum_{i=1}^{M} \{P(t|t, \theta_i) &+ [\hat{x}(t|t, \theta_i) - \hat{x}(t|t)] \\ &\times [\hat{x}(t|t, \theta_i) - \hat{x}(t|t)]^{\mathrm{T}}\}p(\theta_i|t) \end{aligned} \tag{3.26}$$

where

$$\begin{aligned} \Lambda(t|\theta_i) = \exp\bigg\{\int_{t_0}^{t} &\langle H(\tau, \theta_i)\hat{x}(\tau|\tau, \theta_i), R^{-1}(\tau)\, dy(\tau)\rangle \\ &- \frac{1}{2}\int_{t_0}^{t} \|H(\tau, \theta_i)\hat{x}(\tau|\tau, \theta_i)\|^2_{R^{-1}(\tau)}\, d\tau\bigg\} \end{aligned} \tag{3.27}$$

and M denotes the total number of values resulting from discretizing the continuous space of θ.

Proof

This proof is based on substituting $p(\theta) = \sum_{i=1}^{M} p(\theta_i)\delta(\theta - \theta_i)$ or

$p(\theta|\mathscr{Y}_t) = \sum_{i=1}^{M} p(\theta_i|\mathscr{Y}_t)\delta(\theta - \theta_i)$ into equations (3.9), (3.10) and (3.16b) and performing the indicated integrations. □

Note that Corollary 3.1 is often called the *multiple model adaptive filter* (MMAF) in continuous time.

If it is desired to produce the MMSE estimate of the unknown parameter vector itself, the conditional mean of θ at time t, $\hat{\theta}(t) \triangleq E[\theta|\mathscr{Y}_t]$, is

$$\begin{aligned}\hat{\theta}(t) &= E[\theta|\mathscr{Y}_t] = \int \theta p(\theta|\mathscr{Y}_t)\,\mathrm{d}\theta \\ &= \int \theta\left[\sum_{i=1}^{M} p(\theta_i|\mathscr{Y}_t)\delta(\theta - \theta_i)\right]\mathrm{d}\theta \\ &= \sum_{i=1}^{M} \theta_i p(\theta_i|\mathscr{Y}_t)\end{aligned} \tag{3.28}$$

and the estimation error covariance

$$\begin{aligned}&E\{[\theta - \hat{\theta}(t)][\theta - \hat{\theta}(t)]^{\mathrm{T}}|\mathscr{Y}_t\} \\ &= \int [\theta - \hat{\theta}(t)][\theta - \hat{\theta}(t)]^{\mathrm{T}} p(\theta|\mathscr{Y}_t)\,\mathrm{d}\theta \\ &= \sum_{i=1}^{M} [\theta_i - \hat{\theta}(t)][\theta_i - \hat{\theta}(t)]^{\mathrm{T}} p(\theta_i|\mathscr{Y}_t)\end{aligned} \tag{3.29}$$

Example 3.1 Estimation with non-Gaussian initial state vector

This problem has been solved by Lainiotis [1, 2]. We suppose that the dynamic and statistical parameters of the system, i.e. $A(\cdot)$, $B(\cdot)$, $H(\cdot)$, $Q(\cdot)$ and $R(\cdot)$, are completely known, and that the non-Gaussian density $p(x(t_0))$ of $x(t_0)$ can be approximated as the Gaussian sum expression, as follows;

$$\begin{aligned}p(x(t_0)) \simeq \sum_{i=1}^{M} p(\theta_i)&\frac{1}{(2\pi)^{n/2}|P_i(0)|^{1/2}} \\ &\times \exp\{-\tfrac{1}{2}\|x(t_0) - \hat{x}_i(0)\|^2_{P_i^{-1}(0)}\}\end{aligned}$$

where $\hat{x}_i(0)$ and $P_i(0)$ are the mean and covariance of the ith Gaussian (or normal) density, and $p(\theta_i)$ is the a priori probability. Then, the *filtered* estimate $\hat{x}(t|t)$ is given by the following *elemental* Kalman filter matched to θ_i:

$$\begin{aligned}\mathrm{d}\hat{x}(t|t, \theta_i) &= A(t)\hat{x}(t|t, \theta_i)\,\mathrm{d}t \\ &\quad + P(t|t, \theta_i)H^{\mathrm{T}}(t)R^{-1}(t)[\mathrm{d}y(t) - H(t)\hat{x}(t|t, \theta_i)\,\mathrm{d}t]\end{aligned}$$

with the initial condition $\hat{x}(t_0|t_0, \theta_i) = \hat{x}_i(0)$, and where

$$\begin{aligned}\mathrm{d}P(t|t, \theta_i)/\mathrm{d}t = {} & A(t)P(t|t, \theta_i) + P(t|t, \theta_i)A^{\mathrm{T}}(t) \\ & + B(t)Q(t)B^{\mathrm{T}}(t) \\ & - P(t|t, \theta_i)H^{\mathrm{T}}(t)R^{-1}(t)H(t)P(t|t, \theta_i)\end{aligned}$$

with the initial condition $P(t_0|t_0, \theta_i) = P_i(0)$, for $i = 1, \ldots, M$.

The initial *smoothed* estimate $\hat{x}(t_0|t) \triangleq E[x(t_0)|\mathcal{Y}_t]$ is also given by

$$\hat{x}(t_0|t) = \sum_{i=1}^{M} \hat{x}(t_0|t, \theta_i)p(\theta_i|t)$$

where the parameter-conditional smoothed estimate $\hat{x}(t_0|t, \theta_i)$ of the initial state $x(t_0)$ is given by the *fixed-point smoother* due to Kailath and Frost [8] as follows:

$$\begin{aligned}\hat{x}(t_0|t, \theta_i) = {} & \hat{x}_i(0) + \int_{t_0}^{t} P_i(0)\tilde{\Phi}^{\mathrm{T}}(s, t_0)H^{\mathrm{T}}(s)R^{-1}(s) \\ & \times [\mathrm{d}y(s) - H(s)\hat{x}(s|s,\theta_i)\,\mathrm{d}s]\end{aligned}$$

$$\mathrm{d}\tilde{\Phi}(t, s)/\mathrm{d}t = [A(t) - P(t|t, \theta_i)H^{\mathrm{T}}(t)R^{-1}(t)H(t)]\tilde{\Phi}(t, s), \qquad \tilde{\Phi}(s, s) = I$$

The associated estimation error covariance matrix is given by

$$\begin{aligned}P(t_0|t) = {} & \sum_{i=1}^{M} \{P(t_0|t, \theta_i) + [\hat{x}(t_0|t, \theta_i) - \hat{x}(t_0|t)] \\ & \times [\hat{x}(t_0|t, \theta_i) - \hat{x}(t_0|t)]^{\mathrm{T}}\}p(\theta_i|t)\end{aligned}$$

where the parameter-conditional smoothed error covariance is given by

$$\begin{aligned}P(t_0|t, \theta_i) = {} & P_i(0) - \int_{t_0}^{t} P_i(0)\tilde{\Phi}^{\mathrm{T}}(s, t_0)H^{\mathrm{T}}(s)R^{-1}(s) \\ & \times \tilde{\Phi}(s, t_0)P_i(0)\,\mathrm{d}s\end{aligned}$$

In several applications, there may be a nonzero probability that the observations contain noise only. One such application is the tracking of a target trajectory in space where the target may or may not be present, so that the target must be detected as well as its trajectory tracked. This kind of problem is called a *joint detection–estimation* problem [2, 9, 10].

Example 3.2 Joint detection–estimation

The observation equation is now rewritten as

$$\mathrm{d}y(t) = \theta H(t)x(t)\,\mathrm{d}t + \mathrm{d}\eta(t)$$

where θ is taken from the set $\{0, 1\}$.

To facilitate the presentation of the results, two new random processes are introduced:

$$x_0(t) = x(t) - \Phi(t,t_0)\hat{x}(t_0|t_0)$$

$$y_0(t) = y(t) - H(t)\Phi(t,t_0)\hat{x}(t_0|t_0)$$

where $\Phi(\cdot)$ denotes the state transition matrix for the system (3.1) with completely known parameters, and it is easy to see that $x_0(t)$ and $y_0(t)$ satisfy

$$\mathrm{d}x_0(t) = A(t)x_0(t)\,\mathrm{d}t + B(t)\,\mathrm{d}\xi(t)$$

$$\mathrm{d}y_0(t) = H(t)x_0(t)\,\mathrm{d}t + \mathrm{d}\eta(t)$$

with the initial state $x_0(t_0)$, $\hat{x}_0(t_0|t_0) \equiv 0$, and the covariance matrix $P_0(t_0|t_0) \equiv P(0)$. From the equation for $x_0(t)$ we have

$$\hat{x}(t|t) = \Phi(t, t_0)\hat{x}(t_0|t_0) + \hat{x}_0(t|t)$$

In addition, since $p(\theta) = \sum_{i=0}^{1} p(\theta_i)\delta(\theta - i)$, we have from Corollary 3.1

$$\begin{aligned}\hat{x}_0(t|t) &= p(\theta_0|t)\hat{x}_0(t|t, \theta_0) + p(\theta_1|t)\hat{x}_0(t|t, \theta_1)\\ &\equiv p(\theta_1|t)\hat{x}_0(t|t, \theta_1)\end{aligned}$$

because the signal-absent conditional estimate is $\hat{x}_0(t|t, \theta_0) = E[x_0(t)|\mathcal{Y}_t, \theta_0] \equiv 0$. Combining the above two equations yields

$$\hat{x}(t|t) = \Phi(t,t_0)\hat{x}(t_0|t_0) + p(\theta_1|t)\hat{x}_0(t|t, \theta_1)$$

where $\hat{x}_0(t|t, \theta_1) \triangleq E[x_0(t)|\mathcal{Y}_t, \theta_1]$ is given by

$$\begin{aligned}\mathrm{d}\hat{x}_0(t|t, \theta_1) = A(t)&\hat{x}_0(t|t, \theta_1)\,\mathrm{d}t\\ &+ P_0(t|t, \theta_1)H^{\mathrm{T}}(t)R^{-1}(t)\\ &\times [\mathrm{d}y_0(t) - H(t)\hat{x}_0(t|t, \theta_1)\,\mathrm{d}t]\end{aligned}$$

with the initial condition $\hat{x}_0(t_0|t_0, \theta_1) \equiv 0$, in which $P_0(t|t, \theta_1)$ is an error covariance which will be presented later.

The estimation error covariance matrix $P(t|t) \equiv P_0(t|t)$ is obtained from equation (3.26) as follows:

$$P(t|t) = p(\theta_0|t)P_0(t|t, \theta_0) + p(\theta_1|t)P_0(t|t, \theta_1) + \sum_{i=0}^{1} [\hat{x}_0(t|t, \theta_i) - \hat{x}_0(t|t)][\hat{x}_0(t|t, \theta_i) - \hat{x}_0(t|t)]^{\mathrm{T}} p(\theta_i|t)$$

But, substituting equation (3.4) into the third term of the above equation, and noting that $\hat{x}_0(t|t, \theta_0) = 0$ and $p(\theta_0|t) + p(\theta_1|t) = 1$, we have

$$\begin{aligned}
&\sum_{i=0}^{1} [\hat{x}_0(t|t, \theta_i) - \hat{x}_0(t|t)][\hat{x}_0(t|t, \theta_i) - \hat{x}_0(t|t)]^{\mathrm{T}} p(\theta_i|t) \\
&= p^2(\theta_1|t)\hat{x}_0(t|t, \theta_1)\hat{x}_0^{\mathrm{T}}(t|t, \theta_1)p(\theta_0|t) \\
&\quad + [1 - p(\theta_1|t)]^2\hat{x}_0(t|t, \theta_1)\hat{x}_0^{\mathrm{T}}(t|t, \theta_1)p(\theta_1|t) \\
&= \hat{x}_0(t|t, \theta_1)\hat{x}_0^{\mathrm{T}}(t|t, \theta_1)[p(\theta_0|t)p(\theta_1|t)\{p(\theta_0|t) + p(\theta_1|t)\}] \\
&= \hat{x}_0(t|t, \theta_1)\hat{x}_0^{\mathrm{T}}(t|t, \theta_1)p(\theta_0|t)p(\theta_1|t)
\end{aligned}$$

Therefore, it follows that

$$\begin{aligned}
P(t|t) &= p(\theta_0|t)P_0(t|t, \theta_0) + p(\theta_1|t)P_0(t|t, \theta_1) \\
&\quad + p(\theta_0|t)p(\theta_1|t)\hat{x}_0(t|t, \theta_1)\hat{x}_0^{\mathrm{T}}(t|t, \theta_1)
\end{aligned}$$

where the parameter-conditional error covariance matrices are, respectively, given by

$$\begin{aligned}
\mathrm{d}P_0(t|t, \theta_0)/\mathrm{d}t &= A(t)P_0(t|t, \theta_0) + P_0(t|t, \theta_0)A^{\mathrm{T}}(t) \\
&\quad + B(t)Q(t)B^{\mathrm{T}}(t), \qquad P_0(t_0|t_0) = P(0)
\end{aligned}$$

and

$$\begin{aligned}
\mathrm{d}P_0(t|t, \theta_1)/\mathrm{d}t &= A(t)P_0(t|t, \theta_1) + P_0(t|t, \theta_1)A^{\mathrm{T}}(t) \\
&\quad + B(t)Q(t)B^{\mathrm{T}}(t) \\
&\quad - P_0(t|t, \theta_1)H^{\mathrm{T}}(t)R^{-1}(t)H(t)P_0(t|t, \theta_1), \\
&\qquad P_0(t_0|t_0, \theta_1) = P(0)
\end{aligned}$$

Furthermore, from equations (3.25) and (3.27), we find that

$$\begin{aligned}
&\sum_{i=0}^{1} p(\theta_i)\exp\left\{\int_{t_0}^{t} \langle iH(\tau)\hat{x}(\tau|\tau, \theta_i), R^{-1}(\tau)\,\mathrm{d}y(\tau)\rangle \right. \\
&\left. - \frac{1}{2}\int_{t_0}^{t} \|iH(\tau)\hat{x}(\tau|\tau, \theta_i)\|^2_{R^{-1}(\tau)}\,\mathrm{d}\tau\right\} \\
&= p(\theta_0)\left[1 + \frac{p(\theta_1)}{p(\theta_0)}\exp\left\{\int_{t_0}^{t} \langle H(\tau)\hat{x}(\tau|\tau, \theta_1), R^{-1}(\tau)\,\mathrm{d}y(\tau)\rangle \right.\right. \\
&\left.\left. - \frac{1}{2}\int_{t_0}^{t} \|H(\tau)\hat{x}(\tau|\tau, \theta_1)\|^2_{R^{-1}(\tau)}\,\mathrm{d}\tau\right\}\right]
\end{aligned}$$

Hence, we obtain

$$p(\theta_1|t) = 1 - p(\theta_0|t) = \frac{\rho\Lambda(t|\theta_1)}{1 + \rho\Lambda(t|\theta_1)}$$

where $\rho = p(\theta_1)/p(\theta_0)$ and

$$\Lambda(t|\theta_1) \triangleq \exp\left[\int_{t_0}^{t} \langle H(\tau)\hat{x}(\tau|\tau, \theta_1), R^{-1}(\tau)\,\mathrm{d}y(\tau)\rangle - \frac{1}{2}\int_{t_0}^{t} \|H(\tau)\hat{x}(\tau|\tau, \theta_1)\|^2_{R^{-1}(\tau)}\,\mathrm{d}\tau\right]$$

in which

$$\hat{x}(\tau|\tau, \theta_1) = \Phi(\tau, t_0)\hat{x}(t_0|t_0) + \hat{x}_0(\tau|\tau, \theta_1)$$

Example 3.3 System parameter adaptation

The problem of system parameter adaptation involves identifying uncertain elements in the $A(\cdot)$, $B(\cdot)$ and $H(\cdot)$ matrices. This problem can be solved as previously described. For an example involving $A(\cdot)$, see Section 3.5.

Example 3.4 Noise parameter adaptation

This is the adaptation of the first- and second-order statistics of the Gaussian random variables or noises, i.e. $x(0)$, $w(t)$ and $v(t)$. For example, consider a case in which the system noise variance Q is unknown to have one of M possible values $\{Q_i; i = 1, \ldots, M\}$. The solution to this problem is, of course, obtained using Corollary 3.1.

Example 3.5 System structure adaptation

So far, the order of the system, n, is assumed to be fixed and known. However, in practical situations this may not be the case. A typical example can be seen in the classification of cardiac arrhythmias [11, 12] described in Example 1.1 of Chapter 1. In this case, the order of the system model n is known only to have one of three possible values, $\{1, 2, 3\}$, and the problem is regarded as a detection or classification problem instead of as a state-estimation problem. For more detailed discussion of this type of problem, refer to Lainiotis [1, 13] and Chin [14].

3.3 STRUCTURE AND PARAMETER ADAPTIVE ESTIMATION IN DISCRETE TIME

The discrete-time linear system is described by the vector difference equation

$$x(k+1) = \Phi(k+1, k; \theta)x(k) + B(k,\theta)w(k) \tag{3.30}$$

$$z(k) = H(k, \theta)x(k) + v(k), \qquad k = 0, 1, 2, \ldots \tag{3.31}$$

where $x(k)$ is the n-dimensional state vector, $\Phi(\cdot)$ is the $n \times n$ state transition matrix, $B(\cdot)$ is an $n \times p$ matrix, and $\{w(k), k = 1, 2, \ldots\}$ is a p-vector Gaussian white noise sequence, $w(k) \sim N(0, Q(k, \theta))$. $z(k)$ is the m-vector observation, $H(\cdot)$ is an $m \times n$ matrix function, and $\{v(k), k = 1, 2, \ldots\}$ is an m-vector Gaussian white noise sequence, $v(k) \sim N(0, R(k, \theta))$. The distribution of $x(0)$ is Gaussian, $x(0) \sim N(\hat{x}(0, \theta), P(0, \theta))$, and $x(0)$, $\{w(k)\}$ and $\{v(k)\}$ are assumed to be mutually independent. The unknown parameter θ has an a priori probability $p(\theta)$.

The discrete-time version of Lainiotis' partition theorem is presented in the following theorem.

Theorem 3.3

Given the observation data $Z_k \triangleq \{z(l), l = 1, 2, \ldots, k\}$, the MMSE estimate $\hat{x}(k|k)$ of $x(k)$ is given by

$$\hat{x}(k|k) = \int \hat{x}(k|k, \theta)p(\theta|k)\,\mathrm{d}\theta \tag{3.32}$$

where $\hat{x}(k|k) = E[x(k)|Z_k]$ and $\hat{x}(k|k, \theta) \triangleq E[x(k)|Z_k, \theta]$. The a posteriori probability of θ given Z_k, $p(\theta|Z_k) \equiv p(\theta|k)$, is provided by

$$p(\theta|k) = \frac{L(k|\theta)p(\theta|k-1)}{\int L(k|\theta)p(\theta|k-1)\,\mathrm{d}\theta} \tag{3.33}$$

where

$$L(k|\theta) = |\tilde{P}(k|k-1, \theta)|^{-1/2} \exp[-\tfrac{1}{2}\|\nu(k, \theta)\|^2_{\tilde{P}^{-1}(k|k-1,\theta)}] \tag{3.34}$$

Here, the innovation process $\nu(k, \theta) \triangleq z(k) - H(k, \theta)\hat{x}(k|k-1, \theta)$ is a zero-mean white process with covariance matrix $\tilde{P}(k|k-1, \theta) \triangleq H(k, \theta)P(k|k-1, \theta)H^{\mathrm{T}}(k, \theta) + R(k, \theta)$ for $\theta = \theta^*$ (θ^* is the true parameter). Corresponding to equation (3.32), the estimation error covariance matrix $P(k|k) = E\{[x(k) - \hat{x}(k|k)][x(k) - \hat{x}(k|k)]^{\mathrm{T}}|Z_k\}$ is also given by

$$P(k|k) = \int \{P(k|k, \theta) + [\hat{x}(k|k, \theta) - \hat{x}(k|k)] \times [\hat{x}(k|k, \theta) - \hat{x}(k|k)]^{\mathrm{T}}\} p(\theta|k)\, \mathrm{d}\theta \tag{3.35}$$

where $P(k|k, \theta) \triangleq E\{[x(k) - \hat{x}(k|k, \theta)][x(k) - \hat{x}(k|k, \theta)]^{\mathrm{T}} | Z_k, \theta\}$. The parameter-conditional estimates $\hat{x}(k|k, \theta)$, $\hat{x}(k|k-1, \theta)$ and the associated error covariance matrices $P(k|k, \theta)$, $P(k|k-1, \theta)$ are obtained by the Kalman filter matched to the system model with parameter θ.

Proof

The derivations of equations (3.32) and (3.35) are the same as in Theorem 3.1. Therefore, only the proof of equations (3.33) and (3.34) is given [15]. From Bayes' rule,

$$\begin{aligned} p(\theta|Z_k) &= \frac{p(Z_k, \theta)}{p(Z_k)} = \frac{p(z(k), Z_{k-1}, \theta)}{p(z(k), Z_{k-1})} \\ &= \frac{p(z(k), \theta|Z_{k-1})p(Z_{k-1})}{p(z(k)|Z_{k-1})p(Z_{k-1})} = \frac{p(z(k), \theta|Z_{k-1})}{p(z(k)|Z_{k-1})} \\ &= \frac{p(z(k), \theta|Z_{k-1})}{\int p(z(k), \theta|Z_{k-1})\, \mathrm{d}\theta} \end{aligned} \tag{3.36}$$

where the marginal property of a mixed probability/probability density function has been employed in the denominator of equation (3.36). Further, using Bayes' rule again gives

$$\begin{aligned} p(z(k), \theta|Z_{k-1}) &= \frac{p(z(k), \theta, Z_{k-1})}{p(Z_{k-1})} \\ &= \frac{p(z(k)|\theta, Z_{k-1})p(\theta, Z_{k-1})}{p(Z_{k-1})} \\ &= p(z(k)|\theta, Z_{k-1})p(\theta|Z_{k-1}) \end{aligned} \tag{3.37}$$

where $p(z(k)|\theta, Z_{k-1})$ is given, recognizing that it is Gaussian density, by

$$p(z(k)|\theta, Z_{k-1}) = (2\pi)^{-m/2} |\tilde{P}(k|k-1, \theta)|^{-1/2} \times \exp[-\tfrac{1}{2}\|\nu(k,\theta)\|^2_{\tilde{P}^{-1}(k|k-1,\theta)}] \tag{3.38}$$

Substituting equations (3.37), (3.38) into (3.36) proves the theorem. □

Corollary 3.2

If the unknown parameter θ belongs to the discrete space, Theorem 3.3 reduces to

$$\hat{x}(k|k) = \sum_{i=1}^{M} \hat{x}(k|k, \theta_i)p(\theta_i|k) \tag{3.39}$$

$$P(k|k) = \sum_{i=1}^{M}\{P(k|k, \theta_i) + [\hat{x}(k|k, \theta_i) - \hat{x}(k|k)] \times [\hat{x}(k|k, \theta_i) - \hat{x}(k|k)]^{\mathrm{T}}\}p(\theta_i|k) \tag{3.40}$$

$$p(\theta_i|k) = \frac{L(k|\theta_i)p(\theta_i|k-1)}{\sum_{j=1}^{M} L(k|\theta_j)p(\theta_j|k-1)} \tag{3.41}$$

where

$$L(k|\theta_i) = |\tilde{P}(k|k-1, \theta_i)|^{-1/2} \exp[-\tfrac{1}{2}\|\nu(k, \theta_i)\|^2_{\tilde{P}^{-1}(k|k-1,\theta_i)}] \tag{3.42}$$

The block diagram for the MMAF in discrete time is shown in Figure 3.1. As in the continuous-time case, if a specific estimate of θ is desired, it can be formed as a weighted sum of the discrete

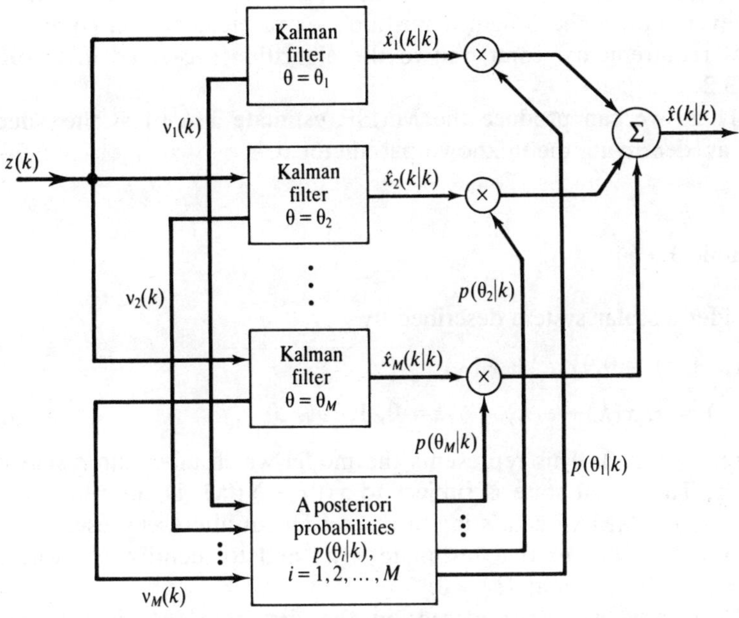

Figure 3.1 Block diagram for the multiple model adaptive filter.

parameter sets:

$$\hat{\theta}(k) = \sum_{i=1}^{M} \theta_i p(\theta_i | Z_k) \tag{3.43}$$

with the estimation error covariance

$$E\{[\theta - \hat{\theta}(k)][\theta - \hat{\theta}(k)]^{\mathrm{T}} | Z_k\}$$
$$= \sum_{i=1}^{M} [\theta_i - \hat{\theta}(k)][\theta_i - \hat{\theta}(k)]^{\mathrm{T}} p(\theta_i | Z_k) \tag{3.44}$$

Magill [16] was first to propose an iterative calculation of the a posteriori probability. That is, he evaluated

$$p(\theta_i | Z_k) = \frac{p(Z_k | \theta_i) p(\theta_i)}{\sum_{j=1}^{M} p(Z_k | \theta_j) p(\theta_j)} \tag{3.45}$$

or, equivalently,

$$p(\theta_i | Z_k) = \frac{\prod_{l=1}^{k} |\tilde{P}(l|l-1, \theta_i)|^{-1/2} \exp[-\frac{1}{2} \sum_{l=1}^{k} \|\nu(l, \theta_i)\|^2_{\tilde{P}^{-1}(l|l-1,\theta_i)}] p(\theta_i)}{\sum_{j=1}^{M} \{\prod_{l=1}^{k} |\tilde{P}(l|l-1, \theta_j)|^{-1/2} \exp[-\frac{1}{2} \sum_{l=1}^{k} \|\nu(l, \theta_j)\|^2_{\tilde{P}^{-1}(l|l-1,\theta_j)}]\} p(\theta_j)} \tag{3.46}$$

where the $p(Z_k | \theta_i)$ are termed *likelihood functions* in discrete time. However, notice the Magill algorithm's extra memory and computational requirements, compared to the algorithm presented in Corollary 3.2.

Thus, we can produce the MMSE estimate $\hat{x}(k|k)$ at the same time as identifying the unknown parameter θ.

Example 3.6

Consider a scalar system described by

$$x(k+1) = 0.9x(k) + w(k)$$
$$z(k) = H_i x(k) + v(k), \qquad k = 0, 1, \ldots, 20$$

where $i = 1, 2, 3$. This represents the model which takes three sensor modes. The initial state is subject to $x(0) \sim N(0.1, 1)$, $Q = 0.4$, and $R = 0.1$. An MMAF consisting of three Kalman filters as presented in Theorem 2.2 was used to estimate $x(k)$ and to identify H_i, where $H_1 = 1$, $H_2 = 0.5$, and $H_3 = 0$.

Two cases were considered: in the first case, H_2 is the true parameter value; for the second case, H_3 is the true parameter value.

The a priori probabilities were set to $p(H_i) = \frac{1}{3}$ for $i = 1, 2, 3$. Figures 3.2 and 3.3 portray the time histories of the a posteriori probabilities for these cases. These show that the algorithm can identify constant unknown parameters with well-behaved convergence.

Example 3.7

We now return to the Kalman filtering problem considered previously in Example 2.2. The problem was solved there using a Kalman filter with a single piece of a priori information on $x(0)$, and we now wish to apply Corollary 3.2 to a case where the initial state is unknown. Let the a priori information on $x(0)$ be given by four hypotheses:

$$h_1\colon \quad \hat{x}^{\mathrm{T}}(0, h_1) = [-\ 14\ \mathrm{km},\ 6\ \mathrm{km},\ 0.014\ \mathrm{km\,s^{-1}},\ -0.006\ \mathrm{km\,s^{-1}}]$$

$$P(0, h_1) = \operatorname{diag}(1\ \mathrm{km}^2,\ 1\ \mathrm{km}^2,\ 1 \times 10^{-6}\ \mathrm{km^2\,s^{-2}},\ 1 \times 10^{-6}\ \mathrm{km^2\,s^{-2}})$$

$$h_2\colon \quad \hat{x}^{\mathrm{T}}(0, h_2) = [-\ 15\ \mathrm{km},\ 7\ \mathrm{km},\ 0.018\ \mathrm{km\,s^{-1}},\ -0.005\ \mathrm{km\,s^{-1}}]$$

$$P(0, h_2) = P(0, h_1)$$

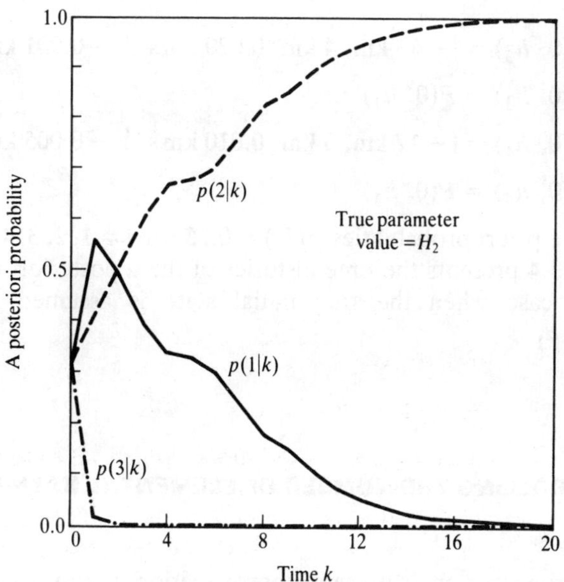

Figure 3.2 A posteriori probabilities for Example 3.6, where H_2 is the true parameter value.

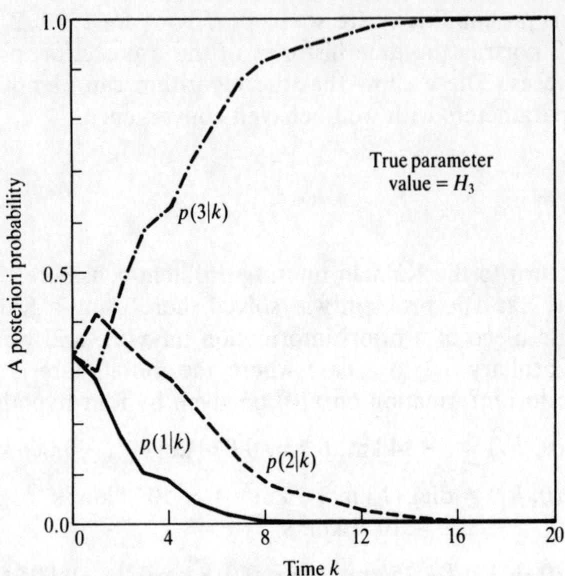

Figure 3.3 A posteriori probabilities for Example 3.6, where H_3 is the true parameter value.

$$h_3: \ \hat{x}^{\mathrm{T}}(0, h_3) = [-13 \text{ km}, 4 \text{ km}, 0.020 \text{ km s}^{-1}, -0.001 \text{ km s}^{-1}]$$

$$P(0, h_3) = P(0, h_1)$$

$$h_4: \ \hat{x}^{\mathrm{T}}(0, h_4) = [-12 \text{ km}, 5 \text{ km}, 0.010 \text{ km s}^{-1}, -0.005 \text{ km s}^{-1}]$$

$$P(0, h_4) = P(0, h_1)$$

with equal a priori probabilities $p(h_i) = 0.25$ for $i = 1, 2, 3, 4$.

Figure 3.4 presents the time histories of the a posteriori probabilities for a case when the true initial state is assumed to be of hypothesis h_1.

3.4 CONTROLLING THE NUMBER OF ELEMENTAL KALMAN FILTERS

In this section, we shall present approximation schemes [17, 18] for use in controlling the number of elemental Kalman filters to a finite or to below a maximum allowable number. Instead of carrying all of

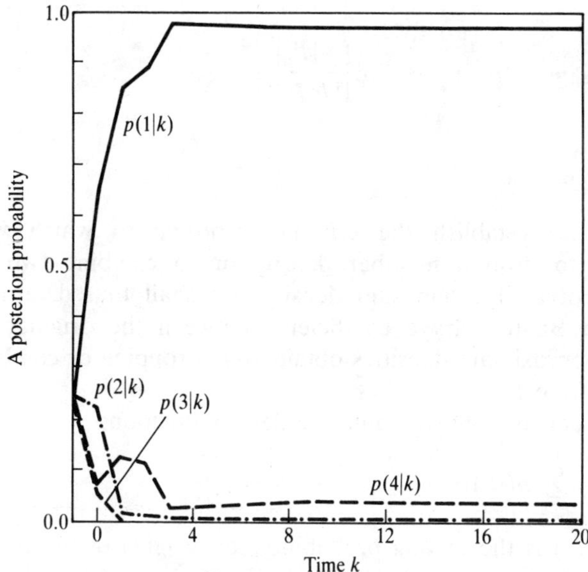

Figure 3.4 A posteriori probabilities for Example 3.7.

the Gaussian member densities, we will either disregard some of the Gaussian densities or combine two or more. The rationale of these approximations lies in the fact that if a Gaussian member density in a Gaussian mixture carries a very small mixing weight (or probability), it carries very little information for detection, estimation and identification, and also if two Gaussian member densities are very close in terms of a given distance measure, the information they carry is similar or indistinguishable for all practical purposes.

The *Bhattacharyya coefficient* [19] may be employed as a measure of distance. The Bhattacharyya coefficient, ρ_{ij}, between two probability densities $p_i(x)$ and $p_j(x)$ is defined as follows:

$$\rho_{ij} = \int [p_i(x)p_j(x)]^{1/2}\,\mathrm{d}x \tag{3.47}$$

Clearly, $0 \leqslant \rho_{ij} \leqslant 1$, and $\rho_{ij} = 1$ if $p_i(x) = p_j(x)$. We can also associate a distance measurement with this coefficient:

$$B_{ij} = -\ln \rho_{ij} \tag{3.48}$$

which is termed the *Bhattacharyya distance*. If $\rho_i(x)$ and $\rho_j(x)$ are Gaussian densities given by $N(m_i, P_i)$ and $N(m_j, P_j)$, respectively, then

$$B_{ij} = \tfrac{1}{8}\|m_i - m_j\|^2_{P_{ij}^{-1}} + \tfrac{1}{2}\ln\left[\frac{|P_{ij}|}{|P_iP_j|^{1/2}}\right] \tag{3.49}$$

where

$$2P_{ij} = P_i + P_j \tag{3.50}$$

In order to establish the criteria according to which we make decisions to drop a member density or to combine two member densities of a Gaussian sum density, we shall first determine how much the Bhattacharyya coefficients between the original densities and the approximate densities obtained by dropping or combining are different from 1.

Consider the following mixture density function:

$$p(x) = \sum_{i=1}^{M} p(\theta_i)p_i(x) \tag{3.51}$$

where $p(\theta_i)$ is the mixing probability (or weight) of the ith member density and $p_i(x)$ is the ith density. Clearly, $\sum_{i=1}^{M} p(\theta_i) = 1$.

The lower bound on the Bhattacharyya coefficient between $p(x)$ and its approximate density $p_{\text{a}}(x)$ is given in the following theorem, where $p_{\text{a}}(x)$ is the probability density of x resulting from dropping a member density.

Theorem 3.4

Let

$$p_{\text{a}}(x) = \sum_{i=1}^{M-1} \tilde{p}(\theta_i)p_i(x) \tag{3.52}$$

be the probability density function approximating the 'true' density function given by equation (3.51), where

$$\tilde{p}(\theta_i) = \frac{p(\theta_i)}{1 - p(\theta_M)}, \qquad \text{for } i = 1, 2, \ldots, M - 1 \tag{3.53}$$

Then the Bhattacharyya coefficient between the two densities, ρ_{a}, is bounded below by

$$\rho_{\text{a}} \geqslant 1 - \tfrac{1}{2}p(\theta_M) \tag{3.54}$$

where

$$\rho_{\text{a}} = \int [p(x)p_{\text{a}}(x)]^{1/2}\,\mathrm{d}x \tag{3.55}$$

Proof

If follows that

$$\int |p(x) - p_a(x)|\,dx \geqslant \int |[p(x)]^{1/2} - [p_a(x)]^{1/2}|^2\,dx$$
$$= 2 - 2\rho_a \tag{3.56}$$

Therefore

$$\rho_a \geqslant 1 - \tfrac{1}{2}\int |p(x) - p_a(x)|\,dx \tag{3.57}$$

However,

$$\int |p(x) - p_a(x)|\,dx = \int \left|\sum_{i=1}^{M} p(\theta_i)p_i(x) - \sum_{i=1}^{M-1} \tilde{p}(\theta_i)p_i(x)\right| dx$$
$$\leqslant \int \left|\sum_{i=1}^{M} p(\theta_i)p_i(x) - \sum_{i=1}^{M-1} p(\theta_i)p_i(x)\right| dx$$
$$= \int p(\theta_M)p_M(x)\,dx \tag{3.58}$$

Therefore,

$$\int |p(x) - p_a(x)|\,dx \leqslant p(\theta_M) \tag{3.59}$$

Combining expressions (3.57) and (3.59) gives the desired result. □

Corollary 3.3

If the density is approximated by dropping l member densities:

$$p_a(x) = \sum_{i=1}^{M-l} \tilde{p}(\theta_i)p_i(x) \tag{3.60}$$

where

$$\tilde{p}(\theta_i) = \frac{p(\theta_i)}{1 - \sum_{i=M-l+1}^{M} p(\theta_i)}, \text{ for } i = 1, 2, \ldots, M-1 \tag{3.61}$$

then

$$\rho_a \geqslant 1 - \frac{1}{2}\sum_{i=M-l+1}^{M} p(\theta_i) \tag{3.62}$$

Proof

The proof is obvious in view of Theorem 3.4. □

Now we consider the case where two close member densities are combined to reduce one member density. From the partition theorem given in Corollary 3.2, it is obvious that the best mean square approximation of two densities with means and covariances given by (m_1, P_1) and (m_2, P_2) with associated probabilities $p(\theta_1)$ and $p(\theta_2)$, respectively, is given by $\tilde{p}(x) = N(\tilde{m}, \tilde{P})$, with weight $\tilde{p}$ as given in the following:

$$\tilde{p} = p(\theta_1) + p(\theta_2) \tag{3.63}$$

$$\tilde{m} = \frac{p(\theta_1)m_1 + p(\theta_2)m_2}{\tilde{p}} \tag{3.64}$$

$$\tilde{P} = \frac{\sum_{i=1}^{2}\{P_i + (m_i - \tilde{m})(m_i - \tilde{m})^{\mathrm{T}}\}p(\theta_i)}{\tilde{p}} \tag{3.65}$$

Theorem 3.5

The Bhattacharyya coefficient between $p(x)$ and $p_{\mathrm{a}}(x)$, which is given by

$$p_{\mathrm{a}}(x) = \tilde{p}\tilde{p}(x) + \sum_{i=1}^{M-2} p(\theta_i)p_i(x) \tag{3.66}$$

is bounded below by

$$\rho_{\mathrm{a}} \geqslant 1 - [p(\theta_{M-1}) + p(\theta_M)](1 - \rho_{M-1,M}^2)^{1/2} \tag{3.67}$$

where

$$\rho_{M-1,M} = \int [p_{M-1}(x)p_M(x)]^{1/2}\,\mathrm{d}x \tag{3.68}$$

For the Gaussian member densities,

$$\rho_{M-1,M}^2 = \exp[-2B_{M-1,M}] \tag{3.69}$$

$$B_{M-1,M} = \tfrac{1}{8}\|m_M - m_{M-1}\|^2_{P_{M-1,M}^{-1}} + \frac{1}{2}\ln\left[\frac{|P_{M-1,M}|}{|P_M P_{M-1}|^{1/2}}\right] \tag{3.70}$$

$$2P_{M-1,M} = P_{M-1} + P_M \tag{3.71}$$

Proof

Define

$$g(x) = p(\theta_M)p_M(x) + p(\theta_{M-1})p_{M-1}(x) \tag{3.72}$$

then

$$\int |p(x) - p_a(x)|\,dx = \int |g(x) - \tilde{p}\tilde{p}(x)|\,dx \tag{3.73}$$

By Schwartz's inequality,

$$\left[\int |g(x) - \tilde{p}\tilde{p}(x)|\,dx\right]^2 \leqslant \left[\int |[g(x)]^{1/2} - [\tilde{p}\tilde{p}(x)]^{1/2}|^2\,dx\right] \times \left[\int |[g(x)]^{1/2} - [\tilde{p}\tilde{p}(x)]^{1/2}|^2\,dx\right] \tag{3.74}$$

The right-hand side of inequality (3.74) reduces to the following after some algebraic manipulation:

$$\left[\int |g(x) - \tilde{p}\tilde{p}(x)|\,dx\right]^2 \leqslant 4[p(\theta_{M-1}) + p(\theta_M)]^2 - 4\left[\int [g(x)\tilde{p}\tilde{p}(x)]^{1/2}\,dx\right]^2 \tag{3.75}$$

By combining expressions (3.57) and (3.75), we have

$$\rho_a \geqslant 1 - \left\{[p(\theta_{M-1}) + p(\theta_M)]^2 - \left[\int [g(x)\tilde{p}\tilde{p}(x)]^{1/2}\,dx\right]^2\right\}^{1/2} \tag{3.76}$$

But it is easy to see that

$$\begin{aligned} &\int [g(x)\tilde{p}\tilde{p}(x)]^{1/2}\,dx \\ &= \int \{[p(\theta_{M-1})p_{M-1}(x) + p(\theta_M)p_M(x)] \\ &\quad \times [p(\theta_{M-1}) + p(\theta_M)]\tilde{p}(x)\}^{1/2}\,dx \\ &\geqslant \int \{[p(\theta_{M-1}) + p(\theta_M)]^2 p_{M-1}(x)p_M(x)\}^{1/2}\,dx \end{aligned} \tag{3.77}$$

We obtain the desired equations (3.67) and (3.68) by substituting (3.77) into (3.76). Equations (3.67)–(3.71) are obvious from (3.48) and (3.49). □

Theorems 3.4 and 3.5, together with Corollary 3.3, determine the maximum possible deviation of the approximate density function from the true density function when the approximation is obtained by either dropping certain member densities or combining member densities that are close together.

Example 3.8

This example illustrates how these bounds may be utilized in approximating the densities involved in processing the MMSE estimate of $x(k)$ to reduce computational and storage requirements. Consider the problem of estimation with non-Gaussian initial state vector as described in Example 3.1, but in the discrete-time case (see also Exercise 3.8). At the beginning, at time $k = 0$, we need to store the initial state vector probability density given by

$$p(x(0)) = \sum_{i=1}^{M} p(\theta_i)N[\hat{x}_i(0), P_i(0)]$$

We consider only the case of disregarding some candidate models. First, determine the a priori threshold δ. Compare $p(\theta_i|k)$, $i = 1, \ldots, M$ to δ at any time $k \geqslant 0$. If some $p(\theta_l|k) < \delta$, let the new a posteriori probability $\tilde{p}(\theta_i|k)$ be

$$\tilde{p}(\theta_i|k) = \frac{p(\theta_i|k)}{1 - \sum p(\theta_l|k)}, \qquad \text{for } i = 1, 2, \ldots, M;\ i \neq l$$

The resulting initial estimate is given approximately by

$$\hat{x}(0|k) \simeq \sum_{\substack{i=1\\ i\neq l}} \hat{x}(0|k, \theta_i)\tilde{p}(\theta_i|k)$$

and the associated estimation error covariance is also given by

$$P(0|k) \simeq \sum_{\substack{i=1\\ i\neq l}} \{P(0|k, \theta_i) + [\hat{x}(0|k, \theta_i) - \hat{x}(0|k)]$$

$$\times [\hat{x}(0|k, \theta_i) - \hat{x}(0|k)]\}\tilde{p}(\theta_i|k)$$

3.5 APPLICATION TO INERTIAL NAVIGATION SYSTEMS

In this section, Corollary 3.2 will be applied to the problem of estimation of inertial navigation system (INS) error states using Doppler information [20]. The results are presented to show that the

MMAF technique is useful for reducing the uncertainties of dynamical parameters due to several flight conditions.

It is well known [20–23] that an integrated aircraft navigation system consisting of a pure inertial navigation system, Doppler radar receiver, Omega receiver and a central computer is a powerful tool for alleviating the rate of error propagation due to gyro drift, mechanization, platform torquing errors, and so on.

A technique for optimally integrating aircraft navigation subsystems usually includes an optimal Kalman filter, which estimates the INS error states. However, such a filter cannot be used in the construction of an integrating INS whose dynamic parameters are partially known. In these circumstances, we must synthesize a practical filter which has an adaptive mechanism to reduce the uncertainties.

Our system model is essentially that of Zimmerman [20]. The formal continuous-time state-space model is

$$\dot{x}(t) = Ax(t) + Bw(t) \tag{3.78}$$

where $x(t) \in \mathbb{R}^7$ is the state vector, $A, B \in \mathbb{R}^{7\times7}$ are time-invariant matrices, and $w(t)$ is a zero-mean Gaussian white noise process. In the case under consideration,

$$x^{\mathrm{T}}(t) = [\phi_x, \phi_y, \phi_z, \delta V_x, \delta V_y, \delta r_x, \delta r_y]$$

$$w^{\mathrm{T}}(t) = [\varepsilon_x, \varepsilon_y, \varepsilon_z, \delta a_x, \delta a_y, \xi_x, \xi_y]$$

$$A =$$

$$\begin{bmatrix}
0 & -(\Omega + \dot{\lambda})\sin L & \dot{L} & 0 & 1/R_r & -\Omega \sin L/R_r & 0 \\
(\Omega + \dot{\lambda})\sin L & 0 & (\Omega + \dot{\lambda})\cos L & -1/R_r & 0 & 0 & 0 \\
-\dot{L} & -(\Omega + \dot{\lambda})\cos L & 0 & 0 & -\tan L/R_r & -(\dot{\lambda}\sec L + \Omega \cos L)/R_r & 0 \\
0 & g & 0 & 0 & -2(\Omega + \dot{\lambda})\sin L & -\dot{\lambda}(2\Omega \cos^2 L + \dot{\lambda}) & 0 \\
-g & 0 & 0 & (2\Omega + \dot{\lambda})\sin L & \dot{L}\tan L & \dot{L}(2\Omega \cos L + \dot{\lambda}\sec L) & 0 \\
0 & 0 & 0 & 1 & 0 & 0 & 0 \\
0 & 0 & 0 & 0 & 1 & (\dot{\lambda}\sin L) & -(\dot{L}\tan L)
\end{bmatrix}$$

$$B = \begin{bmatrix} & 0_{3\times 2} \\ I_{5\times 5} & \\ & I_{2\times 2} \\ 0_{2\times 5} & 0_{2\times 2} \end{bmatrix}$$

where I_{ij} and 0_{ij} are, respectively, $i \times j$ identity and zero matrices. In the above, ϕ, δV and δr are, respectively, misalignment, velocity, and position errors. The subscripts x, y and z denote north, east and down components (see Figure 3.5); R_r is the radial distance from the aircraft to the Earth's centre, i.e. $R_r = R_e + h_a$ (the Earth's radius, $R_e = 2.09 \times 10^7$ ft, and h_a is aircraft altitude) and Ω is the Earth's rate of rotation (0.725×10^{-4} rad s^{-1}); g is acceleration due to gravity (32.2 ft s^{-2}); λ and L are the longitude and latitude angles; and $\dot{\lambda}$ and $\dot{L}$ are their rates of change, i.e. $\dot{L} = V_x/R_r$ and $\dot{\lambda} = (V_y/R_r)\cos L$. Notice that V denotes the nominal vehicle ground velocity while δV denotes the error in the velocity computed by the INS. The gyro drift rate is denoted by ε, δa is the accelerometer error, and ξ denotes the vertical deflection. Although their system noises usually consist of *coloured noises* (or *shaping filters*), they are assumed for the sake of simplicity to be modelled as zero-mean Gaussian white noises with spectral density matrix Q.

It is assumed that the adaptive filter will be implemented in discrete time. That is, an appropriate discrete-time representation of equation (3.78) is derived, though the system (3.78) may be handled in continuous time using various integration routines. To this end, under the condition that the ground velocity V is constant, we

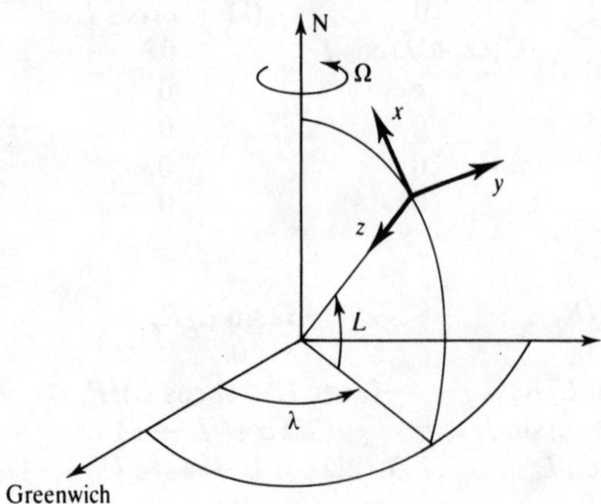

Figure 3.5 Earth coordinate and local frame coordinate systems.

introduce a state transition matrix $\Phi(t, \tau) \in \mathbb{R}^{7\times 7}$, $t \geqslant \tau$,

$$\Phi(t, \tau) = \exp[A(t - \tau)] = \sum_{i=0}^{\infty} \frac{1}{i!}[A(t - \tau)]^i \tag{3.79}$$

which is the solution to

$$\mathrm{d}\Phi(t, \tau)/\mathrm{d}t = A\Phi(t, \tau), \qquad \Phi(\tau, \tau) = I \tag{3.80}$$

Then, the system (3.78) can be rewritten as

$$\begin{aligned} x(t_{k+1}) &= \Phi(t_{k+1}, t_k)x(t_k) + \int_{t_k}^{t_{k+1}} \Phi(t_{k+1}, \tau)Bw(\tau)\mathrm{d}\tau \\ &\simeq \Phi(t_{k+1}, t_k)x(t_k) + \Phi(t_{k+1}, t_k)B\int_{t_k}^{t_{k+1}} w(\tau)\mathrm{d}\tau \end{aligned} \tag{3.81}$$

where t_k indicates the sampling instant at iteration k, $k = 0, 1, 2 \ldots$. Therefore, we obtain

$$x(k + 1) = \Phi x(k) + B_\mathrm{d} w_\mathrm{d}(k) \tag{3.82}$$

where

$$x(k + 1) \triangleq x(t_{k+1})$$

$$\Phi \triangleq \Phi(t_{k+1}, t_k) \equiv \sum_{i=0}^{N_\mathrm{s}} \frac{1}{i!}[A\Delta t]^i$$

$$B_\mathrm{d} \triangleq \Phi(t_{k+1}, t_k)B$$

$$w_\mathrm{d}(k) \triangleq \int_{t_k}^{t_{k+1}} w(\tau)\,\mathrm{d}\tau$$

in which N_s is the truncated number of the matrix exponential series expansion, and $\Delta t = t_{k+1} - t_k$ is a constant sampling interval. The covariance matrix of $w_\mathrm{d}(k)$ can be easily derived:

$$\begin{aligned} Q_\mathrm{d} \triangleq E[w_\mathrm{d}(k)w_\mathrm{d}^\mathrm{T}(k)] &= \int_{t_k}^{t_{k+1}} \int_{t_k}^{t_{k+1}} E[w(t)w^\mathrm{T}(\tau)]\,\mathrm{d}t\,\mathrm{d}\tau \\ &= \int_{t_k}^{t_{k+1}} Q\,\mathrm{d}\tau = Q\Delta t \end{aligned} \tag{3.83}$$

The two-dimensional observation vector $z(t)$ associated with equation (3.78) is of the form:

$$z(t) = Hx(t) - v(t) \tag{3.84}$$

where $v(t)$ is a zero-mean Gaussian white noise process with spectral density matrix R, which represents the Doppler/inertial system error source, and this noise process is assumed to be independent of both

$w(t)$ and $x(0)$, in which $x(0) \sim N(0, P(0))$. It is readily seen that the measurement matrix H reduces to

$$H = [0_{2\times3} \quad I_{2\times2} \quad 0_{2\times2}] \tag{3.85}$$

for this Doppler/inertial mode. The discrete-time model corresponding to equation (3.84) can be written, in a way similar to that used in the derivation of equation (3.85), as follows:

$$\begin{aligned} z(t_k) &= \frac{1}{t_k - t_{k-1}} \int_{t_{k-1}}^{t_k} [Hx(\tau) - v(\tau)]\,\mathrm{d}\tau \\ &\simeq Hx(t_k) - \frac{1}{t_k - t_{k-1}} \int_{t_{k-1}}^{t_k} v(\tau)\,\mathrm{d}\tau \end{aligned} \tag{3.86}$$

Equation (3.86) can also be rewritten in the following compact form:

$$z(k) = Hx(k) - v_{\mathrm{d}}(k) \tag{3.87}$$

where

$$z(k) \triangleq z(t_k)$$

$$v_{\mathrm{d}}(k) \triangleq \frac{1}{\Delta t} \int_{t_{k-1}}^{t_k} v(\tau)\,\mathrm{d}\tau$$

Additionally, the covariance matrix of $v_{\mathrm{d}}(k)$ is obtained by

$$\begin{aligned} R_{\mathrm{d}} \triangleq E[v_{\mathrm{d}}(k)v_{\mathrm{d}}^{\mathrm{T}}(k)] &= \frac{1}{(\Delta t)^2} \int_{t_{k-1}}^{t_k} \int_{t_{k-1}}^{t_k} E[v(t)v^{\mathrm{T}}(\tau)]\,\mathrm{d}t\,\mathrm{d}\tau \\ &= \frac{1}{(\Delta t)^2} \int_{t_{k-1}}^{t_k} R\,\mathrm{d}\tau = R/\Delta t \end{aligned} \tag{3.88}$$

The system (3.78) with uncertain dynamical parameters caused by several flight conditions can consequently be modelled as a set of candidate models, where each candidate has one possible flight condition for the system. That is, the models are described by the following discrete-time forms:

$$\theta_i{:}x(k+1) = \Phi_i x(k) + B_{\mathrm{d}_i} w_{\mathrm{d}}(k) \tag{3.89}$$

$$z(k) = Hx(k) - v_{\mathrm{d}}(k),\ i = 1, 2, \ldots, M \tag{3.90}$$

where it is assumed that the statistics Q, R and $P(0)$ are completely known and the flight conditions, consisting of the aircraft's altitude h_{a}, the nominal ground velocity V and the operating latitude L, are given; θ_i indicates the ith candidate model and M is the number of candidate models.

The adaptive or MMSE estimate for $x(k)$ is, of course, given by equation (3.39). The θ_i-conditional estimate of $x(k)$ is also obtained by the Kalman filter in discrete time:

$$\hat{x}_i(k|k-1) = \Phi_i \hat{x}_i(k-1|k-1), \qquad x_i(0|0) = 0 \tag{3.91}$$

$$P_i(k|k-1) = \Phi_i P_i(k-1|k-1)\Phi_i^{\mathrm{T}} + B_{\mathrm{d}_i} Q_{\mathrm{d}} B_{\mathrm{d}_i}^{\mathrm{T}}, \qquad P_i(0|0) = P(0) \tag{3.92}$$

$$\hat{x}_i(k|k) = \hat{x}_i(k|k-1) + K_i(k)\nu_i(k) \tag{3.93}$$

$$P_i(k|k) = [I - K_i(k)H]P_i(k|k-1) \tag{3.94}$$

for $i = 1, 2, \ldots, M$, where the innovation process $\nu_i(k)$ is given by

$$\nu_i(k) = z(k) - H\hat{x}_i(k|k-1) \tag{3.95}$$

and $K_i(k)$ is the Kalman filter gain for the ith model, which is subject to

$$K_i(k) = P_i(k|k-1)H^{\mathrm{T}}\tilde{P}_i^{-1}(k|k-1) \tag{3.96}$$

$$\tilde{P}_i(k|k-1) = HP_i(k|k-1)H^{\mathrm{T}} + R_{\mathrm{d}} \tag{3.97}$$

Simulations have been performed using the seven hypothetical sets of flight conditions tabulated in Table 3.1. Flight condition 2 was adopted as an actual one. It was assumed that the sampling interval was $\Delta t = 0.04$ hr, the 'reset interval' was equal to it, and the truncated number $N_{\mathrm{s}} = 7$ for the series expanded transition matrix. The root mean square (rms) values of INS and Doppler radar error sources considered are presented in Table 3.2.

Four simulation results are illustrated in Figures 3.6–3.12. All simulations are evaluated in terms of the rms estimation error of 50 Monte Carlo runs. For example, for the north tilt rms error of ϕ_x at time k was calculated as:

$$\text{rms error of } \phi_x = \left\{\frac{1}{N}\sum_{i=1}^{N}[\phi_x^i(k) - \hat{\phi}_x^i(k|k)]^2\right\}^{1/2}$$

where $N = 50$, $\phi_x^i(k)$ denotes ϕ_x at time k in run i and $\hat{\phi}_x^i(k|k)$ is the filtered value. The first simulation gives the results of the optimal Kalman filter matched to flight condition 2, which cannot be realized for the present situation. The second and third simulations describe the results of MMAFs based on two-candidate and three-candidate models, respectively, in which they both contain the true flight condition as one of their hypotheses. The fourth simulation presents the result of an MMAF which does not contain the true flight condition as one of three hypotheses. Notice that the solid curves are theoretical rms trajectories obtained by the square root of the diagonal elements of the covariance matrix associated with the optimal Kalman filter.

Table 3.1 Flight conditions

Flight condition number	*Altitude (ft)*	*Velocity V_y (knots)*	*Velocity V_x (knots)*	*Operating latitude (deg)*
1	3000	300	100	35
2	3000	400	100	35
3	3000	450	100	35
4	3000	500	100	35
5	3000	600	50	20
6	3000	800	10	10
7	6000	800	10	10

Table 3.2 INS error model

System error sources	*rms magnitude*
Gyros	
X random	0.316×10^{-2} (rad hr^{-1})
Y random	0.316×10^{-2} (rad hr^{-1})
Z random	0.707×10^{-2} (rad hr^{-1})
Accelerometers	
X random	20 (knots hr^{-1})
Y random	20 (knots hr^{-1})
Deflection from the vertical	
X random	0.316×10^{-4} (rad)
Y random	0.316×10^{-4} (rad)
Initial conditions	
Attitude (tilts)	0.87×10^{-2} (rad)
Azimuth misalignment	0.87×10^{-1} (rad)
Velocity (N,E)	1.0 (knots)
Latitude, longitude	0.25 (naut. mile)
Doppler/inertial system error sources	
North, East	5 (knots)

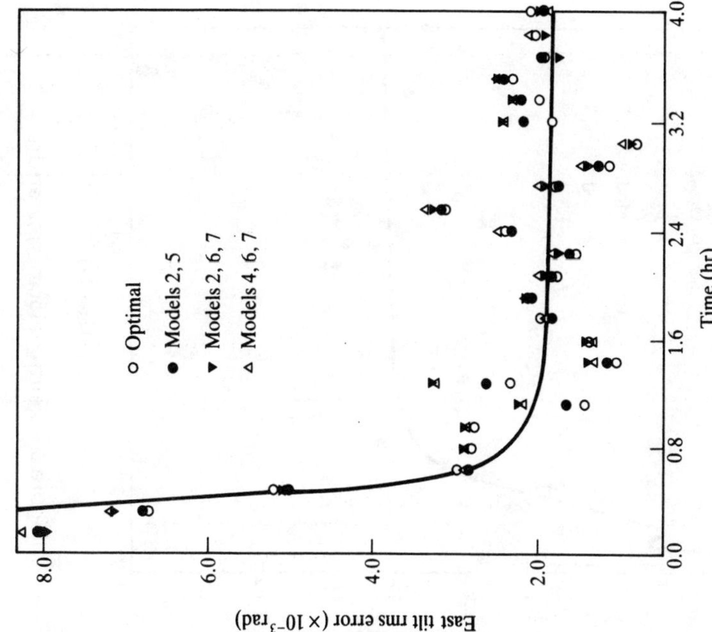

Figure 3.7 East tilt rms errors.

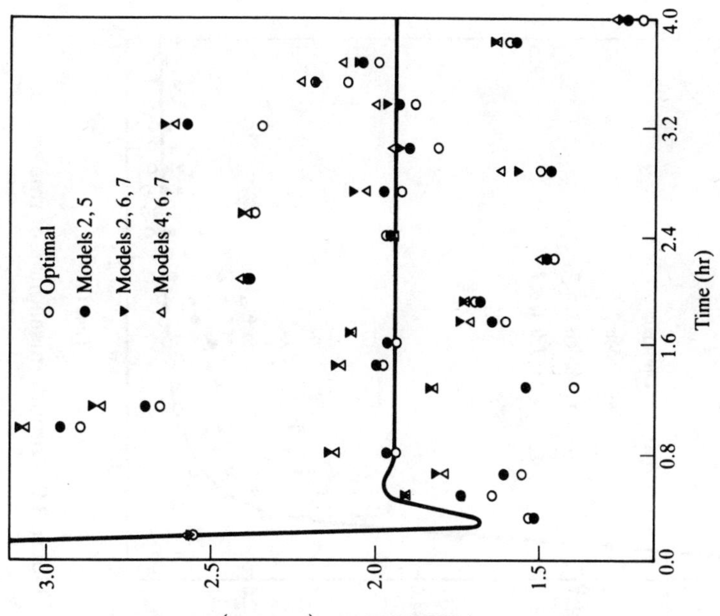

Figure 3.6 North tilt rms errors.

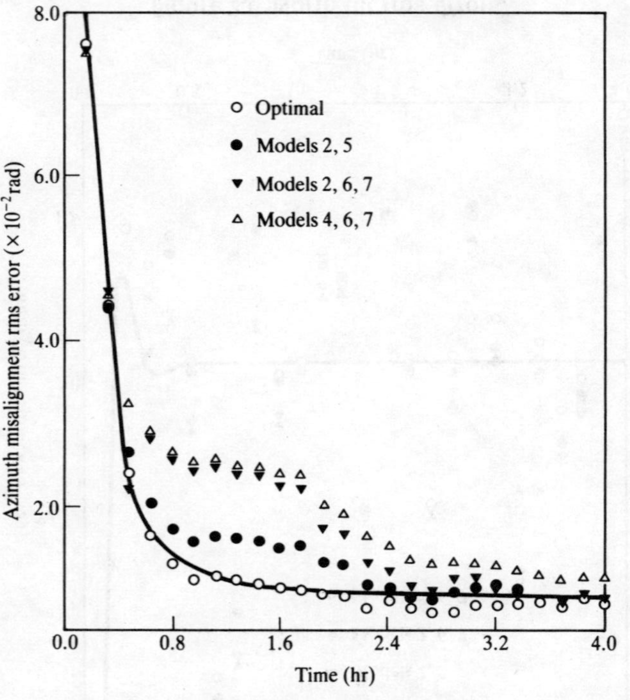

Figure 3.8 Azimuth misalignment rms errors.

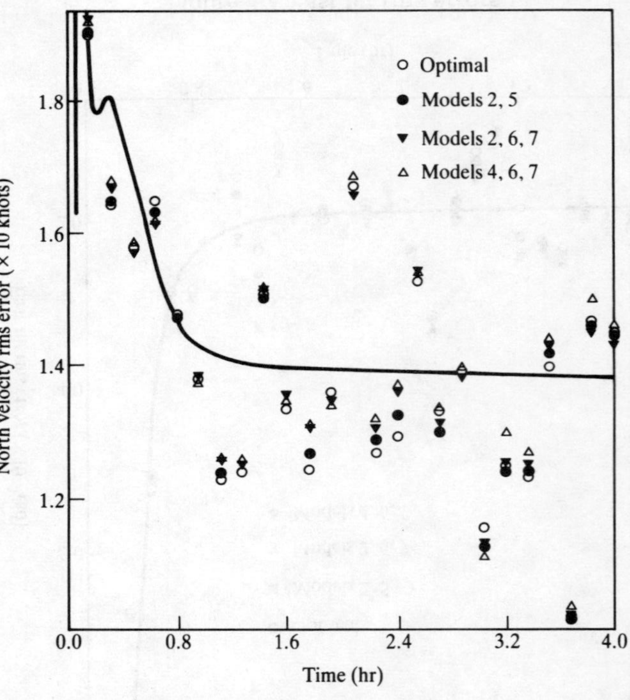

Figure 3.9 North velocity rms errors.

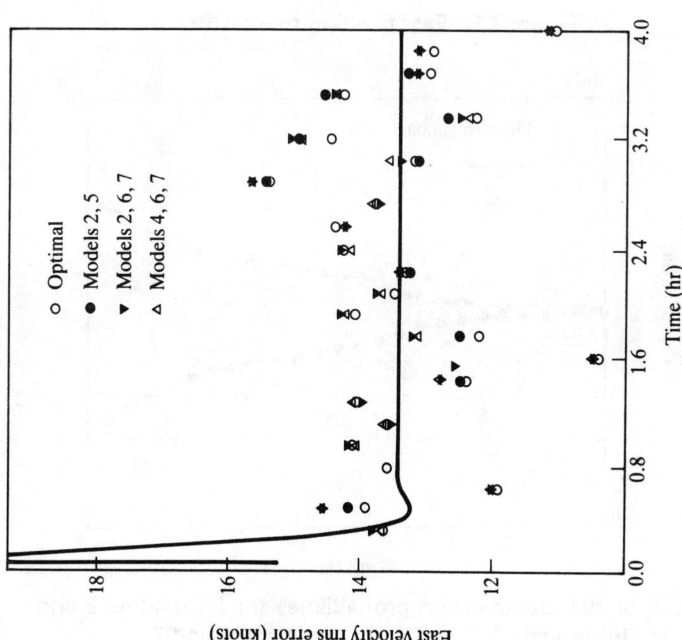

Figure 3.10 East velocity rms errors.

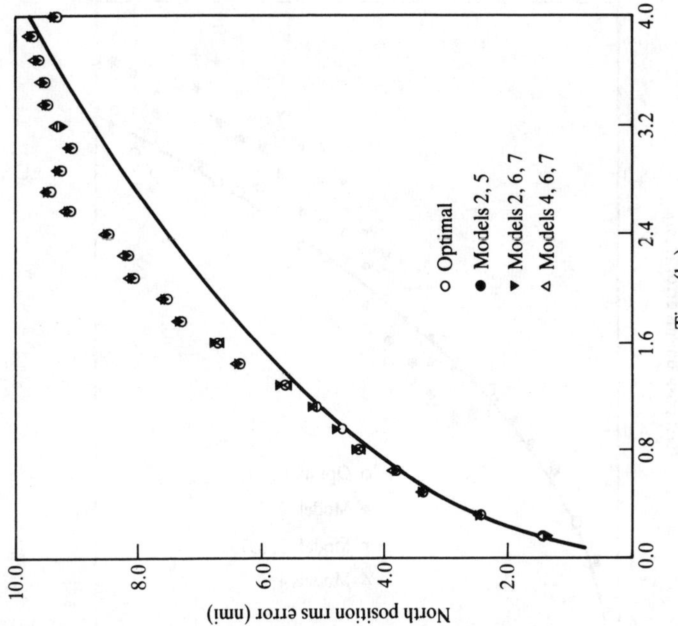

Figure 3.11 North position rms errors.

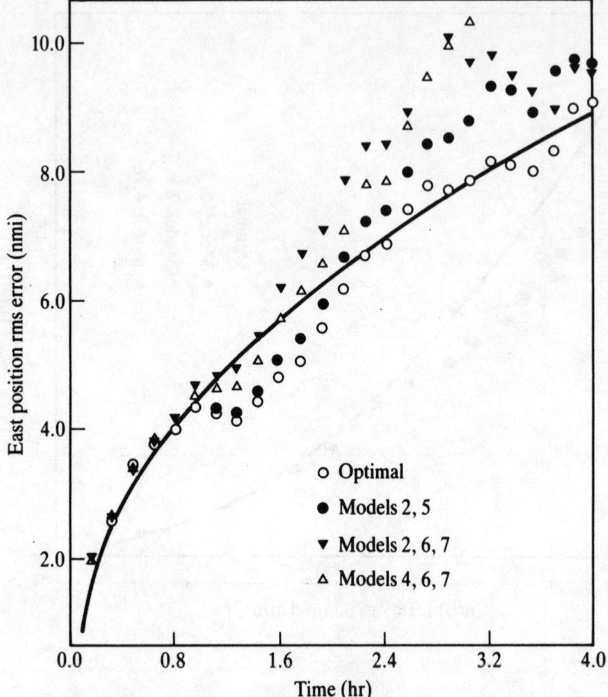

Figure 3.12 East position rms errors.

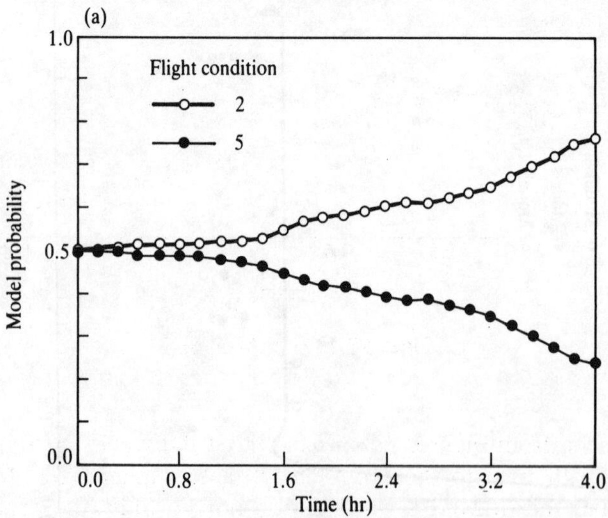

Figure 3.13 Model identification probabilities for (a) models 2 and 5; (b) models 2, 6 and 7; (c) models 4, 6 and 7.

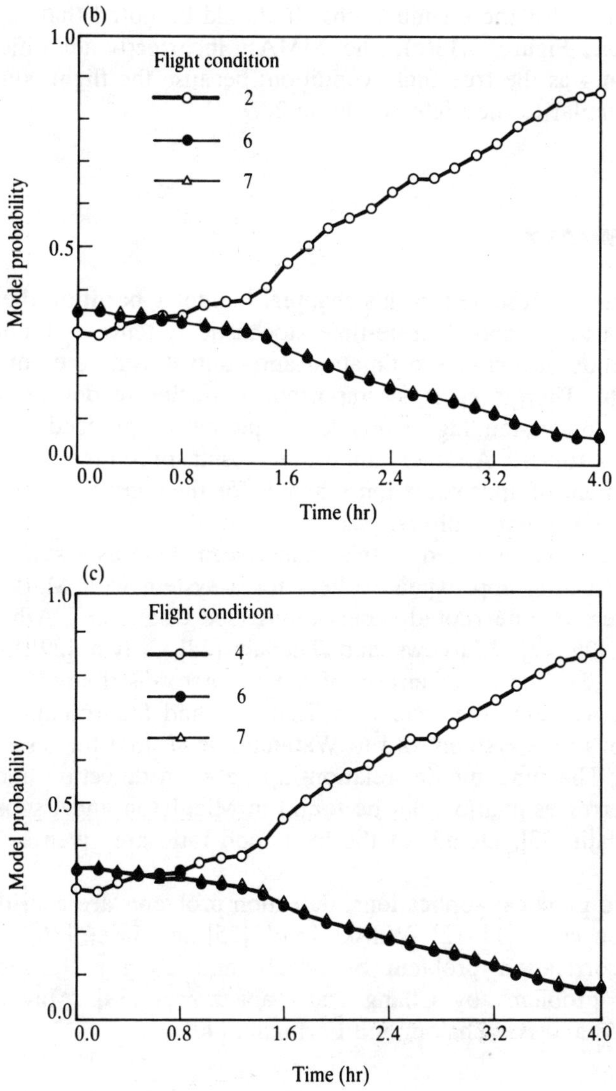

Figure 3.13 (*continued*)

It is seen from Figures 3.8 and 3.12 that the results of the MMAF converge with time to those of the optimal Kalman filter. It is also found that the MMAF, which does not include the true flight condition as a hypothesis, gives the robust response.

Figure 3.13 depicts the trajectories of the model identification probabilities for these simulations. It should be noted that, as can be seen from Figure 3.13(c), the MMAF incorrectly identifies flight condition 4 as the true flight condition; because the flight condition 4 is very similar to the flight condition 2.

3.6 SUMMARY

The material presented in this chapter, Lainiotis' partition theorem in continuous-time and discrete-time stochastic systems, is fundamental to the study of adaptive estimation and control using the partitioning approach. Therefore, it is important that the reader possesses a complete understanding of this core topic before proceeding to more advanced topics. Another important result presented here is the development of approximation schemes for the control of the number of elemental Kalman filters.

The results discussed in this chapter can serve as a starting point for constructing approximate filters for a system with Markov jump parameters or interrupted observations (see Chang and Athans [24], Tugnait [25–27], Mathews and Tugnait [28], Yavin [29]; see also Chapter 12). For an extension of the result to distributed parameter systems, we refer the reader to Tzafestas and Stavroulakis [30] for continuous-time systems and to Watanabe *et al.* [31] for discrete-time systems. The fundamental relationship between detection and mean-square error estimation can be found in Middleton and Esposito [32] and Kailath [33]. Details of the likelihood ratio are given by Kailath [5, 34].

As to practical applications, detection problems are considered by Gustafson *et al.* [11, 12], Willsky *et al.* [35] and Tylee [36]; a terrain height correlation problem by Mealy and Tang [37]; and target tracking problems by Chang and Tabaczynski [38], Maybeck and Suizu [39] and Bar-Shalom and Fortmann [40].

APPENDIX 3A SOME PROPERTIES OF INNOVATION PROCESSES

To supplement the rigorous proof of the likelihood ratio which appears in the partition theorem of continuous-time systems, the concept of the innovation process [5] is reviewed in this appendix.

The innovation process which is suitable for the system of equations (3.1) and (3.2) with $\theta = \theta^*$, is defined by

$$\mathrm{d}\nu(t) \triangleq \mathrm{d}y(t) - \hat{X}(t|t)\,\mathrm{d}t \tag{3A.1a}$$

$$= \tilde{X}(t|t)\,\mathrm{d}t + \mathrm{d}\eta(t) \tag{3A.1b}$$

where

$$\hat{X}(t|t) = H(t)\hat{x}(t|t)$$

$$\tilde{X}(t|t) = H[x(t) - \hat{x}(t|t)]$$

Let $X(t)$ be a second-order process (i.e. $E[\|X(t)\|^2] < \infty$) and

$$\int_{t_0}^{t_f} E[\|X(t)\|^2]\,\mathrm{d}t < \infty \tag{3A.2}$$

Then the following lemma is obtained for the innovation process.

Lemma 3A.1

Let $\{\nu(t), \mathcal{Y}_t, t_0 \leq t \leq t_f\}$ be a real stochastic process, where the $\{\mathcal{Y}_t\}$ are the increasing Borel fields generated by the observations $\{y(s), s \leq t\}$. Then the following conditions are obtained for the innovation process.

(i) Almost all sample functions of the process are continuous.

(ii) The process is a *martingale*, that is,

$$E[\nu(t)|\mathcal{Y}_s, s < t] = \nu(s) \tag{3A.3}$$

holds with probability 1.

(iii)

$$E\{[\nu(t) - \nu(s)][\nu(t) - \nu(s)]^{\mathrm{T}}|\mathcal{Y}_s, s < t\}$$

$$= \int_s^t R(\tau)\,\mathrm{d}\tau \tag{3A.4}$$

holds with probability 1.

Proof

For the sake of future developments, we introduce the Borel field $\mathcal{B}_t$, which is generated by the stochastic process $\{X(s), \eta(s), s \leq t\}$. Clearly, $\mathcal{Y}_t \subseteq \mathcal{B}_t$. The proof of Lemma 3A.1 proceeds as follows:

(i) Integrating equation (3A.1b) from $t = s$ to $t = t$ yields

$$\nu(t) = \nu(s) + \int_s^t \tilde{X}(\sigma|\sigma)\,\mathrm{d}\sigma + \eta(t) - \eta(s) \tag{3A.5}$$

Since $X(t)$ is a second-order process, $\hat{X}(t|t)$ and $\tilde{X}(t|t)$ are second-order processes by Jensen's inequality. Therefore, the integral in equation (3A.5) is a continuous function of its upper limit. Further, since $\eta(\cdot)$ is also continuous, the process $\nu(\cdot)$ must be continuous.

(ii) The fact that $\nu(s)$ is $\mathcal{Y}_s$-measurable gives

$$E[\nu(t)|\mathcal{Y}_s] - \nu(s) = E[\nu(t) - \nu(s)|\mathcal{Y}_s] \tag{3A.6}$$

and substituting equation (3A.5) into (3A.6) yields

$$E[\nu(t)|\mathcal{Y}_s] - \nu(s) = E[\eta(t) - \eta(s)|\mathcal{Y}_s] + \int_s^t E[\tilde{X}(\sigma|\sigma)|\mathcal{Y}_s]\,d\sigma \tag{3A.7}$$

Here, notice that the first term of the right-hand side of equation (3A.7) is zero because the Wiener process increment is independent of $\mathcal{Y}_s$. The interchange of expectation and integration in equation (3A.7) is readily justified by Fubini's theorem. Using these results, making use of the smoothing property of conditional expectations, i.e.

$$E[E[\cdot|\mathcal{Y}_\tau]|\mathcal{Y}_s] = E[\cdot|\mathcal{Y}_s], \qquad \mathcal{Y}_s \subseteq \mathcal{Y}_\tau \tag{3A.8}$$

we obtain

$$\begin{aligned} E[\nu(t)|\mathcal{Y}_s] - \nu(s) &= \int_s^t E[X(\sigma) - E\{X(\sigma)|\mathcal{Y}_\sigma\}|\mathcal{Y}_s]\,d\sigma \\ &= \int_s^t [E\{X(\sigma)|\mathcal{Y}_s\} \\ &\qquad - E\{X(\sigma)|\mathcal{Y}_s\}]\,d\sigma = 0 \end{aligned} \tag{3A.9}$$

(iii) The left-hand side of equation (3A.4), from substitution of (3A.5), becomes

$$\begin{aligned} &E\{[\nu(t) - \nu(s)][\nu(t) - \nu(s)]^{\mathrm{T}}|\mathcal{Y}_s\} \\ &= E\{[\eta(t) - \eta(s)][\eta(t) - \eta(s)]^{\mathrm{T}}|\mathcal{Y}_s\} \\ &\quad + \int_s^t E\{[\eta(t) - \eta(s)]\tilde{X}^{\mathrm{T}}(\sigma|\sigma)|\mathcal{Y}_s\}\,d\sigma \\ &\quad + \int_s^t E\{\tilde{X}(\sigma|\sigma)[\eta(t) - \eta(s)]^{\mathrm{T}}|\mathcal{Y}_s\}\,d\sigma \\ &\quad + E\left[\int_s^t \int_s^t \tilde{X}(\sigma|\sigma)\tilde{X}^{\mathrm{T}}(\tau|\tau)\,d\sigma\,d\tau|\mathcal{Y}_s\right] \end{aligned} \tag{3A.10}$$

Using equation (3A.8), the integrand of the second term of the right-hand side in equation (3A.10) can be rewritten as

$$\begin{aligned}
&E\{E\{[\eta(t) - \eta(\sigma)]\widetilde{X}^{\mathrm{T}}(\sigma|\sigma)|\mathcal{B}_\sigma\}|\mathcal{Y}_s\} \\
&\quad + E\{[\eta(\sigma) - \eta(s)]\widetilde{X}^{\mathrm{T}}(\sigma|\sigma)|\mathcal{Y}_s\} \\
&= E\{E\{[\eta(t) - \eta(\sigma)]|\mathcal{B}_\sigma\}\widetilde{X}^{\mathrm{T}}(\sigma|\sigma)|\mathcal{Y}_s\} \\
&\quad + E\{[\eta(\sigma) - \eta(s)]\widetilde{X}^{\mathrm{T}}(\sigma|\sigma)|\mathcal{Y}_s\} \\
&= 0 + E\{[\eta(\sigma) - \eta(s)]\widetilde{X}^{\mathrm{T}}(\sigma|\sigma)|\mathcal{Y}_s\}
\end{aligned} \tag{3A.11}$$

Using the symmetry of the integrand of the double integral in the fourth term of equation (3A.10) gives

$$\begin{aligned}
\int_s^t \int_s^t \widetilde{X}(\sigma|\sigma)\widetilde{X}^{\mathrm{T}}(\tau|\tau)\,\mathrm{d}\sigma\,\mathrm{d}\tau &= \int_s^t \widetilde{X}(\sigma|\sigma)\int_s^\sigma \widetilde{X}^{\mathrm{T}}(\tau|\tau)\,\mathrm{d}\sigma\,\mathrm{d}\tau \\
&\quad + \int_s^t \int_s^\sigma \widetilde{X}(\tau|\tau)\,\mathrm{d}\tau\widetilde{X}^{\mathrm{T}}(\sigma|\sigma)\,\mathrm{d}\sigma
\end{aligned} \tag{3A.12}$$

Therefore, we can write

$$\begin{aligned}
&\int_s^t E\{[\eta(t) - \eta(s)]\widetilde{X}^{\mathrm{T}}(\sigma|\sigma)|\mathcal{Y}_s\}\,\mathrm{d}\sigma \\
&\quad + \int_s^t E\{\widetilde{X}(\sigma|\sigma)[\eta(t) - \eta(s)]^{\mathrm{T}}|\mathcal{Y}_s\}\,\mathrm{d}\sigma \\
&\quad + E\left[\int_s^t \int_s^t \widetilde{X}(\sigma|\sigma)\widetilde{X}^{\mathrm{T}}(\tau|\tau)\,\mathrm{d}\sigma\,\mathrm{d}\tau|\mathcal{Y}_s\right] \\
&= \int_s^t E\left\{\left[\eta(\sigma) - \eta(s) + \int_s^\sigma \widetilde{X}(\tau|\tau)\,\mathrm{d}\tau\right]\widetilde{X}^{\mathrm{T}}(\sigma|\sigma)|\mathcal{Y}_s\right\}\mathrm{d}\sigma \\
&\quad + \int_s^t E\left\{\widetilde{X}(\sigma|\sigma)\left[\eta(\sigma) - \eta(s)\right.\right. \\
&\quad \left.\left. + \int_s^\sigma \widetilde{X}(\tau|\tau)\,\mathrm{d}\tau\right]^{\mathrm{T}}|\mathcal{Y}_s\right\}\mathrm{d}\sigma
\end{aligned} \tag{3A.13}$$

Then, the first term in equation (3A.13) becomes

$$\begin{aligned}
&\int_s^t E\left\{\left[\eta(\sigma) - \eta(s) + \int_s^\sigma \widetilde{X}\,(\tau|\tau)\,\mathrm{d}\tau\right]\widetilde{X}^{\mathrm{T}}(\sigma|\sigma)|\mathcal{Y}_s\right\}\mathrm{d}\sigma \\
&= \int_s^t E\{E\{[\nu(\sigma) - \nu(s)]\widetilde{X}^{\mathrm{T}}(\sigma|\sigma)|\mathcal{Y}_\sigma\}|\mathcal{Y}_s\}\,\mathrm{d}\sigma \\
&= \int_s^t E\{[\nu(\sigma) - \nu(s)][E\{\widetilde{X}^{\mathrm{T}}(\sigma|\sigma)|\mathcal{Y}_\sigma\}]|\mathcal{Y}_s\}\,\mathrm{d}\sigma \\
&= 0
\end{aligned} \tag{3A.14}$$

where in the above derivation the fact that $\nu(\sigma) - \nu(s)$ is $\mathcal{Y}_\sigma$-measurable has been used, and the last equality is apparent from the martingale property in condition (ii). Similarly, the second term of the right-hand side of equation (3A.13) reduces to zero, and hence we have

$$
\begin{aligned}
E\{[\nu(t) - \nu(s)][\nu(t) - \nu(s)]^{\mathrm{T}} | \mathcal{Y}_s\}
&= E\{[\eta(t) - \eta(s)][\eta(t) - \eta(s)]^{\mathrm{T}} | \mathcal{Y}_s\} \\
&= E\{[\eta(t) - \eta(s)][\eta(t) - \eta(s)]^{\mathrm{T}}\} \\
&= \int_s^t R(\sigma)\,\mathrm{d}\sigma
\end{aligned}
\tag{3A.15}
$$

This completes the proof. □

The following lemma which corresponds to the main theorem of Girsanov [5,41] is restated in the innovation theory.

Lemma 3A.2

If in the detection problem given in equation (3.13),

(i) $\hat{X}_1(t|t, \omega)$ is measurable in t and ω;
(ii) $\hat{X}_1(t|t, \omega)$, for every t, is measurable with respect to the past $\{y(s), s \leq t\}$;
(iii) $\int_{t_0}^{t_f} \|\hat{X}_1(t|t, \omega)\|^2\,\mathrm{d}t < \infty$, for almost all ω;
(iv) $\tilde{\mu}_1 \ll \tilde{\mu}_0$, i.e. $\tilde{\mu}_1$ is absolutely continuous with respect to $\tilde{\mu}_0$, or equivalently $\mu_1 \ll \mu_0$;

then the likelihood ratio function is given by

$$
\begin{aligned}
\mathrm{d}\tilde{\mu}_1/\mathrm{d}\tilde{\mu}_0 = \exp\Bigg[&\int_{t_0}^{t_f} \langle \hat{X}_1(t|t),\, R^{-1}(t)\,\mathrm{d}y(t) \rangle \\
&- \frac{1}{2}\int_{t_0}^{t_f} \|\hat{X}_1(t|t)\|^2_{R^{-1}(t)}\,\mathrm{d}t \Bigg]
\end{aligned}
\tag{3A.16}
$$

Proof

First, condition (i) has been assumed for $X(t, \omega)$ and will be inherited by the conditional expectation $\hat{X}_1(t|t, \omega)$. Condition (ii) is obvious from the definition $\hat{X}_1(t|t, \omega) = E[X(t, \omega)|y(s, \omega),$

$s \leqslant t, h_1] = E[X(t, \omega)|\mathcal{Y}_t, h_1]$. Condition (iii) follows from the integrability of the mean of $||X(t)||^2$. That is, we note that by Jensen's inequality [42] (see also Appendix B)

$$||\hat{X}_1(t|t)||^2 \triangleq ||E[X(t)|\mathcal{Y}_t, h_1]||^2 \leqslant E\{||X(t)||^2|\mathcal{Y}_t, h_1\} \quad \text{(3A.17)}$$

Therefore, taking the h_1-conditional expectation of equation (3A.17) gives

$$\begin{aligned} E\{||\hat{X}_1(t|t)||^2|h_1\} &\leqslant E\{E[||X(t)||^2|\mathcal{Y}_t, h_1]|h_1\} \\ &= E[||X(t)||^2|h_1] \end{aligned} \quad \text{(3A.18)}$$

Integrating both sides of equation (3A.18), using condition (3A.2), yields

$$\begin{aligned} \int_{t_0}^{t_f} E\{||\hat{X}_1(t|t)||^2|h_1\}\,dt &= E\left[\int_{t_0}^{t_f} ||\hat{X}_1(t|t)||^2\,dt|h_1\right] \\ &\leqslant \int_{t_0}^{t_f} E[||X(t)||^2|h_1]\,dt \\ &= E\left[\int_{t_0}^{t_f} ||X(t)||^2\,dt|h_1\right] < \infty \end{aligned} \quad \text{(3A.19)}$$

where Fubini's theorem has been employed. Condition (iv) is obtained by Lemma 3A.3 without proof. □

Lemma 3A.3

Consider the detection problem given in equation (3.12).

(i) If $X(t)$ is deterministic, a necessary and sufficient condition for $\mu_1 \ll \mu_0$ is that $\int_{t_0}^{t_f}||X(t)||^2\,dt < \infty$. Furthermore, $\mu_1 \ll \mu_0$ implies $\mu_0 \ll \mu_1$.

(ii) If $X(t)$ is a zero-mean random process, independent of $\eta(t)$, then the finiteness of $\int_{t_0}^{t_f}||X(t, \omega)||^2\,dt$, for almost all ω, is a sufficient condition for $\mu_1 \ll \mu_0$. However, it is not necessary; furthermore, unless X is Gaussian, $\mu_1 \ll \mu_0$ does not generally imply $\mu_0 \ll \mu_1$.

EXERCISES

3.1 Consider the result of Example 3.2.

(a) Show that

$$\hat{\theta}(t) = E[\theta|\mathcal{Y}_t] = \frac{\rho\Lambda(t|\theta_1)}{1 + \rho\Lambda(t|\theta_1)}$$

(b) Verify that the stochastic evolution for $\hat{\theta}(t)$ can be written by

$$\mathrm{d}\hat{\theta}(t) = \hat{\theta}(t)[1 - \hat{\theta}(t)]\hat{x}_1^{\mathrm{T}}(t|t)H^{\mathrm{T}}(t)R^{-1}(t) \times [\mathrm{d}y(t) - \hat{\theta}(t)H(t)\hat{x}_1(t|t)\,\mathrm{d}t], \qquad \hat{\theta}(t_0) = p(\theta_1)$$

where

$$\hat{x}_1(t|t) = E[x(t)|\mathcal{Y}_t, \theta_1]$$

3.2 (Joint Detection, Estimation and Systems Identification [9]) Consider the following nonlinear system described by

$$\frac{\mathrm{d}x(t)}{\mathrm{d}t} = f(x(t), t; \theta) + g(x(t), t; \theta)w(t)$$

$$z(t) = s(t) + v(t)$$

$$s(t) = \beta y(t)$$

$$y(t) = h(x(t), t; \theta)$$

where $x \in \mathbb{R}^n$ is the state vector; $w \in \mathbb{R}^p$ is a zero-mean Gaussian white noise process with identity covariance matrix; $z \in \mathbb{R}^m$ is the measurement vector; and $v \in \mathbb{R}^m$ is a zero-mean Gaussian white noise process with covariance matrix $R(t)$. β is the so-called indicator variable which takes values 1 or 0 depending on whether H_1 (signal present) or H_0 (signal absent) is true, with a priori probability p_1, p_0, where $p_1 + p_0 = 1$. θ is a time-invariant parameter vector; if it is known, the above model is completely specified. However, θ is considered to be unknown, i.e. it is considered a random variable with known a priori probability density $p(\theta)$. Note that the functions f, g and h are time-varying nonlinear functions of the state vector $x(t)$, and of θ. Furthermore, it is assumed that

$$E[w(t)v^{\mathrm{T}}(s)] = 0$$

$$E[x(t_0)w^{\mathrm{T}}(t)] = E[x(t_0)v^{\mathrm{T}}(t)] = 0$$

$$x(t_0) \sim N(\hat{x}(t_0, \theta), P(t_0, \theta))$$

(a) Apply Theorems 3.1 and 3.2 to derive the following:

$$\hat{\beta}(t) = \frac{\rho\int\Lambda(t|\theta)p(\theta)\,\mathrm{d}\theta}{1 + \rho\int\Lambda(t|\theta)p(\theta)\,\mathrm{d}\theta}$$

$$\hat{s}(t|t) = \hat{\beta}(t)\hat{s}_1(t|t)$$

$$\hat{\theta}(t) = \hat{\theta}(t_0)p(\beta = 0|Z_t) + \frac{\rho\int\theta\Lambda(t|\theta)p(\theta)\,\mathrm{d}\theta}{1 + \rho\int\Lambda(t|\theta)p(\theta)\,\mathrm{d}\theta}$$

where

$$\hat{\beta}(t) \triangleq E[\beta | Z_t]$$
$$= p(\beta = 1 | Z_t) = 1 - p(\beta = 0 | Z_t)$$
$$\rho \triangleq p_1/p_0$$
$$\hat{s}(t|t) \triangleq E[s(t) | Z_t]$$
$$\hat{s}_1(t|t) \triangleq E[s(t) | Z_t, \beta = 1]$$
$$\hat{\theta}(t) \triangleq E[\theta | Z_t]$$

and where

$$p(\theta | Z_t) = \frac{1 + \rho\Lambda(t|\theta)}{1 + \rho\int\Lambda(t|\theta)p(\theta)\,d\theta}\; p(\theta)$$

$\hat{\theta}(t_0) = E[\theta]$ is the a priori mean

and

$$\Lambda(t|\theta) = \exp\left[\int_{t_0}^{t} \hat{h}_1^{T}(\sigma|\sigma, \theta)R^{-1}(\sigma)z(\sigma)\,d\sigma - \frac{1}{2}\int_{t_0}^{t} ||\hat{h}_1(\sigma|\sigma, \theta)||^2_{R^{-1}(\sigma)}\,d\sigma\right]$$

where $\hat{h}_1(\sigma|\sigma, \theta) \triangleq E[h(x(\sigma), \sigma; \theta)|\beta = 1, \theta, Z_\sigma]$.

(b) Prove that

$$\hat{s}(t|t) = \hat{\beta}(t)\hat{h}_1(t|t)$$
$$P_s(t|t) = \hat{\beta}(t)P_{h_1}(t|t)$$

where

$$\hat{h}_1(t|t) \triangleq E[h(x(t), t; \theta)|\beta = 1, Z_t]$$
$$P_s(t|t) \triangleq E\{[s(t) - \hat{s}(t|t)][s(t) - \hat{s}(t|t)]^T | Z_t\}$$

and

$$P_{h_1}(t|t) \triangleq E\{[h(x(t), t; \theta) - \hat{h}_1(t|t)] \times [h(x(t), t; \theta) - \hat{h}_1(t|t)]^T | \beta = 1, Z_t\}$$

$$\hat{h}_1(t|t) = \int \hat{h}_1(t|t, \theta)p_1(\theta|Z_t)\,d\theta$$

where $p_1(\theta|Z_t)$ is given by

$$p_1(\theta|Z_t) \triangleq p(\theta|\beta = 1, Z_t) = \frac{\Lambda(t|\theta)p(\theta)}{\int\Lambda(t|\theta)p(\theta)\,d\theta}$$

The conditional estimation error covariance matrix $P_{h_1}(t|t)$ is given by

$$P_{h_1}(t|t) = \int \{P_{h_1}(t|t, \theta) + [\hat{h}_1(t|t, \theta) - \hat{h}_1(t|t)][\hat{h}_1(t|t, \theta) - \hat{h}_1(t|t)]^{\mathrm{T}}\} p_1(\theta|Z_t)\,\mathrm{d}\theta$$

where

$$P_{h_1}(t|t, \theta) \triangleq E\{[h(x(t), t; \theta) - \hat{h}_1(t|t, \theta)] \times [h(x(t), t; \theta) - \hat{h}_1(t|t, \theta)]^{\mathrm{T}} | \beta = 1, \theta, Z_t\}$$

3.3 Consider again the problem posed in Exercise 3.2, but for the linear dynamic system case:

$$f(x(t), t; \theta) \equiv A(t, \theta)x(t)$$

$$g(x(t), t; \theta) \equiv B(t, \theta)$$

$$h(x(t), t; \theta) \equiv H(t, \theta)x(t)$$

Show that

$$\hat{s}(t|t) = \hat{\beta}(t)\hat{y}_1(t|t)$$

$$P_s(t|t) = \hat{\beta}(t)\Big[\int \{H(t, \theta)P_1(t|t, \theta)H^{\mathrm{T}}(t, \theta) + [H(t, \theta)\hat{x}_1(t|t, \theta) - \hat{y}_1(t|t)] \times [H(t, \theta)\hat{x}_1(t|t, \theta) - \hat{y}_1(t|t)]^{\mathrm{T}}\} p_1(\theta|Z_t)\,\mathrm{d}\theta\Big]$$

where

$$\hat{y}_1(t|t) = \int H(t, \theta)\hat{x}_1(t|t, \theta)p_1(\theta|Z_t)\,\mathrm{d}\theta$$

and $\hat{\beta}(t)$ and $p_1(\theta|Z_t)$ are given by the stochastic differential equations:

$$\mathrm{d}\hat{\beta}(t) = \hat{\beta}(t)[1 - \hat{\beta}(t)]\hat{y}_1^{\mathrm{T}}(t|t)R^{-1}(t) \times [\mathrm{d}z(t) - \hat{\beta}(t)\hat{y}_1(t|t)\,\mathrm{d}t], \qquad \hat{\beta}(t_0) = p_1$$

$$\mathrm{d}p_1(\theta|Z_t) = p_1(\theta|Z_t)[H(t, \theta)\hat{x}_1(t|t, \theta) - \hat{y}_1(t|t)] \times R^{-1}(t)[\mathrm{d}z(t) - \hat{y}_1(t|t)\,\mathrm{d}t], \qquad p_1(\theta|Z_{t_0}) = p(\theta)$$

$p(\theta|Z_t)$ is also subject to the following stochastic differential equation:

$$\mathrm{d}p(\theta|Z_t) = \hat{\beta}(t)[p_1(\theta|Z_t)H(t, \theta)\hat{x}_1(t|t, \theta) - p(\theta|Z_t)\hat{y}_1(t|t)]R^{-1}(t) \times [\mathrm{d}z(t) - \hat{\beta}(t)\hat{y}_1(t|t)\,\mathrm{d}t], \qquad p(\theta|Z_{t_0}) = p(\theta)$$

$\Lambda(t|\theta)$ can be written

$$\Lambda(t|\theta) = \exp\left[\int_{t_0}^{t} \hat{x}_1^{\mathrm{T}}(\sigma|\sigma, \theta)H^{\mathrm{T}}(\sigma, \theta)R^{-1}(\sigma)z(\sigma)\,\mathrm{d}\sigma - \frac{1}{2}\int_{t_0}^{t} \|H(\sigma, \theta)\hat{x}_1(\sigma|\sigma, \theta)\|^2_{R^{-1}(\sigma)}\,\mathrm{d}\sigma\right]$$

Here, $\{\hat{x}_1(t|t, \theta), P_1(t|t, \theta)\}$ are now given by the Kalman filter matched to θ:

$$\mathrm{d}\hat{x}_1(t|t, \theta) = A(t, \theta)\hat{x}_1(t|t, \theta)\,\mathrm{d}t + P_1(t|t, \theta)H^{\mathrm{T}}(t, \theta)R^{-1}(t)[\mathrm{d}z(t) - H(t, \theta)\hat{x}_1(t|t, \theta)\,\mathrm{d}t],$$

$$\hat{x}_1(t_0|t_1, \theta) = \hat{x}(t_0, \theta)$$

$$\frac{\mathrm{d}P_1(t|t, \theta)}{\mathrm{d}t} = A(t, \theta)P_1(t|t, \theta) + P_1(t|t, \theta)A^{\mathrm{T}}(t, \theta) + B(t, \theta)B^{\mathrm{T}}(t, \theta) - P_1(t|t, \theta)H^{\mathrm{T}}(t, \theta)R^{-1}(t)H(t, \theta)P_1(t|t, \theta),$$

$$P_1(t_0|t_0, \theta) = P(t_0, \theta)$$

3.4 Consider expression (3.41).

(a) Show that

$$p(\theta_i|k) = \left[1 + \sum_{\substack{j=1\\ j\neq i}}^{M} \frac{L(k|\theta_j)p(\theta_j|k-1)}{L(k|\theta_i)p(\theta_i|k-1)}\right]^{-1}$$

where $L(k|\theta_i)$ is defined in equation (3.42).

(b) Defining a conditional likelihood ratio function in discrete time as

$$L_{ji}(k) = \frac{p(z(k)|Z_{k-1}, \theta_j)}{p(z(k)|Z_{k-1}, \theta_i)}$$

show that $p(\theta_i|k)$ in part (a) can be rewritten as

$$p(\theta_i|k) = \left[1 + \sum_{\substack{j=1\\ j\neq i}}^{M} L_{ji}(k)\frac{p(\theta_j|k-1)}{p(\theta_i|k-1)}\right]^{-1}$$

where

$$L_{ji}(k) = \left\{\frac{|\tilde{P}(k|k-1, \theta_i)|}{|\tilde{P}(k|k-1, \theta_j)|}\right\}^{1/2} \times \exp\{-\tfrac{1}{2}[\|\nu(k, \theta_j)\|^2_{\tilde{P}^{-1}(k|k-1), \theta_j)} - \|\nu(k, \theta_i)\|^2_{\tilde{P}^{-1}(k|k-1, \theta_i)}]\}$$

3.5 Consider the single-axis motion of a vehicle affected by control input u

and drag a [7]. Letting x_1 denote vehicle position, and x_2 velocity, a continuous-time system of the motion is described by

$$\begin{bmatrix} \dot{x}_1(t) \\ \dot{x}_2(t) \end{bmatrix} = \begin{bmatrix} 0 & 1 \\ 0 & -a \end{bmatrix} \begin{bmatrix} x_1(t) \\ x_2(t) \end{bmatrix} + \begin{bmatrix} 0 \\ 1 \end{bmatrix} u(t)$$

Let radar equipment be located at the origin, and take measurements every $\Delta t = 0.1$ s:

$$z(k) = [1 \quad 0]x(k) + v(k)$$

with $R(k) = 1$. The equivalent discrete-time model of the system is

$$\begin{bmatrix} x_1(k+1) \\ x_2(k+1) \end{bmatrix} = \begin{bmatrix} 1 & \frac{1}{a}(1 - \exp(-a\Delta t)) \\ 0 & \exp(-a\Delta t) \end{bmatrix} \begin{bmatrix} x_1(k) \\ x_2(k) \end{bmatrix}$$
$$+ \begin{bmatrix} 0 \\ \frac{1}{a}(1 - \exp(-a\Delta t)) \end{bmatrix} u(k) + \begin{bmatrix} w_1(k) \\ w_2(k) \end{bmatrix}$$

where noises $\{w_1(k), w_2(k)\}$ with independent covariances Q_1 and Q_2 are added to reflect uncertainty in the model. The initial states are $x_1(t_0) = 100$ and $x_2(t_0) = 50$.

(a) It is assumed that a takes only the values $a_1 = 0$, $a_2 = 0.5$ or $a_3 = 1$. Holding the parameter a constant in the real system, and equal to one of the three possible values, implement the MMAF in discrete time for this system. Here, suppose that the elemental filters are initiated with $\hat{x}_i(0|0)$ set to the correct $x(t_0)$ values, $P_i(0|0) = I$, $u \equiv 0$, $Q_1 = Q_2 = 0$, and the initial probabilities uniformly distributed: $p(\theta_i) = \frac{1}{3}$ for $i = 1$, 2, 3.

(b) Let the true parameter value undergo jump changes so that it equals $a_1 = 0$ for $t \in [0, 2)$, $a_2 = 0.5$ for $t \in [2, 4)$, and $a_3 = 1$ for $t \in [4, 6]$. Assume that the control force u is equal to 50 for all time, and known to the estimator. By setting a lower bound $p_L = 0.0005$ for $p(\theta_i|k)$, $i = 1$, 2, 3, and adding a pseudonoise of strength ($Q_1 = Q_2 = 1$) to the filter models, run the MMAF in discrete time for this system.

3.6 Consider the following nonlinear tracking system [43] (see also Chapter 7):

$$x(k+1) = x(k) + \Delta t v_x$$

$$y(k+1) = y(k) + \Delta t v_y$$

$$V_x(k+1) = V_x(k)$$

$$V_y(k+1) = V_y(k)$$

$$\beta(k) = \beta_r(k) + v(k)$$

$$\beta_r(k) \triangleq \tan^{-1}\left[\frac{y(k) - y_s(k)}{x(k) - x_s(k)}\right]$$

where $x_s(k)$ and $y_s(k)$ are the two components of the sensor position at time k, which are known; Δt is the sampling interval; and $\{v(k)\}$ is a Gaussian white noise process with mean m and variance R. $x(k)$ and $y(k)$ are the two components of the vehicle at time k, and $V_x(k)$ and $V_y(k)$ are the corresponding velocities. $\beta(k)$ is the bearing measurement.

(a) Referring $x(k)$, $y(k)$ to $x(0)$, $y(0)$ and taking into account that $\beta_r(k) \gg v(k)$ and $v(k)$ is small, show that

$$z(k) = -x(0)\tan\beta(k) + y(0) - k\Delta t V_x \tan\beta(k) + k\Delta t V_y + \tilde{v}(k)$$

where

$$z(k) \triangleq y_s(k) - x_s(k)\tan\beta(k)$$

$$\tilde{v}(k) \triangleq v(k)[k\Delta t V_x + x(0) - x_s(k)]$$

(Hint: When $\beta_r(k) \gg v(k)$, it follows that $\beta_r(k) \simeq \tan\beta(k) - v(k)$, where $v(k)$ is in radians.)

(b) Defining the state vector $X(k) = [y(0) \quad V_y]^T$ and assuming that the parameter vector $\theta = [x(0) \quad V_x]^T$ takes on values $\{\theta_i = [x^i(0) \quad V_x^i]^T\}$, $i = 1, \ldots, M$, derive the following equalities:

$$X(k+1) = X(k)$$

$$z(k) = H(k)X(k) + f(k, \theta) + \tilde{v}(k, \theta)$$

where

$$H(k) = [1 \quad k\Delta t]$$

$$f(k, \theta) = [-\tan\beta(k) \quad -k\Delta t \tan\beta(k)]\theta$$

$$E[\tilde{v}(k, \theta)] = m[H(k)\theta - x_s(k)]$$

$$\sigma^2(k, \theta) = \mathrm{Cov}(\tilde{v}(k, \theta), \tilde{v}(k, \theta)) = R[H(k)\theta - x_s(k)]^2$$

where R is measured in square radians.

(c) Develop an MMAF for the model given above.

3.7 Consider the estimation problem with non-Gaussian initial vector for the system given by equations (3.30) and (3.31) by applying the Gaussian sum approximation as used in Example 3.1. Derive the expressions for $\hat{x}(k|k)$, $P(k|k)$, $\hat{x}(0|k)$ and $P(0|k)$.

3.8 Consider the joint detection–estimation problem in discrete time for the system given by equations (3.30) and (3.31), in which (3.31) can be represented by

$$z(k) = \theta H(k)x(k) + v(k)$$

where θ takes values from the set $\{0, 1\}$. Obtain the minimum mean square estimate $\hat{x}(k|k)$ and the associated estimation error covariance $P(k|k)$ by applying Theorem 3.3. Further, express $\hat{\theta}(k)$ in terms of the discrete-time likelihood ratio, $\Lambda(k) = p(Z_k|\theta_1)/p(Z_k|\theta_0)$ (see also Chapter 11).

3.9 Consider the formal backward-time model of equations (3.30) and (3.31), which is described by

$$\begin{aligned} x(k) = \Phi^{-1}&(k+1, k; \theta)x(k+1) \\ &- \Phi^{-1}(k+1, k; \theta)G(k, \theta)u(k) \\ &- \Phi^{-1}(k+1, k; \theta)B(k, \theta)w(k) \end{aligned}$$

$$z(k) = H(k, \theta)x(k) + v(k)$$

where Φ is assumed to be nonsingular, and $u(k)$ is a known input vector. It is assumed that the predictive information on the final state $x(N)$ is given by

$$x(N) \sim N(\hat{x}(N, \theta), P(N, \theta))$$

Supposing the unknown parameter belongs to the discrete parameter space, $\{\theta_1, \ldots, \theta_M\}$, develop the backward-time MMAF for this model. (This problem was studied in Watanabe [44].)

3.10 Verify equation (3.49).

3.11 Let the likelihood ratio be defined as follows:

$$\Lambda(x) = \frac{p_1(x)}{p_2(x)}$$

The *divergence* is defined as the difference in the mean values of the log-likelihood ratio under the two hypotheses:

$$J = E_1[\ln \Lambda(x)] - E_2[\ln \Lambda(x)]$$

where

$$E_i[\ln \Lambda(x)] = \int [\ln \Lambda(x)] p_i(x)\, dx, \qquad i = 1, 2$$

which are called the *Kullback* (or Kullback–Leibler) *information measures* (or numbers) and written

$$I(1, 2) = E_1[\ln \Lambda(x)]$$
$$I(2, 1) = -E_2[\ln \Lambda(x)]$$

In general

$$I(1, 2) \neq I(2, 1)$$

The divergence is a symmetrized form of the Kullback information measures,

$$J = I(1, 2) + I(2, 1)$$

(a) Show that if $p_i(x) = N(m_i, P_i)$, then

$$I(1, 2) = \frac{1}{2} \ln \left[\frac{|P_2|}{|P_1|}\right] + \frac{1}{2} \operatorname{tr} P_1[P_2^{-1} - P_1^{-1}] + \frac{1}{2} \operatorname{tr} P_2^{-1}[m_1 - m_2][m_1 - m_2]^T$$

and

$$J = \tfrac{1}{2} \operatorname{tr}[P_1 - P_2][P_2^{-1} - P_1^{-1}] + \tfrac{1}{2} \operatorname{tr}[P_1^{-1} + P_2^{-1}][m_1 - m_2][m_1 - m_2]^T$$

(b) Show that if $P_1 = P_2$, then

$$B_{12} = J/8$$

3.12 Consider the problem of Example 3.2. When defining the Bhattacharyya coefficient

$$\rho = \int [p(Z_t|\theta_1)p(Z_t|\theta_0)]^{1/2} \, dZ_t$$

show that ρ is given by the average value of the conditional standard deviation of the estimation error of θ. (See Lainiotis [45].)

REFERENCES

[1] LAINIOTIS, D. G., Optimal Adaptive Estimation: Structure and Parameter Adaptation, *IEEE Trans. Aut. Control*, vol. AC-16, no. 2, 1971, pp. 160–70.

[2] LAINIOTIS, D. G., Optimal Nonlinear Estimation, *Int. J. Control*, vol. 14, 1971, pp. 1137–48.

[3] BUCY, R. S. and JOSEPH, P. D., *Filtering for Stochastic Processes with Application to Guidance*, Wiley, New York, 1968.

[4] KAILATH, T., An Innovation Approach to Least-Squares Estimation Part I: Linear Filtering in Additive White Noise, *IEEE Trans. Aut. Control*, vol. AC-13, no. 6, 1968, pp. 646–55.

[5] KAILATH, T., A Generalized Likelihood-Ratio Formula for Random Signals in Gaussian Noise, *IEEE Trans. Inf. Theory*, vol. IT-15, no. 3, 1969, pp. 350–61.

[6] WANG, E. and HAJEK, B., *Stochastic Processes in Engineering Systems*, Springer-Verlag, New York, 1985.

[7] MAYBECK, P., *Stochastic Models, Estimation, and Control*, Vols 1–3, Academic Press, New York, 1979, 1982.

[8] KAILATH, T. and FROST, P., An Innovation Approach to Least-Squares Estimation Part II: Linear Smoothing in Additive White Noise, *IEEE Trans. Aut. Control*, vol. AC-13, no. 6, 1968, pp. 655–60.

[9] LAINIOTIS, D. G., Joint Detection, Estimation and Systems Identification, *Inf. and Control*, vol. 19, 1971, pp. 75–92.

[10] PARK, S. K. and LAINIOTIS, D. G., Joint Detection–Estimation of Gaussian Signals in White Gaussian Noise, *Inf. Sci.*, vol. 4, 1972, pp. 315–25.

[11] GUSTAFSON, D. E., WILLSKY, A. S., WANG, J. Y., LANCASTER, M. C. and TRIEBWASSER, J. H., ECG/VCG Rhythm Diagnosis Using Statistical Signal Analysis—I. Identification of Persistent Rhythms, *IEEE Trans. Biomedical Engng.*, vol. BME-25, no. 4, 1978, pp. 344–53.

[12] GUSTAFSON, D. E., WILLSKY, A. S., WANG, J. Y., LANCASTER, M. C. and TRIEBWASSER, J. H., ECG/VCG Rhythm Diagnosis Using Statistical Signal Analysis—II. Identification of Transient Rhythms, *IEEE Trans. Biomedical Engng.*, vol. BME-25, no. 4, 1978, pp. 353–61.

[13] LAINIOTIS, D. G., Partitioning: A Unifying Framework for Adaptive Systems, I: Estimation, *Proc. IEEE*, vol. 64, no. 8, 1976, pp. 1126–43.

[14] CHIN, L., Advances in Adaptive Filtering. In C. T. Leondes (ed.), *Control and Dynamic Systems*, Vol. 15, Academic Press, New York, 1979, pp. 277–356.

[15] SIMS, F. L., LAINIOTIS, D. S. and MAGILL, D. T., Recursive Algorithm for the Calculation of the Sample Stochastic Processes with an Unknown Parameter, *IEEE Trans. Aut. Control*, vol. AC-14, 1969, pp. 767–70.

[16] MAGILL, D. T., Optimal Adaptive Estimation of Sampled Stochastic Processes, *IEEE Trans. Aut. Control*, vol. AC-10, 1965, pp. 434–9.

[17] LAINIOTIS, D. G. and PARK, S. K., On Joint Detection, Estimation and System Identification: Discrete Data Case, *Int. J. Control*, vol. 17, no. 3, 1973, pp. 609–33.

[18] TUGNAIT, J. K. and HADDAD, A. H., A Detection–Estimation Scheme for State Estimation in Switching Environments, *Automatica*, vol. 15, 1979, pp. 477–81.

[19] KAILATH, T., The Divergence and Bhattacharyya Distance Measures in Signal Selection, *IEEE Trans. Communication Technology*, vol. COM-15, no. 1, 1967, pp. 52–60.

[20] ZIMMERMAN, W., Optimal Integration of Aircraft Navigation Systems, *IEEE Trans. Aero. and Electron. Syst.*, vol. AES-5, 1969, pp. 737–47.

[21] ERZBERGER, H., Application of Kalman Filtering to Error Correction of Inertial Navigator, *NASA Technical Note*, NASA TN D-3874, 1967.

[22] MAYBECK, P. S., Performance Analysis of a Particularly Simple Kalman Filter, *J. Guidance and Control*, vol. 1, 1978, pp. 391–6.

[23] BAR-ITZHACK, I. Y., Minimal Order Time Sharing Filters for INS In-Flight Alignment, *J. Guidance, Control and Dynamics*, vol. 5, 1982, pp. 396–402.

[24] CHANG, C. B. and ATHANS, M., State Estimation for Discrete Systems with Switching Parameters, *IEEE Trans. Aero. and Electron. Syst.*, vol. AES-14, no. 3, 1978, pp. 418–64.

[25] TUGNAIT, J. K., Detection and Estimation for Abruptly Changing Systems, *Automatica*, vol. 18, no. 5, 1982, pp. 607–15.

[26] TUGNAIT, J. K., Adaptive Estimation and Identification for Discrete Systems with Markov Jump Parameters, *IEEE Trans. Aut. Control*, vol. AC-27, no. 5, October 1982, pp. 1054–65.

[27] TUGNAIT, J. K., On Identification and Adaptive Estimation for Systems with Interrupted Observations, *Automatica*, vol. 19, no. 1, 1983, pp. 61–73.

[28] MATHEWS, V. J. and TUGNAIT, J. K., Detection and Estimation with Fixed Lag for Abruptly Changing Systems, *IEEE Trans. Aero. and Electron. Syst.*, vol. AES-19, no. 5, 1983, pp. 730–9.

[29] YAVIN, Y., *Numerical Studies in Nonlinear Filtering*, Lecture Notes in Control and Information Sci., Vol. 65, Springer-Verlag, Berlin, 1985.

[30] TZAFESTAS, S. G. and STAVROULAKIS, P., Partitioning Approach to Adaptive Distributed-Parameter Filtering, *Ricerche di Automatica*, vol. 11, no. 1, 1980, pp. 51–71.

[31] WATANABE, K., YOSHIMURA, T. and SOEDA, T., A Discrete-Time Adaptive Filter for Stochastic Distributed Parameter Systems, *Trans. ASME, J. Dynamic Syst. Meas. Control*, vol. 103, no. 3, 1981, pp. 266–78.

[32] MIDDLETON, D. and ESPOSITO, R., Simultaneous Optimal Detection and Estimation of Signal in Noise, *IEEE Trans. Inf. Theory*, vol. IT-14, no. 3, 1968, pp. 434–44.

[33] KAILATH, T., A Note on Least-Squares Estimation from Likelihood Ratio, *Inf. and Control*, vol. 13, 1968, pp. 534–40.

[34] KAILATH, T., Likelihood Ratios for Gaussian Processes, *IEEE Trans. Inf. Theory*, vol. IT-16, No. 3, 1970, pp. 276–88.

[35] WILLSKY, A. S., CHOW, E. Y., GERSHWIN, S. B., GREENE, C. S., HOUPT, P. K. and KURKJIAN, A. L., Dynamic Model-Based Techniques for the Detection of Incidents on Freeways, *IEEE Trans. Aut. Control*, vol. AC-25, no. 3, 1980, pp. 347–60.

[36] TYLEE, J. L., On-Line Failure Detection in Nuclear Power Plant Instrumentation, *IEEE Trans. Aut. Control*, vol. AC-28, no. 3, 1983, pp. 406–15.

[37] MEALY, G. L. and TANG, W., Application of Multiple Model Estimation to a Recursive Terrain Height Correlation System, *IEEE Trans. Aut. Control*, vol. AC-28, no. 3, 1983, pp. 323–31.

[38] CHANG, C. B. and TABACZYNSKI, J. K., Application of State Estimation to Target Tracking, *IEEE Trans. Aut. Control*, vol. AC-29, no. 2, 1984, pp. 98–109.

[39] MAYBECK, P. S. and SUIZU, R. I., Adaptive Tracker Field-of-View Variation via Multiple Model Filtering, *IEEE Trans. Aero. and Electron. Syst.*, vol. AES-21, no. 4, 1985, pp. 529–39.

[40] BAR-SHALOM, Y. and FORTMANN, T. E., *Tracking and Data Association*, Academic Press, New York, 1988.

[41] SUNAHARA, Y., *Stochastic Systems Theory*, Institute of Electronics and Communications Engineers of Japan, Tokyo, 1979 (in Japanese).

[42] KUNITA, H., *Estimation of Stochastic Processes*, Sangyo Tosho, Tokyo, 1976 (in Japanese).

[43] PETRIDIS, V., A Method for Bearing-Only Velocity and Position Estimation, *IEEE Trans. Aut. Control*, vol. AC-26, no. 2, 1981, pp. 488–93.

[44] WATANABE, K., Backward-Pass Multiple Model Adaptive Filtering for a

Fixed-Interval Smoother, *Int. J. Control*, vol. 49, no. 2, 1989, pp. 385–97.

[45] LAINIOTIS, D. G., On a General Relationship between Estimation, Detection, and the Bhattacharyya Coefficient, *IEEE Trans. Inf. Theory*, vol. IT-15, no. 4, 1969, pp. 504–5.

4

Asymptotic and Convergence Properties of Partitioned Adaptive Filters

4.1 INTRODUCTION

In this chapter we discuss the asymptotic and convergence properties of the partitioned adaptive filters discussed in Chapter 3. It is important to understand in advance how the partitioned adaptive filter behaves as time passes, according to whether the true parameter θ^* is included in the prespecified parameter set or not.

In Section 4.2, we give the result for discrete-time Gauss–Markov systems with finite unknown parameter set [1]. In Section 4.3 we present the result for sampled stochastic processes, where the unknown parameter vector is assumed to be continuous and to belong to a compact subset of a metric space [2]. Finally, in Section 4.4 we analyze the asymptotic behaviour of the adaptive filter for continuous-time linear dynamic Gauss–Markov systems with a finite unknown parameter set [3,4].

4.2 DISCRETE TIME AND DISCRETE PARAMETER SETS

In this section we investigate two convergence properties of partitioned adaptive filters presented in Corollary 3.2, in which the unknown parameter belongs to a finite set of quantified points, and the dynamic systems are of discrete time [1].

By Bayesian learning of an unknown parameter, one means that a sequence of conditional probabilities $p(\theta_i|k)$, $p(\theta_i|k+1)$, ... converges to 1 or 0 according to whether or not $\theta^* = \theta_i$. More precisely, let I^i be a random variable defined as the indicator of the events $\theta^* = \theta_i$; that is

$$I^i = \begin{cases} 1, & \theta^* = \theta_i \\ 0, & \theta^* \neq \theta_i \end{cases} \tag{4.1}$$

Then by Bayesian learning of the parameter θ, we imply that

$$\lim_{k\to\infty} p(\theta_i|k) = I^i \tag{4.2}$$

with probability one (w.p.1) (see also Appendix B).

Lemma 4.1

Let $X(k)$ and $Y(k)$ be sequences of positive random variables such that $0 \leqslant X(k) \leqslant 1$ and $E[Y^r(k)]$ exists and is bounded by $M_1 < \infty$ for $k = 1, 2, \ldots$, for some $r > 1$. Then, if $\lim_{k\to\infty} X(k) = 0$ (w.p.1),

$$\lim_{k\to\infty} E[X(k)Y(k)] = 0 \tag{4.3}$$

Proof

By Hölder's inequality,

$$E[X(k)Y(k)] \leqslant E[X^s(k)]^{1/s} E[Y^r(k)]^{1/r} \tag{4.4}$$

where $1/r + 1/s = 1$, $r > 1$. Then for some $r > 1$,

$$E[X(k)Y(k)] \leqslant E[X^s(k)]^{1/s} M_1^{1/r} \tag{4.5}$$

But, since $X(k) \leqslant 1$,

$$E[X^s(k)]^{1/s} \leqslant E[X(k)]^{1/s} \tag{4.6}$$

By the dominated convergence theorem (see Appendix B), $\lim_{k\to\infty} X(k) = 0$ (w.p.1) means

$$\lim_{k\to\infty} E[X(k)] = 0 \tag{4.7}$$

Hence

$$\lim_{k\to\infty} E[X(k)Y(k)] = 0 \tag{4.8}$$

□

This leads to the following theorem.

Theorem 4.1 Quadratic mean convergence

Suppose that, given a positive definite symmetric matrix F, there exists a bound $M_1 < \infty$ such that

$$E\{[\hat{x}^{\mathrm{T}}(k|k, \theta_i)F\hat{x}(k|k, \theta_j)]^r\} \leqslant M_1 \tag{4.9}$$

for $k = 1, 2, \ldots$, for some $r > 1$. Then if $p(\theta_i|k) \to I^i$ (w.p.1),

$$\lim_{k\to\infty} E\{[\hat{x}(k|k) - \hat{x}(k|k, \theta^*)]^{\mathrm{T}} F[\hat{x}(k|k) - \hat{x}(k|k, \theta^*)]\} = 0 \tag{4.10}$$

where $\hat{x}(k|k)$ is provided by equation (3.39) and $\hat{x}(k|k, \theta^*)$ is given by the Kalman filter matched to $\theta = \theta^*$.

Proof

Substituting equation (3.39) into (4.10) gives

$$\begin{aligned}
&E\{[\hat{x}(k|k) - \hat{x}(k|k, \theta^*)]^{\mathrm{T}} F[\hat{x}(k|k) - \hat{x}(k|k, \theta^*)]\}\\
&\qquad = E\left\{\sum_{i=1}^{M} [I^i - p(\theta_i|k)]\hat{x}^{\mathrm{T}}(k|k, \theta_i)F \right.\\
&\qquad\qquad \left. \times \sum_{j=1}^{M} [I^j - p(\theta_j|k)]\hat{x}(k|k, \theta_j)\right\}\\
&\qquad = E\left\{\sum_{i=1}^{M}\sum_{j=1}^{M} [I^i - p(\theta_i|k)] \right.\\
&\qquad\qquad \left. \times [I^j - p(\theta_j|k)]\hat{x}^{\mathrm{T}}(k|k, \theta_i)F\hat{x}(k|k, \theta_j)\right\}\\
&\qquad \leqslant \sum_{i=1}^{M}\sum_{j=1}^{M} E\{|[I^i - p(\theta_i|k)]\\
&\qquad\qquad \times [I^j - p(\theta_j|k)]|\hat{x}^{\mathrm{T}}(k|k, \theta_i)F\hat{x}(k|k, \theta_j)\}
\end{aligned} \tag{4.11}$$

Convergence to zero follows by application of Lemma 4.1 □

One would expect that, in addition to Theorem 4.1, the *optimal quadratic performance* or *risk* $J^*(k)$ would converge to the average optimal risk for *unknown* parameter optimal estimate $J^{**}(k)$, where

$$J^*(k) \triangleq E\{[x(k) - \hat{x}(k|k)]^{\mathrm{T}} F[x(k) - \hat{x}(k|k)]\} \tag{4.12}$$

and

$$J^{**}(k) \triangleq E\{[x(k) - \hat{x}(k|k, \theta^*)]^{\mathrm{T}} F[x(k) - \hat{x}(k|k, \theta^*)]\} \tag{4.13}$$

Theorem 4.2 Convergence in performance

Suppose that, given a positive definite symmetric matrix F, there exists a bound $M_1 < \infty$ such that

$$E\{[\hat{x}^{\mathrm{T}}(k|k, \theta_i)F\hat{x}(k|k, \theta_j)]^r\} \leqslant M_1 \tag{4.14}$$

for $k = 1, 2, \ldots$, and for some $r > 1$. Then if $\lim_{k\to\infty} p(\theta_i|k) = I^i$ (w.p.1),

$$\lim_{k\to\infty} [J^*(k) - J^{**}(k)] = 0. \tag{4.15}$$

Proof

Noting that $x(k)$ is measurable with respect to $\{z(l), i = 1, \ldots, k\}$ and $\{z(i), \theta^*, i = 1, \ldots, k\}$ yields

$$\begin{aligned}
J^*(k) - J^{**}(k) &= E\{x^{\mathrm{T}}(k)Fx(k) + \hat{x}^{\mathrm{T}}(k|k)F\hat{x}(k|k) \\
&\quad - x^{\mathrm{T}}(k)F\hat{x}(k|k) \\
&\quad -\hat{x}^{\mathrm{T}}(k|k)Fx(k) - x^{\mathrm{T}}(k)Fx(k) \\
&\quad - \hat{x}^{\mathrm{T}}(k|k, \theta^*)F\hat{x}(k|k, \theta^*) \\
&\quad +x^{\mathrm{T}}(k)F\hat{x}(k|k, \theta^*) + \hat{x}^{\mathrm{T}}(k|k, \theta^*)Fx(k)\} \\
&= E\{\hat{x}^{\mathrm{T}}(k|k)F\hat{x}(k|k) \\
&\quad - E[x^{\mathrm{T}}(k)|Z_k]F\hat{x}(k|k) \\
&\quad -\hat{x}^{\mathrm{T}}(k|k)FE[x(k)|Z_k] \\
&\quad - \hat{x}^{\mathrm{T}}(k|k, \theta^*)F\hat{x}(k|k, \theta^*) \\
&\quad + E[x^{\mathrm{T}}(k)|Z_k, \theta^*]F\hat{x}(k|k, \theta^*) \\
&\quad + \hat{x}^{\mathrm{T}}(k|k, \theta^*)FE[x(k)|Z_k, \theta^*]\} \\
&= E\{\hat{x}^{\mathrm{T}}(k|k, \theta^*)F\hat{x}(k|k, \theta^*) \\
&\quad - \hat{x}^{\mathrm{T}}(k|k)F\hat{x}(k|k)\} \\
&= E\{[\hat{x}(k|k) - \hat{x}(k|k, \theta^*)]^{\mathrm{T}} \\
&\quad \times F[\hat{x}(k|k) - \hat{x}(k|k, \theta^*)] \\
&\quad + \hat{x}^{\mathrm{T}}(k|k)F[\hat{x}(k|k, \theta^*) - \hat{x}(k|k)] \\
&\quad + [\hat{x}(k|k, \theta^*) - \hat{x}(k|k)]^{\mathrm{T}}F\hat{x}(k|k)\}
\end{aligned} \tag{4.16}$$

Then, by the same manner as used in the proof of Theorem 4.1, we obtain

$$
\begin{aligned}
J^*(k) - J^{**}(k) = E &\left\{ \sum_{i=1}^{M} \sum_{j=1}^{M} [I^i - p(\theta_i|k)][I^j - p(\theta_j|k)] \right. \\
&\times \hat{x}^{\mathrm{T}}(k|k, \theta_i) F \hat{x}^{\mathrm{T}}(k|k, \theta_j) \\
&+ \sum_{i=1}^{M} \sum_{j=1}^{M} p(\theta_i|k) \\
&\times [I^j - p(\theta_j|k)] \hat{x}^{\mathrm{T}}(k|k, \theta_i) F \hat{x}^{\mathrm{T}}(k|k, \theta_j) \\
&+ \sum_{i=1}^{M} \sum_{j=1}^{M} [I^i - p(\theta_i|k)] p(\theta_j|k) \hat{x}^{\mathrm{T}}(k|k, \theta_i) \\
&\left. \times F \hat{x}^{\mathrm{T}}(k|k, \theta_j) \right\} \\
= E &\left\{ \sum_{i=1}^{M} \sum_{j=1}^{M} [I^i I^j - p(\theta_i|k) p(\theta_j|k)] \right. \\
&\left. \times \hat{x}^{\mathrm{T}}(k|k, \theta_i) F \hat{x}(k|k, \theta_j) \right\} \qquad (4.17)
\end{aligned}
$$

Since, from equations (4.1) and (4.2), $\lim_{k\to\infty}[I^i I^j - p(\theta_i|k)\, p(\theta_j|k)] = 0$ (w.p.1), convergence of $J^*(k) - J^{**}(k)$ to zero follows by application of Lemma 4.1. □

4.3 DISCRETE TIME AND CONTINUOUS PARAMETER SETS

In this section, we analyze the asymptotic behaviour of the partitioned adaptive filtering algorithm given by Theorem 3.3. In particular, it is assumed that the unknown parameter vector θ belongs to a compact subset Θ of an appropriate metric space.

The following conditions are imposed on the processes under consideration [2].

C1. Let $S(\theta^*, \delta)$ be an open sphere with radius $\delta > 0$ and centre at $\theta = \theta^*$, and let $\bar{S}(\theta^*, \delta)$ indicate the complement of $S(\theta^*, \delta)$ in Θ. Then the following holds:

$$\lim_{k\to\infty} \sup_{\theta \in \bar{S}(\theta^*, \delta)} p(\theta|k) = 0 \qquad (4.18)$$

almost everywhere (a.e.).

C2. There exists a bound $M_1 < \infty$ such that

$$E\{|\hat{x}^{\mathrm{T}}(k|k, \theta_i) F \hat{x}(k|k, \theta_j)|^r\} \leqslant M_1, \qquad \forall \theta_i, \theta_j \in \Theta \qquad (4.19)$$

with some $r > 1$.

C3. For some $\delta > 0$ and $r > 1$, there exists a bound $M_2 < \infty$ such that

$$E\left\{\left[\sup_{\theta \in S(\theta^*, \delta)} \|\hat{x}(k|k, \theta) - \hat{x}(k|k, \theta^*)\|\right]^{2r}\right\} \leq M_2 \delta \quad (4.20)$$

These three conditions are sufficient for assuring convergence of the partitioned adaptive estimator $\hat{x}(k|k)$ to the optimal estimator matched to true parameter θ^*, in quadratic mean.

Theorem 4.3

Suppose that conditions C1–C3 hold. Then

$$\lim_{k\to\infty} E\{[\hat{x}(k|k) - \hat{x}(k|k, \theta^*)]^{\mathrm{T}} F[\hat{x}(k|k) - \hat{x}(k|k, \theta^*)]\} = 0 \quad (4.21)$$

Proof

We have

$$\hat{x}(k|k) - \hat{x}(k|k, \theta^*) = \int_\Theta [\hat{x}(k|k, \theta) - \hat{x}(k|k, \theta^*)] p(\theta|k)\,\mathrm{d}\theta \quad (4.22)$$

because $\int_\Theta p(\theta|k)\,\mathrm{d}\theta \equiv 1$. Now

$$\begin{aligned} &E\{[\hat{x}(k|k) - \hat{x}(k|k, \theta^*)]^{\mathrm{T}} F[\hat{x}(k|k) - \hat{x}(k|k, \theta^*)]\} \\ &= \left\{\int_{\bar{S}}\int_{\bar{S}} \tilde{L}(k; \theta_1, \theta_2)\,\mathrm{d}\theta_1\,\mathrm{d}\theta_2 + \int_{\bar{S}}\int_{S} \tilde{L}(k; \theta_1, \theta_2)\,\mathrm{d}\theta_1\,\mathrm{d}\theta_2 \right. \\ &\quad \left. + \int_{S}\int_{\bar{S}} \tilde{L}(k; \theta_1, \theta_2)\,\mathrm{d}\theta_1\,\mathrm{d}\theta_2 + \int_{S}\int_{S} \tilde{L}(k; \theta_1, \theta_2)\,\mathrm{d}\theta_1\,\mathrm{d}\theta_2 \right\} \quad (4.23) \end{aligned}$$

where

$$\begin{aligned} \tilde{L}(k; \theta_1, \theta_2) \triangleq\; &[\hat{x}(k|k, \theta_1) - \hat{x}(k|k, \theta^*)]^{\mathrm{T}} \\ &\times F[\hat{x}(k|k, \theta_2) - \hat{x}(k|k, \theta^*)] \\ &\times p(\theta_1|k) p(\theta_2|k) \quad (4.24) \end{aligned}$$

and $S \triangleq S(\theta^*, \delta)$ and $\bar{S} \triangleq \bar{S}(\theta^*, \delta)$. Moreover, using Minkowski's inequality (see Appendix B) and condition C2, it follows that

$$E\{\|[\hat{x}(k|k, \theta_1) - \hat{x}(k|k, \theta^*)]^{\mathrm{T}} F[\hat{x}(k|k, \theta_2) - \hat{x}(k|k, \theta^*)]\|^r\}$$

$$\leqslant E\{|\hat{x}^{\mathrm{T}}(k|k, \theta_1)F\hat{x}(k|k, \theta_2)|^r + |\hat{x}^{\mathrm{T}}(k|k, \theta_1)F\hat{x}(k|k, \theta^*)|^r$$
$$+ |\hat{x}^{\mathrm{T}}(k|k, \theta^*)F\hat{x}(k|k, \theta_2)|^r + |\hat{x}(k|k, \theta^*)F\hat{x}(k|k, \theta^*)|^r\}$$
$$\leqslant 4M_1 \tag{4.25}$$

On the other hand, from condition C1, we have

$$\lim_{k\to\infty} \sup_{\theta_1,\theta_2\in\bar{S}} [p(\theta_1|k)p(\theta_2|k)] = 0 \quad \text{a.e.} \tag{4.26}$$

Therefore, from expressions (4.25) and (4.26) and Lemma 4.1, we have

$$\lim_{k\to\infty} \int_{\bar{S}}\int_{\bar{S}} E\{\tilde{L}(k; \theta_1, \theta_2)\}\, \mathrm{d}\theta_1\, \mathrm{d}\theta_2 = 0 \tag{4.27}$$

Now consider the second term on the right-hand side of equation (4.23):

$$\left|E\left\{\int_{\bar{S}}\int_{S} \tilde{L}(k; \theta_1, \theta_2)\, \mathrm{d}\theta_1\, \mathrm{d}\theta_2\right\}\right|$$
$$\leqslant E\left\{\int_{\bar{S}} \sup_{\theta_2\in S} |[\hat{x}(k|k, \theta_1) - \hat{x}(k|k, \theta^*)]^{\mathrm{T}}\right.$$
$$\times F[\hat{x}(k|k, \theta_2) - \hat{x}(k|k, \theta^*)]|$$
$$\left.\times \left[\int_{S} p(\theta_2|k)\, \mathrm{d}\theta_2\right] p(\theta_1|k)\, \mathrm{d}\theta_1\right\} \tag{4.28}$$

Since $\int_{\Theta} p(\theta|k)\, \mathrm{d}\theta \equiv 1$ for all k, it follows from condition C1 that

$$\lim_{k\to\infty} \int_{S} p(\theta|k)\, \mathrm{d}\theta = 1 \quad \text{a.e.} \tag{4.29}$$

From this fact, we have, for $k \geqslant k_0(\varepsilon)$,

$$\int_{S} p(\theta_2|k)\, \mathrm{d}\theta_2 \leqslant 1 + \varepsilon \quad \text{a.e. for some } \varepsilon > 0 \tag{4.30}$$

Moreover, from Hölder's inequality, it follows that

$$E\left\{\left[\sup_{\theta_2\in S} |[\hat{x}(k|k, \theta_1) - \hat{x}(k|k, \theta^*)]^{\mathrm{T}} F[\hat{x}(k|k, \theta_2) - \hat{x}(k|k, \theta^*)]|\right.\right.$$
$$\left.\left.\times \int_{S} p(\theta_2|k)\, \mathrm{d}\theta_2\right]\right\}^r \leqslant \left[E\left\{\left[\sup_{\theta_2\in S} |[\hat{x}(k|k, \theta_1) - \hat{x}(k|k, \theta^*)]^{\mathrm{T}}\right.\right.\right.$$
$$\left.\left.\left.\times F[\hat{x}(k|k, \theta_2) - \hat{x}(k|k, \theta^*)]|\right]^{rr_1}\right\}\right]^{1/r_1}\left[E\left\{\int_{S} p(\theta_2|k)\, \mathrm{d}\theta_2\right\}^{rr_2}\right]^{1/r_2} \tag{4.31}$$

where $1/r_1 + 1/r_2 = 1$ and $r_1 > 1$. From conditions C1 and C2, the first term on the right-hand side of expression (4.31) is bounded from above uniformly in k and in $\theta_1 \in \bar{S}$. From expression (4.30), the second term on the right-hand side of inequality (4.31) is also bounded from above uniformly in k. Since

$$\lim_{k\to\infty} \sup_{\theta_1 \in \bar{S}} p(\theta_1|k) = 0 \quad \text{a.e.} \tag{4.32}$$

it follows from Lemma 4.1 that

$$\lim_{k\to\infty} E \int_{\bar{S}} \int_{S} \tilde{L}(k;\, \theta_1,\, \theta_2)\, \mathrm{d}\theta_1\, \mathrm{d}\theta_2 = 0 \tag{4.33}$$

Similarly, it can be proved that

$$\lim_{k\to\infty} E \int_{S} \int_{\bar{S}} \tilde{L}(k;\, \theta_1,\, \theta_2)\, \mathrm{d}\theta_1\, \mathrm{d}\theta_2 = 0 \tag{4.34}$$

Finally, it remains to be proved that the last term on the right-hand side of equation (4.23) vanishes as $k \to \infty$. We have

$$\begin{aligned}
&\left| E\left\{ \int_S \int_S \tilde{L}(k;\, \theta_1,\, \theta_2)\, \mathrm{d}\theta_1\, \mathrm{d}\theta_2 \right\} \right| \\
&\leqslant E\left\{ \sup_{\theta_1, \theta_2 \in S} |[\hat{x}(k|k,\, \theta_1) - \hat{x}(k|k,\, \theta^*)]^{\mathrm{T}} \right. \\
&\quad \times F[\hat{x}(k|k,\, \theta_2) - \hat{x}(k|k,\, \theta^*)]| \\
&\quad \left. \times \int_S p(\theta_1|k)\, \mathrm{d}\theta_1 \int_S p(\theta_2|k)\, \mathrm{d}\theta_2 \right\} \\
&\leqslant \left[E\left\{ \sup_{\theta_1, \theta_2 \in S} |[\hat{x}(k|k,\, \theta_1) - \hat{x}(k|k,\, \theta^*)]^{\mathrm{T}} \right.\right. \\
&\quad \left.\left. \times F[\hat{x}(k|k,\, \theta_2) - \hat{x}(k|k,\, \theta^*)]|^{r_1} \right\} \right]^{1/r_1} \\
&\quad \times \left[E\left\{ \left[\int_S p(\theta_1|k)\, \mathrm{d}\theta_1 \int_S p(\theta_2|k)\, \mathrm{d}\theta_2 \right]^{r_2} \right\} \right]^{1/r_2}
\end{aligned} \tag{4.35}$$

where we have employed Hölder's inequality, and $1/r_1 + 1/r_2 = 1$, $r_1 > 1$. It follows, from condition C3 and Hölder's inequality, that

$$\begin{aligned}
&E\left\{ \sup_{\theta_1, \theta_2 \in S} |[\hat{x}(k|k,\, \theta_1) - \hat{x}(k|k,\, \theta^*)]^{\mathrm{T}} \right. \\
&\quad \left. \times F[\hat{x}(k|k,\, \theta_2) - \hat{x}(k|k,\, \theta^*)]|^{r_1} \right\} \\
&\leqslant \left[E\left\{ \sup_{\theta_1 \in S} [\|\hat{x}(k|k,\, \theta_1) - \hat{x}(k|k,\, \theta^*)\|]^{2r_1} \right\} \right]^{1/2} \|F\|
\end{aligned}$$

$$\times \left[E\left\{ \sup_{\theta_2 \in S} [\|\hat{x}(k|k, \theta_2) - \hat{x}(k|k, \theta^*)\|]^{2r_1} \right\} \right]^{1/2} \tag{4.36}$$

$$\leqslant M_2 \|F\| \delta \tag{4.37}$$

where, in expressions (4.35) and (4.36), r_1 is chosen appropriately. From equation (4.29), it follows that

$$E\left\{ \left[\int_S p(\theta_1|k)\, \mathrm{d}\theta_1 \right]^{2r_1} \right\} \leqslant M_3 < \infty \tag{4.38}$$

for any finite r_1. Therefore, from expressions (4.35), (4.37) and (4.38), we have, for $k \geqslant k_0(\delta)$,

$$\left| E\left\{ \int_S \int_S \tilde{L}(k;\, \theta_1, \theta_2)\, \mathrm{d}\theta_1\, \mathrm{d}\theta_2 \right\} \right| \leqslant M\delta \tag{4.39}$$

for some positive $M < \infty$. Hence, from expressions (4.27), (4.33), (4.34) and (4.39), we obtain the desired result. □

We now claim that the optimal quadratic risk $J^*(k)$ converges to the optimal risk $J^{**}(k)$ for known parameter estimator $\hat{x}(k|k, \theta^*)$, in which $J^*(k)$ and $J^{**}(k)$ are as defined in Section 4.2.

Theorem 4.4

Suppose that conditions C1–C3 are satisfied. Then it follows that

$$\lim_{k \to \infty} [J^*(k) - J^{**}(k)] = 0 \tag{4.40}$$

Proof

We can derive an expression identical to equation (4.16). Hence, following the procedure given in the proof of Theorem 4.3, we obtain the desired result. □

4.4 CONTINUOUS TIME AND DISCRETE PARAMETER SETS

We discuss here the asymptotic behaviour of the partitioned adaptive filter algorithm in continuous-time linear time-invariant systems, presented in Corollary 3.1, where the unknown system parameters are assumed to belong to a finite set.

Let

$$q(t|\theta) = -(2/t)\ln\Lambda(t|\theta). \tag{4.41}$$

Since we are interested in asymptotic properties of the functional, conditions C4 and C5 are imposed on the system of equations (3.1) and (3.2) to ensure convergence of certain expressions as t becomes very large.

C4. For any θ in $\Theta \cup \{\theta^*\}$, where Θ contains a finite number of elements, $A(\theta)$ is a stable matrix, that is all the eigenvalues of $A(\theta)$ have negative real parts.

C5. For any $\theta \in \Theta \cup \{\theta^*\}$, the pair $(A(\theta), D)$ is stabilizable and the pair $(C(\theta), A(\theta))$ is detectable [5] (see also Appendix C), where $BQB^{\mathrm{T}} = DD^{\mathrm{T}}$, $H^{\mathrm{T}}(\theta)R^{-1}H(\theta) = C^{\mathrm{T}}(\theta)C(\theta)$, rank D = rank(BQB^{T}) and rank $C(\theta)$ = rank$(H^{\mathrm{T}}(\theta)R^{-1}H(\theta))$.

Condition C4 means that the initial condition of the state has no effect on the asymptotic behaviour of the system, and this justifies setting the initial condition to zero in equation (3.1), i.e. $x(0) \sim N(0, 0)$. Condition C5 means that $\lim_{t\to\infty} P(t|t, \theta)$ exists and is unique; we denote this limit by $P_\theta \geqslant 0$, which satisfies the algebraic Riccati equation (ARE) [5,6]:

$$A(\theta)P_\theta + P_\theta A^{\mathrm{T}}(\theta) + BQB^{\mathrm{T}} - P_\theta H^{\mathrm{T}}(\theta)R^{-1}H(\theta)P_\theta = 0 \tag{4.42}$$

and furthermore, the closed-loop matrix $[A(\theta) - P_\theta H^{\mathrm{T}}(\theta)R^{-1}H(\theta)]$ is asymptotically stable. In addition, let

$$\bar{q}(t|\theta) = E[q(t|\theta)] \tag{4.43}$$

and define

$$\bar{q}(\theta) \triangleq \lim_{t\to\infty} \bar{q}(t|\theta) \tag{4.44}$$

Before proceeding to the presentation of convergence of the partitioned adaptive estimator, we examine sufficient conditions for almost sure convergence of the maximum likelihood (ML) and the maximum a posteriori probability (MAP) estimates of θ on Θ (see also Appendix B.4). We consider the two cases when $\theta^* \in \Theta$ and when $\theta^* \notin \Theta$. We shall use a stationary version of the likelihood function, which is given by using the steady-state version of the continuous-time Kalman filter given by equations (2.77) and (2.79) matched to θ, but with $\hat{x}(0, \theta) \equiv 0$ and $P(0, \theta) \equiv 0$ for all $\theta \in \Theta$, and by assuming that the system given by equations (3.1) and (3.2) started in the remote past (i.e. initial time $t_0 = -\infty$ instead of $t_0 = 0$).

From equations (3.1), (3.2), (2.77), and (2.79) matched to θ,

$$\begin{bmatrix} x(t) \\ \hat{x}(t|t, \theta) \end{bmatrix} = \int_0^t \widetilde{A} \begin{bmatrix} x(\tau) \\ \hat{x}(\tau|\tau, \theta) \end{bmatrix} d\tau + \int_0^t \widetilde{B} \begin{bmatrix} d\xi(\tau) \\ d\eta(\tau) \end{bmatrix} \tag{4.45}$$

where

$$\widetilde{A} \triangleq \begin{bmatrix} A(\theta^*) & 0 \\ P_\theta H^T(\theta) R^{-1} H(\theta) & A(\theta) - P_\theta H^T(\theta) R^{-1} H(\theta) \end{bmatrix}$$

$$\widetilde{B} \triangleq \begin{bmatrix} B(\theta^*) & 0 \\ 0 & P_\theta H^T(\theta) R^{-1} \end{bmatrix}$$

in which P_θ is obtained by the positive semidefinite solution of equation (4.42). Now let

$$S(t, \theta, \theta^*) \triangleq E\left\{ \begin{bmatrix} x(t) \\ \hat{x}(t|t, \theta) \end{bmatrix} [x^T(t) \quad \hat{x}^T(t|t, \theta)] \right\} \tag{4.46}$$

Then, from expressions (4.45) and (4.46) we have

$$dS(t, \theta, \theta^*)/dt = \widetilde{A} S(t, \theta, \theta^*) + S(t, \theta, \theta^*) \widetilde{A}^T + \widetilde{B} \widetilde{Q} \widetilde{B}^T \tag{4.47}$$

where

$$\widetilde{Q} \triangleq \begin{bmatrix} Q & 0 \\ 0 & R \end{bmatrix}$$

Now $S(\theta, \theta^*) \triangleq \lim_{t\to\infty} S(t, \theta, \theta^*)$ exists if $\widetilde{A}$ is stable. This is true for all $\theta \in \Theta$ because of conditions C4 and C5. Therefore, $S(\theta, \theta^*)$ satisfies the algebraic Lyapunov equation (ALE):

$$\widetilde{A} S(\theta, \theta^*) + S(\theta, \theta^*) \widetilde{A}^T + \widetilde{B} \widetilde{Q} \widetilde{B}^T = 0 \tag{4.48}$$

At this time, defining

$$\hat{X}_1(t|t, \theta) \triangleq H(\theta) \hat{x}(t|t, \theta) \tag{4.49}$$

$$X_1(t) \triangleq H(\theta^*) x(t) \tag{4.50}$$

$$S_1(t, \theta) \triangleq E\{[X_1(t) - \hat{X}_1(t|t, \theta)][X_1(t) - \hat{X}_1(t|t, \theta)]^T\} \tag{4.51}$$

we note that

$$\lim_{t\to\infty} S_1(t, \theta) \triangleq S_1(\theta) = \widetilde{H} S(\theta, \theta^*) \widetilde{H}^T \tag{4.52}$$

where $\widetilde{H} \triangleq [H(\theta^*) - H(\theta)]$. Notice that if we take initial time to be t_0 instead of zero, then expression (4.52) also implies that

$$S_1(\theta) = \lim_{t_0 \to -\infty} S_1(t, \theta) \tag{4.53}$$

Lemma 4.2

Suppose that conditions C4 and C5 are satisfied and that the system given by equations (3.1) and (3.2) starts in the remote past. The process $\{X_1(t) - \hat{X}_1(t|t, \theta)\}$ is ergodic for $t \geq 0$ and for every $\theta \in \Theta \cup \{\theta^*\}$.

Proof

The stochastic process $\{X_1(t) - \hat{X}_1(t|t, \theta)\}$ is a zero-mean Gaussian process. Furthermore, existence of $S_1(\theta)$ implies that it is also stationary. Therefore, it follows that $\{X_1(t) - \hat{X}_1(t|t, \theta)\}$ is ergodic [7] (see also Appendix B) if

$$\int_0^\infty \|\Sigma(\tau)\| \, \mathrm{d}\tau < \infty \tag{4.54}$$

where

$$\begin{aligned}\Sigma(\tau) \triangleq E\{&[X_1(t+\tau) - \hat{X}_1(t+\tau|t+\tau, \theta)] \\ &\times [X_1(t) - \hat{X}_1(t|t, \theta)]^{\mathrm{T}}\}\end{aligned} \tag{4.55}$$

From expressions (4.45) and (4.52), we have

$$\left.\begin{aligned}\Sigma(\tau) &= S_1(\theta) \qquad \text{for } \tau = 0 \\ &= \widetilde{H}[\exp(\widetilde{A}\tau)]S(\theta, \theta^*)\widetilde{H}^{\mathrm{T}} \qquad \text{for } \tau > 0\end{aligned}\right\} \tag{4.56}$$

For any $\tau > 0$, we obtain

$$\|\Sigma(\tau)\| \leq \|\widetilde{H}\|^2 \|S(\theta, \theta^*)\| \; \|\exp(\widetilde{A}\tau)\| \tag{4.57}$$

Since $\widetilde{A}$ is stable $\forall \theta \in \Theta$

$$\|\exp(\widetilde{A}\tau)\| < \exp(\lambda\tau), \qquad \forall \theta \in \Theta \tag{4.58}$$

where $\lambda < 0$ is the largest eigenvalue of $[\widetilde{A} + \widetilde{A}^{\mathrm{T}}]$. Hence, it follows that

$$\int_0^\infty \|\Sigma(\tau)\| \, \mathrm{d}\tau \leq \|\widetilde{H}\|^2 \|S(\theta, \theta^*)\| \int_0^\infty \exp(\lambda\tau) \, \mathrm{d}\tau < \infty \tag{4.59}$$

□

From expressions (3.18), (4.41) and (4.49), we have

$$q(t|\theta) = \frac{1}{t}\left[-2\int_0^t \langle \hat{X}_1(s|s, \theta), R^{-1}\,\mathrm{d}y(s)\rangle + \int_0^t \|\hat{X}_1(s|s, \theta)\|^2_{R^{-1}} \mathrm{d}s\right] \tag{4.60}$$

Substituting expression (2.2) with $\theta = \theta^*$ into equation (4.60) and using expression (4.50) gives, after some algebra,

$$\begin{aligned} q(t|\theta) = \frac{1}{t}\bigg[& \int_0^t \operatorname{tr} R_r[X_1(s) - \hat{X}_1(s|s,\theta)] \\ & \times [X_1(s) - \hat{X}_1(s|s,\theta)]^{\mathrm{T}} R_r \, \mathrm{d}s \\ & - \int_0^t \langle X_1(s), R^{-1} X_1(s) \rangle \, \mathrm{d}s \\ & - 2\int_0^t \langle \hat{X}_1(s|s,\theta), R^{-1} \, \mathrm{d}\eta(s) \rangle \bigg] \end{aligned} \tag{4.61}$$

where R_r indicates the positive square root of R^{-1}, i.e. $R_r R_r = R^{-1}$ and R_r is symmetric positive definite. Using Fubini's theorem, we obtain

$$\begin{aligned} E[q(t|\theta)] = \frac{1}{t}\bigg[& \int_0^t \operatorname{tr} R_r S_1(s,\theta) R_r \, \mathrm{d}s \\ & - \int_0^t E[\langle X_1(s), R^{-1} X_1(s) \rangle] \, \mathrm{d}s \bigg] \end{aligned} \tag{4.62}$$

where we have used the fact that

$$E\left[\int_0^t \langle \hat{X}_1(s|s,\theta), R^{-1} \, \mathrm{d}\eta(s) \rangle\right] = 0 \tag{4.63}$$

because $\eta(s)$ is an independent-increment zero-mean process. Now let $t \to \infty$ and notice that all processes involved are asymptotically stationary. Then we have

$$\begin{aligned} \bar{q}(\theta) &= \lim_{t\to\infty} E[q(t|\theta)] \\ &= \operatorname{tr} R_r S_1(\theta) R_r - \lim_{t\to\infty} E[\langle X_1(t), R^{-1} X_1(t) \rangle] \\ &= \operatorname{tr} R_r S_1(\theta) R_r - \operatorname{tr} R_r [H(\theta^*) \quad 0] S(\theta, \theta^*) [H(\theta^*) \quad 0]^{\mathrm{T}} R_r \end{aligned} \tag{4.64}$$

Lemma 4.3

Under the hypothesis of Lemma 4.2,

$$\lim_{t\to\infty} q(t|\theta) = \bar{q}(\theta) \quad \text{a.e.} \quad \forall \theta \in \Theta \cup \{\theta^*\} \tag{4.65}$$

Proof

From the ergodic theorem and Lemma 4.2, we obtain

$$\lim_{t\to\infty} \frac{1}{t}\int_0^t \operatorname{tr} R_r[X_1(s) - \hat{X}_1(s|s,\theta)][X_1(s) - \hat{X}_1(s|s,\theta)]^{\mathrm{T}} R_r \,\mathrm{d}s$$
$$= \operatorname{tr} R_r S_1(\theta) R_r \tag{4.66}$$

almost surely. Similarly, noting that the processes $X_1(t)$ and $\hat{X}_1(t|t,\theta)$ are also shown to be ergodic for $t \geq 0$ and for every $\theta \in \Theta \cup \{\theta^*\}$ by applying Lemma 4.2, we have

$$\lim_{t\to\infty} \frac{1}{t}\int_0^t \langle X_1(s), R^{-1}X_1(s)\rangle \,\mathrm{d}s$$
$$= \operatorname{tr} R_r[H(\theta^*) \quad 0]S(\theta,\theta^*)[H(\theta^*) \quad 0]^{\mathrm{T}} R_r \tag{4.67}$$

almost surely. Since $\|E[\hat{X}_1(t|t,\theta)\hat{X}_1^{\mathrm{T}}(t|t,\theta)]\| < \infty$, $\forall \theta \in \Theta \cup \{\theta^*\}$, it follows that

$$\lim_{t\to\infty} \frac{1}{t}\int_0^t \langle \hat{X}_1(s|s,\theta), R^{-1}\,\mathrm{d}\eta(s)\rangle = 0 \tag{4.68}$$

almost surely. From equations (4.64) and (4.66)–(4.68) the desired result follows. □

The following lemma defines the limit set of the ML estimate as $t \to \infty$.

Lemma 4.4

Suppose that conditions C4 and C5 are satisfied. Then the ML estimate of θ on Θ belongs to a set Θ_l w.p.1 as $t \to \infty$, where

$$\Theta_l = \{\theta : \theta \in \Theta, \bar{q}(\theta) = \min_{\theta\in\Theta} \bar{q}(\theta)\} \tag{4.69}$$

Proof

We have, for any $\theta' \in \Theta_l$,

$$\lim_{t\to\infty} [\Lambda(t;\omega|\theta)/\Lambda(t;\omega|\theta')]$$
$$= \lim_{t\to\infty} \exp\left[-\frac{t}{2}\{q(t;\omega|\theta) - q(t;\omega|\theta')\}\right]$$

$$= \exp\left[-\lim_{t\to\infty}\frac{t}{2}\{q(t;\omega|\theta) - q(t;\omega|\theta')\}\right] \tag{4.70}$$

where ω is used to denote explicitly the probability space variable. From Lemma 4.3, given any arbitrarily small positive number δ_1, for every sample path ω in a set Ω of probability measure one, and for every θ in the finite set Θ, we obtain

$$|q(t;\omega|\theta) - \bar{q}(\theta)| < \delta_1 \qquad \forall t > t(\delta_1). \tag{4.71}$$

Therefore, it follows that, for all $\theta \in \Theta$,

$$-2\delta_1 + \bar{q}(\theta) - \bar{q}(\theta') < q(t;\omega|\theta) - q(t;\omega|\theta') < 2\delta_1 + \bar{q}(\theta) - \bar{q}(\theta') \tag{4.72}$$

for all $t > t(\delta_1)$ and for all $\omega \in \Omega$. Now let

$$\delta_2 = \min_{\theta\notin\Theta_l,\theta\in\Theta} [\bar{q}(\theta) - \bar{q}(\theta')] \tag{4.73}$$

Apparently, $\delta_2 > 0$; choose $\delta_1 < \delta_2/2$. Then from expressions (4.72) and (4.73), we have

$$q(t;\omega|\theta) - q(t;\omega|\theta') > \delta_2 - 2\delta_1 > 0 \tag{4.74}$$

for all $t > t(\delta_1)$, for all $\omega \in \Omega$, and for all $\theta \notin \Theta_l$, $\theta \in \Theta$. Therefore, from expressions (4.70) and (4.74), it follows that, for any $\theta' \in \Theta_l$

$$\lim_{t\to\infty} \max_{\theta\notin\Theta_l,\theta\in\Theta} [\Lambda(t;\omega|\theta)/\Lambda(t;\omega|\theta')] = 0 \quad \text{w.p.1} \tag{4.75}$$

However,

$$\lim_{t\to\infty} \max_{\theta\in\Theta} [\Lambda(t;\omega|\theta)/\Lambda(t;\omega|\theta')] = 1 \quad \text{w.p.1} \tag{4.76}$$

Hence, from expressions (4.75) and (4.76), the ML estimate of θ on Θ belongs to Θ_l w.p.1 as $t \to \infty$. □

It should be noted that Lemma 4.4 does not mean convergence of the ML estimate of θ to some θ' in Θ_l. It is possible that the ML estimate might oscillate among various θ' in Θ_l.

Now we turn to the question of whether the set Θ_l contains the true parameter vector value θ^* when $\theta^* \in \Theta$. According to Lemma 4.5, $\bar{q}(\theta) \geq \bar{q}(\theta^*)$. Therefore, $\theta^* \in \Theta_l$ whenever $\theta^* \in \Theta$.

Lemma 4.5

Suppose that conditions C4 and C5 are satisfied. Then $\bar{q}(\theta) \geq \bar{q}(\theta^*)$ for all $\theta \in \Theta$.

Proof

From equation (4.62) we have

$$E[q(t|\theta) - q(t|\theta^*)] = \frac{1}{t}\int_0^t \operatorname{tr} R_r[S_1(s, \theta) - S_1(s, \theta^*)]R_r \,\mathrm{d}s \tag{4.77}$$

Since the Kalman filter matched to the true model θ^* is the minimum mean-square error (MMSE) estimator, we obtain $\operatorname{tr} R_r[S_1(s, \theta) - S_1(s, \theta^*)]R_r \geq 0$ for all $\theta \in \Theta$ and for all t. Therefore, letting $t \to \infty$, we conclude that $\bar{q}(\theta) \geq \bar{q}(\theta^*)$ for all $\theta \in \Theta$. □

In light of Lemma 4.5, Lemma 4.4 can now be modified as follows for the case $\theta^* \in \Theta$:

Lemma 4.6

Suppose that conditions C4 and C5 are satisfied and $\theta^* \in \Theta$. Then, the ML estimate of θ on Θ belongs to a set Θ_l^* w.p.1 as $t \to \infty$, where

$$\Theta_l^* \triangleq \{\theta : \theta \in \Theta \text{ and } \bar{q}(\theta) = \bar{q}(\theta^*)\} \tag{4.78}$$

In order to ensure consistency of the ML estimate we need an identifiability condition [8, 9].

C6. (Identifiability Condition) The system model given by equations (3.1) and (3.2) is such that

$$\bar{q}(\theta) \neq \bar{q}(\theta^*) \quad \text{for all } \theta \in \Theta,\ \theta \neq \theta^* \tag{4.79}$$

Notice that from equations (4.64) and (4.77), the identifiability condition (4.79) can also be written as

$$\operatorname{tr} S_1(\theta) \neq \operatorname{tr} S_1(\theta^*) \quad \text{for all } \theta \in \Theta,\ \theta \neq \theta^* \tag{4.80}$$

We now return to the consistency of the MAP estimate of θ on Θ.

Lemma 4.7

Suppose that conditions C4–C6 are satisfied. Then the MAP estimate of θ on Θ belongs to a set Θ_l w.p.1 as $t \to \infty$.

Proof

From expression (3.25), we have

$$p(\theta|t) \leqslant \left[\frac{p(\theta)}{p(\theta')}\right]\left[\frac{\Lambda(t|\theta)}{\Lambda(t|\theta')}\right] \tag{4.81}$$

for every θ in Θ and every θ' in Θ_l. From expressions (4.75) and (4.81), we obtain (w.p.1)

$$\lim_{t\to\infty} p(\theta|t) = 0, \qquad \forall\theta \notin \Theta_l,\ \theta \in \Theta \tag{4.82}$$

Hence (w.p.1)

$$\lim_{t\to\infty} \sum_{\theta\in\Theta_l} p(\theta|t) = 1 \tag{4.83}$$

yielding the desired result. □

We require two more auxiliary results before we present the main theorems. The next lemma follows from Lemma 4.1.

Lemma 4.8

Let $X(t)$ be a positive stochastic process, $t \in [0, \infty)$, such that $E[X^r(t)] \leqslant M_1$ for all $t \in [0, \infty)$ for some $r > 1$. Then if $\lim_{t\to\infty} p(\theta|t) = 0$, w.p.1,

$$\lim_{t\to\infty} E[p(\theta|t)X(t)] = 0 \tag{4.84}$$

Proof

The details are omitted. See the proof of Lemma 4.1. □

Lemma 4.9

Suppose that conditions C4–C6 are satisfied. Then, given a positive definite symmetric matrix F, there exists a bound $M_1 < \infty$ such that

$$E\{[\hat{x}^{\mathrm{T}}(t|t, \theta_i)F\hat{x}(t|t, \theta_j)]^2\} \leqslant M_1 \tag{4.85}$$

for all $t \in [0, \infty)$ and for all $\theta_i, \theta_j \in \Theta$.

Proof

From expressions (4.46) and (4.47), we can show that $S(t, \theta, \theta^*)$ is uniformly bounded by a positive definite matrix, uniformly in t and in $\theta \in \Theta$, by virtue of conditions C4 and C5. The desired result then follows by noting that only Gaussian processes are involved, so that it is sufficient to bound second-order moments. □

We now return to the main results. Theorem 4.5 implies convergence, in the quadratic mean sense, of the partitioned adaptive filter $\hat{x}(t|t)$ to the optimal MMSE estimator $\hat{x}(t|t, \theta^*)$ matched to the true parameter θ^*.

Theorem 4.5

Suppose that conditions C4–C6 hold. Then

$$\lim_{t\to\infty} E\{[\hat{x}(t|t) - \hat{x}(t|t, \theta^*)]^{\mathrm{T}} F[\hat{x}(t|t) - \hat{x}(t|t, \theta^*)]\} = 0 \tag{4.86}$$

where $\hat{x}(t|t)$ is given by Corollary 3.1.

Proof

The proof proceeds in basically the same way as that of Theorem 4.1. The details are therefore omitted here. □

Next we establish that, under the same condition as for Theorem 4.5, the optimal quadratic risk $J^*(t)$ converges to the optimal risk $J^{**}(t)$ corresponding to the true parameter estimator, where

$$J^*(t) \triangleq E\{[x(t) - \hat{x}(t|t)]^{\mathrm{T}} F[x(t) - \hat{x}(t|t)]\} \tag{4.87}$$

and

$$J^{**}(t) \triangleq E\{[x(t) - \hat{x}(t|t, \theta^*)]^{\mathrm{T}} F[x(t) - \hat{x}(t|t, \theta^*)]\} \tag{4.88}$$

Theorem 4.6

Suppose that conditions C4–C6 are satisfied. Then

$$\lim_{t\to\infty} [J^*(t) - J^{**}(t)] = 0 \tag{4.89}$$

Proof

Like that of Theorem 4.5, the proof of this theorem follows from an application of Lemmas 4.8 and 4.9 just as in the proof of Theorem 4.2 for an analogous discrete-time result; we omit the details. □

Note that for the case when $\theta^* \notin \Theta$, the limit set Θ_l defined in Lemma 4.4 may be redefined as follows:

$$\Theta_l = \{\theta:\theta \in \Theta, \bar{q}(\theta) = \min_{\theta\in\Theta} \bar{q}(\theta) - \bar{q}(\theta^*)\} \tag{4.90}$$

Following Theorems 4.5 and 4.6, it is easy to show that, as $t \to \infty$, $\hat{x}(t|t)$ converges to $\hat{x}(t|t, \Theta_l)$ in the quadratic mean sense and in performance, in which

$$\hat{x}(t|t, \Theta_l) \triangleq \sum_{\theta\in\Theta_l} \hat{x}(t|t, \theta)p(\theta|t) \tag{4.91}$$

Finally, Θ_l is a singleton if we have

$$\bar{q}(\theta_1) \neq \bar{q}(\theta_2) \qquad \forall\theta_1, \theta_2 \in \Theta, \theta_1 \neq \theta_2.$$

4.5 SUMMARY

In this chapter we have examined the convergence properties of partitioned adaptive filters. In particular, we first discussed the case of discrete time and discrete parameter sets. Next, the case of discrete time and continuous parameter sets was analyzed. We further presented the case of continuous time and discrete parameter sets, though consideration of continuous time and continuous parameter sets was omitted. This can be found in Tugnait [10], in which the parameter set is assumed to be a compact set in a real Euclidean space as in Section 4.3 and the stationary version of the likelihood functional used in Section 4.4 is applied to obtain the convergence results.

Theorems 4.1 and 4.2 based on Hilborn and Lainiotis [1] are the fundamental results for the convergence of the partitioned adaptive filter. It should be noted that these results motivate the problems of Sections 4.3 and 4.4, which are mainly due to Tugnait [2–4]. An alternative approach, based on the Kullback information measure, can be found in Hawkes and Moore [11] (see also Exercises 4.2 and 4.7). Other related matters can be seen in Tugnait [10], Hawkes and Moore [12] and Anderson and Moore [13].

EXERCISES

4.1 Consider the linear discrete-time system given in equations (3.30) and (3.31), where θ is assumed to belong to the discrete space $\{\theta_1, \ldots, \theta_M\}$. Suppose that $\nu(t, \theta_i)$ is asymptotically ergodic in the autocorrelation function; suppose that $\tilde{P}(k|k-1, \theta_i) \to \tilde{P}(\theta_i)$ as $k \to \infty$, with $\tilde{P}(\theta_i) > 0$; and denote the actual limiting covariance of the innovations, i.e.

$$\lim_{n\to\infty} \frac{1}{n} \sum_{j=k}^{k+n-1} \nu(j, \theta_i)\nu^{\mathrm{T}}(j, \theta_i)$$

by $\Sigma(i)$. Further, define

$$\beta_i = \ln|\tilde{P}(\theta_i)| + \mathrm{tr}\,(\tilde{P}(\theta_i)\Sigma(i))$$

and assume that for some i and all $j \neq i$, we have

$$\beta_i < \beta_j$$

Then, show that $p(\theta_i|k) \to 1$ as $k \to \infty$ and $p(\theta_j|k) \to 0$ as $k \to \infty$ for all $j \neq i$, convergence being exponentially fast. (Hint: Defining $L(k|i) \triangleq p(\theta_i|Z_k)/p(\theta^*|Z_k)$, which is the unconditional likelihood ratio function, where θ^* denotes the true parameter, derive $2n^{-1}\ln(\frac{L(k+n-1|j)}{L(k|j)} \times \frac{L(k|i)}{L(k+n-1|i)}) \to \beta_i - \beta_j$.) (See Anderson and Moore [13].)

4.2 The Kullback information measure in discrete time is defined for a finite measurement sequence as

$$I(\theta_i; \theta_j, k) = E\left\{\ln \frac{p(Z_k|\theta_i)}{p(Z_k|\theta_j)}\middle|\theta_i\right\}$$

and for an infinite measurement sequence, we have an asymptotic per-sample information measure

$$I(\theta_i; \theta_j) = \lim_{k\to\infty} \frac{1}{k} I(\theta_i; \theta_j, k)$$

(a) Show that

$$I(\theta_i; \theta_j) = -\lim_{n\to\infty} \frac{1}{n}\left\{\ln \frac{p(\theta_j|Z_{k+n-1})p(\theta_i|Z_k)}{p(\theta_i|Z_{k+n-1})p(\theta_j|Z_k)}\middle|\theta_i\right\}$$

(b) Show that

$$I(\theta^*; \theta_j) = \tfrac{1}{2}[\beta_j - \ln|\tilde{P}(\theta^*)| - \mathrm{tr}\, I]$$

4.3 Complete the proof of Theorem 4.4 by applying the procedure given in the proof of Theorem 4.3.

4.4 Verify expression (4.52).

4.5 Verify expression (4.75).

4.6 Let $\mu_\theta(t)$ denote the probability measure induced by $\{y(\cdot)\}$ on the space of continuous functions, assuming that $y(s)$, $t_0 \leq s \leq t$, results from the model given by equations (3.1) and (3.2) with θ as the model parameter. Suppose that conditions C4 and C5, presented in Section 4.4, are satisfied. When defining

$$d\mu_{\theta}(t)/d\mu_{\theta^*}(t) = \Lambda(t|\theta)/\Lambda(t|\theta^*) \triangleq \exp(\zeta(t))$$

show that if

$$E[\zeta(t)] \leq \ln E[\exp(\zeta(t))], \qquad \forall t < \infty$$

then

$$\bar{q}(\theta) \geq \bar{q}(\theta^*) \qquad \forall \theta \in \Theta$$

which is an alternative proof of Lemma 4.5. (Hint: Use the fact that $E[\exp(\zeta(t))] = \int (d\mu_{\theta}(t)/d\mu_{\theta^*}(t))\, d\mu_{\theta^*}(t) = 1\ \forall t < \infty$.)

4.7 (Model Approximation via ML Identification) Let us consider the case when $\theta^* \notin \Theta$. The limit set Θ_l may be redefined as follows:

$$\Theta_l = \{\theta : \theta \in \Theta, \bar{q}(\theta) = \min_{\theta \in \Theta} \bar{q}(\theta) - \bar{q}(\theta^*)\}$$

The ML identification technique leads to the selection of a model $\theta \in \Theta$ which minimizes the asymptotic per-unit time Kullback information measure:

$$I(\theta^*; \theta, t) = E\{\ln[d\mu_{\theta^*}(t)/d\mu_{\theta}(t)]\}$$

Show that

$$\lim_{t\to\infty} \frac{1}{t} I(\theta^*; \theta, t) = \frac{1}{2}(\bar{q}(\theta) - \bar{q}(\theta^*))$$

$$= \frac{1}{2} \operatorname{tr} R_r[S_1(\theta) - S_1(\theta^*)]R_r$$

4.8 Verify expression (4.81).

REFERENCES

[1] HILBORN, C. G. and LAINIOTIS, D. G., Optimal Estimation in the Presence of Unknown Parameters, *IEEE Trans. Syst. Sci. and Cyber.*, vol. SSC-5, no. 1, 1969, pp. 38–43.

[2] TUGNAIT, J. K., Convergence Analysis of Partitioned Adaptive Estimator Under Continuous Parameter Uncertainty, *IEEE Trans. Aut. Control*, vol. AC-25, no. 3, 1980, pp. 569–73.

[3] TUGNAIT, J. K., Identification and Model Approximation for Continuous-Time Systems on Finite Parameter Sets, *IEEE Trans. Aut. Control*, vol. AC-25, no. 6, 1980, pp. 1202–6.

[4] TUGNAIT, J. K., Convergence of Continuous-Time Partitioned Adaptive State Estimator, *Int. J. Control*, vol. 33, no. 5, 1981, pp. 923–33.

[5] KUČERA, V., A Contribution to Matrix Quadratic Equations, *IEEE Trans. Aut. Control*, vol. AC-17, no. 3, 1972, pp. 344–7.

[6] WONHAM, W. M., On a Matrix Riccati Equation of Stochastic Control, *SIAM J. Control*, vol. 6, no. 4, 1968, pp. 681–97.

[7] WONG, E. and HAJEK, B., *Stochastic Processes in Engineering Systems*, Springer-Verlag, New York, 1985.

[8] TSE, E. and ANTON, J. J., On the Identifiability of Parameters, *IEEE Trans. Aut. Control*, vol. AC-17, October 1972, pp. 637–46.

[9] TSE, E. and WEINERT, H., Correction and Extension of 'On the Identifiability of Parameters', *IEEE Trans. Aut. Control*, vol. AC-18, December 1973, pp. 687–8.

[10] TUGNAIT, J. K., Continuous-Time System Identification on Compact Parameter Sets, *IEEE Trans. Inf. Theory*, vol. IT-31, no. 5, 1985, pp. 652–9.

[11] HAWKES, R. M. and MOORE, J. B., Performance Bounds for Adaptive Estimation, *Proc. IEEE*, vol. 64, 1976, pp. 1143–50.

[12] HAWKES, R. M. and MOORE, J. B., Performance of Bayesian Parameter Estimators for Linear Signal Models, *IEEE Trans. Aut. Control*, vol. AC-21, no. 4, 1976, pp. 523–7.

[13] ANDERSON, B. D. O. and MOORE, J. B., *Optimal Filtering*, Prentice Hall, Englewood Cliffs, New Jersey, 1979.

5

The Partitioning Filter: A Probabilistic Approach

5.1 INTRODUCTION

In this chapter the partition theorem discussed in Chapter 3 will be utilized to develop the partitioning or Lainiotis filter and multipartitioning filter for both continuous-time and discrete-time systems. A technique for solving the associated Riccati equations is also presented.

In Section 5.2, we present the partitioning filter, which is based on the partitioning of the initial state vector into two statistically dependent or independent random vectors, in continuous-time systems [1]. In section 5.3 we develop the multipartitioning filter, which is based on dividing the initial state vector into many parts, which may be statistically dependent or independent [2, 3].

In Section 5.4, we formulate the discrete-time partitioning filter [4, 5], and in Section 5.5, we apply the results of Section 5.3 to the multipartitioned filtering problem in discrete time [6]. Finally, in Section 5.6, we demonstrate the partitioned numerical algorithms for solving the nonstationary and stationary solutions to Riccati differential equations (RDEs).

5.2 THE PARTITIONING FILTER (CONTINUOUS-TIME SYSTEMS)

In this section we consider the continuous-time dynamical system which is characterized by Langevin's relations:

$$\frac{\mathrm{d}x(t)}{\mathrm{d}t} = A(t)x(t) + B(t)w(t) \tag{5.1}$$

$$z(t) = H(t)x(t) + v(t) \tag{5.2}$$

for $t \geqslant t_0$, where $x(t)$ is an n-vector, $w(t)$ is a p-vector, and $z(t)$ and

$v(t)$ are m-vectors. The matrices $A(t)$, $B(t)$, and $H(t)$ are continuous in time, and of size $n \times n$, $n \times p$, and $m \times n$ respectively.

The stochastic processes $\{w(t), t \geqslant t_0\}$ and $\{v(t), t \geqslant t_0\}$ are zero-mean Gaussian white noises with covariance matrices

$$E[w(t)w^{\mathrm{T}}(s)] = Q(t)\delta(t - s) \tag{5.3}$$

$$E[v(t)v^{\mathrm{T}}(s)] = R(t)\delta(t - s) \tag{5.4}$$

respectivey, for all t, $s \geqslant t_0$. The $p \times p$ matrix $Q(t)$ is continuous in time and positive semidefinite for $t \geqslant t_0$, while the $m \times m$ matrix $R(t)$ is continuous in time and positive definite for $t \geqslant t_0$.

It is assumed that the above two stochastic processess are independent of each other, so that

$$E[w(t)v^{\mathrm{T}}(s)] = 0 \tag{5.5}$$

for all t, $s \geqslant t_0$. Furthermore, the initial state $x(t_0)$ is assumed to be a Gaussian n-vector with mean $\hat{x}(t_0)$; it is independent of $\{w(t), t \geqslant t_0\}$ and $\{v(t), t \geqslant t_0\}$; and its covariance matrix $E\{[x(t_0) - \hat{x}(t_0)][x(t_0) - \hat{x}(t_0)]^{\mathrm{T}}\} = P(t_0)$ is positive semidefinite.

The following lemmas are useful for our future development of the a posteriori probability density function [6].

Lemma 5.1 Sum of two quadratic forms in a Gaussian density

We define

$$a \triangleq \|X - \mu_1\|^2_{P_1^{-1}} + \|X - \mu_2\|^2_{P_2^{-1}} \tag{5.6}$$

which we can re-express as

$$a = \|X - \mu\|^2_{P^{-1}} + \|\mu_1 - \mu_2\|^2_{(P_1 + P_2)^{-1}} \tag{5.7}$$

and

$$\frac{\exp a}{\int \exp a \,\mathrm{d}X} = \frac{\exp[\|X - \mu\|^2_{P^{-1}}]}{\int \exp[\|X - \mu\|^2_{P^{-1}}]\,\mathrm{d}X} \tag{5.8}$$

where

$$\mu \triangleq P[P_1^{-1}\mu_1 + P_2^{-1}\mu_2] \tag{5.9}$$

$$P \triangleq [P_1^{-1} + P_2^{-1}]^{-1} \tag{5.10}$$

Proof

Expanding expression (5.6), we obtain

$$a = X^{\mathrm{T}}(P_1^{-1} + P_2^{-1})X - 2X^{\mathrm{T}}(P_1^{-1}\mu_1 + P_2^{-1}\mu_2) + \mu_1^{\mathrm{T}}P_1^{-1}\mu_1 + \mu_2^{\mathrm{T}}P_2^{-1}\mu_2 \tag{5.11}$$

By substituting expressions (5.9) and (5.10) into (5.11), we have

$$a = (X - \mu)^{\mathrm{T}}P^{-1}(X - \mu) + \mu_1^{\mathrm{T}}P_1^{-1}\mu_1 + \mu_2^{\mathrm{T}}P_2^{-1}\mu_2 - \mu^{\mathrm{T}}P^{-1}\mu \tag{5.12}$$

Further, with the aid of a matrix inversion lemma given in equation (5A.1) (see Appendix 5A) we note that

$$\mu^{\mathrm{T}}P^{-1}\mu = \mu_1^{\mathrm{T}}P_1^{-1}\mu_1 + \mu_2^{\mathrm{T}}P_2^{-1}\mu_2 - (\mu_1 - \mu_2)^{\mathrm{T}}(P_1 + P_2)^{-1}(\mu_1 + \mu_2) \tag{5.13}$$

Therefore, substituting equation (5.13) into (5.12) yields the desired result,

$$a = (X - \mu)^{\mathrm{T}}P^{-1}(X - \mu) + (\mu_1 - \mu_2)^{\mathrm{T}}(P_1 + P_2)^{-1}(\mu_1 - \mu_2) \tag{5.14}$$

Equation (5.8) is apparent from the substitution of expression (5.14) into the left-hand side of equation (5.8). □

Lemma 5.2 Moments for the Gaussian a posteriori probability density function: continuous-time case

The a posteriori probability density function of the random vector X conditioned on data $Z_t \triangleq \{z(s),\ t_0 \leqslant s \leqslant t\}$ and random vector θ is given by

$$p(X|Z_t, \theta) = \frac{\Lambda(Z_t|X, \theta)p(X|\theta)}{\int \Lambda(Z_t|X, \theta)p(X|\theta)\mathrm{d}X} \tag{5.15}$$

where the X- and θ-conditional likelihood ratio function $\Lambda(t|X, \theta) \triangleq \Lambda(Z_t|X, \theta)$ is given by

$$\Lambda(t|X, \theta) = \exp\left\{\int_{t_0}^{t} \langle H(s)\hat{x}(s|s;\, X, \theta),\, R^{-1}(s)z(s)\rangle\,\mathrm{d}s - \frac{1}{2}\int_{t_0}^{t} \|H(s)\hat{x}(s|s;\, X, \theta)\|^2_{R^{-1}(s)}\,\mathrm{d}s\right\} \tag{5.16}$$

It is here assumed that $\hat{x}(s|s; X, \theta)$ is the X- and θ-conditional optimal filtered estimate of $x(s)$, and that it can be divided into two parts, one of which is not a function of X while the other is a linear function of X, i.e.

$$\hat{x}(s|s;\, X,\, \theta) = \hat{x}(s|s,\, \theta) + \phi(s,\, t_0)X \tag{5.17}$$

where $\hat{x}(s|s, \theta)$ is the θ-conditional optimal filtered estimate of $x(s)$ and ϕ is any function, and that the a priori probability density function $p(X|\theta)$ is Gaussian, i.e.

$$p(X|\theta) = C\exp\left[-\tfrac{1}{2}\|X - \hat{X}_\theta\|^2_{P^{-1}(X|\theta)}\right] \tag{5.18}$$

Suppose that the assumptions specified above are satisfied. Then $p(X|Z_t, \theta)$ is a Gaussian density function

$$p(X|Z_t,\, \theta) = C'\exp\left[-\tfrac{1}{2}\|X - E(X|Z_t,\, \theta)\|^2_{P^{-1}(X|Z_t,\theta)}\right] \tag{5.19}$$

where

$$E(X|Z_t,\, \theta) \triangleq P(X|Z_t,\, \theta)[\lambda(t,\, t_0) + P^{-1}(X|\theta)\hat{X}_\theta] \tag{5.20}$$

$$P(X|Z_t,\, \theta) \triangleq [M(t,\, t_0) + P^{-1}(X|\theta)]^{-1} \tag{5.21}$$

in which

$$\lambda(t,\, t_0) = \int_{t_0}^{t} \phi^{\mathrm{T}}(s,\, t_0)H^{\mathrm{T}}(s)R^{-1}(s)[z(s) - H(s)\hat{x}(s|s,\, \theta)]\,\mathrm{d}s \tag{5.22}$$

$$M(t,\, t_0) = \int_{t_0}^{t} \phi^{\mathrm{T}}(s,\, t_0)H^{\mathrm{T}}(s)R^{-1}(s)H(s)\phi(s,\, t_0)\,\mathrm{d}s \tag{5.23}$$

Proof

The derivation of expressions (5.15) and (5.16) can be found in Appendix 5B. To prove expressions (5.19)–(5.23), substitute equation (5.17) into (5.16) to obtain

$$\begin{aligned}\Lambda(t|X,\, \theta) = \exp\Bigg\{&\int_{t_0}^{t} \hat{x}^{\mathrm{T}}(s|s,\, \theta)H^{\mathrm{T}}(s)R^{-1}(s)z(s)\,\mathrm{d}s \\ &- \frac{1}{2}\int_{t_0}^{t} \hat{x}^{\mathrm{T}}(s|s,\, \theta)H(s)R^{-1}(s)H(s)\hat{x}(s|s,\, \theta)\,\mathrm{d}s \\ &+ X^{\mathrm{T}}\int_{t_0}^{t} \phi^{\mathrm{T}}(s,\, t_0)H^{\mathrm{T}}(s)R^{-1}(s) \\ &\times [z(s) - H(s)\hat{x}(s|s,\, \theta)]\,\mathrm{d}s \\ &- \frac{1}{2}X^{\mathrm{T}}\int_{t_0}^{t} \phi^{\mathrm{T}}(s,\, t_0)H^{\mathrm{T}}(s)R^{-1}(s)H(s)\phi(s,\, t_0)\,\mathrm{d}sX\Bigg\}\end{aligned} \tag{5.24}$$

The first and second terms in the exponential function are not functions of X and can therefore be excluded from the evaluation of equation (5.16). Denoting $\Lambda(t|X, \theta)$ without their terms as $\Lambda'(t|X, \theta)$, and using equations (5.22) and (5.23), it follows that

$$\Lambda'(t|X, \theta) = \exp[X^{\mathrm{T}}\lambda(t, t_0) - \tfrac{1}{2}X^{\mathrm{T}}M(t, t_0)X] \tag{5.25}$$

It is easy to verify the following identity:

$$u^{\mathrm{T}}Fv - \tfrac{1}{2}u^{\mathrm{T}}Fu = -\tfrac{1}{2}(u - v)^{\mathrm{T}}F(u - v) + \tfrac{1}{2}v^{\mathrm{T}}Fv \tag{5.26}$$

which can be utilized to rewrite $\Lambda'(t|X, \theta)$ as

$$\begin{aligned}\Lambda'(t|X, \theta) = \exp\{&-\tfrac{1}{2}[X - M^{-1}(t, t_0)\lambda(t, t_0)]^{\mathrm{T}}M(t, t_0) \\ &\times [X - M^{-1}(t, t_0)\lambda(t, t_0)] \\ &+ \tfrac{1}{2}\lambda^{\mathrm{T}}(t, t_0)M^{-1}(t, t_0)\lambda(t, t_0)\}\end{aligned} \tag{5.27}$$

Since the second term in the exponential function of equation (5.27) is not a function of X, $\Lambda'(t|X, \theta)$ can also be redefined as

$$\Lambda''(t|X, \theta) = \exp[-\tfrac{1}{2}||X - M^{-1}(t, t_0)\lambda(t, t_0)||^2_{M(t,t_0)}] \tag{5.28}$$

By making use of equations (5.18) and (5.28), equation (5.15) can now be rewritten as

$$p(X|Z_t, \theta) = \frac{\exp[-\tfrac{1}{2}||X - M^{-1}(t, t_0)\lambda(t, t_0)||^2_{M(t,t_0)} - \tfrac{1}{2}||X - \hat{X}_\theta||^2_{P^{-1}(X|\theta)}]}{\int \exp[-\tfrac{1}{2}||X - M^{-1}(t, t_0)\lambda(t, t_0)||^2_{M(t,t_0)} - \tfrac{1}{2}||X - \hat{X}_\theta||^2_{P^{-1}(X|\theta)}]\mathrm{d}X} \tag{5.29}$$

Applying Lemma 5.1 to equation (5.29) gives the desired results (5.19)–(5.21). □

Consider the estimation problem defined in equations (5.1)–(5.5) with the initial condition partitioned into the sum of two dependent Gaussian random vectors:

$$x(t_0) = x_{\mathrm{n}} + x_{\mathrm{r}} \tag{5.30}$$

where the joint probability density function of random vectors x_{n} and x_{r} is given by

$$\begin{bmatrix} x_{\mathrm{n}} \\ x_{\mathrm{r}} \end{bmatrix} \sim N\left\{\begin{bmatrix} \hat{x}_{\mathrm{n}} \\ \hat{x}_{\mathrm{r}} \end{bmatrix}, \begin{bmatrix} P_{\mathrm{n}} & P_{\mathrm{nr}} \\ P^{\mathrm{T}}_{\mathrm{nr}} & P_{\mathrm{r}} \end{bmatrix}\right\} \tag{5.31}$$

Therefore, the moments of $x(t_0)$ reduce to

$$\hat{x}(t_0) = \hat{x}_{\mathrm{n}} + \hat{x}_{\mathrm{r}} \tag{5.32}$$

$$P(t_0) = P_{\mathrm{n}} + P_{\mathrm{r}} + P_{\mathrm{nr}} + P^{\mathrm{T}}_{\mathrm{nr}} \tag{5.33}$$

The dependent partitioning filter based on the above initial conditions is summarized in the following theorem.

Theorem 5.1 Dependent partitioning filter

When the initial conditions are partitioned as in equations (5.32) and (5.33), the optimal filtered estimate $\hat{x}(t|t) \triangleq E[x(t)|Z_t]$ of $x(t)$ and its error covariance matrix $P(t|t) \triangleq E\{[x(t) - \hat{x}(t|t)][x(t) - \hat{x}(t|t)]^{\mathrm{T}}|Z_t\}$ are given by

$$\hat{x}(t|t) = \hat{x}_{\mathrm{n}}(t|t) + \Phi_{\mathrm{n}}(t, t_0)[I + P_{\mathrm{nr}}P_{\mathrm{r}}^{-1}]\hat{x}_{\mathrm{r}}(t_0|t) \tag{5.34}$$

and

$$P(t|t) = P_{\mathrm{n}}(t|t) + \Phi_{\mathrm{n}}(t, t_0)[I + P_{\mathrm{nr}}P_{\mathrm{r}}^{-1}] \times P_{\mathrm{r}}(t_0|t)[I + P_{\mathrm{nr}}P_{\mathrm{r}}^{-1}]^{\mathrm{T}}\Phi_{\mathrm{n}}^{\mathrm{T}}(t, t_0) \tag{5.35}$$

where $\hat{x}_{\mathrm{n}}(t|t)$ and $P_{\mathrm{n}}(t|t)$ are provided by the nominal Kalman filter as follows:

$$\frac{\mathrm{d}\hat{x}_{\mathrm{n}}(t|t)}{\mathrm{d}t} = [A(t) - K_{\mathrm{n}}(t)H(t)]\hat{x}_{\mathrm{n}}(t|t) + K_{\mathrm{n}}(t)z(t) \tag{5.36}$$

$$\frac{\mathrm{d}P_{\mathrm{n}}(t|t)}{\mathrm{d}t} = A(t)P_{\mathrm{n}}(t) + P_{\mathrm{n}}(t)A^{\mathrm{T}}(t) + B(t)Q(t)B^{\mathrm{T}}(t) - P_{\mathrm{n}}(t|t)H^{\mathrm{T}}(t)R^{-1}(t)H(t)P_{\mathrm{n}}(t|t) \tag{5.37}$$

$$K_{\mathrm{n}}(t) \triangleq P_{\mathrm{n}}(t|t)H^{\mathrm{T}}(t)R^{-1}(t) \tag{5.38}$$

with the initial conditions

$$\hat{x}_{\mathrm{n}}(t_0|t_0) = \hat{x}_{\mathrm{n}} - P_{\mathrm{nr}}P_{\mathrm{r}}^{-1}\hat{x}_{\mathrm{r}} \tag{5.39}$$

$$P_{\mathrm{n}}(t_0|t_0) = P_{\mathrm{n}} - P_{\mathrm{nr}}P_{\mathrm{r}}^{-1}P_{\mathrm{nr}}^{\mathrm{T}} \tag{5.40}$$

The fixed-point smoothed estimate $\hat{x}_{\mathrm{r}}(t_0|t) \triangleq E[x_{\mathrm{r}}|Z_t]$ of x_{r} and its error covariance matrix $P_{\mathrm{r}}(t_0|t) \triangleq E\{[x_{\mathrm{r}} - \hat{x}_{\mathrm{r}}(t_0|t)][x_{\mathrm{r}} - \hat{x}_{\mathrm{r}}(t_0|t)]^{\mathrm{T}}|Z_t\}$ are given by

$$\hat{x}_{\mathrm{r}}(t_0|t) = P_{\mathrm{r}}(t_0|t)[(I + P_{\mathrm{nr}}P_{\mathrm{r}}^{-1})^{\mathrm{T}}\lambda_{\mathrm{n}}(t, t_0) + P_{\mathrm{r}}^{-1}\hat{x}_{\mathrm{r}}] \tag{5.41}$$

$$P_{\mathrm{r}}(t_0|t) = [I + P_{\mathrm{r}}(I + P_{\mathrm{nr}}P_{\mathrm{r}}^{-1})^{\mathrm{T}}M_{\mathrm{n}}(t, t_0)(I + P_{\mathrm{nr}}P_{\mathrm{r}}^{-1})]^{-1}P_{\mathrm{r}} \tag{5.42}$$

where $M_{\mathrm{n}}(t, t_0)$, $\lambda_{\mathrm{n}}(t, t_0)$ and $\Phi_{\mathrm{n}}(t, t_0)$ are evaluated as follows:

$$M_{\mathrm{n}}(t, t_0) = \int_{t_0}^{t} \Phi_{\mathrm{n}}^{\mathrm{T}}(s, t_0)H^{\mathrm{T}}(s)R^{-1}(s)H(s)\Phi_{\mathrm{n}}(s, t_0)\,\mathrm{d}s \tag{5.43}$$

$$\lambda_n(t, t_0) = \int_{t_0}^{t} \Phi_n^T(s, t_0) H^T(s) R^{-1}(s)[z(s) - H(s)\hat{x}_n(s|s)]\,ds \quad (5.44)$$

$$\frac{d\Phi_n(t, t_0)}{dt} = [A(t) - K_n(t)H(t)]\Phi_n(t, t_0), \qquad \Phi_n(t_0, t_0) = I \quad (5.45)$$

Proof

Conditioned on $\theta = x_r$, the problem is a standard Gaussian estimation problem, the solution to which is the Kalman filter with initial conditions [7]:

$$\begin{aligned} E[x(t_0)|x_r] &= E[x_n|x_r] + x_r \\ &= \hat{x}_n + P_{nr}P_r^{-1}[x_r - \hat{x}_r] + x_r \\ &= [\hat{x}_n - P_{nr}P_r^{-1}\hat{x}_r] + [I + P_{nr}P_r^{-1}]x_r \end{aligned} \quad (5.46)$$

$$\begin{aligned} P(x(t_0)|x_r) &\triangleq E\{[x_n - E(x_n|x_r)][x_n - E(x_n|x_r)]^T\} \\ &= P_n - P_{nr}P_r^{-1}P_{nr}^T \end{aligned} \quad (5.47)$$

The x_r-conditional estimate $\hat{x}(t|t, x_r) \triangleq E[x(t)|Z_t, x_r]$ and the error covariance $P(t|t, x_r) \triangleq E\{[x(t) - \hat{x}(t|t, x_r)][x(t) - \hat{x}(t|t, x_r)]^T|Z_t, x_r\}$ $\equiv P_n(t|t)$ are given by the following Kalman filter equations:

$$\frac{d\hat{x}(t|t, x_r)}{dt} = [A(t) - K_n(t)H(t)]\hat{x}(t|t, x_r) + K_n(t)z(t) \quad (5.48)$$

where $K_n(t)$, which is called the nominal filter gain, is obtained by expressions (5.37) and (5.38), but with the initial condition (5.46). Equation (5.48) can also be expressed, by using the closed-loop transition matrix $\Phi_n(t, t_0)$ which is provided by equation (5.45), as

$$\begin{aligned} \hat{x}(t|t, x_r) &= \Phi_n(t, t_0)[\hat{x}_n - P_{nr}P_r^{-1}\hat{x}_r] + \Phi_n(t, t_0)[I + P_{nr}P_r^{-1}]x_r \\ &\quad + \int_{t_0}^{t} \Phi_n(s, t_0)K_n(s)z(s)\,ds \\ &= \hat{x}_n(t|t) + \Phi_n(t, t_0)[I + P_{nr}P_r^{-1}]x_r \end{aligned} \quad (5.49)$$

where $\hat{x}_n(t|t)$ is the solution of equations (5.36) and (5.39). Substituting equation (5.49) into (3.7) with $\theta = x_r$ of the partition theorem gives equation (5.34), where $\hat{x}_r(t_0|t) \triangleq \int x_r p(x_r|Z_t)\,dx_r$. Furthermore, substituting $P(t|t, x_r) = P_n(t)$, along with expressions (5.34) and (5.49), into (3.8) with $\theta = x_r$ yields equation (5.35), where $P_r(t_0|t) \triangleq \int [x_r - \hat{x}_r(t_0|t)][x_r - \hat{x}_r(t_0|t)]^T p(x_r|Z_t)dx_r$. All that remains is to find $\hat{x}_r(t_0|t)$ and $P_r(t_0|t)$. Equations (3.16b) and (3.17) in Theorem 3.2 can now be rewritten as

$$p(x_r|Z_t) = \frac{\Lambda(t|x_r)p(x_r)}{\int \Lambda(t|x_r)p(x_r)\,dx_r} \tag{5.50}$$

where

$$\Lambda(t|x_r) = \exp\left\{\int_{t_0}^{t} \langle H(s)\hat{x}(s|s, x_r), R^{-1}(s)z(s)\rangle\,ds - \frac{1}{2}\int_{t_0}^{t} \|H(s)\hat{x}(s|s, x_r)\|^2_{R^{-1}(s)}\,ds\right\} \tag{5.51}$$

Notice that since the a priori density, $p(x_r)$, is Gaussian, i.e.

$$p(x_r) = C\exp\{-\tfrac{1}{2}\|x_r - \hat{x}_r\|^2_{P_r^{-1}}\} \tag{5.52}$$

the a posteriori probability density $p(x_r|Z_t)$ is also Gaussian. Hence substituting equation (5.49) into (5.51) and applying Lemma 5.2 to equations (5.50)–(5.52) yields the desired results (5.41)–(5.44). □

It is interesting to note that the matrix $M_n(t, t_0)$ can constitute the generalized observability Gramian matrix or Fisher's information matrix in the sense that it constitutes a partial observability matrix for the 'unknown' initial state x_r. Furthermore, note that $\lambda_n(t, t_0)$ is the Lagrange multiplier vector associated with the nominal Kalman filter (see also Chapter 6).

If $P_{nr} = 0$ in Theorem 5.1, the following corollary is obtained.

Corollary 5.1 Independent partitioning filter

If the initial Gaussian state $x(t_0)$ is decomposed into the sum of two statistically independent Gaussian vectors x_n and x_r, the optimal filtered estimate $\hat{x}(t|t)$ and the error covariance matrix $P(t|t)$ are given by

$$\hat{x}(t|t) = \hat{x}_n(t|t) + \Phi_n(t, t_0)\hat{x}_r(t_0|t) \tag{5.53}$$

$$P(t|t) = P_n(t|t) + \Phi_n(t, t_0)P_r(t_0|t)\Phi_n^T(t, t_0) \tag{5.54}$$

where $\hat{x}_n(t|t)$ and $P_n(t|t)$ are obtained from equations (5.36)–(5.38), but $\hat{x}_n(t_0|t_0) = \hat{x}_n$ and $P_n(t_0|t_0) = P_n$. Further, $\hat{x}_r(t_0|t)$ and $P_r(t_0|t)$ are modified as follows:

$$\hat{x}_r(t_0|t) = P_r(t_0|t)[\lambda_n(t, t_0) + P_r^{-1}\hat{x}_r] \tag{5.55}$$

$$P_r(t_0|t) = [I + P_rM_n(t, t_0)]^{-1}P_r \tag{5.56}$$

where $M_n(t, t_0)$, $\lambda_n(t, t_0)$ and $\Phi_n(t, t_0)$ are respectively given by equations (5.43), (5.44) and (5.45).

5.3 THE MULTIPARTITIONING FILTER (CONTINUOUS-TIME SYSTEMS)

The idea of the dependent partitioning filter can now be extended to the case when the initial state $x(t_0)$ is partitioned into M statistically dependent jointly Gaussian random vectors, i.e.

$$x(t_0) = \sum_{i=1}^{M} x_i \tag{5.57}$$

where the x_i have the following Gaussian probability density function:

$$\begin{bmatrix} x_1 \\ x_2 \\ \vdots \\ x_M \end{bmatrix} \sim N\left\{\begin{bmatrix} \hat{x}_1 \\ \hat{x}_2 \\ \vdots \\ \hat{x}_M \end{bmatrix}, \begin{bmatrix} P_{11} & P_{12} & \cdots & P_{1M} \\ P_{12}^{\mathrm{T}} & P_{22} & \cdots & P_{2M} \\ \vdots & \vdots & \ddots & \vdots \\ P_{1M}^{\mathrm{T}} & P_{2M}^{\mathrm{T}} & \cdots & P_{MM} \end{bmatrix}\right\} \tag{5.58}$$

Notice that for each i, $2 \leqslant i \leqslant M$, the conditional probability densities can be obtained from

$$p(x_i | x_{i+1}, x_{i+2}, \ldots, x_M) = N\{\hat{x}_{i|\{i\}}, P_{i|\{i\}}\} \tag{5.59}$$

where

$$\hat{x}_{i|\{i\}} = \hat{x}_i + \sum_{j=i+1}^{M} P_{ij|\{i\}}(x_j - \hat{x}_j) \tag{5.60}$$

$$P_{i|\{i\}} = P_{ii} - \sum_{j=i+1}^{M} P_{ij|\{i\}} P_{ij}^{\mathrm{T}} \tag{5.61}$$

in which $\{i\}$ denotes the set of integers $\{i+1, i+2, \ldots, M\}$. Furthermore, if we define

$$\bar{x}_{i|\{i\}} \triangleq \hat{x}_i - \sum_{j=i+1}^{M} P_{ij|\{i\}} \hat{x}_j \tag{5.62}$$

then

$$\hat{x}_{i|\{i\}} = \bar{x}_{i|\{i\}} + \sum_{j=i+1}^{M} P_{ij|\{i\}} x_j \tag{5.63}$$

Note that $\bar{x}_{i|\{i\}} = E[x_i | x_{i+1} = 0, \ldots, x_M = 0]$.

The solution to this multipartitioning estimation problem is given by the following theorem.

Theorem 5.2 Dependent multipartitioning filter

Assume that the initial state vector is partitioned as in expressions (5.58) and (5.59). The optimal filtered estimate, $\hat{x}(t|t)$, and its error covariance matrix, $P(t|t)$, are given by

$$\hat{x}(t|t) = \hat{x}_1(t|t) + \sum_{i=2}^{M} \Psi_{ii}(t, t_0)\hat{x}_i(t_0|t) \tag{5.64}$$

$$P(t|t) = P_1(t|t) + \sum_{i=2}^{M} \Psi_{ii}(t, t_0)P_i(t_0|t)\Psi_{ii}^{\mathrm{T}}(t, t_0) \tag{5.65}$$

where the conditional fixed-point smoothed estimate $\hat{x}_i(t_0|t) \triangleq E[x_i|Z_t, x_{i+1} = 0, \ldots, x_M = 0]$ and the corresponding error covariance matrix $P_i(t_0|t) \triangleq E\{[x_i - \hat{x}_i(t_0|t)][x_i - \hat{x}_i(t_0|t)]^{\mathrm{T}}|Z_t, x_{i+1}, \ldots, x_M\}$ are given for $2 \leqslant i \leqslant M$ by

$$\hat{x}_i(t_0|t) = P_i(t_0|t)[\lambda_i(t, t_0) + P_{i|\{i\}}^{-1}\bar{x}_{i|\{i\}}] \tag{5.66}$$

$$P_i(t_0|t) = [I + P_{i|\{i\}}M_{ii}(t, t_0)]^{-1}P_{i|\{i\}} \tag{5.67}$$

The $\lambda_i(t, t_0)$, $M_{ij}(t, t_0)$ and $\Psi_{ij}(t, t_0)$ are evaluated as follows:

$$\begin{aligned}\lambda_i(t, t_0) = \int_{t_0}^{t} \Psi_{ii}^{\mathrm{T}}(s, t_0)H^{\mathrm{T}}(s)R^{-1}(s)[z(s) - H(s)\{\hat{x}_1(s|s) \\ + \sum_{j=2}^{i-1} \Psi_{jj}(s, t_0)\hat{x}_j(t_0|s)\}]\,\mathrm{d}s\end{aligned} \tag{5.68}$$

where $2 \leqslant i \leqslant M$;

$$M_{ij}(t, t_0) = \int_{t_0}^{t} \Psi_{ii}^{\mathrm{T}}(s, t_0)H^{\mathrm{T}}(s)R^{-1}(s)H(s)\Psi_{ij}(s, t_0)\,\mathrm{d}s \tag{5.69}$$

where $2 \leqslant i \leqslant M$ and $i \leqslant j \leqslant M$;

$$\begin{aligned}\Psi_{ij}(t, t_0) = \Psi_{i-1,j}(t, t_0) + \Psi_{i-1,i-1}(t, t_0)P_{i-1}(t_0|t) \\ \times [P_{i-1|\{i-1\}}^{-1}P_{i-1,j|\{i-1\}} - M_{i-1,j}(t, t_0)]\end{aligned} \tag{5.70}$$

where $3 \leqslant i \leqslant M$ and $i \leqslant j \leqslant M$; and

$$\Psi_{2j}(t, t_0) = \Phi_2(t, t_0)[I + P_{1j|\{1\}}] \tag{5.71}$$

where $2 \leqslant j \leqslant M$. Finally, the quantities $\hat{x}_1(t|t) \triangleq E[x(t)|Z_t, x_2 = 0, \ldots, x_M = 0]$, $P_1(t|t) \triangleq E\{[x(t) - \hat{x}_1(t|t)][x(t) - \hat{x}_1(t|t)]^{\mathrm{T}}|Z_t, x_2, \ldots, x_M\}$ and $\Phi_2(t, t_0)$ are given by

$$\begin{aligned}\frac{\mathrm{d}\hat{x}_1(t|t)}{\mathrm{d}t} = [A(t) - K_1(t)H(t)]\hat{x}_1(t|t) + K_1(t)z(t), \\ \hat{x}_1(t_0|t_0) = \bar{x}_{1|\{1\}}\end{aligned} \tag{5.72}$$

$$K_1(t) \triangleq P_1(t|t)H^{\mathrm{T}}(t)R^{-1}(t) \tag{5.73}$$

$$\frac{\mathrm{d}P_1(t|t)}{\mathrm{d}t} = A(t)P_1(t|t) + P_1(t|t)A^{\mathrm{T}}(t) + B(t)Q(t)B^{\mathrm{T}}(t) - P_1(t|t)H^{\mathrm{T}}(t)R^{-1}(t)H(t)P_1(t|t),$$

$$P_1(t_0|t_0) = P_{1|\{1\}} \tag{5.74}$$

and

$$\frac{\mathrm{d}\Phi_2(t, t_0)}{\mathrm{d}t} = [A(t) - K_1(t)H(t)]\Phi_2(t, t_0), \qquad \Phi_2(t_0, t_0) = I \tag{5.75}$$

Proof

This proof is quite lengthy. Therefore, a proof for $M = 4$ can be found in Appendix 5C. □

The optimal unconditional smoothed estimates $\hat{x}_i^*(t_0|t) \triangleq E[x_i|Z_t]$ of x_i are given by

$$\hat{x}_i^*(t_0|t) = \hat{x}_i(t_0|t) + P_i(t_0|t) \sum_{j=i+1}^{M} [P_{i|\{i\}}^{-1} P_{ij|\{i\}} - M_{ij}(t, t_0)]\hat{x}_j^*(t_0|t) \tag{5.76}$$

where notice that $\hat{x}_M^*(t_0|t) \equiv \hat{x}_M(t_0|t)$ so that the recursive relation above can be started at $i = M - 1$. The total state filtered estimate can be expressed in terms of the optimal unconditional smoothed estimates as follows:

$$\hat{x}(t|t) = \hat{x}_1(t|t) + \sum_{j=2}^{M} \Psi_{2j}(t, t_0)\hat{x}_j^*(t_0|t) \tag{5.77}$$

In the case where $x_1, \ldots, x_M$ are mutually independent (i.e. $P_{ij} = 0, i \neq j$), Theorem 5.2 reduces to the following corollary:

Corollary 5.2 Independent multipartitioning filter

Assume that the random initial condition $x(t_0)$ is decomposed into M independent Gaussian random vectors. Then the optimal filtered estimate of $x(t)$ and its error covariance matrix are given by

$$\hat{x}(t|t) = \hat{x}_1(t|t) + \sum_{i=2}^{M} \Phi_i(t, t_0)\hat{x}_i(t_0|t) \tag{5.78}$$

$$P(t|t) = P_1(t|t) + \sum_{i=2}^{M} \Phi_i(t, t_0) P_i(t_0|t) \Phi_i^{\mathrm{T}}(t, t_0) \tag{5.79}$$

where the conditional fixed-point smoothed estimate and the corresponding error matrix are given, for $i \geq 2$, by

$$\hat{x}_i(t_0|t) = P_i(t_0|t)[\lambda_i(t, t_0) + P_{ii}^{-1}\hat{x}_i] \tag{5.80}$$

$$P_i(t_0|t) = [I + P_{ii}M_i(t, t_0)]^{-1} P_{ii} \tag{5.81}$$

The $\lambda_i(t, t_0)$, $M_i(t, t_0)$ and $\Phi_i(t, t_0)$, for $i \geq 3$, are evaluated using the following recursive relations:

$$\lambda_i(t, t_0) = [I - M_{i-1}(t, t_0)P_{i-1}(t_0|t)][\lambda_{i-1}(t, t_0) - M_{i-1}(t, t_0)\hat{x}_{i-1}] \tag{5.82}$$

$$M_i(t, t_0) = M_{i-1}(t, t_0)[I - P_{i-1}(t_0|t)M_{i-1}(t, t_0)] \tag{5.83}$$

$$\Phi_i(t, t_0) = \Phi_{i-1}(t, t_0)[I - P_{i-1}(t_0|t)M_{i-1}(t, t_0)] \tag{5.84}$$

and their equations for $i = 2$ are given by

$$\lambda_2(t, t_0) = \int_{t_0}^{t} \Phi_2^{\mathrm{T}}(s, t_0) H^{\mathrm{T}}(s) R^{-1}(s)[z(s) - H(s)\hat{x}_1(s|s)]\, \mathrm{d}s \tag{5.85}$$

$$M_2(t, t_0) = \int_{t_0}^{t} \Phi_2^{\mathrm{T}}(s, t_0) H^{\mathrm{T}}(s) R^{-1}(s) H(s) \Phi_2(s, t_0)\, \mathrm{d}s \tag{5.86}$$

$$\Phi_2(t, t_0) = I + \int_{t_0}^{t} [A(s) - K_1(s)H(s)]\Phi_2(s, t_0)\, \mathrm{d}s \tag{5.87}$$

Finally, the nominal Kalman filter is given by the following equations:

$$\frac{\mathrm{d}\hat{x}_1(t|t)}{\mathrm{d}t} = [A(t) - K_1(t)H(t)]\hat{x}_1(t|t) + K_1(s)z(t), \qquad \hat{x}_1(t_0|t_0) = \hat{x}_1 \tag{5.88}$$

$$\frac{\mathrm{d}P_1(t|t)}{\mathrm{d}t} = A(t)P_1(t|t) + P_1(t|t)A^{\mathrm{T}}(t) + B(t)Q(t)B^{\mathrm{T}}(t) - P_1(t|t)H^{\mathrm{T}}(t)R^{-1}(t)H(t)P_1(t|t), \qquad P_1(t_0|t_0) = P_{11} \tag{5.89}$$

$$K_1(t) \triangleq P_1(t|t)H^{\mathrm{T}}(t)R^{-1}(t) \tag{5.90}$$

Proof

From Theorem 5.2, we obtain, for $P_{ij} = 0$, $i \neq j$, the following equalities:

$$\Psi_{2j}(t, t_0) = \Phi_2(t, t_0), \qquad 2 \leqslant j \leqslant M \tag{5.91}$$

$$\Psi_{ij}(t, t_0) = \Phi_i(t, t_0) = \Phi_{i-1}(t, t_0)[I - P_{i-1}(t_0|t)M_{i-1}(t, t_0)] \tag{5.92}$$

because $P_{1j|\{1\}} \equiv 0$ for $2 \leqslant j \leqslant M$, and $\Psi_{i-1,j}(t, t_0) \equiv \Psi_{i-1,i-1}(t, t_0)$ and $P_{i-1,j|\{i-1\}} \equiv 0$ for $3 \leqslant i \leqslant M$ and $i \leqslant j \leqslant M$. Furthermore, setting $\Psi_{ii}(t, t_0) = \Phi_i(t, t_0)$, $P_{i|\{i\}} = P_{ii}$, $\bar{x}_{i|\{i\}} = \hat{x}_i$ in equations (5.64)–(5.67) gives expressions (5.78)–(5.81). All that remains is to find expressions (5.81)–(5.83). Note that $M_i(t, t_0)$ and $M_{i-1}(t, t_0)$ differ only in that their respective Kalman filters have different initial conditions. As is shown by Lainiotis [8] and Watanabe [9], $\lambda_i(t, t_0)$ and $M_i(t, t_0)$ can be expressed as

$$\lambda_i(t, t_0) = [I + M_{i-1}(t, t_0)P_{i-1,i-1}]^{-1}[\lambda_{i-1}(t, t_0) - M_{i-1}(t, t_0)\hat{x}_{i-1}] \tag{5.93}$$

$$M_i(t, t_0) = M_{i-1}(t, t_0)[I + P_{i-1,i-1}M_{i-1}(t, t_0)]^{-1} \tag{5.94}$$

By virtue of the fact that

$$\begin{aligned}[I + P_{i-1,i-1}M_{i-1}(t, t_0)]^{-1} &= I - [I + P_{i-1,i-1}M_{i-1}(t, t_0)]^{-1} \\ &\quad \times P_{i-1,i-1}M_{i-1}(t, t_0) \\ &= I - P_{i-1}(t_0|t)M_{i-1}(t, t_0)\end{aligned} \tag{5.95}$$

it is easy to obtain the desired results (5.82) and (5.83). □

Although the usual partitioning and multipartitioning filters were derived with the assumption that P_r and P_{ii}, $2 \leqslant i \leqslant M$, were covariance matrices (nonnegative definite symmetric matrices), this restrictive assumption is not necessary. However, note that as a result the quantities $\hat{x}_i(t_0|t)$ and $P_i(t_0|t)$ have no statistical meaning.

The dependent partitioning filter derived in Theorem 5.1 can now be approached by means of Corollary 5.2. That is, it is assumed that

$$\hat{x}_1 = \hat{x}_n, \qquad \hat{x}_2 = \hat{x}_r, \qquad \hat{x}_3 = 0 \tag{5.96}$$

$$P_{11} = P_n, \qquad P_{22} = P_r, \qquad P_{33} = P_{nr} + P_{nr}^T \tag{5.97}$$

The resulting algorithm is presented in the following theorem:

Theorem 5.3 Dependent partitioning filter: independent multipartitioning approach

The dependent partitioning filter shown in expressions (5.34)–(5.45) can be re-expressed by

$$\hat{x}(t|t) = \hat{x}_1(t|t) + \Phi_2(t, t_0)\hat{x}_2(t_0|t) + \Phi_3(t, t_0)\hat{x}_3(t_0|t) \tag{5.98}$$

$$P(t|t) = P_1(t|t) + \Phi_2(t, t_0)P_2(t_0|t)\Phi_2^{\mathrm{T}}(t, t_0) + \Phi_3(t, t_0)P_3(t_0|t)\Phi_3^{\mathrm{T}}(t, t_0) \tag{5.99}$$

where

$$\hat{x}_3(t_0|t) = P_3(t_0|t)\lambda_3(t, t_0) \tag{5.100}$$

$$P_3(t_0|t) = [I + (P_{\mathrm{nr}} + P_{\mathrm{nr}}^{\mathrm{T}})M_3(t, t_0)]^{-1}(P_{\mathrm{nr}} + P_{\mathrm{nr}}^{\mathrm{T}}) \tag{5.101}$$

$$\hat{x}_2(t_0|t) = P_2(t_0|t)[\lambda_2(t, t_0) + P_{\mathrm{r}}^{-1}\hat{x}_{\mathrm{r}}] \tag{5.102}$$

$$P_2(t_0|t) = [I + P_{\mathrm{r}}M_2(t, t_0)]^{-1}P_{\mathrm{r}} \tag{5.103}$$

The auxiliary equations are given by

$$\lambda_3(t, t_0) = [I - M_2(t, t_0)P_2(t_0|t)][\lambda_2(t, t_0) - M_2(t, t_0)\hat{x}_{\mathrm{r}}] \tag{5.104}$$

$$M_3(t, t_0) = M_2(t, t_0)[I - P_2(t_0|t)M_2(t, t_0)] \tag{5.105}$$

$$\Phi_3(t, t_0) = \Phi_2(t, t_0)[I - P_2(t_0|t)M_2(t, t_0)] \tag{5.106}$$

$$\lambda_2(t, t_0) = \int_{t_0}^{t} \Phi_2^{\mathrm{T}}(s, t_0)H^{\mathrm{T}}(s)R^{-1}(s)[z(s) - H(s)\hat{x}_1(s|s)]\,\mathrm{d}s \tag{5.107}$$

$$M_2(t, t_0) = \int_{t_0}^{t} \Phi_2^{\mathrm{T}}(s, t_0)H^{\mathrm{T}}(s)R^{-1}(s)H(s)\Phi_2(s, t_0)\,\mathrm{d}s \tag{5.108}$$

$$\Phi_2(t, t_0) = I + \int_{t_0}^{t} [A(s) - K_1(s)H(s)]\Phi_2(s, t_0)\,\mathrm{d}s \tag{5.109}$$

Furthermore, the quantities $\hat{x}_1(t|t)$, $P_1(t|t)$ and $K_1(t)$ are given by expressions (5.36), (5.37) and (5.38), respectively, but with the initial conditions $\hat{x}_1(t_0|t) = \hat{x}_{\mathrm{n}}$ and $P_1(t_0|t_0) = P_{\mathrm{n}}$.

Proof

The proof of this theorem consists of partitioning the initial condition into three parts as in equations (5.96) and (5.97) and applying Corollary 5.2 with $M = 3$. □

5.4 THE PARTITIONING FILTER (DISCRETE-TIME SYSTEMS)

In this section we shall deal with the discrete-time dynamical system which is given by the following difference equations:

$$x(k+1) = \Phi(k+1, k)x(k) + B(k)w(k) \tag{5.110}$$

$$z(k) = H(k)x(k) + v(k) \tag{5.111}$$

where $x(k)$ is an n-dimensional state vector; $w(k)$ is a p-dimensional disturbance vector; $z(k)$ is an m-dimensional observation vector; $v(k)$ is an m-dimensional measurement error vector; and $k = l, l+1, \ldots,$ is the discrete-time index. In addition, $\Phi(k+1, k)$ is an $n \times n$ matrix, the state transition matrix (or system matrix in discrete time); $B(k)$ is an $n \times p$ matrix, the distribution matrix of system noise (or disturbance transition matrix); and $H(k)$ is an $m \times n$ matrix, the observation matrix.

The processes $\{w(k), k = l, l+1, \ldots\}$ and $\{v(k), k = l, l+1, \ldots\}$ are mutually independent white Gaussian sequences for which

$$E[w(k)] = E[v(k)] = 0 \tag{5.112}$$

$$E\left\{\begin{bmatrix} w(k) \\ v(k) \end{bmatrix} [w^{\mathrm{T}}(j) \quad v^{\mathrm{T}}(j)]\right\} = \begin{bmatrix} Q(k) & 0 \\ 0 & R(k) \end{bmatrix} \delta_{kj} \tag{5.113}$$

for all $j, k = l, l+1, \ldots,$ where $Q(k)$ and $R(k)$ are, respectively, positive semidefinite $p \times p$ and $m \times m$ matrices, and δ_{kj} is the Kronecker delta function. We also assume that the initial state $x(l)$ is a Gaussian random n-vector with

$$E[x(l)] = \hat{x}(l) \tag{5.114}$$

$$E\{[x(l) - \hat{x}(l)][x(l) - \hat{x}(l)]^{\mathrm{T}}\} = P(l), \qquad P(l) \geqslant 0 \tag{5.115}$$

and that $x(l)$, $\{w(k), k = l, l+1, \ldots\}$ and $\{v(k), k = l, l+1, \ldots\}$ are independent of each other.

The following lemma will be useful for future developments.

Lemma 5.3 Moments for Gaussian a posteriori probability density function: discrete-time case

The a posteriori probability density function of the random vector X, conditioned on data $Z_k \triangleq \{z(i), l \leqslant i \leqslant k\}$ and random vector θ, is given by

$$p(X|Z_k, \theta) = \frac{p(Z_k|X, \theta)p(X|\theta)}{\int p(Z_k|X, \theta)p(X|\theta)\,\mathrm{d}X} \tag{5.116}$$

$$= \frac{p(z(k)|Z_{k-1}, X, \theta)p(X|Z_{k-1}, \theta)}{\int p(z(k)|Z_{k-1}, X, \theta)p(X|Z_{k-1}, \theta)\,\mathrm{d}X} \tag{5.117}$$

where

$$p(Z_k|X,\theta) = \prod_{i=l+1}^{k} p(z(i)|Z_{i-1},X,\theta) \tag{5.118}$$

$$p(z(i)|Z_{i-1},X,\theta) = C\exp\{-\tfrac{1}{2}\|\nu(i,X,\theta)\|^2_{\tilde{P}^{-1}(i|i-1,X,\theta)}\} \tag{5.119}$$

in which $\nu(i, X, \theta)$ is the X- and θ-conditional innovation process of a linear Gaussian discrete-time state estimation problem. Here, $\tilde{P}(i|i-1, X, \theta)$ is the covariance of the innovation process, and is not explicitly a function of X. Suppose that $\nu(i, X, \theta)$ can be divided into two parts, one of which is not a function of X while the other is a linear function of X, i.e.

$$\nu(i,X,\theta) = \nu(i,\theta) - \phi(i,l)X \tag{5.120}$$

where $\nu(i, \theta)$ is the θ-conditional innovation process; suppose also that the a priori probability density function of X is given by

$$p(X|\theta) = C'\exp\{-\tfrac{1}{2}\|X - \hat{X}_\theta\|^2_{P^{-1}(X|\theta)}\} \tag{5.121}$$

Then, if the above assumptions are satisfied, $p(X|Z_k, \theta)$ is a Gaussian density function,

$$p(X|Z_k,\theta) = C''\exp\{-\tfrac{1}{2}\|X - E(X|Z_k,\theta)\|^2_{P^{-1}(X|Z_k,\theta)}\} \tag{5.122}$$

where

$$E[X|Z_k,\theta] \triangleq P(X|Z_k,\theta)[\lambda(k,l) + P^{-1}(X|\theta)\hat{X}_\theta] \tag{5.123}$$

$$P(X|Z_k,\theta) \triangleq [M(k,l) + P^{-1}(X|\theta)]^{-1} \tag{5.124}$$

in which

$$\lambda(k,l) = \sum_{i=l+1}^{k} \phi^{\mathrm{T}}(i,l)\tilde{P}^{-1}(i|i-1,\theta)\nu(i,\theta) \tag{5.125}$$

$$M(k,l) = \sum_{i=l+1}^{k} \phi^{\mathrm{T}}(i,l)\tilde{P}^{-1}(i|i-1,\theta)\phi(i,l) \tag{5.126}$$

Proof

Substituting equation (5.120) into (5.119) gives

$$\begin{aligned} p(z(i)|Z_{i-1},X,\theta) = C\exp\{-\tfrac{1}{2}[&\|\nu(i,\theta)\|^2_{\tilde{P}^{-1}(i|i-1,\theta)} \\ &+ \|\phi(i,l)X\|^2_{\tilde{P}^{-1}(i|i-1,\theta)} \\ &- 2X^{\mathrm{T}}\phi^{\mathrm{T}}(i,l)\tilde{P}^{-1}(i|i-1,\theta)\nu(i,\theta)]\} \end{aligned} \tag{5.127}$$

Since the first term in the exponential function of equation (5.127) is not a function of X, it can be cancelled out in the evaluation of equation (5.117). Now, define

$$\tilde{\lambda}(i, l) \triangleq \phi^{\mathrm{T}}(i, l)\tilde{P}^{-1}(i|i-1, \theta)\nu(i, \theta) \tag{5.128}$$

$$\widetilde{M}(i, l) \triangleq \phi^{\mathrm{T}}(i, l)\tilde{P}^{-1}(i|i-1, \theta)\phi(i, l) \tag{5.129}$$

Substitution of these equations into (5.127) without the first term of the exponential function yields

$$p(X|Z_k, \theta) = \frac{\exp[-\frac{1}{2}X^{\mathrm{T}}\widetilde{M}(k, l)X + X^{\mathrm{T}}\tilde{\lambda}(k, l)]p(X|Z_{k-1}, \theta)}{\int \exp[-\frac{1}{2}X^{\mathrm{T}}\widetilde{M}(k, l)X + X^{\mathrm{T}}\tilde{\lambda}(k, l)]p(X|Z_{k-1}, \theta)\,\mathrm{d}X} \tag{5.130}$$

A similar expression can be obtained for $p(X|Z_{k-1}, \theta)$. After repeating this manipulation, it follows that

$$p(X|Z_k, \theta) = \frac{\exp[-\frac{1}{2}X^{\mathrm{T}}M(k, l)X + X^{\mathrm{T}}\lambda(k, l)]p(X|\theta)}{\int \exp[-\frac{1}{2}X^{\mathrm{T}}M(k, l)X + X^{\mathrm{T}}\lambda(k, l)]p(X|\theta)\,\mathrm{d}X} \tag{5.131}$$

where $\lambda(k, l)$ and $M(k, l)$ are defined as in equations (5.125) and (5.126). Furthermore, from equation (5.26), we obtain

$$\begin{aligned} &-\tfrac{1}{2}X^{\mathrm{T}}M(k, l)X + X^{\mathrm{T}}\lambda(k, l) \\ &= -\tfrac{1}{2}[X - M^{-1}(k, l)\lambda(k, l)]^{\mathrm{T}}M(k, l)[X - M^{-1}(k, l)\lambda(k, l)] \\ &\quad + \tfrac{1}{2}\lambda^{\mathrm{T}}(k, l)M^{-1}(k, l)M(k, l)M^{-1}(k, l)\lambda(k, l) \end{aligned} \tag{5.132}$$

This results in

$$p(X|Z_k, \theta) = \frac{\exp[-\frac{1}{2}\|X - M^{-1}(k, l)\lambda(k, l)\|^2_{M(k,l)}]p(X|\theta)}{\int \exp[-\frac{1}{2}\|X - M^{-1}(k, l)\lambda(k, l)\|^2_{M(k,l)}]p(X|\theta)\,\mathrm{d}X} \tag{5.133}$$

Expressions (5.121) and (5.133) are now substituted into Lemma 5.1 to obtain the desired results (5.122)–(5.124). □

Consider the discrete-time state estimation problem set out in equations (5.110)–(5.115) with the initial condition partitioned into two dependent jointly Gaussian random vectors,

$$x(l) = x_n + x_r \tag{5.134}$$

$$\begin{bmatrix} x_n \\ x_r \end{bmatrix} \sim N\left\{\begin{bmatrix} \hat{x}_n \\ \hat{x}_r \end{bmatrix}, \begin{bmatrix} P_n & P_{nr} \\ P_{nr}^T & P_r \end{bmatrix}\right\} \tag{5.135}$$

The results of the application of the discrete-time partition theorem given in Theorem 3.3 with $\theta = x_r$ is obtained in the following theorem:

Theorem 5.4 Dependent partitioning filter in discrete time

When the initial condition is partitioned as in expressions (5.134) and (5.135), the optimal filtered state estimate $\hat{x}(k|k) \triangleq E[x(k)|Z_k]$ of $x(k)$ and its error covariance matrix $P(k|k) \triangleq E\{[x(k) - \hat{x}(k|k)][x(k) - \hat{x}(k|k)]^T|Z_k\}$ are given by the following relations:

$$\hat{x}(k|k) = \hat{x}_n(k|k) + \Phi_n(k, l)[I + P_{nr}P_r^{-1}]\hat{x}_r(l|k) \tag{5.136}$$

$$\begin{aligned} P(k|k) = P_n(k|k) &+ \Phi_n(k, l)[I + P_{nr}P_r^{-1}] \\ &\times P_r(l|k)[I + P_{nr}P_r^{-1}]^T\Phi_n^T(k, l) \end{aligned} \tag{5.137}$$

where $\hat{x}_n(k|k) \triangleq E[x(k)|Z_k, x_r \equiv 0]$ and $P_n(k|k) \triangleq E\{[x(k) - \hat{x}_n(k|k)][x(k) - \hat{x}_n(k|k)]^T|Z_k, x_r\}$ are provided by the following nominal Kalman filter:

$$\begin{aligned} \hat{x}_n(k|k) = [\Phi(k, k-1) &- K_n(k)H(k)\Phi(k, k-1)] \\ &\times \hat{x}_n(k-1|k-1) \\ &+ K_n(k)z(k), \qquad \hat{x}_n(l|l) = \hat{x}_n - P_{nr}P_r^{-1}\hat{x}_r \end{aligned} \tag{5.138}$$

$$K_n(k) \triangleq P_n(k|k-1)H^T(k)\tilde{P}_n^{-1}(k|k-1) \tag{5.139}$$

$$\begin{aligned} P_n(k|k-1) = \Phi(k, k-1)&P_n(k-1|k-1)\Phi^T(k, k-1) \\ &+ B(k-1)Q(k-1)B^T(k-1), \\ &P_n(l|l) = P_n - P_{nr}P_r^{-1}P_{nr}^T \end{aligned} \tag{5.140}$$

$$\tilde{P}_n(k|k-1) = H(k)P_n(k|k-1)H^T(k) + R(k) \tag{5.141}$$

$$P_n(k|k) = [I - K_n(k)H(k)]P_n(k|k-1) \tag{5.142}$$

The fixed-point smoothed estimate $\hat{x}_r(l|k) \triangleq E[x_r|Z_k]$ of x_r and its error covariance matrix $P_r(l|k) \triangleq E\{[x_r - \hat{x}_r(l|k)][x_r - \hat{x}_r(l|k)]^T|Z_k\}$ are given by

$$\hat{x}_r(l|k) = P_r(l|k)[(I + P_{nr}P_r^{-1})^T\lambda_n(k, l) + P_r^{-1}\hat{x}_r] \tag{5.143}$$

$$P_r(l|k) = [I + P_r(I + P_{nr}P_r^{-1})^T M_n(k, l)(I + P_{nr}P_r^{-1})]^{-1} P_r \tag{5.144}$$

where the auxiliary quantities $M_n(k, l)$, $\lambda_n(k, l)$ and $\Phi_n(k, l)$ are evaluated as follows:

$$\begin{aligned} M_n(k, l) = \sum_{j=l+1}^{k} &\Phi_n^T(j-1, l)\Phi^T(j, j-1)H^T(j) \\ &\times \tilde{P}^{-1}(j, j-1)H(j) \\ &\times \Phi(j, j-1)\Phi_n(j-1, l) \end{aligned} \tag{5.145}$$

$$\begin{aligned} \lambda_n(k, l) = \sum_{j=l+1}^{k} &\Phi_n^T(j-1, l)\Phi^T(j, j-1)H^T(j)\tilde{P}^{-1}(j|j-1) \\ &\times [z(j) - H(j)\Phi(j, j-1)\hat{x}_n(j-1|j-1)] \end{aligned} \tag{5.146}$$

$$\Phi_n(k, k-1) = [I - K_n(k)H(k)]\Phi(k, k-1) \tag{5.147}$$

Proof

Conditioned on $\theta = x_r$, the estimation problem is a standard Gaussian estimation problem, the solution to which is the following Kalman filter with initial conditions:

$$E[x(l)|x_r] = [\hat{x}_n - P_{nr}P_r^{-1}\hat{x}_r] + [I + P_{nr}P_r^{-1}]x_r \tag{5.148}$$

$$\begin{aligned} P(x(l)|x_r) &\triangleq E\{[x_n - E(x_n|x_r)][x_n - E(x_n|x_r)]^T\} \\ &= P_n - P_{nr}P_r^{-1}P_{nr}^T \end{aligned} \tag{5.149}$$

The x_r-conditional estimate $\hat{x}(j|j, x_r) \triangleq E[x(j)|Z_j, x_r]$ and the error covariance $P(j|j, x_r) \triangleq E\{[x(j) - \hat{x}(j|j, x_r)][x(j) - \hat{x}(j|j, x_r)]^T|Z_j, x_r\} \equiv P_n(j|j)$ are given by the following Kalman filter equations:

$$\begin{aligned} \hat{x}(j|j, x_r) = \Phi(j, j-1)\hat{x}(j-1|j-1, x_r) &+ K_n(j)[z(j) \\ &- H(j)\Phi(j, j-1)\hat{x}(j-1|j-1, x_r)] \end{aligned} \tag{5.150}$$

where $K_n(j)$ is given by expressions (5.139)–(5.142), but with initial condition (5.148). By using the closed-loop state transition matrix of equation (5.147), expression (5.150) can also be rewritten as

$$\begin{aligned} \hat{x}(j|j, x_r) &= \Phi_n(j, l)[\hat{x}_n - P_{nr}P_r^{-1}\hat{x}_r] + \Phi_n(j, l)[I + P_{nr}P_r^{-1}]x_r \\ &\quad + \sum_{i=l+1}^{j} \Phi_n(j, i)K_n(i)z(i) \\ &= \hat{x}_n(j|j) + \Phi_n(j, l)[I + P_{nr}P_r^{-1}]x_r \end{aligned} \tag{5.151}$$

where $\hat{x}_n(j|j)$ is the unique solution of equation (5.138). Substituting expression (5.151) into (3.55) with $\theta = x_r$ gives equation (5.136), where $\hat{x}_r(l|k) \triangleq \int x_r p(x_r|Z_k) dx_r$. Moreover, substituting $P(k|k, x_r) \equiv P_n(k|k)$, along with expressions (5.143) and (5.151), into expression (3.58) with $\theta = x_r$ provides equation (5.144), where $P_r(l|k) \triangleq \int [x_r - \hat{x}_r(l|k)][x_r - \hat{x}_r(l|k)]^T p(x_r|Z_k)\, dx_r$.

Equations (3.33) and (3.34) in Theorem 3.3 can now be represented as

$$p(x_r|Z_j) = \frac{L(j|x_r)p(x_r|Z_{j-1})}{\int L(j|x_r)p(x_r|Z_{j-1})\, dx_r} \tag{5.152}$$

where

$$L(j|x_r) = C_1 \exp\{-\tfrac{1}{2}\|v(j, x_r)\|^2_{\tilde{P}_n^{-1}(j|j-1)}\} \tag{5.153}$$

in which

$$\begin{aligned} v(j, x_r) &= z(j) - H(j)\Phi(j, j-1)\hat{x}(j-1|j-1, x_r) \\ &= z(j) - H(j)\Phi(j, j-1)\hat{x}_n(j-1|j-1) \\ &\quad - H(j)\Phi(j, j-1)\Phi_n(j, l)[I + P_{nr}P_r^{-1}]x_r \end{aligned} \tag{5.154}$$

We now write

$$v(j, x_r) = v_n(j) - H(j)\Phi(j, j-1)\Phi_n(j, l)[I + P_{nr}P_r^{-1}]x_r \tag{5.155}$$

where

$$v_n(j) \triangleq z(j) - H(j)\Phi(j, j-1)\hat{x}_n(j-1|j-1) \tag{5.156}$$

Recall that the a priori probability density function of x_r is given by

$$p(x_r) = C_2 \exp[-\tfrac{1}{2}\|x_r - \hat{x}_r\|_{P_r^{-1}}] \tag{5.157}$$

Hence, substitution of equation (5.155) into (5.153) and application of Lemma 5.3 to equations (5.152), (5.153) and (5.157) give the desired results (5.143)–(5.146). □

Corollary 5.3 Independent partitioning filter in discrete time

If the initial Gaussian state $x(l)$ is decomposed into two statistically independent Gaussian random vectors x_n and x_r, the optimal filtered estimate $\hat{x}(k|k)$ and the error covariance matrix $P(k|k)$ are given by

$$\hat{x}(k|k) = \hat{x}_n(k|k) + \Phi_n(k, l)\hat{x}_r(l|k) \tag{5.158}$$

$$P(k|k) = P_n(k|k) + \Phi_n(k, l)P_r(l|k)\Phi_n^T(k, l) \tag{5.159}$$

where $\hat{x}_n(k|k)$ and $P_n(k|k)$ are given by equations (5.138)–(5.142), but with initial conditions $\hat{x}_n(l|l) = \hat{x}_n$ and $P_n(l|l) = P_n$. Furthermore, $\hat{x}_r(l|k)$ and $P_r(l|k)$ are obtained by

$$\hat{x}_r(l|k) = P_r(l|k)[\lambda_n(k, l) + P_r^{-1}\hat{x}_r] \tag{5.160}$$

$$P_r(l|k) = [I + P_r M_n(k, l)]^{-1} P_r \tag{5.161}$$

where $M_n(k, l)$, $\lambda_n(k, l)$ and $\Phi_n(k, k-1)$ are given by equations (5.145), (5.146) and (5.147), respectively.

Proof

The proof is apparent from the substitution of $P_{nr} = 0$ into Theorem 5.4. □

Example 5.1 Improvement of relative observability

The smoothed covariance in the partitioning filter (or any multipartitioning filters) can be written as

$$P_2(l|k) = [M_2(k, l) + P_2^{-1}]^{-1}$$

If $P_2^{-1} = 0$, the question arises as to how difficult will it be to invert $M_2(k, l)$. A measure of the degree of invertibility of $M_2(k, l)$ can be determined by investigating the eigenvalues of $M_2(k, l)$. We define the observability index as the square root of the ratio of the smallest eigenvalue of $M_2(k, l)$ to the largest.† The greater the observability index, the more observable the nominal system. The smaller the observability index, the more one eigenvalue of $M_2(k, l)$ approaches zero relative to the largest eigenvalue.

Partitioning the initial state in different manners provides some control over the observability index of the nominal system. Assume that a system initially contains a group of state variables, x_{small}, which have small initial covariances, and another group, x_{large}, which have large initial covariances, i.e.

$$x(l) = \begin{bmatrix} x_{\text{small}} \\ x_{\text{large}} \end{bmatrix}$$

Suppose also that x_{small} and x_{large} are uncorrelated. If the initial state

† This is not the conventional definition. See Appendix C for the familiar definition in linear system theory.

is partitioned as follows:

$$x_1 = \begin{bmatrix} x_{\text{small}} \\ 0 \end{bmatrix}, \qquad x_2 = \begin{bmatrix} 0 \\ x_{\text{large}} \end{bmatrix}$$

then the nominal Kalman filter based on x_1 will have an increased observability index.

The observability index can also be controlled when certain states, denoted x_{fast}, have fast dynamic response while other states, x_{slow}, have slower dynamic response. Again assume that x_{fast} and x_{slow} are uncorrelated. When the initial condition

$$x(l) = \begin{bmatrix} x_{\text{slow}} \\ x_{\text{fast}} \end{bmatrix}$$

is partitioned as follows:

$$x_1 = \begin{bmatrix} x_{\text{slow}} \\ 0 \end{bmatrix}, \qquad x_2 = \begin{bmatrix} 0 \\ x_{\text{fast}} \end{bmatrix}$$

the nominal Kalman filter has an improved observability index.

Example 5.2 Parameter identification [2, 3]

The determination of unknown system parameters from observations (and input data) is called parameter *identification* or *estimation*. One approach to this problem is to augment the state vector, $x(k)$, with a q-vector of constant unknowns to be identified, denoted by θ; that is, the following augmented state vector:

$$y(k) = \begin{bmatrix} x(k) \\ \theta \end{bmatrix}$$

is used. The augmented model is then linearized around a nominal trajectory

$$y_{\text{n}}(k) = \begin{bmatrix} x_{\text{n}}(k) \\ \theta_{\text{n}} \end{bmatrix}$$

and linear estimation or extended Kalman filter (see Chapter 2) techniques are often applied to small perturbed state variables and parameters. It is assumed that the a priori estimates of $x(0)$ and θ are uncorrelated. Then, the initial statistics of such a linearized model can be partitioned as follows:

$$\delta\hat{y}(0) = \delta\hat{y}_{\text{n}}(0) + \delta\hat{y}_{\text{r}}(0)$$

$$P(0) = P_{\mathrm{n}}(0) + P_{\mathrm{r}}(0)$$

where

$$\delta\hat{y}(0) = \begin{bmatrix} \hat{x}(0) - x_{\mathrm{n}}(0) \\ \hat{\theta}(0) - \theta_{\mathrm{n}} \end{bmatrix}, \qquad P(0) = \begin{bmatrix} P_x(0) & 0 \\ 0 & P_\theta(0) \end{bmatrix}$$

$$\delta\hat{y}_{\mathrm{n}}(0) = \begin{bmatrix} \hat{x}(0) - x_{\mathrm{n}}(0) \\ 0 \end{bmatrix}, \qquad \delta\hat{y}_{\mathrm{r}}(0) = \begin{bmatrix} 0 \\ \hat{\theta}(0) - \theta_{\mathrm{n}} \end{bmatrix}$$

$$P_{\mathrm{n}}(0) = \begin{bmatrix} P_x(0) & 0 \\ 0 & 0 \end{bmatrix}, \qquad P_{\mathrm{r}}(0) = \begin{bmatrix} 0 & 0 \\ 0 & P_\theta(0) \end{bmatrix}$$

The application of Corollary 5.3 to this problem yields a parameter estimation problem with the following properties:

(i) The nominal Kalman filter is of order n, whereas the usual extended Kalman filter approach [10–12] requires a filter with order $(n + q)$ (see also Goodwin and Sin [13] for a simpler approach).

(ii) The matrix to be inverted in the solution for the fixed-point smoothed covariance of the parameters $P_{\mathrm{r}}(0)$ is of order $q \times q$ rather than $(n + q) \times (n + q)$.

(iii) The resulting algorithm is of iterative or batch processing type, so that the data are processed several times in order to improve the nominal trajectory around which the problem has been linearized, as in the maximum likelihood (ML) identification technique.

More detailed algorithms and an application to aircraft flight data for this approach can be found in Eulrich *et al.* [14, 15] and Govindaraj *et al.* [16]. In a similar way, two-stage bias correction estimators based on the partitioning approach will be developed in Chapter 8. □

5.5 THE MULTIPARTITIONING FILTER (DISCRETE-TIME SYSTEMS)

The concept of Theorem 5.2 also applies in the discrete-time case. That is, when the initial state $x(l)$ is partitioned into M statistically dependent jointly Gaussian random vectors,

$$x(l) = \sum_{i=1}^{M} x_i \tag{5.162}$$

The discrete-time dependent multipartitioning filter is summarized in the following theorem.

Theorem 5.5 Dependent multipartitioning filter in discrete time

The optimal multipartitioning filter for the discrete-time system given by equations (5.110)–(5.115) consists of the following relations for the filtered state and covariance matrix:

$$\hat{x}(k|k) = \hat{x}_1(k|k) + \sum_{i=2}^{M} \Psi_{ii}(k, l)\hat{x}_i(l|k) \tag{5.163}$$

$$P(k|k) = P_1(k|k) + \sum_{i=2}^{M} \Psi_{ii}(k, l)P_i(l|k)\Psi_{ii}^{\mathrm{T}}(k, l) \tag{5.164}$$

where $\hat{x}_1(k|k) \triangleq E[x(k)|Z_k, x_2 = 0, \ldots, x_M = 0]$ and $P_1(k|k) \triangleq E\{[x(k) - \hat{x}_1(k|k)][x(k) - \hat{x}_1(k|k)]^{\mathrm{T}}|Z_k, x_2, \ldots, x_M\}$ are given by the following Kalman filter:

$$\hat{x}_1(k|k) = \Phi(k, k-1)\hat{x}_1(k-1|k-1) + K_1(k)\nu_1(k), \quad \hat{x}_1(l|l) = \bar{x}_{1|\{1\}} \tag{5.165}$$

$$\nu_1(k) \triangleq z(k) - H(k)\Phi(k, k-1)\hat{x}_1(k-1|k-1) \tag{5.166}$$

$$K_1(k) \triangleq P_1(k|k-1)H^{\mathrm{T}}(k)\tilde{P}_1^{-1}(k|k-1) \tag{5.167}$$

$$P_1(k|k) = \Phi(k, k-1)P_1(k-1|k-1)\Phi^{\mathrm{T}}(k, k-1) + B(k-1)Q(k-1)B(k-1), \quad P_1(l|l) = P_{1|\{1\}} \tag{5.168}$$

$$\tilde{P}_1(k|k-1) = H(k)P_1(k|k-1)H^{\mathrm{T}}(k) + R(k) \tag{5.169}$$

$$P_1(k|k) = [I - K_1(k)H(k)]P_1(k|k-1) \tag{5.170}$$

The conditional fixed-point smoothed estimate $\hat{x}_i(l|k) \triangleq E[x_i|Z_k, x_{i+1} = 0, \ldots, x_M = 0]$ and the associated error covariance matrix $P_i(l|k) \triangleq E\{[x_i - \hat{x}_i(l|k)][x_i - \hat{x}_i(l|k)]^{\mathrm{T}}|Z_k, x_{i+1}, \ldots, x_M\}$ are given for $2 \leqslant i \leqslant M$ by:

$$\hat{x}_i(l|k) = P_i(l|k)[\lambda_i(k, l) + P_{i|\{i\}}^{-1}\bar{x}_{i|\{i\}}] \tag{5.171}$$

$$P_i(l|k) = [I + P_{i|\{i\}}M_{ii}(k, l)]^{-1}P_{i|\{i\}} \tag{5.172}$$

The auxiliary quantities are evaluated as follows:

$$\lambda_i(k, l) = \sum_{j=l+1}^{k} \Psi_{ii}^{\mathrm{T}}(j-1, l)\Phi^{\mathrm{T}}(j, j-1)H^{\mathrm{T}}(j)\tilde{P}_i^{-1}(j|j-1)\nu_i(j) \tag{5.173}$$

$$\nu_i(j) \triangleq z(j) - H(j)\Phi(j, j-1)[\hat{x}_1(j-1|j-1) + \sum_{h=2}^{i-1} \Psi_{hh}(j-1, l)\hat{x}_h(l|j-1)] \tag{5.174}$$

$$\tilde{P}_i(j|j-1) \triangleq \tilde{P}_1(j|j-1) + H(j)\Phi(j, j-1) \times \left[\sum_{h=2}^{i-1} \Psi_{hh}(j-1, l)P_h(l|j-1)\Psi_{hh}^{\mathrm{T}}(j-1, l)\right] \times \Phi^{\mathrm{T}}(j, j-1)H^{\mathrm{T}}(j) \tag{5.175}$$

$$M_{im}(k, l) = \sum_{j=l+1}^{k} \Psi_{ii}^{\mathrm{T}}(j-1, l)\Phi^{\mathrm{T}}(j, j-1)H^{\mathrm{T}}(j)\tilde{P}_i^{-1}(j|j-1) \times H(j)\Phi(j, j-1)\Psi_{im}(j-1, l) \tag{5.176}$$

where $2 \leqslant i \leqslant M$ and $i \leqslant m \leqslant M$. Moreover,

$$\Psi_{im}(k, l) = \Psi_{i-1,m}(k, l) + \Psi_{i-1,i-1}(k, l)P_{i-1}(l|k) \times [P_{i-1|\{i-1\}}^{-1}P_{i-1,m|\{i-1\}} - M_{i-1,m}(k, l)] \tag{5.177}$$

where $3 \leqslant i \leqslant M$ and $i \leqslant m \leqslant M$, and, for $i = 2$,

$$\Psi_{2m}(k, l) = \Phi_2(k, l)[I + P_{1j|\{1\}}] \tag{5.178}$$

$$\Phi_2(k, k-1) = [I - K_1(k)H(k)]\Phi(k, k-1) \tag{5.179}$$

Proof

The proof proceeds in the same way as Theorem 5.2 but with $M = 4$. Therefore, it is omitted here. □

Example 5.3 Block diagonal initial covariance matrix

Suppose that the initial state $x(l)$ has moments given by

$$\hat{x}(l) = \begin{bmatrix} m_1 \\ m_2 \\ \vdots \\ m_M \end{bmatrix}, \qquad P(l) = \begin{bmatrix} V_1 & 0 & \cdots & 0 \\ 0 & V_2 & \cdots & 0 \\ \vdots & \vdots & \ddots & \vdots \\ 0 & 0 & \cdots & V_M \end{bmatrix}$$

where the V_i, $i = 1, \ldots, M$, are symmetric nonnegative definite matrices with dimensions consistent with the vectors m_i. This initial condition can be partitioned as follows:

$$\hat{x}_1 = \begin{bmatrix} m_1 \\ 0 \\ 0 \\ \vdots \\ 0 \end{bmatrix}, \quad \hat{x}_2 = \begin{bmatrix} 0 \\ m_2 \\ 0 \\ \vdots \\ 0 \end{bmatrix}, \ldots, \quad \hat{x}_M = \begin{bmatrix} 0 \\ 0 \\ 0 \\ \vdots \\ m_M \end{bmatrix}$$

$$P_1 = \begin{bmatrix} V_1 & 0 & 0 & \cdots & 0 \\ 0 & 0 & 0 & \cdots & 0 \\ 0 & 0 & 0 & \cdots & 0 \\ \vdots & \vdots & \vdots & \ddots & \vdots \\ 0 & 0 & 0 & \cdots & 0 \end{bmatrix},$$

$$P_2 = \begin{bmatrix} 0 & 0 & 0 & \cdots & 0 \\ 0 & V_2 & 0 & \cdots & 0 \\ 0 & 0 & 0 & \cdots & 0 \\ \vdots & \vdots & \vdots & \ddots & \vdots \\ 0 & 0 & 0 & \cdots & 0 \end{bmatrix}, \ldots,$$

$$P_M = \begin{bmatrix} 0 & 0 & 0 & \cdots & 0 \\ 0 & 0 & 0 & \cdots & 0 \\ 0 & 0 & 0 & \cdots & 0 \\ \vdots & \vdots & \vdots & \ddots & \vdots \\ 0 & 0 & 0 & \cdots & V_M \end{bmatrix}$$

As a result, the effects of each block of the initial state can be separated from the others. Moreover, considerable computational benefit is obtained in the matrix inversion for $P_i(l|k)$. For instance, to compute $P_2(l|k)$ would normally require the inversion of an $n \times n$ matrix, but when P_2 is defined as above, the matrix to be inverted is only the size of V_2, a substantial computational saving. This results because

$$P_2(l|k) = [I + P_2 M_2(k, l)]^{-1} P_2$$
$$= \begin{bmatrix} 0 & 0 & 0 & \cdots & 0 \\ 0 & [I + V_2 M_{2,22}(k, l)]^{-1} V_2 & 0 & \cdots & 0 \\ 0 & 0 & 0 & \cdots & 0 \\ \vdots & \vdots & \vdots & \ddots & \vdots \\ 0 & 0 & 0 & \cdots & 0 \end{bmatrix}$$

where $M_{2,22}(k, l)$ denotes the $(2, 2)$ block of the $M_2(k, l)$ matrix corresponding in size and location to V_2. In addition, when the m_i and V_i are scalars, the matrix inversions are totally eliminated.

Corollary 5.4 Independent multipartitioning filter in discrete time

If the random initial state $x(l)$ is decomposed into M independent Gaussian random vectors, the optimal filtered estimate of $x(k)$ and its error covariance matrix are given by

$$\hat{x}(k|k) = \hat{x}_1(k|k) + \sum_{i=2}^{M} \Phi_i(k, l)\hat{x}_i(l|k) \tag{5.180}$$

$$P(k|k) = P_1(k|k) + \sum_{i=2}^{M} \Phi_i(k, l)P_i(l|k)\Phi_i^{\mathrm{T}}(k, l) \tag{5.181}$$

where $\hat{x}_1(k|k)$ and $P_1(k|k)$ are given by equations (5.165)–(5.170), but with the initial conditions $\hat{x}_1(l|l) = \hat{x}_1$ and $P_1(l|l) = P_{11}$. The conditional fixed-point smoothed estimate $\hat{x}_i(l|k)$ and the error covariance matrix $P_i(l|k)$ are given, for $2 \leqslant i \leqslant M$, by:

$$\hat{x}_i(l|k) = P_i(l|k)[\lambda_i(k, l) + P_{ii}^{-1}\hat{x}_i] \tag{5.182}$$

$$P_i(l|k) = [I + P_{ii}M_i(k, l)]^{-1}P_{ii} \tag{5.183}$$

The auxiliary equations for $i \geqslant 3$ are evaluated using the following recursive expressions:

$$\lambda_i(k, l) = [I - M_{i-1}(k, l)P_{i-1}(l|k)][\lambda_{i-1}(k, l) - M_{i-1}(k, l)\hat{x}_{i-1}] \tag{5.184}$$

$$M_i(k, l) = M_{i-1}(k, l)[I - P_{i-1}(l|k)M_{i-1}(k, l)] \tag{5.185}$$

$$\Phi_i(k, l) = \Phi_{i-1}(k, l)[I - P_{i-1}(l|k)M_{i-1}(k, l)] \tag{5.186}$$

For $i = 2$ the auxiliary equations are obtained by

$$\lambda_2(k, l) = \sum_{i=l+1}^{k} \Phi_2^{\mathrm{T}}(i-1, l)\Phi^{\mathrm{T}}(i, i-1)H^{\mathrm{T}}(i)\tilde{P}_1^{-1}(i|i-1)\nu_1(i) \tag{5.187}$$

$$M_2(k, l) = \sum_{i=l+1}^{k} \Phi_2^{\mathrm{T}}(i-1, l)\Phi^{\mathrm{T}}(i, i-1)H^{\mathrm{T}}(i) \times \tilde{P}_1^{-1}(i|i-1)H(i)\Phi(i, i-1)\Phi_2^{\mathrm{T}}(i-1, l) \tag{5.188}$$

$$\Phi_2(k, k-1) = [I - K_1(k)H(k)]\Phi(k, k-1) \tag{5.189}$$

Proof

Although this theorem can be proved as a special case of Theorem 5.5, an alternative proof will be presented here, involving repeated use of the discrete-time independent partitioning filter shown in Corollary 5.3. The initial condition is first partitioned as follows:

$$x(l) = x_{\mathrm{n}}^{(1)} + x_{\mathrm{r}}^{(1)} \tag{5.190}$$

where

$$\left.\begin{aligned} x_{\mathrm{n}}^{(1)} &= x_1 + x_2 + \ldots + x_{M-1} \\ x_{\mathrm{r}}^{(1)} &= x_M \end{aligned}\right\} \tag{5.191}$$

The resulting independent partitioning (or Lainiotis) filter will contain an $M_M(k, l)$, $\lambda_M(k, l)$ and $\Phi_M(k, l)$ given by equations (5.145), (5.146) and (5.147), respectively, but based on an embedded Kalman filter with initial conditions $\hat{x}_{\mathrm{n}}^{(1)} = \hat{x}_1 + \hat{x}_2 + \ldots + \hat{x}_{M-1}$ and $P_{\mathrm{n}}^{(1)} = P_{11} + P_{22} + \ldots + P_{M-1,M-1}$. The independent partitioning filter is again used to solve for the embedded Kalman filter by repartitioning as follows:

$$\left.\begin{aligned} x_{\mathrm{n}}^{(2)} &= x_1 + x_2 + \ldots + x_{M-2} \\ x_{\mathrm{r}}^{(2)} &= x_{M-1} \end{aligned}\right\} \tag{5.192}$$

What results will be equations for $M_{M-1}(k, l)$, $\lambda_{M-1}(k, l)$ and $\Phi_{M-1}(k, l)$, and an embedded Kalman filter with initial conditions $\hat{x}_{\mathrm{n}}^{(2)} = \hat{x}_1 + \hat{x}_2 + \ldots + \hat{x}_{M-2}$ and $P_{\mathrm{n}}^{(2)} = P_{11} + P_{22} + \ldots + P_{M-2,M-2}$. This process is continued until the final embedded Kalman filter has initial conditions $\hat{x}_1$ and P_{11}. Equations (5.184)–(5.186) are derived as for Corollary 5.2. □

Theorem 5.6 Dependent partitioning filter in discrete time: independent multipartitioning approach

The dependent partitioning filter presented in equations (5.136)–(5.147) can also be expressed as follows:

$$\hat{x}(k|k) = \hat{x}_1(k|k) + \Phi_2(k, l)\hat{x}_2(l|k) + \Phi_3(k, l)\hat{x}_3(l|k) \tag{5.193}$$

$$\begin{aligned} P(k|k) = {} & P_1(k|k) + \Phi_2(k, l)P_2(l|k)\Phi_2^{\mathrm{T}}(k, l) \\ & + \Phi_3(k, l)P_3(l|k)\Phi_3^{\mathrm{T}}(k, l) \end{aligned} \tag{5.194}$$

where

$$\hat{x}_3(l|k) = P_3(l|k)\lambda_3(k, l) \tag{5.195}$$

$$P_3(l|k) = [I + (P_{\mathrm{nr}} + P_{\mathrm{nr}}^{\mathrm{T}})M_3(k, l)]^{-1}(P_{\mathrm{nr}} + P_{\mathrm{nr}}^{\mathrm{T}}) \tag{5.196}$$

$$\hat{x}_2(l|k) = P_2(l|k)[\lambda_2(k, l) + P_{\mathrm{r}}^{-1}\hat{x}_{\mathrm{r}}] \tag{5.197}$$

$$P_2(l|k) = [I + P_r M_2(k, l)]^{-1} P_r \tag{5.198}$$

The auxiliary equations are obtained by

$$\lambda_3(k, l) = [I - M_2(k, l)P_2(l|k)][\lambda_2(k, l) - M_2(k, l)\hat{x}_r] \tag{5.199}$$

$$M_3(k, l) = M_2(k, l)[I - P_2(l|k)M_2(k, l)] \tag{5.200}$$

$$\Phi_3(k, l) = \Phi_2(k, l)[I - P_2(l|k)M_2(k, l)] \tag{5.201}$$

$$\lambda_2(k, l) = \sum_{j=l+1}^{k} \Phi_2^{\mathrm{T}}(j-1, l)\Phi^{\mathrm{T}}(j, j-1)H^{\mathrm{T}}(j)\tilde{P}_1^{-1}(j|j-1)\nu_1(j) \tag{5.202}$$

$$M_2(k, l) = \sum_{j=l+1}^{k} \Phi_2^{\mathrm{T}}(j-1, l)\Phi^{\mathrm{T}}(j, j-1)H^{\mathrm{T}}(j) \times \tilde{P}_1^{-1}(j|j-1)H(j)\Phi(j, j-1)\Phi_2(j-1, l) \tag{5.203}$$

$$\Phi_2(j, j-1) = [I - K_1(j)H(j)]\Phi(j, j-1) \tag{5.204}$$

Moreover, the quantities $\hat{x}_1(k|k)$, $P_1(k|k)$ and $K_1(k)$ are given by equations (5.138)–(5.142), but with initial conditions $\hat{x}_1(l|l) = \hat{x}_n$ and $P_1(l|l) = P_n$.

Proof

The proof simply consists of partitioning the initial condition into three parts as in expressions (5.96) and (5.97) and applying Corollary 5.4 with $M = 3$. □

5.6 PARTITIONED NUMERICAL ALGORITHMS FOR RICCATI EQUATIONS

In this section, we apply the independent partitioning filter method presented in Section 5.2 to develop two numerical algorithms for solving the nonsteady-state and steady-state solutions of the Riccati differential equations (RDEs).

Before presenting the main algorithms, we shall study the relationships between the partitioning filter discussed in Corollary 5.1 and the usual Kalman filter from the viewpoint of the well-known fixed-point smoother, already cited in Chapter 3. Recalling such a smoother, we obtain the following equations (see also Sage and Melsa [17]):

$$\frac{d\hat{x}(\tau|t)}{dt} = P(\tau|\tau)\Phi_K^T(t, \tau)H^T(t)R^{-1}(t) \times [z(t) - H(t)\hat{x}(t|t)], \qquad \hat{x}(\tau|t)|_{t=\tau} = \hat{x}(\tau|\tau) \quad (5.205)$$

$$\frac{dP(\tau|t)}{dt} = -P(\tau|\tau)\Phi_K^T(t, \tau)H^T(t)R^{-1}(t)H(t)\Phi_K(t, \tau)P(\tau|\tau), \qquad P(\tau|t)|_{t=\tau} = P(\tau|\tau) \quad (5.206)$$

where τ indicates a fixed time $\tau \geqslant t_0$, $\hat{x}(t|t)$ and $P(t|t)$ are the Kalman filtered estimate and the associated error covariance matrix, respectively, and $\Phi_K(t, \tau)$ is the closed-loop transition matrix such that

$$\frac{d\Phi_K(t, \tau)}{dt} = A_K(t)\Phi_K(t, \tau), \qquad \Phi_K(\tau, \tau) = I \quad (5.207)$$

$$A_K(t) = A(t) - K(t)H(t) \quad (5.208)$$

in which $K(t) \triangleq P(t|t)H^T(t)R^{-1}(t)$. If this smoother is used to estimate x_r at $t_0 = \tau$, we have the following lemmas [9, 18].

Lemma 5.4

The relationships of the closed-loop transition matrices of the partitioning and Kalman filters, and of the corresponding (dynamic) system matrices, are respectively given by

$$\Phi_K(t, t_0) = \Phi_n(t, t_0) - \Phi_n(t, t_0)P_r(t_0|t)M_n(t, t_0) \quad (5.209a)$$

or

$$\Phi_K(t, t_0) = \Phi_n(t, t_0)[M_n(t, t_0) + P_r^{-1}(t_0|t_0)]^{-1}P_r^{-1}(t_0|t_0) \quad (5.209b)$$

and

$$A_K(t) = A_n(t) - \tilde{A}_n(t) \quad (5.210)$$

where

$$A_n(t) = A(t) - K_n(t)H(t) \quad (5.211)$$

$$\tilde{A}_n(t) = \Phi_n(t, t_0)P_r(t_0|t)\Phi_n^T(t, t_0)H^T(t)R^{-1}(t)H(t) \quad (5.212)$$

Lemma 5.5

The relationship between the observability matrix for the partitioning filter, $M_{\mathrm{n}}(t, t_0)$, and that for the usual Kalman filter, $M_{\mathrm{K}}(t, t_0)$, is given by

$$M_{\mathrm{K}}(t, t_0) = M_{\mathrm{n}}(t, t_0)P_{\mathrm{r}}(t_0|t)P_{\mathrm{r}}^{-1}(t_0|t_0) \tag{5.213}$$

where

$$M_{\mathrm{K}}(t, t_0) = \int_{t_0}^{t} \Phi_{\mathrm{K}}^{\mathrm{T}}(s, t_0)H^{\mathrm{T}}(s)R^{-1}(s)H(s)\Phi_{\mathrm{K}}(s, t_0)\,\mathrm{d}s \tag{5.214}$$

Lemma 5.6

The relationship between the Lagrange multiplier vector for the partitioning filter, $\lambda_{\mathrm{n}}(t, t_0)$, and that for the usual Kalman filter, $\lambda_{\mathrm{K}}(t, t_0)$, is given by

$$\lambda_{\mathrm{K}}(t, t_0) = P_{\mathrm{r}}^{-1}(t_0|t_0)P_{\mathrm{r}}(t_0|t)[\lambda_{\mathrm{n}}(t, t_0) - M_{\mathrm{n}}(t, t_0)\hat{x}_{\mathrm{r}}(t_0|t_0)] \tag{5.215}$$

where

$$\lambda_{\mathrm{K}}(t, t_0) = \int_{t_0}^{t} \Phi_{\mathrm{K}}^{\mathrm{T}}(s, t_0)H^{\mathrm{T}}(s)R^{-1}(s)[z(s) - H(s)\hat{x}_{\mathrm{r}}(s|s)]\,\mathrm{d}s \tag{5.216}$$

We now return to the discussion of partitioned numerical algorithms. The Riccati differential equation associated with the usual Kalman filter is given by

$$\begin{aligned}\frac{\mathrm{d}P(t|t)}{\mathrm{d}t} &= A(t)P(t|t) + P(t|t)A^{\mathrm{T}}(t) + B(t)Q(t)B^{\mathrm{T}}(t)\\ &\quad - P(t|t)H^{\mathrm{T}}(t)R^{-1}(t)H(t)P(t|t), \qquad P(t_0|t_0) = P(t_0)\end{aligned} \tag{5.217}$$

A great number of methods are known for solving nonsteady-state RDEs. Among these, as discussed in Chapter 2, are numerical integration routines, the Hamiltonian matrix method, the negative exponential method and the Chandrasekhar-type algorithm (see also Kortüm [19]).

It is seen that equation (5.54) can be used as an alternative method incorporated in the initialization technique for solving the Riccati equation (5.217). An effective numerical algorithm, called the *partitioned numerical algorithm* or *lambda algorithm* [9, 20, 21], is summarized in the following theorem.

Theorem 5.7 Partitioned numerical algorithm, or lambda algorithm

If the chosen partitioning of computation interval $\{t_0, t_f\}$ into the nonoverlapping subintervals (t_i, t_{i+1}), $i = 0, 1, \ldots, N-1$ is given (where $t_N \equiv t_f$), then:

(i) The most general partitioned numerical algorithm is given by

$$\begin{aligned} P(i+1, 0) = P_n(i+1, i) &+ \Phi_n(i+1, i) \\ &\times [I + P_r(i, 0) M_n(i+1, i)]^{-1} \\ &\times P_r(i, 0) \Phi_n^T(i+1, i) \end{aligned} \tag{5.218}$$

where $P(i+1, 0) \equiv P(t_{i+1}|t_{i+1})$ is the solution of equation (5.217) at t_{i+1} with the initial condition $P(t_0|t_0) = P(t_0)$ at t_0; $P_n(i+1, i) \equiv P_n(t_{i+1}|t_{i+1})$ is the solution of equation (5.37), but with the nominal initial condition $P_n(t_i|t_i) = P_{ni}$ at t_i; the remainder matrix $P_r(i, 0) \equiv P_r(t_i|t_i)$ is given by

$$P_r(i, 0) = P(i, 0) - P_{ni} \tag{5.219}$$

and $\Phi_n(i+1, i) \equiv \Phi_n(t_{i+1}, t_i)$ is the transition matrix at t_{i+1} corresponding to the (dynamic) system matrix

$$A(t) - P_n(t|t) H^T(t) R^{-1}(t) H(t) \tag{5.220}$$

with the initial condition I at t_i, where $P_n(t|t)$ is the solution of equation (5.37) with the initial condition P_{ni} at t_i.

(ii) For a particularly convenient set of nominal initial conditions, $P_{ni} = 0$ for all i, the partitioned numerical algorithm (5.218) reduces to

$$\begin{aligned} P(i+1, 0) = P_0(i+1, i) &+ \Phi_0(i+1, i) \\ &\times [I + P(i, 0) M_0(i+1, i)]^{-1} \\ &\times P(i, 0) \Phi_0^T(i+1, i) \end{aligned} \tag{5.221}$$

Proof

Substituting equation (5.56) into (5.54), we have

$$P(t|t) = P_n(t|t) + \Phi_n(t, t_0)[I + P_r M_n(t, t_0)]^{-1} P_r \Phi_n^T(t, t_0) \tag{5.222}$$

Applying equation (5.222) to every subinterval, we obtain the desired result (5.218). The proof of equation (5.221) is apparent from equation (5.219) because $P_{ni} \equiv 0$. □

It should be noted that the partitioned numerical algorithm gives the relationship of the time progress at the final point of the subinterval. It is also found that each elemental solution is completely decoupled from the others and is computable by means of parallel processing, with the consequent advantages of fast operation and reduced storage requirements.

In particular situations, these algorithms may be effective for widespread time-invariant systems. In such systems, it will be convenient for the algorithm given in equation (5.221) to take the constant step size $\Delta = t_{i+1} - t_i$. Hence, we obtain the following corollary without proof.

Corollary 5.5 Partitioned numerical algorithm for time-invariant systems

The partitioned numerical algorithm for time-invariant systems is given by

$$P[(i+1)\Delta] = P_0(\Delta) + \Phi_0(\Delta)\{I + P[i\Delta]M_0(\Delta)\}^{-1} \times P[i\Delta]\Phi_0^{\mathrm{T}}(\Delta), \qquad P[0] = P(t_0) \tag{5.223}$$

for $i = 0, 1, \ldots$, where the quantities $P_0(\Delta)$, $\Phi_0(\Delta)$ and $M_0(\Delta)$ need only be computed for the first subinterval by using equations (5.37), (5.45) and (5.43) with the condition $P_{\mathrm{n}}(t_0|t_0) = 0$, and remain the same thereafter.

Actually, if the constant step size Δ is determined, the following approximations to $P_0(\Delta)$, $\Phi_0(\Delta)$ and $M_0(\Delta)$ can be obtained from equations (5.37), (5.45) and (5.43):

$$P_0(\Delta) \simeq BQB^{\mathrm{T}}\Delta, \; \Phi_0(\Delta) \simeq I + A\Delta, \; M_0(\Delta) \simeq H^{\mathrm{T}}R^{-1}H\Delta \tag{5.224}$$

Despite the fact that the Riccati equation for time-invariant systems can be solved by using a partitioned numerical algorithm (5.223), it is difficult to obtain the steady-state solution without extensive computation. If the steady-state solution is required, the solution to the ARE:

$$AP + PA^{\mathrm{T}} + BQB^{\mathrm{T}} - PH^{\mathrm{T}}R^{-1}HP = 0 \tag{5.225}$$

or its discrete equivalent, has to be sought. We can use various techniques for the solution of equation (5.225), among them long time-integration (Runge–Kutta or Adams–Bashforth) methods; iterative methods (such as Kleinman's iterative scheme [19, 22, 23]); the eigenvalue–eigenvector method [19, 24–27]; and the matrix sign function method [23] (cf. Section 2.8).

We can now present a *doubling* partitioned numerical algorithm [18, 21], which is effective for time-invariant systems, by applying relationships between the partitioning filter and the usual Kalman filter. This algorithm is sometimes called the *delta* algorithm, and the form is similar to that used in the partitioned numerical algorithm addressed in Theorem 5.7 or Corollary 5.5.

Theorem 5.8 Doubling partitioned numerical algorithm or delta algorithm

If the system given by equations (5.1) and (5.2) is time-invariant, i.e. $A(t) = A$, $B(t) = B$, $H(t) = H$, $Q(t) = Q$, and $R(t) = R$, and the preselected initial time step $\Delta = t_1 - t_0$ is given, then the solution to equation (5.225) can be obtained by the following recursive equation:

$$P(2^{n+1}\Delta) = P_0(2^n\Delta) + \Phi_0(2^n\Delta)[I + P(2^n\Delta)M_0(2^n\Delta)]^{-1} \times P(2^n\Delta)\Phi_0^{\mathrm{T}}(2^n\Delta) \tag{5.226}$$

where $P_0(2^n\Delta)$, $\Phi_0(2^n\Delta)$ and $M_0(2^n\Delta)$ are given by the following iterative equations:

$$P_0(2^{n+1}\Delta) = P_0(2^n\Delta) + \Phi_0(2^n\Delta)[I + P_0(2^n\Delta)M_0(2^n\Delta]^{-1} \times P_0(2^n\Delta)\Phi_0^{\mathrm{T}}(2^n\Delta) \tag{5.227}$$

$$\Phi_0(2^{n+1}\Delta) = \Phi_0(2^n\Delta)[I + P_0(2^n\Delta)M_0(2^n\Delta)]^{-1}\Phi_0(2^n\Delta) \tag{5.228}$$

$$M_0(2^{n+1}\Delta) = M_0(2^n\Delta) + \Phi_0^{\mathrm{T}}(2^n\Delta)M_0(2^n\Delta) \times [I + P_0(2^n\Delta)M_0(2^n\Delta)]^{-1}\Phi_0(2^n\Delta) \tag{5.229}$$

in which $P_0(\Delta)$, $\Phi_0(\Delta)$ and $M_0(\Delta)$ can be obtained by integrating equations (5.37), (5.45) and (5.43) with $P_{\mathrm{n}}(t_0|t_0) = 0$ at initial time t_0, or obtained approximately from equation (5.224).

Proof

The result of this theorem can be derived by applying Lemmas 5.4 and 5.5. See Watanabe [9], Watanabe and Iwasaki [18] and Lainiotis [21] for more details. □

Note that, as can be seen from equation (5.225), the steady-state solution is independent of its initial condition $P(t_0)$. However, the solution method presented in Theorem 5.8 is dependent on its initial condition, i.e. $P(2^0\Delta) = P(t_0)$. Taking account of this fact, setting $P(t_0) \equiv 0$, it follows that

$$\begin{aligned}\lim_{t\to\infty} P(t|t) &= \lim_{t\to\infty} P_0(t|t) \\ &= \lim_{n\to\infty} P_0(2^n\Delta) \end{aligned} \tag{5.230}$$

if the system given by equations (5.1) and (5.2) is stabilizable and detectable (cf. Chapter 2). Therefore, we obtain the following corollary without proof:

Corollary 5.6 Doubling algorithm

If the system is time-invariant, stabilizable and detectable, the solution $P \geqslant 0$ to equation (5.225) is given by the following steps:

Step 1: For a preselected $\Delta > 0$, compute $P_0(\Delta)$, $\Phi_0(\Delta)$ and $M_0(\Delta)$ from equations (5.37), (5.45) and (5.43).

Step 2: Then recursively compute $P_0(2^n\Delta)$, $\Phi_0(2^n\Delta)$ and $M_0(2^n\Delta)$ by means of the recursions

$$\begin{aligned}P_0(2^{n+1}\Delta) = {} & P_0(2^n\Delta) + \Phi_0(2^n\Delta) \\ & \times [I + P_0(2^n\Delta)M_0(2^n\Delta)]^{-1} \\ & \times P_0(2^n\Delta)\Phi_0^{\mathrm{T}}(2^n\Delta)\end{aligned} \tag{5.231}$$

$$\begin{aligned}\Phi_0(2^{n+1}\Delta) = {} & \Phi_0(2^n\Delta) \\ & \times [I + P_0(2^n\Delta)M_0(2^n\Delta)]^{-1}\Phi_0(2^n\Delta)\end{aligned} \tag{5.232}$$

$$\begin{aligned}M_0(2^{n+1}\Delta) = {} & M_0(2^n\Delta) + \Phi_0(2^n\Delta)M_0(2^n\Delta) \\ & \times [I + P_0(2^n\Delta)M_0(2^n\Delta)]^{-1}\Phi_0(2^n\Delta)\end{aligned} \tag{5.233}$$

for $n = 0, 1, \ldots$. If

$$||P_0(2^{n+1}\Delta) - P_0(2^n\Delta)|| \leqslant \varepsilon \tag{5.234}$$

then the recursion is stopped and we have $P = P_0(2^{n+1}\Delta)$, where ε is preselected to give the steady-state solution to the accuracy desired.

It is interesting to note [28] that the above algorithm holds without change for the ARE in discrete-time systems,

$$P - \Phi P\Phi^{\mathrm{T}} + \Phi PH^{\mathrm{T}}[HPH^{\mathrm{T}} + R]^{-1}HP\Phi^{\mathrm{T}} - BQB^{\mathrm{T}} = 0 \tag{5.235}$$

In this case, we have $\Delta = 1$ and the doubling algorithm in discrete time is initialized at Step 1 by

$$P_0(1) = BQB^{\mathrm{T}}, \qquad \Phi_0(1) = \Phi, \qquad M_0(1) = H^{\mathrm{T}}R^{-1}H \qquad (5.236)$$

Example 5.4

Consider the motion of an antenna described by the differential equation

$$J\ddot{\theta}(t) + B\dot{\theta}(t) = \tau(t) + \tau_{\mathrm{d}}(t)$$

where J is the moment of inertia of all the rotating parts, including the antenna, B is the coefficient of viscous friction, $\tau(t)$ is the torque applied by the motor, and $\tau_{\mathrm{d}}(t)$ is the disturbance torque caused by the wind. The motor torque is assumed to be proportional to $\mu(t)$, the input voltage to the motor, so that

$$\tau(t) = k\mu(t)$$

Defining the state variables $x_1(t) = \theta(t)$ and $x_2(t) = \dot{\theta}(t)$, the differential equation of the system becomes

$$\dot{x}(t) = \begin{bmatrix} 0 & 1 \\ 0 & -\alpha \end{bmatrix} x(t) + \begin{bmatrix} 0 \\ \kappa \end{bmatrix} \mu(t) + \begin{bmatrix} 0 \\ \gamma \end{bmatrix} \tau_{\mathrm{d}}(t)$$

where

$$\alpha = \frac{B}{J}, \ \kappa = \frac{k}{J}, \ \gamma = \frac{1}{J}$$

If the fluctuations of the disturbance torque are fast compared to the motion of the system itself, the assumption might be justified that $\tau_{\mathrm{d}}(t)$ is white noise with variance W_{d}. Let us further assume that the observed variable is given by

$$\eta(t) = [1 \quad 0]x(t) + v_{\mathrm{m}}(t)$$

where $v_{\mathrm{m}}(t)$ is white noise with variance V_{m}.

Plotted in Figure 5.1 is a numerical solution of the associated filtering error covariance matrix for the case $\kappa = 0.787$ rad V^{-1} s^{-2}, $\alpha = 4.6$ s^{-1}, $\gamma = 0.1$ kg^{-1} m^{-2}, $W_{\mathrm{d}} = 10$ N^2 m^2 s, $V_{\mathrm{m}} = 10^{-7}$ rad^2 s and $P(0) = \mathrm{diag}(1\ \mathrm{rad}^2, 1\ \mathrm{rad}^2\ \mathrm{s}^{-2})$. Here, the partitioned numerical algorithm given by equations (5.223) and (5.224) with $\Delta = 0.001$ s was implemented in single precision to obtain the result. The steady-state error covariance matrix which is proposed by using the doubling algorithm given by equations (5.224) and (5.231)–(5.234) with

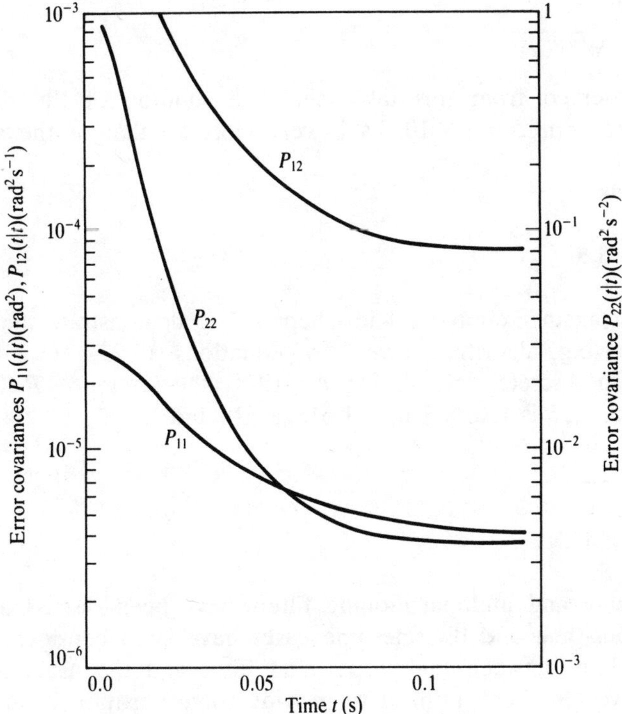

Figure 5.1 History of the estimation error covariance matrix for the motion of an antenna.

$\varepsilon = 10^{-7}$ is also tabulated in Table 5.1, where the analytic solution is as follows [29]:

$$P_{11} = V_m(-\alpha + \sqrt{\alpha^2 + 2\beta})$$
$$P_{12} = V_m(\alpha^2 + \beta - \alpha\sqrt{\alpha^2 + 2\beta})$$
$$P_{22} = V_m(-\alpha^3 - 2\alpha\beta + (\alpha^2 + \beta)\sqrt{\alpha^2 + 2\beta})$$

Table 5.1 Steady-state solutions of the estimation error covariance matrix for the motion of an antenna

	$P_{11}(rad^2)$	$P_{12}(rad^2\,s^{-1})$	$P_{22}(rad^2\,s^{-2})$	*Iteration*
Analytic	4.036×10^{-6}	8.143×10^{-5}	3.661×10^{-3}	*
$\Delta = 1 \times 10^{-3}$(s)	4.123×10^{-6}	8.308×10^{-5}	3.740×10^{-3}	10
$\Delta = 1 \times 10^{-4}$(s)	4.044×10^{-6}	8.160×10^{-5}	3.669×10^{-3}	13
$\Delta = 1 \times 10^{-5}$(s)	4.036×10^{-6}	8.145×10^{-5}	3.662×10^{-3}	16

in which

$$\beta = \gamma\sqrt{W_{\rm d}/V_{\rm m}}$$

It is observed from this table that the solution of the doubling algorithm with $\Delta = 1 \times 10^{-5}$ s is very close to that of the analytic approach.

Example 5.5

Consider again Example 2.4 in Chapter 2 to demonstrate the use of the doubling algorithm given by equations (5.231)–(5.234) and (5.236) in discrete time. Using $\varepsilon = 10^{-7}$, the sequence $P_0(2^n)$ for $n = 0, 1, \ldots, 5$ is 1, 1.5, 1.615, 1.618, 1.618 and 1.618.

5.7 SUMMARY

Partitioning and multipartitioning filters have been discussed. Both continuous-time and discrete-time cases have been considered. The relationships between the partitioning filter and the usual Kalman filter have also been utilized to provide some partitioned numerical algorithms suitable for solving the nonsteady-state or steady-state Riccati equation.

Most of the material of this chapter is drawn from the fundamental studies of Lainiotis and Andrisani [2, 3] and Andrisani [6]. Further information on the numerical methods for RDEs (or AREs) can be found in Kenney and Leipnik [30] and Jamshidi [31]. Some convergence results on the doubling algorithm can be seen in Kimura [32]. A direct extension of the results presented here to distributed parameter systems is made by Watanabe *et al.* [20, 33, 34] and Stavroulakis and Tzafestas [35]. A more generalized theory for evolution systems characterized by an evolution operator includes Watanabe [9, 18].

APPENDIX 5A MATRIX INVERSION LEMMA

If, for any $n \times n$ nonsingular matrix A and any two $n \times m$ matrices B and C, the two matrices $(A + BC^{\rm T})$ and $(I + C^{\rm T}A^{-1}B)$ are nonsingular, then the matrix identity

$$(A + BC^{\mathrm{T}})^{-1} = A^{-1} - A^{-1}B(I + C^{\mathrm{T}}A^{-1}B)^{-1}C^{\mathrm{T}}A^{-1} \tag{5A.1}$$

is valid [17].

Furthermore, let P, R and M be $n \times n$, $m \times m$ and $m \times n$ matrices, respectively. Suppose $P \geqslant 0$ and $R > 0$. Then the following additional equalities are useful [7]:

$$(I + PM^{\mathrm{T}}R^{-1}M)^{-1} = I - PM^{\mathrm{T}}(MPM^{\mathrm{T}} + R)^{-1}M \tag{5A.2}$$

$$(I + PM^{\mathrm{T}}R^{-1}M)^{-1}P = P - PM^{\mathrm{T}}(MPM^{\mathrm{T}} + R)^{-1}MP \tag{5A.3}$$

$$(I + PM^{\mathrm{T}}R^{-1}M)^{-1}PM^{\mathrm{T}}R^{-1} = PM^{\mathrm{T}}(MPM^{\mathrm{T}} + R)^{-1} \tag{5A.4}$$

If, in addition, $P > 0$, then

$$(I + PM^{\mathrm{T}}R^{-1}M)^{-1}P = (P^{-1} + M^{\mathrm{T}}R^{-1}M)^{-1} \tag{5A.5}$$

$$(P^{-1} + M^{\mathrm{T}}R^{-1}M)^{-1} = P - PM^{\mathrm{T}}(MPM^{\mathrm{T}} + R)^{-1}MP \tag{5A.6}$$

$$(P^{-1} + M^{\mathrm{T}}R^{-1}M)^{-1}M^{\mathrm{T}}R^{-1} = PM^{\mathrm{T}}(MPM^{\mathrm{T}} + R)^{-1} \tag{5A.7}$$

APPENDIX 5B PROOF OF THE CONDITIONAL PARTITION THEOREM

From the marginal property of the probability density function and Bayes' rule, we obtain

$$\Lambda(Z_t) = \int \Lambda(Z_t, \theta)\mathrm{d}\theta = \int \Lambda(Z_t|\theta)p(\theta)\mathrm{d}\theta \tag{5B.1}$$

$$= \int \Lambda(Z_t|X, \theta)p(X, \theta)\mathrm{d}X\mathrm{d}\theta \tag{5B.2}$$

$$p(X|Z_t, \theta) = \frac{p(X, Z_t, \theta)}{p(Z_t, \theta)} = \frac{p(X, \theta|Z_t)}{p(\theta|Z_t)} \tag{5B.3}$$

Moreover, notice that the usual partition theorem gives the following two a posteriori probability density functions:

$$p(X, \theta|Z_t) = \frac{\Lambda(Z_t|X, \theta)p(X, \theta)}{\int \Lambda(Z_t|X, \theta)p(X, \theta)\mathrm{d}X\mathrm{d}\theta} \tag{5B.4}$$

$$p(\theta|Z_t) = \frac{\Lambda(Z_t|\theta)p(\theta)}{\int \Lambda(Z_t|\theta)p(\theta)\mathrm{d}\theta} \tag{5B.5}$$

Therefore, substituting equations (5B.4) and (5B.5) into the second equality of (5B.3) gives

$$p(X|Z_t, \theta) = \frac{\Lambda(Z_t|X, \theta)p(X|\theta)}{\Lambda(Z_t|\theta)}$$

$$= \frac{\Lambda(Z_t|X, \theta)p(X|\theta)}{\int \Lambda(X, Z_t|\theta)\,\mathrm{d}X} \tag{5B.6}$$

But notice that

$$\int \Lambda(X, Z_t|\theta)\mathrm{d}X = \int \Lambda(X, Z_t, \theta)/p(\theta)\,\mathrm{d}X$$

$$= \int \Lambda(Z_t|X, \theta)p(X|\theta)\,\mathrm{d}X \tag{5B.7}$$

Hence, we have the desired result (5.15).

APPENDIX 5C PROOF OF THE CONTINUOUS-TIME DEPENDENT MULTIPARTITIONING FILTER

For simplicity's sake, it is assumed that the initial state is partitioned into four components, i.e. $M = 4$.

Step 1: form the estimate of $x(t)$ conditioned on x_2, x_3, x_4. Conditioned on x_2, x_3, x_4, the estimation is of the standard Gaussian state type with initial estimates

$$E[x(t_0)|x_2, x_3, x_4] = E[x_1|x_2, x_3, x_4] + x_2 + x_3 + x_4 \tag{5C.1}$$

$$\mathrm{Cov}\,[x(t_0)|x_2, x_3, x_4] = \mathrm{Cov}\,[x_1|x_2, x_3, x_4] \tag{5C.2}$$

The moments of $p(x_1|x_2, x_3, x_4)$ can be written as

$$E[x_1|x_2, x_3, x_4] \triangleq \hat{x}_{1|234} = \hat{x}_1 + P_{12|234}[x_2 - \hat{x}_2] + P_{13|234}[x_3 - \hat{x}_3] + P_{14|234}[x_4 - \hat{x}_4] \tag{5C.3}$$

$$\mathrm{Cov}\,[x_1|x_2, x_3, x_4] \triangleq P_{1|234} = P_{11} - P_{12|234}P_{12}^{\mathrm{T}} - P_{13|234}P_{13}^{\mathrm{T}} - P_{14|234}P_{14}^{\mathrm{T}} \tag{5C.4}$$

Note that expression (5C.3) can also be rewritten as

$$E[x_1|x_2, x_3, x_4] = \bar{x}_{1|234} + P_{12|234}x_2 + P_{13|234}x_3 + P_{14|234}x_4 \tag{5C.5}$$

where

$$\bar{x}_{1|234} \triangleq \hat{x}_1 - P_{12|234}\hat{x}_2 - P_{13|234}\hat{x}_3 - P_{14|234}\hat{x}_4 \tag{5C.6}$$

From this point of view, equation (5C.1) can be expressed as

$$E[x(t_0)|x_2, x_3, x_4] = \bar{x}_{1|234} + [I + P_{12|234}]x_2 + [I + P_{13|234}]x_3 + [I + P_{14|234}]x_4 \qquad (5C.7)$$

A Kalman filter solution with the initial conditions given by expressions (5C.7) and (5C.4) is given by

$$\hat{x}(t|t, x_2, x_3, x_4) = \hat{x}_1(t|t) + \Psi_{22}(t, t_0)x_2 + \Psi_{23}(t, t_0)x_3 + \Phi_{24}(t, t_0)x_4 \qquad (5C.8)$$

$$P(t|t, x_2, x_3, x_4) = P_1(t|t) \qquad (5C.9)$$

where $\hat{x}_1(t|t) \triangleq E[x(t)|Z_t, x_2 = 0, x_3 = 0, x_4 = 0]$ and $P_1(t|t) \triangleq E\{[x(t) - \hat{x}_1(t|t)][x(t) - \hat{x}_1(t|t)]^{\mathrm{T}}|Z_t, x_2, x_3, x_4\}$ are given by expressions (5.72)–(5.74), but with the initial conditions $\hat{x}_1(t_0|t_0) = \bar{x}_{1|234}$ and $P_1(t_0|t_0) = P_{1|234}$, and

$$\Psi_{2j}(t, t_0) = \Phi_2(t, t_0)[I + P_{1j|234}], \qquad j = 2, 3, 4 \qquad (5C.10)$$

in which $\Phi_2(t, t_0)$ is obtained from equation (5.75). Then, the optimal state estimate of $x(t)$ is given by

$$\begin{aligned}\hat{x}(t|t) &= \int \hat{x}(t|t, x_2, x_3, x_4)p(x_2, x_3, x_4|Z_t)\,\mathrm{d}x_2\,\mathrm{d}x_3\,\mathrm{d}x_4 \\ &= \hat{x}_1(t|t) + \Psi_{22}(t, t_0)\hat{x}_2^*(t_0|t) + \Psi_{23}(t, t_0)\hat{x}_3^*(t_0|t) \\ &\quad + \Psi_{24}(t, t_0)\hat{x}_4^*(t_0|t) \qquad (5C.11)\end{aligned}$$

where

$$\hat{x}_i^*(t_0|t) = \int x_i\, p(x_i|Z_t)\,\mathrm{d}x_i, \qquad i = 2, 3, 4 \qquad (5C.12)$$

Step 2: remove the conditioning on x_2. The partition theorem again provides the following equations:

$$\hat{x}(t|t, x_3, x_4) = \int \hat{x}(t|t, x_2, x_3, x_4)p(x_2|Z_t, x_3, x_4)\,\mathrm{d}x_2 \qquad (5C.13)$$

$$\begin{aligned}P(t|t, x_3, x_4) = \int &\{P(t|t, x_2, x_3, x_4) + [\hat{x}(t|t, x_3, x_4) \\ &- \hat{x}(t|t, x_2, x_3, x_4)] \\ &\times [\hat{x}(t|t, x_3, x_4) - \hat{x}(t|t, x_2, x_3, x_4)]^{\mathrm{T}}\} \\ &\times p(x_2|Z_t, x_3, x_4)\,\mathrm{d}x_2 \qquad (5C.14)\end{aligned}$$

Substituting equation (5C.8) into (5C.13), and noting that $\hat{x}_1(t|t)$ is not a function of x_2, gives

$$\hat{x}(t|t, x_3, x_4) = \hat{x}_1(t|t) + \Psi_{22}(t, t_0)\hat{x}_2(t_0|t, x_3, x_4) + \Psi_{23}(t, t_0)x_3 + \Psi_{24}(t, t_0)x_4 \tag{5C.15}$$

where

$$\hat{x}_2(t_0|t, x_3, x_4) = \int x_2 p(x_2|Z_t, x_3, x_4)\,\mathrm{d}x_2 \tag{5C.16}$$

Furthermore, substituting equations (5C.8), (5C.9) and (5C.15) into (5C.14), and taking account that $P_1(t|t)$ is not a function of x_2, yields

$$P(t|t, x_3, x_4) = P(t|t, x_2, x_3, x_4) + \Psi_{22}(t, t_0)P_2(t_0|t, x_3, x_4)\Psi_{22}^{\mathrm{T}}(t, t_0) \tag{5C.17}$$

where

$$P_2(t_0|t, x_3, x_4) = \int [x_2 - \hat{x}_2(t_0|t, x_3, x_4)][x_2 - \hat{x}_2(t_0|t, x_3, x_4)]^{\mathrm{T}} \times p(x_2|Z_t, x_3, x_4)\,\mathrm{d}x_2 \tag{5C.18}$$

Step 3: find the a posteriori probability density of x_2, and separate the x_3 and x_4 dependence. From equation (5.15) of Lemma 5.2 we obtain

$$p(x_2|Z_t, x_3, x_4) = \frac{\Lambda(Z_t|x_2, x_3, x_4)p(x_2|x_3, x_4)}{\int \Lambda(Z_t|x_2, x_3, x_4)p(x_2|x_3, x_4)\,\mathrm{d}x_2} \tag{5C.19}$$

where the a priori conditional probability density function $p(x_2|x_3, x_4)$ has moments

$$E[x_2|x_3, x_4] = \bar{x}_{2|34} + P_{23|34}x_3 + P_{24|34}x_4 \tag{5C.20}$$

$$\mathrm{Cov}[x_2|x_3, x_4] = P_{22} - P_{23|34}P_{23}^{\mathrm{T}} - P_{24|34}P_{24}^{\mathrm{T}} \triangleq P_{2|34} \tag{5C.21}$$

in which

$$\bar{x}_{2|34} = \hat{x}_2 - P_{23|34}\hat{x}_3 - P_{24|34}\hat{x}_4 \tag{5C.22}$$

The $\{x_2, x_3, x_4\}$-conditional likelihood function is also given by

$$\Lambda(Z_t|x_2, x_3, x_4) = \exp\left\{\int_{t_0}^{t} \langle H(s)\hat{x}(s|s, x_2, x_3, x_4), R^{-1}(s)z(s)\rangle \mathrm{d}s - \frac{1}{2}\|H(s)\hat{x}(s|s, x_2, x_3, x_4)\|_{R^{-1}(s)}^2\,\mathrm{d}s\right\} \tag{5C.23}$$

Then, from Lemma 5.2, we have

$$\hat{x}_2(t_0|t, x_3, x_4) = P_2(t_0|t, x_3, x_4)[\tilde{\lambda}_2(t, t_0) + P_{2|34}^{-1} \times (\bar{x}_{2|34} + P_{23|34}x_3 + P_{24|34}x_4)] \tag{5C.24}$$

$$P_2(t_0|t, x_3, x_4) = [M_{22}(t, t_0) + P_{2|34}^{-1}]^{-1} \tag{5C.25}$$

where

$$\begin{aligned}\tilde{\lambda}_2(t, t_0) = \int_{t_0}^{t} &\Psi_{22}^{\mathrm{T}}(s, t_0) H^{\mathrm{T}}(s) R^{-1}(s) \\ &\times [z(s) - H(s)\{\hat{x}_1(s|s) + \Psi_{23}(s, t_0)x_3 \\ &+ \Psi_{24}(s, t_0)x_4\}]\,\mathrm{d}s\end{aligned} \tag{5C.26}$$

$$M_{22}(t, t_0) = \int_{t_0}^{t} \Psi_{22}^{\mathrm{T}}(s, t_0) H^{\mathrm{T}}(s) R^{-1}(s) H(s) \Psi_{22}(s, t_0)\,\mathrm{d}s \tag{5C.27}$$

If we define

$$\hat{x}_2(t_0|t) \triangleq P_2(t_0|t, x_3, x_4)[\lambda_2(t, t_0) + P_{2|34}^{-1}\bar{x}_{2|34}] \tag{5C.28}$$

$$\lambda_2(t, t_0) \triangleq \int_{t_0}^{t} \Psi_{22}^{\mathrm{T}}(s, t_0) H^{\mathrm{T}}(s) R^{-1}(s)[z(s) - H(s)\hat{x}_1(s|s)]\,\mathrm{d}s \tag{5C.29}$$

$$M_{2j}(t, t_0) \triangleq \int_{t_0}^{t} \Psi_{22}^{\mathrm{T}}(s, t_0) H^{\mathrm{T}}(s) R^{-1}(s) H(s) \Psi_{2j}(s, t_0)\,\mathrm{d}s, \quad j = 3, 4 \tag{5C.30}$$

then

$$\tilde{\lambda}_2(t, t_0) = \lambda_2(t, t_0) - M_{23}(t, t_0)x_3 - M_{24}(t, t_0)x_4 \tag{5C.31}$$

$$\begin{aligned}\hat{x}_2(t_0|t, x_3, x_4) = \hat{x}_2(t_0|t) &+ P_2(t_0|t, x_3, x_4) \\ &\times \{[P_{2|34}^{-1}P_{23|34} - M_{23}(t, t_0)]x_3 \\ &+ [P_{2|34}^{-1}P_{24|34} - M_{24}(t, t_0)]x_4\}\end{aligned} \tag{5C.32}$$

Substitution of equation (5C.32) into (5C.15) gives

$$\begin{aligned}\hat{x}(t|t, x_3, x_4) = \hat{x}_1(t|t) &+ \Psi_{22}(t, t_0)\hat{x}_2(t_0|t) \\ &+ \Psi_{33}(t, t_0)x_3 + \Psi_{34}(t, t_0)x_4\end{aligned} \tag{5C.33}$$

where

$$\begin{aligned}\Psi_{33}(t, t_0) = \Psi_{23}(t, t_0) &+ \Psi_{22}(t, t_0)P_2(t_0|t, x_3, x_4) \\ &\times [P_{2|34}^{-1}P_{23|34} - M_{23}(t, t_0)]\end{aligned} \tag{5C.34}$$

$$\begin{aligned}\Psi_{34}(t, t_0) = \Psi_{24}(t, t_0) &+ \Psi_{22}(t, t_0)P_2(t_0|t, x_3, x_4) \\ &\times [P_{2|34}^{-1}P_{24|34} - M_{24}(t, t_0)]\end{aligned} \tag{5C.35}$$

Step 4: remove the conditioning on x_3. The partition theorem provides the following equalities:

$$\hat{x}(t|t, x_4) = \int \hat{x}(t|t, x_3, x_4)p(x_3|Z_t, x_4)\,dx_3 \tag{5C.36}$$

$$P(t|t, x_4) = \int \{P(t|t, x_3, x_4) + [\hat{x}(t|t, x_4) - \hat{x}(t|t, x_3, x_4)] \times [\hat{x}(t|t, x_4) - \hat{x}(t|t, x_3, x_4)]^{T}\}p(x_3|Z_t, x_4)\,dx_3 \tag{5C.37}$$

Combining equations (5C.33) and (5C.36) gives

$$\hat{x}(t|t, x_4) = \hat{x}_1(t|t) + \Psi_{22}(t, t_0)\hat{x}_2(t_0|t) + \Psi_{33}(t, t_0)\hat{x}_3(t_0|t, x_4) + \Psi_{34}(t, t_0)x_4 \tag{5C.38}$$

where

$$\hat{x}_3(t_0|t, x_4) = \int x_3 p(x_3|Z_t, x_4)\,dx_3 \tag{5C.39}$$

Substitution of expressions (5C.17), (5C.33) and (5C.38) into (5C.37) yields

$$P(t|t, x_4) = P(t|t, x_2, x_3, x_4) + \Psi_{22}(t, t_0)P_2(t_0|t, x_3, x_4)\Psi_{22}^{T}(t, t_0) + \Psi_{33}(t, t_0)P_3(t_0|t, x_4)\Psi_{33}^{T}(t, t_0) \tag{5C.40}$$

where

$$P_3(t_0|t, x_4) = \int [x_3 - \hat{x}_3(t_0|t, x_4)] \times [x_3 - \hat{x}_3(t_0|t, x_4)]^{T} p(x_3|Z_t, x_4)\,dx_3 \tag{5C.41}$$

Step 5: find the a posteriori probability density of x_3, and separate out the dependence on x_4. From the partition theorem, it follows that

$$p(x_3|Z_t, x_4) = \frac{\Lambda(Z_t|x_3, x_4)p(x_3|x_4)}{\int \Lambda(Z_t|x_3, x_4)p(x_3|x_4)\,dx_3} \tag{5C.42}$$

where

$$\Lambda(Z_t|x_3, x_4) = \exp\left\{\int_{t_0}^{t} \langle H(s)\hat{x}(s|s, x_3, x_4), R^{-1}(s)z(s)\rangle\,ds - \frac{1}{2}\int_{t_0}^{t} \|H(s)\hat{x}(s|s, x_3, x_4)\|^2_{R^{-1}(s)}\,ds\right\} \tag{5C.43}$$

and

$$p(x_3|x_4) = N(\bar{x}_{3|4} + P_{34|4}x_4, P_{3|4}) \tag{5C.44}$$

in which

$$\bar{x}_{3|4} \triangleq \hat{x}_3 - P_{34|4}\hat{x}_4 \tag{5C.45}$$

$$P_{3|4} \triangleq P_{33} - P_{34|4}P_{34}^{\mathrm{T}} \tag{5C.46}$$

Substituting equation (5C.33) into (5C.43), Lemma 5.2 gives the following result:

$$\hat{x}_3(t_0|t, x_4) = P_3(t_0|t, x_4)[\tilde{\lambda}_3(t, t_0) + P_{3|4}^{-1}(\bar{x}_{3|4} + P_{34|4}x_4)] \tag{5C.47}$$

$$P_3(t_0|t, x_4) = [M_{33}(t, t_0) + P_{3|4}^{-1}]^{-1} \tag{5C.48}$$

where

$$\begin{aligned}\tilde{\lambda}_3(t, t_0) = \int_{t_0}^{t} &\Psi_{33}^{\mathrm{T}}(s, t_0)H^{\mathrm{T}}(s)R^{-1}(s)\{z(s) \\ &- H(s)[\hat{x}_1(s|s) + \Psi_{22}(s, t_0)\hat{x}_2(t_0|s) + \Psi_{34}(s, t_0)x_4]\}\,\mathrm{d}s\end{aligned} \tag{5C.49}$$

$$M_{33}(t, t_0) = \int_{t_0}^{t} \Psi_{33}^{\mathrm{T}}(s, t_0)H^{\mathrm{T}}(s)R^{-1}(s)H(s)\Psi_{33}(s, t_0)\,\mathrm{d}s \tag{5C.50}$$

Defining

$$\hat{x}_3(t_0|t) \triangleq P_3(t_0|t, x_4)[\lambda_3(t, t_0) + P_{3|4}^{-1}\bar{x}_{3|4}] \tag{5C.51}$$

$$\begin{aligned}\lambda_3(t, t_0) \triangleq \int_{t_0}^{t} &\Psi_{33}^{\mathrm{T}}(s, t_0)H^{\mathrm{T}}(s)R^{-1}(s)\{z(s) \\ &- H(s)[\hat{x}_1(s|s) + \Psi_{22}(s, t_0)\hat{x}_2(t_0|s)]\}\,\mathrm{d}s\end{aligned} \tag{5C.52}$$

$$M_{34}(t, t_0) \triangleq \int_{t_0}^{t} \Psi_{33}^{\mathrm{T}}(s, t_0)H^{\mathrm{T}}(s)R^{-1}(s)H(s)\Psi_{34}(s, t_0)\,\mathrm{d}s \tag{5C.53}$$

gives

$$\hat{x}_3(t_0|t, x_4) = \hat{x}_3(t_0|t) + P_3(t_0|t, x_4)[P_{3|4}^{-1}P_{34|4} - M_{34}(t, t_0)]x_4 \tag{5C.54}$$

Substituting equation (5C.54) into (5C.38) gives

$$\begin{aligned}\hat{x}(t|t, x_4) = \hat{x}_1(t_0|t) &+ \Psi_{22}(t, t_0)\hat{x}_2(t_0|t) + \Psi_{33}(t, t_0)\hat{x}_3(t_0|t) \\ &+ \Psi_{44}(t, t_0)x_4\end{aligned} \tag{5C.55}$$

where

$$\begin{aligned}\Psi_{44}(t, t_0) = \Psi_{34}(t, t_0) &+ \Psi_{33}(t, t_0)P_3(t_0|t, x_4) \\ &\times [P_{3|4}^{-1}P_{34|4} - M_{34}(t, t_0)]\end{aligned} \tag{5C.56}$$

Step 6: remove the conditioning on x_4. From the partition theorem, we obtain that

$$\hat{x}(t|t) = \int \hat{x}(t|t, x_4)p(x_4|Z_t)\,\mathrm{d}x_4 \tag{5C.57}$$

$$P(t|t) = \int \{P(t|t, x_4) + [\hat{x}(t|t) - \hat{x}(t|t, x_4)] \times [\hat{x}(t|t) - \hat{x}(t|t, x_4)]^{\mathrm{T}}\}p(x_4|Z_t)\,\mathrm{d}x_4 \tag{5C.58}$$

Substitution of equation (5C.55) into (5C.57) yields

$$\hat{x}(t|t) = \hat{x}_1(t|t) + \Psi_{22}(t, t_0)\hat{x}_2(t_0|t) + \Psi_{33}\hat{x}_3(t_0|t) + \Psi_{44}(t, t_0)\hat{x}_4(t_0|t) \tag{5C.59}$$

where

$$\hat{x}_4(t_0|t) \triangleq \int x_4 p(x_4|Z_t)\,\mathrm{d}x_4 \tag{5C.60}$$

Combine expressions (5C.40), (5C.55), (5C.58) and (5C.59) to get

$$P(t|t) = P(t|t, x_2, x_3, x_4) + \Psi_{22}(t, t_0)P_2(t_0|t, x_3, x_4)\Psi_{22}^{\mathrm{T}}(t, t_0) + \Psi_{33}(t, t_0)P_3(t_0|t, x_4)\Psi_{33}^{\mathrm{T}}(t, t_0) + \Psi_{44}(t, t_0)P_4(t_0|t)\Psi_{44}^{\mathrm{T}}(t|t_0) \tag{5C.61}$$

where

$$P_4(t_0|t) = \int [x_4 - \hat{x}_4(t_0|t)][x_4 - \hat{x}_4(t_0|t)]^{\mathrm{T}}p(x_4|Z_t)\,\mathrm{d}x_4 \tag{5C.62}$$

Step 7: find the a posteriori probability density of x_4. From the partition theorem, it is apparent that

$$p(x_4|Z_t) = \frac{\Lambda(Z_t|x_4)p(x_4)}{\int \Lambda(Z_t|x_4)p(x_4)\,\mathrm{d}x_4} \tag{5C.63}$$

where

$$\Lambda(Z_t|x_4) = \exp\left\{\int_{t_0}^{t} \langle H^{\mathrm{T}}(s)\hat{x}^{\mathrm{T}}(s|s, x_4), R^{-1}(s)z(s)\rangle\,\mathrm{d}s - \frac{1}{2}\,||H(s)\hat{x}(s|s, x_4)||^2_{R^{-1}(s)}\,\mathrm{d}s||\right\} \tag{5C.64}$$

$$p(x_4) = N(\hat{x}_4, P_{44}) \tag{5C.65}$$

Substituting expression (5C.55) into (5C.64) and applying Lemma 5.2 gives the final results:

$$\hat{x}_4(t_0|t) = P_4(t_0|t)[\lambda_4(t, t_0) + P_{44}^{-1}\hat{x}_4] \quad (5C.66)$$

$$P_4(t_0|t) = [M_{44}(t, t_0) + P_{44}^{-1}]^{-1} \quad (5C.67)$$

where

$$\begin{aligned}\lambda_4(t, t_0) = \int_{t_0}^{t} \Psi_{22}^{\mathrm{T}}(s, t_0)H^{\mathrm{T}}(s)R^{-1}(s)\{z(s) \\ - H(s)[\hat{x}_1(s|s) + \Psi_{22}(s, t_0)\hat{x}_2(t_0|s) \\ + \Psi_{33}(s, t_0)\hat{x}_3(t_0|s)]\}\,\mathrm{d}s\end{aligned} \quad (5C.68)$$

$$M_{44}(t, t_0) = \int_{t_0}^{t} \Psi_{44}^{\mathrm{T}}(s, t_0)H^{\mathrm{T}}(s)R^{-1}(s)H(s)\Psi_{44}(s, t_0)\,\mathrm{d}s \quad (5C.69)$$

This completes the proof.

EXERCISES

5.1 Let $x \in \mathbb{R}^n$ and $y \in \mathbb{R}^m$ be jointly Gaussian random vectors, in which x is a Gaussian n-vector of mean $\hat{x}$ and covariance P_{xx}, and y is a Gaussian m-vector of mean $\hat{y}$ and covariance P_{yy}, so that $p(x, y)$ can be written as

$$p(x, y) = \left[(2\pi)^{(n+m)/2}\left|\begin{bmatrix} P_{xx} & P_{xy} \\ P_{yx} & P_{yy}\end{bmatrix}\right|^{1/2}\right]^{-1} \times \exp\left\{-\frac{1}{2}\begin{bmatrix} x - \hat{x} \\ y - \hat{y}\end{bmatrix}^{\mathrm{T}}\begin{bmatrix} P_{xx} & P_{xy} \\ P_{yx} & P_{yy}\end{bmatrix}^{-1}\begin{bmatrix} x - \hat{x} \\ y - \hat{y}\end{bmatrix}\right\}$$

where the covariance matrix

$$\begin{bmatrix} P_{xx} & P_{xy} \\ P_{yx} & P_{yy}\end{bmatrix}$$

is assumed to be positive definite. Apply Bayes' rule

$$p(x|y) = \frac{p(x, y)}{p(y)}$$

to obtain

$$p(x|y) = \frac{1}{(2\pi)^{n/2}|P_{x|y}|^{1/2}} \times \exp\{-\tfrac{1}{2}[x - E[x|y]]^{\mathrm{T}}P_{x|y}^{-1}[x - E[x|y]]\}$$

where

$$E[x|y] = \hat{x} + P_{xy}P_{yy}^{-1}(y - \hat{y})$$

$$P_{x|y} \triangleq E\{[x - E[x|y]][x - E[x|y]]^{\mathrm{T}}\}$$

$$= P_{xx} - P_{xy}P_{yy}^{-1}P_{yx}$$

(Hint: Use the inversion rule for a block matrix given in Appendix A.5 to expand $p(x, y)$.)

5.2 Show that the differential forms of equations (5.41) and (5.42) can be represented as

$$\frac{\mathrm{d}}{\mathrm{d}t}\hat{x}(t_0|t) = P_{\mathrm{r}}(t_0|t)[I + P_{\mathrm{nr}}P_{\mathrm{r}}^{-1}]^{\mathrm{T}}\Phi_{\mathrm{n}}^{\mathrm{T}}(t, t_0)H^{\mathrm{T}}(t)R^{-1}(t)$$

$$\times [z(t) - H(t)\hat{x}(t|t)], \qquad \hat{x}(t_0|t_0) = \hat{x}_{\mathrm{r}}$$

$$\frac{\mathrm{d}}{\mathrm{d}t}P(t_0|t) = -P_{\mathrm{r}}(t_0|t)[I + P_{\mathrm{nr}}P_{\mathrm{r}}^{-1}]^{\mathrm{T}}\Phi_{\mathrm{n}}^{\mathrm{T}}(t, t_0)H^{\mathrm{T}}(t)R^{-1}(t)$$

$$\times H(t)\Phi_{\mathrm{n}}(t, t_0)[I + P_{\mathrm{nr}}P_{\mathrm{r}}^{-1}]P_{\mathrm{r}}(t_0|t),$$

$$P(t_0|t_0) = P_{\mathrm{r}}$$

5.3 Prove that, for $M = 3$, equations (5.60) and (5.61) yield the following:

$$P_{12|\{1\}} = P_{12}A_{11} + P_{13}A_{12}^{\mathrm{T}}$$

$$P_{13|\{1\}} = P_{12}A_{12} + P_{13}A_{22}$$

$$P_{23|\{2\}} = P_{23}P_{33}^{-1}$$

where

$$A_{11} = [P_{22} - P_{23}P_{33}^{-1}P_{23}^{\mathrm{T}}]^{-1}$$

$$A_{22} = [P_{33} - P_{23}^{\mathrm{T}}P_{22}^{-1}P_{23}]^{-1}$$

$$A_{12} = -A_{11}P_{23}P_{33}^{-1}$$

5.4 Verify the result of Theorem 5.3.

5.5 Consider the exponentially correlated acceleration tracking problem described by

$$\dot{x}(t) = \begin{bmatrix} 0 & 1 & 0 \\ 0 & 0 & 1 \\ 0 & 0 & -1/\tau \end{bmatrix} x(t) + \begin{bmatrix} 0 \\ 0 \\ 1 \end{bmatrix} w(t)$$

$$z(t) = [1 \quad 0 \quad 0]x(t) + v(t)$$

where x is the state vector $[y \quad \dot{y} \quad \ddot{y}]^{\mathrm{T}}$, w and v are Gaussian white noises with zero means and spectral densities Q and R, respectively, and τ is a correlation time. Suppose that $x(0) \sim N\{\hat{x}(0), P(0)\}$, where

$$\hat{x}(0) = \begin{bmatrix} \hat{y}(0) \\ \hat{\dot{y}}(0) \\ \hat{\ddot{y}}(0) \end{bmatrix}, \; P(0) = \begin{bmatrix} P_{11}(0) & P_{12}(0) & P_{13}(0) \\ P_{12}(0) & P_{22}(0) & P_{23}(0) \\ P_{13}(0) & P_{23}(0) & P_{33}(0) \end{bmatrix}$$

(a) Setting

$$\hat{x}_1 = \begin{bmatrix} \hat{y}(0) \\ 0 \\ 0 \end{bmatrix}, \hat{x}_2 = \begin{bmatrix} 0 \\ \dot{\hat{y}}(0) \\ 0 \end{bmatrix}, \hat{x}_3 = \begin{bmatrix} 0 \\ 0 \\ \ddot{\hat{y}}(0) \end{bmatrix}$$

$$P_{11} = \begin{bmatrix} P_{11}(0) & 0 & 0 \\ 0 & 0 & 0 \\ 0 & 0 & 0 \end{bmatrix}$$

$$P_{22} = \begin{bmatrix} 0 & 0 & 0 \\ 0 & P_{22}(0) & 0 \\ 0 & 0 & 0 \end{bmatrix}$$

$$P_{33} = \begin{bmatrix} 0 & 0 & 0 \\ 0 & 0 & 0 \\ 0 & 0 & P_{33}(0) \end{bmatrix}$$

$$P_{12} = \begin{bmatrix} 0 & P_{12}(0) & 0 \\ 0 & 0 & 0 \\ 0 & 0 & 0 \end{bmatrix}$$

$$P_{13} = \begin{bmatrix} 0 & 0 & P_{13}(0) \\ 0 & 0 & 0 \\ 0 & 0 & 0 \end{bmatrix}$$

$$P_{23} = \begin{bmatrix} 0 & 0 & 0 \\ 0 & 0 & P_{23}(0) \\ 0 & 0 & 0 \end{bmatrix}$$

develop a dependent multipartitioning filter with $M = 3$ for the system above by using Theorem 5.2.

(b) Develop a dependent partitioning filter for the system above by applying Theorem 5.3, and by setting

$$\hat{x}_1 = \begin{bmatrix} \hat{y}(0) \\ 0 \\ 0 \end{bmatrix}, \hat{x}_2 = \begin{bmatrix} 0 \\ \dot{\hat{y}}(0) \\ \ddot{\hat{y}}(0) \end{bmatrix}, \hat{x}_3 = \begin{bmatrix} 0 \\ 0 \\ 0 \end{bmatrix}$$

$$P_{11} = \begin{bmatrix} P_{11}(0) & 0 & 0 \\ 0 & 0 & 0 \\ 0 & 0 & 0 \end{bmatrix}$$

$$P_{22} = \begin{bmatrix} 0 & 0 & 0 \\ 0 & P_{22}(0) & P_{23}(0) \\ 0 & P_{23}(0) & P_{33}(0) \end{bmatrix}$$

$$P_{33} = \begin{bmatrix} 0 & P_{12}(0) & P_{33}(0) \\ P_{12}(0) & 0 & 0 \\ P_{13}(0) & 0 & 0 \end{bmatrix}$$

5.6 Consider the system given in Exercise 5.5.

(a) Show that, for a sampling interval Δt, the state transition matrix Φ can be represented as

$$\Phi = \begin{bmatrix} 1 & \tau\theta & \tau^2 a \\ 0 & 1 & \tau(1 - \exp(-\theta)) \\ 0 & 0 & \exp(-\theta) \end{bmatrix}$$

where $\theta = \Delta t/\tau$ and $a = \theta - 1 + \exp(-\theta)$.

(b) Consider the following discrete-time system described by

$$x(k+1) = \Phi x(k) + \begin{bmatrix} 0 \\ 0 \\ 1 \end{bmatrix} w(k)$$

$$z(k) = [1 \quad 0 \quad 0]x(k) + v(k)$$

where w and v are white Gaussian noises with zero means and spectral densities $\bar{Q}$ and $\bar{R}$ respectively. Decomposing the initial conditions as in Exercise 5.5(a), develop a dependent multipartitioning filter for this system by using Theorem 5.5.

(c) Develop a dependent partitioning filter using Theorem 5.6 for the system above by defining the initial conditions as shown in Exercise 5.5(b).

5.7 When the system noise vector is further decomposed into statistically independent parts,

$$w(k) = w_n(k) + w_r(k)$$

how can Theorem 5.4 be amended? (See Andrisani and Gau [36].)

5.8 Prove the result of Lemma 5.4.

5.9 Prove the result of Lemma 5.5.

5.10 Prove the result of Lemma 5.6.

5.11 Complete the proof of Theorem 5.8

5.12 An alternative derivation of Corollary 5.6 was considered by Anderson [37] and Anderson and Moore [38]. Consider the following RDE:

$$\begin{aligned} P(k+1) = {} & FP(k)F^T - FP(k)H^T \\ & \times [HP(k)H^T + R]^{-1} HP(k)F^T + BQB^T \end{aligned}$$

By rearranging, and assuming that F is nonsingular, we have

$$\begin{aligned} P(k+1) &= [BQB^T F^{-T} + (F + BQB^T F^{-T} H^T R^{-1} H)P(k)] \\ &\quad \times [F^{-T} + F^{-T} H^T R^{-1} HP(k)]^{-1} \\ &= [C + DP(k)][A + BP(k)]^{-1} \end{aligned}$$

where

$$A = F^{-T}, \; B = F^{-T} H^T R^{-1} H$$

$$C = BQB^T F^{-T}, \; D = F + BQB^T F^{-T} H^T R^{-1} H$$

Consider also the linear equation

$$\begin{bmatrix} X(k+1) \\ Y(k+1) \end{bmatrix} = \begin{bmatrix} A & B \\ C & D \end{bmatrix} \begin{bmatrix} X(k) \\ Y(k) \end{bmatrix} \triangleq \Phi \begin{bmatrix} X(k) \\ Y(k) \end{bmatrix}$$

(a) Show that if $X(k)$ and $Y(k)$ are such that $Y(k)X^{-1}(k) = P(k)$, then we have $Y(k+1)X^{-1}(k+1) = P(k+1)$ and generally $Y(m)X^{-1}(m) = P(m)$.

(b) Show that Φ^{2^k}, for $k = 0, 1, \ldots$, is symplectic, i.e. that

$$[\Phi^{2^k}]^{\mathrm{T}} J \Phi^{2^k} = J$$

where

$$J = \begin{bmatrix} 0 & -I \\ I & 0 \end{bmatrix}, \qquad J^{-1} = -J$$

(c) Prove that with initial conditions $\alpha_0 = F^{\mathrm{T}}$, $\beta_0 = H^{\mathrm{T}} R^{-1} H$ and $\gamma_0 = BQB^{\mathrm{T}}$, $\gamma_k = P(2^k)$ can be updated by the following doubling algorithm:

$$\alpha_{k+1} = \alpha_k (I + \beta_k \gamma_k)^{-1} \alpha_k$$
$$\beta_{k+1} = \beta_k + \alpha_k (I + \beta_k \gamma_k)^{-1} \beta_k \alpha_k^{\mathrm{T}}$$
$$\gamma_{k+1} = \gamma_k + \alpha_k^{\mathrm{T}} \gamma_k (I + \beta_k \gamma_k)^{-1} \alpha_k$$

(Hint: The fact that Φ^{2^k} is symplectic for $k = 0, 1, \ldots$ gives

$$\Phi = \begin{bmatrix} \alpha_0^{-1} & \alpha_0^{-1} \beta_0 \\ \gamma_0 \alpha_0^{-1} & \alpha_0^{\mathrm{T}} + \gamma_0 \alpha_0^{-1} \beta_0 \end{bmatrix}$$

so that

$$\Phi^{2^k} = \begin{bmatrix} \alpha_k^{-1} & \alpha_k^{-1} \beta_k \\ \gamma_k \alpha_k^{-1} & \alpha_k^{\mathrm{T}} + \gamma_k \alpha_k^{-1} \beta_k \end{bmatrix}$$

assuming α_k^{-1} exists. To verify that $\gamma_k = P(2^k)$, use the relation

$$\begin{bmatrix} X(2^k) \\ Y(2^k) \end{bmatrix} = \Phi^{2^k} \begin{bmatrix} X(0) \\ Y(0) \end{bmatrix} = \Phi^{2^k} \begin{bmatrix} I \\ 0 \end{bmatrix}$$

and to derive the updating algorithm, calculate $\Phi^{2^{k+1}} = \Phi^{2^k} \times \Phi^{2^k}$.)

(d) Show that if β_k, γ_k are symmetric matrices, so are $(I + \beta_k \gamma_k)^{-1} \beta_k$ and $\gamma_k (I + \beta_k \gamma_k)^{-1}$, assuming the inverse exists.

5.13 Write down the doubling algorithm in Corollary 5.6 using FORTRAN77 or BASIC.

(a) Implement the program above for the continuous-time system considered in Exercise 5.5, but using $Q = R = \tau = 1$, $\Delta = 1 \times 10^{-3}$ and $\varepsilon = 1 \times 10^{-7}$.

(b) Implement the program above for the discrete-time system given in Exercise 5.6, but using $Q = R = \tau = 1$, $\Delta t = 10$, and $\varepsilon = 1 \times 10^{-7}$.

5.14 Consider the following algebraic Lyapunov equation (ALE) in discrete time:

$$P - \Phi P \Phi^{T} - BQB^{T} = 0$$

where it is assumed that Φ is asymptotically stable and (Φ, D) is reachable, where $DD^{T} = BQB^{T}$. Modify the doubling algorithm in Corollary 5.6 to obtain the solution for $P > 0$. (Hint: For this case, $R \equiv \infty$.)

REFERENCES

[1] LAINIOTIS, D. G., Partitioned Estimation Algorithms, II: Linear Estimation, *Inf. Sci.*, vol. 7, 1974, pp. 317–40.

[2] LAINIOTIS, D. G. and ANDRISANI II, D., Multipartitioning Solutions for State and Parameter Estimation: Continuous Systems, *Proc. 1979 Joint Aut. Control Conf.*, Philadelphia, 1978, pp. 215–46.

[3] LAINIOTIS, D. G. and ANDRISANI II, D., Multipartitioning Linear Estimation Algorithms: Continuous Systems, *IEEE Trans. Aut. Control,* vol. AC–24, no. 6, 1979, pp. 937–44.

[4] GOVINDARAJ, K. S. and LAINIOTIS, D. G., A Unifying Framework for Discrete Linear Estimation: Generalized Partitioned Algorithms, *Int. J. Control*, vol. 28, no. 4, 1978, pp. 571–88.

[5] LAINIOTIS, D. G., Partitioning Filters, *Inf. Sci.*, vol. 17, 1979, pp. 177–93.

[6] ANDRISANI II, D., Multipartitioned Estimation Algorithms, Ph.D. dissertation, State University of New York at Buffalo, NY, December 1978.

[7] JAZWINSKI, A. H., *Stochastic Processes and Filtering Theory*, Academic Press, New York, 1970.

[8] LAINIOTIS, D. G., A Unifying Framework for Linear Estimation: Generalized Partitioned Algorithms, *Inf. Sci.*, vol. 10, 1976, pp. 243–78.

[9] WATANABE, K., Optimal Partitioned Filter of Stochastic Distributed-Parameter Dynamical Systems with Unknown Initial State, *J. Franklin Inst.*, vol. 315, nos 5–6, 1983, pp. 347–85.

[10] YOSHIMURA, T., WATANABE, K., KONISHI, K. and SOEDA, T., A Sequential Failure Detection Approach and the Identification of Failure Parameters, *Int. J. Syst. Sci.*, vol. 10, no. 7, 1979, pp. 827–36.

[11] WATANABE, K., YOSHIMURA, T. and SOEDA, T., A Diagnosis System Design for a Parametric Failure, *Trans. Soc. Instrum. Control Engrs*, vol. 15, no. 7, 1979, pp. 901–6 (in Japanese).

[12] WATANABE, K., YOSHIMURA, T. and SOEDA, T., A Diagnosis Method for Linear Stochastic Systems with Parametric Failures, *Trans. ASME, J. Dynamic Syst., Measure. and Control*, vol. 103, no. 1, 1981, pp. 28–35.

[13] GOODWIN, G. C. and SIN, K. S., *Adaptive Filtering Prediction and Control*, Prentice Hall, Englewood Cliffs, New Jersey, 1984.

[14] EULRICH, B. J., ANDRISANI II, D. and LAINIOTIS, D. G., New Identification Algorithms and Their Relationships to Maximum-Likelihood Methods: The Partitioned Approach, *Proc. 1978 Joint Aut. Control Conf.*, Philadelphia, 1978, pp. 231–46.

[15] EULRICH, B. J., ANDRISANI II, D. and LAINIOTIS, D. G., Partitioning Identification Algorithms, *IEEE Trans. Aut. Control,* vol. AC–25, no. 3, 1980, pp. 521–8.

[16] GOVINDARAJ, K. S., LEBACQZ, J. V. and EULRICH, B. J., Application of Identification Algorithms to Aircraft Flight Data, *Proc. 1978 Joint Aut. Control Conf.*, Philadelphia, 1978, pp. 247–58.

[17] SAGE, A. P. and MELSA, J. L., *Estimation Theory with Applications to Communications and Control*, Robert E. Krieger, New York, 1979.

[18] WATANABE, K. and IWASAKI, M., A Fast Computational Approach in Optimal Distributed-Parameter State Estimation, *Trans. ASME, J. Dynamic Syst. Measure. and Control,* vol. 105, no. 1, 1983, pp. 1–10.

[19] KORTÜM, W., Computational Techniques in Optimal State-Estimation—A Tutorial Review, *Trans. ASME, J. Dynamic Syst. Measure. and Control*, vol. 101, no. 2, 1979, pp. 99–107.

[20] WATANABE, K., Partitioned Estimation Problem for Linear Distributed Parameter Systems, *Trans. Soc. Instrum. Control Engrs.*, vol. 18, no. 7, 1982, pp. 677–84 (in Japanese).

[21] LAINIOTIS, D. G., Partitioned Riccati Solutions and Integration-Free Doubling Algorithms, *IEEE Trans. Aut. Control,* vol. AC–21, no. 5, 1976, pp. 677–89.

[22] KLEINMAN, D., On an Iterative Technique for Riccati Equation Computations, *IEEE Trans. Aut. Control*, vol. AC–13, no. 1, 1968, pp. 114–15.

[23] CASTI, J. L., *Dynamical Systems and Their Applications: Linear Theory*, Academic Press, New York, 1977.

[24] POTTER, J. F., Matrix Quadratic Solutions, *J. SIAM Appl. Math.*, vol. 14, May 1966, pp. 496–501.

[25] KUČERA, V., A Contribution to Matrix Quadratic Equations, *IEEE Trans. Aut. Control*, vol. AC–17, no. 3, 1972, pp. 344–7.

[26] KUČERA, V., On Nonnegative Definite Solutions to Matrix Quadratic Equations, *Automatica*, vol. 8, 1972, pp. 413–23.

[27] WATANABE, K., Algebraic Solution of a Forward-Pass Fixed-Interval Smoother: Continuous-Time Systems, *Int. J. Syst. Sci.*, vol. 16, no. 10, 1985, pp. 1217–27.

[28] SIDHU, G. S. and BIERMAN, G. J., Integration-Free Interval Doubling for Riccati Equation Solutions, *IEEE Trans. Aut. Control*, vol. AC–22, no. 5, 1977, pp. 831–4.

[29] KWAKERNAAK, H. and SIVAN, R., *Linear Optimal Control Systems*, Wiley-Interscience, New York, 1972.

[30] KENNEY, C. S. and LEIPNIK, R. B., Numerical Integration of the Differential Matrix Riccati Equation, *IEEE Trans. Aut. Control*, vol. AC–30, no. 10, 1985, pp. 962–70.

[31] JAMSHIDI, M., An Overview on the Solutions of the Algebraic Matrix Riccati Equation and Related Problems, *Large Scale Syst.*, vol. 1, 1980, pp. 167–92.

[32] KIMURA, M., Convergence of the Doubling Algorithm for the Discrete-Time Algebraic Riccati Equation, *Int. J. Syst. Sci.*, vol. 19, no. 5, 1988, pp. 701–11.

[33] WATANABE, K., YOSHIMURA, T. and SOEDA, T., Optimal Nonlinear Estimation for Distributed-Parameter Systems via the Partition Theorem, *Int. J. Syst. Sci.*, vol. 11, 1980, pp. 1113–30.

[34] WATANABE, K., An Alternative Approach to the Derivation of Distributed-Type Partitioned Filters, *Int. J. Syst. Sci.*, vol. 12, no. 3, 1981, pp. 351–6.

[35] STAVROULAKIS, P. and TZAFESTAS, S. G., Multipartitioning in Distributed Parameter Adaptive Estimation, *Int. J. Syst. Sci.*, vol. 13, no. 3, 1982, pp. 301–15.

[36] ANDRISANI II, D. and GAU, C. F., Multistage Linear Estimation Using Partitioning, *IEEE Trans. Aut. Control*, vol. AC–30, no. 2, 1985, pp. 182–5.

[37] ANDERSON, B. D. O., Second-Order Algorithms for the Steady-State Riccati Equation, *Int. J. Control*, vol. 28, no. 2, 1978, pp. 295–306.

[38] ANDERSON, B. D. O. and MOORE, J. B., *Optimal Filtering*, Prentice Hall, Englewood Cliffs, New Jersey, 1979.

6

Partitioning Estimators: The Scattering Approach

6.1 INTRODUCTION

In the previous chapter, we derived partitioning filters by applying the probabilistic approach. In this chapter, we further demonstrate the partitioning estimators and other related matters by means of scattering theory [1–4].

In Section 6.2 we introduce Redheffer's scattering theory [1], the star product and a general scattering calculus [5, 6]. We then formulate the partitioning filter and smoother algorithms in Sections 6.3 and 6.4, respectively.

In Section 6.5, we show the backward partitioning filter based on forward-time models and prove the Weinert–Desai smoothing algorithm [7] by a simple method. Furthermore, we present a computationally advantageous approach, known as the $X-Y$ or Chandrasekhar algorithm [8–10], to the solution of matrix Riccati equations. Additionally, we prove the so-called Mayne–Fraser or Mehra two-filter smoothing algorithm [11–13].

6.2 REDHEFFER'S SCATTERING THEORY

Consider the scattering section provided by cascading two other sections, as in Figure 6.1, with scattering matrices (or operators)

$$S_1 = \begin{bmatrix} a & b \\ c & d \end{bmatrix}, \qquad S_2 = \begin{bmatrix} A & B \\ C & D \end{bmatrix} \tag{6.1}$$

By tracing paths through Figure 6.1, it is easy to see that the composite scattering matrix is

$$S = \begin{bmatrix} a & b \\ c & d \end{bmatrix} * \begin{bmatrix} A & B \\ C & D \end{bmatrix} \tag{6.2a}$$

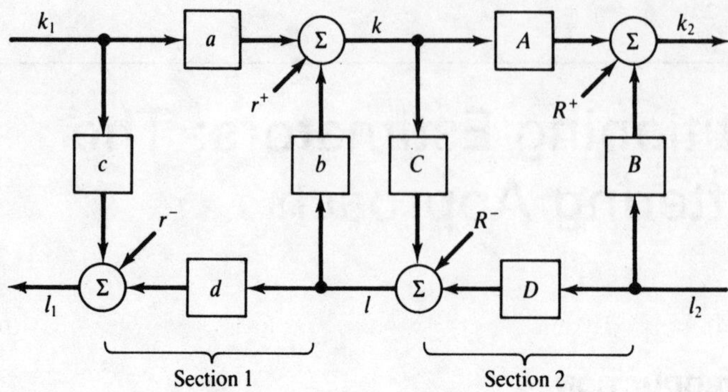

Figure 6.1 Cascade scattering layers.

$$\triangleq \begin{bmatrix} A(I - bC)^{-1}a & B + Ab(I - Cb)^{-1}D \\ c + dC(I - bC)^{-1}a & d(I - Cb)^{-1}D \end{bmatrix} \tag{6.2b}$$

We call S the (Redheffer) *star product* [1] of S_1 and S_2. Furthermore, the internal source contribution vectors are defined as

$$q_1 = \begin{bmatrix} r^+ \\ r^- \end{bmatrix}, \qquad q_2 = \begin{bmatrix} R^+ \\ R^- \end{bmatrix} \tag{6.3}$$

The waves emerging from section 1 of Figure 6.1 are related to those incident on the section as follows:

$$\begin{bmatrix} k \\ l_1 \end{bmatrix} = S_1 \begin{bmatrix} k_1 \\ l \end{bmatrix} + q_1 \tag{6.4}$$

Then, the source vector for the combined sections can be written

$$q = q_1 \bullet q_2$$

$$= \begin{bmatrix} r^+ \\ r^- \end{bmatrix} \bullet \begin{bmatrix} R^+ \\ R^- \end{bmatrix} \triangleq \begin{bmatrix} R^+ \\ r^- \end{bmatrix} + \left(\begin{bmatrix} I & b \\ 0 & d \end{bmatrix} * \begin{bmatrix} A & 0 \\ C & I \end{bmatrix} \right) \begin{bmatrix} r^+ \\ R^- \end{bmatrix} \tag{6.5a}$$

$$= \begin{bmatrix} R^+ \\ r^- \end{bmatrix} + \begin{bmatrix} A(I - bC)^{-1}(r^+ + bR^-) \\ d(I - Cb)^{-1}(R^- + Cr^+) \end{bmatrix} \tag{6.5b}$$

We shall call q the *dot sum* [5] of q_1 and q_2.

Notice that, if S is invertible, the following significant property holds:

$$S^{-1}*S = I = S*S^{-1} \tag{6.6}$$

Therefore if, for instance,

$$\begin{aligned} S_1 &= S_3*S \\ S_2 &= S_4*S \end{aligned} \tag{6.7}$$

then we could compute S_2 from S_1:

$$S_2 = S_4*S_3^{-1}*S_1 \tag{6.8}$$

To examine briefly a calculus for such operators, consider a scattering matrix S_0 such that

$$S_0(\tau, t) = \begin{bmatrix} a_0(t, \tau) & \rho_0(t, \tau) \\ r_0(t, \tau) & \alpha_0(t, \tau) \end{bmatrix}, \qquad S_0(\tau, \tau) = I,\ \tau \leqslant t \tag{6.9}$$

Suppose that

$$S_0(t, t + \Delta t) = S_0(t, t) + M(t)\Delta t + o(\Delta t), \qquad S_0(t, t) \equiv I \tag{6.10}$$

and

$$M(t) = \lim_{\Delta t \to 0} \frac{S_0(t, t + \Delta t) - I}{\Delta t} \tag{6.11a}$$

$$\triangleq \begin{bmatrix} f(t) & g(t) \\ h(t) & e(t) \end{bmatrix} \tag{6.11b}$$

which is referred to as the *infinitesimal generator* of S. Now note that

$$\begin{aligned} S_0(\tau, t + \Delta t) &= S_0(\tau, t)*S_0(t, t + \Delta t) \\ &= S_0(\tau, t)*[I + M(t)\Delta t + o(\Delta t)] \\ &= S_0(\tau, t) + S_0(\tau, t)*M(t)\Delta t + o(\Delta t) \end{aligned} \tag{6.12}$$

After some simple algebra on $S_0(\tau, t)*M(t)$, we obtain the forward evolution equation for $S_0(\tau, t)$:

$$\frac{\partial S_0(\tau, t)}{\partial t} = \begin{bmatrix} [f(t) + \rho_0(t, \tau)h(t)]a_0(t, \tau) & \begin{array}{l} g(t) + f(t)\rho_0(t, \tau) + \rho_0(t, \tau)e(t) \\ \quad + \rho_0(t, \tau)h(t)\rho_0(t, \tau) \end{array} \\ \alpha_0(t, \tau)h(t)a_0(t, \tau) & \alpha_0(t, \tau)[e(t) + h(t)\rho_0(t, \tau)] \end{bmatrix},$$

$$S_0(\tau, \tau) = I \tag{6.13}$$

Similarly, by noting that we can write

$$
\begin{aligned}
S_0(s - \mathrm{d}s, t) &= S_0(s - \mathrm{d}s, s) * S_0(s, t) \\
&= [I + M(s - \mathrm{d}s)\,\mathrm{d}s + o(\mathrm{d}s)] * S_0(s, t) \\
&= [I + M(s)\,\mathrm{d}s + o(\mathrm{d}s)] * S_0(s, t) \\
&= S_0(s, t) + M(s) * S_0(s, t)\,\mathrm{d}s + o(\mathrm{d}s) \qquad (6.14)
\end{aligned}
$$

we obtain the backward evolution equation for $S_0(s, t)$:

$$
-\frac{\partial S_0(s, t)}{\partial s} =
\begin{bmatrix}
a_0(t, s)[f(s) + g(s) r_0(t, s)] & a_0(t, s) g(s) \alpha_0(t, s) \\
h(s) + e(s) r_0(t, s) + r_0(t, s) f(s) & [e(s) + r_0(t, s) g(s)] \alpha_0(t, s) \\
+ r_0(t, s) g(s) r_0(t, s) &
\end{bmatrix},
$$

$$
S_0(t, t) = I \qquad (6.15)
$$

Note that there is a certain interesting 'duality' between equations (6.13) and (6.15). Furthermore, if the forward and backward projection operators are defined by

$$
P_+ = \begin{bmatrix} I & 0 \\ 0 & 0 \end{bmatrix}, \qquad P_- = \begin{bmatrix} 0 & 0 \\ 0 & I \end{bmatrix} \qquad (6.16)
$$

then equations (6.13) and (6.15) can be rewritten more compactly [6] as follows:

$$
\frac{\partial S_0(\tau, t)}{\partial t} = (P_+ + S_0(\tau, t) P_-) M(t) (P_+ S_0(\tau, t) + P_-) \qquad (6.17)
$$

$$
-\frac{\partial S_0(s, t)}{\partial s} = (P_- + S_0(s, t) P_+) M(s) (P_- S_0(s, t) + P_+) \qquad (6.18)
$$

In addition, if the source vector corresponding to equation (6.9) is defined as

$$
q_0(\tau, t) = \begin{bmatrix} q^+(\tau, t) \\ q^-(\tau, t) \end{bmatrix}, \qquad q_0(\tau, \tau) = 0 \qquad (6.19)
$$

it can be shown that

$$
\frac{\partial q_0(\tau, t)}{\partial t} = [P_+ + S_0(\tau, t) P_-][q_0(\tau, t) + M(t) P_+ q_0(\tau, t)] \qquad (6.20)
$$

and

$$-\frac{\partial q_0(s, t)}{\partial s} = [P_- + S_0(s, t)P_+][q_0(s, t) + M(s)P_- q_0(s, t)] \tag{6.21}$$

6.3 THE FORWARD PARTITIONING FILTER

Consider the problem of the least-squares (LS) estimate of the state $x(s)$, $\tau \leqslant s \leqslant t$, given by equations (5.1)–(5.5), but with $x(\tau) \sim N(\hat{x}(\tau|\tau), P(\tau))$. A well-known approach to this problem is the variational procedure described by Bryson and Ho [14] (see also Rauch *et al.* [15] and Jazwinski [16]). That is, the smoothed estimate $\hat{x}(s|t) \triangleq E[x(s)|Z_t]$ of $x(s)$ is determined by the solution of the following Hamiltonian equations:

$$\frac{\mathrm{d}}{\mathrm{d}s}\begin{bmatrix} \hat{x}(s|t) \\ -\lambda(s|t) \end{bmatrix} = \begin{bmatrix} A(s) & B(s)Q(s)B^{\mathrm{T}}(s) \\ -H^{\mathrm{T}}(s)R^{-1}(s)H(s) & A^{\mathrm{T}}(s) \end{bmatrix} \times \begin{bmatrix} \hat{x}(s|t) \\ \lambda(s|t) \end{bmatrix} + \begin{bmatrix} 0 \\ H^{\mathrm{T}}(s)R^{-1}(s)z(s) \end{bmatrix} \tag{6.22}$$

with boundary conditions

$$\hat{x}(\tau|t) = \hat{x}(\tau|\tau) + P(\tau)\lambda(\tau|t), \qquad \lambda(t|t) = 0 \tag{6.23}$$

where $\lambda(\cdot)$ denotes the adjoint state (or Lagrange multiplier) vector. A problem involving expressions (6.22) and (6.23) is called the *two-point boundary-value problem* (TPBVP).

Suppose that the moments of $x(\tau)$ are decomposed such that

$$\hat{x}(\tau|\tau) = \hat{x}_{\mathrm{n}}(\tau|\tau) + \hat{x}_{\mathrm{r}}(\tau|\tau) \tag{6.24a}$$

$$P(\tau) = P_{\mathrm{n}}(\tau) + P_{\mathrm{r}}(\tau) + P_{\mathrm{nr}}(\tau) + P_{\mathrm{nr}}^{\mathrm{T}}(\tau) \tag{6.24b}$$

and define the following:

$$\hat{m}_{\mathrm{n}}(\tau) \triangleq \hat{x}_{\mathrm{n}}(\tau|\tau) - P_{\mathrm{nr}}(\tau)P_{\mathrm{r}}^{-1}(\tau)\hat{x}_{\mathrm{r}}(\tau|\tau) \tag{6.25a}$$

$$\Pi_{\mathrm{n}}(\tau) \triangleq P_{\mathrm{n}}(\tau) - P_{\mathrm{nr}}(\tau)P_{\mathrm{r}}^{-1}(\tau)P_{\mathrm{nr}}^{\mathrm{T}}(\tau) \tag{6.25b}$$

$$\begin{aligned} \hat{m}_{\mathrm{r}}(\tau) &\triangleq \hat{x}(\tau|\tau) - \hat{m}_{\mathrm{n}}(\tau) \\ &= [I + P_{\mathrm{nr}}(\tau)P_{\mathrm{r}}^{-1}(\tau)]\hat{x}_{\mathrm{r}}(\tau|\tau) \end{aligned} \tag{6.26a}$$

$$\begin{aligned} \Pi_{\mathrm{r}}(\tau) &\triangleq P(\tau) - \Pi_{\mathrm{n}}(\tau) \\ &= P_{\mathrm{nr}}(\tau) + P_{\mathrm{nr}}^{\mathrm{T}}(\tau) + P_{\mathrm{r}}(\tau) + P_{\mathrm{nr}}(\tau)P_{\mathrm{r}}^{-1}(\tau)P_{\mathrm{nr}}^{\mathrm{T}}(\tau) \\ &= [I + P_{\mathrm{nr}}(\tau)P_{\mathrm{r}}^{-1}(\tau)]P_{\mathrm{r}}(\tau)[I + P_{\mathrm{nr}}(\tau)P_{\mathrm{r}}^{-1}(\tau)]^{\mathrm{T}} \end{aligned} \tag{6.26b}$$

Now consider the finite scattering section illustrated in Figure 6.2, where $\Phi_0(t, \tau)$, $\Phi_0^T(t, \tau)$ are forward and backward transmission operators; $-W_0(t, \tau)$, $P_0(t, \tau)$ are left and right reflection operators; and $q_0^+(\tau, t)$, $q_0^-(\tau, t)$ are forward and backward source contribution vectors. When the composite and nominal scattering matrices are defined by

$$S(\tau, t) = \begin{bmatrix} \Phi(t, \tau) & P(t, \tau) \\ -W(t, \tau) & \Phi^T(t, \tau) \end{bmatrix} \tag{6.27}$$

$$S_n(\tau, t) = \begin{bmatrix} \Phi_n(t, \tau) & P_n(t, \tau) \\ -W_n(t, \tau) & \Phi_n^T(t, \tau) \end{bmatrix} \tag{6.28}$$

the following theorem is obtained.

Theorem 6.1

The following identity holds:

$$\begin{bmatrix} \Phi(t, \tau) & P(t, \tau) \\ -W(t, \tau) & \Phi^T(t, \tau) \end{bmatrix}$$

$$= \begin{bmatrix} \Phi_n(t, \tau)[I + \Pi_r(\tau)W_n(t, \tau)]^{-1} & P_n(t, \tau) + \Phi_n(t, \tau)\Pi_r(\tau)[I + W_n(t, \tau)\Pi_r(\tau)]^{-1}\Phi_n^T(t, \tau) \\ -W_n(t, \tau)[I + \Pi_r(\tau)W_n(t, \tau)]^{-1} & [I + W_n(t, \tau)\Pi_r(\tau)]^{-1}\Phi_n^T(t, \tau) \end{bmatrix} \tag{6.29}$$

Proof

From the decomposition rule for the initial layer, it follows that

$$S(\tau, t) = \overline{\Pi}(\tau) * S_0(\tau, t) \tag{6.30a}$$

$$= \overline{\Pi}_r(\tau) * \overline{\Pi}_n(\tau) * S_0(\tau, t) \tag{6.30b}$$

$$= \overline{\Pi}_r(\tau) * S_n(\tau, t) \tag{6.30c}$$

where

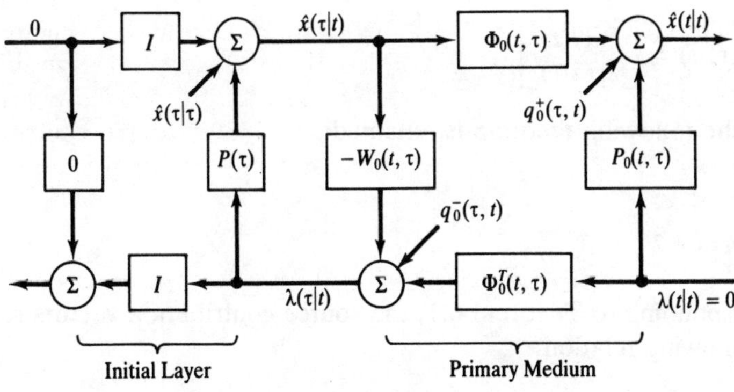

Figure 6.2 A finite scattering section.

$$\bar{\Pi}(\tau) \triangleq \begin{bmatrix} I & \Pi_n(\tau) + \Pi_r(\tau) \\ 0 & I \end{bmatrix}, \qquad \bar{\Pi}_r(\tau) \triangleq \begin{bmatrix} I & \Pi_r(\tau) \\ 0 & I \end{bmatrix}$$

$$\bar{\Pi}_n(\tau) \triangleq \begin{bmatrix} I & \Pi_n(\tau) \\ 0 & I \end{bmatrix}, \qquad S_n(\tau, t) \triangleq \bar{\Pi}_n(\tau) * S_0(\tau, t)$$

Hence, applying the star product rule given by expression (6.2) to (6.30c) gives the desired result. □

Figure 6.3 illustrates the decomposition of the initial layer. In addition, if the source vectors corresponding to $S(\tau, t)$ and $S_n(\tau, t)$ are defined as

$$q(\tau, t) \triangleq \begin{bmatrix} \hat{x}(t|t) \\ \lambda(\tau|t) \end{bmatrix} \tag{6.31}$$

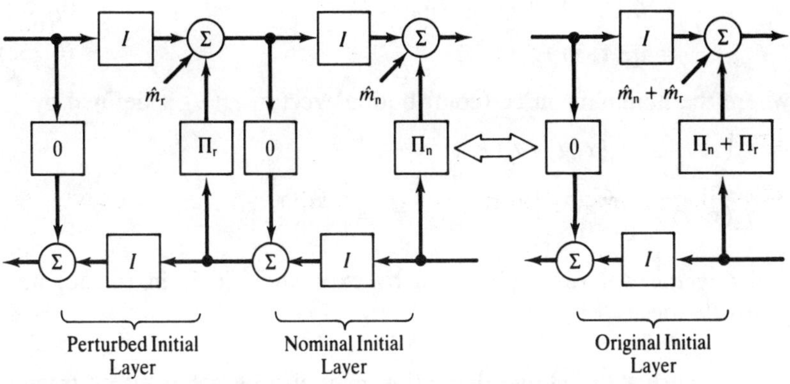

Figure 6.3 Decomposition of initial layer.

$$q_n(\tau, t) \triangleq \begin{bmatrix} \hat{x}_n(t|t) \\ \lambda_n(\tau|t) \end{bmatrix} \tag{6.32}$$

then the following theorem is obtained:

Theorem 6.2

Corresponding to Theorem 6.1, the source contribution vectors satisfy the following relation:

$$\begin{bmatrix} \hat{x}(t|t) \\ \lambda(\tau|t) \end{bmatrix} = \begin{bmatrix} \hat{x}_n(t|t) + \Phi_n(t, \tau)\hat{x}_r(\tau|t) \\ \{I + W_n(t, \tau)\Pi_r(\tau)\}^{-1}\{\lambda_n(\tau|t) - W_n(t, \tau)\hat{m}_r(\tau)\} \end{bmatrix} \tag{6.33}$$

where

$$\hat{x}_r(\tau|t) = [I + \Pi_r(\tau)W_n(t, \tau)]^{-1}[\hat{m}_r(\tau) + \Pi_r(\tau)\lambda_n(\tau|t)] \tag{6.34}$$

Proof

From the decomposition of the initial source vector

$$q(\tau, \tau) = \begin{bmatrix} \hat{x}(\tau|\tau) \\ 0 \end{bmatrix}$$

it is easy to see that

$$q(\tau, t) = q(\tau, \tau)\bullet q_0(\tau, t) \tag{6.35a}$$

$$= q_r(\tau, \tau)\bullet q_n(\tau, \tau)\bullet q_0(\tau, t) \tag{6.35b}$$

$$= q_r(\tau, \tau)\bullet q_n(\tau, t) \tag{6.35c}$$

where the nominal source (contribution) vector $q_n(\cdot)$ is defined by

$$q_n(\tau, t) \triangleq q_n(\tau, \tau)\bullet q_0(\tau, t)$$

$$q_n(\tau, \tau) = \begin{bmatrix} \hat{m}_n(\tau) \\ 0 \end{bmatrix}, \qquad q_r(\tau, \tau) = \begin{bmatrix} \hat{m}_r(\tau) \\ 0 \end{bmatrix}$$

Utilizing the dot sum rule given by expression (6.5) in (6.35c) now yields the desired result. □

Notice that the above derivation may also be achieved by tracing the flow-graph relationships in Figure 6.4. That is, the upper identity

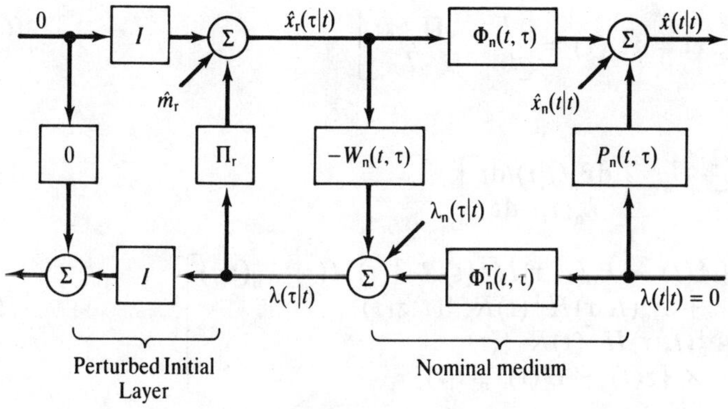

Figure 6.4 A scattering layer with a perturbed initial layer.

of expression (6.33) is followed by the right-hand summation in Figure 6.4. Furthermore, from the middle and left-hand summations, the following expressions are obtained:

$$\lambda(\tau|t) = \lambda_n(\tau|t) - W_n(t, \tau)\hat{x}_r(\tau|t) \tag{6.36}$$

$$\hat{x}_r(\tau|t) = \hat{m}_r(\tau) + \Pi_r(\tau)\lambda(\tau|t) \tag{6.37}$$

Using these relations yields the lower identity of expression (6.33) as well as equation (6.34).

Theorem 6.3

The forward evolution equations of $S_n(\tau, t)$ and $q_n(\tau, t)$ are given by

$$\frac{dS_n(\tau, t)}{dt} = \begin{bmatrix} d\Phi_n(t, \tau)/dt & dP_n(t, \tau)/dt \\ -dW_n(t, \tau)/dt & d\Phi_n^T(t, \tau)/dt \end{bmatrix}$$

$$= \begin{bmatrix} \{A(t) - P_n(t, \tau)H^T(t) \\ \quad \times R^{-1}(t)H(t)\}\Phi_n(t, \tau) & A(t)P_n(t, \tau) + P_n(t, \tau)A^T(t, \tau)A^T(t) \\ & \quad + B(t)Q(t)B^T(t) \\ & \quad - P_n(t, \tau)H^T(t)R^{-1}(t)H(t)P_n(t, \tau) \\ -\Phi_n^T(t, \tau)H^T(t)R^{-1}(t)H(t)\Phi_n(t, \tau) & \Phi_n^T(t, \tau)\{A(t) - P_n(t, \tau)H^T(t) \\ & \quad \times R^{-1}(t)H(t)\}^T \end{bmatrix},$$

$$S_n(\tau, \tau) = \bar{\Pi}_n(\tau) = \begin{bmatrix} I & \Pi_n(\tau) \\ 0 & I \end{bmatrix} \tag{6.38}$$

and

$$\frac{dq_n(\tau, t)}{dt} = \begin{bmatrix} d\hat{x}_n(t|t)/dt \\ \lambda_n(\tau|t)dt \end{bmatrix}$$

$$= \begin{bmatrix} \{A(t) - P_n(t, \tau)H^T(t)R^{-1}(t)H(t)\}\hat{x}_n(t|t) \\ \quad + P_n(t, \tau)H^T(t)R^{-1}(t)z(t) \\ \Phi_n^T(t, \tau)H^T(t)R^{-1}(t) \\ \quad \times \{z(t) - H(t)\hat{x}_n(t|t)\} \end{bmatrix},$$

$$q_n(\tau, \tau) = \begin{bmatrix} \hat{m}_n(\tau) \\ 0 \end{bmatrix} \tag{6.39}$$

Proof

From the decomposition rule for $S_n(\tau, t + \Delta t)$, we obtain

$$\begin{aligned} S_n(\tau, t + \Delta t) &= \bar{\Pi}_n(\tau) * S_0(\tau, t + \Delta t) \\ &= \bar{\Pi}_n(\tau) * S_0(\tau, t) * S_0(t, t + \Delta t) \\ &= S_n(\tau, t) * S_0(t, t + \Delta t) \end{aligned} \tag{6.40}$$

Taking account of the fact that

$$S_0(t, t + \Delta t) = I + M(t)\Delta t + o(\Delta t) \tag{6.41}$$

$$M(t) \triangleq \lim_{\Delta t \to \infty} \frac{S_0(t, t + \Delta t) - I}{\Delta t}$$

$$= \begin{bmatrix} A(t) & B(t)Q(t)B^T(t) \\ -H^T(t)R^{-1}(t)H(t) & A^T(t) \end{bmatrix} \tag{6.42}$$

it follows that

$$S_n(\tau, t + \Delta t) = \begin{bmatrix} \Phi_n(t, \tau) & P_n(t, \tau) \\ -W_n(t, \tau) & \Phi_n^T(t, \tau) \end{bmatrix}$$

$$* \begin{bmatrix} I + A(t)\Delta t & B(t)Q(t)B^T(t)\Delta t \\ -H^T(t)R^{-1}(t)H(t)\Delta t & I + A^T(t)\Delta t \end{bmatrix}$$

$$+ o(\Delta t) \tag{6.43}$$

Using the star product rule given by expression (6.2) on (6.43), and dropping terms higher than $(\Delta t)^2$ yields

$$S_n(\tau, t + \Delta t) = S_n(\tau, t) + o(\Delta t) + \Delta t \begin{bmatrix} \{A(t) - P_n(t, \tau)H^T(t)R^{-1}(t)H(t)\} \\ \quad \times \Phi_n(t, \tau) & A(t)P_n(t, \tau) + P_n(t, \tau)A^T(t) \\ & \quad + B(t)Q(t)B^T(t) - P_n(t, \tau) \\ & \quad \times H^T(t)R^{-1}(t)H(t)P_n(t, \tau) \\ -\Phi_n^T(t, \tau)H^T(t)R^{-1}(t)H(t)\Phi_n(t, \tau) & \Phi_n^T(t, \tau)\{A(t) - P_n(t, \tau) \\ & \quad \times H^T(t)R^{-1}(t)H(t)\}^T \end{bmatrix} \tag{6.44}$$

Therefore, equation (6.38) is apparent from (6.44).

On the other hand, by the decomposition rule for $q_n(\tau, t + \Delta t)$,

$$\begin{aligned} q_n(\tau, t + \Delta t) &= q_n(\tau, \tau)\bullet q_0(\tau, t + \Delta t) \\ &= q_n(\tau, \tau)\bullet q_0(\tau, t)\bullet q_0(t, t + \Delta t) \\ &= q_n(\tau, t)\bullet q_0(t, t + \Delta t) \end{aligned} \tag{6.45}$$

Using the fact that

$$q_0(t, t + \Delta t) = F(t)\Delta t + o(\Delta t) \tag{6.46}$$

$$F(t) \triangleq \lim_{\Delta t \to \infty} \frac{q_0(t, t + \Delta t)}{\Delta t} \equiv \begin{bmatrix} 0 \\ H^T(t)R^{-1}(t)z(t) \end{bmatrix} \tag{6.47}$$

expression (6.45) reduces to

$$q_n(\tau, t + \Delta t) = q_n(\tau, t) + o(\Delta t) + \Delta t \begin{bmatrix} \{A(t) - P_n(t, \tau)H^T(t)R^{-1}(t)H(t)\}\hat{x}_n(t|t) \\ \quad + P_n(t, \tau)H^T(t)R^{-1}(t)z(t) \\ \Phi_n^T(t, \tau)H^T(t)R^{-1}(t)\{z(t) - H(t)\hat{x}_n(t|t)\} \end{bmatrix} \tag{6.48}$$

Equation (6.39) then follows from (6.48). □

Thus, Theorems 6.1–6.3 constitute the dependent partitioning filter described in Theorem 5.1 of Chapter 5, though there are some notational discrepancies. Therefore, it is interesting to note that $\Phi_n(t, \tau)$ represents the closed-loop state transition matrix for the nominal Kalman filter with state $\hat{x}_n(t|t)$; $P_n(t, \tau)$ the nominal filtered error covariance matrix; and $W_n(t, \tau)$ the partial observability Gramian.

6.4 THE FORWARD PARTITIONING SMOOTHER

In this section, we shall derive a set of forward partitioning smoother algorithms. Utilizing a method due to Zachrisson [17], we shall consider $x(\tau)$, $\tau \leqslant t$, as an unknown parameter, and transform the smoothing problem into an extended Kalman filtering problem.

Let

$$\mathscr{X}^T(t) \triangleq [x^T(t) \quad x^T(\tau)] \tag{6.49}$$

Then we can write

$$\frac{d\mathscr{X}(t)}{dt} = \mathscr{A}(t)\mathscr{X}(t) + \mathscr{B}(t)w(t) \tag{6.50}$$

$$z(t) = \mathscr{H}(t)\mathscr{X}(t) + v(t) \tag{6.51}$$

where

$$\mathscr{A}(t) \triangleq \begin{bmatrix} A(t) & 0 \\ 0 & 0 \end{bmatrix}, \qquad \mathscr{B}(t) \triangleq \begin{bmatrix} B(t) \\ 0 \end{bmatrix} \tag{6.52}$$

$$\mathscr{H}(t) \triangleq [H(t) \quad 0] \tag{6.53}$$

For the problem of the least-squares estimate of equations (6.49)–(6.53), the Hamiltonian equations are obtained as follows:

$$\frac{d}{ds}\begin{bmatrix} \hat{\mathscr{X}}(s|t) \\ -\Lambda(s|t) \end{bmatrix} = \begin{bmatrix} \mathscr{A}(s) & \mathscr{B}(s)\mathscr{Q}(s)\mathscr{B}^T(s) \\ -\mathscr{H}^T(s)R^{-1}(s)\mathscr{H}(s) & \mathscr{A}^T(s) \end{bmatrix}\begin{bmatrix} \hat{\mathscr{X}}(s|t) \\ \Lambda(s|t) \end{bmatrix} + \begin{bmatrix} 0 \\ \mathscr{H}^T(s)R^{-1}(s)z(s) \end{bmatrix} \tag{6.54}$$

with boundary conditions

$$\hat{\mathscr{X}}(\tau|t) = \hat{\mathscr{X}}(\tau|\tau) + \mathscr{P}(\tau)\Lambda(\tau|t), \qquad \Lambda(t|t) = 0, \quad \tau \leqslant s \leqslant t \tag{6.55}$$

Now assume that the moments of $\mathscr{X}(\tau)$ are partitioned as follows:

$$\hat{\mathscr{X}}(\tau|\tau) = \hat{\mathscr{X}}_n(\tau|\tau) + \hat{\mathscr{X}}_r(\tau|\tau) \tag{6.56a}$$

$$\mathcal{P}(\tau) = \mathcal{P}_n(\tau) + \mathcal{P}_r(\tau) + \mathcal{P}_{nr}(\tau) + \mathcal{P}_{nr}^T(\tau) \tag{6.56b}$$

where

$$\hat{\mathscr{X}}_n(\tau|\tau) = [\hat{x}_n^T(\tau|\tau) \quad \hat{x}_n^T(\tau|\tau)]^T, \qquad \hat{\mathscr{X}}_r(\tau|\tau) = [\hat{x}_r^T(\tau|\tau) \quad \hat{x}_r^T(\tau|\tau)]$$

$$\mathcal{P}_n(\tau) = \begin{bmatrix} P_n(\tau) & P_n(\tau) \\ P_n(\tau) & P_n(\tau) \end{bmatrix}, \qquad \mathcal{P}_r(\tau) = \begin{bmatrix} P_r(\tau) & P_r(\tau) \\ P_r(\tau) & P_r(\tau) \end{bmatrix}$$

$$\mathcal{P}_{nr}(\tau) = \begin{bmatrix} P_{nr}(\tau) & P_{nr}(\tau) \\ P_{nr}(\tau) & P_{nr}(\tau) \end{bmatrix}$$

Furthermore, it is assumed that

$$\hat{\xi}_n(\tau) \triangleq \hat{\mathscr{X}}_n(\tau|\tau) - \mathcal{P}_{nr}(\tau)\mathcal{P}_r^{-1}(\tau)\hat{\mathscr{X}}_r(\tau|\tau) \tag{6.57a}$$

$$\tilde{\Pi}_n(\tau) \triangleq \mathcal{P}_n(\tau) - \mathcal{P}_{nr}(\tau)\mathcal{P}_r^{-1}(\tau)\mathcal{P}_{nr}^T(\tau) \tag{6.57b}$$

$$\hat{\xi}_r(\tau) \triangleq [I + \mathcal{P}_{nr}(\tau)\mathcal{P}_r^{-1}]\hat{\mathscr{X}}_r(\tau|\tau) \tag{6.58a}$$

$$\tilde{\Pi}_r(\tau) \triangleq [I + \mathcal{P}_{nr}(\tau)\mathcal{P}_r^{-1}(\tau)]\mathcal{P}_r(\tau)[I + \mathcal{P}_{nr}(\tau)\mathcal{P}_r^{-1}(\tau)]^T \tag{6.58b}$$

Note that since

$$\mathcal{P}_r(\tau) = \begin{bmatrix} P_r(\tau) & P_r(\tau) \\ P_r(\tau) & P_r(\tau) \end{bmatrix},$$

$\mathcal{P}_r^{-1}(\tau)$ does not exist. However, $\mathcal{P}_{nr}(\tau)\mathcal{P}_r^{-1}(\tau)$ can then be replaced by

$$\mathcal{P}_{nr}(\tau)\mathcal{P}_r^{-1}(\tau) = \begin{bmatrix} P_{nr}(\tau)P_r^{-1}(\tau) & 0 \\ 0 & P_{nr}(\tau)P_r^{-1}(\tau) \end{bmatrix} \tag{6.59}$$

because

$$\mathcal{P}_{nr}(\tau) = \begin{bmatrix} P_{nr}(\tau)P_r^{-1}(\tau) & 0 \\ 0 & P_{nr}(\tau)P_r^{-1}(\tau) \end{bmatrix}\mathcal{P}_r(\tau)$$

$$= \begin{bmatrix} P_{nr}(\tau) & P_{nr}(\tau) \\ P_{nr}(\tau) & P_{nr}(\tau) \end{bmatrix} \tag{6.60}$$

If the nominal scattering matrix for equations (6.54) and (6.55) is defined by

$$\mathcal{S}_n(\tau, t) = \begin{bmatrix} \tilde{\Phi}_n(t, \tau) & \mathcal{P}_n(t, \tau) \\ -\mathcal{W}_n(t, \tau) & \tilde{\Phi}_n^T(t, \tau) \end{bmatrix} \tag{6.61}$$

then the composite scattering matrix for the same equations (6.54) and (6.55), $\mathcal{S}(\tau, t)$, can be expressed as

$$\mathcal{S}(\tau, t) = \begin{bmatrix} \widetilde{\Phi}(t, \tau) & \mathcal{P}(t, \tau) \\ -\mathcal{W}(t, \tau) & \widetilde{\Phi}^{\mathrm{T}}(t, \tau) \end{bmatrix}$$

$$= \begin{bmatrix} \widetilde{\Phi}_{\mathrm{n}}(t, \tau)\{I + \widetilde{\Pi}_{\mathrm{r}}(\tau)\mathcal{W}_{\mathrm{n}}(t, \tau)\}^{-1} & \mathcal{P}_{\mathrm{n}}(t, \tau) + \widetilde{\Phi}_{\mathrm{n}}(t, \tau)\widetilde{\Pi}_{\mathrm{r}}(\tau)\{I + \mathcal{W}_{\mathrm{n}}(t, \tau)\widetilde{\Pi}_{\mathrm{r}}(\tau)\}^{-1}\widetilde{\Phi}_{\mathrm{n}}^{\mathrm{T}}(t, \tau) \\ -\mathcal{W}_{\mathrm{n}}(t, \tau)\{I + \widetilde{\Pi}_{\mathrm{r}}(\tau)\mathcal{W}_{\mathrm{n}}(t, \tau)\}^{-1} & \{I + \mathcal{W}_{\mathrm{n}}(t, \tau)\widetilde{\Pi}_{\mathrm{r}}(\tau)\}^{-1}\widetilde{\Phi}_{\mathrm{n}}^{\mathrm{T}}(t, \tau) \end{bmatrix} \tag{6.62}$$

Additionally, the corresponding internal source vector to (6.62) reduces to

$$\begin{bmatrix} \hat{\mathcal{X}}(t|t) \\ \Lambda(\tau|t) \end{bmatrix} = \begin{bmatrix} \hat{\mathcal{X}}_{\mathrm{n}}(t|t) + \widetilde{\Phi}_{\mathrm{n}}(t, \tau)\hat{\mathcal{X}}_{\mathrm{r}}(\tau|t) \\ \{I + \mathcal{W}_{\mathrm{n}}(t, \tau)\widetilde{\Pi}_{\mathrm{r}}(\tau)\}^{-1}\{\Lambda_{\mathrm{n}}(\tau|t) - \mathcal{W}_{\mathrm{n}}(t, \tau)\hat{\xi}_{\mathrm{r}}(\tau)\} \end{bmatrix} \tag{6.63}$$

where

$$\hat{\mathcal{X}}_{\mathrm{r}}(\tau|t) \triangleq [I + \widetilde{\Pi}_{\mathrm{r}}(\tau)\mathcal{W}_{\mathrm{n}}(t, \tau)]^{-1}[\hat{\xi}_{\mathrm{r}}(\tau) + \widetilde{\Pi}_{\mathrm{r}}(\tau)\Lambda_{\mathrm{n}}(\tau|t)] \tag{6.64}$$

in which

$$\begin{bmatrix} \hat{\mathcal{X}}_{\mathrm{n}}(t|t) \\ \Lambda(\tau|t) \end{bmatrix}$$

denotes the nominal source vector associated with equation (6.61). Furthermore, the forward evolution equations for $\mathcal{S}_{\mathrm{n}}(\tau, t)$ and

$$\begin{bmatrix} \hat{\mathcal{X}}_{\mathrm{n}}(t|t) \\ \Lambda_{\mathrm{n}}(\tau|t) \end{bmatrix}$$

can also be derived, from the results of the previous section, as follows:

$$\frac{\mathrm{d}\mathcal{S}_{\mathrm{n}}(\tau, t)}{\mathrm{d}t} = \begin{bmatrix} \mathrm{d}\widetilde{\Phi}_{\mathrm{n}}(t, \tau)/\mathrm{d}t & \mathrm{d}\mathcal{P}_{\mathrm{n}}(t, \tau)/\mathrm{d}t \\ -\mathrm{d}\mathcal{W}_{\mathrm{n}}(t, \tau)/\mathrm{d}t & \mathrm{d}\widetilde{\Phi}_{\mathrm{n}}^{\mathrm{T}}(t, \tau)/\mathrm{d}t \end{bmatrix}$$

$$= \begin{bmatrix} \{\mathcal{A}(t) - \mathcal{P}_{\mathrm{n}}(t, \tau)\mathcal{H}^{\mathrm{T}}(t) \times R^{-1}(t)\mathcal{H}(t)\}\widetilde{\Phi}_{\mathrm{n}}(t, \tau) \\ -\widetilde{\Phi}_{\mathrm{n}}^{\mathrm{T}}(t, \tau)\mathcal{H}^{\mathrm{T}}(t)R^{-1}(t)\mathcal{H}(t)\widetilde{\Phi}_{\mathrm{n}}(t, \tau) \end{bmatrix}$$

$$\left.\begin{matrix} \mathcal{A}(t)\mathcal{P}_{\mathrm{n}}(t,\tau) + \mathcal{P}_{\mathrm{n}}(t,\tau)\mathcal{A}^{\mathrm{T}}(t) \\ \quad + \mathcal{B}(t)Q(t)\mathcal{B}^{\mathrm{T}}(t) \\ \quad - \mathcal{P}_{\mathrm{n}}(t,\tau)\mathcal{H}^{\mathrm{T}}(t)R^{-1}(t)\mathcal{H}(t)\mathcal{P}_{\mathrm{n}}(t,\tau) \\ \widetilde{\Phi}_{\mathrm{n}}^{\mathrm{T}}(t,\tau)\{\mathcal{A}(t) - \mathcal{P}_{\mathrm{n}}(t,\tau)\mathcal{H}^{\mathrm{T}}(t) \\ \quad \times R^{-1}(t)\mathcal{H}(t)\}^{\mathrm{T}} \end{matrix}\right],$$

$$\mathcal{S}_{\mathrm{n}}(\tau,\tau) = \begin{bmatrix} I & \widetilde{\Pi}_{\mathrm{n}}(\tau) \\ 0 & I \end{bmatrix} \tag{6.65}$$

and

$$\begin{bmatrix} \mathrm{d}\hat{\mathcal{X}}_{\mathrm{n}}(t|t)/\mathrm{d}t \\ \mathrm{d}\Lambda_{\mathrm{n}}(\tau|t)/\mathrm{d}t \end{bmatrix}$$

$$= \begin{bmatrix} \{\mathcal{A}(t) - \mathcal{P}_{\mathrm{n}}(t,\tau)\mathcal{H}^{\mathrm{T}}(t)R^{-1}(t)\mathcal{H}(t)\}\hat{\mathcal{X}}_{\mathrm{n}}(t|t) \\ \quad + \mathcal{P}_{\mathrm{n}}(t,\tau)\mathcal{H}^{\mathrm{T}}(t)R^{-1}(t)z(t) \\ \widetilde{\Phi}_{\mathrm{n}}^{\mathrm{T}}(t,\tau)\mathcal{H}^{\mathrm{T}}(t)R^{-1}(t)\{z(t) - \mathcal{H}(t)\hat{\mathcal{X}}_{\mathrm{n}}(t|t)\} \end{bmatrix},$$

$$\begin{bmatrix} \hat{\mathcal{X}}_{\mathrm{n}}(\tau|\tau) \\ \Lambda_{\mathrm{n}}(\tau|\tau) \end{bmatrix} = \begin{bmatrix} \hat{\xi}_{\mathrm{n}}(\tau) \\ 0 \end{bmatrix} \tag{6.66}$$

The following lemma is then valid:

Lemma 6.1

The nominal right reflection operator $\mathcal{P}_{\mathrm{n}}(t,\tau)$ for the augmented system given by expressions (6.49)–(6.53) is obtained by

$$\mathcal{P}_{\mathrm{n}}(t,\tau) = \begin{bmatrix} P_{\mathrm{n}}(t,\tau) & \Phi_{\mathrm{n}}(t,\tau)\Pi_{\mathrm{n}}(\tau) \\ \Pi_{\mathrm{n}}(\tau)\Phi_{\mathrm{n}}^{\mathrm{T}}(t,\tau) & \Pi_{\mathrm{n}}(\tau) - \Pi_{\mathrm{n}}(\tau)W_{\mathrm{n}}(t,\tau)\Pi_{\mathrm{n}}(\tau) \end{bmatrix} \tag{6.67}$$

and the corresponding forward transmission operator $\widetilde{\Phi}_{\mathrm{n}}(t,\tau)$ is given by

$$\widetilde{\Phi}_{\mathrm{n}}(t,\tau) = \begin{bmatrix} \Phi_{\mathrm{n}}(t,\tau) & 0 \\ -\Pi_{\mathrm{n}}(\tau)W_{\mathrm{n}}(t,\tau) & I \end{bmatrix} \tag{6.68}$$

Proof

From the (1,2) entry of the matrix in equation (6.65), if we define

$$\mathcal{P}_n(t, \tau) \triangleq \begin{bmatrix} P_{11n}(t, \tau) & P_{12n}(t, \tau) \\ P_{12n}^T(t, \tau) & P_{22n}(t, \tau) \end{bmatrix} \tag{6.69}$$

then each element becomes

$$\frac{dP_{11n}(t, \tau)}{dt} = A(t)P_{11n}(t, \tau) + P_{11n}(t, \tau)A^T(t) + B(t)Q(t)B^T(t) - P_{11n}(t, \tau)H^T(t)R^{-1}(t)H(t)P_{11n}(t, \tau),$$

$$P_{11n}(\tau, \tau) = \Pi_n(\tau) \tag{6.70}$$

$$\frac{dP_{12n}(t, \tau)}{dt} = [A(t) - P_{11n}(t, \tau)H^T(t)R^{-1}(t)H(t)]P_{12n}(t, \tau),$$

$$P_{12n}(\tau, \tau) = \Pi_n(\tau) \tag{6.71}$$

$$\frac{dP_{22n}(t, \tau)}{dt} = -P_{12n}^T(t, \tau)H^T(t)R^{-1}(t)H(t)P_{12n}(t, \tau),$$

$$P_{22}(\tau, \tau) = \Pi_n(\tau) \tag{6.72}$$

Comparing the (1, 2) and (1, 1) matrix entries of equation (6.38) with equations (6.70) and (6.71), we have

$$P_{11n}(t, \tau) = P_n(t, \tau) \tag{6.73}$$

$$P_{12n}(t, \tau) = \Phi_n(t, \tau)\Pi_n(\tau) \tag{6.74}$$

Therefore from equation (6.72)

$$P_{22n}(t, \tau) = \Pi_n(\tau) - \Pi_n(\tau)W_n(t, \tau)\Pi_n(\tau) \tag{6.75}$$

Hence, equation (6.67) is apparent. Furthermore, utilizing (6.67) in the (1,1) matrix entry of equation (6.65) gives the desired result (6.68). □

Theorem 6.4

The augmented estimate $\hat{\mathcal{X}}(t|t)$ and its right reflection operator $\mathcal{P}(t, \tau)$ are obtained as follows:

$$\hat{\mathcal{X}}(t|t) = \begin{bmatrix} \hat{x}(t|t) \\ \hat{x}(\tau|t) \end{bmatrix} = \begin{bmatrix} \hat{x}_n(t|t) + \Phi_n(t, \tau)\hat{x}_r(\tau|t) \\ \hat{x}_n(\tau|t) + [I - \Pi_n(\tau)W_n(t, \tau)]\hat{x}_r(\tau|t) \end{bmatrix} \tag{6.76}$$

and

$$\mathcal{P}(t, \tau) = \begin{bmatrix} P_n(t, \tau) + \Phi_n(t, \tau)P_r(\tau|t)\Phi_n^T(t, \tau) & \Phi_n(t, \tau)\Pi_n(\tau) + \Phi_n(t, \tau)P_r(\tau|t) \times [I - \Pi_n(\tau)W_n(t, \tau)] \\ \Pi_n(\tau)\Phi_n(t, \tau) + [I - \Pi_n(\tau) \times W_n(t, \tau)]P_r(\tau|t)\Phi_n^T(t, \tau) & \Pi_n(\tau) - \Pi_n(\tau)W_n(t, \tau)\Pi_n(\tau) + [I - \Pi_n(\tau)W_n(t, \tau)]P_r(\tau|t)[I - \Pi_n(\tau)W_n(t, \tau)]^T \end{bmatrix} \tag{6.77}$$

where

$$\begin{aligned} \hat{x}_n(\tau|t) &= \hat{m}_n(\tau) + \Pi_n(\tau)\lambda_n(\tau|t) \\ &= \hat{m}_n(\tau) + \Pi_n(\tau) \int_\tau^t \Phi_n(s, \tau)H^T(s)R^{-1}(s) \\ &\quad \times [z(s) - H(s)\hat{x}_n(s|s)]\,\mathrm{d}s \end{aligned} \tag{6.78}$$

and

$$\begin{aligned} P_r(\tau|t) &\triangleq \Pi_r(\tau)[I + W_n(t, \tau)\Pi_r(\tau)]^{-1} \\ &= [\Pi_r^{-1}(\tau) + W_n(t, \tau)]^{-1} \end{aligned} \tag{6.79}$$

Proof

Taking account of equation (6.68), from the (2,1) matrix equation entry of (6.65),

$$\mathcal{W}_n(t, \tau) = \begin{bmatrix} W_n(t, \tau) & 0 \\ 0 & 0 \end{bmatrix} \tag{6.80}$$

and from equation (6.66),

$$\Lambda_n(\tau|t) = \begin{bmatrix} \lambda_n(\tau|t) \\ 0 \end{bmatrix} \tag{6.81}$$

Substituting equations (6.80) and (6.81) into (6.64), we obtain

$$\hat{\mathscr{X}}_{\mathrm{r}}(\tau|t) = \begin{bmatrix} \hat{x}_{\mathrm{r}}(\tau|t) \\ \hat{x}_{\mathrm{r}}(\tau|t) \end{bmatrix} \tag{6.82}$$

Hence, from equations (6.63), (6.68) and (6.82), we may obtain the desired result (6.76). Similarly, from the (1,2) matrix equation entry of (6.62), we also get equation (6.77). In addition, equation (6.78) follows from the upper identity of equation (6.66). □

Equation (6.78) and the (2,2) entry of equation (6.67) constitute the so-called Kailath–Frost innovation smoother [18]. It may be noted that these partitioning filter and smoother algorithms based on Redheffer's scattering theory are closely related to those based on the perturbed Kalman filter equations [19, 20]. Additionally, the three kinds of smoothers—the fixed-interval, fixed-point and fixed-lag partitioning smoothers—are readily obtained from these results (see Watanabe [20]). In the remaining part of this section, we shall present the partitioning predictor based on such equations.

For some fixed time $t_1 \geqslant \tau$, consider the problem of determining the optimal predicted value $\hat{x}(t|t_1) \triangleq E[x(t)|Z_{t_1}]$, $t \geqslant t_1$, in a partitioned form. As in the previous filtering or smoothing problems, it is assumed that at time t_1 the optimal filtered estimate $\hat{x}(t_1|t_1)$ and its error covariance matrix $P(t_1)$ are, respectively, decomposed as follows:

$$\hat{x}(t_1|t_1) = \hat{x}_{\mathrm{n}}(t_1|t_1) + \hat{x}_{\mathrm{r}}(t_1|t_1) \tag{6.83a}$$

$$P(t_1) = P_{\mathrm{n}}(t_1) + P_{\mathrm{r}}(t_1) + P_{\mathrm{nr}}(t_1) + P_{\mathrm{nr}}^{\mathrm{T}}(t_1) \tag{6.83b}$$

and that

$$\hat{m}_{\mathrm{n}}(t_1) \triangleq \hat{x}_{\mathrm{n}}(t_1|t_1) - P_{\mathrm{nr}}(t_1)P_{\mathrm{r}}^{-1}(t_1)\hat{x}_{\mathrm{r}}(t_1|t_1) \tag{6.84a}$$

$$\Pi_{\mathrm{n}}(t_1) \triangleq P_{\mathrm{n}}(t_1) - P_{\mathrm{nr}}(t_1)P_{\mathrm{r}}^{-1}(t_1)P_{\mathrm{nr}}^{\mathrm{T}}(t_1) \tag{6.84b}$$

$$\hat{m}_{\mathrm{n}}(t_1) \triangleq [I + P_{\mathrm{nr}}(t_1)P_{\mathrm{r}}^{-1}(t_1)]\hat{x}_{\mathrm{r}}(t_1|t_1) \tag{6.85a}$$

$$\Pi_{\mathrm{r}}(t_1) \triangleq [I + P_{\mathrm{nr}}(t_1)P_{\mathrm{r}}^{-1}(t_1)]P_{\mathrm{r}}(t_1)[I + P_{\mathrm{nr}}(t_1)P_{\mathrm{r}}^{-1}(t_1)]^{\mathrm{T}} \tag{6.85b}$$

With these preliminaries completed, we state and prove the following result:

Theorem 6.5

The optimal predicted estimate $\hat{x}(t|t_1)$ and the associated error covariance matrix $P(t)$, $t \geqslant t_1$, are partitioned as follows:

$$\hat{x}(t|t_1) = \hat{x}_n(t|t_1) + \Phi(t, t_1)\hat{m}_r(t_1) \tag{6.86}$$

$$P(t|t_1) = P_n(t) + \Phi(t, t_1)\Pi_r(t_1)\Phi^T(t, t_1) \tag{6.87}$$

where $\Phi(t, t_1)$ denotes the state transition matrix for $A(t)$, and

$$\frac{d\hat{x}_n(t|t_1)}{dt} = A(t)\hat{x}_n(t|t_1), \qquad \hat{x}_n(t_1|t_1) = \hat{m}_n(t_1) \tag{6.88}$$

$$\frac{dP_n(t)}{dt} = A(t)P_n(t) + P_n(t)A^T(t) + B(t)Q(t)B^T(t), \qquad P_n(t_1) = \Pi_n(t_1) \tag{6.89}$$

Proof

The optimal predictor, with initial conditions given by equations (6.83), is obtained from [21]

$$\frac{d\hat{x}(t|t_1)}{dt} = A(t)\hat{x}(t|t_1), \qquad \hat{x}(t_1|t_1) = \hat{m}_n(t_1) + \hat{m}_r(t_1) \tag{6.90}$$

with error covariance matrix $P(t) \triangleq E\{[x(t) - \hat{x}(t|t_1)][x(t) - \hat{x}(t|t_1)]^T | Z_{t_1}\}$ as follows:

$$\frac{dP(t)}{dt} = A(t)P(t) + P(t)A^T(t) + B(t)Q(t)B^T(t), \qquad P(t_1) = P_n(t_1) + P_r(t_1) + P_{nr}(t_1) + P_{nr}^T(t_1) \tag{6.91}$$

Similarly, we have the nominal predictor given by equations (6.88) and (6.89). With equations (6.88)–(6.91), it is found that the differential equations of $\delta\hat{x}(t|t_1) \triangleq \hat{x}(t|t_1) - \hat{x}_n(t|t_1)$ and $\delta P(t) \triangleq P(t) - P_n(t)$ are subject to the following:

$$\frac{d\delta\hat{x}(t|t_1)}{dt} = A(t)\delta\hat{x}(t|t_1), \qquad \delta\hat{x}(t_1|t_1) = \hat{m}_r(t_1) \tag{6.92}$$

$$\frac{d\delta P(t)}{dt} = A(t)\delta P(t) + \delta P(t)A^T(t), \qquad \delta P(t_1) = \Pi_r(t_1) \tag{6.93}$$

Therefore, we obtain the following solutions to equations (6.92) and (6.93):

$$\delta\hat{x}(t|t_1) = \Phi(t, t_1)\hat{m}_r(t_1) \tag{6.94}$$

$$\delta P(t) = \Phi(t, t_1)\Pi_r(t_1)\Phi^T(t, t_1) \tag{6.95}$$

These complete the proof. □

6.5 THE BACKWARD PARTITIONING FILTER

Since in the backward filtered estimate $\hat{x}(t|t; s)$ of $x(s)$ the data sequence is given as $\{z(s'), s' \in (t, s)\}$, $\tau \leqslant s \leqslant t$, the only information given in the interval (τ, s) is summarized in the a priori information, $\{\hat{x}(\tau|\tau), P(\tau)\}$. That is, the forward estimate at time s is given by the optimal predictor:

$$\frac{d\hat{x}(s; \tau)}{ds} = A(s)\hat{x}(s; \tau), \qquad \hat{x}(\tau; \tau) = \hat{x}(\tau|\tau) \tag{6.96}$$

$$\frac{dP(s; \tau)}{ds} = A(s)P(s; \tau) + P(s; \tau)A^{\mathrm{T}}(s) + B(s)Q(s)B^{\mathrm{T}}(s), \qquad P(\tau; \tau) = P(\tau) \tag{6.97}$$

Now suppose that $\hat{x}(s; \tau)$ and $P(s; \tau)$ are decomposed into

$$\hat{x}(s; \tau) = \hat{x}_{\mathrm{n}}(s; \tau) + \hat{x}_{\mathrm{r}}(s; \tau) \tag{6.98a}$$

$$P(s; \tau) = P_{\mathrm{n}}(s; \tau) + P_{\mathrm{r}}(s; \tau) + P_{\mathrm{nr}}(s; \tau) + P_{\mathrm{nr}}^{\mathrm{T}}(s; \tau) \tag{6.98b}$$

and define

$$\hat{m}_{\mathrm{n}}(s; \tau) \triangleq \hat{x}_{\mathrm{n}}(s; \tau) - P_{\mathrm{nr}}(s; \tau)P_{\mathrm{r}}^{-1}(s; \tau)\hat{x}_{\mathrm{r}}(s; \tau) \tag{6.99a}$$

$$\Pi_{\mathrm{n}}(s; \tau) \triangleq P_{\mathrm{n}}(s; \tau) - P_{\mathrm{nr}}(s; \tau)P_{\mathrm{r}}^{-1}(s; \tau)P_{\mathrm{nr}}^{-1}(s; \tau) \tag{6.99b}$$

$$\hat{m}_{\mathrm{n}}(s; \tau) \triangleq [I + P_{\mathrm{nr}}(s; \tau)P_{\mathrm{r}}^{-1}(s; \tau)]\hat{x}_{\mathrm{r}}(s; \tau) \tag{6.100a}$$

$$\Pi_{\mathrm{r}}(s; \tau) \triangleq [I + P_{\mathrm{nr}}(s; \tau)P_{\mathrm{r}}^{-1}(s; \tau)]P_{\mathrm{r}}(s; \tau) \times [I + P_{\mathrm{nr}}(s; \tau)P_{\mathrm{r}}^{-1}(s; \tau)]^{\mathrm{T}} \tag{6.100b}$$

Introducing the fictitious initial layer,

$$\bar{\Pi}(s; \tau) = \begin{bmatrix} I & \Pi_{\mathrm{n}}(s; \tau) + \Pi_{\mathrm{r}}(s; \tau) \\ 0 & I \end{bmatrix} \tag{6.101}$$

and decomposing it, we find that

$$S(s, t) \triangleq \begin{bmatrix} \Phi(t, s) & P(t, s) \\ -W(t, s) & \Phi^{\mathrm{T}}(t, s) \end{bmatrix} = \bar{\Pi}_{\mathrm{r}}(s; \tau) * S_{\mathrm{n}}(s, t) \tag{6.102}$$

where

$$\bar{\Pi}_r(s;\tau) = \begin{bmatrix} I & \Pi_r(s;\tau) \\ 0 & I \end{bmatrix} \tag{6.103}$$

$$S_n(s;\tau) \triangleq \bar{\Pi}_n(s;\tau) * S_0(s,t) \equiv \begin{bmatrix} \Phi_n(t,s) & P_n(t,s) \\ -W_n(t,s) & \Phi_n^T(t,s) \end{bmatrix} \tag{6.104}$$

$$\bar{\Pi}_n(s;\tau) = \begin{bmatrix} I & \Pi_n(s;\tau) \\ 0 & I \end{bmatrix} \tag{6.105}$$

Then, from expression (6.102)

$$\begin{bmatrix} \Phi(t,s) & P(t,s) \\ -W(t,s) & \Phi^T(t,s) \end{bmatrix}$$

$$= \begin{bmatrix} \Phi_n(t,s)[I + \Pi_r(s;\tau)W_n(t,s)]^{-1} & P_n(t,s) + \Phi_n(t,s)\Pi_r(s;t)[I + W_n(t,s)\Pi_r(s;\tau)]^{-1}\Phi_n^T(t,s) \\ -W_n(t,s)[I + \Pi_r(s;\tau)W_n(t,s)]^{-1} & [I + W_n(t,s)\Pi_r(s;\tau)]^{-1}\Phi_n(t,s) \end{bmatrix} \tag{6.106}$$

Furthermore, we can see that the source vector corresponding to the scattering matrix of expression (6.102) is given by

$$q(s,t) \triangleq \begin{bmatrix} \hat{x}(t|t;s) \\ \lambda(s|t) \end{bmatrix} = q_r(s,s;\tau) \bullet q_n(s,t) \tag{6.107}$$

where

$$q_r(s,s;\tau) = \begin{bmatrix} \hat{m}_r(s;\tau) \\ 0 \end{bmatrix} \tag{6.108}$$

$$q_n(s,t) = q_n(s,s;\tau) \bullet q_0(s,t) = \begin{bmatrix} \hat{x}_n(t|t,s) \\ \lambda_n(s|t) \end{bmatrix} \tag{6.109}$$

$$q_n(s,s;\tau) = \begin{bmatrix} \hat{m}_n(s;\tau) \\ 0 \end{bmatrix} \tag{6.110}$$

Therefore, by applying the dot sum rule of expression (6.5) to (6.107), we get

$$\begin{bmatrix} \hat{x}(t|t; s) \\ \lambda(s|t) \end{bmatrix} = \begin{bmatrix} \hat{x}_{\mathrm{n}}(t|t; s) + \Phi_{\mathrm{n}}(t, s)\hat{x}_{\mathrm{r}}(s|t) \\ [I + W_{\mathrm{n}}(t, s)\Pi_{\mathrm{r}}(s; \tau)]^{-1}[\lambda_{\mathrm{n}}(s|t) - W_{\mathrm{n}}(t, s)\hat{m}_{\mathrm{r}}(s; \tau)] \end{bmatrix} \tag{6.111}$$

where

$$\begin{aligned} \hat{x}_{\mathrm{r}}(s|t) &= [I + \Pi_{\mathrm{r}}(s; \tau)W_{\mathrm{n}}(t, s)]^{-1}[\hat{m}_{\mathrm{r}}(s; \tau) + \Pi_{\mathrm{r}}(s; \tau)\lambda(s|t)] \\ &= [W_{\mathrm{n}}(t, s) + \Pi_{\mathrm{r}}^{-1}(s; \tau)]^{-1}[\lambda_{\mathrm{n}}(s|t) + \Pi_{\mathrm{r}}^{-1}(s; \tau)\hat{m}_{\mathrm{r}}(s; \tau)] \end{aligned} \tag{6.112}$$

It should be noted that equations (6.106) and (6.111) are the same as the equalities given in Theorems 6.1 and 6.2 but with $\tau = s$.

Theorem 6.6

The nominal scattering matrix $S_{\mathrm{n}}(s, t)$ can be represented by the backward-time (or reverse-time) evolution equation:

$$-\frac{\mathrm{d}S_{\mathrm{n}}(s, t)}{\mathrm{d}s} = \begin{bmatrix} -\mathrm{d}\Phi_{\mathrm{n}}(t, s)/\mathrm{d}s & -\mathrm{d}P_{\mathrm{n}}(t, s)/\mathrm{d}s \\ -\mathrm{d}W_{\mathrm{n}}(t, s)/\mathrm{d}s & -\mathrm{d}\Phi_{\mathrm{n}}^{\mathrm{T}}(t, s)/\mathrm{d}s \end{bmatrix}$$

$$= \begin{bmatrix} \Phi_{\mathrm{n}}(t, s)[A_{\mathrm{b}}(s) - Q_{\mathrm{b}}(s)W_{\mathrm{n}}(t, s)] & \Phi_{\mathrm{n}}(t, s)Q_{\mathrm{b}}(s)\Phi_{\mathrm{n}}^{\mathrm{T}}(t, s) \\ A_{\mathrm{b}}^{\mathrm{T}}(s)W_{\mathrm{n}}(t, s) + W_{\mathrm{n}}(t, s)A_{\mathrm{b}}(s) - W_{\mathrm{n}}(t, s)Q_{\mathrm{b}}(s)W_{\mathrm{n}}(t, s) + H_{\mathrm{b}}(s) & [A_{\mathrm{b}}(s) - Q_{\mathrm{b}}(s)W_{\mathrm{n}}(t, s)]^{\mathrm{T}}\Phi_{\mathrm{n}}^{\mathrm{T}}(t, s) \end{bmatrix},$$

$$S_{\mathrm{n}}(t, t) = \bar{\Pi}_{\mathrm{n}}(t; \tau) \tag{6.113}$$

where

$$A_{\mathrm{b}}(s) = A(s) - \Pi_{\mathrm{n}}(s; \tau)H^{\mathrm{T}}(s)R^{-1}(s)H(s) \tag{6.114}$$

$$\begin{aligned} Q_{\mathrm{b}}(s) = {} & A(s)\Pi_{\mathrm{n}}(s; \tau) + \Pi_{\mathrm{n}}(s; \tau)A^{\mathrm{T}}(s) + B(s)Q(s)B^{\mathrm{T}}(s) \\ & - \Pi_{\mathrm{n}}(s; \tau)H^{\mathrm{T}}(s)R^{-1}(s)H(s)\Pi_{\mathrm{n}}(s; \tau) \end{aligned} \tag{6.115}$$

$$H_{\mathrm{b}}(s) = H^{\mathrm{T}}(s)R^{-1}(s)H(s) \tag{6.116}$$

Proof

The incremental form of the backward-time nominal scattering matrix can be written as follows:

$$\begin{aligned} S_n(s - \Delta s, t) &= \bar{\Pi}_n(s - \Delta s; \tau) * S_0(s - \Delta s, t) \\ &= \bar{\Pi}_n(s - \Delta s; \tau) * S_0(s - \Delta s, s) \\ &\quad * \bar{\Pi}_n^{-1}(s; \tau) * \bar{\Pi}_n(s; \tau) * S_0(s, \tau) \\ &= \bar{\Pi}_n(s - \Delta s; \tau) * S_0(s - \Delta s, s) * \bar{\Pi}_n^{-1}(s; \tau) * S_n(s, t) \end{aligned} \tag{6.117}$$

Then, in terms of the forward-time generator of expressions (6.41) and (6.42),

$$\bar{\Pi}_n(s - \Delta s; \tau) * S_0(s - \Delta s, s) * \bar{\Pi}_n^{-1}(s; \tau)$$

$$= \bar{\Pi}_n(s - \Delta s; \tau) * [I + M(s)\Delta s + o(\Delta s)] * \bar{\Pi}_n^{-1}(s; \tau) \tag{6.118a}$$

$$\triangleq I + \mathcal{M}(s)\Delta s + o(\Delta s) \tag{6.118b}$$

where

$$\mathcal{M}(s) \triangleq \begin{bmatrix} A_b(s) & Q_b(s) \\ -H_b(s) & A_b^T(s) \end{bmatrix} \tag{6.119}$$

Therefore, from equation (6.117)

$$S_n(s - \Delta s, t) = \begin{bmatrix} I + A_b(s)\Delta s & Q_b(s)\Delta s \\ -H_b(s)\Delta s & I + A_b^T(s)\Delta s \end{bmatrix}$$

$$* \begin{bmatrix} \Phi_n(t, s) & P_n(t, s) \\ -W_n(t, s) & \Phi_n^T(t, s) \end{bmatrix} + o(\Delta s)$$

$$= \begin{bmatrix} \Phi_n(t, s)[I + Q_b(s)W_n(t, s)\Delta s]^{-1} \times [I + A_b(s)\Delta s] & P_n(t, s) + \Phi_n(t, s)Q_b(s)\Delta s[I + W_n(t, s)Q_b(s)\Delta s]^{-1}\Phi_n^T(t, s) \\ -H_b(s)\Delta s - [I + A_b^T(s)\Delta s] \times W_n(t, s)[I + Q_b(s)W_n(t, s)\Delta s]^{-1} \times [I + A_b(s)\Delta s] & [I + A_b^T(s)\Delta s][I + W_n(t, s) \times Q_b(s)\Delta s]^{-1}\Phi_n^T(t, s) \end{bmatrix}$$

$$+ o(\Delta s)$$

$$
= \begin{bmatrix} \Phi_n(t,s)[I + A_b(s)\Delta s] - \Phi_n(t,s)Q_b(s)W_n(t,s)\Delta s & P_n(t,s) + \Phi_n(t,s)Q_b(s)\Phi_n^T(t,s)\Delta s \\ -H_b(s)\Delta s - W_n(t,s) + W_n(t,s)Q_b(s)W_n(t,s)\Delta s - A_b^T(s)W_n(t,s)\Delta s - W_n(t,s)A_b(s)\Delta s & [I - W_n(t,s)Q_b(s)\Delta s + A_b^T(s)\Delta s]\Phi_n^T(t,s) \end{bmatrix} + o(\Delta s) \tag{6.120}
$$

Equation (6.120) can be rewritten as

$$
\begin{bmatrix} \Phi_n(t, s-\Delta s) & P_n(t, s-\Delta s) \\ W_n(t, s-\Delta s) & \Phi_n^T(t, s-\Delta s) \end{bmatrix} - \begin{bmatrix} \Phi_n(t,s) & P_n(t,s) \\ W_n(t,s) & \Phi_n^T(t,s) \end{bmatrix}
$$

$$
= \begin{bmatrix} \Phi_n(t,s)[A_b(s) - Q_b(s)W_n(r,s)]\Delta s & \Phi_n(t,s)Q_b(s)\Phi_n^T(t,s)\Delta s \\ [A_b^T(s)W_n(t,s) + W_n(t,s)A_b(s) - W_n(t,s)Q_b(s)W_n(t,s) + H_b(s)]\Delta s & [A_b^T(s) - W_n(t,s)Q_b(s)] \times \Phi_n^T(t,s)\Delta s \end{bmatrix} + o(\Delta s) \tag{6.121}
$$

from which equation (6.113) immediately follows.

The backward-time generator of equations (6.114)–(6.116) can be constructed by directly expanding expression (6.118a). That is,

$$
\begin{bmatrix} I & \Pi_n(s;\tau) \\ 0 & I \end{bmatrix} * \begin{bmatrix} I + A(s)\Delta s & B(s)Q(s)B^T(s)\Delta s \\ -H^T(s)R^{-1}(s)H(s)\Delta s & I + A^T(s)\Delta s \end{bmatrix}
$$

$$
* \begin{bmatrix} I & -\Pi_n(s;\tau) \\ 0 & I \end{bmatrix} + o(\Delta s)
$$

$$
= \begin{bmatrix} [I + A(s)\Delta s][I + \Pi_n(s;\tau) \times H^T(s)R^{-1}(s)H(s)\Delta s]^{-1} \\ -H^T(s)R^{-1}(s)H(s)\Delta s[I + \Pi_n(s;\tau)H^T(s)R^{-1}(s) \times H(s)\Delta s]^{-1} \end{bmatrix}
$$

$$\left.\begin{matrix} B(s)Q(s)B^{\mathrm{T}}(s)\Delta s + [I + A(s)\Delta s]\Pi_{\mathrm{n}}(s;\,\tau) \\ \quad \times\, [I + H^{\mathrm{T}}(s)R^{-1}(s)H(s)\Pi_{\mathrm{n}}(s;\,\tau)\Delta s]^{-1} \\ \quad \times\, [I + A^{\mathrm{T}}(s)\Delta s] \\ [I + H^{\mathrm{T}}(s)R^{-1}(s)H(s)\Pi_{\mathrm{n}}(s;\,\tau)\Delta s]^{-1} \\ \quad \times\, [I + A^{\mathrm{T}}(s)\Delta s] \end{matrix}\right]$$

$$* \begin{bmatrix} I & -\Pi_{\mathrm{n}}(s;\,\tau) \\ 0 & I \end{bmatrix} + o(\Delta s)$$

$$= \begin{bmatrix} I + [A(s) - \Pi_{\mathrm{n}}(s;\,\tau) \\ \quad \times\, H^{\mathrm{T}}(s)R^{-1}(s)H(s)]\Delta s & \begin{matrix} B(s)Q(s)B^{\mathrm{T}}(s)\Delta s + \Pi_{\mathrm{n}}(s;\,\tau) \\ \quad - \Pi_{\mathrm{n}}(s;\,\tau)H^{\mathrm{T}}(s)R^{-1}(s)H(s)\Pi_{\mathrm{n}}(s;\,\tau)\Delta s \\ \quad + A(s)\Pi_{\mathrm{n}}(s;\,\tau)\Delta s + \Pi_{\mathrm{n}}(s;\,\tau)A^{\mathrm{T}}(s)\Delta s \end{matrix} \\ -H^{\mathrm{T}}(s)R^{-1}(s)H(s)\Delta s & \begin{matrix} I + A^{\mathrm{T}}(s)\Delta s \\ \quad - H^{\mathrm{T}}(s)R^{-1}(s)H(s)\Pi_{\mathrm{n}}(s;\,\tau)\Delta s \end{matrix} \end{bmatrix}$$

$$* \begin{bmatrix} I & -\Pi_{\mathrm{n}}(s;\,\tau) \\ 0 & I \end{bmatrix} + o(\Delta s) \tag{6.122a}$$

$$= \begin{bmatrix} I + [A(s) - \Pi_{\mathrm{n}}(s;\,\tau)H^{\mathrm{T}}(s)R^{-1}(s) \\ \quad \times\, H(s)]\Delta s & \begin{matrix} B(s)Q(s)B^{\mathrm{T}}(s)\Delta s - \Pi_{\mathrm{n}}(s;\,\tau)H^{\mathrm{T}}(s) \\ \quad \times\, R^{-1}(s)H(s)\Pi_{\mathrm{n}}(s;\,\tau)\Delta s \\ \quad + A(s)\Pi_{\mathrm{n}}(s;\,\tau)\Delta s + \Pi_{\mathrm{n}}(s;\,\tau)A^{\mathrm{T}}(s)\Delta s \end{matrix} \\ -H^{\mathrm{T}}(s)R^{-1}(s)H(s)\Delta s & \begin{matrix} I + [A(s) - \Pi_{\mathrm{n}}(s;\,\tau)H^{\mathrm{T}}(s)R^{-1}(s) \\ \quad \times\, H(s)]^{\mathrm{T}}\Delta s \end{matrix} \end{bmatrix}$$

$$+\, o(\Delta s) \tag{6.122b}$$

The desired results (6.114)–(6.116) are provided by comparing equations (6.118b) and (6.122b).

Theorem 6.7

The nominal source vector $q_n(s, t)$, corresponding to the scattering matrix $S_n(s, t)$, satisfies the backward-time evolution equation:

$$-\frac{dq_n(s, t)}{ds} = \begin{bmatrix} -d\hat{x}_n(t|t; s)/ds \\ -d\lambda_n(s|t)/ds \end{bmatrix}$$

$$= \begin{bmatrix} \Phi_n(t, s)[\hat{x}_b(s) + Q_b(s)\lambda_n(s|t)] \\ [A_b(s) - Q_b(s)W_n(t, s)]^T\lambda_n(s|t) + \lambda_b(s) - W_n(t, s)\hat{x}_b(s) \end{bmatrix},$$

$$q_n(t|t) = \begin{bmatrix} \hat{m}_n(t; \tau) \\ 0 \end{bmatrix} \tag{6.123}$$

where

$$\hat{x}_b(s) = [A(s) - \Pi_n(s; \tau)H^T(s)R^{-1}(s)H(s)]\hat{m}_n(s; \tau) + \Pi_n(s; \tau)H^T(s)R^{-1}(s)z(s) \tag{6.124}$$

$$\lambda_b(s) = H^T(s)R^{-1}(s)z(s) - H^T(s)R^{-1}(s)H(s)\hat{m}_n(s; \tau) \tag{6.125}$$

Proof

The incremental form of the nominal source vector in backward time can be represented by

$$\begin{aligned} q_n(s - \Delta s, t) &= q_n(s - \Delta s, s - \Delta s)\bullet q_0(s - \Delta s, t) \\ &= q_n(s - \Delta s, s - \Delta s)\bullet q_0(s - \Delta s, s) \\ &\quad \bullet q_n'(s, s)\bullet q_n(s, s)\bullet q_0(s, t) \\ &= q_n(s - \Delta s, s - \Delta s)\bullet q_0(s - \Delta s, s) \\ &\quad \bullet q_n'(s, s)\bullet q_n(s, t) \end{aligned} \tag{6.126}$$

where $q_n'(s, s)$, which is given by

$$q_n'(s, s) = \begin{bmatrix} -\hat{m}_n(s; \tau) \\ 0 \end{bmatrix}$$

denotes the internal source vector associated with the inverse scattering matrix $\Pi_n^{-1}(s; \tau)$. Furthermore, from expressions (6.46) and (6.47) it is easy to derive the following:

$$q_n(s - \Delta s, s - \Delta s)\bullet q_0(s - \Delta s, s)\bullet q_n'(s, s)$$

$$= q_n(s - \Delta s, s - \Delta s)\bullet[F(s)\Delta s + o(\Delta s)]\bullet q'_n(s, s) \tag{6.127a}$$

$$\triangleq F_b(s)\Delta s + o(\Delta s) \tag{6.127b}$$

where

$$F_b(s) \triangleq \begin{bmatrix} \hat{x}_b(s) \\ \lambda_b(s) \end{bmatrix} \tag{6.128}$$

It can also be seen that

$$q_n(s - \Delta s, s - \Delta s)\bullet F(s)\Delta s =$$

$$\begin{bmatrix} \hat{m}_n(s;\tau) \\ 0 \end{bmatrix} \bullet \begin{bmatrix} 0 \\ H^T(s)R^{-1}(s)z(s)\Delta s \end{bmatrix} = \begin{bmatrix} 0 \\ 0 \end{bmatrix}$$

$$+ \begin{bmatrix} [I + A(s)\Delta s][I + \Pi_n(s;\tau)H^T(s)R^{-1}(s)H(s)\Delta s]^{-1} \\ \quad \times [\hat{m}_n(s;\tau) + \Pi_n(s;\tau)H^T(s)R^{-1}(s)z(s)\Delta s] \\ [I + H^T(s)R^{-1}(s)H(s)\Pi_n(s;\tau)\Delta s]^{-1} \\ \quad \times [H^T(s)R^{-1}(s)z(s)\Delta s - H^T(s)R^{-1}(s)H(s)\hat{m}_n(s;\tau)\Delta s \end{bmatrix}$$

$$= \begin{bmatrix} [I + A(s)\Delta s - \Pi_n(s;\tau)H^T(s)R^{-1}(s)H(s)\Delta s]\hat{m}_n(s;\tau) \\ \quad + \Pi_n(s;\tau)H^T(s)R^{-1}(s)z(s)\Delta s \\ H^T(s)R^{-1}(s)z(s)\Delta s \\ \quad - H(s)R^{-1}(s)H(s)\hat{m}_n(s;\tau)\Delta s \end{bmatrix}$$

$$+ o(\Delta s) \tag{6.129}$$

Using equations (6.122a) and (6.129), and applying the dot sum rule of expression (6.5), it can be proved that

$$\begin{bmatrix} \hat{m}_n(s;\tau) \\ 0 \end{bmatrix} \bullet \begin{bmatrix} 0 \\ H^T(s)R^{-1}(s)z(s)\Delta s \end{bmatrix} \bullet \begin{bmatrix} -\hat{m}_n(s;\tau) \\ 0 \end{bmatrix} + o(\Delta s)$$

$$= \begin{bmatrix} -\hat{m}_n(s;\tau) \\ H^T(s)R^{-1}(s)z(s)\Delta s - H^T(s)R^{-1}(s)H(s)\hat{m}_n(s;\tau)\Delta s \end{bmatrix}$$

$$+ \begin{bmatrix} [I + A(s)\Delta s - \Pi_n(s;\tau)H^T(s)R^{-1}(s)H(s)\Delta s]\hat{m}_n(s;\tau) \\ \quad + \Pi_n(s;\tau)H^T(s)R^{-1}(s)z(s)\Delta s \\ 0 \end{bmatrix}$$

$$+ o(\Delta s) \tag{6.130}$$

Now, comparing (6.127b) with (6.130) yields equations (6.124) and (6.125).

In addition, by straightforwardly applying the dot sum rule of expression (6.5) to (6.126), we can see that

$$
\begin{aligned}
q_n(s - \Delta s, t) &= [F_b(s)\Delta s]\bullet q_n(s, t) + o(\Delta s) \\
&= \begin{bmatrix} \hat{x}_b(s)\Delta s \\ \lambda_b(s)\Delta s \end{bmatrix} \bullet \begin{bmatrix} \hat{x}_n(t|t;\, s) \\ \lambda_n(s|t) \end{bmatrix} + o(\Delta s) \\
&= \begin{bmatrix} \hat{x}_n(t|t;\, s) \\ \lambda_b(s)\Delta s \end{bmatrix} + \begin{bmatrix} \Phi_n(t, s)[I + Q_b(s)W_n(t, s)\Delta s]^{-1} \\ \times\, [\hat{x}_b(s)\Delta s + Q_b(s)\lambda_n(s|t)\Delta s] \\ [I + A_b^{\mathrm{T}}(s)\Delta s][I + W_n(t, s)Q_b(s)\Delta s]^{-1} \\ \times\, [\lambda_n(s|t) - W_n(t, s)\hat{x}_b(s)\Delta s] \end{bmatrix} \\
&\quad + o(\Delta s) \\
&= \begin{bmatrix} \hat{x}_n(t|t;\, s) \\ \lambda_b(s)\Delta s \end{bmatrix} + \begin{bmatrix} \Phi_n(t, s)[\hat{x}_b(s)\Delta s + Q_b(s)\lambda_n(s|t)\Delta s] \\ [I - W_n(t, s)Q_b(s)\Delta s + A_b^{\mathrm{T}}(s)\Delta s] \\ \times\, \lambda_n(s|t) - W_n(t, s)\hat{x}_b(s)\Delta s \end{bmatrix} + o(\Delta s) \\
&= \begin{bmatrix} \hat{x}_n(t|t;\, s) \\ \lambda_n(s|t) \end{bmatrix} + \begin{bmatrix} \Phi_n(t, s)[\hat{x}_b(s) + Q_b(s)\lambda_n(s|t)]\Delta s \\ [A_b^{\mathrm{T}}(s) - W_n(t, s)Q_b(s)]\lambda_n(s|t)\Delta s \\ +\, \lambda_b(s)\Delta s - W_n(t, s)\hat{x}_b(s)\Delta s \end{bmatrix} + o(\Delta s)
\end{aligned} \tag{6.131}
$$

Equation (6.131) gives the desired result (6.123) □

If in Theorems 6.6 and 6.7, $\hat{x}_r(s;\tau) = 0$, $P_r(s;\tau) = P_{nr}(s;\tau) = 0$, then we have the following corollary:

Corollary 6.1

The filtered estimate $\hat{x}(t|t;s)$ and the corresponding error covariance matrix $P(t, s)$ satisfy the backward-time evolution equations:

$$
-\frac{d\hat{x}(t|t;\, s)}{ds} = \Phi(t, s)[\hat{x}_b(s) + Q_b(s)\lambda(s|t)], \qquad \hat{x}(t|t;\, t) = \hat{x}(t;\, \tau) \tag{6.132}
$$

$$
-\frac{dP(t, s)}{ds} = \Phi(t, s)Q_b(s)\Phi^{\mathrm{T}}(t, s), \qquad P(t, t) = P(t;\, \tau) \tag{6.133}
$$

where

$$
-\frac{d\lambda(s|t)}{ds} = [A_b(s) - Q_b(s)W(t, s)]^{\mathrm{T}}\lambda(s|t) + \lambda_b(s) - W(t, s)\hat{x}_b(s), \qquad \lambda(t|t) = 0 \tag{6.134}
$$

$$-\frac{\mathrm{d}\Phi(t, s)}{\mathrm{d}s} = \Phi(t, s)[A_{\mathrm{b}}(s) - Q_{\mathrm{b}}(s)W(t, s)], \qquad \Phi(t, t) = I \tag{6.135}$$

$$\begin{aligned}-\frac{\mathrm{d}W(t, s)}{\mathrm{d}s} &= A_{\mathrm{b}}^{\mathrm{T}}(s)W(t, s) + W(t, s)A_{\mathrm{b}}(s) \\ &\quad - W(t, s)Q_{\mathrm{b}}(s)W(t, s) \\ &\quad + H_{\mathrm{b}}(s), \qquad W(t, t) = 0\end{aligned} \tag{6.136}$$

in which the quantities $A_{\mathrm{b}}(s)$, $Q_{\mathrm{b}}(s)$, $H_{\mathrm{b}}(s)$, $\hat{x}_{\mathrm{b}}(s)$ and $\lambda_{\mathrm{b}}(s)$ are as follows:

$$A_{\mathrm{b}}(s) = A(s) - P(s; \tau)H^{\mathrm{T}}(s)R^{-1}(s)H(s) \tag{6.137}$$

$$\begin{aligned}Q_{\mathrm{b}}(s) &= A(s)P(s; \tau) + P(s; \tau)A^{\mathrm{T}}(s) + B(s)Q(s)B^{\mathrm{T}}(s) \\ &\quad - P(s; \tau)H^{\mathrm{T}}(s)R^{-1}(s)H(s)P(s; \tau)\end{aligned} \tag{6.138}$$

$$H_{\mathrm{b}}(s) = H^{\mathrm{T}}(s)R^{-1}(s)H(s) \tag{6.139}$$

$$\begin{aligned}\hat{x}_{\mathrm{b}}(s) &= [A(s) - P(s; \tau)H^{\mathrm{T}}(s)R^{-1}(s)H(s)]\hat{x}(s; \tau) \\ &\quad + P(s; \tau)H^{\mathrm{T}}(s)R^{-1}(s)z(s)\end{aligned} \tag{6.140}$$

$$\lambda_{\mathrm{b}}(s) = H^{\mathrm{T}}(s)R^{-1}(s)[z(s) - H(s)\hat{x}(s; \tau)] \tag{6.141}$$

Proof

For this case, since $\Pi_{\mathrm{r}}(s; \tau) = 0$ and $\hat{m}_{\mathrm{r}}(s; \tau) = 0$ from the above assumptions, we have $\hat{x}_{\mathrm{n}}(t|t; s) \to \hat{x}(t|t; s)$, $\lambda_{\mathrm{n}}(s|t) \to \lambda(s|t)$, $\Phi_{\mathrm{n}}(t, s) \to \Phi(t, s)$, $P_{\mathrm{n}}(t, s) \to P(t, s)$, $W_{\mathrm{n}}(t, s) \to W(t, s)$, $\hat{m}_{\mathrm{n}}(s; \tau) \to \hat{x}(s; \tau)$ and $\Pi_{\mathrm{n}}(s; \tau) \to P(s; \tau)$. □

Moreover, we note that when $\Pi_{\mathrm{n}}(s; \tau) \equiv 0$ and $\hat{m}_{\mathrm{n}}(s; \tau) \equiv 0$ in Theorem 6.7 or $P(s; \tau) \equiv 0$ and $\hat{x}(s; \tau) \equiv 0$ in Corollary 6.1, the following corollary is also valid:

Corollary 6.2

If $\Pi_{\mathrm{n}}(s; \tau) = 0$ and $\hat{m}_{\mathrm{n}}(s; \tau) = 0$ in Theorem 6.7 or $P(s; \tau) = 0$ and $\hat{x}(s; \tau) = 0$ in Corollary 6.1, then $\hat{x}_0(t|t; s)$, which is the filtered estimate $\hat{x}_n(t|t; s)$ with $\Pi_n(s; \tau) = \hat{m}_n(s; \tau) = 0$, or $\hat{x}(t|t; s)$ with $P(s; \tau) = \hat{x}(s; \tau) = 0$, satisfies

$$-\frac{d\hat{x}_0(t|t; s)}{ds} = \Phi_0(t, s)B(s)Q(s)B^{T}(s)\lambda_0(s|t), \qquad \hat{x}_0(t|t; t) = 0 \tag{6.142}$$

$$-\frac{dP_0(t, s)}{ds} = \Phi_0(t, s)B(s)Q(s)B^{T}(s)\Phi_0^{T}(t, s), \qquad P_0(t, t) = 0 \tag{6.143}$$

where

$$\begin{aligned} -\frac{d\lambda_0(s|t)}{ds} &= [A(s) - B(s)Q(s)B^{T}(s)W_0(t, s)]^{T}\lambda_0(s|t) \\ &\quad + H^{T}(s)R^{-1}(s)z(s), \qquad \lambda_0(t|t) = 0 \end{aligned} \tag{6.144}$$

$$-\frac{d\Phi_0(t, s)}{ds} = \Phi_0(t, s)[A(s) - B(s)Q(s)B^{T}(s)W_0(t, s)], \qquad \Phi_0(t, t) = I \tag{6.145}$$

and

$$\begin{aligned} -\frac{dW_0(t, s)}{ds} &= A^{T}(s)W_0(t, s) + W_0(t, s)A(s) \\ &\quad + H^{T}(s)R^{-1}(s)H(s) \\ &\quad - W_0(t, s)B(s)Q(s)B^{T}(s)W_0(t, s), \qquad W(t, t) = 0 \end{aligned} \tag{6.146}$$

Proof

In this case, it is readily seen that $A_b(s) = A(s)$, $Q_b(s) = B(s)Q(s)B^{T}(s)$, $H_b(s) = H^{T}(s)R^{-1}(s)H(s)$, $\hat{x}_b(s) = 0$ and $\lambda_b(s) = H^{T}(s)R^{-1}(s)z(s)$. Thus, equations (6.142)–(6.146) are apparent. □

Note that these results are entirely the same as those given by Lainiotis [13] based on the backward differentiation of the forward Kalman filter algorithms. In addition, it should be noted that, as can be found in later developments, this 'zero-nominal Kalman filter' plays an important role in smoothing problems.

6.5.1 Generalized Two-Filter Smoothing Algorithms

We now consider the smoothing problem at $s = \tau$. First, we obtain a backward-time Kalman filter in the following corollary.

Corollary 6.3

If, in Theorems 6.6 and 6.7, $\hat{x}_n(s;\tau) = P_n(s;\tau) = P_{nr}(s;\tau) = 0$, then equation (6.122) satisfies (see also Figure 6.5)

$$\hat{x}_B(s|t) = P_B(s|t)[\lambda_0(s|t) + P^{-1}(s;\tau)\hat{x}(s;\tau)] \tag{6.147}$$

$$P_B(s|t) = [W_0(t,s) + P^{-1}(s;\tau)]^{-1} \tag{6.148}$$

where $\lambda_0(s|t)$ and $W_0(t,s)$ are given by expressions (6.144) and (6.146), respectively.

Proof

It is readily seen from expressions (6.98)–(6.100) that $\hat{m}_n(s;\tau) = 0$, $\Pi_n(s;\tau) = 0$, $\hat{m}_r(s;\tau) = \hat{x}_r(s;\tau) = \hat{x}(s;\tau)$ and $\Pi_r(s;\tau) = P_r(s;\tau) = P(s;\tau)$. Hence using Corollary 6.2, equation (6.112) reduces to the above result. □

Equations (6.144) and (6.146)–(6.148) constitute the so-called Mayne–Fraser or Mehra two-filter formula for the smoothing problem [11, 12]. To see this more clearly, substitute $P^{-1}(s;\tau) = 0$ into equations (6.147) and (6.148) and define, at $s = \tau$,

$$P_b(t,\tau) \triangleq W_0^{-1}(t,\tau) \tag{6.149}$$

$$\hat{x}_b(\tau|t) \triangleq W_0^{-1}(t,\tau)\lambda_0(\tau|t) = P_b(t,\tau)\lambda_0(\tau|t) \tag{6.150}$$

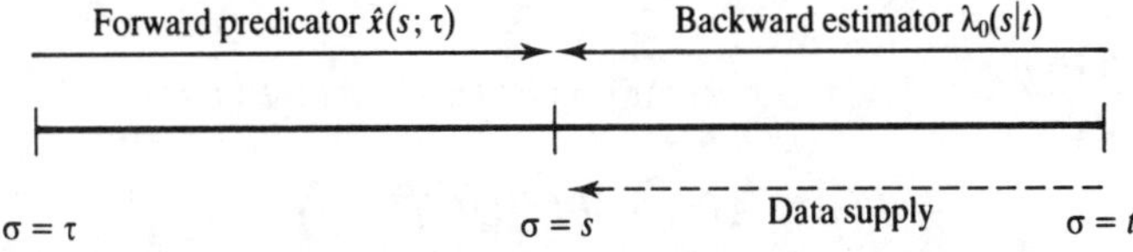

Figure 6.5 The strict backward-time Kalman filter.

From these relations, we obtain, at time $s = \tau$, the following expressions:

$$\hat{x}(\tau|t) = P(\tau|t)[\lambda_0(\tau|t) + P^{-1}(\tau)\hat{x}(\tau|\tau)] \tag{6.151a}$$

$$= P(\tau|t)[P_b^{-1}(t, \tau)\hat{x}_b(\tau|t) + P^{-1}(\tau)\hat{x}(\tau|\tau)] \tag{6.151b}$$

$$P(\tau|t) = [W_0(t, \tau) + P^{-1}(\tau)]^{-1} \tag{6.152a}$$

$$= [P_b^{-1}(t, \tau) + P^{-1}(\tau)]^{-1} \tag{6.152b}$$

where $\{\hat{x}(\tau|\tau), P(\tau)\}$ denote the forward-time Kalman filtered values, and from expressions (6.144), (6.146), (6.149) and (6.150), $\hat{x}_b(\tau|t)$ and $P_b(t, \tau)$ are subject to the following relations:

$$-\frac{d\hat{x}_b(\tau|t)}{d\tau} = -[A(\tau) + P_b(t, \tau)H^T(\tau)R^{-1}(\tau)H(\tau)]\hat{x}_b(\tau|t) + P_b(t, \tau)H^T(\tau)R^{-1}(\tau)z(\tau), \quad \hat{x}_b(t|t) \text{ arbitrary} \tag{6.153}$$

$$-\frac{dP_b(t, \tau)}{d\tau} = -A(\tau)P_b(t, \tau) - P_b(t, \tau)A^T(\tau) + B(\tau)Q(\tau)B^T(\tau) - P_b(t, \tau)H^T(\tau)R^{-1}(\tau)H(\tau)P_b(t, \tau), \quad P_b(t, t) = \infty \tag{6.154}$$

Thus, equations (6.151)–(6.154) are actually the Mayne–Fraser two-filter smoother.

It is also interesting to note that the so-called Rauch–Tung–Striebel (RTS) smoother [15] can be readily derived by using these results. That is, by differentiating equations (6.151a) and (6.152a) with respect to τ it follows that

$$-\frac{d\hat{x}(\tau|t)}{d\tau} = -[A(\tau) + B(\tau)Q(\tau)B^T(\tau)P^{-1}(\tau)]\hat{x}(\tau|t) + B(\tau)Q(\tau)B^T(\tau)P^{-1}(\tau)\hat{x}(\tau|\tau), \quad \hat{x}(\tau|t)|_{\tau=t} = \hat{x}(t|t) \tag{6.155}$$

$$-\frac{dP(\tau|t)}{d\tau} = -[A(\tau) + B(\tau)Q(\tau)B^T(\tau)P^{-1}(\tau)]P(\tau|t) - P(\tau|t)[A(\tau) + B(\tau)Q(\tau)B^T(\tau)P^{-1}(\tau)]^T + B(\tau)Q(\tau)B^T(\tau), \quad P(t|t) = P(t) \tag{6.156}$$

Although there remain certain questions regarding the final conditions, equations (6.153) and (6.154) are usually regarded as a 'formal (or classical) backward-time Kalman filter' [22], solved from $\tau = t$, while equations (6.147) and (6.148) are regarded as a 'strict (or modern) backward-time Kalman filter' [23]. Note that equation (6.152b) indicates that $P(\tau|t) \leq P(\tau)$, which means that the smoothed estimate is always as good as or better than its filtered estimate. This is depicted graphically in Figure 6.6 for a typical fixed-interval smoothing problem, where the initial and final times 0 and T are fixed, and τ and t in equation (6.152b) are replaced with t and T, respectively. With similar replacement of time variables, we can obtain, from equations (6.155) and (6.156), the backward-pass fixed-interval smoother in continuous-time systems.

A strict interpretation of the backward-time Kalman filter can be seen from the direct differentiation of $\hat{x}_{\mathrm{B}}(s|t)$ and $P_{\mathrm{B}}(s|t)$ with respect to s [13]. This result is summarized in the following corollary.

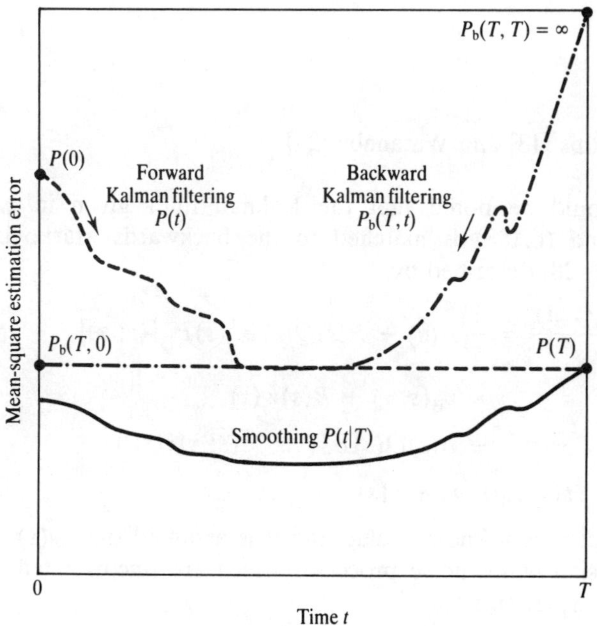

Figure 6.6 Mean-square estimation errors for the Mayne–Fraser two-filter smoother in the fixed-interval smoothing problem.

Corollary 6.4

The smoothed estimate $\hat{x}_B(s|t)$ and the corresponding error covariance matrix $P_B(s|t)$ satisfy the following backward-time evolution equations:

$$\begin{aligned}
-\frac{d\hat{x}_B(s|t)}{ds} &= -[A(s) + B(s)Q(s)B^T(s)P^{-1}(s;\tau)]\hat{x}_B(s|t) \\
&\quad + P_B(s|t)H^T(s)R^{-1}(s)[z(s) - H(s)\hat{x}_B(s|t)] \\
&\quad + B(s)Q(s)B^T(s)P^{-1}(s;\tau)\hat{x}(s;\tau), \\
&\qquad \hat{x}_B(t|t) = \hat{x}(t;\tau)
\end{aligned} \tag{6.157}$$

and

$$\begin{aligned}
-\frac{dP_B(s|t)}{ds} &= -[A(s) + B(s)Q(s)B^T(s)P^{-1}(s;\tau)]P_B(s|t) \\
&\quad - P_B(s|t)[A(s) + B(s)Q(s)B^T(s)P^{-1}(s;\tau)]^T \\
&\quad - P_B(s|t)H^T(s)R^{-1}(s)H(s)P_B(s|t) \\
&\quad + B(s)Q(s)B^T(s), \qquad P_B(t|t) = P(t;\tau)
\end{aligned} \tag{6.158}$$

Proof

See Lainiotis [13] and Watanabe [24]. □

It should be noted that the Kalman filter given in equations (6.157) and (6.158) is matched to the backwards Markovian state model [25–28] described by

$$\begin{aligned}
-\frac{dx_B(s,t)}{ds} &= -[A(s) + B(s)Q(s)B^T(s)P^{-1}(s;\tau)] \\
&\quad \times x_B(s,t) + B(s)w(s) \\
&\quad + B(s)Q(s)B^T(s)P^{-1}(s;\tau)\hat{x}(s;\tau)
\end{aligned} \tag{6.159}$$

$$z(s) = H(s)x_B(s,t) + v(s) \tag{6.160}$$

where $\hat{x}(s;\tau)$ is a known value and it is assumed that $w(s)$ and $v(s)$ are Gaussian white noise processes which are uncorrelated with the final state $x_B(t,t)$, i.e.

$$E\left[\begin{pmatrix} w(s) \\ v(s) \end{pmatrix}\right] = 0, \qquad E = \left[\begin{pmatrix} w(s) \\ v(s) \end{pmatrix} x_B^T(t,t)\right] = 0 \tag{6.161}$$

and

$$E\left[\begin{pmatrix} w(s) \\ v(s) \end{pmatrix} (w^{\mathrm{T}}(s') \quad v^{\mathrm{T}}(s'))\right] = \begin{bmatrix} Q(s) & 0 \\ 0 & R(s) \end{bmatrix} \delta(s - s') \quad (6.162)$$

Moreover, the final statistics on the final state are given by $x_{\mathrm{B}}(t, t) \sim N(\hat{x}(t; \tau), P(t; \tau))$.

Example 6.1 [29]

A spacecraft is falling radially away from the Earth at an almost constant speed, and is subject to small, random, high-frequency disturbance accelerations of spectral density Q. Determine the accuracy to which the vehicle velocity can be estimated, using ground-based Doppler radar with spectral density error R.

Let x denote the deviation from the predicted nominal spacecraft velocity, using available gravitational models. We then have

$$\dot{x}(t) = w(t), \qquad w(t) \sim N(0, Q)$$

$$z(t) = x(t) + v(t), \qquad v(t) \sim N(0, R)$$

We desire to examine the steady-state, fixed-interval smoother in both (a) its Mayne–Fraser two-filter form, and (b) its RTS form.

(i) The estimation error covariance associated with the forward-time Kalman filter is

$$\dot{P} = Q - P^2/R$$

which, in the steady state ($\dot{P} = 0$), yields $P = \sqrt{RQ} \triangleq \alpha$. The estimation error covariance associated with the backward-time Kalman filter is, from equation (6.154),

$$\frac{\mathrm{d}P_{\mathrm{b}}}{\mathrm{d}\tau} = Q - P_{\mathrm{b}}^2/R$$

which has the steady state $P_{\mathrm{b}} = \sqrt{RQ} = \alpha$. Thus, we find, for the smoothed covariance,

$$\begin{aligned} P(t|T) &= [P_{\mathrm{b}}^{-1}(T, t) + P^{-1}(t)]^{-1} \\ &= \frac{\alpha}{2} \end{aligned}$$

which is half the error covariance associated with the

optimal Kalman filter. Consequently, from equation (6.151b),

$$\hat{x}(t|T) = P(t|T)[P_b^{-1}(T, t)\hat{x}_b(t|T) + P^{-1}(t)\hat{x}(t|t)]$$
$$= \tfrac{1}{2}[\hat{x}_b(t|T) + \hat{x}(t|t)] \tag{6.163}$$

The smoothed estimate of x is the average of the forward and backward estimates, in steady state.

(ii) In the RTS form, the steady-state smoothed covariance differential equation (6.156) is

$$\dot{P}(t|T) = \frac{2Q}{\alpha} P(t|T) - Q$$

for which the solution is ($Q/\alpha \triangleq \beta$, $P(T|T) = \alpha$)

$$P(t|T) = \frac{\alpha}{2}\{1 + \exp[-2\beta(T - t)]\}, \quad t \leq T$$

For $T - t$ sufficiently large (i.e. $T - t > 2/\beta$), the backward sweep is in steady state. In this case, we obtain $P(t|T) = \alpha/2$, as before. The corresponding differential equation for the smoothed state estimation, from equation (6.155), is

$$\frac{\mathrm{d}\hat{x}(t|T)}{\mathrm{d}t} = \beta\hat{x}(t|T) - \beta\hat{x}(t|t), \qquad \hat{x}(t|T)|_{t=T} = \hat{x}(T|T)$$

This can be shown to be identical to equation (6.163) by direct differentiation of the latter with respect to time.

If $\lambda_0(s|t)$ and $W_0(t, s)$ are excluded from equations (6.147), (6.148), (6.151a) and (6.152a), one obtains

$$\hat{x}(s|t) = P(s|t)[P_B^{-1}(s|t)\hat{x}_B(s|t) + P^{-1}(s)\hat{x}(s|s) - P^{-1}(s;\tau)\hat{x}(s;\tau)] \tag{6.164}$$

$$P(s|t) = [P_B^{-1}(s|t) + P^{-1}(s) - P^{-1}(s;\tau)]^{-1} \tag{6.165}$$

This result can be seen as a generalized two-filter smoother formula which represents the smoothed estimate as a combination of one forward predictor and two optimal linear filters, one of which works forward over the data and the other of which runs backward over the interval. A similar result to this for the special case of initial condition $\hat{x}(\tau|\tau) = 0$ can be found in Wall *et al*. [23]. The graphic concept of the generalized two-filter smoother is shown in Figure 6.7 for the fixed-interval smoothing problem.

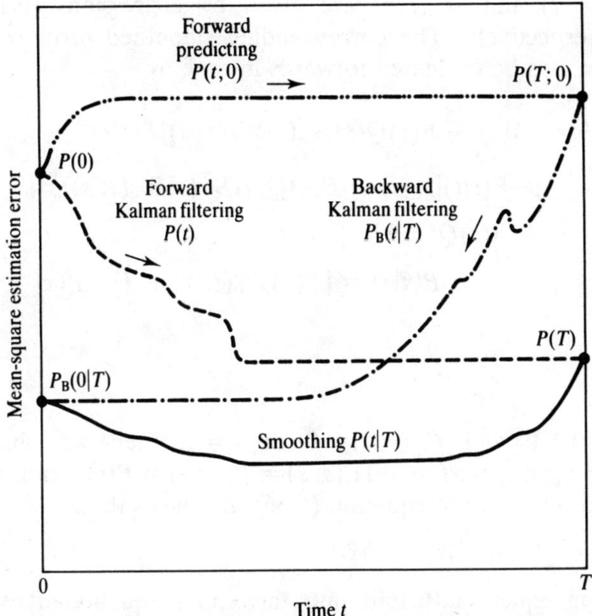

Figure 6.7 Mean-square estimation errors for a generalized two-filter smoother in the fixed-interval smoothing problem.

6.5.2 Weinert–Desai Smoothing Algorithm

It is of interest to note that equation (6.36), together with equations (6.144) and (6.146), gives directly the forward-pass smoothing algorithm due to Weinert and Desai [7], in which they utilize the concept of 'complementary models' to derive the desired formulas. The following derivation was first suggested by Verghese *et al*. [5], but the result was not given in a complete form.

Corollary 6.5 Weinert–Desai smoothing algorithm

The smoothed estimate can be processed forwards in time by

$$\frac{\mathrm{d}\hat{x}(s|t)}{\mathrm{d}s} = [A(s) - B(s)Q(s)B^{\mathrm{T}}(s)W_0(t, s)]\hat{x}(s|t) + B(s)Q(s)B^{\mathrm{T}}(s)\lambda_0(s|t)$$

$$\hat{x}(\tau|t) = [P(\tau)W_0(t, \tau) + I]^{-1}[P(\tau)\lambda_0(\tau|t) + \hat{x}(\tau|\tau)] \qquad (6.166)$$

where $\lambda_0(s|t)$ and $W_0(t, s)$ are given by expressions (6.144) and (6.146), respectively. The corresponding smoothed error covariance matrix can also be evaluated forwards in time by

$$\begin{aligned}\frac{\mathrm{d}P(s|t)}{\mathrm{d}s} &= [A(s) - B(s)Q(s)B^{\mathrm{T}}(s)W_0(t, s)]P(s|t) \\ &\quad + P(s|t)[A(s) - B(s)Q(s)B^{\mathrm{T}}(s)W_0(t, s)]^{\mathrm{T}} \\ &\quad + B(s)Q(s)B^{\mathrm{T}}(s), \\ P(\tau|\tau) &= [P(\tau)W_0(t, \tau) + I]^{-1}P(\tau) \qquad (6.167)\end{aligned}$$

Proof

If $\Pi_0(s; \tau) = 0$ and $\hat{m}_{\mathrm{n}}(s; \tau) = 0$ at $s = \tau$, then we obtain that $\hat{m}_{\mathrm{r}}(s; s) = \hat{x}_{\mathrm{r}}(s; s) \equiv \hat{x}(s|s)$, $\Pi_{\mathrm{r}}(s; s) = P_{\mathrm{r}}(s; s) \equiv P(s)$, and $\hat{x}_{\mathrm{r}}(s|t) \equiv \hat{x}(s|t)$. Therefore, from equation (6.36), it follows that

$$\lambda(s|t) = \lambda_0(s|t) - W_0(t, s)\hat{x}(s|t) \qquad (6.168)$$

Substituting equation (6.168) into the upper equality of expression (6.22) gives (6.166), where the initial condition was derived from equation (6.151a). By differentiating equation (6.152a) with respect to $\tau(= s)$, utilizing equation (6.146) and noting that

$$\begin{aligned}\frac{\mathrm{d}P^{-1}(s)}{\mathrm{d}s} &= -P^{-1}(s)A(s) - A^{\mathrm{T}}(s)P^{-1}(s) + H^{\mathrm{T}}(s)R^{-1}(s)H(s) \\ &\quad - P^{-1}(s)B(s)Q(s)B^{\mathrm{T}}(s)P^{-1}(s) \qquad (6.169)\end{aligned}$$

we have the desired equation (6.167), where the initial condition was derived from equation (6.152a). □

Example 6.2

Determine the structure of the (Weinert–Desai) forward-pass fixed-interval smoother for the system considered in Example 2.5. The differential equation for $W_0(t, s)$ is governed, from equation (6.146), by

$$-\frac{\mathrm{d}W_0(T, s)}{\mathrm{d}s} = -2aW_0(T, s) + \frac{1}{R} - QW_0^2(T, s), \qquad 0 \leq s \leq T$$

Setting $T - s \triangleq \tau$, we find

$$\frac{\mathrm{d}W_0}{\mathrm{d}\tau} = -2aW_0 + \frac{1}{R} - QW_0^2$$

Applying the result of Example 2.5 and letting γ_1 and γ_2 denote the roots of $W_0^2 + 2aW_0/Q - 1/(QR) = 0$ we have

$$W_0(T, s) = \frac{\gamma_1 - \gamma_2 c \exp[-2\mu(T - s)]}{1 - c\exp[-2\mu(T - s)]}, \qquad W(T, T) = 0$$

where

$$\gamma_{1,2} = \frac{1}{Q}\left[-a \pm \sqrt{a^2 \pm \frac{Q}{R}}\right]$$

$$c = \gamma_1/\gamma_2, \qquad \mu = \sqrt{a^2 + \frac{Q}{R}}$$

We also have, from equations (6.144) and (6.166),

$$-\frac{d\lambda_0(s|T)}{ds} = [-a - QW_0(T, s)]\lambda(s|T) + \frac{1}{R}z(s),$$

$$\lambda_0(T|T) = 0$$

and

$$\frac{d\hat{x}(s|T)}{ds} = [-a - QW_0(T, s)]\hat{x}(s|T) + Q(s)\lambda_0(s|T)$$

$$\hat{x}(0|T) = [P(0)W_0(T, 0) + I]^{-1}[P(0)\lambda_0(0|T) + \hat{x}(0)]$$

When $T - s$ is sufficiently large, it follows, as suggested in Watanabe [30], that

$$W_0(T, s) = \gamma_1$$
$$= -\frac{1}{\rho_1}$$

(see Example 2.5).

An inspection of equations (6.166) and (6.167) shows that the a priori statistics $\{\hat{x}(\tau|\tau), P(\tau)\}$ will affect only the linear equations. From this fact, the effects of changes in $\{\hat{x}(\tau|\tau), P(\tau)\}$ on $\{\hat{x}(s|t), P(s|t)\}$, $\tau \leqslant s \leqslant t$, can be evaluated immediately. This result is summarized in the following corollary.

Corollary 6.6

It is assumed that $\hat{x}_1(s|t)$ and $P_1(s|t)$ are the smoothed estimate and the error covariance for a system with initial statistics

$\{\hat{x}_1(\tau|\tau), P_1(\tau)\}$. When $\{\hat{x}_1(\tau|\tau), P_1(\tau)\}$ changes into $\{\hat{x}_2(\tau|\tau), P_2(\tau)\}$, the smoothed estimate $\hat{x}_2(s|t)$ and the error covariance matrix $P_2(s|t)$ corresponding to $\{\hat{x}_2(\tau|\tau), P_2(\tau)\}$ can be obtained by

$$\hat{x}_2(s|t) = \hat{x}_1(s|t) + \delta\hat{x}(s|t) \tag{6.170}$$

$$P_2(s|t) = P_1(s|t) + \delta P(s|t) \tag{6.171}$$

where $\hat{x}_1(s|t)$ and $P_1(s|t)$ are given by equations (6.166) and (6.167) with initial conditions $\{\hat{x}_1(\tau|\tau), P(\tau)\}$, and $\delta\hat{x}(s|t)$ and $\delta P(s|t)$ are subject to

$$\begin{aligned}\frac{\mathrm{d}\delta\hat{x}(s|t)}{\mathrm{d}s} &= [A(s) - B(s)Q(s)B^{\mathrm{T}}(s)W_0(t, s)]\delta\hat{x}(s|t),\\ \delta\hat{x}(\tau|t) &= \hat{x}_2(\tau|t) - \hat{x}_1(\tau|t)\\ &= \delta P(\tau|t)\lambda_0(\tau|t) + [I + P_2(\tau)W_0(t, \tau)]^{-1}\hat{x}_2(\tau|\tau)\\ &\quad -[I + P_1(\tau)W_0(t, \tau)]^{-1}\hat{x}_1(\tau|\tau)\end{aligned} \tag{6.172}$$

and

$$\begin{aligned}\frac{\mathrm{d}\delta P(s|t)}{\mathrm{d}s} &= [A(s) - B(s)Q(s)B^{\mathrm{T}}(s)W_0(t, s)]\delta P(s|t)\\ &\quad + \delta P(s|t)[A(s) - B(s)Q(s)B^{\mathrm{T}}(s)W_0(t, s)]^{\mathrm{T}},\\ \delta P(\tau|t) &= P_2(\tau|t) - P_1(\tau|t)\\ &= [I + P_2(\tau)W_0(t, \tau)]^{-1}P_2(\tau)\\ &\quad - [I + P_1(\tau)W_0(t, \tau)]^{-1}P_1(\tau)\end{aligned} \tag{6.173}$$

Proof

Subtracting the solutions of equations (6.166) and (6.167) with $\{\hat{x}_1(\tau|\tau), P_1(\tau)\}$ from the solutions of the same equations with $\{\hat{x}_2(\tau|\tau), P_2(\tau)\}$ leads to the desired results. □

6.5.3 Generalized Chandrasekhar Algorithms

A computationally advantageous approach, known as *X–Y functions* or *Chandrasekhar algorithms* [8], is derived in this subsection. The preliminary results are given in the following two corollaries.

Corollary 6.7

The generalized Chandrasekhar algorithms, which are applicable to time-varying models with arbitrary initial conditions, are given by [31]

$$-\frac{\mathrm{d}P_{\mathrm{n}}(t,\tau)}{\mathrm{d}\tau}=\Phi_{\mathrm{n}}(t,\tau)G_{\mathrm{n}}(\tau)\Sigma_{\mathrm{n}}(\tau)G_{\mathrm{n}}^{\mathrm{T}}(\tau)\Phi_{\mathrm{n}}^{\mathrm{T}}(t,\tau),$$

$$P_{\mathrm{n}}(t,t)=\Pi_{\mathrm{n}}(t;\tau) \qquad (6.174)$$

$$\frac{\mathrm{d}\Phi_{\mathrm{n}}(t,\tau)}{\mathrm{d}t}=[A(t)-P_{\mathrm{n}}(t,\tau)H^{\mathrm{T}}(t)R^{-1}(t)H(t)]\Phi_{\mathrm{n}}(t,\tau),$$

$$\Phi_{\mathrm{n}}(\tau,\tau)=I \qquad (6.175)$$

where

$$\begin{aligned}&G_{\mathrm{n}}(\tau)\Sigma_{\mathrm{n}}(\tau)G_{\mathrm{n}}^{\mathrm{T}}(\tau)\\&\triangleq A(\tau)\Pi_{\mathrm{n}}(\tau;\tau)+\Pi_{\mathrm{n}}(\tau;\tau)A^{\mathrm{T}}(\tau)+B(\tau)Q(\tau)B^{\mathrm{T}}(\tau)\\&\quad-\Pi_{\mathrm{n}}(\tau;\tau)H^{\mathrm{T}}(\tau)R^{-1}(\tau)H(\tau)\Pi_{\mathrm{n}}(\tau;\tau)\end{aligned} \qquad (6.176)$$

and $G_{\mathrm{n}}(\tau)$ is a full rank $n\times\alpha$ matrix, in which α indicates the rank of $G_{\mathrm{n}}(\tau)\Sigma_{\mathrm{n}}(\tau)G_{\mathrm{n}}^{\mathrm{T}}(\tau)$.

Proof

These results follow from Theorem 6.3 and Theorem 6.6 with $s=\tau$.□

Corollary 6.8

The 'dual' version of the generalized Chandrasekhar algorithms is obtained by

$$\begin{aligned}-\frac{\mathrm{d}\Phi_{\mathrm{n}}^{\mathrm{T}}(t,\tau)}{\mathrm{d}\tau}&=[A(\tau)-\Pi_{\mathrm{n}}(\tau;\tau)H^{\mathrm{T}}(\tau)R^{-1}(\tau)H(\tau)\\&\quad-G_{\mathrm{n}}(\tau)\Sigma_{\mathrm{n}}(\tau)G_{\mathrm{n}}^{\mathrm{T}}(\tau)\\&\quad\times W_{\mathrm{n}}(t,\tau)]^{\mathrm{T}}\Phi_{\mathrm{n}}^{\mathrm{T}}(t,\tau),\qquad \Phi_{\mathrm{n}}^{\mathrm{T}}(t,t)=I\end{aligned} \qquad (6.177)$$

$$\frac{\mathrm{d}W_{\mathrm{n}}(t,\tau)}{\mathrm{d}t}=\Phi_{\mathrm{n}}^{\mathrm{T}}(t,\tau)H^{\mathrm{T}}(t)R^{-1}(t)H(t)\Phi_{\mathrm{n}}(t,\tau),$$

$$W_{\mathrm{n}}(\tau,\tau)=0 \qquad (6.178)$$

Proof

This corollary is derived in the same way as Corollary 6.7. □

Assuming the system matrices A, B, H, Q and R are constant, we have the following corollaries.

Corollary 6.9

The nominal Kalman filter can, for a time-invariant system, be written by using Chandrasekhar algorithms as follows:

$$\frac{\mathrm{d}\hat{x}_{\mathrm{n}}(t|t,\tau)}{\mathrm{d}t} = [A - K_{\mathrm{n}}(t,\tau)H]\hat{x}_{\mathrm{n}}(t|t,\tau) + K_{\mathrm{n}}(t,\tau)z(t), \qquad \hat{x}_{\mathrm{n}}(\tau|\tau,\tau) = \hat{m}_{\mathrm{n}}(\tau;\tau) \tag{6.179}$$

$$\frac{\mathrm{d}K_{\mathrm{n}}(t,\tau)}{\mathrm{d}t} = L_{\mathrm{n}}(t,\tau)\Sigma_{\mathrm{n}}(\tau)L_{\mathrm{n}}^{\mathrm{T}}(t,\tau)H^{\mathrm{T}}R^{-1}, \qquad K_{\mathrm{n}}(\tau,\tau) = \Pi_{\mathrm{n}}(\tau;\tau)H^{\mathrm{T}}R^{-1} \tag{6.180}$$

$$\frac{\mathrm{d}L_{\mathrm{n}}(t,\tau)}{\mathrm{d}t} = [A - K_{\mathrm{n}}(t,\tau)H]L_{\mathrm{n}}(t,\tau), \qquad L_{\mathrm{n}}(\tau,\tau) = G_{\mathrm{n}}(\tau) \tag{6.181}$$

Proof

It is noted that for the time-invariant system the so-called 'Stokes identity' [9] holds:

$$\frac{\partial}{\partial t}S_{\mathrm{n}}(\tau,t) = -\frac{\partial}{\partial \tau}S_{\mathrm{n}}(\tau,t) \tag{6.182}$$

Therefore, noting that

$$\frac{\mathrm{d}P_{\mathrm{n}}(t,\tau)}{\mathrm{d}t} = \Phi_{\mathrm{n}}(t,\tau)G_{\mathrm{n}}(\tau)\Sigma_{\mathrm{n}}(\tau)G_{\mathrm{n}}^{\mathrm{T}}(\tau)\Phi_{\mathrm{n}}^{\mathrm{T}}(t,\tau), \qquad P_{\mathrm{n}}(\tau,\tau) = \Pi_{\mathrm{n}}(\tau;\tau) \tag{6.183}$$

and defining

$$K_{\mathrm{n}}(t,\tau) \triangleq P_{\mathrm{n}}(t,\tau)H^{\mathrm{T}}R^{-1} \tag{6.184}$$

$$L_{\mathrm{n}}(t,\tau) \triangleq \Phi_{\mathrm{n}}(t,\tau)G_{\mathrm{n}}(\tau) \tag{6.185}$$

equation (6.174) gives the desired results. □

It should be remarked that equations (6.180) and (6.181) do not directly provide the error covariance matrix $P_n(t, \tau)$, which may be of interest in certain applications. However, it can be readily obtained by means of a quadrature:

$$P_n(t, \tau) = \Pi_n(\tau; \tau) + \int_\tau^t L_n(s, \tau)\Sigma_n(\tau)L_n^T(s, \tau)\,ds \tag{6.186}$$

Example 6.3

The example considered here is often treated as a test example [8, 32]. The system is assumed to be

$$A = \begin{bmatrix} -0.0297 & -1 & 0 & 0.0438 & 0 \\ 0.331 & -0.0042 & -0.0461 & 0 & 0 \\ -1.13 & 0.128 & -0.803 & 0 & 0 \\ 0 & 0 & 1 & 0 & 0 \\ 0 & 1 & 0 & 0 & 0 \end{bmatrix}$$

$$B = \begin{bmatrix} -0.0297 & 0 & 0 \\ 0.331 & 0.381 & 0.040 \\ -1.13 & 0.067 & 1.59 \\ 0 & 0 & 0 \\ 0 & 0 & 0 \end{bmatrix}$$

$$H = \begin{bmatrix} 0 & 0 & 0 & 1 & 0 \\ 0 & 0 & 0 & 0 & 1 \end{bmatrix}$$

$$Q = \begin{bmatrix} 0.01 & 0 & 0 \\ 0 & 0.001 & 0 \\ 0 & 0 & 0.001 \end{bmatrix}$$

$$R = \begin{bmatrix} 0.001 & 0 \\ 0 & 0.001 \end{bmatrix}$$

In Figure 6.8, we have plotted the values of some typical elements of $K(t)(K_n(t, \tau)$ for $\tau = 0$, $\Pi_n(0; 0) \equiv 0)$ computed by solving the Chandrasekhar algorithms (6.180) and (6.181). The Runge–Kutta–Gill method was adopted to obtain the numerical solution, where the numerical step interval was 0.0001 s. The steady-state value of $K(\cdot)$ was also computed by applying the doubling

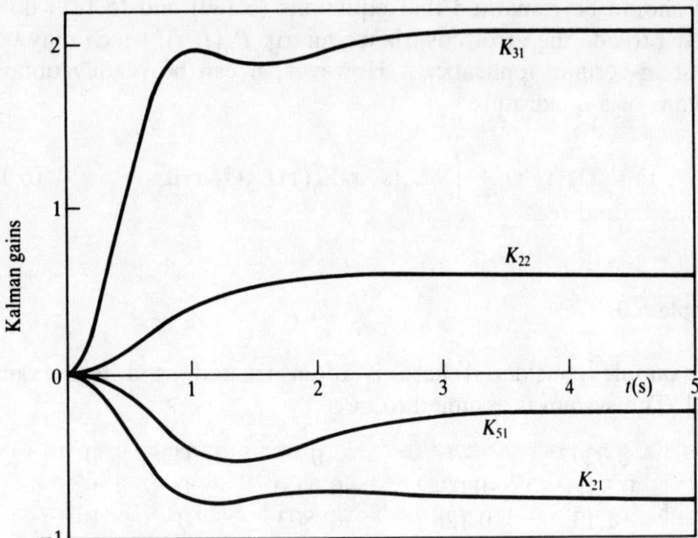

Figure 6.8 History of Kalman gain matrix in Example 6.3.

algorithm with $\Delta = 0.0001$, $\varepsilon = 1 \times 10^{-7}$, as shown in Chapter 5. The resultant values were $K_{21} = -0.7844$, $K_{31} = 2.0974$, $K_{51} = -0.2462$, $K_{22} = 0.5964$ and $n = 22$. It is seen that the solutions of the Chandrasekhar algorithms run quite smoothly into steady state.

Corollary 6.10

For a time-invariant system, the nominal adjoint vector can be expressed by using dual Chandrasekhar algorithms as follows:

$$\frac{d\lambda_n(\tau|t)}{dt} = F_n(t, \tau)[z(t) - H\hat{x}_n(t|t, \tau)], \qquad \lambda_n(\tau|\tau) = 0 \tag{6.187}$$

$$\frac{dF_n(t, \tau)}{dt} = [A - \Pi_n(\tau; \tau)H^T R^{-1} H - G_n(\tau)\Sigma_n(\tau)N_n(t, \tau)]^T F_n(t, \tau), \qquad F_n(\tau, \tau) = H^T R^{-1} \tag{6.188}$$

$$\frac{dN_n(t, \tau)}{dt} = G_n^T(\tau)F_n(t, \tau)RF_n^T(t, \tau), \qquad N_n(\tau, \tau) = 0 \tag{6.189}$$

Proof

In the same manner as Corollary 6.9, defining

$$F_n(t, \tau) \triangleq \Phi_n^T(t, \tau) H^T R^{-1} \tag{6.190}$$

$$N_n(t, \tau) \triangleq G_n^T(\tau) W_n(t, \tau) \tag{6.191}$$

gives the desired results. □

Note that the partial observability matrix $W_n(t, \tau)$ can also be evaluated by using

$$W_n(t, \tau) = \int_\tau^t F_n(s, \tau) R F_n^T(s, \tau)\, ds \tag{6.192}$$

Since many observations on the nature, computational advantages, and use of the above Chandrasekhar algorithms have been made in several papers (see, for example, Kailath [8], Kailath and Ljung [10], Ljung and Kailath [19], Lainiotis [31], Brammer [33] and Kasti [34]), we state only the principal properties:

- Assume that $n = \dim x(t)$, $p = \dim w(t)$ and $m = \dim z(t)$. While the solution of the original Riccati equation, $P_n(t, \tau)$, requires solution of a system of $n(n+1)/2$ simultaneous nonlinear differential equations, the Chandrasekhar algorithm requires the solution of the $n(m+\alpha)$ simultaneous nonlinear equations given by expressions (6.180) and (6.181). Thus, if $m + \alpha < (n+1)/2$, there then may be a significant computational advantage in using the Chandrasekhar algorithm instead of the Riccati equation. This feature also holds for the dual Chandrasekhar algorithm.
- Note, however, that the computational advantages are usually more closely related to the number of arithmetic operations to be performed than to the number of simultaneous equations to be solved. Corollaries 6.9 and 6.10 imply that the number of operations per time iteration required by the Chandrasekhar algorithms is proportional to $n^2(m+\alpha)$, whereas the number of operations per time iteration required by the straightforward calculations of $P_n(t, \tau)$ and $\Phi_n^T(t, \tau)$ is proportional to n^3.
- It is found from the above observations that the computational advantages of the Chandrasekhar algorithms depend upon the low-rank property of $G_n(\tau)\Sigma_n(\tau)G_n^T(\tau)$. In particular, for the zero-nominal initial condition case, i.e. $\Pi_n(\tau; \tau) = 0$, we obtain $\alpha = p$ so that $G_n(\tau) = B(\tau)$ and $\Sigma_n(\tau) = Q(\tau)$.

6.6 SUMMARY

In this chapter partitioning filters and smoothers have been derived from Redheffer's scattering theory. Also, popular and useful algorithms such as generalized two-filter smoothing algorithms, the Rauch–Tung–Striebel smoothing algorithm, the Weinert–Desai smoothing algorithm, and generalized Chandrasekhar algorithms have been shown to be derived by applying the partitioning filtering and smoothing results.

The Chandrasekhar (or $X-Y$) algorithm originated in S. Chandrasekhar's investigations [35] on radiative transfer in plane parallel atmospheres. Relationships between the star product of Redheffer and the theory of Riccati matrix differential equations are also found in the work by Reid [36]. Other applications of scattering theory can be found in the work of Verriest *et al*. [37], Levey *et al*. [38] and Watanabe [39] for the estimation problem for distributed sensor networks and in Bruckstein and Kailath [40] for limited memory filtering.

Where predictive information on the final state is available in addition to the usual a priori information on the initial state, generalized forward and backward Markovian models are developed in Watanabe and Tzafestas [41] for discrete-time systems. Further work on 'complementary models' is reported by Adams *et al*. [42, 43] and Ackner and Kailath [44] for continuous-time systems, and by Ackner and Kailath [45] for discrete-time systems.

The Chandrasekhar algorithms for the linear quadratic Gaussian (LQG) control problems are detailed in Casti [3] and Gorez [46]. Chandrasekhar algorithms in discrete-time systems are developed in Morf *et al*. [47]. A successful result using such algorithms is reported by Mahalanabis and Ray [48]. The distributed-type Chandrasekhar algorithm can also be seen in Casti and Ljung [49], Baras and Lainiotis [50] and Watanabe [51].

EXERCISES

6.1 Suppose that A, B, H, Q and R are time-invariant. Show that

$$S(0, 2^p t) = [S(0, t)]^{*2^p}$$

where $*2^p$ denotes powers with respect to the star product. This relation is an interesting doubling formula, and in fact it gives a fast doubling algorithm for computing the steady-state solution of $P(t, 0)$ (cf. Chapter 5).

6.2 Suppose again that A, B, H, Q and R are time-invariant, and that (A, D) is stabilizable and (C, A) is detectable, where $BQB^{\mathrm{T}} = DD^{\mathrm{T}}$ and $H^{\mathrm{T}}R^{-1}H = C^{\mathrm{T}}C$.

(a) Show that the limit of $S_{\mathrm{n}}(\tau, t)$ exists as

$$\lim_{t\to\infty} S_0(\tau, t) = \begin{bmatrix} 0 & \bar{P}_0 \\ \bar{W}_0 & 0 \end{bmatrix}$$

where

$$\lim_{t\to\infty} P_0(t, \tau) = \bar{P}_0$$

$$\lim_{t\to\infty} W_0(t, \tau) = \bar{W}_0$$

(b) Show that, for $P(\tau) \geqslant 0$,

$$P(t, \tau) \to \bar{P}_0 \text{ as } t \to \infty$$

6.3 Complete the proof of Theorem 6.2 by applying equations (6.36) and (6.37).

6.4 Consider the following second-order system:

$$\dot{x}(t) = \begin{bmatrix} 0 & 1 \\ 0 & 0 \end{bmatrix} x(t) + \begin{bmatrix} 0 \\ 1 \end{bmatrix} w(t)$$

$$z(t) = [1 \quad 0]x(t) + v(t)$$

where

$$w(t) \sim N(0, 2), \qquad v(t) \sim N(0, 1)$$

$$P(0) \sim N\left(\begin{bmatrix} 0 \\ 0 \end{bmatrix}, \begin{bmatrix} 4 & 0 \\ 0 & 1 \end{bmatrix}\right)$$

Apply Theorem 6.4 to derive a partitioning fixed-point smoother with respect to time 0.

6.5 Complete the derivation of equations (6.155) and (6.156).

6.6 Verify the result of Corollary 6.4.

6.7 Consider again the system given in Exercise 6.4. By using the generalized two-filter smoother described in equations (6.164) and (6.165), develop a fixed-interval smoother for time interval $[0, T]$.

6.8 Apply the Hamiltonian system given by equations (6.22) and (6.23) to show that

$$\hat{x}(s|T) = \hat{x}(s|s) + P(s)\lambda(s|T)$$

$$P(s|T) = P(s) - P(s)\Lambda(s|T)P(s)$$

$$-\frac{\mathrm{d}}{\mathrm{d}s}\lambda(s|T) = [A(s) - P(s)H^{\mathrm{T}}(s)R^{-1}(s)H(s)]^{\mathrm{T}}\lambda(s|T) + H^{\mathrm{T}}(s)R^{-1}(s)[z(s) - H(s)\hat{x}(s|s)], \qquad \lambda(T|T) = 0$$

$$-\frac{\mathrm{d}}{\mathrm{d}s}\Lambda(s|T) = [A(s) - P(s)H^{\mathrm{T}}(s)R^{-1}(s)H(s)]^{\mathrm{T}}\Lambda(s|T)$$
$$+ \Lambda(s|T)[A(s) - P(s)H^{\mathrm{T}}(s)R^{-1}(s)H(s)]$$
$$- H^{\mathrm{T}}(s)R^{-1}(s)H(s), \qquad \Lambda(T|T) = 0$$

where $\Lambda(s|T) = E[\lambda(s|t)\lambda^{\mathrm{T}}(s|T)]$, $P(s|T) = E[\tilde{x}(s|T)\tilde{x}^{\mathrm{T}}(s|T)]$, $\tilde{x}(s|T) = x(s) - \hat{x}(s|T)$ and $E[\tilde{x}(s|T)\lambda^{\mathrm{T}}(s|T)] = 0$. This is called the Bryson–Frazier smoother in continuous time (see Bryson and Ho [7]).

6.9 Apply again the Hamiltonian system given by equations (6.22) and (6.23) to derive the following:

$$\hat{x}(s|T) = \hat{x}_{\mathrm{b}}(s|s) - P_{\mathrm{b}}(T, s)\lambda(s|T)$$

$$P(s|T) = P_{\mathrm{b}}(T, s) - P_{\mathrm{b}}(T, s)\Lambda(s|T)P_{\mathrm{b}}(T, s)$$

$$\frac{\mathrm{d}}{\mathrm{d}s}\lambda(s|T) = -[A(s) + P_{\mathrm{b}}(T, s)H^{\mathrm{T}}(s)R^{-1}(s)H(s)]^{\mathrm{T}}\lambda(s|T)$$
$$- H^{\mathrm{T}}(s)R^{-1}(s)[z(s) - H(s)\hat{x}_{\mathrm{b}}(s|T)],$$
$$\lambda(0|T) = [P_{\mathrm{b}}(t, 0) + P(0)]^{-1}[\hat{x}_{\mathrm{b}}(0|T) - \hat{x}(0)]$$

$$\frac{\mathrm{d}}{\mathrm{d}s}\Lambda(s|T) = -[A(s) + P_{\mathrm{b}}(T, s)H^{\mathrm{T}}(s)R^{-1}(s)H(s)]^{\mathrm{T}}\Lambda(s|T)$$
$$- \Lambda(s|T)[A(s) + P_{\mathrm{b}}(T, s)H^{\mathrm{T}}(s)R^{-1}(s)H(s)]$$
$$+ H^{\mathrm{T}}(s)R^{-1}(s)H(s),$$
$$\Lambda(0|T) = [P_{\mathrm{b}}(T, 0) + P(0)]^{-1}$$

where $\{\hat{x}_{\mathrm{b}}(s|T), P_{\mathrm{b}}(T, s)\}$ are given by equations (6.153) and (6.154). This can be seen as a dual version (in time) of the Bryson–Frazier smoother in continuous time.

6.10 Apply the Hamiltonian system given by equations (6.22) and (6.23) to derive the RTS smoother given in equations (6.155) and (6.156).

6.11 Consider the results of Exercise 6.8.

(a) Prove that

$$\lambda(t|T) = \int_t^T \Phi^{\mathrm{T}}(s, t)H^{\mathrm{T}}(s)R^{-1}(s) \times [z(s) - H(s)\hat{x}(s|s)]\,\mathrm{d}s$$

$$\Lambda(t|T) = -\int_t^T \Phi^{\mathrm{T}}(s, t)H^{\mathrm{T}}(s)R^{-1}(s)H(s)\Phi(s, t)\,\mathrm{d}s$$

where $t \leqslant s$ is current time, T is fixed and $\Phi(s, t)$ is subject to

$$\frac{\mathrm{d}\Phi(s, t)}{\mathrm{d}t} = -\Phi(s, t)[A(t) - P(t)H^{\mathrm{T}}(t)R^{-1}(t)H(t)],$$
$$\Phi(s, s) = I$$

Hint: Use Leibniz' rule

$$\frac{\mathrm{d}}{\mathrm{d}t}\int_{a(t)}^{b(t)} f(t, \tau)\,\mathrm{d}\tau =$$

$$f(t, b(t))\,\frac{\mathrm{d}b(t)}{\mathrm{d}t} - f(t, a(t))\,\frac{\mathrm{d}a(t)}{\mathrm{d}t} + \int_{a(t)}^{b(t)} \frac{\partial}{\partial t} f(t, \tau)\,\mathrm{d}\tau$$

(b) Apply the results of Exercise 6.8 and part (a) to derive the RTS smoother given in equations (6.155) and (6.156).

(c) Use the expressions

$$\lambda(t_*|t) = \int_{t_*}^{t} \Phi^{\mathrm{T}}(s, t_*)H^{\mathrm{T}}(s)R^{-1}(s)[z(s) - H(s)\hat{x}(s|s)]\,\mathrm{d}s$$

$$\Lambda(t_*|t) = -\int_{t_*}^{t} \Phi^{\mathrm{T}}(s, t_*)H^{\mathrm{T}}(s)R^{-1}(s)H(s)\Phi(s, t_*)\,\mathrm{d}s$$

to derive the *usual* fixed-point smoother:

$$\frac{\mathrm{d}}{\mathrm{d}t}\hat{x}(t_*|t) =$$

$$P(t_*)\Phi^{\mathrm{T}}(t, t_*)H^{\mathrm{T}}(t)R^{-1}(t)[z(t) - H(t)\hat{x}(t|t)],$$

$$\hat{x}(t_*|t)|_{t=t_*} = \hat{x}(t_*|t_*)$$

$$\frac{\mathrm{d}}{\mathrm{d}t}P(t_*|t) =$$

$$-P(t_*)\Phi^{\mathrm{T}}(t, t_*)H^{\mathrm{T}}(t)R^{-1}(t)H(t)\Phi(t, t_*)P(t_*),$$

$$P(t_*|t)|_{t=t_*} = P(t_*)$$

where t_* is a constant.

(d) Using the expressions

$$\lambda(t|t+\Delta) = \int_{t}^{t+\Delta} \Phi^{\mathrm{T}}(s, t)H^{\mathrm{T}}(s)R^{-1}(s) \times [z(s) - H(s)\hat{x}(s|s)]\,\mathrm{d}s$$

$$\Lambda(t|t+\Delta) = -\int_{t}^{t+\Delta} \Phi^{\mathrm{T}}(s, t)H^{\mathrm{T}}(s)R^{-1}(s)H(s)\Phi(s, t)\,\mathrm{d}s$$

where Δ is a fixed constant, derive the formula for the fixed-lag smoother:

$$\begin{aligned}\frac{\mathrm{d}}{\mathrm{d}t}\hat{x}(t|t+\Delta) &= A(t)\hat{x}(t|t+\Delta)\\ &+ \widetilde{Q}(t)P^{-1}(t)[\hat{x}(t|t+\Delta) - \hat{x}(t|t)]\\ &+ P(t)\Phi^{\mathrm{T}}(t+\Delta, t)\\ &\times H^{\mathrm{T}}(t+\Delta)R^{-1}(t+\Delta)\\ &\times [z(t+\Delta) - H(t+\Delta)\hat{x}(t+\Delta|t+\Delta)]\end{aligned}$$

$$\frac{\mathrm{d}}{\mathrm{d}t} P(t|t+\Delta) = [A(t) + \tilde{Q}(t)P^{-1}(t)]P(t|t+\Delta) + P(t|t+\Delta)[A(t) + \tilde{Q}(t)P^{-1}(t)]^{\mathrm{T}} - \tilde{Q}(t) - J(t)$$

where

$$\tilde{Q}(t) = B(t)Q(t)B^{\mathrm{T}}(t)$$

$$J(t) = \Phi^{\mathrm{T}}(t+\Delta, t)H^{\mathrm{T}}(t+\Delta)R^{-1}(t+\Delta) \times H(t+\Delta)\Phi(t+\Delta, t)$$

and the initial conditions $\{\hat{x}(0|\Delta), P(0|\Delta)\}$ are obtained from the fixed-point smoother given in part (c).

6.12 Verify the result of Corollary 6.7.

6.13 Complete the proof of Corollary 6.8.

6.14 Consider the partitioning filtering problem for the system given in Exercise 6.4.

(a) Setting $P_{\mathrm{n}}(0) = 0$, develop a nominal Kalman filter by applying the Chandrasekhar algorithm described in Corollary 6.9.

(b) Write down a nominal adjoint vector by using the dual Chandrasekhar algorithm given in Corollary 6.10 and applying the result of part (a).

6.15 Let us consider the discrete-time version of the scattering structure [52]. If we define the generator of the scattering matrix as

$$M(s) = S_0(s, s+1) \triangleq \begin{bmatrix} f(s) & g(s) \\ h(s) & e(s) \end{bmatrix}$$

for any integer s, then the forward and backward evolution equations become

$$S_0(\tau, s+1) = S_0(\tau, s) * M(s), \qquad S_0(\tau, \tau) = I$$

$$S_0(s-1, t) = M(s-1) * S_0(s, t), \qquad S_0(t, t) = I$$

Let the generator be

$$M(t) = \begin{bmatrix} A(t) & B(t)Q(t)B^{\mathrm{T}}(t) \\ -H^{\mathrm{T}}(t)R^{-1}(t)H(t) & A^{\mathrm{T}}(t) \end{bmatrix}$$

and the initial condition matrix $\bar{\Pi}$ be

$$\bar{\Pi}(\tau) = \begin{bmatrix} I & \Pi \\ 0 & I \end{bmatrix}$$

Further, denote the entries of $S(\tau, t; \bar{\Pi}) = \bar{\Pi} * S_0(\tau, t)$ by

$$S(\tau, t; \bar{\Pi}) = \begin{pmatrix} \Phi(t, \tau; \Pi) & P(t, \tau; \Pi) \\ W(t, \tau; \Pi) & \Phi^{\mathrm{T}}(t, \tau; \Pi) \end{pmatrix}$$

(a) Show that we can write the forward evolution equations as follows:

$$P(t+1, \tau; \bar{\Pi}) = B(t)Q(t)B^{\mathrm{T}}(t) + A(t)P(t, \tau; \bar{\Pi}) \times [I + H^{\mathrm{T}}(t)R^{-1}(t)H(t) \times P(t, \tau; \bar{\Pi})]^{-1}A^{\mathrm{T}}(t),$$

$$P(\tau, \tau; \bar{\Pi}) = \Pi$$

$$\Phi(t+1, \tau; \bar{\Pi}) = A(t)[I + P(t, \tau; \bar{\Pi})H^{\mathrm{T}}(t)R^{-1}(t)H(t)]^{-1} \times \Phi(t, \tau; \bar{\Pi}), \qquad \Phi(\tau, \tau; \bar{\Pi}) = I$$

$$W(t+1, \tau; \bar{\Pi}) = W(t, \tau; \bar{\Pi}) - \Phi^{\mathrm{T}}(t, \tau; \bar{\Pi})H^{\mathrm{T}}(t)R^{-1}(t)H(t) \times [I + P(t, \tau; \bar{\Pi})H^{\mathrm{T}}(t)R^{-1}(t)H(t)]^{-1} \times \Phi(t, \tau; \bar{\Pi}), \qquad W(\tau, \tau; \bar{\Pi}) = 0$$

(b) Show that we can identify $P(t, \tau; \bar{\Pi})$ as $P(t|t-1)$ and $\Phi(t, \tau; \bar{\Pi})$ as the state transition matrix of the closed-loop system for the predictive-type Kalman filter in the case of uncorrelated system and measurement noises. (Hint: Use the matrix inversion lemma given in equation (5A.2).)

6.16 Consider again the discrete-time version of the scattering structure as described in Exercise 6.15.

(a) Show that the backward evolution equation can be rewritten as

$$S(s-1, t) = \mathcal{M}(s-1) * S(s, t), \qquad s \leqslant t$$

where

$$\mathcal{M}(s) = \bar{\Pi}(s) * M(s) * \bar{\Pi}^{-1}(s+1)$$

(b) When defining

$$\mathcal{M}(s) \triangleq \begin{bmatrix} \tilde{A}(s) & \tilde{Q}(s) \\ -\tilde{H}^{\mathrm{T}}(s)\tilde{H}(s) & \tilde{A}^{\mathrm{T}}(s) \end{bmatrix}$$

show that the backward evolution equations become

$$P(t, s-1; \bar{\Pi}) = P(t, s; \bar{\Pi}) + \Phi(t, s; \bar{\Pi})\tilde{Q} \times [I - W(t, s; \bar{\Pi})\tilde{Q}]^{-1}\Phi^{\mathrm{T}}(t, s; \bar{\Pi}),$$

$$P(t, t; \bar{\Pi}) = \Pi$$

$$\Phi(t, s-1; \bar{\Pi}) = \Phi(t, s; \bar{\Pi})[I - \tilde{Q}W(t, s; \bar{\Pi})]^{-1}\tilde{A}(t),$$

$$\Phi(t, t; \bar{\Pi}) = I$$

$$W(t, s-1; \bar{\Pi}) = \widetilde{H}^{\mathrm{T}}\widetilde{H} + \Phi(t, s; \bar{\Pi})W(t, s; \bar{\Pi}) \times [I - \widetilde{Q}W(t, s; \bar{\Pi})]^{-1}\Phi(t, s; \bar{\Pi}), \qquad W(t, t; \bar{\Pi}) = 0$$

(c) Show that the following expression for $\widetilde{Q}(\tau)$ can be obtained:

$$\begin{aligned}\widetilde{Q}(\tau) &= B(\tau)Q(\tau)B^{\mathrm{T}}(\tau) + A(\tau)\Pi \\ &\quad \times [I + H^{\mathrm{T}}(\tau)R^{-1}(\tau)H(\tau)\Pi]^{-1} \\ &\quad \times A^{\mathrm{T}}(\tau) - \Pi \\ &= P(\tau+1, \tau; \Pi) - \Pi\end{aligned}$$

(d) By decomposing

$$\widetilde{Q}(\tau) = \widetilde{G}\,\Sigma\widetilde{G}^{\mathrm{T}}$$

where $\widetilde{G}$ is a full-rank matrix and Σ is the signature matrix of $\widetilde{Q}$, show that the backward evolution equation for $P(t, s; \bar{\Pi})$ can be rewritten as

$$\begin{aligned}P(t, s-1; \bar{\Pi}) = P(t, s; \bar{\Pi}) &+ \Phi(t, s; \bar{\Pi})\widetilde{G} \\ &\times [I - \Sigma\widetilde{G}^{\mathrm{T}}W(t, s; \bar{\Pi})\widetilde{G}\Sigma]^{-1} \\ &\times \widetilde{G}^{\mathrm{T}}\Phi^{\mathrm{T}}(t, s; \bar{\Pi})\end{aligned}$$

6.17 Let the system considered be time-invariant. Applying the backward evolution equation for $P(t, s; \bar{\Pi})$ given in Exercise 6.16(d) and the forward evolution equations for $\Phi(s, \tau; \bar{\Pi})$ and $W(s, \tau; \bar{\Pi})$, and defining

$$Y(s, \tau) \triangleq \Phi(s, \tau; \bar{\Pi})\widetilde{G}$$

$$K(s, \tau) \triangleq AP(s, \tau; \bar{\Pi})H^{\mathrm{T}}[R + HP(s, \tau; \bar{\Pi})H^{\mathrm{T}}]^{-1}$$

$$M(s, \tau) \triangleq [I - \Sigma\widetilde{G}W(s, \tau; \bar{\Pi})\widetilde{G}\Sigma]^{-1}$$

$$\Omega(s, \tau) \triangleq R + HP(s, \tau; \bar{\Pi})H^{\mathrm{T}}$$

show that

$$\Omega(s+1, \tau) = \Omega(s, \tau) + HY(s, \tau)M(s, \tau)Y^{\mathrm{T}}(s, \tau)H^{\mathrm{T}}, \qquad \Omega(\tau, \tau) = R + H\Pi H^{\mathrm{T}}$$

$$K(s+1, \tau) = [K(s, \tau)\Omega(s, \tau) + AY(s, \tau)M(s, \tau)Y^{\mathrm{T}}(s, \tau)H^{\mathrm{T}}]\Omega^{-1}(s+1, \tau), \qquad K(\tau, \tau) = A\Pi H^{\mathrm{T}}\Omega^{-1}(\tau, \tau)$$

$$Y(s+1, \tau) = [A - K(s, \tau)H]Y(s, \tau), \qquad Y(\tau, \tau) = \widetilde{G}$$

$$M(s+1, \tau) = M(s, \tau) + M(s, \tau)Y^{\mathrm{T}}(s, \tau)H^{\mathrm{T}}\Omega^{-1}(s, \tau)HY(s, \tau)M(s, \tau), \qquad M(\tau, \tau) = I$$

which is the discrete-time version of the Chandrasekhar-type equations (cf. Chapter 2). (Hint: The system is time-invariant, i.e. $\Omega(s-1, t) = \Omega(s, t+1)$ and $K(s-1, t) = K(s, t+1)$, so that $\Omega(s-1, t) = \Omega(s+1, \tau)$ and $K(s-1, t) = K(s+1, \tau)$.)

REFERENCES

[1] REDHEFFER, R. M., On the Relation of Transmission-Line Theory to Scattering and Transfer, *J. Math. Phys.*, vol. 41, 1962, pp. 1–41.

[2] NICHOLSON, H., Concepts of Scattering in the Linear Optimal-Control Problem, *Proc. IEE*, vol. 118, December 1971, pp. 1823–9.

[3] NICHOLSON, H., Structure of Kron's Polyhedron Model and the Scattering Problem, *Proc. IEE*, vol. 120, December 1973, pp. 1545–57.

[4] NICHOLSON, H., System Concepts in Kron's Polyhedron Model and the Scattering Problem, *Int. J. Control*, vol. 20, no. 4, 1974, pp. 529–55.

[5] VERGHESE, G., FRIEDLANDER, B. and KAILATH, T., Scattering Theory and Linear Least-Squares Estimation, Part III: The Estimates, *IEEE Trans. Aut. Control*, vol. AC-25, no. 4, August 1980, pp. 794–802.

[6] KAILATH, T., Redheffer Scattering Theory and Linear State-Space Estimation Problems, *Ricerche di Automatica*, vol. 10, no. 2, December 1979, pp. 136–62.

[7] WEINERT, H. L. and DESAI, U. B., On Complementary Models and Fixed-Interval Smoothing, *IEEE Trans. Aut. Control*, vol. AC-26, no. 4, August 1981, pp. 863–7.

[8] KAILATH, T., Some New Algorithms for Recursive Estimation in Constant Linear Systems, *IEEE Trans. Inf. Theory*, vol. IT-19, no. 6, November 1973, pp. 750–60.

[9] LJUNG, L., KAILATH, T. and FRIEDLANDER, B., Scattering Theory and Linear Least Squares Estimation, Part I: Continuous-Time Problems, *Proc. IEEE*, vol. 64, January 1976, pp. 131–8.

[10] KAILATH, T. and LJUNG, L., A Scattering Theory and Framework for Fast Least-Squares Algorithms. In *Multivariate Analysis—VI*, North-Holland, Amsterdam, 1977, pp. 387–406.

[11] LAINIOTIS, D. G., Optimal Linear Smoothing: Continuous Data Case, *Int. J. Control*, vol. 17., no. 5, 1973, pp. 921–30.

[12] LJUNG, L. and KAILATH, T., A Unified Approach to Smoothing Formulas, *Automatica*, vol. 12, 1976, pp. 147–57.

[13] LAINIOTIS, D. G., General Backwards Markov Models, *IEEE Trans. Aut. Control*, vol. AC-21, no. 4, 1976, pp. 595–8.

[14] BRYSON, A. E. and HO, Y.-H., *Applied Optimal Control*, Hemisphere, New York, 1975.

[15] RAUCH, H. E., TUNG, F. and STRIEBEL, C. T., Maximum Likelihood Estimates of Linear Dynamic Systems, *AIAA J.*, vol. 3, no. 8, August 1965, pp. 1445–50.

[16] JAZWINSKI, A. H., *Stochastic Processes and Filtering Theory*, Academic Press, New York, 1970.

[17] ZACHRISSON, L., On Optimal Smoothing of Continuous Time Kalman Processes, *Inf. Sci.*, vol. 1, April 1969, pp. 143–72.

[18] KAILATH, T. and FROST, P., An Innovation Approach to Least-Squares Estimation Part II: Linear Smoothing in Additive White Noise, *IEEE Trans. Aut. Control*, vol. AC-13, no. 6, December 1968, pp. 655–60.

[19] LJUNG, L. and KAILATH, T., Efficient Change of Initial Conditions, Dual Chandrasekhar Equations, and Some Applications, *IEEE Trans. Aut. Control*, vol. AC-22, June 1977, pp. 443–7.

[20] WATANABE, K., Partitioned Estimators Based on the Perturbed Kalman Filter Equations, *Int. J. Syst. Sci.*, vol. 14, no. 9, 1983, pp. 1115–28.

[21] MEDITCH, J. S., *Stochastic Optimal Linear Estimation and Control*, McGraw-Hill, New York, 1969.

[22] FRASER, D. C. and POTTER, J. E., The Optimum Linear Smoother as a Combination of Two Optimum Linear Filters, *IEEE Trans. Aut. Control*, vol. AC-14, 1969, pp. 387–90.

[23] WALL, J. E., WILLSKY, A. S. and SANDELL, N. R., On The Fixed-Interval Smoothing Problem, *Stochastics*, vol. 5, 1981, pp. 1–42.

[24] WATANABE, K., Scattering Framework for Backwards Partitioned Estimators, *Int. J. Syst. Sci.*, vol. 16, no. 5, 1985, pp. 553–72.

[25] LJUNG, L. and KAILATH, T., Backwards Markovian Models for Second-Order Stochastic Processes, *IEEE Trans. Inf. Theory*, vol. IT-22, July 1976, pp. 488–91.

[26] VERGHESE, G. and KAILATH, T., A Further Note on Backward Markovian Models, *IEEE Trans. Inf. Theory*, vol. IT-25, 1979, pp. 121–4; Correction to 'A Further Note on Backward Markovian Models', *IEEE Trans. Inf. Theory*, vol. IT-25, July 1979.

[27] SIDHU, G. S. and DESAI, U. B., New Smoothing Algorithms Based on Reversed-Time Lumped Models, *IEEE Trans. Aut. Control*, vol. AC-21, no. 4, 1976, pp. 538–41.

[28] SOLO, V., Smoothing Estimation of Stochastic Processes: Two-Filter Formulas, *IEEE Trans. Aut. Control*, vol. AC-27, no. 2, April 1982, pp. 473–6.

[29] GELB, A., *Applied Optimal Estimation*, MIT Press, Cambridge, Massachusetts, 1974.

[30] WATANABE, K., Algebraic Solution of a Forward-Pass Fixed-Interval Smoother: Continuous-Time Systems, *Int. J. Syst. Sci.*, vol. 16, no. 10, 1985, pp. 1217–27.

[31] LAINIOTIS, D. G., Generalized Chandrasekhar Algorithms: Time Varying Models, *IEEE Trans. Aut. Control*, vol. AC-21, no. 5, 1976, pp. 728–32.

[32] SIDHU, G. S. and BIERMAN, G. J., Integration-Free Integral Doubling for Riccati Equation Solutions, *IEEE Trans. Aut. Control*, vol. AC-22, no. 5, 1977, pp. 831–4.

[33] BRAMMER, R. F., A Note on the Use of Chandrasekhar Equations for the Calculation of the Kalman Gain Matrix, *IEEE Trans. Inf. Theory*, vol. IT-21, 1975, pp. 334–6.

[34] CASTI, J. L., *Dynamical Systems and Their Applications: Linear Theory*, Academic Press, New York, 1977.

[35] CHANDRASEKHAR, S., *Radiative Transfer*, Dover Publications, New York, 1960.

[36] REID, W. T., *Riccati Differential Equations*, Academic Press, New York, 1972.

[37] VERRIEST, E., FRIEDLANDER, B. and MORF, M., Distributed Processing in Estimation and Detection, *Proc. 1979 IEEE Symp. Dec. Contr.*, Florida, December 1979, pp. 153–8.

[38] LEVEY, B. C., CASTANON, D. A., VERGHESE, G. C. and WILLSKY, A. S., A Scattering Framework for Decentralized Estimation Problems, *Automatica*, vol. 19, no. 4, 1983, pp. 373–84.

[39] WATANABE, K., Decentralized Fixed-Interval Smoothing Algorithms, *Trans. ASME, J. Dynamic Syst. Meas. Control*, vol. 108, no. 1, 1986, pp. 86–9.

[40] BRUCKSTEIN, A. and KAILATH, T., Recursive Limited Memory Filtering and Scattering Theory, *IEEE Trans. Inf. Theory*, vol. IT-31, no. 3, May 1985, pp. 440–3.

[41] WATANABE, K. and TZAFESTAS, S. G., Generalized Fixed-Interval Smoothers for Discrete-Time Systems with Predictive Information, *Preprint of 8th IFAC/IFROS Symposium on Identification and System Parameter Estimation*, Beijing, 27–31 August 1988, Vol. 2, pp. 872–7.

[42] ADAMS, M. B., WILLSKY, A. S. and LEVY, B. C., Linear Estimation of Boundary Value Stochastic Processes—Part I: The Role and Construction of Complementary Models, *IEEE Trans. Aut. Control*, vol. AC-29, no. 9, September 1984, pp. 803–11.

[43] ADAMS, M. B., WILLSKY, A. S. and LEVY, B. C., Linear Estimation of Boundary Value Stochastic Processes—Part II: 1-D Smoothing Problems, *IEEE Trans. Aut. Control*, vol. AC-29, no. 9, September 1984, pp. 811–21.

[44] ACKNER, R. and KAILATH, T., Complementary Models and Smoothing, *IEEE Trans. Aut. Control*, vol. AC-34, no. 9, 1989, pp. 963–9.

[45] ACKNER, R. and KAILATH, T., Discrete-Time Complementary Models and Smoothing, *Int. J. Control*, vol. 49, no. 5, 1989, pp. 1665–82.

[46] GOREZ, R., Matrix Fraction and Chandrasekhar Equation Techniques in the Design of Linear Quadratic Optimal Control Systems, *Int. J. Syst. Sci.*, vol. 12, no. 8, 1981, pp. 907–15.

[47] MORF, M., SIDHU, G. S. and KAILATH, T., Some New Algorithms for Recursive Estimation in Constant Linear, Discrete-Time Systems, *IEEE Trans. Aut. Control*, vol. AC-19, no. 4, 1974, pp. 315–23.

[48] MAHALANABIS, A. K. and RAY, G., Efficient Algorithm for the Optimal Control of Interconnected Stochastic Systems, *Proc. IEE*, vol. 131, pt. D, no. 3, 1984, pp. 99–102.

[49] CASTI, J. and LJUNG, L., Some New Analytic and Computational Results for Operator Riccati Equations, *SIAM J. Control*, vol. 13, no. 4, July 1975, pp. 817–26.

[50] BARAS, J. S. and LAINIOTIS, D. G., Chandrasekhar Algorithms for Linear Time Varying Distributed Systems, *Inf. Sci.*, vol. 17, 1979, pp. 153–67.

[51] WATANABE, K., Generalized Chandrasekhar Algorithms for Distributed-Parameter Filtering Problem with Pointwise Coloured Measurement Noise, *Int. J. Syst. Sci.*, vol. 13, no. 6, 1982, pp. 619–37.

[52] FRIEDLANDER, B., KAILATH, T. and LJUNG, L., Scattering Theory and Linear Least Squares Estimation—II. Discrete-Time Problems, *J. Franklin Inst.*, vol. 301, nos 1/2, 1976, pp.71–82.

7

The Pseudolinear Partitioning Filter and Tracking Motion Analysis

7.1 INTRODUCTION

The problem of estimating the position and velocity of a vehicle travelling on a straight-line course at constant speed from measurements of bearings is known as *passive bearing-only tracking*. It has already been shown by Aidala [1] and Lindgren and Gong [2] that the extended Kalman filter often exhibits unstable behaviours when applied to such bearing-only tracking motion analysis (TMA). They have therefore proposed a *pseudolinear tracking filter* algorithm as an alternative approach. This algorithm is known to be stable and computationally simple to implement, though it has the undesirable property of generating biased estimates whenever noisy measurements are processed [3].

This chapter presents a *pseudolinear partitioned tracking filter* [4] by applying the discrete-time partitioning filter technique presented in Chapter 5. After transforming the nonlinear observation system into a linear one, we make an initialization for a pseudolinear filter. We then derive a pseudolinear partitioning filter in a form suitable for *recursive* processing, since the previous partitioning filters discussed in Chapter 5 (cf. Park and Lainiotis [5] and Govindaraj and Lainiotis [6]) are given in a form suitable for *batch* processing. In addition, with an eye on the parallel processing mechanism, we construct a data compression algorithm to average the bearing data contaminated by random measurement noise.

7.2 PROBLEM FORMULATION

The bearing-only TMA problem is described by a linear state model and a nonlinear scalar measurement equation [1–3]:

$$x(k+1) = \Phi(k+1, k)x(k) - M(k) \tag{7.1}$$

$$\beta(k+1) = h[x(k+1)] + v(k+1), \qquad k = 0, 1, \ldots \tag{7.2}$$

where the target velocity is assumed to be constant, and $x(k)$ is a four-dimensional state vector at time $k\Delta t$ (where Δt is the sampling interval) given in partitioned form by

$$x(k) = \begin{bmatrix} r(k) \\ V(k) \end{bmatrix} \tag{7.3}$$

Here, $r(k) = [r_x(k) \quad r_y(k)]^{\mathrm{T}}$ (where $r_x(k) \triangleq r_{\mathrm{T}_x}(k) - r_{\mathrm{S}_x}(k)$, $r_y(k) \triangleq r_{\mathrm{T}_y}(k) - r_{\mathrm{S}_y}(k)$) and $V(k) = [V_x(k) \quad V_y(k)]^{\mathrm{T}}$ (where $V_x(k) \triangleq V_{\mathrm{T}_x}(k) - V_{\mathrm{S}_x}(k)$, $V_y(k) \triangleq V_{\mathrm{T}_y}(k) - V_{\mathrm{S}_y}(k)$) represent the relative target position and velocity vectors, respectively. The geometric configuration of this two-dimensional bearing-only TMA is depicted in Figure 7.1.

$\Phi(k+1, k)$, which is a 4×4 deterministic transition matrix, is given by

$$\Phi(k+1, k) = \begin{bmatrix} 1 & 0 & \Delta t & 0 \\ 0 & 1 & 0 & \Delta t \\ 0 & 0 & 1 & 0 \\ 0 & 0 & 0 & 1 \end{bmatrix} \tag{7.4}$$

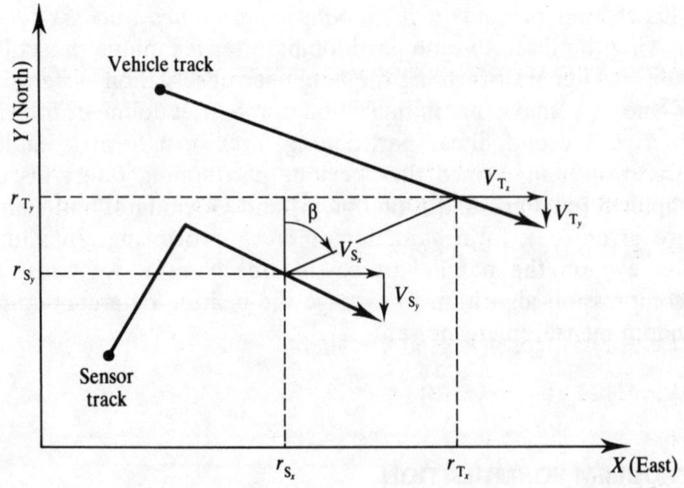

Figure 7.1 Geometrical configuration for two-dimensional bearing-only tracking motion analysis.

and $M(k)$ is a four-dimensional vector of deterministic inputs described by (cf. Appendix 7A):

$$M(k) = \begin{bmatrix} \int_0^{\Delta t} t a_s(k\Delta t - t)\,\mathrm{d}t \\ \int_0^{\Delta t} a_s(k\Delta t - t)\,\mathrm{d}t \end{bmatrix} \tag{7.5}$$

where $a_s(\cdot) = [a_{s_x}(\cdot) \quad a_{s_y}(\cdot)]^{\mathrm{T}}$ specifies own-sensor acceleration. The measured bearing $\beta(k+1)$ is contaminated by a zero-mean Gaussian white noise $v(k+1)$ with variance $\sigma_v^2(k+1)$ and the nonlinear function $h[x(k+1)]$ is defined by

$$h[x(k+1)] = \tan^{-1}[r_x(k+1)/r_y(k+1)] \tag{7.6}$$

Moreover, it is assumed that the initial state $x(0) \sim N(\hat{x}(0),\ P(0))$ is independent of $v(\cdot)$.

Because of nonlinearities in equation (7.6), direct use of the usual Kalman filter is not applicable to equations (7.1) and (7.2). Although the approximate filters, such as the extended Kalman filter or first-order filter (see Chapter 2) and the second-order filter [7–10] must be employed for this purpose, it should be noted that the extended Kalman filter presents various instances of premature covariance collapse and solution divergence in these kinds of problems [1]. To circumvent these difficulties, pseudolinear filtering algorithms [1–3] are proposed as an alternative method of tracking.

7.3 DERIVATION OF THE PSEUDOLINEAR PARTITIONING FILTER

In this section, a pseudolinear partitioning filter will be derived by directly manipulating equation (7.2). That is, we first note that

$$0 = H(k+1)x(k+1) + \varepsilon(k+1) \tag{7.7}$$

where

$$H(k+1) \triangleq [\cos\beta(k+1) \quad -\sin\beta(k+1) \quad 0 \quad 0] \tag{7.8}$$

$$\varepsilon(k+1) \triangleq r(k+1)\sin v(k+1) \tag{7.9}$$

$$r(k+1) \triangleq [r_x^2(k+1) + r_y^2(k+1)]^{1/2} \tag{7.10}$$

If we define

$$\sin v(k) \sim N(0,\ \sigma^2(k)) \tag{7.11}$$

then $\varepsilon(k)$ in expression (7.9) has the following statistical properties [3]:

$$\varepsilon(k) \sim N(0, r^2(k)\sigma^2(k)) \tag{7.12a}$$

where

$$\sigma^2(k) = \tfrac{1}{2}[1 - \exp\{-2\sigma_v^2(k)\}] \tag{7.12b}$$

However, in order to construct a pseudolinear filter, we have to know the noise variance $r^2(k)\sigma^2(k)$. To overcome this inconvenience, it is here assumed that

$$r^2(k)\sigma^2(k) = r^2(k-1)\sigma^2(k-1) \tag{7.13}$$

At the same time, normalizing equation (7.9) and the initial error covariance $P(0)$ by means of equation (7.13), it is found that

$$\varepsilon(k) \sim N(0, 1) \tag{7.14}$$

$$P(0) \sim N(\hat{x}(0), I) \tag{7.15}$$

For the initialization of the state vector, the following range-parameterized solution [2] may be adopted:

$$\hat{x}(0) = [\hat{r}(0)\sin\beta(0) \quad \hat{r}(0)\cos\beta(0)\ 0\ 0]^{\mathrm{T}} \tag{7.16}$$

where

$$\hat{r}(0) \triangleq [\hat{r}_x^2(0) + \hat{r}_y^2(0)]^{1/2} \tag{7.17}$$

We can now state the following lemma.

Lemma 7.1 Pseudolinear partitioning filter in batch-processing form

The pseudolinear partitioning filter for equations (7.1) and (7.7) and its associated error covariance are given by

$$\hat{x}(k|k) = \hat{x}_{\mathrm{n}}(k|k) + \Phi_{\mathrm{n}}(k, 0)\hat{x}_{\mathrm{r}}(0|k) \tag{7.18}$$

$$P(k|k) = P_{\mathrm{n}}(k|k) + \Phi_{\mathrm{n}}(k, 0)P_r(0|k)\Phi_{\mathrm{n}}^{\mathrm{T}}(k, 0) \tag{7.19}$$

where $\hat{x}_{\mathrm{r}}(0|k)$ and $P_{\mathrm{r}}(0|k)$ are the partitioning fixed-point smoothed value and the corresponding estimation error covariance, respectively, obtained from the following equalities:

$$\hat{x}_{\mathrm{r}}(0|k) = P_{\mathrm{r}}(0|k)[\lambda_{\mathrm{n}}(k, 0) + P_{\mathrm{r}}^{-1}\hat{x}_{\mathrm{r}}] \tag{7.20}$$

$$P_{\mathrm{r}}(0|k) = [M_{\mathrm{n}}(k, 0) + P_{\mathrm{r}}^{-1}]^{-1} \tag{7.21}$$

where

$$\lambda_n(k, 0) = \sum_{j=1}^{k} \Phi_n^T(j-1, 0)\Phi^T(j, j-1)H^T(j)[H(j)P_n(j|j-1) \times H^T(j) + 1]^{-1}[0 - H(j)\hat{x}_n(j|j-1)] \tag{7.22}$$

$$M_n(k, 0) = \sum_{j=1}^{k} \Phi_n^T(j-1, 0)\Phi^T(j, j-1)H^T(j)[H(j)P_n(j|j-1) \times H^T(j) + 1]^{-1}H(j)\Phi(j, j-1)\Phi_n(j-1, 0) \tag{7.23}$$

Furthermore, the nominal filtered value $\hat{x}_n(k|k)$ is also given by

$$\hat{x}_n(k|k) = \hat{x}_n(k|k-1) - K_n(k)H(k)\hat{x}_n(k|k-1) \tag{7.24a}$$

$$\hat{x}_n(0|0) = \hat{x}_n \tag{7.24b}$$

where

$$\hat{x}_n(k|k-1) = \Phi(k, k-1)\hat{x}_n(k-1|k-1) - M(k-1) \tag{7.25}$$

$$K_n(k) = P_n(k|k-1)H^T(k)[H(k)P_n(k|k-1)H^T(k) + 1]^{-1} \tag{7.26}$$

$$\Phi_n(k, k-1) = [I - K_n(k)H(k)]\Phi(k, k-1) \tag{7.27}$$

$$P_n(k|k-1) = \Phi(k, k-1)P_n(k-1|k-1)\Phi^T(k, k-1) \tag{7.28}$$

$$P_n(k|k) = [I - K_n(k)H(k)]P_n(k|k-1) \tag{7.29a}$$

$$P_n(0|0) = P_n \tag{7.29b}$$

Proof

Using equations (7.7) and (7.14) for the independent partitioning filter algorithm developed in Corollary 5.3 with $l = 0$, we immediately obtain these equations. □

When estimates of relative position and velocity at every sampling instant are needed, the aforementioned algorithms are unsuccessful in the processing of bearing data because of the 4×4 inversion matrix which has to be calculated in expression (7.21). For this reason, we construct the following recursive-type pseudolinear partitioning filter.

Theorem 7.1 Pseudolinear partitioning filter in recursive processing form

The pseudolinear partitioning filter for equations (7.1) and (7.7) and its error covariance are given by

$$\hat{x}(k|k) = \hat{x}_n(k|k) + \Phi_n(k, 0)\hat{x}_r(0|k) \tag{7.18}$$

$$P(k|k) = P_n(k|k) + \Phi_n(k, 0)P_r(0|k)\Phi_n^T(k, 0) \tag{7.19}$$

where $\hat{x}_n(k|k)$ and $P_n(k|k)$ are as obtained in equations (7.24)–(7.26), (7.28) and (7.29). However, the pseudolinear partitioning fixed-point smoothed value $\hat{x}_r(0|k)$, the corresponding error covariance $P_r(0|k)$ and the closed-loop transition matrix $\Phi_n(k, 0)$ can be expressed, in recursive processing form, as follows:

$$\hat{x}_r(0|k) = \hat{x}_r(0|k-1) - K_r(k)H(k)[\hat{x}_n(k|k-1) + \hat{\Phi}_n(k, 0)\hat{x}_r(0|k-1)], \qquad \hat{x}_r(0|0) = \hat{x}_r \tag{7.30}$$

$$K_r(k) = P_r(0|k-1)\hat{\Phi}_n^T(k, 0)H^T(k)[H(k)\hat{\Phi}_n(k, 0)P_r(0|k-1) \times \hat{\Phi}_n^T(k, 0)H^T(k) + H(k)P_n(k|k-1)H^T(k) + 1]^{-1} \tag{7.31}$$

$$P_r(0|k) = [I - K_r(k)H(k)\hat{\Phi}_n(k, 0)]P_r(0|k-1), \qquad P_r(0|0) = P_r \tag{7.32}$$

$$\hat{\Phi}_n(k, 0) = \Phi(k, k-1)\Phi_n(k-1, 0) \tag{7.33}$$

$$\Phi_n(k, 0) = [I - K_n(k)H(k)]\hat{\Phi}_n(k, 0) \tag{7.34}$$

Proof

Using equation (7.27), we can write

$$\Phi_n(k, 0) = \Phi_n(k, k-1)\Phi_n(k-1, k-2)\ldots\Phi_n(1, 0) = [I - K_n(k)H(k)]\Phi(k, k-1)\Phi_n(k-1, 0) \tag{7.35}$$

which we can re-express as

$$\Phi_n(k, 0) = [I - K_n(k)H(k)]\hat{\Phi}_n(k, 0) \tag{7.36}$$

where

$$\hat{\Phi}_n(k, 0) \triangleq \Phi(k, k-1)\Phi_n(k-1, 0) \tag{7.37}$$

Making use of expression (7.37), we find that $M_n(k, 0)$ of equation (7.23) reduces to

$$M_n(k, 0) = M_n(k-1, 0) + \hat{\Phi}_n^T(k, 0)H^T(k) \times [H(k)P_n(k|k-1)H^T(k) + 1]^{-1}H(k)\hat{\Phi}_n(k, 0) \tag{7.38}$$

Substituting equation (7.38) into (7.21) gives

$$P_r(0|k) = \{P_r(0|k-1)^{-1} + \hat{\Phi}_n^T(k,0)H^T(k) \times [H(k)P_n(k|k-1)H^T(k)+1]^{-1}H(k)\hat{\Phi}_n(k,0)\}^{-1} \tag{7.39}$$

and further recalling expression (5A.6), which is given in the discussion of matrix inversion in Appendix 5A, yields equations (7.31) and (7.32).

Similarly, $\lambda_n(k, 0)$ from equation (7.22) can be rewritten, in recursive form, as follows:

$$\lambda_n(k,0) = \lambda_n(k-1,0) + \hat{\Phi}_n^T(k,0)H^T(k) \times [H(k)P_n(k|k-1)H^T(k)+1]^{-1} \times [0 - H(k)\hat{x}_n(k|k-1)] \tag{7.40}$$

and therefore equation (7.20) can be re-expressed by

$$\begin{aligned}\hat{x}_r(0|k) &= P_r(0|k)\{P_r^{-1}(0|k-1)\hat{x}_r(0|k-1) + \hat{\Phi}_n^T(k,0)H^T(k) \\ &\quad \times [H(k)P_n(k|k-1)H^T(k)+1]^{-1} \\ &\quad \times [0 - H(k)\hat{x}_n(k|k-1)]\} \\ &= P_r(0|k)P_r^{-1}(0|k-1)\hat{x}_r(0|k-1) \\ &\quad + P_r(0|k)\hat{\Phi}_n^T(k,0)H^T(k) \\ &\quad \times [H(k)P_n(k|k-1)H^T(k)+1]^{-1} \\ &\quad \times [0 - H(k)\hat{x}_n(k|k-1)]\end{aligned} \tag{7.41}$$

Using equation (7.32), the first term of the right-hand side of equation (7.41) becomes

$$\begin{aligned}&P_r^{-1}(0|k)P_r^{-1}(0|k-1)\hat{x}_r(0|k-1) \\ &= \hat{x}_r(0|k-1) - K_r(k)H(k)\hat{\Phi}_n(k,0)\hat{x}_r(0|k-1)\end{aligned} \tag{7.42}$$

Using equation (7.39), it follows that

$$\begin{aligned}&P_r(0|k)\hat{\Phi}_n^T(k,0)H^T(k)[H(k)P_n(k|k-1)H^T(k)+1]^{-1} \\ &\quad \times [0 - H(k)\hat{x}_n(k|k-1)] \\ &= \{P_r(0|k-1)^{-1} + \hat{\Phi}_n^T(k,0)H^T(k) \\ &\quad \times [H(k)P_n(k|k-1)H^T(k)+1]^{-1}H(k) \\ &\quad \times \hat{\Phi}_n(k,0)\}^{-1}\hat{\Phi}_n^T(k,0)H^T(k)[H(k)P_n(k|k-1)H^T(k)+1]^{-1} \\ &\quad \times [0 - H(k)\hat{x}_n(k|k-1)]\end{aligned} \tag{7.43}$$

Moreover, applying equation (5A.7), once again from the discussion of matrix inversion in Appendix 5A, to (7.43), results in

$$\begin{aligned}
&\{P_r(0|k-1) + \hat{\Phi}_n^T(k,0)H^T(k) \\
&\quad \times [H(k)P_n(k|k-1)H^T(k)+1]^{-1}H(k) \\
&\quad \times \hat{\Phi}_n(k,0)\}^{-1}\hat{\Phi}_n^T(k,0)H^T(k)[H(k)P_n(k|k-1)H^T(k)+1]^{-1} \\
&\quad \times [0 - H(k)\hat{x}_n(k|k-1)] \\
&= K_r(k)[0 - H(k)\hat{x}_n(k|k-1)]
\end{aligned} \tag{7.44}$$

Combining equations (7.41), (7.42) and (7.44) gives the desired result (7.30). □

7.4 APPLICATION TO TRACKING MOTION ANALYSIS

Now let us apply the preceding results to the problem of estimation of target states. We first select the following initial conditions:

$$\hat{x}_n = 0, \qquad P_n = 0 \tag{7.45a}$$

$$\hat{x}_r = [\hat{r}(0)\sin\beta(0) \quad \hat{r}(0)\cos\beta(0)\ 0\ 0]^T \tag{7.45b}$$

$$P_r = I \tag{7.45c}$$

It is interesting to note that, since the present system has no random noise, the nominal predicted error covariance $P_n(k|k-1)$, $k = 1, 2, \ldots$, reduces to zero under initial condition (7.45a). From this point of view, the nominal filter becomes a predictor driven by the deterministic own-sensor manoeuvre input $M(k-1)$. This is called a *manoeuvre predictor*, and is given by

$$\hat{x}_n(k|k) = \Phi(k, k-1)\hat{x}_n(k-1|k-1) - M(k-1), \qquad \hat{x}_n(0|0) = 0 \tag{7.46}$$

Moreover, noting the fact that the 'blending matrix' becomes $\Phi_n(k, 0) \equiv \hat{\Phi}_n(k\ 0)$, because $K_n(k) \equiv 0$ in the nominal filter, we find that the set of equations (7.30)–(7.34) can be slightly simplified as follows:

$$\begin{aligned}
\hat{x}_r(0|k) = \hat{x}_r(0|k-1) &- K_r(k)H(k) \\
&\times [\hat{x}_n(k|k) + \Phi_n(k,0)\hat{x}_r(0|k-1)]
\end{aligned} \tag{7.47a}$$

$$\hat{x}_r(0|0) = [\hat{r}(0)\sin\beta(0) \quad \hat{r}(0)\cos\beta(0)\ 0\ 0]^T \tag{7.47b}$$

$$K_r(k) = P_r(0|k-1)\Phi_n^T(k, 0)H^T(k)S^{-1}(k) \tag{7.48}$$

$$S(k) = H(k)\Phi_n(k, 0)P_r(0|k-1)\Phi_n^T(k, 0)H^T(k) + 1 \tag{7.49}$$

$$P_r(0|k) = [I - K_r(k)H(k)\Phi_n(k, 0)]P_r(0|k-1) \tag{7.50a}$$

$$P_r(0|0) = I \tag{7.50b}$$

$$\Phi_n(k, 0) = \Phi(k, k-1)\Phi_n(k-1, 0), \qquad \Phi_n(0, 0) = I \tag{7.51}$$

It should be noted that these estimation problems remain unobservable prior to an own-sensor manoeuvre [1–3, 11]. That is, it can be seen from equations (7.46) and (7.47) that if $\hat{x}_r(0|0) = 0$, then expression (7.18) generates the trivial solution $\hat{x}(k|k) = 0$, for all k on the first leg.

Additionally, the present partitioning approach can quantitatively assess the effect that the deterministic own-sensor manoeuvre contributes to the total state estimate $\hat{x}(k|k)$, and explicitly replace the filtering problem by the parameter estimation problem.

Noting the parallel mechanism of the partitioning filter, we may further construct the so-called data compression mechanism. This mechanism averages the m measurements as shown in Figure 7.2, so that the mean value $\bar{\beta}(ml)$ can be fed into the fixed-point smoother at time ml, where the manoeuvre predictor is propagated over such circumstances, and $m \geqslant 1$ denotes the compressing or renovating interval. We summarize this concept in the following corollary.

Corollary 7.1

The renovated state estimate $\hat{x}(k|k)$ and the associated error covariance $P(k|k)$ are, respectively, determined by

$$\hat{x}(k|k) = \hat{x}_n(k|k) + \Phi_n(ml, 0)\hat{x}_r(0|ml) \tag{7.52}$$

$$P(k|k) = \Phi_n(ml, 0)P_r(0|ml)\Phi_n^T(ml, 0), \qquad l = 1, 2, \ldots \tag{7.53}$$

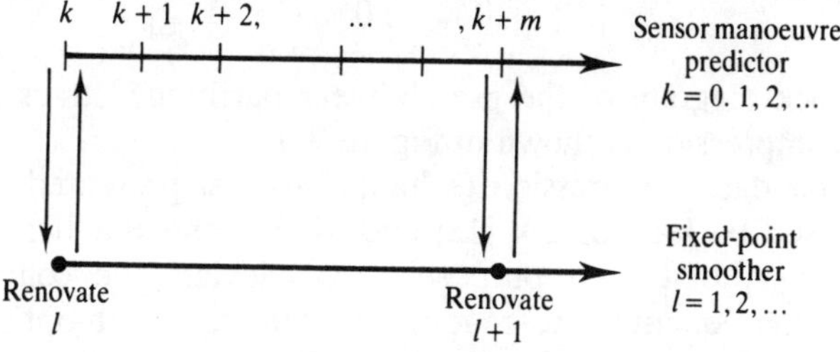

Figure 7.2 Data compression time sequence.

where $\hat{x}_n(k|k)$ is generated by the manoeuvre predictor:

$$\hat{x}_n(k|k) = \Phi(k, k-1)\hat{x}_n(k-1|k-1) - M(k-1),$$
$$\hat{x}_n(0|0) = 0 \qquad (7.54)$$

The pseudolinear partitioning fixed-point smoother at the instant ml is governed by the relation

$$\hat{x}_r(0|ml) = \hat{x}_r(0|m(l-1)) - K_r(ml)H(ml)[\hat{x}_n(k|k) + \Phi_n(ml, 0)\hat{x}_r(0|m(l-1))] \qquad (7.55a)$$

$$\hat{x}_r(0|0) = [\hat{r}(0)\sin\beta(0) \quad \hat{r}(0)\cos\beta(0) \quad 0 \quad 0]^T \qquad (7.55b)$$

where $K_r(ml)$ is the smoothing gain obtained from

$$K_r(ml) = P_r(0|m(l-1))\Phi_n^T(ml, 0)H^T(ml)S(ml)^{-1} \qquad (7.56)$$

$$S(ml) \triangleq H(ml)\Phi_n(ml, 0)P_r(0|m(l-1))\Phi_n^T(ml, 0)H^T(ml) + 1 \qquad (7.57)$$

$$H(ml) \triangleq [\cos\bar{\beta}(ml) \quad -\sin\bar{\beta}(ml) \quad 0 \ \ 0]^T \qquad (7.58)$$

$$\bar{\beta}(ml) = \frac{1}{m}\sum_{i=m(l-1)+1}^{ml} \beta(i) \qquad (7.59)$$

The smoothed error covariance $P_r(0|ml)$ and the blending matrix $\Phi_n(ml, 0)$ are, respectively, given by

$$P_r(0|ml) = [I - K_r(ml)H(ml)\Phi_n(ml, 0)]P_r(0|m(l-1)) \qquad (7.60)$$

$$\Phi_n(ml, 0) = \Phi_m(ml, m(l-1))\Phi_n(m(l-1), 0) \qquad (7.61)$$

where

$$\Phi_m(ml, m(l-1)) = \begin{bmatrix} 1 & 0 & m\Delta t & 0 \\ 0 & 1 & 0 & m\Delta t \\ 0 & 0 & 1 & 0 \\ 0 & 0 & 0 & 1 \end{bmatrix} \qquad (7.62)$$

The block diagram of the pseudolinear partitioned tracking filter with data compression is shown in Figure 7.3.

A similar data compression technique to that presented here has been proposed by Bar-Itzhack [12] and Medan and Bar-Itzhack [13]. Note, however, that their purpose is to alleviate the computation loads, and consequently the renovated estimate is sub-optimal. On the other hand, the present approach is aimed at averaging the measurement data.

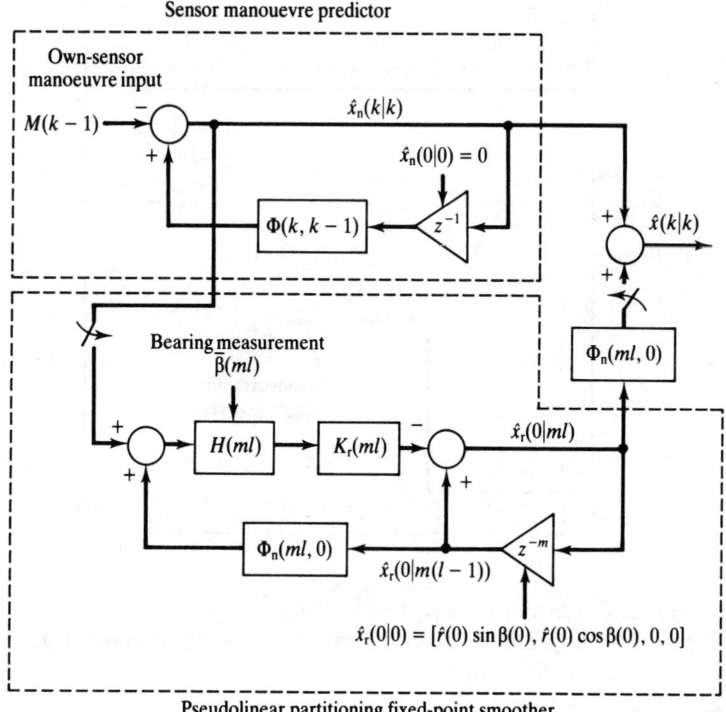

Figure 7.3 Block diagram for pseudolinear partitioned tracking filter with data compression.

Example 7.1

The effectiveness of the pseudolinear partitioned tracking filter discussed here is demonstrated by computer simulations. A representative result is obtained for the sensor manoeuvre described in Figure 7.4, in which the sampling interval is $\Delta t = 1$ s, the initial range estimate $\hat{r}(0) = 6000$ m, the related bearing $\beta(0) = 0$ deg, and the rms measurement noise level $\sigma_\eta = 1.0$ deg. It is further assumed that both vehicles are travelling at a constant speed of 5 m s^{-1} and the sensor traverses three 420 s legs involving course changes of 90 deg, where a point manoeuvre, in which the course change is made in an instant, is considered.

Figures 7.5–7.8 illustrate a typical sample result for the above condition, but with the renovating interval $m = 20$. It is understood from these figures that in this example the manoeuvre predicted value

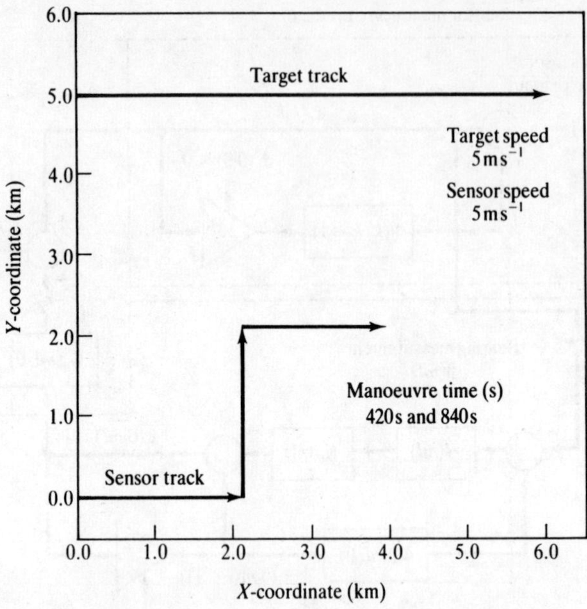

Figure 7.4 Target and sensor tracks for a simulation geometry.

is dominant in the renovated estimate, except for the estimate of relative Y position. In addition, it is seen from Figure 7.8 that the fixed-point smoother rapidly converges to the true solution after the first own-sensor manoeuvre.

Figures 7.9–7.12 show the rms estimation errors for three renovating intervals. Notice that for the estimate of position, remarkable differences appear in this example. In particular, a filter with renovating interval $m = 40$ exhibits divergence in the estimation of X position (Figure 7.9). However, in the simulations with $m = 10$ and $m = 20$ increasing the renovating interval significantly improves the estimates of X position and speed.

7.5 SUMMARY

In this chapter the pseudolinear partitioned tracking filter has been developed in recursive form. In addition, a data compression algorithm suitable for parallel processing has been presented.

It is remarked that estimates cannot necessarily be improved in proportion to the renovating or compressing interval; rather, the

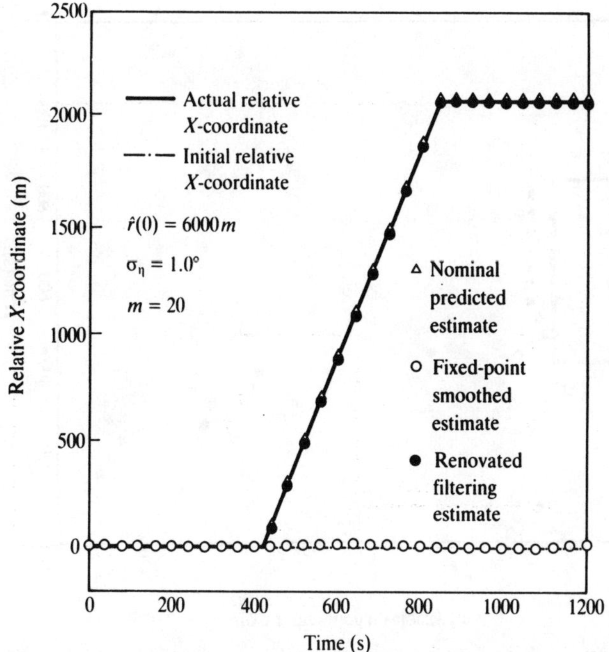

Figure 7.5 Estimate of relative X position for one sample path.

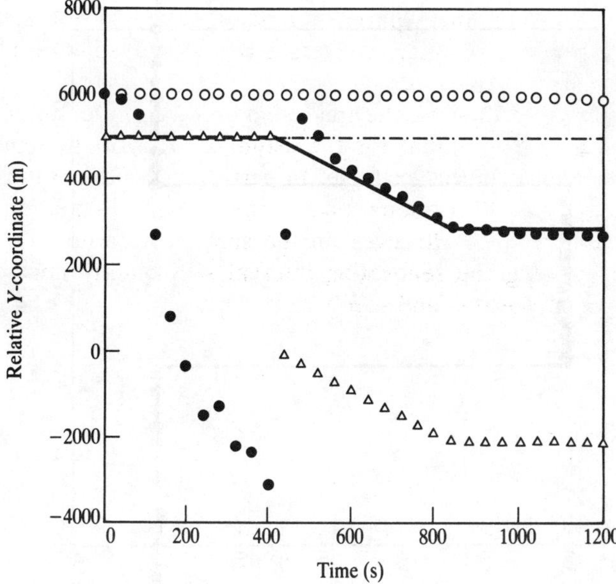

Figure 7.6 Estimate of relative Y position for one sample path.

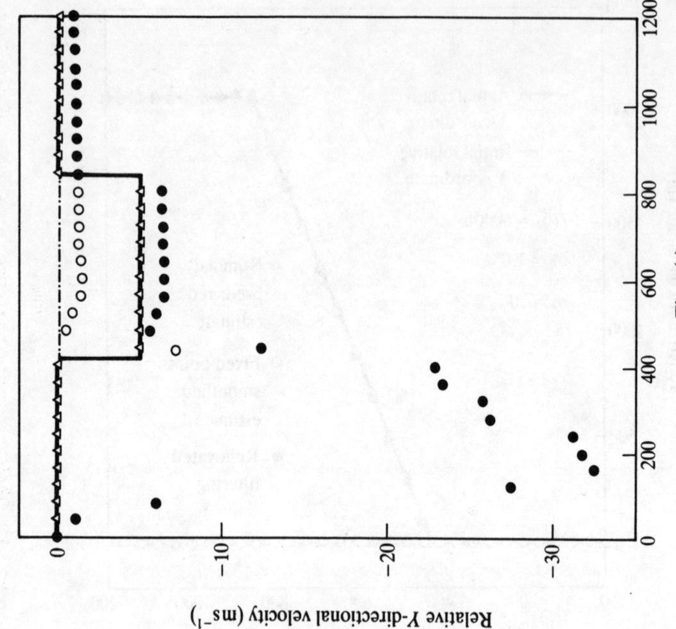

Figure 7.8 Estimate of relative velocity along Y-axis for one sample path.

Figure 7.7 Estimate of relative velocity along X-axis for one sample path.

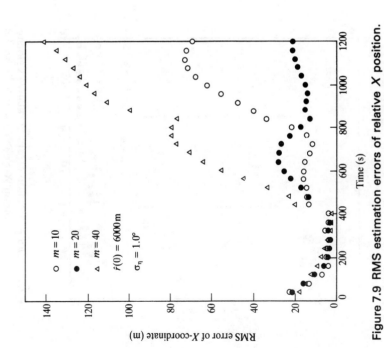

Figure 7.9 RMS estimation errors of relative X position.

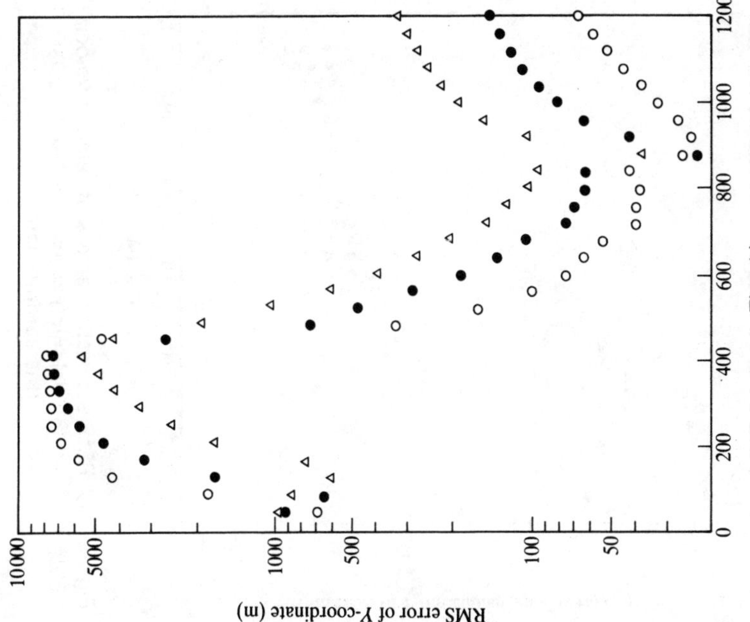

Figure 7.10 RMS estimation errors of relative Y position.

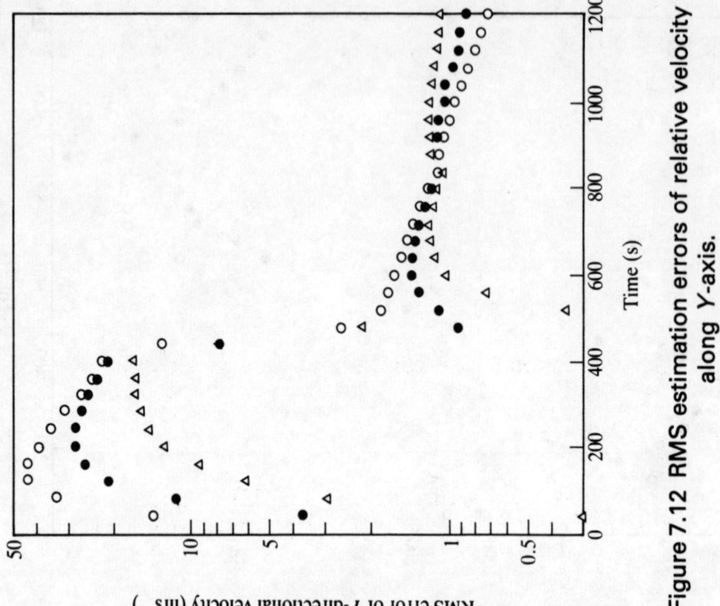

Figure 7.12 RMS estimation errors of relative velocity along Y-axis.

Figure 7.11 RMS estimation errors of relative velocity along X-axis.

convergence of the estimator depends heavily upon the sensor–target range and on the target velocity [14, 15]. It is demonstrated by Aidala and Hammel [16] that a shortcoming of the Cartesian-coordinate extended Kalman filter, namely that it exhibits unstable behaviour characteristics when utilized in the TMA problem, can be overcome by using modified polar coordinates. For bearing-only TMA problems in three dimensions, see Hammel and Aidala [17] on observability requirements and Song and Speyer [18, 19] on a modified gain extended Kalman filter.

The extension of the results to the case where the target manoeuvres are random [20, 21] and work on the optimal geometries of own-sensor manoeuvres [22] and their instances are open problems.

APPENDIX 7A EQUATION OF MOTION FOR BEARING-ONLY TRACKING MOTION ANALYSIS

Consider the geometry shown in Figure 7.1, with target and own-sensor confined to the same horizontal plane. If constant target velocity is assumed, then the equations of motion for the two-dimensional configuration (see Hassab [23] for the three-dimensional case) may be described as follows:

$$\dot{r}(t) = V(t) \tag{7A.1}$$

$$\dot{V}(t) = -a_s(t) \tag{7A.2}$$

where

$$r(t) = r_T(t) - r_S(t) \tag{7A.3}$$

$$V(t) = V_T(t) - V_S(t) \tag{7A.4}$$

and note that $a_s(t)$ denotes the only own-sensor acceleration. Here, $r^T(t) = [r_x(t) \quad r_y(t)]$, where $r_x(t)$ is the x-component of relative range, and $r_y(t)$ is the y-component of relative range; $V^T(t) = [V_x(t) \quad V_y(t)]$, where $V_x(t)$ is the x-component of relative velocity, and $V_y(t)$ is the y-component of relative velocity; and $a_s^T(t) = [a_{s_x}(t) \quad a_{s_y}(t)]$, where $a_{s_x}(t)$ is the x-component of own-sensor acceleration, and $a_{s_y}(t)$ is the y-component of own-sensor acceleration. Furthermore, in equations (7A.3) and (7A.4) the subscripts 'T' and 'S' indicate the target and own-sensor components, respectively.

Integrating equations (7A.1) and (7A.2) gives the following expressions:

$$r(t) = r(t^*) + (t - t^*)V(t^*) - \int_0^{(t-t^*)} \tau a_0(t - \tau)\, d\tau \tag{7A.5}$$

$$V(t) = V(t^*) - \int_0^{(t-t^*)} a_0(t - \tau)\, d\tau \tag{7A.6}$$

where t^* denotes any arbitrarily fixed time. The discrete-time state equations for bearing-only TMA, taking $t = k\Delta t$ and $t^* = (k-1)\Delta t$ in equations (7A.5) and (7A.6), reduce to

$$r(k) = r(k-1) + \Delta t V(k-1) - \int_0^{\Delta t} \tau a_0(k\Delta t - \tau)\, d\tau \tag{7A.7}$$

$$V(k) = V(k-1) - \int_0^{\Delta t} a_0(k\Delta t - \tau)\, d\tau, \qquad k = 1, 2, \ldots \tag{7A.8}$$

where $r(k\Delta t)$ and $V(k\Delta t)$ are newly defined as $r(k)$ and $V(k)$, respectively.

EXERCISES

7.1 Develop an extended Kalman filter for the system given by equations (7.1)–(7.6).

7.2 Show that if $\hat{x}_r = 0$ and $P_r = 0$ in Theorem 7.1, then $\hat{x}(k|k) = \hat{x}_n(k|k)$ and $P(k|k) = P_n(k|k)$.

7.3 Suppose that $x(0)$ is deterministic and $\hat{x}_n = \hat{x}_r = 0$. Show that before the first own-sensor

$$E[\tilde{x}(0|k)] = x(0)$$

and

$$E[\tilde{x}(k|k)] = x(k)$$

where $\tilde{x}(0|k) = x(0) - \hat{x}_r(0|k)$ and $\tilde{x}(k|k) = x(k) - \hat{x}(k|k)$.

7.4 Suppose that $P_n = 0$ in Lemma 7.1. Prove that

$$H(j)\Phi(j, j-1)\Phi_n(j-1, 0) \equiv H(j)\Phi(j, 0) \triangleq \widetilde{H}(j)$$
$$= [\cos\beta(j) \quad -\sin\beta(j) \quad j\Delta t\cos\beta(j) \quad -j\Delta t\sin\beta(j)]$$

7.5 Let $\varepsilon(k) \equiv 0$ in equation (7.7).

(a) Show that equation (7.7) can be rewritten as a linear equation:

$$\widetilde{H}(k)x(0) = y(k)$$

where $\widetilde{H}(k)$ is defined in Exercise 7.4 and

$$y(k) = \sum_{j=0}^{k-1} y_x(k, j) - y_y(k, j)$$

$$y_x(k, j) = \cos \beta(k) \int_0^{\Delta t} t a_{s_x}(k\Delta t - t)\, dt$$

$$+ (k - j - 1)\Delta t \cos \beta(k) \int_0^{\Delta t} a_{s_x}(k\Delta t - t)\, dt$$

$$y_y(k, j) = \sin \beta(k) \int_0^{\Delta t} t a_{s_y}(k\Delta t - t)\, dt$$

$$+ (k - j - 1)\Delta t \sin \beta(k) \int_0^{\Delta t} a_{s_y}(k\Delta t - t)\, dt$$

(Hint: Substitute $x(k) = \Phi^k x(0) - \sum_{j=0}^{k-1} \Phi^{k-j-1} M(j)$ into $H(k)x(k) = 0$.)

(b) What is a necessary and sufficient condition for uniquely determining the unknown initial state vector $x(0)$?

7.6 Consider a bearing-only TMA problem in three-dimensional Cartesian coordinates (see Figure 7.13) described by the following model:

$$\dot{x}(t) = Ax(t) - M(t)$$

where $x(t) = [r^{\mathrm{T}}(t)\ V^{\mathrm{T}}(t)]^{\mathrm{T}} = [r_x(t)\ r_y(t)\ r_z(t)\ V_x(t)\ V_y(t)\ V_z(t)]^{\mathrm{T}}$ denotes the system state vector whose components consist of relative vehicle position and velocity, and $M(t) = [0^{\mathrm{T}}\ a_s^{\mathrm{T}}]^{\mathrm{T}} = [0\ 0\ 0\ a_{s_x}\ a_{s_y}\ a_{s_z}]$, in which a_s is own-sensor acceleration. The three-dimensional TMA problem involves simultaneously processing azimuth bearing $\beta(t)$, conical bearing $\theta(t)$, and/or depth/elevation angle $\phi(t)$ measurements in one of the following pairwise combinations [17]: (β, ϕ); (θ, ϕ); (β, θ).

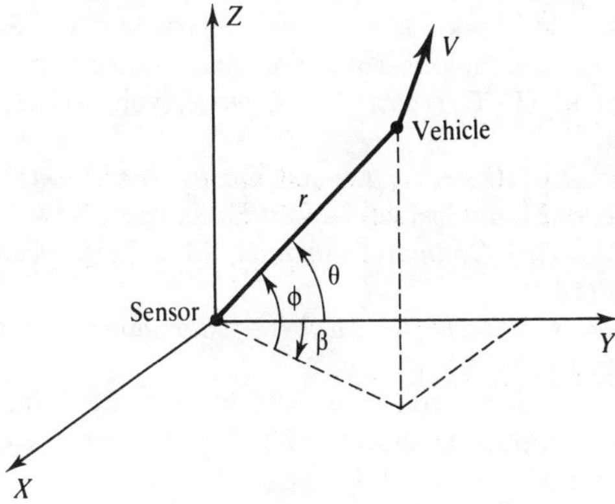

Figure 7.13 Vehicle and sensor geometry in three-dimensional Cartesian coordinates.

(a) Write down β, θ and ϕ in terms of r_x, r_y and r_z.
(b) Apply the representation of β and ϕ given in part (a) to show that

$$0 = H(t)x(t)$$

where

$$H(t) = \begin{bmatrix} 1 & 0 & -\sin\beta(t)\cot\phi(t) & 0 & 0 & 0 \\ 0 & 1 & -\cos\beta(t)\cot\phi(t) & 0 & 0 & 0 \end{bmatrix}$$

REFERENCES

[1] AIDALA, V. J., Kalman Filter Behavior in Bearing-Only Tracking Applications, *IEEE Trans. Aero. and Electron. Syst.*, vol. AES-15, no. 1, January 1979, pp.29–39.

[2] LINDGREN, A. G. and GONG, K. F., Position and Velocity Estimation via Bearing Observations, *IEEE Trans. Aero. and Electron. Syst.*, vol. AES-14, no. 4, July 1978, pp. 564–77.

[3] AIDALA, V. J. and NARDONE, S. C., Biased Estimation Properties of the Pseudolinear Tracking Filter, *IEEE Trans. Aero. and Electron. Syst.*, vol. AES-18, no. 4, July 1982, pp. 432–41.

[4] WATANABE, K., Application of Pseudolinear Partitioned Filter to Passive Vehicle Tracking, *Int. J. Syst. Sci.*, vol. 15, no. 9, 1984, pp. 959–75.

[5] PARK, S. K. and LAINIOTIS, D. G., Monte-Carlo Study of the Optimal Nonlinear Estimator: Linear Systems with the Non-Gaussian Initial State, *Int. J. Control,* vol. 16, no. 6, 1972, pp. 1029–40.

[6] GOVINDARAJ, K. S. and LAINIOTIS, D. G., A Unifying Framework for Discrete Linear Estimation: Generalized Partitioned Algorithms, *Int. J. Control,* vol. 28, no. 4, 1978, pp. 571–88.

[7] ATHANS, M., WISHNER, R. P. and BERTOLINI, A., Suboptimal State Estimation for Continuous-Time Nonlinear Systems from Discrete Noisy Measurements, *IEEE Trans. Aut. Control,* vol. AC-13, no. 5, October 1968, pp. 504–14.

[8] SAGE, A. P. and MELSA, J. L., *Estimation Theory with Applications to Communications and Control,* Robert E. Krieger, New York, 1979.

[9] GELB, A., *Applied Optimal Estimation,* MIT Press, Cambridge, Massachusetts, 1974.

[10] MAYBECK, P. S., *Stochastic Models, Estimation, and Control,* vol. 2, Academic Press, New York, 1982.

[11] NARDONE, S. C. and AIDALA, V. J., Observability Criteria For Bearing-Only Target Motion Analysis. *IEEE Trans. Aero. and Elecıron. Syst.*, vol. AES-17, no. 2, March 1981, pp. 162–6.

[12] BAR-ITZHACK, I. Y., Novel Method for Data Compression in Recursive INS Error Estimation, *J. Guidance and Control,* vol. 3, no. 3, May–June 1980, pp. 245–50.

[13] MEDAN, Y. and BAR-ITZHACK, I. Y., Error and Sensitivity Analysis Scheme of a New Data Compression Technique in Estimation, *J. Guidance and Control*, vol. 4, no. 5, September–October 1981, pp. 510–17.

[14] PETRIDIS, V., A Method for Bearing-Only Velocity and Position Estimation, *IEEE Trans. Aut. Control,* vol. AC-26, no. 2, April 1981, pp. 488–93.

[15] NARDONE, S. C., LINDGREN, A. G. and GONG, K. F., Fundamental Properties and Performance of Conventional Bearing-Only Target Motion Analysis, *IEEE Trans. Aut. Control,* vol. AC-29, no. 9, September 1984, pp. 775–87.

[16] AIDALA, V. J. and HAMMEL, S. E., Utilization of Modified Polar Coordinates for Bearings-Only Tracking, *IEEE Trans. Aut. Control,* vol. AC-28, no. 3, 1983, pp. 283–94.

[17] HAMMEL, S. E. and AIDALA, V. J., Observability Requirements for Three-Dimensional Tracking via Angle Measurements, *IEEE Trans. Aero. and Electron. Syst.,* vol. AES-21, no. 2, 1985, pp. 200–7.

[18] SONG, T. L. and SPEYER, J. L., A Stochastic Analysis of a Modified Gain Extended Kalman Filter with Applications to Estimation with Bearings Only Measurements, *IEEE Trans. Aut. Control,* vol. AC-30, no. 10, 1985, pp. 940–9.

[19] SONG, T. L. and SPEYER, J. L., The Modified Gain Extended Kalman Filter and Parameter Identification in Linear Systems, *Automatica,* vol. 22, no. 1, 1986, pp. 59–75.

[20] MOORE, R. L., An Adaptive State Estimation Solution to the Maneuvering Target Problem, *IEEE Trans. Aut. Control,* vol. AC-20, no. 3, 1975, pp. 359–62.

[21] TENNEY, R. R., HEBBERT, R. S. and SANDELL JR, N. R., A Tracking Filter for Maneuvering Sources, *IEEE Trans. Aut. Control,* vol. AC-22, no. 2, 1977, pp. 246–51.

[22] LIU, P.-T. and BONGIOVANNI, P., On a Passive Vehicle Tracking Problem and Max-Minimization, *IEEE Trans. Aut. Control,* vol. AC-28, no. 2, February 1983, pp. 233–5.

[23] HASSAB, J. C., Passive Tracking of a Moving Source by a Single Observer in Shallow Water, *J. Sound and Vibration*, vol 44, no. 1, 1976, pp. 127–45.

8

Two-Stage Bias Correction Estimators Based on the Generalized Partitioning Filter

8.1 INTRODUCTION

In this chapter, we apply the initial dependent-type partitioning filter developed in Chapter 5 to the problem of synthesizing two-stage bias correction estimators for continuous- or discrete-time systems [1–4].

In Section 8.2, for a continuous-time stochastic system with any system structure, we provide the condition for the feasibility of constructing a structurally partitioning filter [5, 6] consisting of a Kalman-type nominal filter of order n_c and a fixed-point smoother of order $(n - n_c)$, where n is the order of the state and $0 < n_c < n$.

We derive in Section 8.3 a generalized two-stage bias correction filter and predictor in continuous time [7] for the case where the original system state is initially dependent upon the unknown bias state. We then proceed, in Section 8.4, to develop an algorithm for a two-stage bias correction fixed-interval smoother based on a generalized partitioning smoother evolving forwards in time [8], and present two Chandrasekhar algorithms to alleviate the computation burden. In Section 8.5, we further analyze the stability of the two-stage bias correction filter by applying a generalized controllability (or disturbability) matrix [9]. Finally, a generalized two-stage bias correction filter and predictor in discrete time will be discussed.

8.2 A STRUCTURALLY PARTITIONING FILTER FOR ANY SYSTEM

Before proceeding to the derivation of two-stage bias correction estimators based on the generalized partitioning filter, in this section we address structurally partitioning filters [5], which have more extended applicability than the two-stage bias correction filters that will be derived later.

Consider a linear continuous-time stochastic system described by

$$\dot{x}(t) = Ax(t) + Bw(t) \tag{8.1}$$

$$z(t) = Hx(t) + v(t) \tag{8.2}$$

where $x(t) \in \mathbb{R}^n$, $w(t) \in \mathbb{R}^p$, $z(t) \in \mathbb{R}^m$, and A, B and H are matrices with compatible dimensions. Here, $1 < n$, $0 < p \leqslant n$ and $0 < m \leqslant n$. It is assumed that the vectors $\{x(\tau), w(t), v(t); \tau \leqslant t \leqslant t_f\}$ are mutually independent Gaussian random vectors with known statistical properties:

$$\left.\begin{aligned} &x(\tau) \sim N(\hat{x}(\tau), P(\tau)), \qquad P(\tau) > 0 \\ &E[w(t)] = 0,\ E[w(t)w^{\mathrm{T}}(s)] = Q\delta(t-s), \qquad Q \geqslant 0 \\ &E[v(t)] = 0,\ E[v(t)v^{\mathrm{T}}(s)] = R\delta(t-s), \qquad R > 0 \end{aligned}\right\} \tag{8.3}$$

Let the systems $\mathcal{S}_1$ consisting of equations (8.1) and (8.2) be partitioned as follows:

$$\mathcal{P}_1: \left\{ \begin{aligned} &A = \begin{matrix} n_c \\ n-n_c \end{matrix} \overset{\begin{matrix} n_c & \quad n-n_c \end{matrix}}{\begin{bmatrix} A_{11} & A_{12} \\ A_{21} & A_{22} \end{bmatrix}}, \qquad B = \begin{matrix} n_c \\ n-n_c \end{matrix} \begin{bmatrix} B_1 \\ B_2 \end{bmatrix} \\ &x(t) = \begin{matrix} n_c \\ n-n_c \end{matrix} \begin{bmatrix} x_1(t) \\ x_2(t) \end{bmatrix}, \qquad H = m \overset{\begin{matrix} n_c & \quad n-n_c \end{matrix}}{\begin{bmatrix} H_1 & H_2 \end{bmatrix}} \end{aligned} \right.$$

Given the measurement data $Z_t = \{z(s); \tau \leqslant s \leqslant t\}$, the optimal MV estimate $\hat{x}(t|t) = E[x(t)|Z_t]$ is obtained by the following Kalman filter:

$$\begin{aligned} \dot{\hat{x}}(t|t) &= \begin{bmatrix} \dot{\hat{x}}_1(t|t) \\ \dot{\hat{x}}_2(t|t) \end{bmatrix} \\ &= A\hat{x}(t|t) + P(t)H^{\mathrm{T}}R^{-1}[z(t) - H\hat{x}(t|t)] \end{aligned} \tag{8.4a}$$

$$\begin{aligned} \hat{x}(\tau|\tau) &= \hat{x}(\tau) \\ &= \begin{bmatrix} \hat{x}_1(\tau|\tau) \\ \hat{x}_2(\tau|\tau) \end{bmatrix} \end{aligned} \tag{8.4b}$$

Note that $\hat{x}(t|t)$ minimizes the following MSE criterion, subject to constraints (8.1)–(8.3):

$$J = E[\|x(t) - \hat{x}(t|t)\|^2 | Z_t] \tag{8.5}$$

The associated filtering error covariance $P(t) = E\{[x(t) - \hat{x}(t|t)][x(t) - \hat{x}(t|t)]^{\mathrm{T}}|Z_t\}$ is given by the RDE:

$$\dot{P}(t|t) = AP(t) + P(t)A^{\mathrm{T}} - P(t)H^{\mathrm{T}}R^{-1}HP(t) + BQB^{\mathrm{T}} \tag{8.6a}$$

$$P(t)|_{t=\tau} = P(\tau)$$

$$= \begin{bmatrix} P_{11}(\tau) & P_{12}(\tau) \\ P_{21}(\tau) & P_{22}(\tau) \end{bmatrix} \tag{8.6b}$$

In addition, when $P(t)$ is partitioned as follows:

$$P(t) = \begin{bmatrix} P_{11}(t) & P_{12}(t) \\ P_{12}^{\mathrm{T}}(t) & P_{22}(t) \end{bmatrix} \tag{8.7}$$

each block matrix is given as the solution to the following differential equations:

$$\begin{aligned} \dot{P}_{11}(t) = {} & A_{11}P_{11}(t) + A_{12}P_{12}^{\mathrm{T}}(t) + P_{11}(t)A_{11}^{\mathrm{T}} + P_{12}(t)A_{12}^{\mathrm{T}} \\ & - [P_{11}(t)H_1^{\mathrm{T}} + P_{12}(t)H_2^{\mathrm{T}}]R^{-1}[H_1P_{11}(t) + H_2P_{12}^{\mathrm{T}}(t)] \\ & + B_1QB_1^{\mathrm{T}}, \qquad P_{11}(t)|_{t=\tau} = P_{11}(\tau) \end{aligned} \tag{8.8}$$

$$\begin{aligned} \dot{P}_{12}(t) = {} & \{A_{11} - [P_{11}(t)H_1^{\mathrm{T}} + P_{12}(t)H_2^{\mathrm{T}}]R^{-1}H_1\}P_{12}(t) \\ & + \{A_{12} - [P_{11}(t)H_1^{\mathrm{T}} + P_{12}(t)H_2^{\mathrm{T}}]R^{-1}H_2\}P_{22}(t) \\ & + P_{11}(t)A_{21}^{\mathrm{T}} + P_{12}(t)A_{22}^{\mathrm{T}} + B_1QB_2^{\mathrm{T}}, \\ & \qquad P_{12}(t)|_{t=\tau} = P_{12}(\tau) \end{aligned} \tag{8.9}$$

$$\begin{aligned} \dot{P}_{22}(t) = {} & \{A_{22} - [P_{12}(t)H_1^{\mathrm{T}} + P_{22}(t)H_2^{\mathrm{T}}]R^{-1}H_2\}P_{22}(t) \\ & + \{A_{21} - [P_{12}(t)H_1^{\mathrm{T}} + P_{22}(t)H_2^{\mathrm{T}}]R^{-1}H_1\}P_{12}(t) \\ & + P_{12}^{\mathrm{T}}(t)A_{21}^{\mathrm{T}} + P_{22}(t)A_{22}^{\mathrm{T}} + B_2QB_2^{\mathrm{T}}, \\ & \qquad P_{22}(t)|_{t=\tau} = P_{22}(\tau) \end{aligned} \tag{8.10}$$

In what follows, we will present a necessary and sufficient condition for the possibility of constructing an optimal structurally partitioning filter, which minimizes the cost given by equation (8.5) subject to the dynamic constraints (8.1)–(8.3) and solves equations (8.8) and (8.10), by applying the partition theorem shown in Chapter 5.

Let us decompose the initial state $x(\tau)$ as follows:

$$x(\tau) = x_{\mathrm{n}}(\tau) + x_{\mathrm{r}}(\tau) \tag{8.11}$$

where

$$x_{\mathrm{n}}(\tau) = \begin{matrix} n_{\mathrm{c}} \\ n - n_{\mathrm{c}} \end{matrix}\begin{bmatrix} x_1(\tau) \\ 0 \end{bmatrix}, \qquad x_{\mathrm{r}}(\tau) = \begin{matrix} n_{\mathrm{c}} \\ n - n_{\mathrm{c}} \end{matrix}\begin{bmatrix} 0 \\ x_2(\tau) \end{bmatrix},$$

$$0 < n_{\mathrm{c}} < n \qquad (8.12)$$

The mean and covariance values are respectively assumed to be

$$\hat{x}(\tau) = \hat{\mu}_{\mathrm{n}}(\tau) + \hat{\mu}_{\mathrm{r}}(\tau) \qquad (8.13a)$$

$$P(\tau) = V_{\mathrm{n}}(\tau) + V_{\mathrm{nr}}(\tau) + V_{\mathrm{nr}}^{\mathrm{T}}(\tau) + V_{\mathrm{r}}(\tau) \qquad (8.13b)$$

where

$$\hat{\mu}_{\mathrm{n}}(\tau) = \begin{bmatrix} \hat{x}_1(\tau|\tau) \\ 0 \end{bmatrix}, \qquad \hat{\mu}_{\mathrm{r}}(\tau) = \begin{bmatrix} 0 \\ \hat{x}_2(\tau|\tau) \end{bmatrix} \qquad (8.14a)$$

$$V_{\mathrm{n}}(\tau) = \begin{bmatrix} P_{11}(\tau) & 0 \\ 0 & 0 \end{bmatrix}, \qquad V_{\mathrm{nr}}(\tau) = \begin{bmatrix} 0 & P_{12}(\tau) \\ 0 & 0 \end{bmatrix},$$

$$V_{\mathrm{r}}(\tau) = \begin{bmatrix} 0 & 0 \\ 0 & P_{22}(\tau) \end{bmatrix} \qquad (8.14b)$$

From Theorem 5.1, we then obtain the following partitioning filter:

$$\hat{x}(t|t) = \hat{x}_{\mathrm{n}}(t|t) + \Phi_{\mathrm{n}}(t, \tau)\Pi_{\mathrm{r}}(\tau)\hat{x}_{\mathrm{r}}(\tau|t) \qquad (8.15)$$

$$P(t) = P_{\mathrm{n}}(t) + \Phi_{\mathrm{n}}(t, \tau)\Pi_{\mathrm{r}}(\tau)P_{\mathrm{r}}(\tau|t)\Pi_{\mathrm{r}}^{\mathrm{T}}(\tau)\Phi_{\mathrm{n}}^{\mathrm{T}}(t, \tau) \qquad (8.16)$$

where

$$\Pi_{\mathrm{r}}(\tau) = [I + V_{\mathrm{nr}}(\tau)V_{\mathrm{r}}^{-1}(\tau)] \qquad (8.17)$$

$$\dot{\Phi}_{\mathrm{n}}(t, \tau) = [A - K_{\mathrm{n}}(t)H]\Phi_{\mathrm{n}}(t, \tau), \qquad \Phi_{\mathrm{n}}(\tau, \tau) = I \qquad (8.18)$$

$$K_{\mathrm{n}}(t) = P_{\mathrm{n}}(t)H^{\mathrm{T}}R^{-1} \qquad (8.19)$$

The nominal filtered estimate $\hat{x}_{\mathrm{n}}(t|t)$ and its error covariance $P_{\mathrm{n}}(t)$ are given by

$$\dot{\hat{x}}_{\mathrm{n}}(t|t) = A\hat{x}_{\mathrm{n}}(t|t) + K_{\mathrm{n}}(t)[z(t) - H\hat{x}_{\mathrm{n}}(t|t)], \qquad \hat{x}_{\mathrm{n}}(\tau|\tau) = \hat{m}_{\mathrm{n}}(\tau) \qquad (8.20)$$

$$\dot{P}_{\mathrm{n}}(t) = AP_{\mathrm{n}}(t) + P_{\mathrm{n}}(t)A^{\mathrm{T}} + BQB^{\mathrm{T}} - P_{\mathrm{n}}(t)H^{\mathrm{T}}R^{-1}HP_{\mathrm{n}}(t), \qquad P_{\mathrm{n}}(\tau) = \Pi_{\mathrm{n}}(\tau) \qquad (8.21)$$

where

$$\hat{m}_{\mathrm{n}}(\tau) = \hat{\mu}_{\mathrm{n}}(\tau) - V_{\mathrm{nr}}(\tau)V_{\mathrm{r}}^{-1}(\tau)\hat{\mu}_{\mathrm{r}}(\tau) \qquad (8.22)$$

$$\Pi_{\mathrm{n}}(\tau) = V_{\mathrm{n}}(\tau) - V_{\mathrm{nr}}(\tau)V_{\mathrm{r}}^{-1}(\tau)V_{\mathrm{nr}}^{\mathrm{T}}(\tau) \qquad (8.23)$$

Furthermore, the partitioned-type fixed-point smoothed estimate

$\hat{x}_r(\tau|t)$ can be obtained by simply differentiating (5.41)–(5.44):

$$\dot{\hat{x}}_r(\tau|t) = -K_r(t,\tau)H\Phi_n(t,\tau)\Pi_r(\tau)\hat{x}_r(\tau|t) + K_r(t,\tau)[z(t) - H\hat{x}_n(t|t)], \qquad \hat{x}_r(\tau|\tau) = \hat{\mu}_r(\tau) \tag{8.24}$$

$$K_r(t,\tau) = P_r(\tau|t)\Pi_r^T(\tau)\Phi_n^T(t,\tau)H^T R^{-1} \tag{8.25}$$

$$\dot{P}_r(\tau|t) = -K_r(t,\tau)RK_r^T(t,\tau), \qquad P_r(\tau|\tau) = V_r(\tau) \tag{8.26}$$

If we assume that $A_{21} \equiv 0$ and $B_2 \equiv 0$ in the partition $\mathcal{P}_1$, then for initial values such as those given in equations (8.14) $P_n(t)$ can constitute the following block matrix:

$$P_n(t) = \begin{bmatrix} P_1^n(t) & 0 \\ 0 & 0 \end{bmatrix} \tag{8.27}$$

for $t \geqslant \tau$, where $P_1^n(t)$ is the solution of the following RDE:

$$\dot{P}_1^n(t) = A_{11}P_1^n(t) + P_1^n(t)A_{11}^T - P_1^n(t)H_1^T R^{-1}H_1P_1^n(t) + B_1QB_1^T,$$

$$P_1^n(\tau) = P_{11}(\tau) - P_{12}(\tau)P_{22}^{-1}(\tau)P_{21}(\tau) \tag{8.28}$$

From this fact, the nominal filter gain of equation (8.19) can also be rewritten by

$$K_n(t) = \begin{matrix} n_c \\ n - n_c \end{matrix}\overset{m}{\begin{bmatrix} K_1^n(t) \\ 0 \end{bmatrix}} \tag{8.29a}$$

$$K_1^n(t) = P_1^n(t)H_1^T R^{-1} \tag{8.29b}$$

These discussions lead to the following theorem.

Theorem 8.1

A nominal filter of order n_c and a fixed-point smoother of order $(n - n_c)$ can be optimally synthesized for a linear time-invariant system $\mathcal{S}_1$ if and only if the partition $\mathcal{P}_1$ has the condition $A_{21} \equiv 0$ and $B_2Q \equiv 0$.

Proof

To prove the theorem, introduce a set defined by

$$\mathcal{K}_{\mathrm{n}} \triangleq \left\{ K_{\mathrm{n}}(t) = \begin{bmatrix} K_1^{\mathrm{n}}(t) \\ 0 \end{bmatrix} \in \mathbb{R}^{n \times m}, \right.$$

$$\left. t \in [\tau, t_{\mathrm{f}}] \| \quad K_1^{\mathrm{n}}(t) \in \mathbb{R}^{n_{\mathrm{c}} \times m} \text{ satisfies (8.29b)} \right\} \tag{8.30}$$

First, in order to prove the sufficient condition, put $A_{21} = 0$ and $B_2 Q = 0$ in the partition $\mathcal{D}_1$. If the initial condition $x_2(\tau)$ is given, then the nominal initial error covariance becomes $P_{\mathrm{n}}(\tau) = V_{\mathrm{n}}(\tau) - V_{\mathrm{nr}}(\tau) V_{\mathrm{r}}^{-1}(\tau) V_{\mathrm{nr}}^{\mathrm{T}}(\tau)$, and equations (8.9) and (8.10) reduce to homogeneous equations with zero initial conditions. Therefore, the nominal filter gain given in equations (8.29) becomes

$$K_{\mathrm{n}}(t) \in \mathcal{K}_{\mathrm{n}} \tag{8.31}$$

and hence the efficient order of equation (8.20) reduces to n_{c}, where the efficient order is defined as constituting the state of the nominal filter of equation (8.20) in closed loop.

Conversely, to prove the necessary condition, set $A_{21} \neq 0$ or $B_2 Q \neq 0$ in the partition $\mathcal{D}_1$. If the initial condition $x_2(\tau)$ is provided, then the error covariance matrix given by equation (8.21) for the nominal filter is obtained by

$$P_{\mathrm{n}}(t) = \begin{bmatrix} P_{11}(\mathrm{t}) & P_{12}(\mathrm{t}) \\ P_{12}^{\mathrm{T}}(\mathrm{t}) & P_{22}(\mathrm{t}) \end{bmatrix},$$

$$P_{12}(\mathrm{t}) \neq 0,\ P_{22}(\mathrm{t}) \neq 0 \quad \text{for} \quad t \geqslant \tau \tag{8.32}$$

because (8.9) and (8.10) are nonhomogeneous equations, where $P_{ij}(t)$, $i,j = 1,\ 2$, satisfy equation (8.8) with $P_{11}(t)|_{t=\tau} = (P_{11}(\tau) - P_{12}(\tau) P_{22}^{-1}(\tau) P_{21}(\tau))$, equation (8.9) with $P_{12}(t)|_{t=\tau} = 0$, and equation (8.10) with $P_{22}(t)|_{t=\tau} = 0$, respectively. Therefore, the nominal filter gain given in equations (8.29) becomes

$$K_{\mathrm{n}}(t) \notin \mathcal{K}_{\mathrm{n}} \tag{8.33}$$

and hence the filtered estimate $\hat{x}_{\mathrm{n}}(t|t)$ is not of order n_{c}. □

It is easy to see that Theorem 8.1 is equivalent to the following corollary.

Corollary 8.1

A necessary and sufficient condition for the possibility of constructing a nominal filter of order n_c and a fixed-point smoother of order $(n - n_c)$ is that in the nominal filtering error covariance the (1, 2) and (2, 2) block elements satisfy each homogeneous equation for $t \geqslant \tau$.

Thus, in order to synthesize an optimal structurally partitioning filter we require the structural restrictions for the system $\mathcal{S}_1$. In fact, by virtue of Kalman's canonical structural model [10, 11], an optimal structurally partitioning filter, which consists of a nominal filter of order n_c and a fixed-point smoother of order $(n - n_c)$, can always be synthesized [5] for incompletely controllable (or disturbable) and completely observable systems, where n_c is the order of the state for completely controllable and completely observable subsystems. In addition, as a special class of this problem, a partitioning filter based on the Jordan canonical form can be obtained for completely controllable systems, where there exists no system noise.

Example 8.1

Consider the following system with unknown bias-type input:

$$\dot{x}_1 = -x_1 + x_2 + w$$

$$\dot{x}_2 = -2x_2$$

$$z = x_1 + v$$

This system has state variable representation

$$\frac{\mathrm{d}}{\mathrm{d}t}\begin{bmatrix} x_1 \\ x_2 \end{bmatrix} = \begin{bmatrix} -1 & 1 \\ 0 & -2 \end{bmatrix}\begin{bmatrix} x_1 \\ x_2 \end{bmatrix} + \begin{bmatrix} 1 \\ 0 \end{bmatrix} w$$

$$z = \begin{bmatrix} 1 & 0 \end{bmatrix}\begin{bmatrix} x_1 \\ x_2 \end{bmatrix} + v$$

It can be seen that states $x_1(t)$ and $x_2(t)$ can be estimated by using a nominal filter of order 1 and a fixed-point smoother of order 1, respectively, because the above system satisfies Theorem 8.1.

8.3 BIAS CORRECTION FILTER AND PREDICTOR

We now return to the problem of two-stage bias correction estimation. As before, we consider a linear time-invariant stochastic system, such as an inertial navigation system [12] with accelerometer error and gyro drifts, etc., as unknown bias states, described by

$$\dot{x}(t) = Ax(t) + Cw(t) \tag{8.34}$$

$$z(t) = Hx(t) + v(t) \tag{8.35}$$

where $x(t) \in \mathbb{R}^n$, $w(t) \in \mathbb{R}^p$, $z(t) \in \mathbb{R}^m$, and $v(t) \in \mathbb{R}^m$. It is assumed that the above augmented system is decomposed into

$$\left.\begin{aligned} x^{\mathrm{T}}(t) &= [y^{\mathrm{T}}(t) \quad b^{\mathrm{T}}(t)], \qquad A = \begin{bmatrix} F & B \\ 0 & D \end{bmatrix} \\ C &= \begin{bmatrix} G \\ 0 \end{bmatrix}, \qquad H = [H_1 \quad H_2] \end{aligned}\right\} \tag{8.36}$$

where the original state $y(t) \in \mathbb{R}^{n_c}$, the unknown bias state $b(t) \in \mathbb{R}^{n_b}$, and $p \leq n_c$, $m \leq n_c$, and $n = n_c + n_b$. Figure 8.1 shows the block diagram for such a system involving the unknown bias subsystem.

Assume that the zero-mean Gaussian white noise processes $w(t)$ and $v(t)$ are independent of the initial state $x(\tau)$, $\tau \leq t$, and have covariances $Q \geq 0$ and $R > 0$, respectively. Assume, further, that the

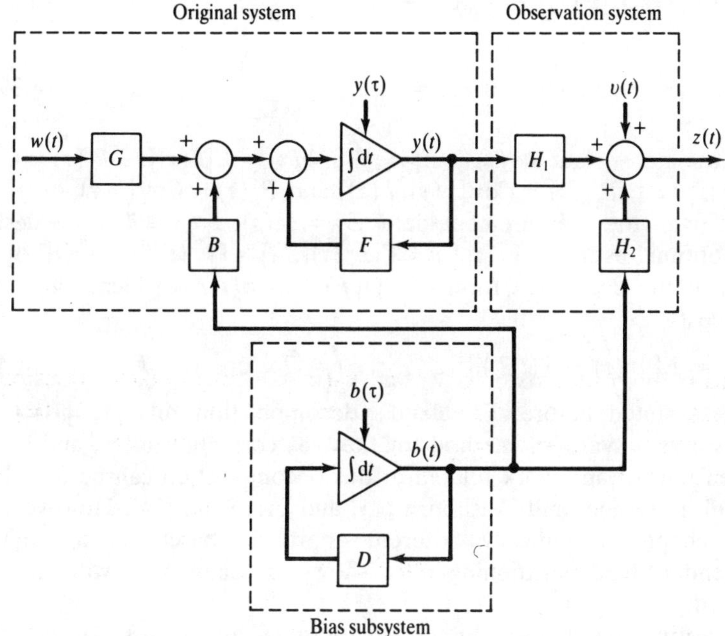

Figure 8.1 Block diagram for an augmented system with a bias subsystem.

Gaussian random initial state $x(\tau)$ can be decomposed as follows:

$$x(\tau) = x_n(\tau) + x_r(\tau) \tag{8.37}$$

but with the moments

$$E[x(\tau)] = \hat{x}(\tau) = \begin{bmatrix} \hat{y}(\tau) \\ \hat{b}(\tau) \end{bmatrix} \tag{8.38a}$$

$$\text{Cov}\,[x(\tau), x(\tau)] = P(\tau) = \begin{bmatrix} P_y(\tau) & P_{yb}(\tau) \\ P_{by}(\tau) & P_b(\tau) \end{bmatrix} \tag{8.38b}$$

and

$$E[x_n(\tau)] = \hat{\mu}_n(\tau) = \begin{bmatrix} \hat{y}(\tau) \\ 0 \end{bmatrix},$$

$$E[x_r(\tau)] = \hat{\mu}_r(\tau) = \begin{bmatrix} 0 \\ \hat{b}(\tau) \end{bmatrix} \tag{8.39a}$$

$$\text{Cov}\,[x_n(\tau), x_n(\tau)] = V_n(\tau) = \begin{bmatrix} P_y(\tau) & 0 \\ 0 & 0 \end{bmatrix} \tag{8.39b}$$

$$\text{Cov}\,[x_n(\tau), x_r(\tau)] = V_{nr}(\tau) = \begin{bmatrix} 0 & P_{yb}(\tau) \\ 0 & 0 \end{bmatrix} \tag{8.39c}$$

$$\text{Cov}\,[x_r(\tau), x_r(\tau)] = V_r(\tau) = \begin{bmatrix} 0 & 0 \\ 0 & P_b(\tau) \end{bmatrix} \tag{8.39d}$$

where note that $\hat{y}(\tau) = E[y(\tau)]$, $\hat{b}(\tau) = E[b(\tau)]$, $P_y(\tau) = \text{Cov}\,[y(\tau), y(\tau)]$, $P_{yb}(\tau) = \text{Cov}\,[y(\tau), b(\tau)]$ and $P_b(\tau) = \text{Cov}\,[b(\tau), b(\tau)]$.

Given the measurement data $Z_T = \{z(s); \tau \leq s \leq T\}$, we desire the optimal estimate $\hat{x}^T(t|T) = E[x^T(t)|Z_T] = [\hat{y}^T(t|T) \quad \hat{b}^T(t|T)]$ of $x(t)$ in the decoupled form of $\hat{y}(t|T)$ and $\hat{b}(t|T)$, which minimizes the cost

$$J = E[\|x(t) - \hat{x}(t|T)\|^2 | Z_T], \qquad t \leq T \quad \text{or} \quad t > T \tag{8.40}$$

As stated before, Friedland's decomposition rule [1, 3] is well known as a synthesis method for the bias correction filter, and some extensions of and work related to this decomposition can be found in detail in Mendel and Washburn [13] and Friedland [14]. However, in this chapter a radically different approach, based on the initial dependent-type partitioning filter shown in Chapter 5, will be presented.

Noting that the system of equations (8.34) and (8.35) with the partition given in equations (8.36) apparently satisfies Theorem 8.1, we obtain the following theorem.

Theorem 8.2 Generalized bias correction filter in continuous time

The filtered estimate of the original state $y(t)$ and the associated error covariance can be obtained in the following two-stage structures:

$$\hat{y}(t|t) = \hat{y}_{\rm n}(t|t) + \widetilde{\Phi}_{y\rm bn}(t, \tau)\hat{b}(\tau|t) \tag{8.41}$$

$$P_y(t) = P_{y\rm n}(t) + \widetilde{\Phi}_{y\rm bn}(t, \tau)P_{\rm b}(\tau|t)\,\widetilde{\Phi}^{\rm T}_{y\rm bn}(t, \tau) \tag{8.42}$$

where

$$\widetilde{\Phi}_{y\rm bn}(t, \tau) = \Phi_{y\rm n}(t, \tau)P_{y\rm b}(\tau)P_{\rm b}^{-1}(\tau) + \Phi_{y\rm bn}(t, \tau) \tag{8.43}$$

which is called a *generalized blending matrix* and $\hat{y}_{\rm n}(t|t)$ and $P_{y\rm n}(t)$ are the bias-free filtered estimate and its error covariance matrix, respectively, governed by the following Kalman filter equations:

$$\dot{\hat{y}}_{\rm n}(t|t) = F\hat{y}_{\rm n}(t|t) + K_{y\rm n}(t)\nu_{y\rm n}(t),$$
$$\hat{y}_{\rm n}(\tau|\tau) = \hat{y}(\tau) - P_{y\rm b}(\tau)P_{\rm b}^{-1}(\tau)\hat{b}(\tau) \tag{8.44}$$

$$\nu_{y\rm n}(t) = z(t) - H_1\hat{y}_{\rm n}(t|t) \tag{8.45}$$

$$K_{y\rm n}(t) = P_{y\rm n}(t)H_1^{\rm T}R^{-1} \tag{8.46}$$

$$\dot{P}_{y\rm n}(t) = FP_{y\rm n}(t) + P_{y\rm n}(t)F^{\rm T} + GQG^{\rm T}$$
$$- P_{y\rm n}(t)H_1^{\rm T}R^{-1}H_1P_{y\rm n}(t), \qquad P_{y\rm n}(\tau) = \widetilde{P}_y(\tau) \tag{8.47}$$

where

$$\widetilde{P}_y(\tau) = P_y(\tau) - P_{y\rm b}(\tau)P_{\rm b}^{-1}(\tau)P^{\rm T}_{y\rm b}(\tau) \tag{8.48}$$

The matrices $\Phi_{y\rm n}(t, \tau) \in \mathbb{R}^{n_{\rm c}\times n_{\rm c}}$ and $\Phi_{y\rm bn}(t, \tau) \in \mathbb{R}^{n_{\rm c}\times n_{\rm b}}$ included in equation (8.43) satisfy the following differential equations:

$$\dot{\Phi}_{y\rm n}(t, \tau) = [F - K_{y\rm n}(t)H_1]\Phi_{y\rm n}(t, \tau), \qquad \Phi_{y\rm n}(\tau, \tau) = I \tag{8.49}$$

$$\dot{\Phi}_{y\rm bn}(t, \tau) = [F - K_{y\rm n}(t)H_1]\Phi_{y\rm bn}(t, \tau)$$
$$+ [B - K_{y\rm n}(t)H_2]\Phi_{\rm bn}(t, \tau), \qquad \Phi_{y\rm bn}(\tau, \tau) = 0 \tag{8.50}$$

where $\Phi_{\rm bn}(t, \tau) \in \mathbb{R}^{n_{\rm b}\times n_{\rm b}}$ is given by

$$\dot{\Phi}_{\rm bn}(t, \tau) = D\Phi_{\rm bn}(t, \tau), \qquad \Phi_{\rm bn}(\tau, \tau) = I \tag{8.51}$$

for $D \neq 0$, and $\Phi_{\rm bn}(t, \tau) \equiv I$ for $D = 0$. Moreover, the smoothed estimate $\hat{b}(\tau|t)$ of $b(\tau)$ is presented by

$$\dot{\hat{b}}(\tau|t) = K_{\rm b}(t, \tau)[z(t) - H_1\hat{y}(t|t) - H_2\hat{b}(t|t)] \tag{8.52a}$$

or

$$\dot{\hat{b}}(\tau|t) = -K_{\mathrm{b}}(t, \tau)[H_1 \widetilde{\Phi}_{y\mathrm{bn}}(t, \tau) + H_2\Phi_{\mathrm{bn}}(t, \tau)]\hat{b}(\tau|t) + K_{\mathrm{b}}(t, \tau)\nu_{y\mathrm{n}}(t), \quad \hat{b}(\tau|\tau) = \hat{b}(\tau) \tag{8.52b}$$

where

$$K_{\mathrm{b}}(t, \tau) = P_{\mathrm{b}}(\tau|t)[H_1 \widetilde{\Phi}_{y\mathrm{bn}}(t, \tau) + H_2\Phi_{\mathrm{bn}}(t, \tau)]^{\mathrm{T}} R^{-1} \tag{8.53}$$

$$\hat{b}(t|t) = \Phi_{\mathrm{bn}}(t, \tau)\hat{b}(\tau|t) \tag{8.54}$$

$$\dot{P}_{\mathrm{b}}(\tau|t) = -K_{\mathrm{b}}(t, \tau)RK_{\mathrm{b}}^{\mathrm{T}}(t, \tau), \quad P_{\mathrm{b}}(\tau|\tau) = P_{\mathrm{b}}(\tau) \tag{8.55}$$

Proof

Applying equations (8.15)–(8.26) to the augmented system of equations (8.34)–(8.36), and decomposing each matrix or vector quantity of their results into 2×2 or 2×1 block elements, we obtain the above results. More precisely, from expressions (8.39b)–(8.39d), we see that

$$P_{\mathrm{n}}(\tau) = \begin{bmatrix} \widetilde{P}_y(\tau) & 0 \\ 0 & 0 \end{bmatrix} \tag{8.56}$$

for $t \geqslant \tau$, so that equation (8.21) can be rewritten as

$$P_{\mathrm{n}}(t) = \begin{bmatrix} P_{y\mathrm{n}}(t) & 0 \\ 0 & 0 \end{bmatrix} \tag{8.57}$$

where $P_{y\mathrm{n}}(t)$ and $\widetilde{P}_y(\tau)$ are given by equations (8.47) and (8.48), respectively. From these facts, it follows that

$$K_{\mathrm{n}}(t) = \begin{bmatrix} K_{y\mathrm{n}}(t) \\ 0 \end{bmatrix}, \quad t \geqslant \tau \tag{8.58}$$

where $K_{y\mathrm{n}}(t)$ is as defined in equation (8.46). Hence equation (8.20) becomes

$$\hat{x}_{\mathrm{n}}(t|t) = \begin{bmatrix} \hat{y}_{\mathrm{n}}(t|t) \\ 0 \end{bmatrix}, \quad t \geqslant \tau \tag{8.59}$$

where $\hat{y}_{\mathrm{n}}(t|t)$ is obtained from equation (8.44), in which the initial condition for $\hat{y}_{\mathrm{n}}(t|t)$ follows from expressions (8.22), (8.39a), (8.39c) and (8.39d).

Substituting equations (8.36) and (8.58) into (8.18) and defining

$$\Phi_{\mathrm{n}}(t, \tau) = \begin{bmatrix} \Phi_{y\mathrm{n}}(t, \tau) & \Phi_{y\mathrm{bn}}(t, \tau) \\ 0 & \Phi_{\mathrm{bn}}(t, \tau) \end{bmatrix} \tag{8.60}$$

we readily obtain equations (8.49)–(8.51). In addition, we see that, under the initial condition $V_r(\tau)$ in equation (8.39d), equation (8.26) becomes

$$P_r(\tau|t) = \begin{bmatrix} 0 & 0 \\ 0 & P_b(\tau|t) \end{bmatrix}, \quad t \geqslant \tau \tag{8.61}$$

where $P_b(\tau|t)$ is governed by equation (8.55). Since equation (8.19) with (8.39c) and (8.39d) reduces to

$$\Pi_r(\tau) = \begin{bmatrix} I & P_{yb}(\tau)P_b^{-1}(\tau) \\ 0 & I \end{bmatrix} \tag{8.62}$$

we find that equation (8.25) can be rewritten as follows:

$$K_r(t, \tau) = \begin{bmatrix} 0 \\ K_b(t, \tau) \end{bmatrix}, \quad t \geqslant \tau \tag{8.63}$$

where $K_b(t, \tau)$ is as given in equation (8.53). Therefore, we can see from equations (8.59), (8.60), (8.62) and (8.63) that equation (8.24) gives

$$\hat{x}_r(\tau|t) = \begin{bmatrix} 0 \\ \hat{b}(\tau|t) \end{bmatrix}, \quad t \geqslant \tau \tag{8.64}$$

where $\hat{b}(\tau|t)$ is obtained from equation (8.52). From equations (8.59), (8.60), (8.62) and (8.64), it is observed that expression (8.15) includes (8.41) and (8.54) as subelements. In the same manner, substituting equations (8.57) and (8.60)–(8.62) into (8.16), we can show that

$$P(t) = \begin{bmatrix} P_y(t) & P_{yb}(t) \\ P_{yb}^T(t) & P_b(t) \end{bmatrix} \tag{8.65}$$

where $P_y(t)$ is given by equation (8.42) and

$$P_{yb}(t) = \tilde{\Phi}_{ybn}(t, \tau)P_b(\tau|t)\Phi_{bn}^T(t, \tau) \tag{8.66}$$

$$P_b(t) = \Phi_{bn}(t, \tau)P_b(\tau|t)\Phi_{bn}^T(t, \tau) \tag{8.67}$$

This completes the proof. □

If the original initial state $y(\tau)$ and the bias state $b(\tau)$ are mutually independent, i.e. $P_{yb}(\tau) \equiv 0$, then the resultant bias correction filter derived here becomes that studied by Tacker and Lee [15]. Ignagni [16] investigated the discrete-time bias correction filter for a certain initial correlation, under the condition that $\Phi_{ybn}(t, \tau)$ is unconstrained initially. However, his result does not necessarily provide a convincing solution for such a situation.

The present approach also provides a new way of synthesizing the two-stage bias correction filter. This aspect can be seen from the interpretation of the blending matrix, $\widetilde{\Phi}_{y\mathrm{bn}}(t, \tau)$. Bierman [17] defined it as a sensitivity matrix, $\widetilde{\Phi}_{y\mathrm{bn}} = \mathrm{grad}_{b(t)}\hat{y}(t|t)$ and Friedland [3] introduced it as $\widetilde{\Phi}_{y\mathrm{bn}} = E[\hat{b}(t|t)\hat{y}_{\mathrm{n}}^{\mathrm{T}}(t|t)]$, which represents the cross-moment between $\hat{b}(t|t)$ and $\hat{y}_{\mathrm{n}}(t|t)$. In the present approach, however, this matrix can be interpreted as the cross-term between $y(t)$ and $b(t)$ of the closed transition matrix in the nominal Kalman filter for the augmented system given by equations (8.34) and (8.35).

The key to solving the bias correction estimators, using partitioned estimation techniques, is to see if the nominal estimator for the augmented systems given by equations (8.34) and (8.35) includes only the estimator for the original system having no bias state.

As can be seen from equations (8.47) and (8.55), this bias correction filtering approach is able to deal with Riccati differential equations of dimension $n(n+1)/2 - n_{\mathrm{c}}n_{\mathrm{b}}$, while the usual Kalman filtering approach needs RDEs of dimension $n(n+1)/2$, though an extra calculation of equations (8.49)–(8.51) is required: this is the price one has to pay for partitioning the filter. Furthermore, when $n_{\mathrm{c}} \geqslant n_{\mathrm{b}}$, the number of operations per time iteration required by equations (8.47), (8.49)–(8.51) and (8.55) is proportional to $2n_{\mathrm{c}}^3 + 2n_{\mathrm{b}}^3 + n_{\mathrm{c}}^2 n_{\mathrm{b}}$,† whereas for equation (8.6) it is proportional to n^3. Therefore, substantial computational savings can be obtained for large n and $n_{\mathrm{c}} \simeq n_{\mathrm{b}}$, compared to the usual Kalman filtering method.

The block diagram of the present bias correction filter is depicted in Figure 8.2.

Example 8.2

Consider the straight-line motion of a spacecraft subject to no random acceleration. The information is assumed to be supplied as range data which are contaminated by zero-mean Gaussian white noise with spectral density R. Let x_1 and x_2 be deviation of position and speed of the spacecraft from the standard trajectory, respectively. Then, the system equations are

$$\frac{\mathrm{d}}{\mathrm{d}t}\begin{bmatrix} x_1 \\ x_2 \end{bmatrix} = \begin{bmatrix} 0 & 1 \\ 0 & 0 \end{bmatrix}\begin{bmatrix} x_1 \\ x_2 \end{bmatrix}$$

† Notice that the number of operations required per time iteration becomes $2n_{\mathrm{c}}^3 + n_{\mathrm{b}}^3 + n_{\mathrm{c}}^2 n_{\mathrm{b}}$ for $D \equiv 0$, $n_{\mathrm{c}}^3 + 2n_{\mathrm{b}}^3 + n_{\mathrm{c}}^2 n_{\mathrm{b}}$ for $P_{y\mathrm{b}}(\tau) \equiv 0$, and $n_{\mathrm{c}}^3 + n_{\mathrm{b}}^3 + n_{\mathrm{c}}^2 n_{\mathrm{b}}$ for $D \equiv 0$ and $P_{y\mathrm{b}}(\tau) \equiv 0$.

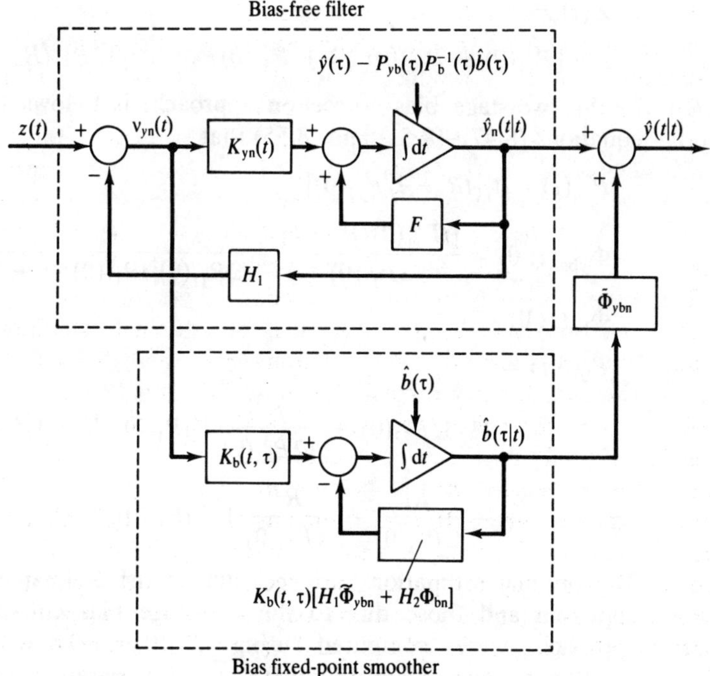

Figure 8.2 Block diagram for a two-stage bias correction filter with initial correlation.

$$z = [1 \quad 0]\begin{bmatrix} x_1 \\ x_2 \end{bmatrix} + v$$

where the a priori covariance of the initial state $x(0)$ is assumed to be $P(0) = \text{diag}\,(P_{11}(0), P_{22}(0))$. Examine the estimation error covariances in terms of (a) the optimal Kalman filter, and (b) the two-stage bias correction filter.

(a) In the Kalman filter approach, the filtering error covariance [18] is given by

$$P_{11}(t) = \frac{1}{Z(t)}\left[P_{11}(0) + P_{22}(0)t^2 + \frac{P_{11}(0)P_{22}(0)t^3}{3R}\right]$$

$$P_{12}(t) = \frac{1}{Z(t)}\,[P_{22}(0)t(1 + P_{11}(0)t/2R)]$$

$$P_{22}(t) = P_{22}(0)(1 + P_{11}(0)t/R)$$

where

$$Z(t) =$$
$$1 + [P_{11}(0) + P_{22}(0)t^2/3 + P_{11}(0)P_{22}(0)t^3/12R]t/R$$

(b) In the two-stage bias correction approach, it follows from equations (8.47)–(8.51) and (8.55) that

$$P_{yn}(t) = 1/[tR + 1/P_{11}(0)]$$

$$\Phi_{ybn}(t, 0) = \frac{[tP_{11}(0)/R + 1]}{2P_{11}(0)} - \frac{R}{2P_{11}(0)[tP_{11}(0)/R + 1]}$$

$$\Phi_{bn}(t, 0) = 1$$

$$P_b(0|t) =$$
$$1\bigg/\left\{1/P_{22}(0) + \frac{R^2}{12P_{11}^3(0)}[(tP_{11}(0)/R + 1)^3 - 1] - \frac{Rt}{2P_{11}^2(0)} - \frac{R^2}{4P_{11}^3(0)}[1/(tP_{11}(0)/R + 1) - 1]\right\}$$

The optimal estimation variances due to the Kalman filter approach and those due to the two-stage bias correction approach were computed using $P_{11}(0) = 1.0 \times 10^4\ \text{m}^2$, $P_{22}(0) = 1.0\ \text{m}^2\,\text{s}^{-2}$ and $R = 1.0\ \text{km}^2\,\text{s}$. Comparative results from the two different approaches are displayed in Figure 8.3, which shows that the computational error covariances due to both methods are completely consistent.

In the remainder of this section, we shall present a bias correction predictor based on the partitioning predictor given in Theorem 6.5 of Chapter 6. That is, given the measurement data Z_{t_1}, $t \geq t_1 \geq \tau$ (where t_1 is fixed), the following corollary is obtained.

Corollary 8.2 Generalized bias correction predictor

The predicted estimate $\hat{y}(t|t_1)$ of the original state $y(t)$ and the predicted error covariance matrix $P_y(t|t_1)$ are obtained by the following relations:

$$\hat{y}(t|t_1) = \hat{y}_n(t|t_1) + \widetilde{\Phi}_{yb}(t, t_1)\hat{b}(t_1) \tag{8.68}$$

$$P_y(t|t_1) = P_{yn}(t|t_1) + \widetilde{\Phi}_{yb}(t, t_1)P_b(t_1)\widetilde{\Phi}_{yb}^{\mathrm{T}}(t, t_1) \tag{8.69}$$

The bias-free predicted estimate $\hat{y}_n(t|t_1)$ and its error covariance matrix $P_{yn}(t|t_1)$ are, respectively, given by

$$\dot{\hat{y}}_{\mathrm{n}}(t|t_1) = F\hat{y}_{\mathrm{n}}(t|t_1), \qquad \hat{y}_{\mathrm{n}}(t_1|t_1) = \hat{y}(t_1) - P_{y\mathrm{b}}(t_1)P_{\mathrm{b}}^{-1}(t_1)\hat{b}(t_1) \tag{8.70}$$

$$\dot{P}_{y\mathrm{n}}(t|t_1) = FP_{y\mathrm{n}}(t|t_1) + P_{y\mathrm{n}}(t|t_1)F^{\mathrm{T}} + GQG^{\mathrm{T}}, \qquad P_{y\mathrm{n}}(t_1|t_1) = \tilde{P}_y(t_1) \tag{8.71}$$

where $\tilde{P}_y(t_1)$ is given by equation (8.48) with $\tau = t_1$. In addition, a blending matrix $\tilde{\Phi}_{y\mathrm{b}}(t, t_1)$ is defined by

$$\tilde{\Phi}_{y\mathrm{b}}(t, t_1) = \Phi_y(t, t_1)P_{y\mathrm{b}}(t_1)P_{\mathrm{b}}^{-1}(t_1) + \Phi_{y\mathrm{b}}(t, t_1) \tag{8.72}$$

and

$$\dot{\Phi}_y(t, t_1) = F\Phi_y(t, t_1), \qquad \Phi_y(t_1, t_1) = I \tag{8.73}$$

$$\dot{\Phi}_{y\mathrm{b}}(t, t_1) = F\Phi_{y\mathrm{b}}(t, t_1) + B\Phi_{\mathrm{b}}(t, t_1), \qquad \Phi_{y\mathrm{b}}(t_1, t_1) = 0 \tag{8.74}$$

$$\dot{\Phi}_{\mathrm{b}}(t, t_1) = D\Phi_{\mathrm{b}}(t, t_1), \qquad \Phi_{\mathrm{b}}(t_1, t_1) = I \tag{8.75}$$

where

$$P_{y\mathrm{b}}(t_1) = \tilde{\Phi}_{y\mathrm{bn}}(t_1, \tau)P_{\mathrm{b}}(\tau|t_1)\tilde{\Phi}_{\mathrm{bn}}^{\mathrm{T}}(t_1, \tau) \tag{8.76}$$

$$P_{\mathrm{b}}(t_1) = \Phi_{\mathrm{bn}}(t_1, \tau)P_{\mathrm{b}}(\tau|t_1)\Phi_{\mathrm{bn}}^{\mathrm{T}}(t_1, \tau) \tag{8.77}$$

Proof

This derivation, together with Theorem 6.5, follows in exactly the same manner as for the filtering problem. But a direct derivation using Theorem 8.2 is useful. The initial conditions for this estimator can be readily obtained from those for the bias correction filter at the time $\tau = t_1$, and the current estimates and the error covariances can also be provided by the results of Theorem 8.2 with $R = \infty$. □

Figure 8.4 shows the block diagram for the bias correction predictor.

8.4 A BIAS CORRECTION SMOOTHER

In this section, we shall present a bias correction fixed-interval smoother based on the partitioning fixed-interval smoother. The partitioning fixed-interval smoothing algorithm can be readily obtained by applying Theorems 6.3 and 6.4, and setting $t = T$ and $\tau = t$ in their results. We summarize this in the following lemma.

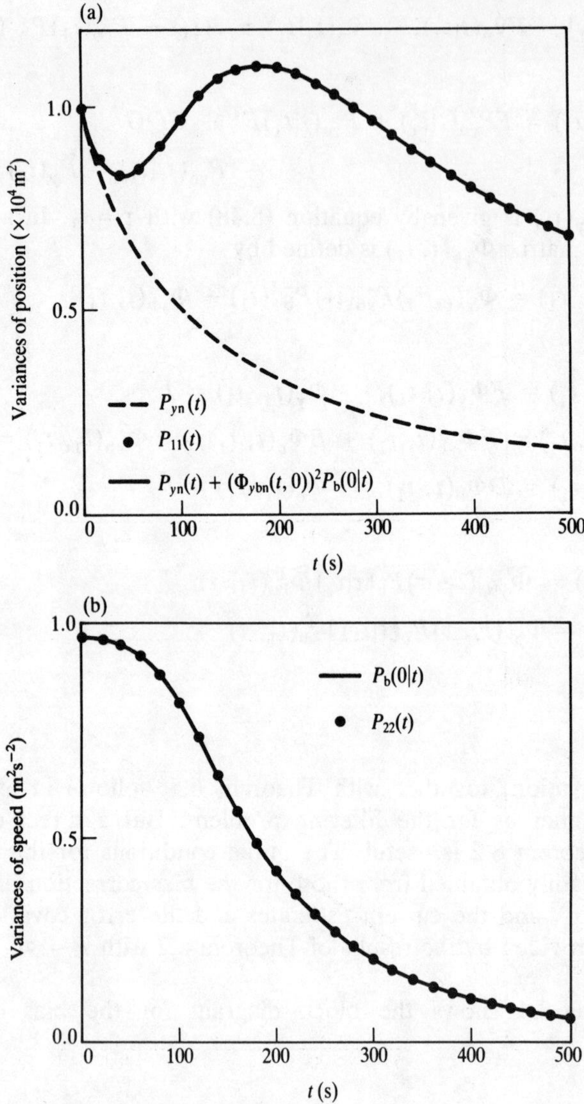

Figure 8.3 (a) Optimal variances of spacecraft position; (b) optimal variances of spacecraft speed; (c) estimator gains of spacecraft speed; (d) blending matrix for a two-stage bias correction filter.

Lemma 8.1 Partitioning fixed-interval smoother

Suppose that the filtered state of $x(t)$ can be obtained by the generalized partitioning filter consisting of equations (8.15)–(8.26).

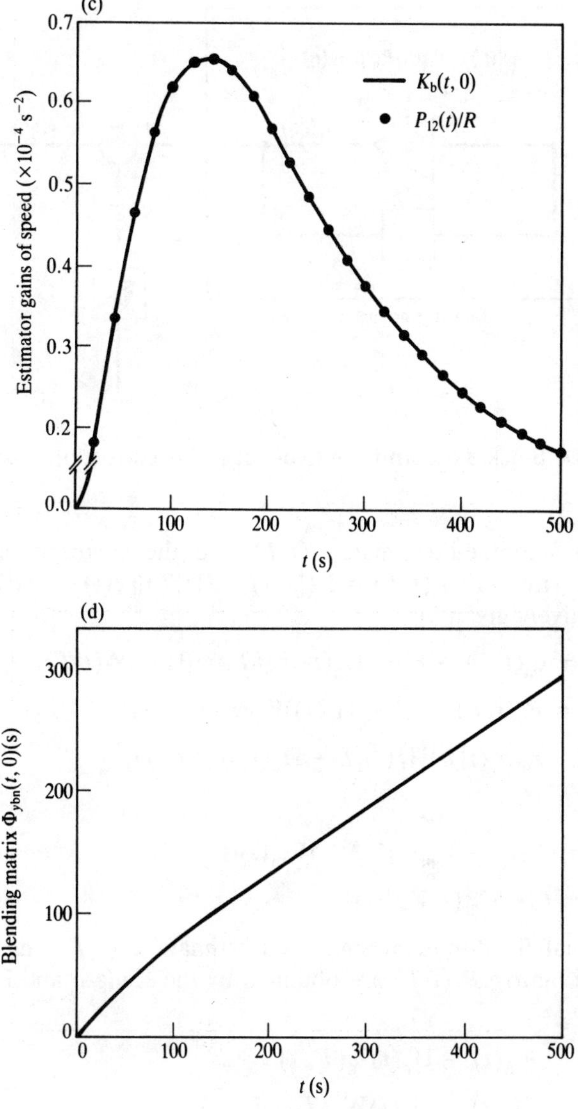

Figure 8.3 (*continued*)

Additionally, the filtered estimate $\hat{x}(t|t)$ and its error covariance $P(t)$ are assumed to be decomposed into

$$\hat{x}(t|t) = \hat{\mu}_n(t) + \hat{\mu}_r(t) \tag{8.78a}$$

$$P(t) = V_n(t) + V_{nr}(t) + V_{nr}^T(t) + V_r(t) \tag{8.78b}$$

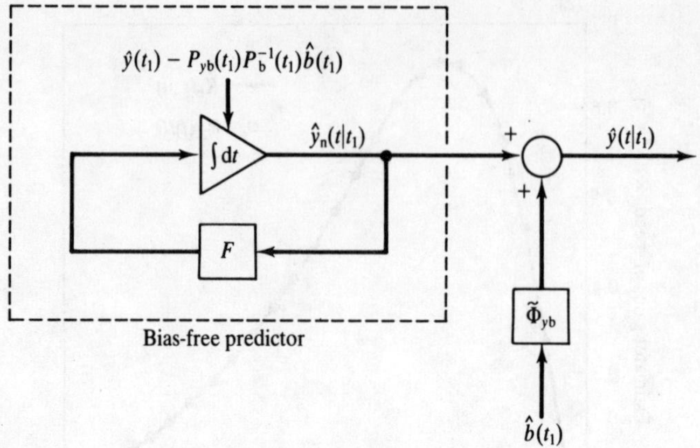

Figure 8.4 Block diagram for a two-stage bias correction predictor.

Then, the smoothed estimate $\hat{x}(t|T)$ and the corresponding error covariance matrix $P(t|T) = E\{[x(t) - \hat{x}(t|T)][x(t) - \hat{x}(t|T)]^{\mathrm{T}}|Z_T\}$ are respectively given by

$$\hat{x}(t|T) = \hat{x}_{\mathrm{n}}(t|T) + [I - \Pi_{\mathrm{n}}(t)W_{\mathrm{n}}(T, t)]\Pi_{\mathrm{r}}(t)\hat{x}_{\mathrm{r}}(t|T) \tag{8.79}$$

$$P(t|T) = P_{\mathrm{n}}(t|T) + [I - \Pi_{\mathrm{n}}(t)W_{\mathrm{n}}(T, t)]\Pi_{\mathrm{r}}(t) \times P_{\mathrm{r}}(t|T)\Pi_{\mathrm{r}}^{\mathrm{T}}(t)[I - \Pi_{\mathrm{n}}(t)W_{\mathrm{n}}(T, t)]^{\mathrm{T}} \tag{8.80}$$

where

$$\Pi_{\mathrm{n}}(t) = V_{\mathrm{n}}(t) - V_{\mathrm{nr}}(t)V_{\mathrm{r}}^{-1}(t)V_{\mathrm{nr}}^{\mathrm{T}}(t) \tag{8.81}$$

$$\Pi_{\mathrm{r}}(t) = [I + V_{\mathrm{nr}}(t)V_{\mathrm{r}}^{-1}(t)] \tag{8.82}$$

The nominal fixed-interval smoothed estimate $\hat{x}_{\mathrm{n}}(t|T)$ and its error covariance matrix $P_{\mathrm{n}}(t|T)$ are obtained by the Kailath and Frost-type smoother:

$$\hat{x}_{\mathrm{n}}(t|T) = \hat{m}_{\mathrm{n}}(t) + \Pi_{\mathrm{n}}(t)\lambda_{\mathrm{n}}(T, t) \tag{8.83}$$

$$P_{\mathrm{n}}(t|T) = \Pi_{\mathrm{n}}(t) - \Pi_{\mathrm{n}}(t)W_{\mathrm{n}}(T, t)\Pi_{\mathrm{n}}(t) \tag{8.84}$$

where

$$\hat{m}_{\mathrm{n}}(t) = \hat{\mu}_{\mathrm{n}}(t) - V_{\mathrm{nr}}(t)V_{\mathrm{r}}^{-1}(t)\hat{\mu}_{\mathrm{r}}(t) \tag{8.85}$$

$$\lambda_{\mathrm{n}}(T, t) = \int_t^T \Phi_{\mathrm{n}}^{\mathrm{T}}(s, t)H^{\mathrm{T}}R^{-1}[z(s) - H\hat{x}_{\mathrm{n}}(s|s)]\,\mathrm{d}s \tag{8.86}$$

$$W_{\mathrm{n}}(T, t) = \int_t^T \Phi_{\mathrm{n}}^{\mathrm{T}}(s, t)H^{\mathrm{T}}R^{-1}H\Phi_{\mathrm{n}}(s, t)\,\mathrm{d}s \tag{8.87}$$

Further, the residual fixed-interval smoothed estimate $\hat{x}_r(t|T)$ and the error covariance matrix $P_r(t|T)$ are derived from equations (8.24)–(8.26):

$$\hat{x}_r(t|T) = \hat{\mu}_r(t) - \int_t^T K_r(s, t)H\Phi_n(s, t)\Pi_r(t)\hat{x}_r(t|s)\,ds + \int_t^T K_r(s, t)[z(s) - H\hat{x}_n(s|s)]\,ds \tag{8.88}$$

$$P_r(t|T) = V_r(t) - \int_t^T K_r(s, t)RK_r^T(s, t)\,ds \tag{8.89}$$

where

$$K_r(s, t) = P_r(t|s)\Pi_r^T(t)\Phi_n^T(s, t)H^T R^{-1} \tag{8.90}$$

Moreover, the nominal filtered estimate $\hat{x}_n(s|s)$, $t \leqslant s \leqslant T$, and its error covariance matrix $P_n(t)$, which are involved in the integrands of equations (8.86)–(8.89), satisfy the following Kalman filter equations:

$$\dot{\hat{x}}_n(s|s) = A\hat{x}_n(s|s) + K_n(s)[z(s) - H\hat{x}_n(s|s)], \qquad \hat{x}_n(t|t) = \hat{m}_n(t) \tag{8.91}$$

$$K_n(s) = P_n(s)H^T R^{-1} \tag{8.92}$$

$$\dot{P}_n(s) = AP_n(s) + P_n(s)A^T + CQC^T - P_n(s)H^T R^{-1} HP_n(s), \qquad P_n(t) = \Pi_n(t) \tag{8.93}$$

and the closed-loop transition matrix $\Phi_n(s, t)$ is described by the following relation:

$$\dot{\Phi}_n(s, t) = [A - K_n(s)H]\Phi_n(s, t), \qquad \Phi_n(t, t) = I \tag{8.94}$$

Now we can state the following theorem on a bias correction fixed-interval smoother.

Theorem 8.3 Bias correction fixed-interval smoother

The fixed-interval smoothed estimate of the original state $y(t)$ and its error covariance matrix are obtained by the following expressions:

$$\hat{y}(t|T) = \hat{y}_n(t|T) + [P_{yb}(t)P_b^{-1}(t) - \tilde{P}_y(t)\tilde{W}_{ybn}(T, t)]\hat{b}(t|T) \tag{8.95}$$

$$P_y(t|T) = P_{yn}(t|T) + [P_{yb}(t)P_b^{-1}(t) - \widetilde{P}_y(t)\widetilde{W}_{ybn}(T, t)] \times P_b(t|T)[P_{yb}(t)P_b^{-1}(t) - \widetilde{P}_y(t)\widetilde{W}_{ybn}(T, t)]^T \tag{8.96}$$

where

$$\widetilde{P}_y(t) = P_y(t) - P_{yb}(t)P_b^{-1}(t)P_{yb}^T(t) \tag{8.97}$$

$$\widetilde{W}_{ybn}(T, t) = W_{yn}(T, t)P_{yb}(t)P_b^{-1}(t) + W_{ybn}(T, t) \tag{8.98}$$

in which $P_y(t)$, $P_{yb}(t)$ and $P_b(t)$ are as given by equations (8.42), (8.66) and (8.67), respectively, and $W_{yn}(T, t)$ and $W_{ybn}(T, t)$ are defined by

$$W_{yn}(T, t) = \int_t^T \Phi_{yn}^T(s, t)H_1^T R^{-1} H_1 \Phi_{yn}(s, t)\,\mathrm{d}s \tag{8.99}$$

$$W_{ybn}(T, t) = \int_t^T \Phi_{yn}^T(s, t)H_1^T R^{-1} \times [H_1\Phi_{ybn}(s, t) + H_2\Phi_{bn}(s, t)]\,\mathrm{d}s \tag{8.100}$$

where $\Phi_{yn}(s, t)$, $\Phi_{ybn}(s, t)$ and $\Phi_{bn}(s, t)$ are subject to equations (8.49), (8.50) and (8.51) with $t = s$ and $\tau = t$, respectively. The bias-free fixed-interval smoothed estimate $\hat{y}_n(t|T)$ and its error covariance matrix $P_{yn}(t|T)$ are obtained from the Kailath and Frost-type smoother evolving forwards in time:

$$\hat{y}_n(t|T) = \hat{y}(t|t) - P_{yb}(t)P_b^{-1}(t)\hat{b}(t|t) + \widetilde{P}_y(t)\int_t^T \Phi_{yn}^T(s, t)H_1^T R^{-1} v_{yn}(s)\,\mathrm{d}s \tag{8.101}$$

$$P_{yn}(t|T) = \widetilde{P}_y(t) - \widetilde{P}_y(t)W_{yn}(T, t)\widetilde{P}_y(t) \tag{8.102}$$

Furthermore, the bias smoothed estimate $\hat{b}(t|T)$ and the error covariance $P_b(t|T)$ are given by the partitioning smoother formulas:

$$\hat{b}(t|T) = \hat{b}(t|t) - \int_t^T K_b(s, t) \times [H_1\widetilde{\Phi}_{ybn}(s, t) + H_2\Phi_{bn}(s, t)]\hat{b}(t|s)\,\mathrm{d}s + \int_t^T K_b(s, t)v_{yn}(s)\,\mathrm{d}s \tag{8.103}$$

$$K_b(s, t) = P_b(t|s)[H_1\widetilde{\Phi}_{ybn}(s, t) + H_2\Phi_{bn}(s, t)]^T R^{-1} \tag{8.104}$$

$$v_{yn}(s) = z(s) - H_1\hat{y}_n(s|s) \tag{8.105}$$

$$P_b(t|T) = P_b(t) - \int_t^T K_b(s, t)RK_b^T(s, t)\,\mathrm{d}s \tag{8.106}$$

where $\widetilde{\Phi}_{ybn}(s, t)$ is obtained by equation (8.43) with $t = s$ and $\tau = t$. The bias-free filtered estimate $\hat{y}_n(s|s)$ and its error covariance $P_{yn}(s)$, which are involved in each integrand of the above equations, are also given by the following Kalman filter equations:

$$\dot{\hat{y}}_n(s|s) = F\hat{y}_n(s|s) + K_{yn}(s)[z(s) - H_1\hat{y}_n(s|s)],$$
$$\hat{y}_n(t|t) = \hat{y}(t|t) - P_{yb}(t)P_b^{-1}(t)\hat{b}(t|t) \qquad (8.107)$$

$$K_{yn}(s) = P_{yn}(s)H_1^T R^{-1} \qquad (8.108)$$

$$\dot{P}_{yn}(s) = FP_{yn}(s) + P_{yn}(s)F^T - P_{yn}(s)H_1^T R^{-1} H_1 P_{yn}(s) + GQG^T, \qquad P_{yn}(t) = \widetilde{P}_y(t) \qquad (8.109)$$

for $\tau \leqslant t \leqslant s \leqslant T$. □

Proof

See Watanabe [7]. □

It should be noted that the blending matrix $[P_{yb}(t)P_b^{-1}(t) - \widetilde{P}_y(t)\widetilde{W}_{ybn}(T, t)]$ for this smoothing problem originates from the closed-loop transition matrix for the smoothing problem as an extended filtering problem, which was discussed in Section 6.4. We also observe that this matrix consists of the partial information matrix of the nominal filter in the partitioning filter problem for equations (8.34) and (8.35), whereas Bierman [17] defines it as $\text{grad}_{b(t)}\hat{y}(t|T)$.

Notice that the dual Chandrasekhar algorithm shown in Corollary 6.10 can be readily applied to this bias correction smoothing formula. That is, if we define

$$L_{yn}(s, t) = R^{-1}H_1\Phi_{yn}(s, t) \in \mathbb{R}^{m\times n_c} \qquad (8.110)$$

$$L_{ybn}(s, t) = R^{-1}H_1\Phi_{ybn}(s, t) \in \mathbb{R}^{m\times n_b} \qquad (8.111)$$

$$L_{bn}(s, t) = R^{-1}H_2\Phi_{bn}(s, t) \in \mathbb{R}^{m\times n_b} \qquad (8.112)$$

where $\tau \leqslant t \leqslant s \leqslant T$, we see that the bias-free smoother mechanism of equation (8.101) can be rewritten as

$$\hat{y}_n(t|T) = \hat{y}(t|t) - P_{yb}(t)P_b^{-1}(t)\hat{b}(t|t) + \widetilde{P}_y(t)\int_t^T L_{yn}^T(s, t)\nu_{yn}(s)\,ds \qquad (8.101')$$

In addition, the bias smoother of equation (8.103) and the gain matrix of equation (8.104) become

$$\hat{b}(t|T) = \hat{b}(t|t) - \int_t^T K_b(s, t)R \times [L_{yn}(s, t)P_{yb}(t)P_b^{-1}(t) + L_{ybn}(s, t) + L_{bn}(s, t)] \times \hat{b}(t|s)\,ds + \int_t^T K_b(s, t)\nu_{yn}(s)\,ds \tag{8.103'}$$

$$K_b(s, t) = P_b(s|t) \times [L_{yn}(s, t)P_{yb}(t)P_b^{-1}(t) + L_{ybn}(s, t) + L_{bn}(s, t)]^T \tag{8.104'}$$

where $L_{yn}(\cdot)$, $L_{ybn}(\cdot)$ and $L_{bn}(\cdot)$ are obtained from Corollary 6.10 as follows:

$$\frac{dL_{yn}(s, t)}{ds} = L_{yn}(s, t)[F - \tilde{P}_y(t)H_1^T R^{-1} H_1 - N_n(t)\Sigma_n(t)N_n^T(t)W_{yn}(s, t)], \quad L_{yn}(t, t) = R^{-1}H_1 \tag{8.113}$$

$$\frac{dL_{ybn}(s, t)}{ds} = L_{ybn}(s, t)D + L_{yn}(s, t)[B - \tilde{P}_y(t)H_1^T R^{-1} H_2 - N_n(t)\Sigma_n(t)N_n^T(t)W_{ybn}(s, t)], \quad L_{ybn}(t, t) = 0 \tag{8.114}$$

$$\frac{dL_{bn}(s, t)}{ds} = L_{bn}(s, t)D, \qquad L_{bn}(t, t) = R^{-1}H_2 \tag{8.115}$$

in which $W_{yn}(s, t)$ and $W_{ybn}(s, t)$ are given by combining equations (8.99), (8.100) and (8.110)–(8.112) as follows:

$$\frac{dW_{yn}(s, t)}{ds} = L_{yn}^T(s, t)RL_{yn}(s, t), \qquad W_{yn}(t, t) = 0 \tag{8.116}$$

$$\frac{dW_{ybn}(s, t)}{ds} = L_{yn}^T(s, t)R[L_{ybn}(s, t) + L_{bn}(s, t)], \quad W_{ybn}(t, t) = 0 \tag{8.117}$$

where

$$N_n(t)\Sigma_n(t)N_n^T(t) = F\tilde{P}_y(t) + \tilde{P}_y(t)F^T + GQG^T - \tilde{P}_y(t)H_1^T R^{-1} H_1 \tilde{P}_y(t) \tag{8.118}$$

and $N_n(t)$ is a full-rank $n_c \times \alpha$ matrix, where

$$\alpha = \text{rank}\,[F\widetilde{P}_y(t) + \widetilde{P}_y(t)F^{\mathrm{T}} + GQG^{\mathrm{T}} - \widetilde{P}_y(t)H_1^{\mathrm{T}}R^{-1}H_1\widetilde{P}_y(t)]$$

Moreover, the gain matrix $K_{yn}(s)$ of the bias-free filter of equation (8.107) can be rewritten by the usual Chandrasekhar equation given in Corollary 6.9 as follows:

$$\frac{\mathrm{d}K_{yn}(s)}{\mathrm{d}s} = \widetilde{L}_{yn}(s, t)\Sigma_n(t)\widetilde{L}_{yn}^{\mathrm{T}}(s, t)H_1^{\mathrm{T}}R^{-1},$$

$$K_{yn}(t) = \widetilde{P}_y(t)H_1^{\mathrm{T}}R^{-1} \qquad (8.119)$$

$$\frac{\mathrm{d}\widetilde{L}_{yn}(s, t)}{\mathrm{d}s} = [F - K_{yn}(s)H_1]\widetilde{L}_{yn}(s, t), \qquad \widetilde{L}_{yn}(t, t) = N_n(t) \qquad (8.120)$$

The block diagram of the bias correction fixed-interval smoother is depicted in Figure 8.5.

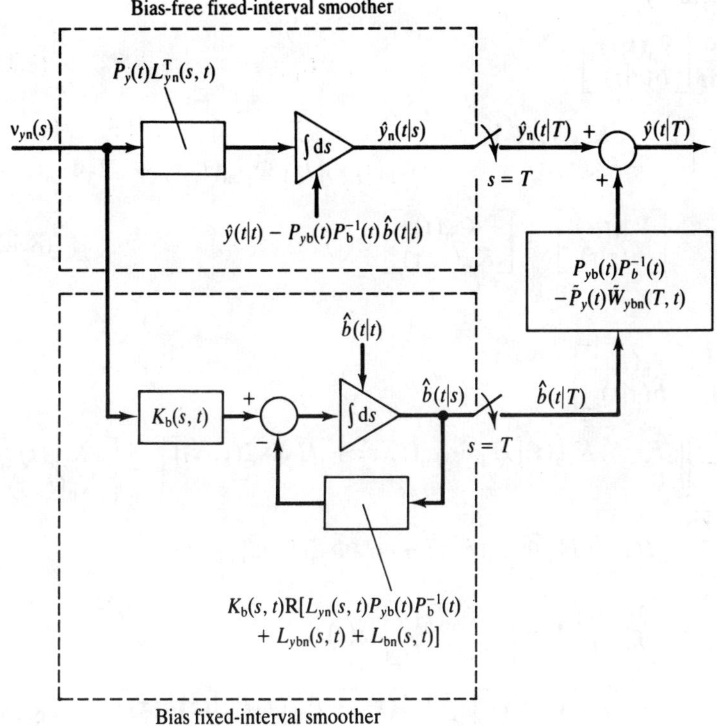

Figure 8.5 Block diagram for a two-stage bias correction fixed-interval smoother evolving forwards in time.

8.5 STABILITY OF THE BIAS CORRECTION FILTER

In this section, we examine the stability of bias correction filters and their asymptotic behaviour [19]. To this end, it is assumed that the original subsystem of the augmented system of equations (8.34) and (8.35), i.e. (F, G, H_1), is completely controllable (or reachable) and observable. It is further assumed that equations (8.49)–(8.51) are asymptotically stable, i.e.

$$\left.\begin{aligned} &\|\Phi_{yn}(t;\tau)\| \to 0, \qquad \|\Phi_{ybn}(t;\tau)\| \to 0 \\ &\|\Phi_{bn}(t;\tau)\| \to 0 \quad (t \to \infty) \end{aligned}\right\} \tag{8.121}$$

where matrix norm $\|\Phi_{yn}(t;\tau)\|$ is given by $\max_{\|x\|=1}\|\Phi_{yn}(t;\tau)x\|$. Under these assumptions, it is sufficient for examining the stability of the bias correction filter to consider those of the n_cth-order bias-free filter given in equation (8.44) and the n_bth-order partitioned-type fixed-point smoother of equation (8.52). We express this augmented system by

$$\begin{aligned} &\frac{d}{dt}\begin{bmatrix} \hat{y}_n(t|t) \\ \hat{b}(\tau|t) \end{bmatrix} \\ &= \begin{bmatrix} [F - K_{yn}(t)H_1] & 0 \\ -K_b(t,\tau) & -K_b(t,\tau)[H_1\widetilde{\Phi}_{ybn}(t,\tau) + H_2\Phi_{bn}(t,\tau)] \end{bmatrix} \\ &\quad \times \begin{bmatrix} \hat{y}_n(t|t) \\ \hat{b}(\tau|t) \end{bmatrix} + \begin{bmatrix} K_{yn}(t) \\ K_b(t,\tau) \end{bmatrix} z(t) \end{aligned} \tag{8.122a}$$

or

$$\begin{aligned} &\frac{d}{dt}\begin{bmatrix} (\hat{y}_n(t|t) \\ \hat{b}(\tau|t) \end{bmatrix} \\ &= \left\{\begin{bmatrix} F & K_{yn}(t)[H_1\widetilde{\Phi}_{ybn}(t,\tau) + H_2\Phi_{bn}(t,\tau)] \\ 0 & 0 \end{bmatrix} - \begin{bmatrix} K_{yn}(t) \\ K_b(t,\tau) \end{bmatrix}\right. \\ &\quad \left.\times [H_1 \quad \{H_1\widetilde{\Phi}_{ybn}(t,\tau) + H_2\Phi_{bn}(t,\tau)\}]\right\} \\ &\quad \times \begin{bmatrix} \hat{y}_n(t|t) \\ \hat{b}(\tau|t) \end{bmatrix} + \begin{bmatrix} K_{yn}(t) \\ K_b(t,\tau) \end{bmatrix} z(t), \\ &\qquad\qquad \begin{bmatrix} \hat{y}_n(\tau|\tau) \\ \hat{b}(\tau|\tau) \end{bmatrix} = \begin{bmatrix} \hat{y}(\tau) - P_{yb}(\tau)P_b^{-1}(\tau)\hat{b}(\tau) \\ \hat{b}(\tau) \end{bmatrix} \end{aligned} \tag{8.122b}$$

where

$$\begin{bmatrix} K_{yn}(t) \\ K_b(t, \tau) \end{bmatrix} =$$

$$\begin{bmatrix} P_{yn}(t) & 0 \\ 0 & P_b(\tau|t) \end{bmatrix} \begin{bmatrix} H_1^T \\ \{H_1 \widetilde{\Phi}_{ybn}(t, \tau) + H_2 \Phi_{bn}(t, \tau)\}^T \end{bmatrix} R^{-1} \tag{8.123}$$

Moreover, if we define the following matrices

$$\widetilde{P}(t) \triangleq \begin{bmatrix} P_{yn}(t) & 0 \\ 0 & P_b(\tau|t) \end{bmatrix} \tag{8.124}$$

$$\widetilde{F}(t) \triangleq \begin{bmatrix} F & K_{yn}(t)[H_1 \widetilde{\Phi}_{ybn}(t, \tau) + H_2 \Phi_{bn}(t, \tau)] \\ 0 & 0 \end{bmatrix} \tag{8.125}$$

$$\widetilde{H}(t) \triangleq [H_1 \quad \{H_1 \widetilde{\Phi}_{ybn}(t, \tau) + H_2 \Phi_{bn}(t, \tau)\}] \tag{8.126}$$

$$\widetilde{G} \triangleq \begin{bmatrix} G \\ 0 \end{bmatrix}, \qquad \widetilde{K}(t) \triangleq \widetilde{P}(t) \widetilde{H}^T(t) R^{-1} \tag{8.127}$$

then $\widetilde{P}(t)$, from equations (8.47) and (8.55), obeys the following time-varying RDE:

$$\begin{aligned} \dot{\widetilde{P}}(t) = \widetilde{F}(t)\widetilde{P}(t) + \widetilde{P}(t)\widetilde{F}^T(t) - \widetilde{P}(t)\widetilde{H}^T(t)R^{-1}\widetilde{H}(t)\widetilde{P}(t) \\ + \widetilde{G}Q\widetilde{G}^T, \qquad \widetilde{P}(\tau) = \operatorname{diag}(\widetilde{P}_y(\tau), P_b(\tau)) \end{aligned} \tag{8.128}$$

where $\widetilde{P}_y(\tau)$ is as defined in equation (8.48). At the same time, consider the following dynamic system:

$$\frac{d}{dt} \begin{bmatrix} y_n(t) \\ b(\tau) \end{bmatrix} = \widetilde{F}(t) \begin{bmatrix} y_n(t) \\ b(\tau) \end{bmatrix} + \widetilde{G} w(t) \tag{8.129}$$

$$z(t) = \widetilde{H}(t) \begin{bmatrix} y_n(t) \\ b(\tau) \end{bmatrix} + v(t) \tag{8.130}$$

with the initial statistical distribution

$$\begin{bmatrix} y_n(\tau) \\ b(\tau) \end{bmatrix} \sim N\left\{ \begin{bmatrix} \hat{y}(\tau) - P_{yb}(\tau) P_b^{-1}(\tau) \hat{b}(\tau) \\ \hat{b}(\tau) \end{bmatrix}, \begin{bmatrix} \widetilde{P}_y(\tau) & 0 \\ 0 & P_b(\tau) \end{bmatrix} \right\} \tag{8.131}$$

as one candidate model for equation (8.122).

8.5.1 Stability Problem

Now returning to the stability problem, we can see that the structure of the bias-free filter is identical to that of the usual Kalman filter, and hence the uniform complete controllability of $(F, GQ^{1/2})$ and the uniform complete observability of $(R^{-1/2}H_1, F)$ are guaranteed by the results of Kalman and Bucy [20]. However, the uniform complete controllability of the partitioned-type fixed-point smoother of equation (8.52) is evidently not assured because it is based on the uncontrollable subsystem. It is concluded from these facts that the augmented estimator in equation (8.131) is not in fact uniformly asymptotically stable, i.e. exponentially asymptotically stable. Therefore, in the following steps, we shall provide a more relaxed set of results than the above conclusion by using the nonsingularity property of $\widetilde{P}(\tau)$.

Let us introduce a generalized form of controllability matrix associated with equation (8.129):

$$\widetilde{W}(t_1, \tau) = \widetilde{\Phi}(t_1, \tau)\widetilde{P}(\tau)\widetilde{\Phi}^{\mathrm{T}}(t_1, \tau) + \int_{\tau}^{t_1} \widetilde{\Phi}(t_1, s)\widetilde{G}Q\widetilde{G}^{\mathrm{T}}\widetilde{\Phi}^{\mathrm{T}}(t_1, s)\,\mathrm{d}s \tag{8.132}$$

where $\widetilde{\Phi}(t_1, \tau)$ is the state transition matrix of equation (8.129) and $\widetilde{W}(t_1, \tau)$ is the second-order moment of the same equation. Then, with the aid of the results in Anderson [9], we obtain a necessary and sufficient condition for the invertibility of $\widetilde{P}(t_1)$.

Lemma 8.2

The null space of $\widetilde{P}(t_1)$, $\mathcal{N}(\widetilde{P}(t_1))$, and the null space of $\widetilde{W}(t_1, \tau)$, $\mathcal{N}(\widetilde{W}(t_1, \tau))$, are the same; in particular, both matrices are either simultaneously singular or simultaneously nonsingular.

Note that if $\widetilde{P}(t_1)$ is invertible, $\widetilde{P}(t)$ is invertible for all $t > t_1$. Specifically, if $\widetilde{P}(\tau)$ is nonsingular, $\widetilde{P}(t)$ is nonsingular for all t. Furthermore, the following lemma is easily obtained for the boundedness of $\widetilde{P}(\cdot)$.

Lemma 8.3

If $\widetilde{F}(t)$, $\widetilde{G}$, Q and R^{-1} are bounded, and if $(\widetilde{H}(t), F(t))$ is uniformly completely observable, then $\widetilde{P}(t)$ is bounded.

From these lemmas, we can state the stability result for the two-stage bias correction filter.

Theorem 8.4

Suppose that $\widetilde{F}(t)$, $\widetilde{G}$, Q, $\widetilde{H}(t)$ and R^{-1} are bounded, that $(R^{-1/2}\widetilde{H}(t), \widetilde{F}(t))$ is uniformly completely observable, and that $\widetilde{W}(t_1, \tau)$ is nonsingular for some t_1. Then the two-stage bias correction filter for equations (8.34) and (8.35) is asymptotically stable *sensu* Lyapunov.

Proof

Taking account of Lemma 8.2, $\widetilde{P}^{-1}(t)$ exists for all $t \geqslant t_1$; and from Lemma 8.3, $\widetilde{P}^{-1}(t) \geqslant \alpha_1 I > 0$ for some positive constant α_1. As a Lyapunov function for the closed-loop system of equation (8.122), i.e.

$$\frac{\mathrm{d}\xi(t)}{\mathrm{d}t} = [\widetilde{F}(t) - \widetilde{P}(t)\widetilde{H}^{\mathrm{T}}(t)R^{-1}\widetilde{H}(t)]\xi(t) \tag{8.133}$$

we adopt $V(\xi, t) = \xi^{\mathrm{T}}(t)\widetilde{P}^{-1}(t)\xi(t)$. Differentiating $V(\cdot)$ with respect to time t yields

$$\begin{aligned}\frac{\mathrm{d}V(\xi, t)}{\mathrm{d}t} &= \frac{\mathrm{d}[\xi^{\mathrm{T}}(t)\widetilde{P}^{-1}(t)\xi(t)]}{\mathrm{d}t} \\ &= -\xi^{\mathrm{T}}(t)[\widetilde{H}^{\mathrm{T}}(t)R^{-1}\widetilde{H}(t) + \widetilde{P}^{-1}(t)\widetilde{G}Q\widetilde{G}^{\mathrm{T}}\widetilde{P}^{-1}(t)]\xi(t) \\ &\leqslant -\xi^{\mathrm{T}}(t)\widetilde{H}^{\mathrm{T}}(t)R^{-1}\widetilde{H}(t)\xi(t)\end{aligned} \tag{8.134}$$

The uniform complete observability of $(R^{-1/2}\widetilde{H}(t), \widetilde{F}(t))$ implies uniform complete observability of $(R^{-1/2}\widetilde{H}(t), \widetilde{F}(t) - \widetilde{P}(t)\widetilde{H}^{\mathrm{T}}(t) \times R^{-1}\widetilde{H}(t))$ [21]. This fact, together with equation (8.134), guarantees the asymptotic stability of equation (8.133) [22]. □

Thus, since the two-stage bias correction filter developed in this chapter is for a system having an uncontrollable bias subsystem, the existence of a lower bound on $\widetilde{P}(\cdot)$ is not ensured. Consequently, this cannot ensure exponential asymptotic stability. Therefore, we can conclude that the two-stage bias correction filter is asymptotically stable, but not necessarily *uniformly* asymptotically stable.

As can be noted from the above results, the fixed-point smoother error covariance $P_{\mathrm{b}}(\tau|t) \to 0$ as $t \to \infty$. Therefore, in practice, the smoother gain $K_{\mathrm{b}}(t, \tau) \to 0$ in such a situation, and the estimator ignores the new measurements. It is well known that such an estimator will almost certainly *diverge* [23–26]. In addition, if the matrices F and D are both stable (i.e. their characteristic roots have negative real parts), then the transition matrices $\Phi_{y\mathrm{n}}(t, \tau)$, $\Phi_{y\mathrm{bn}}(t, \tau)$ and $\Phi_{\mathrm{bn}}(t, \tau)$ all tend to 0 as $t \to \infty$.

8.5.2 A Steady-State Bias Correction Filter

Finally, we shall give a way to construct a two-stage bias correction filter in the steady state. For this purpose, it is assumed that the magnitudes of matrices $P_b(\tau|t)$, $\Phi_{yn}(t, \tau)$ and $\Phi_{bn}(t, \tau)$ are already so low that they cannot be substantially reduced in a time interval $[\tau, t_f]$ of practical significance [4]. That is, it is assumed that there exists P_{bs}, Φ_{ys}, Φ_{bs} such that

$$P_{bs}^{-1} \gg P_b^{-1}(\tau) + \int_{\tau}^{t_f} [H_1 \widetilde{\Phi}_{ybn}(s, \tau) + H_2 \Phi_{bn}(s, \tau)]^T R^{-1} \times [H_1 \widetilde{\Phi}_{ybn}(s, \tau) + H_2 \Phi_{bn}(s, \tau)] \, ds \tag{8.135}$$

$$\Phi_{ys} \ll I + \int_{\tau}^{t_f} [F - K_{yn}(s) H_1] \Phi_{yn}(s, \tau) \, ds \tag{8.136}$$

and

$$\Phi_{bs} \ll I + \int_{\tau}^{t_f} D \Phi_{bn}(s, \tau) \, ds \tag{8.137}$$

for $D \neq 0$, where $\Phi_{bs} = I$ for $D \equiv 0$. Since asymptotic stability for the augmented estimator of equation (8.122) is already assured in Theorem 8.4, the subsystem $d\xi_1(t)/dt = [F - K_{yn}(t) H_1] \xi_1(t)$ is asymptotically stable, and hence it follows from equation (8.50) that

$$\lim_{t \to \infty} [d\Phi_{ybn}(t, \tau)/dt] = 0 \tag{8.138}$$

If we adopt Φ_{bs}, which satisfies equation (8.137), as $\Phi_{bn}(t, \tau)$ for $t \to \infty$, we then get from equation (8.50)

$$\Phi_{ybs} = [F - K_{ys} H_1]^{-1} [K_{ys} H_2 - B] \Phi_{bs} \tag{8.139}$$

where

$$\lim_{t \to \infty} \Phi_{ybn}(t, \tau) = \Phi_{ybs} \tag{8.140}$$

$$K_{ys} = P_{ys} H_1^T R^{-1} \tag{8.141}$$

in which P_{ys} is the unique solution of the following ARE:

$$F P_{ys} + P_{ys} F^T - P_{ys} H_1^T R^{-1} H_1 P_{ys} + G Q G^T = 0 \tag{8.142a}$$

Here, it is assumed that in the above equation there exists P_{ys} such that

$$P_{ys}^{-1} \gg \widetilde{P}_y^{-1}(\tau) + \int_{\tau}^{t_f} [-F^T P_{yn}^{-1}(s) - P_{yn}^{-1}(s) F + H_1^T R^{-1} H_1] \, ds \tag{8.142b}$$

for $Q \equiv 0$ and $F \neq 0$, and that

$$P_{ys} \ll [\tilde{P}_y^{-1}(\tau) + H_1^{\mathrm{T}} R^{-1} H_1 (t_{\mathrm{f}} - \tau)]^{-1} \tag{8.142c}$$

for $Q \equiv 0$ and $F \equiv 0$.

From these results, a two-stage bias correction filter in the steady state can consist of

$$\hat{y}(t|t) = \hat{y}_{\mathrm{n}}(t|t) + \tilde{\Phi}_{y\mathrm{bs}} \hat{b}(\tau|t) \tag{8.143}$$

$$\hat{b}(t|t) = \Phi_{\mathrm{bs}} \hat{b}(\tau|t) \tag{8.144}$$

$$\tilde{\Phi}_{y\mathrm{bs}} = \Phi_{y\mathrm{s}} P_{y\mathrm{b}}(\tau) P_{\mathrm{b}}^{-1}(\tau) + \Phi_{y\mathrm{bs}} \tag{8.145}$$

where $\hat{y}_{\mathrm{n}}(t|t)$ and $\hat{b}(\tau|t)$ are the solutions to the following equations:

$$\dot{\hat{y}}_{\mathrm{n}}(t|t) = [F - K_{y\mathrm{s}} H_1] \hat{y}_{\mathrm{n}}(t|t) + K_{y\mathrm{s}} z(t) \tag{8.146}$$

$$\dot{\hat{b}}(\tau|t) = -K_{\mathrm{bs}}[H_1 \tilde{\Phi}_{y\mathrm{bs}} + H_2 \Phi_{\mathrm{bs}}] \hat{b}(\tau|t) + K_{\mathrm{bs}}[z(t) - H_1 \hat{y}_{\mathrm{n}}(t|t)] \tag{8.147}$$

$$K_{\mathrm{bs}} = P_{\mathrm{bs}}[H_1 \tilde{\Phi}_{y\mathrm{bs}} + H_2 \Phi_{\mathrm{bs}}] R^{-1} \tag{8.148}$$

Note that this estimator is apparently dependent upon the initial statistics given in equations (8.38), whereas the usual steady-state Kalman filter, in the uniformly completely observable and controllable system, is independent of it. The block diagram of the steady-state two-stage bias correction filter is shown in Figure 8.6.

Example 8.3

Consider a biased sensor [2]. Suppose that the quantity y measured by the sensor is governed by

$$\dot{y} = -ay + w$$

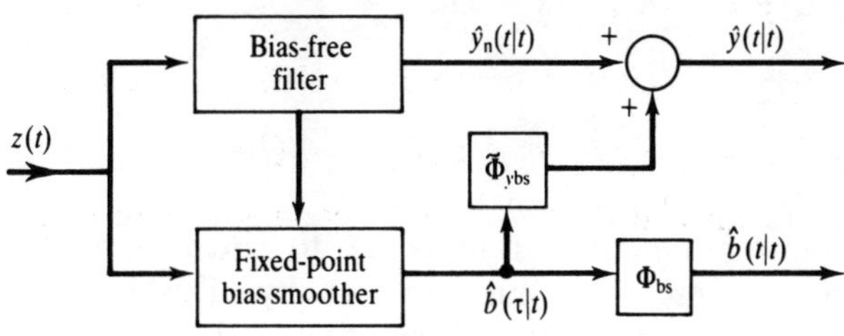

Figure 8.6 Block diagram for a steady-state two-stage bias correction filter.

and that the measured output is

$$z = y + b + v$$

where w and v are independent Gaussian white noises with variances Q and R, and $y(0)$ and $b(0)$ are also independent of each other.

The optimal estimate $\hat{y}(t|t)$ for the case in which the bias uncertainty cannot be reduced is obtained from equations (7.143)–(7.148):

$$\hat{y}(t|t) = \hat{y}_n(t|t) + \Phi_{ybs}\hat{b}(0|t)$$

where

$$\dot{\hat{y}}_n(t|t) = -a\hat{y}_n(t|t) + K_{ys}[z(t) - \hat{y}_n(t|t)]$$

$$\dot{\hat{b}}(0|t) = K_{bs}[-(\Phi_{ybs} + 1)\hat{b}(0|t) + (z(t) - \hat{y}_n(t|t))]$$

The steady-state gains K_{ys} and K_{bs} are

$$K_{ys} = P_{ys}/R$$

$$K_{bs} = P_{bs}(\Phi_{ybs} + 1)/R$$

where

$$P_{ys} = \sqrt{QR + a^2R^2} - aR$$

$$\Phi_{ybs} = -\frac{K_{ys}}{a + K_{ys}} = -\frac{K_{ys}}{\sqrt{\dfrac{Q}{R} + a^2}}$$

The block diagram of the optimal filter is shown in Figure 8.7. The transfer functions of the filter to the state and bias estimates, respectively, are

$$\frac{\hat{Y}(s)}{Z(s)} = \frac{K_{ys}}{s + a + K_{ys}} \frac{(1 - G)s}{s + Ga} \tag{8.149}$$

$$\frac{\hat{B}(s)}{Z(s)} = \frac{s + a}{s + a + K_{ys}} \frac{G(a + K_{ys})}{s + Ga}$$

where

$$G = \frac{P_{bs}}{R} \frac{a}{(a + K_{ys})^2} \tag{8.150}$$

The transfer function from $Z(s)$ to $\hat{B}(s)$ is such that the steady-state response to a unit step is unity. This implies that if a constant bias is present in the observation $z(t)$, it will be estimated perfectly in the steady state.

On the other hand, the first term in the transfer function from

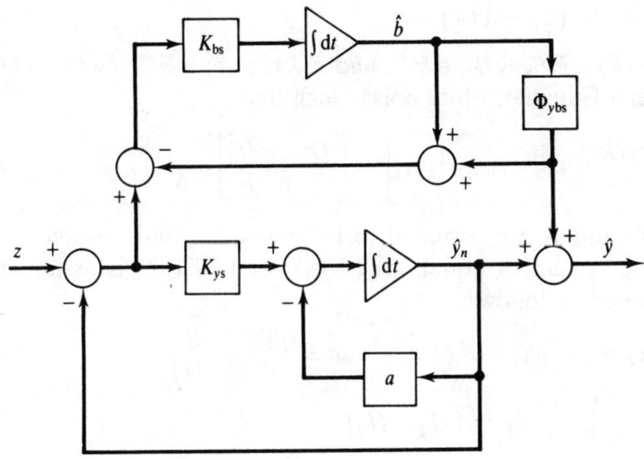

Figure 8.7 Block diagram for a filter with a biased sensor.

$Z(s)$ to $\hat{Y}(s)$ is that of the bias-free filter. The second term $(1 - G)s/(s + Ga)$ is that of a high-pass filter which may be interpreted as preventing the transmission of a constant in the observation to the optimal estimate of the state. For the high-frequency characteristics of the filter in equation (8.149) to be correct, however, it is reasonable to require that the parameter G of equation (8.150) be much smaller than unity, i.e.

$$\frac{P_{bs}}{R} \ll \frac{1}{a}\left(\frac{Q}{R} + a^2\right)$$

If this condition does not hold, the assumption that P_b, the covariance of the bias, cannot be reduced by the operation of the filter is likely not to be valid.

8.6 A BIAS CORRECTION FILTER AND PREDICTOR IN DISCRETE-TIME SYSTEMS

In this final section, we shall briefly discuss the two-stage bias correction filtering and prediction problems for linear discrete-time systems [27].

Consider the following time-invariant discrete-time system:

$$x(k + 1) = \Phi x(k) + Cw(k) \tag{8.151}$$

$$z(k) = Hx(k) + v(k) \tag{8.152}$$

where $x(k) \in \mathbb{R}^n$, $w(k) \in \mathbb{R}^p$, and $z(k)$, $v(k) \in \mathbb{R}^m$. $\{w(k), v(k)\}$ are zero-mean Gaussian white noises such that

$$E\left[\begin{pmatrix} w(k) \\ v(k) \end{pmatrix} \begin{pmatrix} w^{\mathrm{T}}(j) & v^{\mathrm{T}}(j) \end{pmatrix}\right] = \begin{bmatrix} Q & 0 \\ 0 & R \end{bmatrix} \delta_{kj} \tag{8.153}$$

where Q and R are assumed to be symmetric and positive semidefinite. The system of equations (8.151) and (8.152) is assumed to be decomposed as follows:

$$x^{\mathrm{T}}(k) = [y^{\mathrm{T}}(k) \quad b^{\mathrm{T}}(k)], \qquad \Phi = \begin{pmatrix} F & B \\ 0 & D \end{pmatrix}$$
$$C = \begin{pmatrix} G \\ 0 \end{pmatrix}, \qquad H = [H_1 \quad H_2] \tag{8.154}$$

where the original system state $y(k) \in \mathbb{R}^{n_c}$, and the unknown bias state $b(k) \in \mathbb{R}^{n_b}$. In addition, the a priori information of $x(0)$ is given by

$$\hat{x}(0) \sim N(\hat{x}(0), P(0)) \tag{8.155a}$$

$$\hat{x}(0) \triangleq \begin{bmatrix} \hat{y}(0) \\ \hat{b}(0) \end{bmatrix}, \qquad P(0) \triangleq \begin{bmatrix} P_y(0) & P_{yb}(0) \\ P_{by}(0) & P_b(0) \end{bmatrix} \tag{8.155b}$$

where $x(0)$ is assumed to be independent of the sequences $\{w(k), v(k)\}$. It is also assumed that $\hat{x}(0)$ and $P(0)$ are partitioned as follows:

$$x(0) = \hat{\mu}_{\mathrm{n}}(0) + \hat{\mu}_{\mathrm{r}}(0) \tag{8.156a}$$

$$P(0) = V_{\mathrm{n}}(0) + V_{\mathrm{nr}}(0) + V_{\mathrm{nr}}^{\mathrm{T}}(0) + V_{\mathrm{r}}(0) \tag{8.156b}$$

in which

$$\hat{\mu}_{\mathrm{n}}(0) = \begin{pmatrix} \hat{y}(0) \\ 0 \end{pmatrix}, \qquad \hat{\mu}_{\mathrm{r}}(0) = \begin{pmatrix} 0 \\ \hat{b}(0) \end{pmatrix} \tag{8.157a}$$

$$V_{\mathrm{n}}(0) = \begin{pmatrix} P_y(0) & 0 \\ 0 & 0 \end{pmatrix}, \qquad V_{\mathrm{nr}}(0) = \begin{pmatrix} 0 & P_{yb}(0) \\ 0 & 0 \end{pmatrix}$$

$$V_{\mathrm{r}}(0) = \begin{pmatrix} 0 & 0 \\ 0 & P_b(0) \end{pmatrix} \tag{8.157b}$$

In the sense of the discrete-time system discussed in Section 5.4, the nominal initial conditions can be decomposed as follows:

$$\hat{m}_{\mathrm{n}}(0) = \hat{\mu}_{\mathrm{n}}(0) - V_{\mathrm{nr}}(0)V_{\mathrm{r}}^{-1}(0)\hat{\mu}_{\mathrm{r}}(0) \tag{8.158a}$$

$$\Pi_{\mathrm{n}}(0) = V_{\mathrm{n}}(0) - V_{\mathrm{nr}}(0)V_{\mathrm{r}}^{-1}(0)V_{\mathrm{nr}}^{\mathrm{T}}(0) \tag{8.158b}$$

If the measurement data $Z_k \triangleq \{z(i), 1 \leqslant i \leqslant k\}$ are given, then the MV estimate $\hat{x}^{\mathrm{T}}(k|k) \triangleq E[x(k)|Z_k]$ and the corresponding estimation error covariance matrix $P(k|k) \triangleq E\{[x(k) - \hat{x}(k|k)][x(k) - \hat{x}(k|k)]^{\mathrm{T}}|Z_k\}$ are obtained by the dependent partitioning filter with $l = 0$, which was shown in Theorem 5.4 of Chapter 5:

$$\hat{x}(k|k) = \hat{x}_{\mathrm{n}}(k|k) + \Phi_{\mathrm{n}}(k, 0)\Pi_{\mathrm{r}}(0)\hat{x}_{\mathrm{r}}(0|k) \tag{8.159}$$

$$P(k|k) = P_{\mathrm{n}}(k|k) + \Phi_{\mathrm{n}}(k, 0)\Pi_{\mathrm{r}}(0)P_{\mathrm{r}}(0|k)\Pi_{\mathrm{r}}^{\mathrm{T}}(0)\Phi_{\mathrm{n}}^{\mathrm{T}}(k, 0) \tag{8.160}$$

where

$$\Pi_{\mathrm{r}}(0) = [I + V_{\mathrm{nr}}(0)V_{\mathrm{r}}^{-1}(0)] \tag{8.161}$$

$$\Phi_{\mathrm{n}}(k, 0) = [I - K_{\mathrm{n}}(k)H]\hat{\Phi}_{\mathrm{n}}(k, 0) \tag{8.162}$$

$$\hat{\Phi}_{\mathrm{n}}(k, 0) = \Phi\Phi_{\mathrm{n}}(k - 1, 0), \qquad \Phi_{\mathrm{n}}(0, 0) = I \tag{8.163}$$

The nominal filtered estimate $\hat{x}_{\mathrm{n}}(k|k)$ and its error covariance $P_{\mathrm{n}}(k|k)$ are given by

$$\hat{x}_{\mathrm{n}}(k|k) = \hat{x}_{\mathrm{n}}(k|k - 1) + K_{\mathrm{n}}(k)[z(k) - H\hat{x}_{\mathrm{n}}(k|k - 1)] \tag{8.164}$$

$$P_{\mathrm{n}}(k|k) = [I - K_{\mathrm{n}}(k)H]P_{\mathrm{n}}(k|k - 1) \tag{8.165}$$

where

$$\hat{x}_{\mathrm{n}}(k|k - 1) = \Phi\hat{x}_{\mathrm{n}}(k - 1|k - 1), \qquad \hat{x}_{\mathrm{n}}(0|0) = \hat{m}_{\mathrm{n}}(0) \tag{8.166}$$

$$K_{\mathrm{n}}(k) = P_{\mathrm{n}}(k|k - 1)H^{\mathrm{T}}[HP_{\mathrm{n}}(k|k - 1)H^{\mathrm{T}} + R]^{-1} \tag{8.167}$$

$$P_{\mathrm{n}}(k|k - 1) = \Phi P_{\mathrm{n}}(k - 1|k - 1)\Phi^{\mathrm{T}} + CQC^{\mathrm{T}}, \qquad P_{\mathrm{n}}(0|0) = \Pi_{\mathrm{n}}(0) \tag{8.168}$$

The partitioned-type fixed-point smoothed estimate $\hat{x}_{\mathrm{r}}(0|k)$ can be obtained, in recursive form (see also Chapter 7), from the following expression:

$$\begin{aligned}\hat{x}_{\mathrm{r}}(0|k) = \hat{x}_{\mathrm{r}}(0|k - 1) + K_{\mathrm{r}}(k)\{z(k) - H[\hat{x}_{\mathrm{n}}(k|k - 1)\\ + \hat{\Phi}_{\mathrm{n}}(k, 0)\Pi_{\mathrm{r}}(0)\hat{x}_{\mathrm{r}}(0|k - 1)]\}, \qquad \hat{x}_{\mathrm{r}}(0|0) = \hat{\mu}_{\mathrm{r}}(0)\end{aligned} \tag{8.169}$$

where

$$\begin{aligned}K_{\mathrm{r}}(k) = P_{\mathrm{r}}(0|k - 1)\Pi_{\mathrm{r}}^{\mathrm{T}}(0)\hat{\Phi}_{\mathrm{n}}^{\mathrm{T}}(k, 0)H^{\mathrm{T}}\\ \times [H\hat{\Phi}_{\mathrm{n}}(k, 0)\Pi_{\mathrm{r}}(0)P_{\mathrm{r}}(0|k - 1)\\ \times \Pi_{\mathrm{r}}^{\mathrm{T}}(0)\hat{\Phi}_{\mathrm{n}}^{\mathrm{T}}(k, 0)H^{\mathrm{T}} + HP_{\mathrm{n}}(k|k - 1)H^{\mathrm{T}} + R]^{-1}\end{aligned} \tag{8.170}$$

$$P_r(0|k) = [I - K_r(k)H\hat{\Phi}_n(k,0)\Pi_r(0)]P_r(0|k-1), \quad P_r(0|0) = V_r(0) \qquad (8.171)$$

Note that, since in the prediction problem both $K_n(k)$ in equation (8.167) and $K_r(k)$ in equation (8.170) are 0 from the fact $R \equiv \infty$, equations (8.159) and (8.160) reduce to

$$\hat{x}(k|k-1) = \hat{x}_n(k|k-1) + \hat{\Phi}_n(k,0)\Pi_r(0)\hat{x}_r(0|k-1) \qquad (8.172)$$

$$P(k|k-1) = P_n(k|k-1) + \hat{\Phi}_n(k,0)\Pi_r(0)P_r(0|k-1)\Pi_r^T(0)\hat{\Phi}_n^T(k,0) \qquad (8.173)$$

Now, noting that the system structure given by equations (8.154) satisfies the condition for optimally synthesizing a two-stage bias correction filter (i.e. Theorem 8.1), we obtain the generalized two-stage bias correction filter for equations (8.151)–(8.155) in the following theorem.

Theorem 8.5 Generalized bias correction filter in discrete time

The optimal filtered estimate for the discrete-time system of equations (8.151)–(8.155) can be obtained by the following two-stage plan:

$$\hat{y}(k|k) = \hat{y}_n(k|k) + \tilde{\Phi}_{ybn}(k,0)\hat{b}(0|k) \qquad (8.174)$$

$$\hat{b}(k|k) = \Phi_{bn}(k,0)\hat{b}(0|k) \qquad (8.175)$$

where $\tilde{\Phi}_{ybn}(k,0)$ is a generalized blending matrix at the filtering instant, which is given by

$$\tilde{\Phi}_{ybn}(k,0) = \Phi_{yn}(k,0)P_{yb}(0)P_b^{-1}(0) + \Phi_{ybn}(k,0) \qquad (8.176)$$

and the bias-free filtered estimate $\hat{y}_n(k|k)$ is governed by the following Kalman filter:

$$\hat{y}_n(k|k-1) = F\hat{y}_n(k-1|k-1), \quad \hat{y}_n(0|0) = \hat{y}(0) - P_{yb}(0)P_b^{-1}(0)\hat{b}(0) \qquad (8.177)$$

$$P_{yn}(k|k-1) = FP_{yn}(k-1|k-1)F^T + GQG^T, \quad P_{yn}(0|0) = \tilde{P}_y(0) \triangleq P_y(0) - P_{yb}(0)P_b^{-1}(0)P_{by}(0) \qquad (8.178)$$

$$K_{yn}(k) = P_{yn}(k|k-1)H_1^T[H_1P_{yn}(k|k-1)H_1^T + R]^{-1} \qquad (8.179)$$

$$\hat{y}_n(k|k) = \hat{y}_n(k|k-1) + K_{yn}(k)[z(k) - H_1\hat{y}_n(k|k-1)] \qquad (8.180)$$

$$P_{yn}(k|k) = [I - K_{yn}(k)H_1]P_{yn}(k|k-1) \qquad (8.181)$$

The bias smoothed estimate $\hat{b}(0|k)$ is also given by the following fixed-point smoother:

$$\hat{b}(0|k) = \hat{b}(0|k-1) + K_b(k)\{z(k) - [H_1\hat{y}_n(k|k-1) \\ - \bar{H}(k,0)\hat{b}(0|k-1)]\}, \qquad \hat{b}(0|0) = \hat{b}(0) \tag{8.182}$$

$$K_b(k) = P_b(0|k-1)\bar{H}^T(k,0)\{\bar{H}(k,0) \\ \times P_b(0|k-1)\bar{H}^T(k,0) + \tilde{R}(k)\}^{-1} \tag{8.183}$$

$$\bar{H}(k,0) \triangleq H_1\hat{\Psi}_{ybn}(k,0) + H_2\hat{\Psi}_{bn}(k,0) \tag{8.184}$$

$$\tilde{R}(k) \triangleq H_1 P_{yn}(k|k-1)H_1^T + R \tag{8.185}$$

$$P_b(0|k) = [I - K_b(k)\bar{H}(k,0)]P_b(0|k-1), \qquad P_b(0|0) = P_b(0) \tag{8.186}$$

where $\hat{\Psi}_{ybn}(k,0)$ denotes a generalized blending matrix at the predicting instant, defined by

$$\hat{\Psi}_{ybn}(k,0) = \hat{\Phi}_{ybn}(k,0) + \hat{\Phi}_{yn}(k,0)P_{yb}(0)P_b^{-1}(0) \tag{8.187}$$

The generalized blending matrices $\tilde{\Phi}_{ybn}(k,0)$ and $\hat{\Psi}_{ybn}(k,0)$ can be constructed recursively from the following relations:

$$\hat{\Phi}_{yn}(k,0) = F\Phi_{yn}(k-1,0), \qquad \Phi_{yn}(0,0) = I \tag{8.188}$$

$$\hat{\Phi}_{ybn}(k,0) = F\Phi_{ybn}(k-1,0) + B\Phi_{bn}(k-1,0), \\ \Phi_{ybn}(0,0) = I \tag{8.189}$$

$$\hat{\Phi}_{bn}(k,0) = D\Phi_{bn}(k-1,0), \qquad \Phi_{bn}(0,0) = I \tag{8.190}$$

$$\Phi_{yn}(k,0) = [I - K_{yn}(k)H_1]\hat{\Phi}_{yn}(k,0) \tag{8.191}$$

$$\Phi_{ybn}(k,0) = [I - K_{yn}(k)H_1]\hat{\Phi}_{ybn}(k,0) - K_{yn}(k)H_2\hat{\Phi}_{bn}(k,0) \tag{8.192}$$

$$\Phi_{bn}(k,0) = \hat{\Phi}_{bn}(k,0) \tag{8.193}$$

Furthermore, the estimation error covariances associated with equations (8.174) and (8.175) are given by

$$P_y(k|k) = P_{yn}(k|k) + \tilde{\Phi}_{ybn}(k,0)P_b(0|k)\tilde{\Phi}_{ybn}(k,0) \tag{8.194}$$

$$P_{yb}(k|k) = \tilde{\Phi}_{ybn}(k,0)P_b(0|k)\Phi_{bn}^T(k,0) \tag{8.195}$$

$$P_b(k|k) = \Phi_{bn}(k,0)P_b(0|k)\Phi_{bn}^T(k,0) \tag{8.196}$$

Proof

See Watanabe [27]. □

The generalized two-stage bias correction predictor is also obtained by the following corollary.

Corollary 8.3 Generalized bias correction predictor in discrete time

The optimal (one-stage) predicted estimate and the corresponding estimation error covariance can be obtained by the following two-stage plan:

$$\hat{y}(k|k-1) = \hat{y}_{\mathrm{n}}(k|k-1) + \hat{\Psi}_{y\mathrm{bn}}(k,0)\hat{b}(0|k-1) \tag{8.197}$$

$$\hat{b}(k|k-1) = \hat{\Phi}_{\mathrm{bn}}(k,0)\hat{b}(0|k-1) \tag{8.198}$$

and

$$P_y(k|k-1) = P_{y\mathrm{n}}(k|k-1) + \hat{\Psi}_{y\mathrm{bn}}(k,0)P_{\mathrm{b}}(0|k-1)\hat{\Psi}^{\mathrm{T}}_{y\mathrm{bn}}(k,0) \tag{8.199}$$

$$P_{y\mathrm{b}}(k|k-1) = \hat{\Psi}_{y\mathrm{bn}}(k,0)P_{\mathrm{b}}(0|k-1)\hat{\Psi}^{\mathrm{T}}_{\mathrm{bn}}(k,0) \tag{8.200}$$

$$P_{\mathrm{b}}(k|k-1) = \hat{\Phi}_{\mathrm{bn}}(k,0)P_{\mathrm{b}}(0|k-1)\hat{\Phi}^{\mathrm{T}}_{\mathrm{bn}}(k,0) \tag{8.201}$$

Proof

Setting $R = \infty$ in Theorem 8.5 gives the desired results. □

The two-stage bias correction estimators discussed in this chapter have three possible improvements over the well-known extended Kalman filter: first, they provide a potential improvement in numerical accuracy due to the lower order of the matrices involved (cf. Section 8.3); second, the bias-free state form is useful in applications where it is desired to monitor the bias-free state in failure detection problems, in which failures may be modelled as bias jumps [28, 29]; and third, infinite uncertainty in the a priori bias estimate can be readily handled within this framework by applying the concept of the information filter (see Exercise 2.4) [27].

Example 8.4

To illustrate the philosophy of the two-stage bias correction filter, consider the state estimation for the longitudinal dynamics of an aircraft. The system considered here is described by a linear equation of a longitudinal short-period dynamics subject to gust disturbances:

$$\dot{y}(t) = Ay(t) + Mu(t) + Nb(t) + Sn(t) \tag{8.202}$$

$$\dot{b}(t) = Jb(t) \tag{8.203}$$

where

$$y^{\mathrm{T}}(t) \triangleq [w(t) \quad q(t) \quad \theta(t) \quad \xi(t) \quad \eta(t)]$$

$$u^{\mathrm{T}}(t) \triangleq [\delta_{\mathrm{e}}(t) \quad \delta_{\mathrm{f}}(t)]$$

$$b^{\mathrm{T}}(t) \triangleq [b_{\mathrm{w}}(t) \quad b_{\mathrm{q}} \quad b_{\mathrm{a}}]$$

$$A = \begin{bmatrix} Z_{\mathrm{w}} & V & 0 & V^2 Z_{\mathrm{w}}/L^2 & \sqrt{3}VZ_{\mathrm{w}}/L \\ M_{\mathrm{w}} & M_{\mathrm{q}} & 0 & V^2 M_{\mathrm{w}}/L^2 & \sqrt{3}VZ_{\mathrm{w}}/L \\ 0 & 1 & 0 & 0 & 0 \\ 0 & 0 & 0 & 0 & 1 \\ 0 & 0 & 0 & -V^2/L^2 & -2V/L \end{bmatrix}$$

$$M = \begin{bmatrix} Z_{\delta_{\mathrm{e}}} & Z_{\delta_{\mathrm{f}}} \\ M_{\delta_{\mathrm{e}}} & M_{\delta_{\mathrm{f}}} \\ 0 & 0 \\ 0 & 0 \\ 0 & 0 \end{bmatrix}, \qquad N = \begin{bmatrix} 1 & 0 & 0 \\ 0 & 0 & 0 \\ 0 & 0 & 0 \\ 0 & 0 & 0 \\ 0 & 0 & 0 \end{bmatrix}$$

$$S = \begin{bmatrix} 0 \\ 0 \\ 0 \\ 0 \\ 1 \end{bmatrix}, \qquad J = \begin{bmatrix} M_{\mathrm{b_w}} & 0 & 0 \\ 0 & 0 & 0 \\ 0 & 0 & 0 \end{bmatrix}$$

and where $w(t)$ is the normal velocity of an aircraft, $q(t)$ the pitch rate, $\theta(t)$ the pitch angle, $\xi(t)$ and $\eta(t)$ are the states of the random vertical gust which are expressed by the Dryden turbulence model, and $\delta_{\mathrm{e}}(t)$ and $\delta_{\mathrm{f}}(t)$ deflections of the elevator and flap, respectively. In addition, $b_{\mathrm{w}}(t)$ is an unknown time-varying bias, and b_{q} and b_{a} are unknown time-invariant biases. The scalar-valued zero-mean Gaussian white noise $n(t)$ is considered to be subject to

$$E[n(t)n(\tau)] = (\sigma_{w_g}^2 L/\pi V)\delta(t - \tau)$$

where σ_{w_g} and L are the standard deviation and integral scale of the gust, and V is the constant velocity of the aircraft.

The observation equation is expressed by

$$z(k) = H_1 y(k) + H_2 b(k) + v(k), \qquad k = 1, 2, \ldots$$

where

$$H_1 = \begin{bmatrix} 0 & 1 & 0 & 0 & 0 \\ Z_w & 0 & 0 & V^2 Z_w/L^2 & \sqrt{3} V Z_w/L \end{bmatrix}$$

$$H_2 = \begin{bmatrix} 0 & 1 & 0 \\ 1 & 0 & 1 \end{bmatrix}$$

$$v^T(k) = [v_q(k) \quad v_a(k)]$$

and the variances of zero-mean Gaussian white noises $v_q(k)$ and $v_a(k)$ are respectively assumed to be

$$E[v_q(k)v_q(j)] = (\sigma_q^2/\Delta t)\delta_{kj}$$
$$E[v_a(k)v_a(j)] = (\sigma_a^2/\Delta t)\delta_{kj}$$

where Δt denotes the sampling interval.

Table 8.1 Stability derivatives

Stability derivatives	*Values*	*Units*
Z_w	−0.02	s^{-1}
Z_{δ_e}	−0.75	s^{-2}
Z_{δ_f}	−16	$m\ s^{-2}$
M_w	−0.07	$m^{-1}\ s^{-1}$
M_q	−2.0	s^{-1}
M_{δ_e}	−17	s^{-2}
M_{δ_f}	0.59	s^{-2}
M_{b_w}	−0.5	s^{-1}

Table 8.2 Filter initialization

Variable	*Initial state*	*Standard deviation*	*Units*
w	3.0615	1.0	$m\ s^{-1}$
q	-3.5539×10^{-3}	8.9442×10^{-3}	$rad\ s^{-1}$
θ	-4.9793×10^{-3}	1.0×10^{-2}	rad
ξ	−3.3707	10.0	m s
η	2.2884	2.0	m
b_w	7.2187×10^{-2}	3.1305×10^{-1}	$m\ s^{-2}$
b_q	-1.0956×10^{-3}	3.1623×10^{-3}	$rad\ s^{-1}$
b_a	-3.4689×10^{-1}	6.2610×10^{-1}	$m\ s^{-2}$

Simulations were implemented, with stability derivatives as tabulated in Table 8.1 and under the initial conditions given in Table 8.2. The parameters were set as follows: $\hat{y}(0) = \hat{b}(0) = 0$, $\sigma_q = 0.005 \text{ rad s}^{-1}$, $\sigma_a = 0.098 \text{ m s}^{-2}$, $\Delta t = 0.02$ s, $\sigma_{w_g} = 1.7 \text{ m s}^{-1}$, $\delta_e(t)$, $\delta_f(t) \sim N(0,1)$, $V = 67 \text{ m s}^{-1}$ and $L = 533$ m. The continuous-time model given by equations (8.202) and (8.203) was discretized using the state transition matrix expanded by a fifth-order series matrix.

Figure 8.8 shows the estimation errors of the original states: (a) normal velocity w; (b) pitch rate q; (c) pitch angle θ; (d) gust state ξ; and (e) gust state η for a case when the following initial correlation exists:

$$P_{yb}(0) = \begin{bmatrix} 0.31 \times 10^{-4} & 0 & 0.63 \times 10^{-4} \\ 0 & 0.28 \times 10^{-4} & 0 \\ 0 & 0 & 0 \\ 0.31 \times 10^{-4} & 0 & 0.63 \times 10^{-4} \\ 0.64 \times 10^{-4} & 0 & 0.125 \times 10^{-4} \end{bmatrix}$$

The bias-corrected rms value $\sqrt{\text{diag}(P_y(k|k))}$ and the bias-free one $\sqrt{\text{diag}(P_{yn}(k|k))}$ are shown in bold lines and bold broken lines,

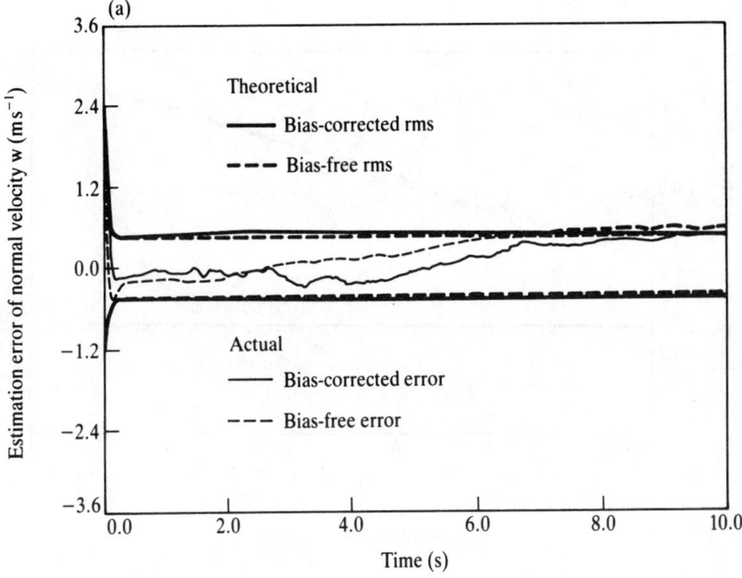

Figure 8.8 Estimation errors, where the initial correlation exists, (a) of the normal velocity $w(t)$; (b) of the pitch rate $q(t)$; (c) of the pitch angle $\theta(t)$; (d) of the gust state $\xi(t)$; (e) of the gust state $\eta(t)$.

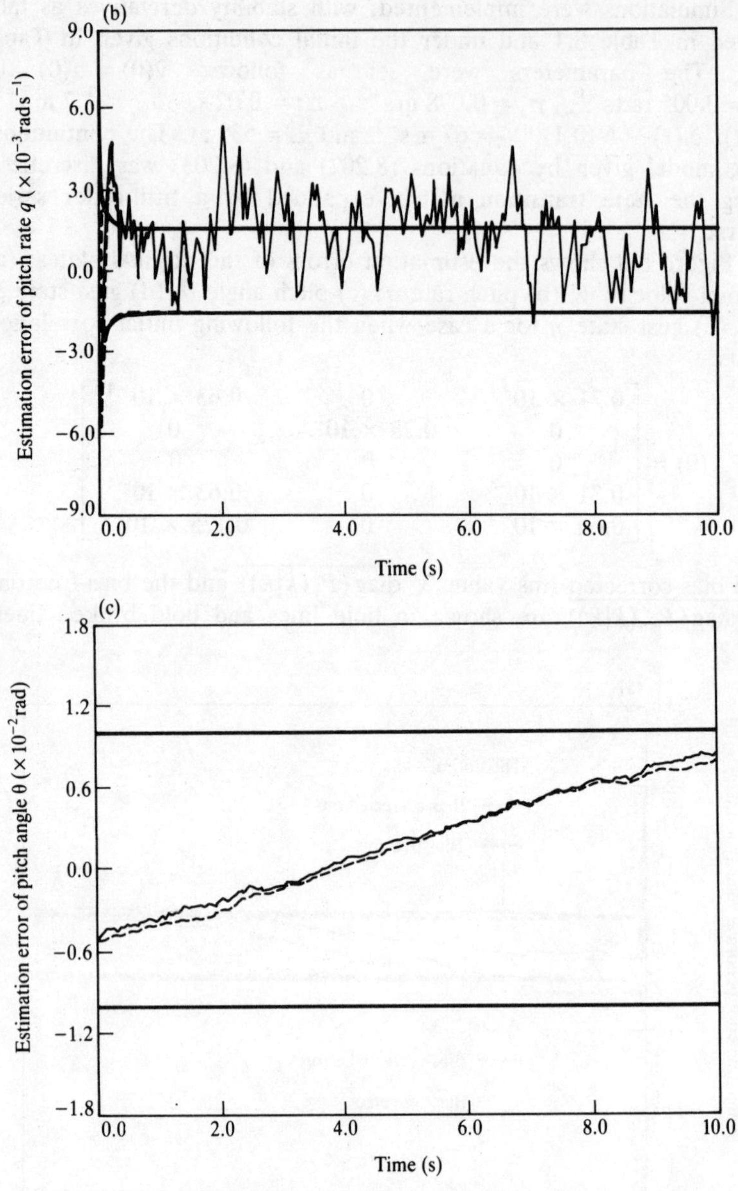

Figure 8.8 (*continued*)

respectively. The actual errors are also shown in fine lines. The corresponding estimation errors of biases are shown in Figure 8.9: plots (a), (b) and (c) depict the errors of b_w, b_q and b_a, respectively.

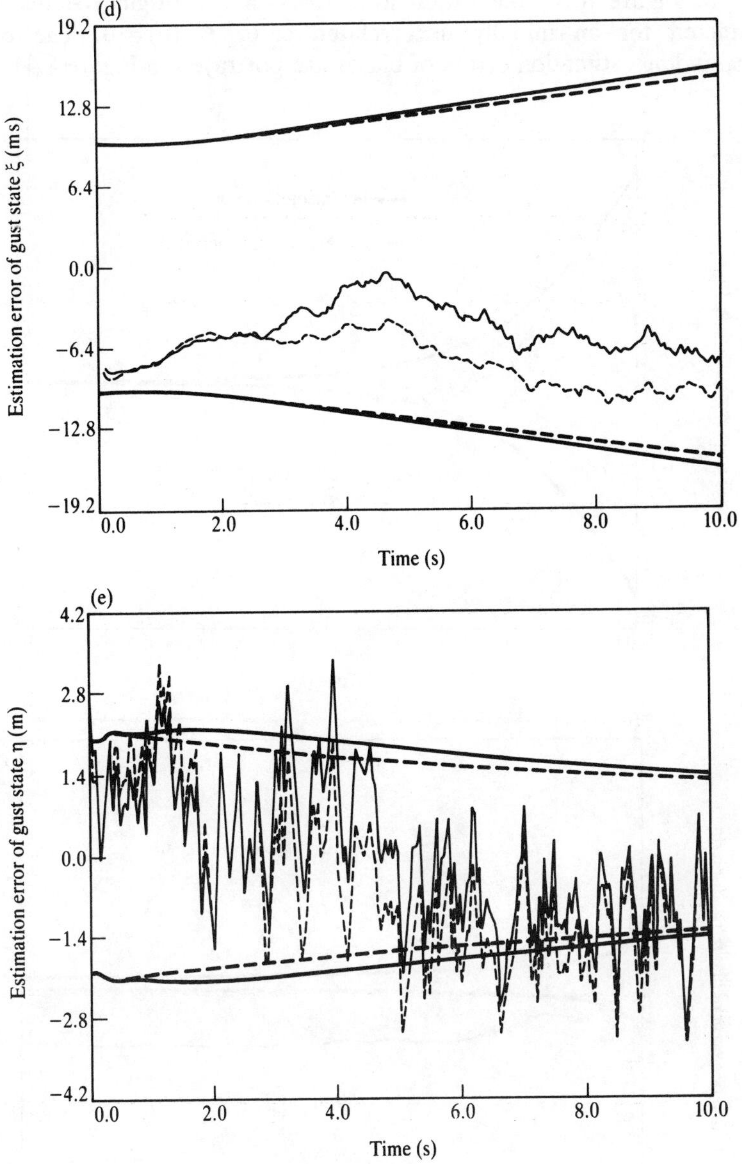

Figure 8.8 (*continued*)

It can be seen from Figures 8.8(a), 8.8(d) and 8.8(e) that the effect of the unknown biases on the estimate of the original states is quantitatively evaluated.

In Figure 8.10, the estimation errors of the original states are depicted for an initially uncorrelated case, $P_{yb}(0) = 0$. The corresponding estimation errors of biases are portrayed in Figure 8.11.

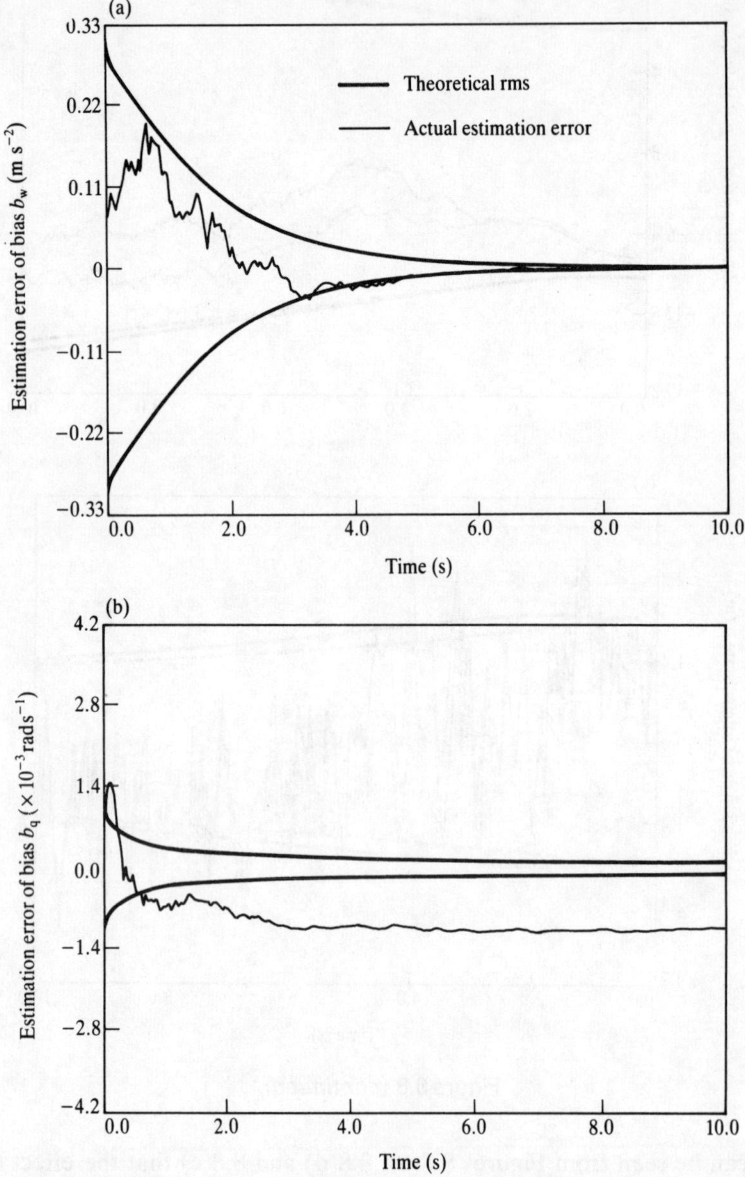

Figure 8.9 Estimation errors where the initial correlation exists, (a) of the bias $b_w(t)$; (b) of the constant bias b_q; (c) of the constant bias b_a.

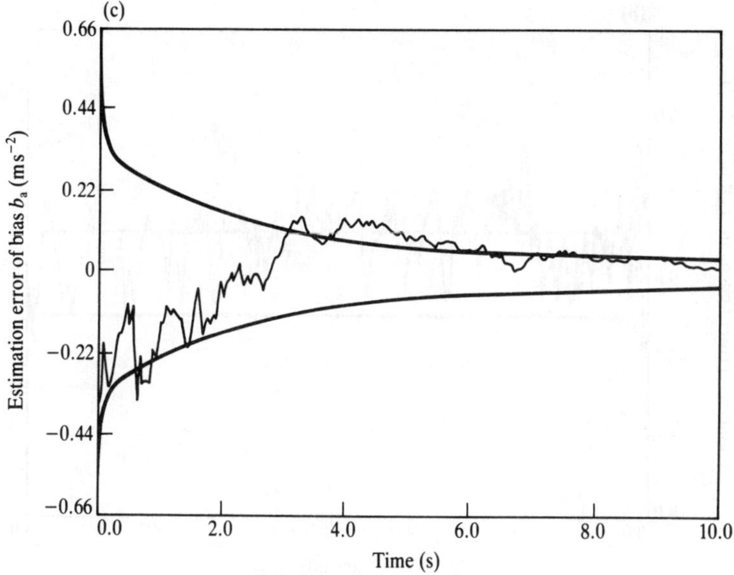

Figure 8.9 (*continued*).

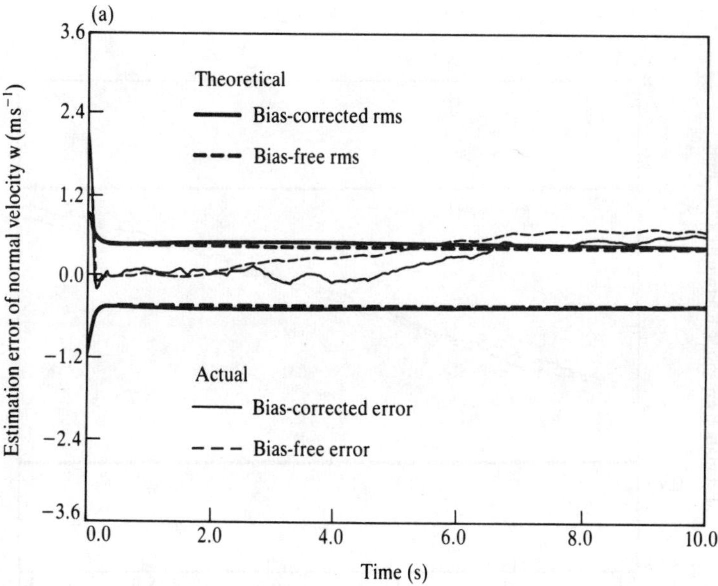

Figure 8.10 Estimation errors, where the initial correlation does not exist, (a) of the normal velocity $w(t)$; (b) of the pitch rate $q(t)$; (c) of the pitch angle $\theta(t)$; (d) of the gust state $\xi(t)$; (e) of the gust state $\eta(t)$.

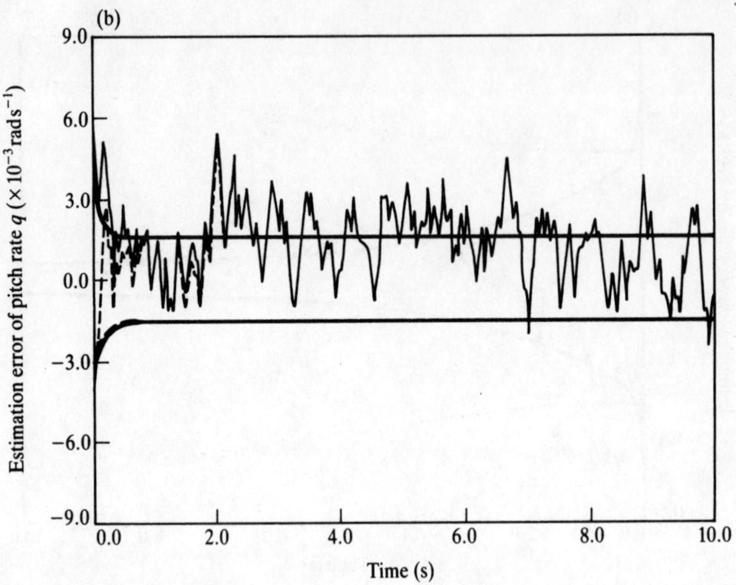

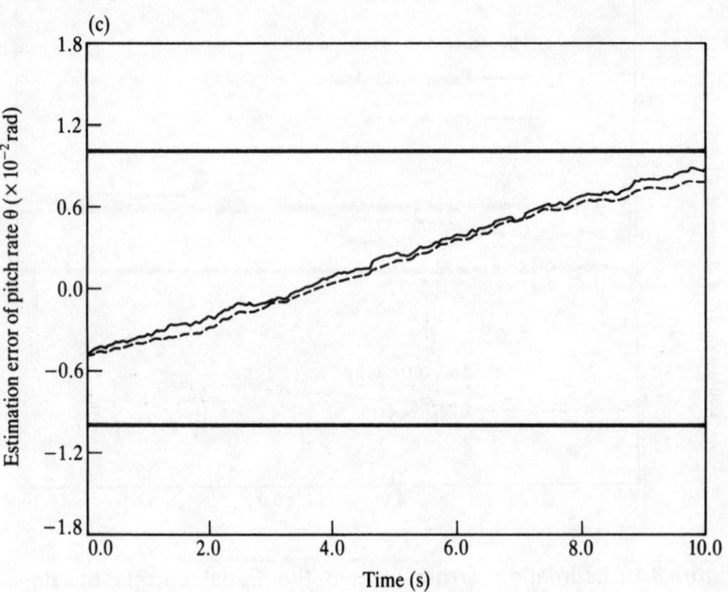

Figure 8.10 (*continued*)

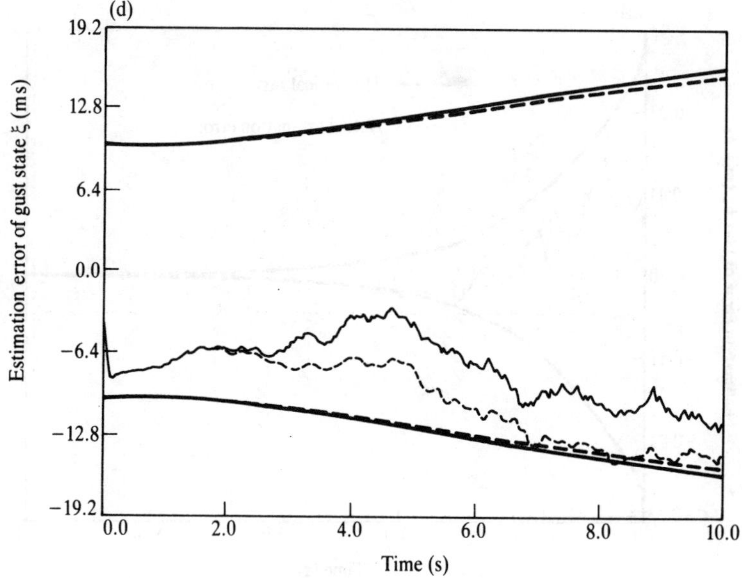

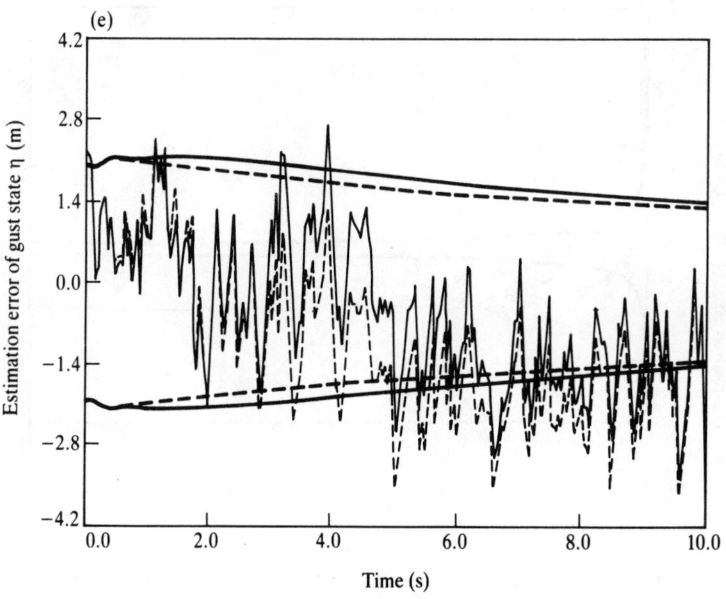

Figure 8.10 (*continued*)

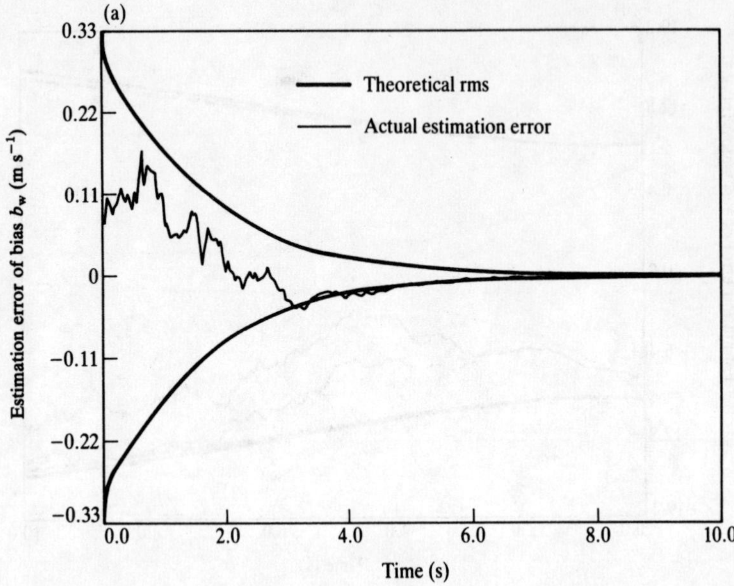

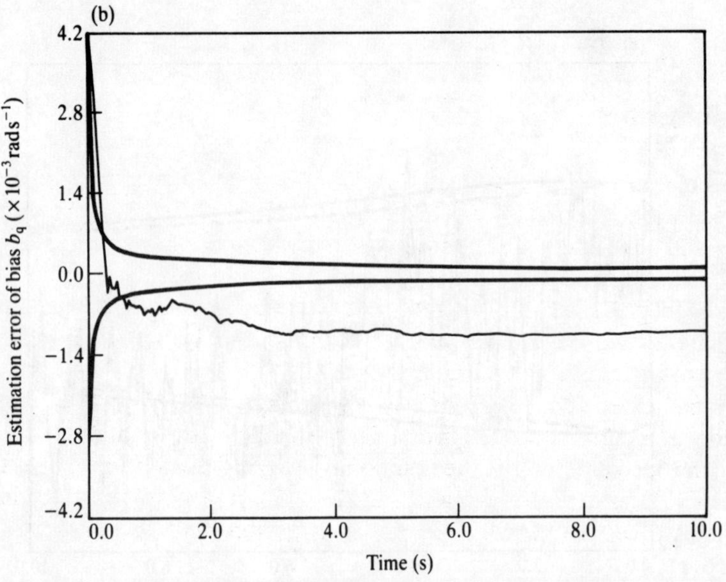

Figure 8.11 Estimation errors where the initial correlation does not exist, (a) of the bias $b_w(t)$; (b) of the constant bias b_q; (c) of the constant bias b_a.

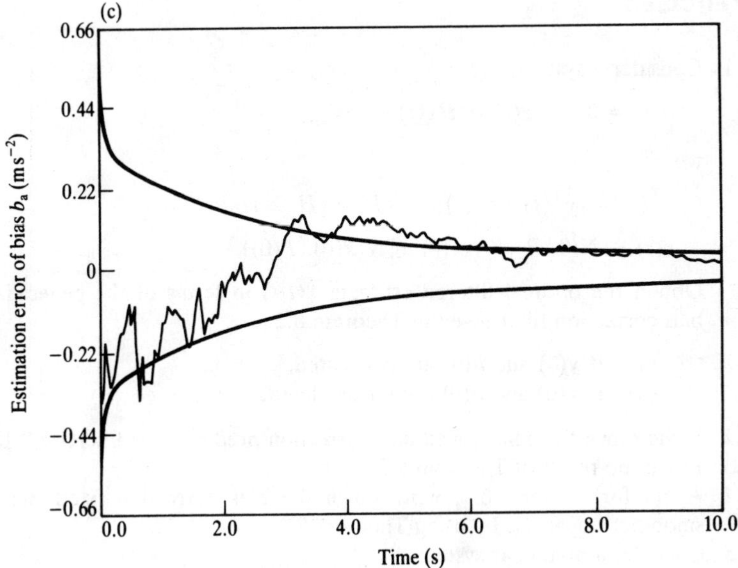

Figure 8.11 (*continued*)

8.7 SUMMARY

In this chapter we have applied the initial dependent-type partitioning filter developed in Chapter 5 to synthesize a two-stage bias correction filter and smoother for continuous-time systems. Using the same approach, a generalized two-stage bias correction filter and predictor have been presented for discrete-time systems. It is then shown that Friedland's two-stage bias correction filter can be derived in the framework of the partitioning approach.

Some extensions of Friedland's two-stage bias correction filter include a recursive second-order filter due to Shreve and Hedrick [30], the decoupling of the randomly varying bias terms due to Tanaka [31] and Washburn and Mendel [32], and nonlinear filtering problems due to Mendel [33] and Caglayan and Lancraft [29]. Moreover, practical applications can be seen in recent work on failure detections or monitoring [28, 34–36]. On other applications to trajectory estimation, added-inertial navigation, calibration, satellite-attitude estimation, etc., see Friedland [14].

EXERCISES

8.1 Consider a system

$$\dot{x}(t) = 0, \qquad z(t) = Hx(t) + v(t)$$

where

$$x^{\mathrm{T}}(t) = [y^{\mathrm{T}}(t) \quad b^{\mathrm{T}}(t)], \qquad H = [H_1 \quad H_2]$$

$$v(t) \sim N(0, R), \qquad x(0) \sim N(\hat{x}(0), P(0))$$

Obtain the optimal filtered estimate $\hat{x}(t|t)$ in terms of the generalized bias correction filter given in Theorem 8.2:

(a) If $y(0)$ and $b(0)$ are correlated.
(b) If $y(0)$ and $b(0)$ are uncorrelated.

8.2 Write down the generalized bias correction predictor for Exercise 8.1.

8.3 Prove the result of Theorem 8.3.

8.4 Again for Exercise 8.1, write down the bias correction fixed-interval smoother equations by using Theorem 8.3.

8.5 Consider a nonlinear system

$$\dot{x}(t) = f(x(t)) + w(t)$$

$$z(t) = h(x(t)) + B(t)b(t) + v(t)$$

where $w(t)$ and $v(t)$ are Gaussian white noise processes with zero mean and covariances $Q(t)$ and $R(t)$. The unknown bias parameter $b(t)$ is assumed to be constant. Apply the extended Kalman filter (see Chapter 2) to derive a generalized bias correction filter as described in Theorem 8.2.

8.6 When defining $e_1(t) \triangleq y(t) - \hat{y}_{\mathrm{n}}(t|t)$, $e_2(t) \triangleq y(t) - \hat{y}(t|t)$ and $e_3(t) \triangleq b(t) - \hat{b}(t)$ in equations (8.34)–(8.36) and (8.143)–(8.148), what type of asymptotic solutions do they have in the average sense?

8.7 Describe how one can apply the algorithms (8.143)–(8.148) to the problem of failure detection and correction in dynamic systems (see Friedland [28] and Friedland and Grabrowsky [36]).

8.8 Prove the result of Theorem 8.5.

8.9 Complete the proof of Corollary 8.3.

8.10 Consider the following simple system:

$$x(k+1) = \begin{bmatrix} a_1 & a_2 \\ 0 & a_3 \end{bmatrix} x(k) + \begin{bmatrix} b \\ 0 \end{bmatrix} w(k)$$

$$z(k) = [h_1 \quad h_2]x(k) + v(k)$$

where a_1, a_2, a_3, b, h_1, h_2 are constant scalar values, and $\{w(k), v(k)\}$ are zero-mean Gaussian white noise processes such that

$$E\left[\begin{pmatrix} w(k) \\ v(k) \end{pmatrix} (w(j) \quad v(j))\right] = \begin{bmatrix} q & 0 \\ 0 & r \end{bmatrix} \delta_{kj}$$

$$E[w(k)x^{\mathrm{T}}(0)] = E[v(k)x^{\mathrm{T}}(0)] = 0$$

and

$$x(0) \sim N\left(0, \begin{bmatrix} P_{11} & P_{12} \\ P_{12} & P_{22} \end{bmatrix}\right)$$

(a) Develop a generalized bias correction filter for $a_1 = a_3 = 1$ and $a_2 = 0$ by applying Theorem 8.5.
(b) Develop a generalized bias correction (one-step) predictor for the case of part (a) by using Corollary 8.3.

REFERENCES

[1] FRIEDLAND, B., Treatment of Bias in Recursive Filtering, *IEEE Trans. Aut. Control*, vol. AC-14, no. 4, August 1969, pp. 359–67.

[2] FRIEDLAND, B., Recursive Filtering in the Presence of Biases with Irreducible Uncertainty, *IEEE Trans. Aut. Control*, vol. AC-21, no. 5, October 1976, pp. 789–90.

[3] FRIEDLAND, B., On the Calibration Problem, *IEEE Trans. Aut. Control*, vol. AC-22, no. 6, December 1977, pp. 899–905.

[4] FRIEDLAND, B., Notes on Separate-Bias Estimation, *IEEE Trans. Aut. Control*, vol. AC-23, no. 4, August 1978, pp. 735–8.

[5] WATANABE, K., Optimal Structurally Partitioned Filter for Undisturbable Stochastic Systems: Part I. Basic Theory, *Int. J. Syst. Sci.*, vol. 14, no. 10, 1983, pp. 1139–58.

[6] WATANABE, K. and KUROSAKI, R., Optimal Structurally Partitioned Filter for Uncontrollable Stochastic Systems, *Syst. and Control*, vol. 27, no. 5, May 1983, pp. 65–76 (in Japanese).

[7] WATANABE, K., Two-Stage Bias Correction Estimators Based on Generalized Partitioning Estimation Method, *Int. J. Control*, vol. 38, no. 3, 1983, pp. 621–37.

[8] WATANABE, K., Partitioned Estimators Based on the Perturbed Kalman Filter Equations, *Int. J. Syst. Sci.*, vol. 14., no. 9, 1983, pp. 1115–28.

[9] ANDERSON, B. D. O., Stability Properties of Kalman–Bucy Filters, *J. Franklin Inst.*, vol. 291, no. 2, February 1971, pp. 137–44.

[10] KALMAN, R. E., Mathematical Description of Linear Dynamical Systems, *SIAM J. Control*, vol. 1, no. 2, 1963, pp. 152–92.

[11] WEISS, L., On the Structure Theory of Linear Differential Systems, *SIAM J. Control*, vol. 6, no. 4, 1968, pp. 659–80.

[12] BROWN, R. G., *Introduction to Random Signal Analysis and Kalman Filtering*, Wiley, New York, 1983.

[13] MENDEL, J. M. and WASHBURN, H. D., Multistage Estimation of Bias States in Linear Systems, *Int. J. Control*, vol. 28, no. 4, 1978, pp. 511–24.

[14] FRIEDLAND, B., Separated-Bias Estimation and Some Applications. In C.

T. Leondes (ed.), *Control and Dynamic Systems, Advances in Theory and Applications*, Vol. 20, Academic Press, New York, 1983, pp. 1–41.

[15] TACKER, E. C. and LEE, C. C., Linear Filtering in the Presence of Time-Varying Bias, *IEEE Trans. Aut. Control*, vol. AC-17, no. 6, 1972, pp. 828–9.

[16] IGNAGNI, M. B., An Alternate Derivation and Extension of Friedland's Two-Stage Kalman Estimator, *IEEE Trans. Aut. Control*, vol. AC-26, no. 3, June 1981, pp. 746–750.

[17] BIERMAN, G. J., The Treatment of Bias in the Square-Root Information Filter/Smoother, *J. Optim. Theory Applic.*, vol. 16, nos 1–2, July 1975, pp. 165–78.

[18] WATANABE, K., Optimal Structurally Partitioned Filter for Undisturbable Stochastic Systems: Part II. Stability and its Asymptotic Behaviour, *Int. J. Syst. Sci.*, vol. 14, no. 10, 1983, pp. 1159–69.

[19] NISHIMURA, T., Error Bounds of Continuous Kalman Filters and the Application to Orbit Determination Problems, *IEEE Trans. Aut. Control*, vol. AC-12, no. 3, June 1967, pp. 268–75.

[20] KALMAN, R. E. and BUCY, R. S., New Results in Linear Filtering and Prediction Theory, *J. Bas. Engng. Trans. ASME, Series D*, vol. 83, no. 1, 1961, pp. 95–108.

[21] SILVERMAN, L. M. and ANDERSON, B. D. O., Controllability, Observability and Stability of Linear Systems, *SIAM J. Control*, vol. 6, no. 1, 1968, pp. 121–30.

[22] ANDERSON, B. D. O. and MOORE, J. B., Time-Varying Version of the Lemma of Lyapunov, *Electronics Letters*, vol. 3, no. 7, July 1967, pp. 293–4.

[23] SCHLEE, F. H., STANDISH, C. J. and TODA, N. F., Divergence in the Kalman Filter, *AIAA J.*, vol. 5, no. 6, 1967, pp. 1114–20.

[24] TARN, T. J. and ZABORSZKY, J., A Practical, Nondiverging Filter, *AIAA J.*, vol. 8, no. 6, 1970, pp. 1127–33.

[25] FITZGERALD, R. J., Divergence of the Kalman Filter, *IEEE Trans. Aut. Control*, vol. AC-16, no. 6, 1971, pp. 736–47.

[26] GELB, A., *Applied Optimal Estimation*, MIT Press, Cambridge, Massachusetts, 1974.

[27] WATANABE, K., General Two-Stage Bias Correction Filter and Predictor for Linear Discrete-Time Systems, *Control-Theory and Advanced Technology*, vol. 1, no. 1, 1985, pp. 87–101.

[28] FRIEDLAND, B., Maximum Likelihood Failure Detection of Aircraft Flight Control Sensors, *J. Guidance and Control*, vol. 5, no. 5, 1982, pp. 498–503.

[29] CAGLAYAN, A. K. and LANCRAFT, R. E., A Separated Bias Identification and State Estimation Algorithm for Nonlinear Systems, *Automatica*, vol. 19, no. 5, 1983, pp. 561–70.

[30] SHREVE, E. L. and HEDRICK, W. R., Separating Bias and State Estimates in a Recursive Second-Order Filter, *IEEE Trans. Aut. Control*, vol. AC-19, no. 5, 1974, pp. 585–6.

[31] TANAKA, A., Parallel Computation in Linear Discrete Filtering, *IEEE Trans. Aut. Control*, vol. AC-20, no. 4, August 1975, pp. 573–5.

[32] WASHBURN, H. D. and MENDEL, J. M., Multistage Estimation of Dynamical and Weakly Coupled States in Continuous-Time Linear Systems, *IEEE Trans. Aut. Control*, vol. AC-25, no. 1, 1980, pp. 71–6.

[33] MENDEL, J. M., Extension of Friedland's Bias Filtering Technique to a Class of Nonlinear Systems, *IEEE Trans. Aut. Control*, vol. AC-21, no. 2, April 1976, pp. 296–8.

[34] CHANG, C. B. and DUNN, K. P., On GLR Detection and Estimation of Unexpected Inputs in Linear Discrete Systems, *IEEE Trans. Aut. Control*, vol. AC-24, no. 3, June 1979, pp. 499–501.

[35] CAGLAYAN, A. K., Necessary and Sufficient Conditions for Detectability of Jumps in Linear Systems, *IEEE Trans. Aut. Control*, vol. AC-25, no. 4, August 1980, pp. 833–4.

[36] FRIEDLAND, B. and GRABROWSKY, S. M., Estimating Sudden Changes of Biases in Linear Dynamic Systems, *IEEE Trans. Aut. Control*, vol. AC-27, no. 1, February 1982, pp. 237–40.

9

Forward-Pass Fixed-Interval Smoothers in Discrete-Time Systems

9.1 INTRODUCTION

This chapter extends the forward-pass fixed-interval smoother discussed in Section 6.5 to linear discrete-time systems, and studies the factorization and the steady-state solution of such a smoother.

In Section 9.2, we clarify the relationship between the Lagrange multiplier method [1–3] and the two-filter smoother due to Mayne [4] by applying simple orthogonal projection lemmas [5, 6]. Based on this, the forward-pass fixed-interval smoother is derived algebraically in Section 9.3. In Section 9.4, such a smoother recursion is used to develop a $U-D$ factorization-based smoother algorithm, which is effective in guaranteeing computational efficiency, reliance on numerically stable matrix modification algorithms, and reduced computer storage.

Section 9.5 discusses the solution of algebraic Lyapunov equations (ALEs) associated with the error covariance of the smoother. An algorithm for solving such equations is also proposed by using the classical eigenvector approach [7].

9.2 THE LAGRANGE MULTIPLIER METHOD AND THE TWO-FILTER SMOOTHER

Consider the following state-variable model for linear systems:

$$x(k+1) = \Phi(k+1, k)x(k) + B(k)w(k) \tag{9.1}$$

$$z(k) = H(k)x(k) + v(k) \tag{9.2}$$

where $x(k) \in \mathbb{R}^n$, $w(k) \in \mathbb{R}^p$, $z(k) \in \mathbb{R}^m$ and $v(k) \in \mathbb{R}^m$.

$\Phi(k+1, k)$, $B(k)$ and $H(k)$ are real matrices of compatible dimensions. The zero-mean Gaussian white noise processes $\{w(k), v(k)\}$ are described by the following second-order statistics:

$$E\left\{\begin{bmatrix} w(k) \\ v(k) \end{bmatrix} [w^{\mathrm{T}}(j) \quad v^{\mathrm{T}}(j)]\right\} = \begin{bmatrix} Q(k) & 0 \\ 0 & R(k) \end{bmatrix} \delta_{kj} \tag{9.3}$$

where $Q(k) = Q^{\mathrm{T}}(k) \geqslant 0$ and $R(k) = R^{\mathrm{T}}(k) > 0$. The distribution of the initial state vector $x(0)$ is assumed to be

$$x(0) \sim N(\hat{x}(0), P(0)), E[w(k)x^{\mathrm{T}}(0)] = E[v(k)x^{\mathrm{T}}(0)] = 0 \tag{9.4}$$

Given the measurement data $Z_N \triangleq \{z(0), \ldots, z(N)\}$, we desire an MV estimate $\hat{x}(k|N) \triangleq E[x(k)|Z_N]$ of $x(k)$. This problem can be expressed as one of minimizing, under constraint (9.1), the following cost functional [8, 9]:

$$J_N = \frac{1}{2}\,||x(0) - \hat{x}(0)||^2_{P^{-1}(0)} + \frac{1}{2}\sum_{k=0}^{N} ||z(k) - H(k)x(k)||^2_{R^{-1}(k)} + \frac{1}{2}\sum_{k=0}^{N-1} ||w(k)||^2_{Q^{-1}(k)} \tag{9.5}$$

If an n-vector Lagrange multiplier $\lambda(k|N)$ is introduced to incorporate the constraint, the following functional may be minimized:

$$J'_N = J_N + \sum_{k=0}^{N-1} \lambda^{\mathrm{T}}(k|N)[x(k+1) - \Phi(k+1, k)x(k) - B(k)w(k)] \tag{9.6}$$

This necessary condition reduces to the two-point boundary-value problem (TPBVP) described by the following discrete-time Hamiltonian system:

$$\begin{bmatrix} \hat{x}(k+1|N) \\ \lambda(k-1|N) \end{bmatrix} = \begin{bmatrix} \Phi(k+1, k) & B(k)Q(k)B^{\mathrm{T}}(k) \\ -H(k)R^{-1}(k)H(k) & \Phi^{\mathrm{T}}(k+1, k) \end{bmatrix} \times \begin{bmatrix} \hat{x}(k|N) \\ \lambda(k|N) \end{bmatrix} + \begin{bmatrix} 0 \\ H^{\mathrm{T}}(k)R^{-1}(k) \end{bmatrix} z(k) \tag{9.7a}$$

$$\hat{x}(0|N) = \hat{x}(0) + P(0)\lambda(-1|N), \qquad \lambda(N|N) = 0 \tag{9.7b}$$

Now express the Rauch–Tung–Striebel (RTS) smoother algorithm [1] (see also Appendix 9D) with respect to the above multiplier as follows:

$$\hat{x}(k|N) = \hat{x}(k|k-1) + P(k|k-1)\lambda(k-1|N) \tag{9.8}$$

and describe the Mayne–Fraser two-filter smoother [4], as discussed in Chapter 6, by the expressions:

$$\hat{x}(k|N) = P(k|N)[P^{-1}(k|k-1)\hat{x}(k|k-1) + q(k)] \tag{9.9}$$

$$P(k|N) = [P^{-1}(k|k-1) + S(k)]^{-1} \tag{9.10}$$

where $\hat{x}(k|k-1) = E[x(k)|Z_{k-1}]$ and $P(k|k-1) = E[\tilde{x}(k|k-1)\tilde{x}^{\mathrm{T}}(k|k-1)]$ denote the one-step predicted estimates and the associated estimation error covariance, respectively, in which $\tilde{x}(k|k-1) \triangleq x(k) - \hat{x}(k|k-1)$. Here, the quantities $q(k)$ and $S(k)$ are an n-dimensional vector and $n \times n$ matrix yet to be determined, and $P(k|N) = E[\tilde{x}(k|N)\tilde{x}^{\mathrm{T}}(k|N)]$, where $\tilde{x}(k|N) \triangleq x(k) - \hat{x}(k|N)$ which denotes the smoothed error.

The following lemmas are useful in deriving a forward-pass fixed-interval smoother algebraically.

Lemma 9.1

The Lagrange multiplier $\lambda(k-1|N)$ is related to $q(k)$ as follows:

$$\lambda(k-1|N) = q(k) - S(k)\hat{x}(k|N) \tag{9.11}$$

Proof

Equating equations (9.8) and (9.9) and rearranging gives the desired result. □

Lemma 9.2

The estimates $\hat{x}(k|k-1)$, $\hat{x}(k|N)$ and the error states $\tilde{x}(k|k-1)$, $\tilde{x}(k|N)$ satisfy the following equalities:

$$E[\hat{x}(k|k-1)\tilde{x}^{\mathrm{T}}(k|k-1)] = 0 \tag{9.12}$$

$$E[\tilde{x}(k|N)\hat{x}^{\mathrm{T}}(k|N)] = 0 \tag{9.13}$$

$$E[\hat{x}(k|k-1)\tilde{x}^{\mathrm{T}}(k|N)] = 0 \tag{9.14}$$

$$E[\tilde{x}(k|k-1)\tilde{x}^{\mathrm{T}}(k|N)] = P(k|N) \tag{9.15}$$

$$E[\tilde{x}(k|k-1)\hat{x}^{\mathrm{T}}(k|N)] = P(k|k-1) - P(k|N) \tag{9.16}$$

Proof

Equations (9.12) and (9.13) can be obtained from the well-known orthogonal projection lemmas [5, 6] (cf. Section 2.2). See Watanabe [10] for the proof of the remainder. □

Rearranging the Mayne–Fraser two-filter smoother given by equations (9.9) and (9.10):

$$P^{-1}(k|N)\hat{x}(k|N) = q(k) + P^{-1}(k|k-1)\hat{x}(k|k-1) \tag{9.17}$$

and subtracting $[S(k) + P^{-1}(k|k-1)]x(k)$ from both sides yields

$$P^{-1}(k|N)\tilde{x}(k|N) = P^{-1}(k|k-1)\tilde{x}(k|k-1) - \tilde{q}(k) \tag{9.18}$$

where

$$\tilde{q}(k) \triangleq q(k) - S(k)x(k) \tag{9.19}$$

Equations (9.18) and (9.19) may be used to establish the following lemma.

Lemma 9.3

The quantities $\tilde{q}(k)$, $x(k)$, $\tilde{x}(k|k-1)$ and $\tilde{x}(k|N)$ satisfy the following relations:

$$E[\tilde{q}(k)x^{\mathrm{T}}(k)] = 0 \tag{9.20}$$

$$E[\tilde{q}(k)\tilde{x}^{\mathrm{T}}(k|k-1)] = 0 \tag{9.21}$$

$$E[\tilde{q}(k)\tilde{x}^{\mathrm{T}}(k|N)] = -S(k)P(k|N) \tag{9.22}$$

$$E[\tilde{q}(k)\tilde{q}^{\mathrm{T}}(k)] = S(k) \tag{9.23}$$

Proof

See Watanabe [10]. □

Using this lemma, some properties can be derived for the Lagrange multiplier. Combining equations (9.11) and (9.19) gives the following equality:

$$\tilde{q}(k) = \lambda(k-1|N) - S(k)\tilde{x}(k|N) \tag{9.24}$$

This equation and Lemma 8.3 together constitute the following lemma.

Lemma 9.4

The Lagrange multiplier $\lambda(k-1|N)$ has the following properties:

$$E[\lambda(k-1|N)x^{\mathrm{T}}(k)] = E[\lambda(k-1|N)\hat{x}^{\mathrm{T}}(k|N)] = S(k)P(k|N) \tag{9.25}$$

$$E[\lambda(k-1|N)\hat{x}^{\mathrm{T}}(k|k-1)] = 0 \tag{9.26}$$

$$E[\lambda(k-1|N)\tilde{x}^{\mathrm{T}}(k|k-1)] = S(k)P(k|N) \tag{9.27}$$

$$E[\lambda(k-1|N)\tilde{x}^{\mathrm{T}}(k|N)] = 0 \tag{9.28}$$

$$E[\lambda(k-1|N)\lambda^{\mathrm{T}}(k-1|N)] = S(k) - S(k)P(k|N)S(k) \tag{9.29}$$

Proof

See Watanabe [10]. □

The expectation results of Lemmas 9.2–9.4 are of interest and may be an aid in interpreting estimation results. In particular, note that Lemma 9.3, together with Lemma 9.1, will play a fundamental role in algebraically deriving a forward-pass fixed-interval smoother in the next section.

9.3 A FORWARD-PASS FIXED-INTERVAL SMOOTHER

In this section, it is shown that the results obtained in the previous section may be used to construct a forward-pass fixed-interval smoother. The fixed-interval smoothed estimate $\hat{x}(k|N)$ is given by the following theorem.

Theorem 9.1

The fixed-interval smoothed estimate for the system of equations (9.1) and (9.2) is governed, in forward time, by the relation

$$\hat{x}(k+1|N) = [I + B(k)Q(k)B^{\mathrm{T}}(k)S(k+1)]^{-1}$$

$$\times [\Phi(k+1, k)\hat{x}(k|N) + B(k)Q(k)B^{\mathrm{T}}(k)q(k+1)],$$

$$\hat{x}(0|N) = [I + P(0)S(0)]^{-1}[P(0)q(0) + \hat{x}(0)] \tag{9.30}$$

for $k = 0, 1, \ldots, N$, where $S(k)$ and $q(k)$, are, in backward time, subject to

$$S(k) = \Phi^{\mathrm{T}}(k+1, k)S(k+1) \times [I + B(k)Q(k)B^{\mathrm{T}}(k)S(k+1)]^{-1}\Phi(k+1, k) + H^{\mathrm{T}}(k)R^{-1}(k)H(k),$$

$$S(N) = H^{\mathrm{T}}(N)R^{-1}(N)H(N) \tag{9.31}$$

$$q(k) = \Phi^{\mathrm{T}}(k+1, k)\{I - S(k+1) \times [I + B(k)Q(k)B^{\mathrm{T}}(k)S(k+1)]^{-1} \times B(k)Q(k)B^{\mathrm{T}}(k)\}q(k+1) + H^{\mathrm{T}}(k)R^{-1}(k)z(k), \qquad q(N) = H^{\mathrm{T}}(N)R^{-1}(N)z(N) \tag{9.32}$$

for $k = N, N-1, \ldots, 0$.

Proof

Substitution of equation (9.11), along with conditions (9.9) and (9.10), into the upper equality of expression (9.7a) gives equation (9.30). The initial condition is obtained by combining equations (9.9) and (9.10) and setting $k \to 0$. Equations (9.31) and (9.32) are proved as follows. Using equation (9.11) in the lower equality of expression (9.7a) yields

$$\lambda(k-1|N) = \Phi^{\mathrm{T}}(k+1, k)q(k+1) - \Phi^{\mathrm{T}}(k+1, k)S(k+1)\hat{x}(k+1|N) - H^{\mathrm{T}}(k)R^{-1}(k)H(k)\hat{x}(k|N) + H^{\mathrm{T}}(k)R^{-1}(k)z(k) \tag{9.33}$$

Further, substituting equation (9.30) into (9.33) and rearranging terms gives

$$\begin{aligned}\lambda(k-1|N) = {} & \Phi^{\mathrm{T}}(k+1, k)q(k+1) - \Phi^{\mathrm{T}}(k+1, k)S(k+1) \\ & \times [I + B(k)Q(k)B^{\mathrm{T}}(k) \\ & \times S(k+1)]^{-1}B(k)Q(k)B^{\mathrm{T}}(k)q(k+1) \\ & + H^{\mathrm{T}}(k)R^{-1}(k)z(k) \\ & - \{\Phi^{\mathrm{T}}(k+1, k)S(k+1) \\ & \times [I + B(k)Q(k)B^{\mathrm{T}}(k)S(k+1)]^{-1}\Phi(k+1, k) \\ & + H^{\mathrm{T}}(k)R^{-1}(k)H(k)\}\hat{x}(k|N) \end{aligned} \tag{9.34}$$

Comparing equations (9.11) and (9.34) leads to the desired results (9.31) and (9.32). The associated final conditions are obvious from equations (9.7) and (9.11), because $q(N+1) = 0$ and $S(N+1) = 0$ for $\hat{x}(N+1|N) \neq 0$. □

The corresponding estimation error covariance matrix is summarized in the following theorem.

Theorem 9.2

The smoothing error covariance $P(k|N)$, $k = 0, 1, \ldots, N$, satisfies the following recursive relation:

$$\begin{aligned}P(k+1|N) = {} & [I + B(k)Q(k)B^{\mathrm{T}}(k)S(k+1)]^{-1} \\ & \times [\Phi(k+1, k)P(k|N)\Phi^{\mathrm{T}}(k+1, k) \\ & + B(k)Q(k)B^{\mathrm{T}}(k) \\ & + B(k)Q(k)B^{\mathrm{T}}(k)S(k+1)B(k)Q(k)B^{\mathrm{T}}(k)] \\ & \times [I + B(k)Q(k)B^{\mathrm{T}}(k)S(k+1)]^{-\mathrm{T}}, \\ P(0|N) = {} & [I + P(0)S(0)]^{-1}P(0) \end{aligned} \tag{9.35}$$

Proof

From equations (9.1) and (9.30), it follows that

$$\begin{aligned}\tilde{x}(k+1|N) = {} & \Phi(k+1, k)\tilde{x}(k|N) + B(k)w(k) \\ & + B(k)Q(k)B^{\mathrm{T}}(k)S(k+1)\hat{x}(k+1|N) \\ & - B(k)Q(k)B^{\mathrm{T}}(k)q(k+1) \end{aligned} \tag{9.36}$$

This equation can also be rearranged as follows:

$$
\begin{aligned}
&[I + B(k)Q(k)B^{\mathrm{T}}(k)S(k+1)]\tilde{x}(k+1|N) \\
&= \Phi(k+1, k)\tilde{x}(k|N) + B(k)w(k) - B(k)Q(k)B^{\mathrm{T}}(k)\tilde{q}(k+1)
\end{aligned}
\tag{9.37}
$$

where $\tilde{q}(k+1)$ is as defined in equation (9.19). If equation (9.1) is used above, then equation (9.37) can be expressed as follows:

$$
\begin{aligned}
&[I + B(k)Q(k)B^{\mathrm{T}}(k)S(k+1)]\tilde{x}(k+1|N) \\
&\quad + B(k)Q(k)B^{\mathrm{T}}(k)\tilde{q}(k+1) - x(k+1) \\
&= -\Phi(k+1, k)\hat{x}(k|N)
\end{aligned}
\tag{9.38}
$$

Then, taking account of the results of Lemma 9.3 and taking the second-order moment of the left-hand side of equation (9.38) leads to

$$
\begin{aligned}
&E[\{[I - B(k)Q(k)B^{\mathrm{T}}(k)S(k+1)]\tilde{x}(k+1|N) \\
&\quad + B(k)Q(k)B^{\mathrm{T}}(k)\tilde{q}(k+1) - x(k+1)]\} \\
&\quad \times \{[I - B(k)Q(k)B^{\mathrm{T}}(k)S(k+1)] \\
&\quad \times \tilde{x}(k+1|N) + B(k)Q(k)B^{\mathrm{T}}(k)\tilde{q}(k+1) - x(k+1)]\}^{\mathrm{T}}] \\
&= -B(k)Q(k)B^{\mathrm{T}}(k)S(k+1)P(k+1|N) - P(k+1|N) \\
&\quad - P(k+1|N)S(k+1)B(k)Q(k)B^{\mathrm{T}}(k) \\
&\quad - B(k)Q(k)B^{\mathrm{T}}(k)S(k+1)P(k+1|N) \\
&\quad \times S(k+1)B(k)Q(k)B^{\mathrm{T}}(k) \\
&\quad + B(k)Q(k)B^{\mathrm{T}}(k)S(k+1)B(k)Q(k)B^{\mathrm{T}}(k) \\
&\quad + E[x(k+1)x^{\mathrm{T}}(k+1)]
\end{aligned}
\tag{9.39}
$$

Furthermore, from the following equalities:

$$
\begin{aligned}
E[x(k+1)x^{\mathrm{T}}(k+1)] &= \Phi(k+1, k)E[x(k)x^{\mathrm{T}}(k)]\Phi^{\mathrm{T}}(k+1, k) \\
&\quad + B(k)Q(k)B^{\mathrm{T}}(k) \\
E[x(k)x^{\mathrm{T}}(k)] &= E[\hat{x}(k|N)\hat{x}^{\mathrm{T}}(k|N)] + P(k|N)
\end{aligned}
$$

it is seen that

$$
\begin{aligned}
&E[x(k+1)x^{\mathrm{T}}(k+1)] \\
&\quad - \Phi(k+1, k)E[\hat{x}(k|N)\hat{x}^{\mathrm{T}}(k|N)]\Phi^{\mathrm{T}}(k+1, k) \\
&= \Phi(k+1, k)P(k|N)\Phi^{\mathrm{T}}(k+1, k) + B(k)Q(k)B^{\mathrm{T}}(k)
\end{aligned}
\tag{9.40}
$$

Therefore, the second-order moment of equation (9.38), along with equations (9.39) and (9.40), yields

$$\begin{aligned}
-[I + B(k)Q(k)B^{\mathrm{T}}(k)S(k+1)]P(k+1|N)& \\
\times [I + B(k)Q(k)B^{\mathrm{T}}(k)S(k+1)]^{\mathrm{T}}& \\
+ B(k)Q(k)B^{\mathrm{T}}(k)S(k+1)B(k)Q(k)B^{\mathrm{T}}(k)& \\
+ \Phi(k+1, k)P(k|N)\Phi^{\mathrm{T}}(k+1, k)& \\
+ B(k)Q(k)B^{\mathrm{T}}(k) = 0&
\end{aligned} \tag{9.41}$$

Since the associated initial condition is apparent from equation (9.10), the proof of the theorem is complete. □

Example 9.1

In order to illustrate forward-pass fixed-interval smoothing in a very simple way, let us consider the following scalar system: $\Phi = 1$, $B = 1$, $H = 1$, $P(0) = 100$, $\hat{x}(0) = 0$, $Q = 25$ and $R = 15$, with $N = 4$.

Since $\Phi = 1$, it follows that equation (9.30) becomes

$$\begin{aligned}
\hat{x}(k+1|4) &= [1 + 25S(k+1)]^{-1}[\hat{x}(k|4) + 25q(k+1)] \\
&= \frac{\hat{x}(k|4) + 25q(k+1)}{1 + 25S(k+1)}
\end{aligned}$$

for $k = 0, 1, 2, 3$ and

$$\hat{x}(0|4) = \frac{100q(0)}{1 + 100S(0)}$$

Also, equations (9.31) and (9.35) become

$$S(k) = \frac{S(k+1)}{1 + 25S(k+1)} + \frac{1}{15}$$

$$S(4) = \frac{1}{R} = \frac{1}{15}$$

and

$$P(k+1|4) = \frac{P(k|4) + 25 + 25^2 S(k+1)}{(1 + 25S(k+1))^2} \quad \text{for } k = 0, 1, 2, 3$$

$$P(0|4) = \frac{100}{1 + 100S(0)}$$

Utilizing these equations, we compute $S(k)$ in backward time for $k = 4, 3, 2, 1, 0$ and $P(k|4)$ in forward time for $k = 0, 1, 2, 3, 4$ in Table 9.1.

It is emphasized that $\{q(k), S(k)\}$ in Theorem 9.1 are just the quantities associated with the backward-time information filter, i.e. $q(k) = P_b^{-1}(k|k)\hat{x}_b(k|k)$ and $S(k) = P_b^{-1}(k|k)$, where $\{\hat{x}_b(k|k), P_b(k|k)\}$ are the quantities due to the formal backward-time Kalman filter (see also Exercises 2.8 and 9.6). Furthermore, if equation (5A.1), presented in the discussion of matrix inversion in Appendix 5A, is utilized in Theorem 9.1, the backward-time information filter can also be expressed as follows:

$$\begin{aligned} q(k) &= \Phi^{\mathrm{T}}(k+1, k)[I + S(k+1)B(k)Q(k)B^{\mathrm{T}}(k)]^{-1}q(k+1) \\ &\quad + H^{\mathrm{T}}(k)R^{-1}(k)z(k), \qquad q(N) = H^{\mathrm{T}}(N)R^{-1}(N)z(N) \end{aligned} \tag{9.42}$$

It can be found, from this fact, that the present smoothing formula with $\hat{x}(0) = 0$ is equivalent to the one suggested by Desai *et al* [11] and Ackner and Kailath [12]. The proof and the tools used by Desai *et al.* and Ackner and Kailath are based on the concepts of complementary model [13–15]. The present derivation differs from their work in that the method used here depends primarily on simple orthogonal conditions. The block diagram of the fixed-interval smoother is presented in Figure 9.1.

As discussed in the continuous-time systems in Chapter 6, the so-called backward-time Kalman filter is realizable using equations (9.31) and (9.32). That is, since in the backward-time Kalman filtering problem the measurement data Z_k are available at the interval $\{z(j), j \in [N, k]\}$ and not at time $j \in [0, k)$, the optimal one-step predictor $\hat{x}(k|k-1)$ and the estimation error covariance

Table 9.1 Computation of $P(k|4)$ via forward-pass fixed-interval smoothing

Time k	*S(k)*	*P(k|4)*
0	0.0948	9.542
1	0.0948	8.261
2	0.0945	8.166
3	0.0917	8.349
4	0.0667	10.549

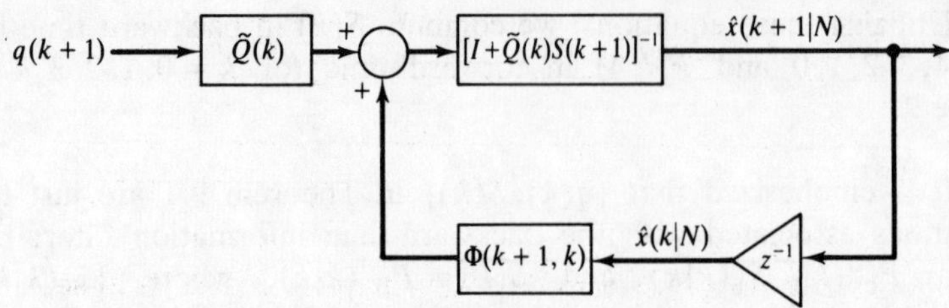

Figure 9.1 Block diagram for a forward-pass fixed-interval smoother with *n*-dimensional inversion, where $\tilde{Q}(k) = B(k)Q(k)B^{\mathrm{T}}(k)$.

$P(k|k-1)$ which are described by

$$\begin{aligned}\hat{x}(k+1|k) = \Phi(k+1, k)\{&\hat{x}(k|k-1) \\ &+ P(k|k-1)H^{\mathrm{T}}(k)[H(k)P(k|k-1)H^{\mathrm{T}}(k) \\ &+ R(k)]^{-1}[z(k) - H(k)\hat{x}(k|k-1)]\}, \\ &\hat{x}(0|-1) = \hat{x}(0) \end{aligned} \tag{9.43}$$

$$\begin{aligned}P(k+1|k) = \Phi(k+1, k)\{&P(k|k-1) \\ &- P(k|k-1)H^{\mathrm{T}}(k)[H(k)P(k|k-1)H^{\mathrm{T}}(k) \\ &+ R(k)]^{-1}H(k)P(k|k-1)\}\Phi^{\mathrm{T}}(k+1, k) \\ &+ B(k)Q(k)B^{\mathrm{T}}(k), \\ &P(0|-1) = P(0)\end{aligned} \tag{9.44}$$

reduce to

$$\hat{m}(k+1) = \Phi(k+1, k)\hat{m}(k), \qquad \hat{m}(0) = \hat{x}(0) \tag{9.45}$$

$$\begin{aligned}\Pi(k+1) = \Phi(k+1, k)&\Pi(k)\Phi^{\mathrm{T}}(k+1, k) \\ &+ B(k)Q(k)B^{\mathrm{T}}(k), \qquad \Pi(0) = P(0)\end{aligned} \tag{9.46}$$

Given this fact and equations (9.9) and (9.10), the strict backward-time Kalman filter (see Figure 9.2) can be expressed by

$$\hat{x}_{\mathrm{B}}(k|k) = P_{\mathrm{B}}(k|k)[\Pi^{-1}(k)\hat{m}(k) + q(k)] \tag{9.47}$$

$$P_{\mathrm{B}}(k|k) = [\Pi^{-1}(k) + S(k)]^{-1} \tag{9.48}$$

Moreover, if equations (9.9), (9.10), (9.47) and (9.48) are combined appropriately, it follows that

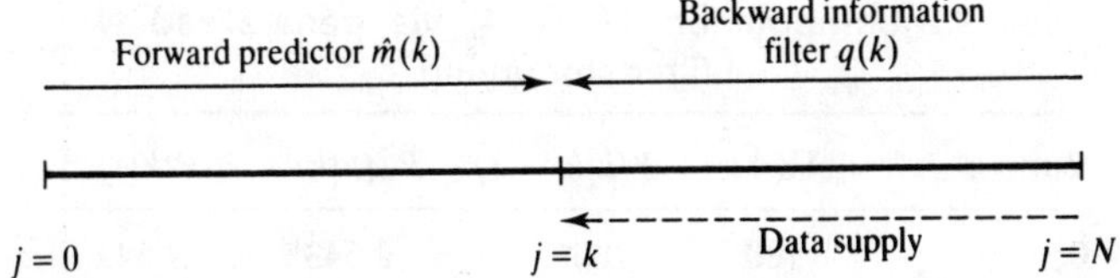

Figure 9.2 A strict backward-time Kalman filter in discrete time.

$$\hat{x}(k|N) = P(k|N)[P_{\mathrm{B}}^{-1}(k|k)\hat{x}_{\mathrm{B}}(k|k) + P^{-1}(k|k-1)\hat{x}(k|k-1) - \Pi^{-1}(k)\hat{m}(k)] \tag{9.49}$$

$$P(k|N) = [P_{\mathrm{B}}^{-1}(k|k) + P^{-1}(k|k-1) - \Pi^{-1}(k)]^{-1} \tag{9.50}$$

which can be seen as a generalized two-filter smoother formula [16]. Figure 9.3 illustrates this concept.

Example 9.2

Consider again the same problem as Example 9.1, but with the generalized two-filter smoother formula given by equations (9.48) and (9.50). The results are tabulated in Table 9.2. The results of backward-pass fixed-interval smoothing due to Rauch *et al*. [1] are presented in Table 9.3. It is clear that the smoothed estimate using forward recursion completely coincides with that using backward recursion.

The algorithms indicated in Theorems 9.1 and 9.2 are not convenient for practical calculation with large state dimension, because the calculation of the smoothed estimate $\hat{x}(k|N)$ requires the inversion of the $n \times n$ matrix $[I + B(k)Q(k)B^{\mathrm{T}}(k)S(k+1)]$ at time k. An expression which requires only two inversions of the $p \times p$ matrix $[I + G^{\mathrm{T}}(k)S(k+1)G(k)]$ and the $m \times m$ matrix $R(k)$, except for the initial instant, is given by the following corollary.

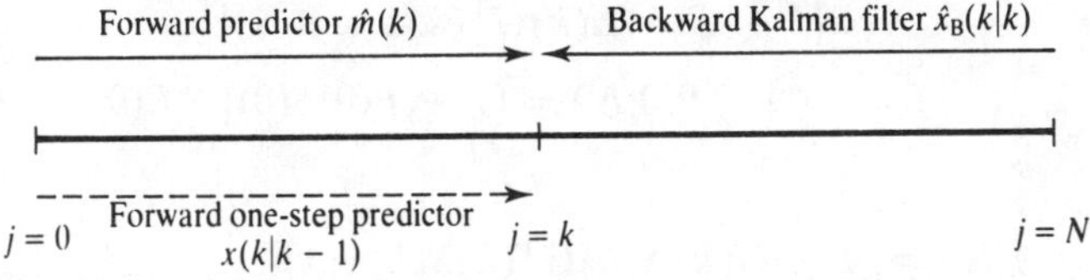

Figure 9.3 A generalized two-filter smoother in discrete time.

Table 9.2 Computation of $P(k|4)$ via generalized two-filter smoothing

Time k	$\Pi(k)$	$P(k\|k-1)$	$P_B(k\|k)$	$P(k\|4)$
0	100	100	9.543	9.542
1	125	38.043	9.730	8.261
2	150	35.758	9.883	8.166
3	175	35.567	10.269	8.349
4	200	35.550	13.953	10.549

Table 9.3 Computation of $P(k|4)$ via backward-pass fixed-interval smoothing

Time k	$P(k\|k)$	$P(k\|4)$
0	13.043	9.542
1	10.758	8.261
2	10.567	8.166
3	10.550	8.349
4	10.549	10.549

Corollary 9.1

The smoothed estimate $\hat{x}(k|N)$ and the associated estimation error covariance matrix $P(k|N)$ can be rewritten as follows:

$$\hat{x}(k+1|N) = [\Phi(k+1, k) + G(k)\beta(k)]\hat{x}(k|N) + G(k)\alpha(k),$$

$$\hat{x}(0|N) = [I + P(0)S(0)]^{-1}[P(0)q(0) + \hat{x}(0)] \tag{9.51}$$

$$P(k+1|N) = [I - G(k)\Delta(k)G^{\mathrm{T}}(k)S(k+1)] \times \{\Phi(k+1, k)P(k|N)\Phi^{\mathrm{T}}(k+1, k) + G(k)[I + G^{\mathrm{T}}(k)S(k+1)G(k)]G^{\mathrm{T}}(k)\} \times [I - G(k)\Delta(k)G^{\mathrm{T}}(k)S(k+1)]^{\mathrm{T}},$$

$$P(0|N) = [I + P(0)S(0)]^{-1}P(0) \tag{9.52}$$

or

$$P(k+1|N) = [I - G(k)\Delta(k)G^{\mathrm{T}}(k)S(k+1)] \times \Phi(k+1, k)P(k|N)\Phi^{\mathrm{T}}(k+1, k)$$

$$\times [I - G(k)\Delta(k)G^{\mathrm{T}}(k)S(k+1)]^{\mathrm{T}}$$
$$+ G(k)\Delta(k)G^{\mathrm{T}}(k),$$
$$P(0|N) = [I + P(0)S(0)]^{-1}P(0) \qquad (9.53)$$

where

$$\alpha(k) \triangleq \Delta(k)G^{\mathrm{T}}(k)q(k+1) \qquad (9.54)$$

$$\beta(k) \triangleq -\Delta(k)G^{\mathrm{T}}(k)S(k+1)\Phi(k+1, k) \qquad (9.55)$$

$$\Delta(k) \triangleq [I + G^{\mathrm{T}}(k)S(k+1)G(k)]^{-1} \qquad (9.56)$$

$$G(k)G^{\mathrm{T}}(k) \triangleq B(k)Q(k)B^{\mathrm{T}}(k) \qquad (9.57)$$

and

$$S(k) = \Phi^{\mathrm{T}}(k+1, k)S(k+1)[I - G(k)\Delta(k)G^{\mathrm{T}}(k)S(k+1)] \times \Phi(k+1, k) + H^{\mathrm{T}}(k)R^{-1}(k)H(k), \qquad S(N) = H^{\mathrm{T}}(N)R^{-1}(N)H(N) \qquad (9.58)$$

$$q(k) = \Phi^{\mathrm{T}}(k+1, k)[I - S(k+1)G(k)\Delta(k)G^{\mathrm{T}}(k)]q(k+1) + H^{\mathrm{T}}(k)R^{-1}(k)z(k), \qquad q(N) = H^{\mathrm{T}}(N)R^{-1}(N)z(N) \qquad (9.59)$$

Proof

Expressing $[I + B(k)Q(k)B^{\mathrm{T}}(k)S(k+1)]$ as $[I + G(k)G^{\mathrm{T}}(k) S(k+1)]$ and applying equation (5A.1), the corollary immediately follows. The proof of the equivalence of equations (9.52) and (9.53) is left as an exercise. □

Example 9.3

Consider the following system:

$$\Phi = \begin{bmatrix} 1 & 2 \\ 0 & 1 \end{bmatrix}, \qquad B = \begin{bmatrix} 0 \\ 1 \end{bmatrix}, \qquad Q = 4$$

$$H = [1 \quad 1], \qquad R = 1, \qquad P(0) = \begin{bmatrix} 1 & 0 \\ 0 & 1 \end{bmatrix}$$

at time $k = 0$ and $k = 1$. The purpose of this example is to derive the information matrix $S(k)$ and the smoothed error covariance $P(k|N)$ algebraically. We first obtain

$$G = \begin{bmatrix} 0 \\ 2 \end{bmatrix}$$

From equation (9.58),

$$S(1) = \begin{bmatrix} 1 \\ 1 \end{bmatrix} [1][1 \quad 1] = \begin{bmatrix} 1 & 1 \\ 1 & 1 \end{bmatrix}$$

$$\Delta(0) = 1 \Big/ \left\{ 1 + [0 \quad 2] \begin{bmatrix} 1 & 1 \\ 1 & 1 \end{bmatrix} \begin{bmatrix} 0 \\ 2 \end{bmatrix} \right\} = 1/5$$

$$G\Delta(0)G^{\mathrm{T}} = \begin{bmatrix} 0 \\ 2 \end{bmatrix} [1/5][0 \quad 2] = \begin{bmatrix} 0 & 0 \\ 0 & 4/5 \end{bmatrix}$$

$$I - G\Delta(0)G^{\mathrm{T}}S(1) = \begin{bmatrix} 1 & 0 \\ 0 & 1 \end{bmatrix} - \begin{bmatrix} 0 & 0 \\ 0 & 4/5 \end{bmatrix} \begin{bmatrix} 1 & 1 \\ 1 & 1 \end{bmatrix}$$

$$= \begin{bmatrix} 1 & 0 \\ -4/5 & 1/5 \end{bmatrix}$$

so that

$$S(0) = \begin{bmatrix} 1 & 0 \\ 2 & 1 \end{bmatrix} \begin{bmatrix} 1 & 1 \\ 1 & 1 \end{bmatrix} \begin{bmatrix} 1 & 0 \\ -4/5 & 1/5 \end{bmatrix}$$

$$\times \begin{bmatrix} 1 & 2 \\ 0 & 1 \end{bmatrix} + \begin{bmatrix} 1 & 1 \\ 1 & 1 \end{bmatrix}$$

$$= \begin{bmatrix} 1/5 & 3/5 \\ 3/5 & 9/5 \end{bmatrix} + \begin{bmatrix} 1 & 1 \\ 1 & 1 \end{bmatrix} = \begin{bmatrix} 6/5 & 8/5 \\ 8/5 & 14/5 \end{bmatrix}$$

By (9.53),

$$P(0|N) = \left[\begin{bmatrix} 1 & 0 \\ 0 & 1 \end{bmatrix} + \begin{bmatrix} 6/5 & 8/5 \\ 8/5 & 14/5 \end{bmatrix} \right]^{-1} = \begin{bmatrix} 11/5 & 8/5 \\ 8/5 & 19/5 \end{bmatrix}^{-1}$$

$$= \begin{bmatrix} 19/29 & -8/29 \\ -8/29 & 11/29 \end{bmatrix}$$

$$[I - G\Delta(0)G^{\mathrm{T}}S(1)]\Phi = \begin{bmatrix} 1 & 0 \\ -4/5 & 1/5 \end{bmatrix}\begin{bmatrix} 1 & 2 \\ 0 & 1 \end{bmatrix}$$

$$= \begin{bmatrix} 1 & 2 \\ -4/5 & -7/5 \end{bmatrix}$$

$$[I - G\Delta(0)G^{\mathrm{T}}S(1)]\Phi P(0|N)\Phi^{\mathrm{T}}[I - G\Delta(0)G^{\mathrm{T}}S(1)]^{\mathrm{T}}$$

$$= \begin{bmatrix} 1 & 2 \\ -4/5 & -7/5 \end{bmatrix}\begin{bmatrix} 19/29 & -8/29 \\ -8/29 & 11/29 \end{bmatrix}\begin{bmatrix} 1 & -4/5 \\ 2 & -7/5 \end{bmatrix}$$

$$= \begin{bmatrix} 31/29 & -22/29 \\ -22/29 & 79/145 \end{bmatrix}$$

Thus

$$P(1|N) = [I - G\Delta(0)G^{\mathrm{T}}S(1)]\Phi P(0|N)\Phi^{\mathrm{T}} \times [I - G\Delta(0)G^{\mathrm{T}}S(1)]^{\mathrm{T}} + G\Delta(0)G^{\mathrm{T}}$$

$$= \begin{bmatrix} 31/29 & -22/29 \\ -22/29 & 79/145 \end{bmatrix} + \begin{bmatrix} 0 & 0 \\ 0 & 4/5 \end{bmatrix}$$

$$= \begin{bmatrix} 31/29 & -22/29 \\ -22/29 & 39/29 \end{bmatrix}$$

It should be noted that the present formulas are equivalent to the smoothing algorithms due to Mayne [4] derived by the dynamic programming (DP) method, except for the smoothing error covariance described in equations (9.52) and (9.53) (see also Figure 9.4).

For the time-invariant system, the algorithm presented in Corollary 9.1, of course, requires only one inversion of the $p \times p$ matrix $[I + G^{\mathrm{T}}(k)S(k+1)G(k)]$, except for the initial condition. Therefore, it is computationally efficient for the case when $p \ll n$, as are most aerospace-related applications. Moreover, we note that the smoothed

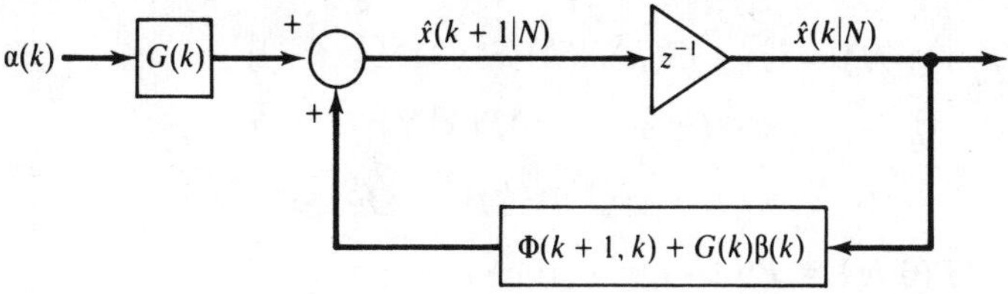

Figure 9.4 Block diagram for a forward-pass fixed-interval smoother with p-dimensional inversion, where $p \leqslant n$.

covariance recursion equation (9.53) is a numerically stable form which is similar to the Joseph form [17] (cf. Exercise 2.1). Hence, the present formulation is at least numerically superior to the best-known RTS formulation, except for some factorization-based smoother algorithms such as the square-root information filter/smoother and $U-D$ covariance factored filter/smoother algorithms [18–20] (see also Section 9.4).

As suggested in Chapter 6, we can see that the a priori statistics $(\hat{x}(0), P(0))$ will affect only the linear equations (9.51) and (9.52) or (9.53). From this point of view, the effect of changes in $(\hat{x}(0), P(0))$ on $\{\hat{x}(k|N), P(k|N)\}$ can be immediately evaluated as the following corollary.

Corollary 9.2

Let $\hat{x}_0(k+1|N)$ and $P_0(k+1|N)$ be the smoothed estimate and the error covariance matrix for the system of equations (9.1) and (9.2) with initial statistics $(\hat{x}(0), P(0))$. When $(\hat{x}(0), P(0))$ changes into $(m, \Pi(0))$, the smoothed estimate and the error covariance matrix corresponding to $(m, \Pi(0))$ can be obtained from the following equalities:

$$\hat{x}(k+1|N) = \hat{x}_0(k+1|N) + \delta\hat{x}(k+1|N) \tag{9.60}$$

$$P(k+1|N) = P_0(k+1|N) + \delta P(k+1|N) \tag{9.61}$$

where $\hat{x}_0(k+1|N)$ and $P_0(k+1|N)$ are given by Corollary 9.1, and $\delta\hat{x}(k+1|N)$, $\delta P(k+1|N)$ are subject to

$$\begin{aligned}
&\delta\hat{x}(k+1|N) = [\Phi(k+1, k) + G(k)\beta(k)]\delta\hat{x}(k|N), \\
&\delta\hat{x}(0|N) = \hat{x}(0|N) - \hat{x}_0(0|N) \\
&= \delta P(0|N)q(0) + [I + \Pi(0)S(0)]^{-1}m - [I + P(0)S(0)]^{-1}\hat{x}(0)
\end{aligned} \tag{9.62}$$

and

$$\begin{aligned}
\delta P(k+1|N) &= [I - G(k)\Delta(k)G^{\mathrm{T}}(k)S(k+1)] \\
&\quad \times \Phi(k+1, k)\delta P(k|N)\Phi^{\mathrm{T}}(k+1, k) \\
&\quad \times [I - G(k)\Delta(k)G^{\mathrm{T}}(k)S(k+1)]^{\mathrm{T}}, \\
\delta P(0|N) &= P(0|N) - P_0(0|N) \\
&= [I + \Pi(0)S(0)]^{-1}\Pi(0) - [I + P(0)S(0)]^{-1}P(0)
\end{aligned} \tag{9.63}$$

Proof

Subtracting equations (9.51) and (9.53) with $(\hat{x}(0), P(0))$ from those with $(m, \Pi(0))$ leads to the desired results. □

It is interesting to note that for any change in initial conditions the smoothed estimate can be computed from the basic estimate $\hat{x}_0(k|N)$ without reprocessing of the data. For the particular case when $\hat{x}(0) = 0$ and $P(0) = 0$, it follows that

$$\delta\hat{x}(0|N) = [I + \Pi(0)S(0)]^{-1}[\Pi(0)q(0) + m] \equiv \hat{x}(0|N) \qquad (9.64)$$

$$\delta P(0|N) = [I + \Pi(0)S(0)]^{-1}\Pi(0) \equiv P(0|N) \qquad (9.65)$$

In this case, $\{\hat{x}(k+1|N), P(k+1|N)\}$ consists of two parts, one the basic quantities $\{\hat{x}_0(k+1|N),\ P_0(k+1|N)\}$, irrespective of the initial conditions, and another the corrections $\{\delta\hat{x}(k+1|N), \delta P(k+1|N)\}$ due to the initial statistics $(m, \Pi(0))$.

Example 9.4

Consider a practical fourth-order satellite tracking system [1]. Let x_1 be the angular position of the satellite, x_2 the angular velocity of the satellite, x_3 constant acceleration, and x_4 stochastic acceleration. Further, we have

$$\Phi = \begin{bmatrix} 1 & 1 & 0.5 & 0.5 \\ 0 & 1 & 1 & 1 \\ 0 & 0 & 1 & 0 \\ 0 & 0 & 0 & 0.606 \end{bmatrix}, \qquad B = \begin{bmatrix} 0 \\ 0 \\ 0 \\ 1 \end{bmatrix}$$

$$H = [1 \ \ 0 \ \ 0 \ \ 0], \qquad w(k) \sim N(0, 0.63 \times 10^{-2})$$

$$v(k) \sim N(0, 1)$$

The main purpose of this example is to give the quantitative evaluation of the effect of the initial statistics on the smoothed estimate, by using Corollary 9.2. It is assumed that $P(0) = 0$, $\Pi(0) = \mathrm{diag}(1, 1, 1, 1 \times 10^{-2})$ for case 1 and $\Pi(0) = \mathrm{diag}(100, 100, 100, 1 \times 10^{-2})$ for case 2. These results are shown in Figures 9.5–9.9. The quantitative effect of the initial conditions can be seen in Figures 9.6 and 9.7, i.e. $\delta P(k|N)$ rapidly dies out. In addition, note that, in Figure 9.8, $P_0(k|N)$ is markedly smaller than $\delta P(k|N)$ for the constant acceleration modelled as a random constant. This means that

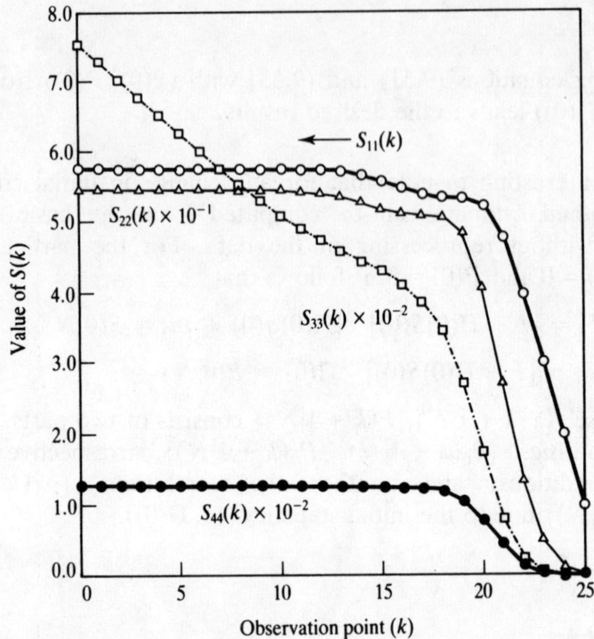

Figure 9.5 History of backward Riccati equation $S(k)$.

state x_3 is unsmoothable. This condition can also be confirmed by calculating the associated controllable (or smoothable) matrix [21].

9.4 *U–D* INFORMATION MATRIX FACTORIZATION

Consider a linear discrete-time stochastic system described by

$$x(k+1) = \Phi(k+1, k)x(k) + Bw(k) \tag{9.66}$$

$$z(k) = H(k)x(k) + v(k) \tag{9.67}$$

where $x(k) \in \mathbb{R}^n$, $w(k) \in \mathbb{R}^p$, $z(k)$, $v(k) \in \mathbb{R}^m$ and $\{w(k), v(k)\}$ are zero-mean, statistically independent Gaussian white noise processes with covariances

$$\begin{aligned} E[w(k)w^{\mathrm{T}}(j)] &= Q\delta_{kj} \\ E[v(k)v^{\mathrm{T}}(j)] &= R\delta_{kj} \end{aligned} \tag{9.68}$$

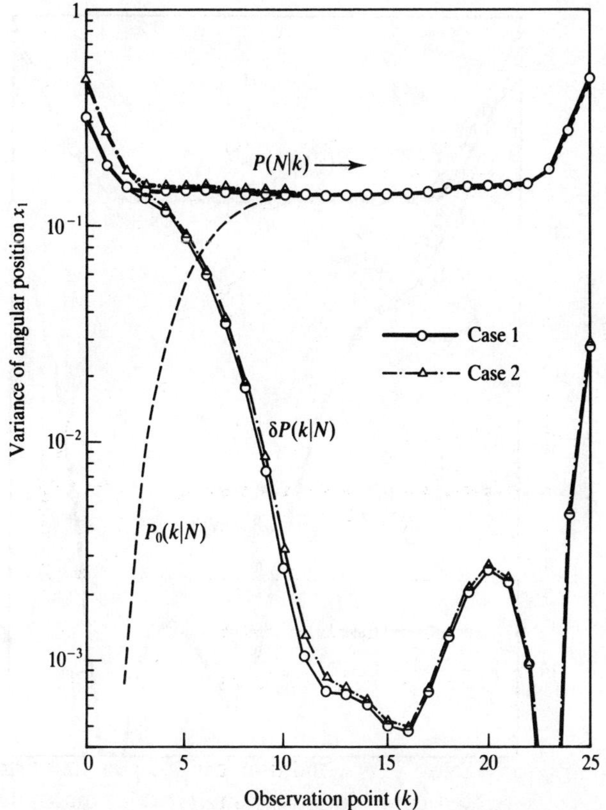

Figure 9.6 Error variance history of angular position x_1.

in which Q and R are assumed to be diagonal.† The initial state $x(0)$ is assumed to be distributed as

$$x(0) \sim N(\hat{x}(0), P(0)), \; E[w(k)x^{\mathrm{T}}(0)] = E[v(k)x^{\mathrm{T}}(0)] = 0 \quad (9.69)$$

Given the observation data Z_N, we seek the MV estimate $\hat{x}(k|N)$ of $x(k)$ to minimize the cost functional as in equation (9.5).

A nonfactorized forward-pass fixed-interval smoother solution to the above problem can be obtained by slightly modifying the results shown in the previous section such that the smoothed estimate is

† The choice of B assures that Q being diagonal involves no loss of generality, and by Cholesky decomposition, as discussed in Bierman [19], we can similarly assume without loss of generality that R is diagonal.

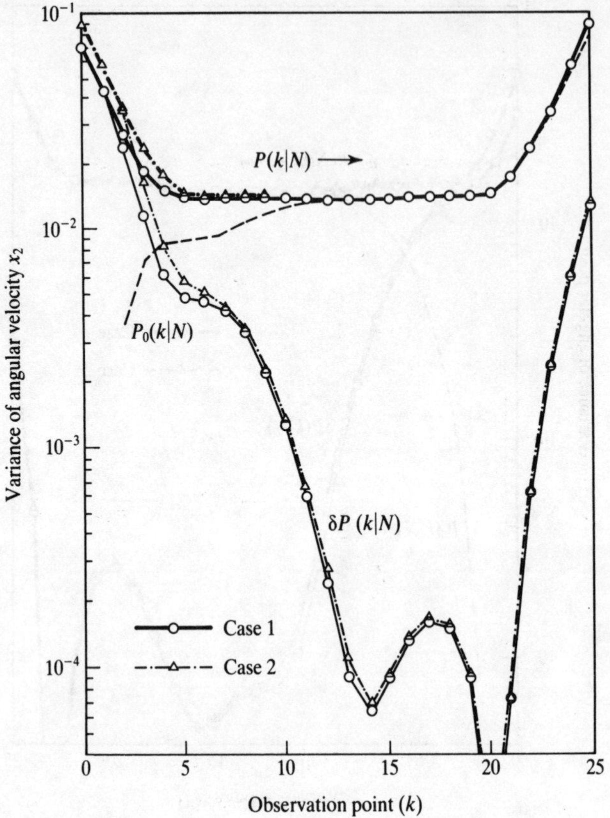

Figure 9.7 Error variance history of angular velocity x_2.

given by

$$\bar{x}(k+1|N) = \Phi(k+1, k)\hat{x}(k|N),$$

$$\hat{x}(0|N) = [I + P(0)S(0)]^{-1}[P(0)q(0) + \hat{x}(0)] \tag{9.70}$$

$$\hat{x}(k+1|N) = [I - GK^{\mathrm{T}}(k+1)]\bar{x}(k+1|N) + G[I + G^{\mathrm{T}}S(k+1)G]^{-1}y(k+1) \tag{9.71}$$

and the smoothed error covariance by

$$\bar{P}(k+1|N) = \Phi(k+1, k)P(k|N)\Phi^{\mathrm{T}}(k+1, k),$$

$$P(0|N) = [I + P(0)S(0)]^{-1}P(0) \tag{9.72}$$

$$P(k+1|N) = [I - GK^{\mathrm{T}}(k+1)]\bar{P}(k+1|N)[I - GK^{\mathrm{T}}(k+1)]^{\mathrm{T}}$$

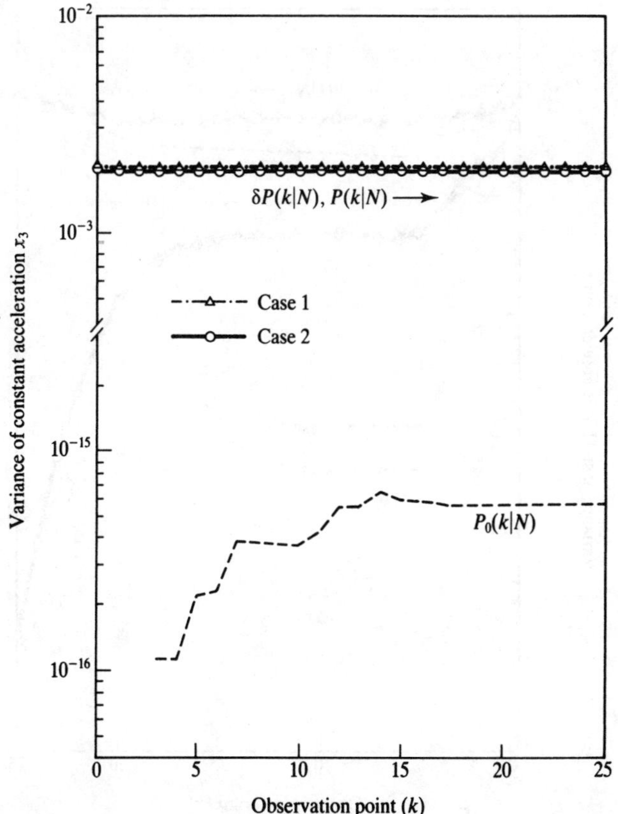

Figure 9.8 Error variance history of constant acceleration x_3.

$$+ G[I + G^{\mathrm{T}}S(k+1)G]^{-1}G^{\mathrm{T}}, \qquad GG^{\mathrm{T}} \triangleq BQB^{\mathrm{T}} \tag{9.73}$$

where G is an $n \times p$ matrix. It should be noted that an 'input' $y(k+1)$ and a 'gain' $K(k+1)$ are newly defined to ease future developments. These quantities are provided by the following backward-time information filter, with measurement update given by

$$K(k+1) = S(k+1)G[I + G^{\mathrm{T}}S(k+1)G]^{-1}, \qquad S(N+1) = 0 \tag{9.74}$$

$$\bar{S}(k+1) = [I - K(k+1)G^{\mathrm{T}}]S(k+1) \tag{9.75}$$

$$\bar{q}(k+1) = [I - K(k+1)G^{\mathrm{T}}]q(k+1), \qquad q(N+1) = 0 \tag{9.76}$$

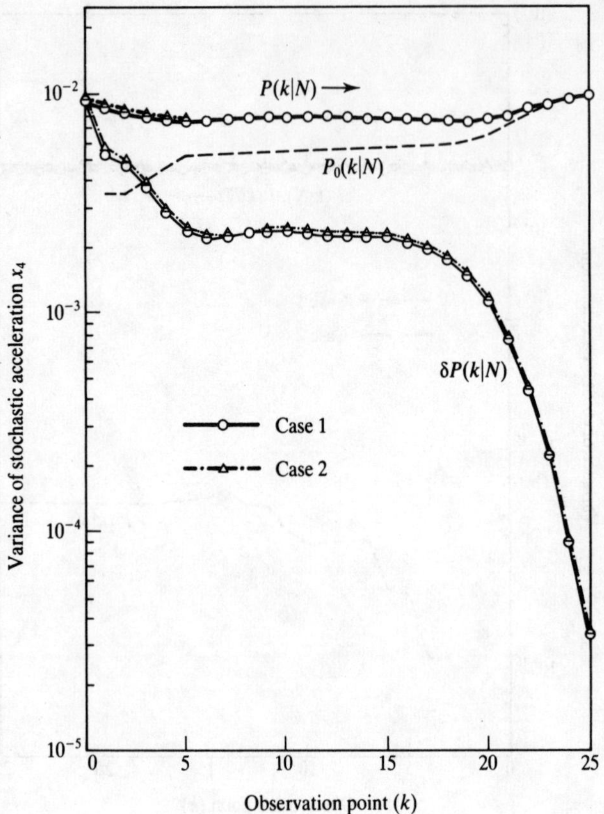

Figure 9.9 Error variance history of stochastic acceleration x_4.

and the time update by

$$S(k) = \Phi^{\mathrm{T}}(k+1, k)\bar{S}(k+1)\Phi(k+1, k) + H^{\mathrm{T}}(k)\bar{R}H(k), \qquad \bar{R} \triangleq R^{-1} \tag{9.77}$$

$$q(k) = \Phi^{\mathrm{T}}(k+1, k)\bar{q}(k+1) + H^{\mathrm{T}}(k)\bar{R}z(k) \tag{9.78}$$

$$y(k) = G^{\mathrm{T}}q(k) \tag{9.79}$$

Note that terms 'measurement update' and 'time update' are simply adopted from the fact that the recursions of information matrices $\bar{S}(k+1)$ and $S(k)$ are identical in form to those of the Kalman filter. As can be observed from the above results, the information matrix recursions (9.74), (9.75) and (9.77) are entirely similar to the matrix Riccati recursions required for the LQG control gain (see Chapter 11), which are *dual* to those required for the Kalman filter gain. Therefore, the $U-D$ filter algorithms with scalar measurements (cf. Appendix 9B) can be applied directly to obtain the gain for the backward information filter or the forward-pass smoother. Thus we naturally note that the system noise must be handled in component-wise fashion. That is, let G be partitioned such that

$$G = [g_1, \ldots, g_i, \ldots, g_p] \tag{9.80}$$

where the g_i are n-dimensional vectors. Then, we have the following $U-D$ factorization of the backward information filter [22].

Theorem 9.3 A U–D factorization backward information filter: measurement update

Suppose that at time $k+1$ the $n \times n$ information matrix $S(k+1)$ is factored such that

$$S(k+1) = U(k+1)D(k+1)U^{\mathrm{T}}(k+1) \tag{9.81}$$

Then the information filter gains $F_1, \ldots, F_i, \ldots, F_p$ (F_i: n-dimensional vector) due to each single-component system noise, the corresponding $U-D$ factors of $\bar{S}(k+1)$:

$$\bar{S}(k+1) = \bar{U}(k+1)\bar{D}(k+1)\bar{U}^{\mathrm{T}}(k+1) \tag{9.82}$$

and $\bar{q}(k+1)$ may be obtained as follows:

For $i = 1, 2, \ldots, p$ recursively cycle through steps (i) and (ii):

(i) Set $A \equiv g_i^{\mathrm{T}}$, $r \equiv 1$, $\tilde{U} \equiv \bar{U}^{(i-1)}(k+1)$, $\tilde{D} \equiv \bar{D}^{(i-1)}(k+1)$ and apply the $U-D$ measurement update of Appendix 9B to obtain the update factors $\bar{U}^{(i)}(k+1)$, $\bar{D}^{(i)}(k+1)$, the gains F_i and the 'innovation variance' $\alpha^{(i)}(k+1)$, where initially $\bar{U}^{(0)}(k+1) \equiv U(k+1)$ and $\bar{D}^{(0)}(k+1) \equiv D(k+1)$.

(ii) Compute

$$\alpha_i^* = 1/\alpha^{(i)}(k+1) \tag{9.83}$$

$$e_i = g_i^{\mathrm{T}}\bar{q}_{i-1} \tag{9.84}$$

$$\bar{q}_i = \bar{q}_{i-1} - F_i e_i \tag{9.85}$$

where initially $\bar{q}_0 = q(k+1)$.

At the conclusion of these recursions, the $U-D$ factors of $\bar{S}(k+1)$ and $\bar{q}(k)$ are given by

$$\bar{U}(k+1) = \bar{U}^{(p)}(k+1), \qquad \bar{D}(k+1) = \bar{D}^{(p)}(k+1),$$
$$\bar{q}(k+1) = \bar{q}_p \tag{9.86}$$

Example 9.5

This example illustrates the measurement update of the $U-D$ factorized backward information filter described in Theorem 9.3. Let

$$S(k+1) = \begin{bmatrix} 1 & 1 \\ 1 & 1 \end{bmatrix}, \qquad G = \begin{bmatrix} 0 \\ 2 \end{bmatrix}$$

as computed in Example 9.3.

From Appendix 9A, the factors of $S(k+1)$ are obtained as

$$d_2 = S(2, 2) = 1$$
$$U(2, 2) = 1$$
$$U(1, 2) = S(1, 2)/d_2 = 1$$
$$S(1, 1) := S(1, 1) - U(1, 2)U(1, 2)d_2$$
$$= 1 - 1 = 0$$
$$U(1, 1) = 1, \qquad d_1 = S(1, 1) = 0$$

Thus

$$U(k+1) = \begin{bmatrix} 1 & 1 \\ 0 & 1 \end{bmatrix}, \qquad D(k+1) = \begin{bmatrix} 0 & 0 \\ 0 & 1 \end{bmatrix}$$

From Appendix 9B.1,

$$f^{\mathrm{T}} = [0 \;\; 2] \begin{bmatrix} 1 & 1 \\ 0 & 1 \end{bmatrix} = [0 \;\; 2]$$
$$v_1 = [0][0] = 0$$
$$v_2 = [1][2] = 2$$
$$\bar{K}_1 = \begin{bmatrix} 0 \\ 0 \end{bmatrix}$$
$$\alpha_1 = 1 + [0][0] = 1$$
$$\hat{d}_1 = [0][1]/1 = 0$$

For $j = 2$, we have

$$\alpha_2 = [1] + [2][2] = 5$$

$$\hat{d}_2 = [1][1]/[5] = 1/5$$

$$\lambda_2 = -[2]/[1] = -2$$

$$\hat{U}_2 = \begin{bmatrix} 1 \\ 1 \end{bmatrix} + [-2]\begin{bmatrix} 0 \\ 0 \end{bmatrix} = \begin{bmatrix} 1 \\ 1 \end{bmatrix}$$

$$\bar{K}_2 = \begin{bmatrix} 0 \\ 0 \end{bmatrix} + [2]\begin{bmatrix} 1 \\ 1 \end{bmatrix} = \begin{bmatrix} 2 \\ 2 \end{bmatrix}$$

$$K = \begin{bmatrix} 2 \\ 2 \end{bmatrix}/[5] = \begin{bmatrix} 2/5 \\ 2/5 \end{bmatrix}$$

so that

$$\bar{U}(k+1) = \begin{bmatrix} 1 & 1 \\ 0 & 1 \end{bmatrix}$$

$$\bar{D}(k+1) = \begin{bmatrix} 0 & 0 \\ 0 & 1/5 \end{bmatrix}, \qquad \alpha_1^* = 1/5, \qquad F_1 = \begin{bmatrix} 2/5 \\ 2/5 \end{bmatrix}$$

Finally, equations (9.84) and (9.85) generate

$$\bar{q}(k+1) = q(k+1) - F_1 G^{\mathrm{T}} q(k+1)$$

Theorem 9.4 A *U–D* factorized backward information filter: time update

Given the $\bar{U}-\bar{D}$ factors of $\bar{S}(k+1)$ at time $k+1$, the $U-D$ factors of $S(k)$ at time k may be obtained by applying the modified weighted Gram–Schmidt (MWGS) time update algorithm of Appendix 9B with W and $\bar{D}$ arrays set to

$$W = [\Phi^{\mathrm{T}}(k+1, k)\bar{U}(k+1) \vdots\, H^{\mathrm{T}}(k)],\ \bar{D} = \operatorname{diag}(\bar{D}(k+1), \bar{R}) \tag{9.87}$$

In addition, the backward information filtered estimate may be computed from equation (9.78) and save the smoother input (9.79).

Example 9.6

Consider the time update of the $U–D$ factorized backward-time information filter given in Theorem 9.4, but using the matrices shown in Examples 9.3 and 9.5:

$$\Phi = \begin{bmatrix} 1 & 2 \\ 0 & 1 \end{bmatrix}, \qquad H = [1 \quad 1], \qquad R = 1$$

$$\bar{U}(k+1) = \begin{bmatrix} 1 & 1 \\ 0 & 1 \end{bmatrix}, \qquad \bar{D}(k+1) = \begin{bmatrix} 0 & 0 \\ 0 & 1/5 \end{bmatrix}$$

First, we compose matrices W and $\bar{D}$

$$W = \left[\begin{bmatrix} 1 & 0 \\ 2 & 1 \end{bmatrix} \begin{bmatrix} 1 & 1 \\ 0 & 1 \end{bmatrix} \vdots \begin{bmatrix} 1 \\ 1 \end{bmatrix}\right] = \begin{bmatrix} 1 & 1 & 1 \\ 2 & 3 & 1 \end{bmatrix}$$

$$\bar{D} = \begin{bmatrix} 0 & 0 & 0 \\ 0 & 1/5 & 0 \\ 0 & 0 & 1 \end{bmatrix}$$

so that

$$w_1^{\mathrm{T}} = [1 \quad 1 \quad 1], \; w_2^{\mathrm{T}} = [2 \quad 3 \quad 1]$$

The first iteration of equations (9B.13)–(9B.15), for $j = 2$, produces

$$\tilde{d}_2 = [2 \quad 3 \quad 1] \begin{bmatrix} 0 & 0 & 0 \\ 0 & 1/5 & 0 \\ 0 & 0 & 1 \end{bmatrix} \begin{bmatrix} 2 \\ 3 \\ 1 \end{bmatrix} = 14/5$$

$$\tilde{U}_{12} = [1 \quad 1 \quad 1] \begin{bmatrix} 0 & 0 & 0 \\ 0 & 1/5 & 0 \\ 0 & 0 & 1 \end{bmatrix} \begin{bmatrix} 2 \\ 3 \\ 1 \end{bmatrix} \Big/ [14/5] = 4/7$$

$$w_1^{(1)} = \begin{bmatrix} 1 \\ 1 \\ 1 \end{bmatrix} - [4/7] \begin{bmatrix} 2 \\ 3 \\ 1 \end{bmatrix} = \begin{bmatrix} -1/7 \\ -5/7 \\ 3/7 \end{bmatrix}$$

$$\tilde{d}_1 = [-1/7 \quad -5/7 \quad 3/7] \begin{bmatrix} 0 & 0 & 0 \\ 0 & 1/5 & 0 \\ 0 & 0 & 1 \end{bmatrix} \begin{bmatrix} -1/7 \\ -5/7 \\ 3/7 \end{bmatrix} = 2/7$$

Thus

$$U(k) = \begin{bmatrix} 1 & 4/7 \\ 0 & 1 \end{bmatrix}, \; D(k) = \begin{bmatrix} 2/7 & 0 \\ 0 & 14/5 \end{bmatrix}$$

Finally, from equations (9.78) and (9.79)

$$q(k) = \begin{bmatrix} 1 & 0 \\ 2 & 1 \end{bmatrix} \bar{q}(k+1) + \begin{bmatrix} 1 \\ 1 \end{bmatrix} z(k)$$

and

$$y(k) = [0 \quad 2]q(k) = 2q_2(k)$$

where $q_2(k)$ denotes the second element of $q(k)$. Note that

$$U(k)D(k)U^{\mathrm{T}}(k) = \begin{bmatrix} 1 & 4/7 \\ 0 & 1 \end{bmatrix} \begin{bmatrix} 2/7 & 0 \\ 0 & 14/5 \end{bmatrix} \begin{bmatrix} 1 & 0 \\ 4/7 & 1 \end{bmatrix}$$

$$= \begin{bmatrix} 6/5 & 8/5 \\ 8/5 & 14/5 \end{bmatrix}$$

which is also consistent with the earlier computations performed in Example 9.3.

Note that in the recursions in Theorem 9.3 the gains $F_i(i = 1, \ldots, p)$ and the quantities α_i^* are assumed to be stored. Using these results, the resultant $U-D$ factored forward-pass fixed-interval smoother is given by the following theorem.

Theorem 9.5 A *U–D* factorized forward-pass fixed-interval smoother: estimates

First, obtain the time update estimate $\bar{x}(k+1|N)$ from equation (9.70). Next, for $i = 1, \ldots, p$, cycle through equations (9.88)–(9.91):

$$\widetilde{h}_i = F_i^{\mathrm{T}}\hat{x}_{i-1} \tag{9.88}$$

$$\beta_i = y_i(k+1)\alpha_i^* \tag{9.89}$$

$$\gamma_i = \beta_i - \widetilde{h}_i \tag{9.90}$$

$$\hat{x}_i = \hat{x}_{i-1} + \gamma_i g_i \tag{9.91}$$

where initially $\hat{x}_0 = \bar{x}(k+1|N)$. Upon completion of this recursion, the desired smoothed estimate is $\hat{x}(k+1|N) \equiv \hat{x}_p$.

Example 9.7

This example illustrates the estimate of the $U-D$ factorized forward-pass fixed-interval smoother given by Theorem 9.5, but using the results of Examples 9.5 and 9.6. First, from equation (9.70), we obtain

$$\bar{x}(k+1|N) = \begin{bmatrix} 1 & 2 \\ 0 & 1 \end{bmatrix} \hat{x}(k|N)$$

Then from equations (9.88)–(9.91), it follows that

$$\widetilde{h}_1 = [2/5 \quad 2/5]\begin{bmatrix} \bar{x}_1(k+1|N) \\ \bar{x}_2(k+1|N) \end{bmatrix}$$
$$= \frac{2}{5}[\bar{x}_1(k+1|N) + \bar{x}_2(k+1|N)]$$
$$\beta_1 = 2q_2(k+1)[1/5] = \tfrac{2}{5}q_2(k+1)$$
$$\gamma_1 = \beta_1 - \widetilde{h}_1 = \tfrac{2}{5}[q_2(k+1) - \bar{x}_1(k+1|N) - \bar{x}_2(k+1|N)]$$
$$\hat{x}(k+1|N) = \begin{bmatrix} \bar{x}_1(k+1|N) \\ \bar{x}_2(k+1|N) \end{bmatrix} + \begin{bmatrix} 0 \\ \tfrac{4}{5}\{q_2(k+1) - \bar{x}_1(k+1|N) - \bar{x}_2(k+1|N)\} \end{bmatrix}$$

where the subscript of $\bar{x}(k+1|N)$ denotes the element number.

Theorem 9.6 A U–D factorized forward-pass fixed-interval smoother: covariance

The computation of the smoothed error covariance $P(k+1|N)$ using the U–D factorization method may be accomplished in two steps as follows:

(i) Given the factorization of $P(k|N)$ at time k such that

$$P(k|N) = U(k|N)D(k|N)U^{\mathrm{T}}(k|N) \tag{9.92}$$

set

$$W = [\Phi(k+1, k)U(k+1)], \ \bar{D} = \operatorname{diag}(D(k|N)) \tag{9.93}$$

and apply the MWGS time update algorithm of Appendix 9B to obtain the factors $\bar{U}(k+1|N)$ and $\bar{D}(k+1|N)$, where

$$\bar{U}(k+1|N)\bar{D}(k+1|N)\bar{U}^{\mathrm{T}}(k+1|N) = \Phi(k+1, k)P(k|N)\Phi^{\mathrm{T}}(k+1, k) \tag{9.94}$$

(ii) For $i = 1, \ldots, p$, recursively evaluate the following:

(a) Set

$$\begin{aligned} W &= [\{I - g_i F_i^{\mathrm{T}}\}U^{(i-1)}(k+1|N)] \\ \bar{D} &= \operatorname{diag}(D^{(i-1)}(k+1|N)) \end{aligned} \tag{9.95}$$

and apply the MWGS time update algorithm of Appendix 9B to obtain the intermediate factors $\widetilde{U}(k+1|N)$ and $\widetilde{D}(k+1|N)$, where $U^{(0)}(k+1|N) = \bar{U}(k+1|N)$ and $D^{(0)}(k+1|N) = \bar{D}(k+1|N)$.

(b) Apply the rank-one update algorithm given in Appendix 9C to compute

$$U^{(i)}(k+1|N)D^{(i)}(k+1|N)U^{(i)^{\mathrm{T}}}(k+1|N)$$
$$= \widetilde{U}(k+1|N)\widetilde{D}(k+1|N)\widetilde{U}^{\mathrm{T}}(k+1|N) + \alpha_i^* g_i g_i^{\mathrm{T}} \tag{9.96}$$

Finally, the desired factors of $P(k+1|N)$ are $U(k+1|N) \equiv U^{(p)}(k+1|N)$ and $D(k+1|N) \equiv D^{(p)}(k+1|N)$.

Example 9.8

Let the smoothed error covariance $P(k|N)$ be given by

$$P(k|N) = \begin{bmatrix} 19/29 & -8/29 \\ -8/29 & 11/29 \end{bmatrix}$$

but using matrices Φ and vectors g_1, F_1 as used (or derived) in Examples 9.5 and 9.6.

The U–D factors of $P(k|N)$ can be computed from Appendix 9A as follows:

$$U(k|N) = \begin{bmatrix} 1 & -8/11 \\ 0 & 1 \end{bmatrix}, \qquad D(k|N) = \begin{bmatrix} 5/11 & 0 \\ 0 & 11/29 \end{bmatrix}$$

so that

$$W = \left[\begin{bmatrix} 1 & 2 \\ 0 & 1 \end{bmatrix}\begin{bmatrix} 1 & -8/11 \\ 0 & 1 \end{bmatrix}\right] = \begin{bmatrix} 1 & 14/11 \\ 0 & 1 \end{bmatrix}$$

$$\bar{D} = \begin{bmatrix} 5/11 & 0 \\ 0 & 11/29 \end{bmatrix}$$

From Appendix 9B.2, $\bar{U}(k+1|N)$ and $\bar{D}(k+1|N)$ have been generated as

$$\bar{U}(k+1|N) = \begin{bmatrix} 1 & 14/11 \\ 0 & 1 \end{bmatrix}, \qquad \bar{D}(k+1|N) = \begin{bmatrix} 5/11 & 0 \\ 0 & 11/29 \end{bmatrix}$$

Setting

$$W = \left[\left\{\begin{bmatrix} 1 & 0 \\ 0 & 1 \end{bmatrix} - \begin{bmatrix} 0 \\ 2 \end{bmatrix} [2/5 \;\; 2/5]\right\}\begin{bmatrix} 1 & 14/11 \\ 0 & 1 \end{bmatrix}\right]$$

$$= \begin{bmatrix} 1 & 14/11 \\ -4/5 & -9/11 \end{bmatrix}$$

$$\bar{D} = \begin{bmatrix} 5/11 & 0 \\ 0 & 11/29 \end{bmatrix}$$

and applying again Appendix 9B.2 to W and $\bar{D}$ above, we have the intermediate factors $\widetilde{U}(k+1|N)$ and $\widetilde{D}(k+1|N)$ such that

$$\widetilde{U}(k+1|N) = \begin{bmatrix} 1 & -110/79 \\ 0 & 1 \end{bmatrix}$$

$$\widetilde{D}(k+1|N) = \begin{bmatrix} 1/79 & 0 \\ 0 & 79/145 \end{bmatrix}$$

Furthermore, it follows from Appendix 9C that

$$\alpha_2 = [1/5][2] = 2/5$$

$$d_2 = [79/145] + [2/5][2] = 39/29$$

$$v_2 = [2/5]/[39/29] = 58/195$$

$$\beta_2 = [79/145]/[39/29] = 79/195 > 0.1$$

$$\lambda_1 := [0] - [2][-110/79] = 220/79$$

$$u_{12} = [-110/79] + [220/79][58/195] = -22/39$$

$$c_1 = [79/195][1/5] = 79/975$$

$$d_1 = [1/79] + [79/975][220/79]^2 = 25/39$$

Thus, we have

$$U(k+1|N) = \begin{bmatrix} 1 & -22/39 \\ 0 & 1 \end{bmatrix}$$

$$D(k+1|N) = \begin{bmatrix} 25/39 & 0 \\ 0 & 39/29 \end{bmatrix}$$

Note that

$$U(k+1|N)D(k+1|N)U^{\mathrm{T}}(k+1|N) = \begin{bmatrix} 31/29 & -22/29 \\ -22/29 & 39/29 \end{bmatrix}$$

as found in the previous Example 9.3.

It is observed that the only large matrix inversion appears in the initial instant for the smoothed estimate and the associated error covariance. Practically, if the initial information $P(0)$ is diagonal, then $P(0|N) = U(0|N)P(0|N)U^{\mathrm{T}}(0|N)$ can start with

$$U(0|N) = L^{-\mathrm{T}}(0|N),\ D(0|N) = \bar{D}^{-1}(0|N) \tag{9.97}$$

where $L(0|N)$ and $\bar{D}(0|N)$ are the elements of square-root-free Cholesky decomposition, i.e. L is lower triangular with unit diagonal elements and $\bar{D}$ is diagonal. Thus, the initial condition for equation (9.70) is given by

$$\hat{x}(0|N) = U(0|N)D(0|N)U^{\mathrm{T}}(0|N)[q(0) + P^{-1}(0)\hat{x}(0)] \tag{9.98}$$

Compared with the Bierman rank-two $U-D$ smoother [20] based on the RTS smoother recursion, the present smoother contains no computationally burdensome matrix inversion except for the initial time, whereas the Bierman smoother requires two inversions in the computations of the transition matrix and the smoother gain (cf. Appendix 9D). Consequently, our $U-D$ smoother may be applied to broader systems which cover time-varying systems.

We now compare the computational requirements of the present algorithm and the Bierman rank-two $U-D$ smoother quoted above. Comparative results for the one measurement update are presented in Table 9.4, where the arithmetic operation count for the present algorithm is given by Appendix 9E, and that due to Bierman is obtained from Bierman [19, p. 107].

Table 9.5 shows the comparative results for the filter time update. The operation count associated with the Watanabe rank-two $U-D$ smoother is given by Appendix 9E, in which the operations involved in computing the smoother input are included. Bierman's algorithm is classified into three types, depending on whether the direct, $L^{\mathrm{T}}U^{\mathrm{T}}$ factorization or truncated $L^{\mathrm{T}}U^{\mathrm{T}}$ factorization method is used to compute the smoother gain, where those operation counts can be found in Appendix 9E.

Table 9.4 Operation counts for filtered estimate only after m observations

Arithmetic operations	*Watanabe rank-two U–D smoother*	*Bierman rank-two U–D smoother*
Addition	$(1.5n^2 + 1.5n)p$	$(1.5n^2 + 1.5n)m$
Multiplication	$(1.5n^2 + 5.5n)p$	$(1.5n^2 + 5.5n)m$
Division	$(n + 1)p$	nm

Table 9.5 Operation counts for one filter time update

Arithmetic operations	*Watanabe rank-two U–D smoother*	*Bierman rank-two U–D smoother*		
		Direct	$L^{T}U^{T}$	$L^{T}U^{T}(b_i = e_{k_i})$†
Addition	$1.5n^3 + 0.5n^2 + (n^2 + n)m + np$	$1.5n^3 + 0.5n^2 + (\frac{1}{6}n^3 + \frac{3}{2}n^2 + \frac{1}{3}n)p$	$1.5n^3 + 0.5n^2 + 2n^2p$	$1.5n^3 + 0.5n^2 + (1.5n^2 + 0.5n)p + \sum_{i=1}^{p} 0.5k_i(k_i - 1)$
Multiplication	$1.5n^3 + 2n^2 - 0.5n + (n^2 + 2n)m + np$	$1.5n^3 + 2n^2 - 0.5n + (\frac{1}{6}n^3 + \frac{5}{2}n^2 + \frac{7}{3}n)p$	$1.5n^3 + 2n^2 - 0.5n + (2n^2 + 2n)p$	$1.5n^3 + 2n^2 - 0.5n + (1.5n^2 + 2.5n)p + \sum_{i=1}^{p} 0.5k_i(k_i - 1)$
Division	$n - 1$	$n - 1 + (3n - 1)p$	$n - 1 + (3n - 1)p$	$n - 1 + (2n - 1)p + \sum_{i=1}^{p} k_i$

† The e_{k_i} denote the ith column which has only a unit value at the k_ith row.

Table 9.6 is composed of computational requirements corresponding to one-step smoothing for each of the two algorithms. Their arithmetic operation counts are given by Appendix 9E. Table 9.7 shows the computational requirements for the smoothed covariance calculations, whose operation counts can also be found in Appendix 9E. It can be seen from Tables 9.6 and 9.7 that the present algorithm is comparable in the computation of smoothed estimate and associated error covariance to Bierman's method. Note here that the number of operations of step (ii) in Theorem 9.6 is overestimated. That is, there is an alternative implementation of step (ii) in Theorem 9.6, which is based on using the U–D measurement update and rank-one update algorithms (Kuga [23]; see also Exercise 9.12).

Table 9.8 presents comparative numerical results for several specific examples. The operation totals represent one full filter step required for the computation of smoothed estimates, which are obtained by summing the entries in Tables 9.4 and 9.5 for each formulation. Examples 1 and 2 represent typical low-order and higher-order examples. Cases a and b illustrate the case of $m = p$; cases c and d represent the case of $n = p$, and cases e and f show the case of $n = m$. On the basis of computation speed, the direct formulation of Bierman's method is less attractive in all cases. Our method is the fastest in all cases; in particular, it is 50% faster than

Table 9.6 Operation counts for one-step smoothed estimate

Arithmetic operations	*Watanabe rank-two U–D smoother*	*Bierman rank-two U–D smoother*
Addition	$n^2 - n + 2np$	$n^2 - n + 3np$
Multiplication	$n^2 + 2np$	$n^2 + 2np$
Division	0	0

Table 9.7 Operation counts for one-step smoothed covariance

Arithmetic operations	*Watanabe rank-two U–D smoother*	*Bierman rank-two U–D smoother*
Addition	$1.5n^3 + 0.5n$ $+ (1.5n^3 + 0.5n^2 + 2n)p$	$1.5n^3 + 0.5n$ $+(1.5n^3 + 0.5n^2 + 2n)p$
Multiplication	$1.5n^3 + 2n^2 - 1.5n$ $+ (1.5n^3 + 3n^2 + 1.5n)p$	$1.5n^3 + 2n^2 - 1.5n$ $+ (1.5n^3 + 3n^2 + 2.5n)p$
Division	$n - 1 + (3n - 3)p$	$n - 1 + (3n - 3)p$

Table 9.8 Arithmetic operations for one full filter step required for the computation of smoothed estimates

						Bierman rank-two U–D smoother		
Case	n	m	p	$k_1, \ldots, k_p$	*Watanabe rank-two U–D smoother*	*Direct*	$L^T U^T$	$L^T U^T (b_i = e_{k_i})$
1a	10	2	2	4, 5	4516	5592	4912	4753
1b	10	6	6	2, 3, 5, . . ., 8	7040	10268	8228	7815
1c	10	2	10	1, . . ., 10	7724	11904	8504	7889
1d	10	6	10	1, . . ., 10	8644	13424	10024	9409
1e	10	10	2	4, 5	6356	8632	7952	7793
1f	10	10	6	2, 3, 5, . . ., 8	7960	11788	9748	9335
2a	30	4	4	7, . . ., 10	102828	146100	110020	106714
2b	30	6	6	7, . . ., 10, 12, 14	112610	177518	123398	118632
2c	30	4	30	1, . . ., 30	180854	478094	207494	189949
2d	30	6	30	1, . . ., 30	184634	483974	213374	195829
2e	30	30	4	7, . . ., 10	151968	222540	186460	183154
2f	30	30	6	7, . . ., 10, 12, 14	157970	248078	193958	189192

Table 9.9 Storage requirements

Smoother formulation	*Variable storage*
RTS smoother	$0.5(n^2 + 3n)$
Bierman rank-two smoother	$(n + 1)p + n$
Watanabe rank-two smoother	$(n + 2)p$

the direct method; and 20% faster than the two $L^T U^T$ methods, where $n \simeq m$ and $p \ll n$.

Next we compare storage requirements. For the smoothing implementation, we must store some estimates and gains (or estimation error covariances) which are generated in the filtering stage. Three algorithms are considered in our comparison: the RTS smoother (cf. Appendix 9D), the Bierman rank-two smoother and the Watanabe rank-two smoother. Comparisons are made for the storage variables associated with one measurement update or one time update, in which no storage is required for $\Phi(k+1, k)$. The RTS smoother is at least required for the storage of the filtered estimates $x_{j|j}$ (n-vector) and the associated estimation error covariances $P_{j|j}$ ($n(n+1)/2$ elements). For the Bierman rank-two $U-D$ smoother, it is necessary to store the one-step predicted estimates $x_{j+1|j}$ (n-vector), (gain) vectors v_i (n-vector) and scalar multipliers λ_i, $i = 1, \ldots, p$, as indicated by Bierman [20]. On the other hand, the Watanabe rank-two $U-D$ smoother requires storage for the gains F_i (n-vector), the scalar factors α_i^* and the smoother inputs $y(k)$ (p-vector), $i = 1, \ldots, p$. These comparisons are summarized in Table 9.9. When $p \ll n$, it is obvious from this table that our rank-two $U-D$ smoother algorithm reduces storage requirements over those required by both the RTS and Bierman's formulations.

9.5 THE STEADY-STATE SOLUTION

We now seek the steady-state solution to the algorithm given by Corollary 9.1, but for time-invariant systems. That is, $\Phi(k+1, k) = \Phi$, $B(k) = B$, $H(k) = H$, $Q(k) = Q$ and $R(k) = R$, where Φ is assumed to be nonsingular, that is Φ does not have zero eigenvalues.

Using equation (9.58), it is easy to see that

$$\begin{aligned} GG^{\mathrm{T}}\Phi^{-\mathrm{T}}(S(k) - C^{\mathrm{T}}C) \\ &= GG^{\mathrm{T}}\Phi^{-\mathrm{T}}\{\Phi^{\mathrm{T}}[I - S(k+1)G\Delta(k)G^{\mathrm{T}}]S(k+1)\Phi\} \\ &= G[I - G^{\mathrm{T}}S(k+1)G\Delta(k)]G^{\mathrm{T}}S(k+1)\Phi \end{aligned} \tag{9.99}$$

where $C^{\mathrm{T}}C \triangleq H^{\mathrm{T}}R^{-1}H$. Applying equation (5A.1), given in the discussion of matrix inversion in Appendix 5A, to equation (9.99) gives the identity

$$GG^{\mathrm{T}}\Phi^{\mathrm{T}}(S(k) - C^{\mathrm{T}}C) = G[I + G^{\mathrm{T}}S(k+1)]^{-1}G^{\mathrm{T}}S(k+1)\Phi \tag{9.100}$$

Therefore, equation (9.59) can be rewritten as follows:

$$\begin{aligned} q(k) = \{\Phi - GG^{\mathrm{T}}\Phi^{-\mathrm{T}}[S(k) - C^{\mathrm{T}}C]\}^{\mathrm{T}}q(k+1) \\ + H^{\mathrm{T}}R^{-1}z(k), \qquad q(N) = H^{\mathrm{T}}R^{-1}z(N) \end{aligned} \tag{9.101}$$

If N is sufficiently large, then the steady-state solution S for the information matrix in equation (9.58) will satisfy the following algebraic Riccati equation (ARE), obtained by setting $S(k+1) = S(k) = S$:

$$S - \Phi^{\mathrm{T}}S\Phi + \Phi^{\mathrm{T}}SG(I + G^{\mathrm{T}}SG)^{-1}G^{\mathrm{T}}S\Phi - C^{\mathrm{T}}C = 0 \tag{9.102}$$

Recalling the discussion on the ARE in Chapter 2, it is easy to see that equation (9.102) is closely related to the following $2n \times 2n$ symplectic matrix†

$$M = \begin{bmatrix} \Phi + GG^{\mathrm{T}}\Phi^{-\mathrm{T}}C^{\mathrm{T}}C & -GG^{\mathrm{T}}\Phi^{-\mathrm{T}} \\ -\Phi^{-\mathrm{T}}C^{\mathrm{T}}C & \Phi^{-\mathrm{T}} \end{bmatrix} \tag{9.103}$$

It is assumed for simplicity that the eigenvalues of M are distinct. Then the eigenvalues of M can be arranged in a diagonal matrix

$$\Lambda = \begin{bmatrix} \Lambda_1^{-1} & 0 \\ 0 & \Lambda_1 \end{bmatrix} \tag{9.104}$$

where Λ_1 is a diagonal matrix of the n eigenvalues outside the unit circle. Therefore, using the right eigenvector ($2n \times 2n$) matrix T (see also Appendix A) corresponding to equation (9.104) gives

$$MT = T\Lambda \tag{9.105}$$

where T is partitioned into four $n \times n$ submatrices

† Note that, in this section, the backward Hamiltonian system matrix M^{-1} is considered, because the Riccati difference equation for $S(k)$ proceeds backwards in time.

$$T = \begin{bmatrix} X & U \\ Y & V \end{bmatrix} \tag{9.106}$$

The following lemmas are immediately obtained ([24]; cf. Section 2.4).

Lemma 9.5

A positive semidefinite solution of equation (9.102) takes the form

$$S = YX^{-1} \tag{9.107}$$

assuming X is nonsingular.

Lemma 9.6

Let the closed-loop system for the backward information filter be

$$\tilde{q}(k) = \tilde{\Phi}^{\mathrm{T}}\tilde{q}(k+1) \tag{9.108}$$

where $\tilde{\Phi}$ denotes the closed-loop transition matrix defined by

$$\tilde{\Phi} \triangleq (I - G\Delta G^{\mathrm{T}}S)\Phi \tag{9.109}$$

$$\equiv \Phi - GG^{\mathrm{T}}\Phi^{-\mathrm{T}}(S - C^{\mathrm{T}}C) \tag{9.110}$$

$$\Delta = [I + G^{\mathrm{T}}SG]^{-1} \tag{9.111}$$

There exists a stabilizing solution† $S \geqslant 0$ to (9.102) if and only if (Φ, G) is stabilizable and $|\lambda_i(M)| \neq 1$, $i = 1, \ldots, 2n$.

Lemma 9.7

The stabilizability of (Φ, G) and detectability of (C, Φ) are necessary and sufficient for equation (9.102) to have a unique positive semidefinite solution which asymptotically stabilizes the closed-loop system given in equation (9.108).

It is useful for future developments to note that the decomposed expression of equation (9.105) can be described, using equations

† See Section 2.4 for the definition of stabilizing solution, as well as that of strong solution [25].

(9.103), (9.104) and (9.106), as follows:

$$[\Phi + GG^{\mathrm{T}}\Phi^{-\mathrm{T}}C^{\mathrm{T}}C]X - GG^{\mathrm{T}}\Phi^{-\mathrm{T}}Y = X\Lambda_1^{-1} \tag{9.112}$$

$$\Phi U + GG^{\mathrm{T}}\Phi^{-\mathrm{T}}[C^{\mathrm{T}}CU - V] = U\Lambda_1 \tag{9.113}$$

$$-\Phi^{-\mathrm{T}}C^{\mathrm{T}}CX + \Phi^{-\mathrm{T}}Y = Y\Lambda_1^{-1} \tag{9.114}$$

$$-\Phi^{-\mathrm{T}}C^{\mathrm{T}}CU + \Phi^{-\mathrm{T}}V = V\Lambda_1 \tag{9.115}$$

Substituting $Y = SX$ into equation (9.112) results in

$$[\Phi - GG^{\mathrm{T}}\Phi^{-\mathrm{T}}(S - C^{\mathrm{T}}C)]X = X\Lambda_1^{-1} \tag{9.116}$$

Hence, it is easy to see that the closed-loop transition matrix can be expressed by

$$\widetilde{\Phi} = X\Lambda_1^{-1}X^{-1} \tag{9.117}$$

Then it is assumed that the smoothed error covariance matrix of expression (9.53) reaches the steady-state solution, i.e. $P(k+1|N) = P(k|N) = P$. That is, expression (9.53) is subject to the following algebraic Lyapunov equation (ALE):

$$P - \widetilde{\Phi}P\widetilde{\Phi}^{\mathrm{T}} - G[I + G^{\mathrm{T}}SG]^{-1}G^{\mathrm{T}} = 0 \tag{9.118}$$

When the $2n \times 2n$ symplectic matrix

$$\hat{M} = \begin{bmatrix} \widetilde{\Phi}^{\mathrm{T}} & 0 \\ -\widetilde{\Phi}^{-1}G[I + G^{\mathrm{T}}SG]^{-1}G^{\mathrm{T}} & \widetilde{\Phi}^{-1} \end{bmatrix} \tag{9.119}$$

associated with equation (9.118) is introduced, we can see that all eigenvalues of $\hat{M}$ are identical with Λ, because $\widetilde{\Phi}$ can be represented by equation (9.117). Let the right eigenvectors matrix associated with equation (9.119) be decomposed into four $n \times n$ submatrices

$$\hat{T} = \begin{bmatrix} \hat{X} & \hat{U} \\ \hat{Y} & \hat{V} \end{bmatrix} \tag{9.120}$$

Then, we obtain

$$\hat{M}\hat{T} = \hat{T}\Lambda \tag{9.121}$$

Theorem 9.7

A solution to (9.118) is given by

$$P = \hat{Y}\hat{X}^{-1} \tag{9.122}$$

as long as $\hat{X}$ is nonsingular.

Proof

See Watanabe [7]. □

In the smoothing problem, as well as the filtering problem, P is a symmetric positive semidefinite matrix. We are particularly interested in the question of existence and uniqueness of such a solution that provides a steady-state fixed-interval smoother having roots inside the unit circle.

Theorem 9.8

$|\lambda_i(M)| \neq 1$, $i = 1, \ldots, 2n$ and stabilizability of (Φ, G) are necessary and sufficient for equation (9.118) to have a unique positive semidefinite stabilizing solution.

Proof

When Lemma 9.7 is applied to the ALE in equation (9.118), the necessary and sufficient condition for the existence of the stabilizing solution of equation (9.118) is that $(\widetilde{\Phi}^{\mathrm{T}}, 0)$ is stabilizable and $(\widetilde{G}^{\mathrm{T}}, \widetilde{\Phi}^{\mathrm{T}})$ is detectable, where $\widetilde{G} \triangleq G[I + G^{\mathrm{T}}SG]^{-1/2}$. In the following we prove that the necessary and sufficient conditions for the above two conditions are that $|\lambda_i(M)| \neq 1$, $i = 1, \ldots, 2n$ and the pair (Φ, G) is stabilizable.

To take sufficiency first: it is evident from Lemma 9.6 that if $|\lambda_i(M)| \neq 1$, $i = 1, \ldots, 2n$ and (Φ, G) is stabilizable, then equation (9.102) has a positive semidefinite solution S which stabilizes the closed-loop system. Therefore, $|\lambda_i(\widetilde{\Phi}^{\mathrm{T}})| < 1$, $i = 1, \ldots, n$. Hence, $(\widetilde{\Phi}^{\mathrm{T}}, 0)$ is stabilizable and $(\widetilde{G}^{\mathrm{T}}, \widetilde{\Phi}^{\mathrm{T}})$ is detectable, if $|\lambda_i(M)| \neq 1$, $i = 1, \ldots, 2n$ and (Φ, G) is stabilizable.

Moving on to necessity: it is known [24] that $|\lambda_i(\hat{M})| \neq 1$, $i = 1, \ldots, 2n$, if $(\widetilde{\Phi}, 0)$ is stabilizable and $(\widetilde{G}^{\mathrm{T}}, \widetilde{\Phi}^{\mathrm{T}})$ is detectable. It is also found that $|\lambda_i(M)| \neq 1$, $i = 1, \ldots, 2n$, since $\lambda(\hat{M}) \equiv \lambda(M)$ from equations (9.117) and (9.119). We must further show that when (Φ, G) is unstabilizable $(\widetilde{G}^{\mathrm{T}}, \widetilde{\Phi}^{\mathrm{T}})$ is undetectable, i.e. $(\widetilde{\Phi}, \widetilde{G})$ is unstabilizable. For the case when (Φ, G) is unstabilizable, there exists an eigenvalue λ and a row vector $y \neq 0$ such that (see Appendix C)

$$y\Phi = \lambda y, \qquad |\lambda| \geqslant 1 \tag{9.123}$$

$$yG = 0 \tag{9.124}$$

Combining equations (9.109) and (9.123) gives

$$y\widetilde{\Phi} = y(\Phi - G\Delta G^{\mathrm{T}} S\Phi) \tag{9.125}$$

$$= \lambda y, \qquad |\lambda| \geq 1. \tag{9.126}$$

From equation (9.124), it follows that

$$y\widetilde{G} = yG[I + G^{\mathrm{T}}SG]^{-1/2} = 0 \tag{9.127}$$

Therefore, equations (9.126) and (9.127) together establish that $(\widetilde{\Phi}, \widetilde{G})$ is unstabilizable. From these results, we can see that $|\lambda_i(M)| \neq 1$, $i = 1, \ldots, 2n$ and (Φ, G) is stabilizable, if (Φ, G) is stabilizable and $(\widetilde{G}^{\mathrm{T}}, \widetilde{\Phi}^{\mathrm{T}})$ is detectable. □

It should be noted that the above condition parallels one derived in continuous-time systems [26]. Note also that Nishimura [27, 28] has obtained a similar result for the RTS fixed-interval smoother in continuous-time systems.

We obtain the following theorem on the one-step predicted error covariance of the Kalman filter.

Theorem 9.9

The one-step predicted error covariance matrix corresponding to the Kalman filter is given by

$$P_{\mathrm{f}} = -UV^{-1} \tag{9.128}$$

if V is nonsingular.

Proof

The theorem can be proved by utilizing the negative semidefinite solution $S_{\mathrm{n}} = VU^{-1}$ of the ARE (9.102) [7], or by directly using equations (9.113) and (9.115). □

Note that the detectability of (C, Φ) is necessary for equation (9.102) to assure the existence of a stabilizing solution $P_{\mathrm{f}} \geq 0$. The following theorem can be derived by using four eigenvector matrices X, Y, U and V.

Theorem 9.10

The solution P of the ALE associated with the forward-pass fixed-interval smoother is given by

$$P = \hat{Y}\hat{X}^{-1} \tag{9.129}$$

where

$$\hat{X} = X^{-\mathrm{T}}(Y^{\mathrm{T}}U - X^{\mathrm{T}}V) \tag{9.130a}$$

$$\equiv SU - V \tag{9.130b}$$

$$\hat{Y} = U \tag{9.131}$$

Proof

Substituting equations (9.130b) and (9.131) into $\hat{X}$ and $\hat{Y}$ of

$$\hat{M}\begin{bmatrix}\hat{X}\\ \hat{Y}\end{bmatrix} = \begin{bmatrix}\hat{X}\\ \hat{Y}\end{bmatrix}\Lambda_1^{-1} \tag{9.132}$$

we have

$$\tilde{\Phi}^{\mathrm{T}}(SU - V) = (SU - V)\Lambda_1^{-1} \tag{9.133}$$

$$-\tilde{\Phi}^{-1}G[I + G^{\mathrm{T}}SG]^{-1}G^{\mathrm{T}}(SU - V) + \tilde{\Phi}^{-1}U = U\Lambda_1^{-1} \tag{9.134}$$

We can prove these two relations by invoking the matrix inversion lemma given in equation (5A.1) and equations (9.112)–(9.115). However, the detailed proof is lengthy and we refer the reader to Watanabe [7]. □

Example 9.9

This problem concerns the estimation of the position and velocity of an aircraft or similar vehicle with random acceleration perturbing its motion from a straight line [29]. We desire to solve equation (9.118) when

$$\Phi = \begin{bmatrix}1 & \Delta t\\ 0 & 1\end{bmatrix}, \qquad B = \begin{bmatrix}(\Delta t)^2/2\\ \Delta t\end{bmatrix}, \qquad H = [1 \quad 0]$$

$$Q = \sigma_{\mathrm{a}}^2, \qquad R = \sigma_x^2$$

where it is assumed that $\Delta t = 0.154$ s, $\sigma_{\mathrm{a}} = 10\ \mathrm{m\,s^{-2}}$, and $\sigma_x = 1$ m.

The eigenvalues of Φ are $\lambda_1(\Phi) = \lambda_2(\Phi) = 1$, but (Φ, G) is completely controllable and (C, Φ) is completely observable. Therefore, they are also stabilizable and detectable. Solving for the symplectic matrix M, we find the four eigenvalues of the problem:

$$\lambda_1(M) = 1.3252 + j0.47785$$

$$\lambda_2(M) = 1.3252 - j0.47785$$

$$\lambda_3(M) = 0.66778 + j0.24080$$

$$\lambda_4(M) = 0.66778 - j0.24080$$

The eigenvectors corresponding to these eigenvalues reduce to complex vectors. To overcome the complex arithmetic, we can introduce a technique similar to that discussed in Fath [30].

Consider the complex eigenvalue $\lambda_1 = a + jb$ and the corresponding complex eigenvector u given as $c + jd$. Then

$$M[u \quad u^*] = [u \quad u^*] \operatorname{diag}(\lambda_1, \lambda_1^*)$$

where the asterisk denotes the complex conjugate and $[u \quad u^*]$ is the $2n \times 2$ matrix with u and u^* as columns. By noting that

$$\begin{bmatrix} a & -b \\ b & a \end{bmatrix} = N^{-1} \operatorname{diag}(\lambda_1, \lambda_1^*) N$$

where

$$N = \frac{1}{2} \begin{bmatrix} 1 & j \\ 1 & -j \end{bmatrix}$$

the desired columns of $T := TN$ can be given as $[u \quad u^*]N$. This means that the first column is simply c, the second $-d$.

Following the above technique, the matrices X, Y, U and V become

$$X = \begin{bmatrix} -0.22690 & 0.22027 \\ 1.0000 & 0.0 \end{bmatrix}$$

$$Y = \begin{bmatrix} -0.13270 & 0.75920 \\ 0.14734 & 0.14303 \end{bmatrix}$$

$$U = \begin{bmatrix} 0.22690 & 0.22027 \\ 1.000 & 0.0 \end{bmatrix}$$

$$V = \begin{bmatrix} 0.094199 & -0.53893 \\ -0.14734 & 0.14303 \end{bmatrix}$$

Then two solutions of equation (9.102) are given by

$$S = YX^{-1} = \begin{bmatrix} 3.4467 & 0.64935 \\ 0.64935 & 0.29467 \end{bmatrix} > 0$$

$$S_n = VU^{-1} = \begin{bmatrix} -2.4467 & 0.64935 \\ 0.64935 & -0.29467 \end{bmatrix} < 0$$

Hence, stabilizing solutions P_f and P are obtained by

$$P_f = -UV^{-1} = \begin{bmatrix} 0.98447 & 2.1694 \\ 2.1694 & 8.1742 \end{bmatrix}$$

$$P = U(SU - V)^{-1} = \begin{bmatrix} 0.16968 & -0.98532 \times 10^{-15} \\ 0.31919 \times 10^{-15} & 1.6968 \end{bmatrix}$$

Note that this solution is not a symmetric matrix, because of the numerical errors of the algorithm, but the analytic solution is expected to be a symmetric matrix with zero off-diagonal elements.

Example 9.10

Consider the lateral dynamics of a VSTOL aircraft [31]. We want to solve equation (9.118) when

$$\Phi = \begin{bmatrix} 0.9949 & 0.0 & -0.1 & 0.0238 \\ -0.0737 & 0.08665 & 0.0369 & 0.001 \\ 0.001 & 0.0107 & 0.9668 & 0.0 \\ 0.0 & 0.1 & 0.0 & 0.0 \end{bmatrix}$$

$$B = \begin{bmatrix} 1 & 0 & 0 \\ 0 & 1 & 0 \\ 0 & 0 & 1 \\ 0 & 0 & 0 \end{bmatrix}, \qquad H = \begin{bmatrix} 0 & 1 & 0 & 0 \\ 0 & 0 & 1 & 0 \\ 0 & 0 & 0 & 1 \end{bmatrix}$$

$$Q = \text{diag}\,(0.01, 0.008, 0.01), \qquad R = \text{diag}\,(0.0001, 0.0001, 0.0003)$$

$$x^T(k) = [\beta(k) \quad p(k) \quad r(k) \quad \phi(k)]$$

where β is the angle of sideslip, p is the roll velocity, r is the yaw velocity and ϕ is the roll angle. We find the eigenvalues of Φ, $\lambda_1(\Phi) = 0.089579$, $\lambda_2(\Phi) = 0.99422$, $\lambda_3(\Phi) = -0.003085$, and $\lambda_4(\Phi) =$

0.96764. Therefore, (Φ, G) is stabilizable and (C, Φ) is detectable. The eigenvalues of M are

$$\lambda_1(M) = 0.92097, \qquad \lambda_5(M) = -0.72096 \times 10^3$$

$$\lambda_2(M) = -0.001387, \qquad \lambda_6(M) = 0.40808 \times 10^3$$

$$\lambda_3(M) = 0.0024505, \qquad \lambda_7(M) = 0.10537 \times 10^3$$

$$\lambda_4(M) = 0.0094907, \qquad \lambda_8(M) = 1.0858$$

The real eigenvector matrices corresponding to these values are

$$X = \begin{bmatrix} 0.10101 & 0.71619 \times 10^{-5} & 0.72046 \times 10^{-5} & -0.89247 \times 10^{-5} \\ -0.91879 \times 10^{-4} & 0.41611 \times 10^{-5} & -0.73516 \times 10^{-5} & -0.44324 \times 10^{-5} \\ 0.10985 \times 10^{-4} & -0.46676 \times 10^{-7} & 0.99195 \times 10^{-7} & -0.99081 \times 10^{-4} \\ -0.99764 \times 10^{-5} & -0.30000 \times 10^{-3} & -0.30000 \times 10^{-3} & -0.46702 \times 10^{-4} \end{bmatrix}$$

$$Y = \begin{bmatrix} 0.81077 & 0.42622 \times 10^{-5} & 0.13359 \times 10^{-4} & 0.21919 \times 10^{-4} \\ -1.0000 & 0.041745 & -0.073776 & -0.044610 \\ 0.010916 & -0.46827 \times 10^{-3} & 0.98762 \times 10^{-3} & -1.0000 \\ -0.016404 & -1.0000 & -1.0000 & -0.15567 \end{bmatrix}$$

$$U = \begin{bmatrix} -0.37189 \times 10^{-3} & 0.66324 \times 10^{-3} & -0.82799 \times 10^{-5} & -0.11908 \\ -0.0040257 & 0.0071240 & 0.12692 \times 10^{-3} & 0.90013 \times 10^{-4} \\ 0.13402 \times 10^{-3} & -0.36568 \times 10^{-3} & 0.010093 & 0.68735 \times 10^{-5} \\ 0.55838 \times 10^{-6} & 0.17457 \times 10^{-5} & 0.12046 \times 10^{-6} & 0.82899 \times 10^{-5} \end{bmatrix}$$

$$V = \begin{bmatrix} 0.037242 & -0.066153 & -0.13769 \times 10^{-3} & 0.99700 \\ 0.50326 & -0.89033 & -0.015410 & 1.0000 \\ -0.013414 & 0.036500 & -1.0000 & -0.010954 \\ -1.0000 & -1.0000 & -0.0015674 & 0.054484 \end{bmatrix}$$

From these results, two (symmetric) solutions $S > 0$ and $S_n < 0$ are obtained (we show only the upper half of each matrix):

$$S = \begin{bmatrix} 8.0257 & -0.77287 & -0.98820 & 0.16682 \\ & 0.10034 \times 10^5 & 1.4176 & 0.011045 \\ & & 0.10093 \times 10^5 & -0.013339 \\ & & & 0.33333 \times 10^4 \end{bmatrix}$$

$$S_n = \begin{bmatrix} -8.3839 & -8.4836 & 0.087011 & -70.853 \\ & -0.12421 \times 10^3 & 0.026914 & 0.11487 \times 10^3 \\ & & -99.083 & 10.426 \\ & & & -0.10125 \times 10^7 \end{bmatrix}$$

Finally, two (symmetric) stabilizing solutions for the Kalman filter and the forward-pass fixed-interval smoother are given by

$$P_f = \begin{bmatrix} 0.12823 & -0.0087674 & 0.10918 \times 10^{-3} & -0.99675 \times 10^{-5} \\ & 0.0086510 & -0.51816 \times 10^{-5} & 0.15950 \times 10^{-5} \\ & & 0.010093 & 0.95702 \times 10^{-7} \\ & & & 0.98857 \times 10^{-6} \end{bmatrix}$$

$$P = \begin{bmatrix} 0.060981 & -0.46336 \times 10^{-4} & 0.64352 \times 10^{-5} & -0.42687 \times 10^{-5} \\ & 0.98475 \times 10^{-4} & -0.18311 \times 10^{-7} & 0.14373 \times 10^{-7} \\ & & 0.98118 \times 10^{-4} & 0.55636 \times 10^{-9} \\ & & & 0.98475 \times 10^{-6} \end{bmatrix}$$

It can be expected from these results that the roll and yaw velocities are more smoothed than other states, if the fixed-interval smoother is applied to the system.

Finally, we obtain a solution method based on the use of eigenvalues of M.

Theorem 9.11

The stabilizing solution to ALE (9.118) can also be obtained by the following algorithm.

Step 1: Using the stable eigenvalue Λ_1^{-1} of M and the corresponding eigenvector matrices X and Y, compute

$$\tilde{X} = X^{-1}[I - X\Lambda_1^{-1}X^{-1}\Phi^{-1}]XY^{-1}X^{-\mathrm{T}} \tag{9.135}$$

Step 2: From $\tilde{X} = \{x_{ij}\}$ and $\Lambda_1^{-1} = \mathrm{diag}(\lambda_1, \ldots, \lambda_n)$, obtain $\tilde{P} = \{\tilde{p}_{ij}\}$, where

$$\tilde{p}_{ij} = \frac{\tilde{x}_{ij}}{1 - \lambda_i \lambda_j} \tag{9.136}$$

Step 3: Compute

$$P = X\tilde{P}X^{\mathrm{T}} \tag{9.137}$$

Proof

Using equation (9.109), (9.118) can be rewritten as

$$P - \tilde{\Phi}P\tilde{\Phi}^{\mathrm{T}} - [I - \tilde{\Phi}\Phi^{-1}]S^{-1} = 0 \tag{9.138}$$

Substituting equations (9.107) and (9.117) into the above equation, premultiplying it by X^{-1} and postmultiplying it by $X^{-\mathrm{T}}$ gives

$$X^{-1}PX^{-\mathrm{T}} - X^{-1}\tilde{\Phi}XX^{-1}PX^{-\mathrm{T}}X^{\mathrm{T}}\tilde{\Phi}^{\mathrm{T}}X^{-\mathrm{T}} - X^{-1}[I - X\Lambda_1^{-1}X^{-1}\Phi^{-1}]XY^{-1}X^{-\mathrm{T}} = 0 \tag{9.139}$$

If we set the third term of the left-hand side of equation (9.139) to $\tilde{X}$, then we get equation (9.135). Defining

$$\tilde{P} \triangleq X^{-1}PX^{-\mathrm{T}} \tag{9.140}$$

equation (9.139) is found to reduce to

$$\tilde{P} - \Lambda_1^{-1}\tilde{P}\Lambda_1^{-1} = \tilde{X} \tag{9.141}$$

which provides equation (9.136). Hence, it is seen from equation (9.140) that P can be computed as in equation (9.137). □

9.6 SUMMARY

We have extended the continuous-time forward-pass smoother described in Chapter 6 to develop forward-pass smoothers for discrete-time systems. We have also presented a $U-D$ factorized forward-pass smoothing algorithm, which has been shown to be effective in assuring numerical stability, computational speed-up, and reduction of computer storage. Lastly, we have presented the steady-state solution to the forward-pass smoother.

A generalized fixed-interval smoothing problem is discussed by Watanabe and Tzafestas [32], in which the predictive information on the final state is available in addition to the usual a priori information on the initial state. The optimal smoothing problem is extended to the case where the system is characterized by uncertain observations [33, 34]. The Bierman $U-D$ smoother is further extended to account for situations with singular estimate error covariances that arise in applications [35]. Numerical results for Example 9.3 are taken from Watanabe [10]. A $U-D$ fixed-interval smoothing algorithm for the backward-pass realization that uses the method described in Section 9.4 is developed in Watanabe and Tzafestas [36].

A similar algebraic solution to the forward-pass fixed-interval smoother discussed in Section 9.5 may be obtained [37] for a fixed-point smoother with an initial state, which consists of the steady-state solution to the Kalman filter.

APPENDIX 9A TRIANGULAR *UDU*^{T} FACTORIZATION

9A.1 Upper Square-Root-Free Cholesky Decomposition

If an $n \times n$ matrix $P > 0$, then $P = UDU^T$, where U is an upper triangular factor with unit diagonal elements and $D = \text{diag}(d_1, \ldots, d_n)$. The elements of U and D are given by the following square-root-free Cholesky decomposition algorithm.

For $j = n, n-1, \ldots, 2$, recursively cycle through the following equations:

$$d_j = P(j, j) \tag{9A.1}$$

$$U(j, j) = 1 \tag{9A.2}$$

$$U(k, j) = P(k, j)/d_j, \qquad k = 1, \ldots, j-1 \tag{9A.3}$$

$$P(i, k) := P(i, k) - U(i, j)U(k, j)d_j \begin{cases} k = 1, \ldots, j-1 \\ i = 1, \ldots, k \end{cases} \tag{9A.4}$$

and then

$$U(1, 1) = 1 \quad \text{and} \quad d_1 = P(1, 1) \tag{9A.5}$$

9A.2 Lower Square-Root-Free Cholesky Decomposition

Lower triangular factorization algorithms can also be obtained by mimicking the developments of the above algorithm. If $P > 0$, then $P = LDL^{\mathrm{T}}$, where L is lower triangular with unit diagonal elements and $D = \operatorname{diag}(d_1, \ldots, d_n)$. The elements of L and D are given by the following algorithm.

For $j = 1, \ldots, n-1$, recursively cycle through the following equations:

$$d_j = P(j, j) \tag{9A.6}$$

$$L(j, j) = 1 \tag{9A.7}$$

$$P(i, k) := P(i, k) - L(i, j)U(k, j) \begin{cases} k = j+1, \ldots, n \\ i = k, \ldots, n \end{cases} \tag{9A.8}$$

$$L(k, j) = P(k, j)/d_j, \qquad k = j+1, \ldots, n \tag{9A.9}$$

and then

$$U(n, n) = 1 \quad \text{and} \quad d_n = P(n, n) \tag{9A.10}$$

APPENDIX 9B *U–D* FILTER ALGORITHMS

This appendix summarizes algorithms for performing measurement and time updating of the $U-D$ factors. These algorithms correspond to the conventional Kalman formulas [38].

9B.1 *U–D* Measurement Update Algorithm

Given a priori covariance factors $\widetilde{U}$ and $\widetilde{D}$ and the scalar measurement $z = Ax + v$, where $E[v^2] = r$, the updated $\hat{U}$–$\hat{D}$ covariance factors and the Kalman gain, K, are obtained as follows:

$$f^{\mathrm{T}} = A\widetilde{U}; \qquad f^{\mathrm{T}} = (f_1, \ldots, f_n) \tag{9B.1}$$

$$v = \widetilde{D}f; \qquad v_j = \tilde{d}_j f_j, \qquad \widetilde{D} = \mathrm{diag}(\tilde{d}_1, \ldots, \tilde{d}_n) \tag{9B.2}$$

$$\overline{K}_1^{\mathrm{T}} = (v_1, \overbrace{0, \ldots, 0}^{n-1}) \tag{9B.3}$$

$$\alpha_1 = r + v_1 f_1 \tag{9B.4}$$

$$\hat{d}_1 = \tilde{d}_1 r / \alpha_1 \tag{9B.5}$$

For $j = 2, \ldots, n$ cycle through equations (9B.6)–(9B.10):

$$\alpha_j = \alpha_{j-1} + v_j f_j \tag{9B.6}$$

$$\hat{d}_j = \tilde{d}_j \alpha_{j-1} / \alpha_j \tag{9B.7}$$

$$\lambda_j = -f_j / \alpha_{j-1} \tag{9B.8}$$

$$\hat{U}_j = \widetilde{U}_j + \lambda_j \overline{K}_{j-1} \tag{9B.9}$$

$$\overline{K}_j = \overline{K}_{j-1} + v_j \widetilde{U}_j \tag{9B.10}$$

where $\hat{U} = [\hat{U}_1, \ldots, \hat{U}_n]$ and $\widetilde{U} = [\widetilde{U}_1, \ldots, \widetilde{U}_n]$. The component $\hat{U}$ vectors have the form $\hat{U}_j^{\mathrm{T}} = (\hat{U}_j(1) \ldots \hat{U}_j(j-1)\, 1\, 0 \ldots 0)$ and $\hat{D} = \mathrm{diag}(\hat{d}_1, \ldots, \hat{d}_n)$. The Kalman gain is given by

$$K = \overline{K}_n / \alpha_n \tag{9B.11}$$

where α_n is the innovation covariance.

Refer to Bierman [19] or Thornton and Bierman [18] for a FORTRAN mechanization incorporating these details.

9B.2 Modified Weighted Gram–Schmidt Time Update Algorithm

Let

$$W = \{[\overbrace{\Phi\hat{U}}^{n} \vdots \overbrace{B}^{p}]\}n \quad \text{with row vectors} \begin{bmatrix} w_1^{\mathrm{T}} \\ \vdots \\ w_n^{\mathrm{T}} \end{bmatrix}$$

$$\bar{D} = \text{diag}(\overbrace{\hat{D}}^{n}, \overbrace{Q}^{p}) = \text{diag}(\bar{d}_1, \ldots, \bar{d}_{n+p})$$

Then the $\tilde{U}$–$\tilde{D}$ factors of $\tilde{P} = W\bar{D}W^{\mathrm{T}} = \Phi P \Phi^{\mathrm{T}} + BQB^{\mathrm{T}}$ may be computed as follows [39].

Consider the recursions

$$w_i^{(0)} = w_i \quad \text{for} \quad i = 1, \ldots, n \tag{9B.12}$$

Evaluate the following equations recursively for $j = n, \ldots, 1$:

$$\tilde{d}_j = (w_j^{(n-j)})^{\mathrm{T}} \bar{D} w_j^{(n-j)} \tag{9B.13}$$

$$\tilde{U}_{ij} = (w_i^{(n-j)})^{\mathrm{T}} \bar{D} w_j^{(n-j)}/\tilde{d}_j, \qquad i = 1, \ldots, j-1 \tag{9B.14}$$

$$w_i^{(n-j+1)} = w_i^{(n-j)} - \tilde{U}_{ij} w_j^{(n-j)}, \qquad i = 1, \ldots, j-1 \tag{9B.15}$$

This FORTRAN code can be found in Bierman [19] or Thornton and Bierman [18].

APPENDIX 9C RANK-ONE FACTORIZATION UPDATE ALGORITHM

Consider the symmetric matrix $\bar{P}$ factored such that $\bar{P} = \bar{U}\bar{D}\bar{U}^{\mathrm{T}}$. Then the corresponding factors of

$$P = \bar{P} + c\lambda\lambda^{\mathrm{T}} \tag{9C.1}$$

for $c > 0$ and vector quantity λ may be obtained by using a modest arrangement of the Agee–Turner rank-one update [19].

Evaluate equations (9C.2)–(9C.10) recursively for $j = n$, $n-1, \ldots, 2$.

$$\alpha_j = c_j\lambda_j; \qquad c_n = c \tag{9C.2}$$

If $\alpha_j = 0$, omit equations (9C.3)–(9C.10).

$$d_j = \bar{d}_j + \alpha_j\lambda_j \tag{9C.3}$$

$$v_j = \alpha_j/d_j \tag{9C.4}$$

$$\beta_j = \bar{d}_j/d_j \tag{9C.5}$$

$$f_i := \lambda_i, \qquad i = 1, \ldots, j-1 \tag{9C.6}$$

$$\lambda_i := \lambda_i - \lambda_j\bar{u}_{ij}, \qquad i = 1, \ldots j-1 \tag{9C.7}$$

$$u_{ij} = \bar{u}_{ij} + \lambda_i v_j, \quad \text{for } \beta_j > 0.1, \quad i = 1, \ldots, j-1 \tag{9C.8}$$

$$u_{ij} = \beta_j \bar{u}_{ij} + f_i v_j, \quad \text{for } \beta_j \leqslant 0.1, \quad i = 1, \ldots, j-1 \tag{9C.9}$$

$$c_{j-1} = \beta_j c_j \tag{9C.10}$$

$$d_1 = \bar{d}_1 + c_1 \lambda_1^2 \tag{9C.11}$$

This FORTRAN mechanization is also presented in Thornton and Bierman [18, 40].

APPENDIX 9D THE BIERMAN RANK-TWO *U–D* SMOOTHER

In this appendix we review the Bierman rank-two $U-D$ smoother [20]. The smoother is based on applying the RTS smoother recursion. The recursions that generate smooth estimates $\hat{x}(k|N)$ and the estimation error covariances $P(k|N)$ are as follows:

$$\hat{x}(k|N) = \hat{x}(k|k) + G^*(k)[\hat{x}(k+1|N) - \hat{x}(k+1|k)] \tag{9D.1}$$

$$P(k|N) = P(k|k) + G^*(k)[P(k+1|N) - P(k+1|k)]G^{*\mathrm{T}}(k) \tag{9D.2}$$

where

$$G^*(k) = P(k|k)\Phi^{\mathrm{T}}(k+1, k)P^{-1}(k+1|k) \tag{9D.3}$$

and the recursion is a backward sweep from $k = N - 1$ down to $k = 0$, so that the RTS smoother is sometimes called a *backward-pass* fixed-interval smoother. Furthermore, $\{\hat{x}(k|k), P(k|k)\}$ denote the Kalman filtered estimate (measurement update) and the associated estimation error covariance. Also $\{\hat{x}(k+1|k), P(k+1|k)\}$ denote the one-step predicted estimate and the associated estimation error covariance (cf. Chapter 2). Some relationships between this RTS smoother and the Bryson–Frazier smoother (cf. Exercise 9.5) can be found in McReynolds [41].

9D.1 Review of the *U–D* Kalman Filter for the Rank-Two Smoother

We first present the $U-D$ Kalman filter algorithm required for the above smoother. The measurement update equation can be obtained by applying the $U-D$ measurement update algorithm given in Appendix 9B.1. Let H and R be partitioned such that

$$H^{\mathrm{T}} = [H_1^{\mathrm{T}}, \ldots, H_i^{\mathrm{T}}, \ldots, H_m^{\mathrm{T}}], \qquad H_i\text{:}n\text{-dimensional row vector}$$

$$R = \operatorname{diag}(r_1, \ldots, r_i, \ldots, r_m), \qquad r_i\text{:scalar}$$

Suppose that the a priori estimate and covariance are given by $\hat{x}(k|k-1)$ and $P(k|k-1)$, where $P(k|k-1) = U(k|k-1) D(k|k-1)U^{\mathrm{T}}(k|k-1)$. Then the *measurement* updated estimate $\hat{x}(k|k)$ and its error covariance $P(k|k) = U(k|k)D(k|k)U^{\mathrm{T}}(k|k)$ may be obtained as follows. For $i = 1, \ldots, m$ recursively cycle through steps (i) and (ii):

(i) Set $A \equiv H_i$, $r \equiv r_i$, $\widetilde{U} \equiv U^{(i-1)}(k|k)$, $\widetilde{D} \equiv D^{(i-1)}(k|k)$ and apply the $U-D$ measurement update of Appendix 9B.1 to obtain the update factors $U^{(i)}(k|k)$, $D^{(i)}(k|k)$ and filter gain F_i due to ith scalar measurement, where initially $U^{(0)}(k|k) \equiv U(k|k-1)$ and $D^{(0)}(k|k) \equiv D(k|k-1)$.

(ii) Compute

$$\hat{x}_i = \hat{x}_{i-1} + F_i[z_i(k) - H_i\hat{x}_{i-1}] \tag{9D.4}$$

where $z_i(k)$ is the ith element of $z(k)$ and initially $\hat{x}_0 = \hat{x}(k|k-1)$.

At the conclusion of these recursions, the $U-D$ factors of $P(k|k)$ and $\hat{x}(k|k)$ are given by

$$U(k|k) = U^{(m)}(k|k), \qquad D(k|k) = D^{(m)}(k|k), \qquad \hat{x}(k|k) = \hat{x}_m \tag{9D.5}$$

We now present the *time* update algorithm which is required for the Bierman rank-two smoother. To this end, let B and Q be partitioned as

$$B = [b_1, \ldots, b_i, \ldots, b_p]$$

where the b_i are n-dimensional vectors, and

$$Q = \operatorname{diag}(q_1, \ldots, q_i, \ldots, q_p)$$

where the q_i are scalar. Suppose that the filtered estimate $\hat{x}(k|k)$ and error covariance $P(k|k) = U(k|k)D(k|k)U^{\mathrm{T}}(k|k)$ are given. Then the time update estimate $\hat{x}(k+1|k)$ and its error covariance $P(k+1|k) = U(k+1|k)D(k+1|k)U^{\mathrm{T}}(k+1|k)$ may be obtained as follows:

(i) Compute the deterministic update state,

$$x^{(D)}(k) = \Phi(k+1, k)\hat{x}(k|k) \tag{9D.6}$$

and set $W = [\Phi(k+1, k)U(k|k)]$ and $\bar{D} = \text{diag}(D(k|k))$ in the MWGS time update algorithm given in Appendix 9B.2 to get the factors $U^{(D)}(k)$ and $D^{(D)}(k)$ of $P^{(D)}(k)$, where

$$P^{(D)}(k) = \Phi(k+1, k)P(k|k)\Phi^{T}(k+1, k) \tag{9D.7}$$

(ii) Next, for $i = 1, \ldots, p$, recursively cycle through

$$\hat{x}_{i+1}^{(w_k)} = \hat{x}_i^{(w_k)} \tag{9D.8}$$

$$v_i = [\hat{P}_i^{(w_k)}]^{-1} b_i = [\hat{U}_i^{(w_k)}\hat{D}_i^{(w_k)}(\hat{U}_i^{(w_k)})^{T}]^{-1} b_i \tag{9D.9}$$

$$\lambda_i = \frac{q_i}{1 + q_i v_i^{T} b_i} \tag{9D.10}$$

$$\hat{U}_{i+1}^{(w_k)}\hat{D}_{i+1}^{(w_k)}(\hat{U}_{i+1}^{(w_k)})^{T} = \hat{U}_i^{(w_k)}\hat{D}_i^{(w_k)}(\hat{U}_i^{(w_k)})^{T} + q_i b_i b_i^{T} \tag{9D.11}$$

by applying the rank-one factorization update algorithm given in Appendix 9C, where initially

$$\hat{x}_1^{(w_k)} = x^{(D)}(k), \qquad \hat{U}_1^{(w_k)} = U^{(D)}(k), \quad \hat{D}_1^{(w_k)} = D^{(D)}(k)$$

Finally,

$$\hat{x}(k+1|k) = \hat{x}_{p+1}^{(w_k)}, \qquad U_{k+1|k} = \hat{U}_{p+1}^{(w_k)}, \qquad D(k+1|k) = \hat{D}_{p+1}^{(w_k)} \tag{9D.12}$$

Note that v_i in equation (9D.9) can be obtained by the direct inversion of a triangular matrix or by the $L^{T}U^{T}$ factorization method. The latter approach consists of two steps: equation (9D.9) can be rewritten as

$$\hat{U}_i^{(w_k)}\hat{D}_i^{(w_k)} b_i^{*} = b_i \tag{9D.13}$$

$$(\hat{U}_i^{(w_k)})^{T} v_i = b_i^{*} \tag{9D.14}$$

where we first solve b_i^{*} in equation (9D.13) and next obtain v_i in equation (9D.14). When $b_i(j) = 0$ for $j > k$, as is most often the situation in applications, we can alleviate the associated computation loads.

9D.2 Review of the Bierman Rank-Two *U–D* Smoother

Using the vector v_i and multiplier λ_i obtained in the filtering step, we have the *smoothed estimate* $\hat{x}(k|N)$. That is, $\hat{x}(k|N)$ is solved by the following recursion. For $i = p, p-1, \ldots, 1$, cycle through equations (9D.15) and (9D.16):

$$\delta_i = \lambda_i v_i^{\mathrm{T}}[\hat{x}(k+1|k) - \hat{x}_{i+1|N}^{(w_k)}] \tag{9D.15}$$

$$\hat{x}_{i|N}^{(w_k)} = \hat{x}_{i+1|N}^{(w_k)} + b_i\delta_i \tag{9D.16}$$

where initially $\hat{x}_{p+1|N}^{(w_k)} = \hat{x}(k+1|N)$. Finally,

$$\hat{x}(k|N) = \Phi^{-1}(k+1, k)\hat{x}_{1|N}^{(w_k)} \tag{9D.17}$$

The *smoothed error covariance* update corresponding to the smoothed state update, (9D.16), follows from the RTS covariance result, equation (9D.2) which can be arranged in the rank-two matrix modifications,

$$P_{i|N}^{(w_k)} = [I - \lambda_i b_i v_i^{\mathrm{T}}]P_{i+1|N}^{(w_k)}[I - \lambda_i b_i v_i^{\mathrm{T}}]^{\mathrm{T}} + \lambda_i b_i b_i^{\mathrm{T}} \tag{9D.18}$$

where the p-step recursion is initialized with $P_{p+1|N}^{(w_k)} = P(k+1|N)$. In addition, we note that

$$P(k|N) = \Phi^{-1}(k+1, k)P_{1|N}^{(w_k)}\Phi^{-\mathrm{T}}(k+1, k) \tag{9D.19}$$

From this fact, we obtain $P(k|N)$ through the following recursion. Given the factors $U(k+1|N)$ and $D(k+1|N)$ of $P(k+1|N)$, the smoothed error covariance $P(k|N)$ may be obtained in two steps as follows:

(i) For $i = p, p-1, \ldots, 1$, recursively evaluate the following

(a) Set

$$W = [\{I - \lambda_i b_i v_i^{\mathrm{T}}\}U_{i+1|N}^{(w_k)}], \qquad \bar{D} = \mathrm{diag}\,\{D_{i+1|N}^{(w_k)}\} \tag{9D.20}$$

and apply the MWGS time update algorithm to obtain the intermediate factors $\tilde{U}_{i|N}^{(w_k)}$ and $\tilde{D}_{i|N}^{(w_k)}$, where initially $U_{p+1|N}^{(w_k)} = U(k+1|N)$ and $D_{p+1|N}^{(w_k)} = D(k+1|N)$.

(b) Apply the rank-one update algorithm to compute

$$U_{i|N}^{(w_k)}D_{i|N}^{(w_k)}(U_{i|N}^{(w_k)})^{\mathrm{T}} = \tilde{U}_{i|N}^{(w_k)}\tilde{D}_{i|N}^{(w_k)}(\tilde{U}_{i|N}^{(w_k)})^{\mathrm{T}} + \lambda_i b_i b_i^{\mathrm{T}} \tag{9D.21}$$

(ii) Set

$$W = [\Phi^{-1}(k+1, k)U_{1|N}^{(w_k)}], \qquad \bar{D} = \mathrm{diag}\,(D_{1|N}^{(w_k)}) \tag{9D.22}$$

and apply the MWGS time update algorithm to obtain factors $U(k|N)$ and $D(k|N)$ of $P(k|N)$.

If we directly need $P(k|N)$ then we compute $U(k|N)$

$D(k|N)U^{\mathrm{T}}(k|N)$. Note that step (i) can also be implemented by a more effective means, which combines the $U-D$ measurement and rank-one update algorithms as discussed in Section 9.4 (cf. Exercise 9.12).

APPENDIX 9E ARITHMETIC OPERATION COUNTS FOR RANK-TWO *U–D* SMOOTHERS

In this appendix we tabulate the number of arithmetic operations involved in our rank-two $U-D$ smoother and that of Bierman. We first describe filtering operation counts. Tables 9E.1 and 9E.2 indicate the number of operations necessary to perform one step of three algorithms: our filter, and the Bierman filter with the computation of v_i by $L^{\mathrm{T}}U^{\mathrm{T}}$ factorization and by truncated $L^{\mathrm{T}}U^{\mathrm{T}}$ factorization. Table 9E.3 also shows the number of operations necessary to compute v_i by using the direct inversion of a triangular matrix. Refer to Bierman [19], or Thornton and Bierman [18, 24], for the operation counts and FORTRAN codes incorporating the $U-D$ measurement update, the MWGS time update and the rank-one factorization update algorithms (cf. Appendix 9D). Tables 9E.4 and 9E.5 present the number of operation counts required for the smoothed estimate and the associated covariance calculations, using our rank-two $U-D$ smoother mechanization and that of Bierman.

Table 9E.1 Operation counts of *U–D* backward-pass information filter for Watanabe rank-two smoother

Computation	*Remarks*	*Addition*†	*Multiplication*†	*Division*
F_i, $\bar{U}^{(i)}(k+1)$ $\bar{D}^{(i)}(k+1)$ $\alpha^{(i)}(k+1)$	*U–D* measurement update algorithm, Eqs (9.B1)–(9.B11)	$1.5n^2 - 0.5n$	$1.5n^2 + 3.5n$	n
$\alpha_i^* = 1/\alpha^{(i)}(k+1)$	Eq. (9.83)	0	0	1
$e_i = g_i^T \bar{q}_{i-1}$		$n - 1$	n	0
$F_i e_i$	Eqs (9.84) and (9.85)	0	n	0
$\bar{q}_i = \bar{q}_{i-1} - F_i e_i$		n	0	0
	Total for p updating cycles	$(1.5n^2 + 1.5n)p$	$(1.5n^2 + 5.5n)p$	$(n+1)p$
$U(k)$, $D(k)$	MWGS time update algorithm Eqs (9B.12)–(9B.15)	$1.5n^3 - 0.5n^2 + n + n^2 m$	$1.5n^3 + n^2 - 0.5n + (n^2 + n)m$	$n - 1$
$\Phi^T \bar{q}(k+1)$		$n^2 - n$	n^2	0
$(H^T \bar{R})z(k)$	Eq. (9.78)	$nm - n$	nm	0
$\Phi^T \bar{q}(k+1) + (H^T \bar{R})z(k)$		n	0	0
$G^T q(k)$	Smoother input (eq. (9.79))	$np - p$	np	0
	Total	$1.5n^3 + 0.5n^2 + (n^2 + n)m + np$	$1.5n^3 + 2n^2 - 0.5n + (n^2 + 2n)m + np$	$n - 1$

† Multiplication and addition are included only up to order n. This condition is continued in the future operation counts.

Table 9E.2 Operation counts of $U\text{–}D$ Kalman filter for Bierman rank-two smoother (cf. Appendix 9D.1)

Computation	Remarks	Addition	Multiplication	Division
$U(k\|k),\ D(k\|k),\ \hat{x}(k\|k)$	$U\text{–}D$ measurement update algorithm, Bierman [19, p. 107]	$(1.5n^2 + 1.5n)m$	$(1.5n^2 + 5.5n)m$	nm
$U^{(\mathrm{D})}(k), D^{(\mathrm{D})}(k)$	Deterministic update, MWGS time update algorithm	$1.5n^3 - 0.5n^2 + n$	$1.5n^3 + n^2 - 0.5n$	$n - 1$
$x^{(\mathrm{D})}(k) = \Phi\hat{x}(k\|k)$	(only once per time update)	$n^2 - n$	n^2	0
$\hat{U}_i^{(w_k)}\hat{D}_i^{(w_k)}$	$L^{\mathrm{T}} = UD$	0	n	0
$(\hat{U}_i^{(w_k)}\hat{D}_i^{(w_k)})b_i^* = b_i$	$L^{\mathrm{T}}b_i^* = b_i$	$0.5n(n-1)$ $\{0.5k_i(k_i-1)\}$	$0.5n(n-1)$ $\{0.5k_i(k_i-1)\}$	n $\{k_i\}$
$(\hat{U}_i^{(w_k)})^{\mathrm{T}}v_i = b_i^*$	Vector v_i	$0.5n(n-1)$	$0.5n(n-1)$	0
$v_i^{\mathrm{T}}b_i$		$n - 1$	n	0
$q_i(v_i^{\mathrm{T}}b_i)$		0	1	0
$1 + q_i(v_i^{\mathrm{T}}b_i)$		1	0	0
$\lambda_i = q_i/(1 + q_iv_i^{\mathrm{T}}b_i)$	Multiplier	0	0	1
$\hat{U}_{i+1}^{(w_k)},\ \hat{D}_{i+1}^{(w_k)}$	Rank-one factorization update (eqs (9C.2)–(9C.11))	n^2	$n^2 + 2n - 1$	$2n - 2$
	Total for p updating cycles†	$1.5n^3 + 0.5n^2$ $+ 2n^2p$ $\{1.5n^3 + 0.5n^2$ $+ (1.5n^2 + 0.5n)p$ $+ \sum_{i=1}^{p} 0.5k_i(k_i - 1)\}$	$1.5n^3 + 2n^2 - 0.5n$ $+ (2n^2 + 2n)p$ $\{1.5n^3 + 2n^2 - 0.5n$ $+ (1.5n^2 + 2.5n)p$ $+ \sum_{i=1}^{p} 0.5k_i(k_i - 1)\}$	$n - 1$ $+ (3n - 1)p$ $\{n - 1$ $+ (2n - 1)p$ $+ \sum_{i=1}^{p} k_i\}$

† The values inside curly brackets are due to the truncated $L^{\mathrm{T}}U^{\mathrm{T}}$ method.

Table 9E.3 Operation counts of the smoother gain via the direct inverse of a triangular matrix (cf. Bierman [19, pp. 65 and 109])

Computation	*Remarks*	*Addition*	*Multiplication*	*Division*
$[\hat{U}_i^{(w_k)}]^{-1}$	Unit upper triangular	$\frac{1}{6}n^3 - \frac{1}{2}n^2 + \frac{1}{3}n$	$\frac{1}{6}n^3 + \frac{1}{2}n^2 - \frac{2}{3}n$	0
$[\hat{D}_i^{(w_k)}]^{-1}$	Diagonal	0	0	n
$[\hat{U}_i^{(w_k)}]^{-1} b_i$		$0.5n(n-1)$	$0.5n(n-1)$	0
$[\hat{D}_i^{(w_k)}]^{-1}\{[\hat{U}_i^{(w_k)}]^{-1} b_i\}$		0	n	0
$v_i = [\hat{U}_i^{(w_k)}]^{-T}\{[\hat{D}_i^{(w_k)}]^{-1} \times [\hat{U}_i^{(w_k)}]^{-1} b_i\}$	Smoother gain	$0.5n(n-1)$	$0.5n(n-1)$	0
	Total for p updating cycles	$(\frac{1}{6}n^3 + \frac{1}{2}n^2 - \frac{2}{3}n)p$	$(\frac{1}{6}n^3 + \frac{3}{2}n^2 - \frac{2}{3}n)p$	np

Table 9E.4 Operation counts for Watanabe rank-two $U-D$ smoother

Computation	*Remarks*	*Addition*	*Multiplication*	*Division*
$\Phi\hat{x}(k\|N)$	(only once per time update)	$n^2 - n$	n^2	0
$\tilde{h}_i = F_i^{\mathrm{T}}\hat{x}_{i-1}$	For $i = 1, \ldots, p$, eqs (9.88)–(9.91)	$n - 1$	n	0
$\beta_i = y_i(k+1)\alpha_i^*$		0	1	0
$\gamma_i = \beta_i - \tilde{h}_i$		1	0	0
$g_i\gamma_i$		0	n	0
$\hat{x}_i = \hat{x}_{i-1} + g_i\gamma_i$		n	0	0
	Total for p updating cycles	$n^2 - n + 2np$	$n^2 + 2np$	0
$\bar{U}(k\|N),\ \bar{D}(k\|N)$	MWGS time update algorithm (only once per time update)	$1.5n^3 - 0.5n^2 + n$	$1.5n^3 + n^2 - 0.5n$	$n - 1$
$g_iF_i^{\mathrm{T}}$	For $i = 1, \ldots, p$	0	n^2	0
$1 - \mathrm{diag}(g_iF_i^{\mathrm{T}})$		n	0	0
$\tilde{U}^{(i)}(k+1\|N)$, $\tilde{D}^{(i)}(k+1\|N)$	MWGS time update algorithm	$1.5n^3 - 0.5n^2 + n$	$1.5n^3 + n^2 - 0.5n$	$n - 1$
$U^{(i)}(k+1\|N)$, $D^{(i)}(k+1\|N)$	Rank-one factorization update	n^2	$n^2 + 2n - 1$	$2n - 2$
$U(k+1\|N)D(k+1\|N) \times U^{\mathrm{T}}(k+1\|N)$	(only one per time update)	$0.5n^2 - 0.5n$	$n^2 - n$	0
	Total for p updating cycles	$1.5n^3 + 0.5n + (1.5n^3 + 0.5n^2 + 2n)p$	$1.5n^3 + 2n^2 - 1.5n + (1.5n^3 + 3n^2 + 1.5n)p$	$n - 1 + (3n - 3)p$

Table 9E.5 Operation counts for Bierman rank-two $U-D$ smoother (cf. Appendix 9D.2)

Computation	*Remarks*	*Addition*	*Multiplication*	*Division*
$\hat{x}(k+1\|k) - \hat{x}_{i+1\|N}^{(w_k)}$		n	0	0
$v_i^{\mathrm{T}}(\hat{x}(k+1\|k) - \hat{x}_{i+1\|N}^{(w_k)})$		$n-1$	n	0
$\delta_i = \lambda_i v_i^{\mathrm{T}} \times (\hat{x}(k+1\|k) - \hat{x}_{i+1\|N}^{(w_k)})$	For $i = p, \ldots, 1$ Eqs (9D.15), (9D.16)	0	1	0
$b_i\delta_i$		0	n	0
$\hat{x}_{i\|N}^{(w_k)} = \hat{x}_{i+1\|N}^{(w_k)} + b_i\delta_i$		n	0	0
$\hat{x}(k\|N) = \Phi^{-1}\hat{x}_{1\|N}^{(w_k)}$	Eq. (9D.17) (only once per time update)	$n^2 - n$	n^2	0
	Total for p updating cycles	$n^2 - n + 3np$	$n^2 + 2np$	0
$b_i\lambda_i$	For $i = p, \ldots, 1$ Eq. (9D.18)	0	n	0
$(b_i\lambda_i)v_i^{\mathrm{T}}$		0	n^2	0
$1 - \mathrm{diag}(b_i\lambda_i v_i^{\mathrm{T}})$		n	0	0
$\tilde{U}_{i\|N}^{(w_k)}, \tilde{D}_{i\|N}^{(w_k)}$	MWGS time update algorithm	$1.5n^3 - 0.5n^2 + n$	$1.5n^3 + n^2 - 0.5n$	$n-1$
$U_{i\|N}^{(w_k)}, D_{i\|N}^{(w_k)}$	Rank-one factorization update	n^2	$n^2 + 2n - 1$	$2n - 2$

Table 9E.5 (*Cont.*)

Computation	*Remarks*	*Addition*	*Multiplication*	*Division*
$U(k\|N)$, $D(k\|N)$	MWGS time update algorithm (only once per time update)	$1.5n^3 - 0.5n^2 + n$	$1.5n^3 + n^2 - 0.5n$	$n - 1$
$U(k\|N)D(k\|N)U^{\mathrm{T}}(k\|N)$	Only once per time update	$0.5n^2 - 0.5n$	$n^2 - n$	0
	Total for p updating cycles	$1.5n^3 + 0.5n + (1.5n^3 + 0.5n^2 + 2n)p$	$1.5n^3 + 2n^2 - 1.5n + (1.5n^3 + 3n^2 + 2.5n)p$	$n - 1 + (3n - 3)p$

EXERCISES

9.1 Prove equation (9.11).

9.2 Prove equations (9.14)–(9.16).

9.3 Complete the proof of Lemma 9.3.

9.4 Complete the proof of Lemma 9.4.

9.5 Using the following equations:

$$\hat{x}(k|N) = \hat{x}(k|k-1) - P(k|k-1)\lambda(k-1|N)$$
$$\hat{x}(k|N) = \hat{x}(k|k) - P(k|k)\Phi^{\mathrm{T}}(k+1, k)\lambda(k|N)$$

show that

$$\begin{aligned}\lambda(k-1|N) = {} & [I - K(k)H(k)]^{\mathrm{T}}\Phi^{\mathrm{T}}(k+1, k)\lambda(k|N) \\ & - H^{\mathrm{T}}(k)V^{-1}(k|k-1) \\ & \times [z(k) - H(k)\hat{x}(k|k-1)],\end{aligned} \qquad \lambda(N|N) = 0$$

$$\begin{aligned}\Lambda(k-1|N) = {} & [I - K(k)H(k)]^{\mathrm{T}}\Phi^{\mathrm{T}}(k+1, k)\Lambda(k|N) \\ & \times \Phi(k+1, k)[I - K(k)H(k)] \\ & + H^{\mathrm{T}}(k)V^{-1}(k|k-1)H(k), \qquad \Lambda(N|N) = 0\end{aligned}$$

and

$$P(k|N) = P(k|k-1) - P(k|k-1)\Lambda(k-1|N)P(k|k-1)$$
$$\begin{aligned}P(k|N) = {} & P(k|k) - P(k|k)\Phi^{\mathrm{T}}(k+1, k)\Lambda(k|N) \\ & \times \Phi(k+1, k)P(k|k)\end{aligned}$$

where $K(k)$ is the Kalman filter gain $K(k) = P(k|k-1)H^{\mathrm{T}}(k)V^{-1}(k|k-1)$ or $P(k|k)H^{\mathrm{T}}(k)R^{-1}(k)$, $V(k|k-1) = H^{\mathrm{T}}(k)P(k|k-1)H(k) + R(k)$ and $\Lambda(k|N) \triangleq E[\lambda(k|N)\lambda^{\mathrm{T}}(k|N)]$. These constitute the Bryson–Frazier smoother algorithm in discrete time [2, 3] (cf. Exercise 10.13).

9.6 Consider the formal backward-time Kalman filter:

$$\hat{x}_{\mathrm{b}}(k|k) = \hat{x}_{\mathrm{b}}(k|k+1) + K_{\mathrm{b}}(k)[z(k) - H(k)\hat{x}_{\mathrm{b}}(k|k+1)],$$
$$P_{\mathrm{b}}^{-1}(N|N+1)\hat{x}_{\mathrm{b}}(N|N+1) = 0$$
$$K_{\mathrm{b}}(k) = P_{\mathrm{b}}(k|k+1)H^{\mathrm{T}}(k)V_{\mathrm{b}}^{-1}(k|k+1)$$
$$V_{\mathrm{b}}(k|k+1) = H(k)P_{\mathrm{b}}(k|k+1)H^{\mathrm{T}}(k) + R(k)$$
$$P_{\mathrm{b}}(k|k) = [I - K_{\mathrm{b}}(k)H(k)]P_{\mathrm{b}}(k|k+1), \qquad P_{\mathrm{b}}^{-1}(N|N+1) = 0$$
$$\hat{x}_{\mathrm{b}}(k-1|k) = \Phi^{-1}(k, k-1)\hat{x}_{\mathrm{b}}(k|k)$$
$$P_{\mathrm{b}}(k-1|k) = \Phi^{-1}(k, k-1)P_{\mathrm{b}}(k|k)\Phi^{-\mathrm{T}}(k, k-1)$$

$$+ \Phi^{-1}(k, k-1)B(k-1)Q(k-1)$$
$$\times B^{\mathrm{T}}(k-1)\Phi^{-\mathrm{T}}(k, k-1)$$

which are based on the formally backward model for equation (9.1),

$$x(k) = \Phi^{-1}(k+1, k)x(k+1) - \Phi^{-1}(k+1, k)B(k)w(k)$$
$$z(k) = H(k)x(k) + v(k)$$

(a) By decomposing $w(k)$ into a predictable part from a set $\mathscr{X} \triangleq \{x(N), \ldots, x(k+1)\}$, i.e. $E[x(k)|\mathscr{X}]$, and a remaining error part, transform this model to the backward Markovian model:

$$x(k) = \Psi_{\mathrm{b}}(k, k+1)x(k+1)$$
$$+ y_{\mathrm{b}}(k+1) + w_{\mathrm{b}}(k+1)$$
$$x(N) \sim N(\hat{m}(N), \Pi(N))$$

where $(\hat{m}(N), \Pi(N))$ are given by equations (9.45) and (9.46), and

$$\Psi_{\mathrm{b}}(k, k+1) = \Pi(k)\Phi^{\mathrm{T}}(k+1, k)\Pi^{-1}(k+1)$$
$$y_{\mathrm{b}}(k+1) = [I - \Psi_{\mathrm{b}}(k, k+1)\Phi(k+1, k)]\hat{m}(k)$$
$$E[w_{\mathrm{b}}(k+1)] = 0$$
$$E[w_{\mathrm{b}}(k+1)w_{\mathrm{b}}^{\mathrm{T}}(k+1)]$$
$$= \Pi(k) - \Pi(k)\Phi^{\mathrm{T}}(k+1, k)\Pi^{-1}(k+1)$$
$$\times \Phi(k+1, k)\Pi(k)$$
$$E[x(N)w_{\mathrm{b}}^{\mathrm{T}}(k+1)] = 0, \qquad k+1 \leqslant N$$

(b) Show that the following identities hold:

$$\Psi_{\mathrm{b}}(k, k+1)$$
$$= \Phi^{-1}(k+1, k)[I - B(k)Q(k)B^{\mathrm{T}}(k)\Pi^{-1}(k+1)]$$
$$y_{\mathrm{b}}(k+1) = \Phi^{-1}(k+1, k)B(k)Q(k)B^{\mathrm{T}}(k)$$
$$\times \Pi^{-1}(k+1)\hat{m}(k+1)$$
$$E[w_{\mathrm{b}}(k+1)w_{\mathrm{b}}^{\mathrm{T}}(k+1)]$$
$$= \Phi^{-1}(k+1, k)B(k)E[\tilde{w}(k)\tilde{w}^{\mathrm{T}}(k)]$$
$$\times B^{\mathrm{T}}(k)\Phi^{-\mathrm{T}}(k+1, k)$$

where

$$E[\tilde{w}(k)\tilde{w}^{\mathrm{T}}(k)]$$
$$= Q(k) - Q(k)B^{\mathrm{T}}(k)\Pi^{-1}(k+1)B(k)Q(k)$$

(c) Verify that equations (9.47) and (9.48) can be rewritten as follows:

$$\hat{x}_B(k|k) = \hat{x}_B(k|k+1) + K_B(k) \times [z(k) - H(k)\hat{x}_B(k|k+1)],$$

$$\hat{x}_B(N|N+1) = \hat{m}(N)$$

$$K_B(k) = P_B(k|k+1)H^T(k) \times [H(k)P_B(k|k+1)H^T(k) + R(k)]^{-1}$$

$$P_B(k|k) = [I - K_B(k)H(k)]P_B(k|k+1),$$

$$P_B(N|N+1) = \Pi(N)$$

$$\hat{x}_B(k-1|k) = \Psi_b(k-1, k)\hat{x}_B(k|k) + y_b(k)$$

$$P_B(k-1|k) = \Psi_b(k-1, k)P_B(k|k)\Psi_b^T(k-1, k) + \widetilde{Q}_b(k)$$

where

$$\widetilde{Q}_b(k) = \Phi^{-1}(k, k-1)B(k-1) \times [Q(k-1) - Q(k-1)B^T(k-1)\Pi^{-1}(k) \times B(k-1)Q(k-1)]B^T(k-1)\Phi^{-T}(k, k-1)$$

(d) Minimizing the following cost function:

$$J = \|\hat{x}(k|N) - \Psi_b(k, k+1)x(k+1) - y_b(k+1)\|^2_{\widetilde{Q}_b^{-1}(k+1)} + \|x(k+1) - \hat{x}_B(k+1|k+1)\|^2_{P_B^{-1}(k+1|k+1)}$$

derive the generalized forward-pass fixed-interval smoother:

$$\hat{x}(k+1|N) = \hat{x}_B(k+1|k+1) + C_b(k+1)[\hat{x}(k|N) - \Psi_b(k, k+1)\hat{x}_B(k+1|k+1) - y_b(k+1)],$$

$$\hat{x}(0|N) = \hat{x}_B(0|0)$$

$$P(k+1|N) = P_B(k+1|k+1) + C_b(k+1)[P(k|N) - P_B(k|k+1)]C_b^T(k+1),$$

$$P(0|N) = P_B(0|0)$$

where

$$C_b(k+1) = P_B(k+1|k+1)\Psi_b^T(k, k+1)[\Psi_b(k, k+1) \times P_B(k+1|k+1)\Psi_b^T(k, k+1) + \widetilde{Q}_b(k+1)]^{-1}$$

9.7 Apply the results of Exercise 9.5 to derive the RTS smoother in discrete time (see also Appendix 9D):

$$\hat{x}(k|N) = \hat{x}(k|k) + G^*(k)[\hat{x}(k+1|N) - \hat{x}(k+1|k)]$$

$$P(k|N) = P(k|k) + G^*(k)[P(k+1|N) - P(k+1|k)]G^{*\mathrm{T}}(k)$$

where

$$G^*(k) = P(k|k)\Phi^{\mathrm{T}}(k+1, k)P^{-1}(k+1|k)$$

9.8 Develop an alternative version of the result in Exercise 9.5, using the relations (9.7) and (9.9)–(9.11). That is, show that

$$\hat{x}(k|N) = \hat{x}_{\mathrm{b}}(k|k) - P_{\mathrm{b}}(k|k)\lambda(k-1|N)$$

$$P(k|N) = P_{\mathrm{b}}(k|k) - P_{\mathrm{b}}(k|k)\Lambda(k-1|N)P_{\mathrm{b}}(k|k)$$

where $\{\hat{x}_{\mathrm{b}}(k|k), P_{\mathrm{b}}(k|k)\}$ are provided by the formal backward-time Kalman filter given in Exercise 9.6, and $\lambda(k-1|N)$ and $\Lambda(k-1|N)$ are subject to the following equations:

$$\lambda(k|N) = \Phi^{-\mathrm{T}}(k+1, k)\{[I - K_{\mathrm{b}}(k)H(k)]^{\mathrm{T}}\lambda(k-|N)$$
$$- H^{\mathrm{T}}(k)V_{\mathrm{b}}^{-1}(k|k+1)[z(k) - H(k)\hat{x}_{\mathrm{b}}(k|k+1)],$$
$$\lambda(-1|N) = [P_{\mathrm{b}}(0|0) + P(0)]^{-1}[\hat{x}_{\mathrm{b}}(0|0) - \hat{x}(0)]$$

$$\Lambda(k|N) = \Phi^{-\mathrm{T}}(k+1, k)\{[I - K_{\mathrm{b}}(k)H(k)]^{\mathrm{T}}\Lambda(k-1|N)$$
$$\times [I - K_{\mathrm{b}}(k)H(k)]$$
$$+ H^{\mathrm{T}}(k)V_{\mathrm{b}}^{-1}(k|k+1)H(k)\}\Phi^{-1}(k+1, k),$$
$$\Lambda(-1|N) = [P_{\mathrm{b}}(0|0) + P(0)]^{-1}$$

which can be seen as the dual version in time of the Bryson–Frazier smoother in discrete time [42] given in Exercise 9.5.

9.9 Prove the equivalence of equations (9.52) and (9.52).

9.10 Consider the system described by the equations

$$\begin{bmatrix} x_1(k+1) \\ x_2(k+1) \end{bmatrix} = \begin{bmatrix} 1 & 1 \\ 0 & 1 \end{bmatrix}\begin{bmatrix} x_1(k) \\ x_2(k) \end{bmatrix} + \begin{bmatrix} 1 & 0 \\ 0 & 1 \end{bmatrix}\begin{bmatrix} w_1(k) \\ w_2(k) \end{bmatrix}$$

$$z(k) = [1 \quad 1]\begin{bmatrix} x_1(k) \\ x_2(k) \end{bmatrix} + v(k)$$

where

$$E\left\{\begin{bmatrix} w_1(k) \\ w_2(k) \\ v(k) \end{bmatrix}[w_1(j) \quad w_2(j) \quad v(j)]\right\} = \begin{bmatrix} 2 & 0 & 0 \\ 0 & 1 & 0 \\ 0 & 0 & 2 \end{bmatrix}\delta_{kj}$$

$$E[w(k)x^{\mathrm{T}}(0)] = E[v(k)x^{\mathrm{T}}(0)] = 0$$

$$x(0) \sim N\left(\begin{bmatrix} 4 \\ 1 \end{bmatrix}, \begin{bmatrix} 1 & 1 \\ 0 & 1 \end{bmatrix}\right)$$

Let $z(0) = 4$ and $z(1) = 6$.

(a) Generate the $U-D$ factorized backward information filter at times $k = 1, 0$, using Theorems 9.3 and 9.4.

(b) Generate the estimates of the $U-D$ factorized forward-pass fixed-interval smoother at times $k = 0, 1$, applying the results of part (a) and Theorem 9.5.

(c) Generate the covariances of the $U-D$ factorized forward-pass fixed-interval smoother at times $k = 0, 1$, using the results of part (a) and Theorem 9.6.

9.11 Consider the matrices given in Example 9.3.

(a) Generate the $U-D$ Kalman filter for the rank-two smoother, applying Appendix 9D.1.

(b) Generate the equations for smoothed estimates and covariances by using the Bierman rank-two $U-D$ smoother given in Appendix 9D.2.

9.12 Consider an alternative $U-D$ implementation of smoothed error covariance.

(a) Show that equation (9.73), which is the Joseph form discussed in Exercise 2.1, can be further modified as follows [23, 40]:

$$P(k+1|N) = [I - \bar{K}K^{\mathrm{T}}(k+1)]\bar{P}(k+1|N) + (G-\bar{K})\hat{R}(G-\bar{K})^{\mathrm{T}}$$

where

$$\hat{R} = K^{\mathrm{T}}(k+1)\bar{P}(k+1|N)K(k+1) + \bar{R}$$

$$\bar{R} = [I + G^{\mathrm{T}}S(k+1)G]^{-1}$$

$$\bar{K} = \bar{P}(k+1|N)K(k+1)\hat{R}^{-1}$$

(b) Prove that, when using the result of part (a), step (ii) of Theorem 9.6 can be written as follows:

For $i = 1, \ldots, p$, recursively evaluate the following:

(i) Set $A \equiv F_i^{\mathrm{T}}$, $r \equiv \alpha_i^*$, $\tilde{U} \equiv U^{(i-1)}(k+1|N)$, $\tilde{D} \equiv D^{(i-1)}(k+1|N)$ and apply the $U-D$ measurement update of Appendix 9B.1 to obtain the update factors $\tilde{U}(k+1|N)$, $\tilde{D}(k+1|N)$, the filter gain $\bar{F}_i$ and the innovation variance $\hat{\alpha}_i$, where $U^{(0)}(k+1|N) \equiv \bar{U}(k+|N)$ and $D^{(0)}(k+1|N) \equiv \bar{D}(k+1|N)$.

(ii) Compute

$$f_i := g_i - \bar{F}_i$$

(iii) Apply the rank-one update algorithm given in Appendix 9C to compute

$$U^{(i)}(k+1|N)D^{(i)}(k+1|N)U^{(i)^{\mathrm{T}}}(k+1|N)$$

$$= \widetilde{U}(k+1|N)\widetilde{D}(k+1|N)\widetilde{U}^{\mathrm{T}}(k+1|N) + \hat{\alpha}_i f_i f_i^{\mathrm{T}}$$

Finally, the desired factors of $P(k+1|N)$ are $U(k+1|N) \equiv U^{(p)}(k+1|N)$ and $D(k+1|N) \equiv D^{(p)}(k+1|N)$.

(c) Show that the operation count of Theorem 9.6 with this algorithm yields about $1.5n^3 + 2.5n^2 p$ flops, whereas that of Theorem 9.6 gives about $1.5n^3 + 1.5n^3 p$ flops, where flops is here defined as operation summation of multiplies and divides.

9.13 Solve the problem of Example 9.9 by using Theorem 9.11.

9.14 Reconsider the system given in Example 9.10. Obtain the steady-state estimation error covariance P_{f} for the Kalman filter and P for the forward-pass fixed-interval smoother by applying Theorems 9.9 and 9.10.

9.15 Consider the discrete-time model given by equations (9.1)–(9.4). It is assumed that $Q(k) > 0$ and the predictive information on the final state is also available as follows:

$$x(N) \sim N(\hat{x}(N), P(N))$$

(a) Show that the optimal a priori information can be modified as follows:

$$x(0) \sim N(\hat{m}(0), \Pi(0))$$

where

$$\hat{m}(0) = \Pi(0)[P^{-1}(0)\hat{x}(0) + \Pi_{\mathrm{b}}^{-1}(0)\hat{m}_{\mathrm{b}}(0)]$$

$$\Pi(0) = [P^{-1}(0) + \Pi_{\mathrm{b}}^{-1}(0)]^{-1}$$

$$\hat{m}_{\mathrm{b}}(k) = \Phi^{-1}(k+1, k)\hat{m}_{\mathrm{b}}(k+1), \qquad \hat{m}_{\mathrm{b}}(N) = \hat{x}(N)$$

$$\Pi_{\mathrm{b}}(k) = \Phi^{-1}(k+1, k)\Pi_{\mathrm{b}}(k+1)\Phi^{-\mathrm{T}}(k+1, k) + \Phi^{-1}(k+1, k)B(k)Q(k)B^{\mathrm{T}}(k)\Phi^{-\mathrm{T}}(k+1, k)$$

$$\Pi_{\mathrm{b}}(N) = P(N)$$

(Hint: Remember the Mayne–Fraser two-filter smoother given only two pieces of information on $x(0) \sim N(\hat{x}(0), P(0))$ and $x(N) \sim N(\hat{x}(N), P(N))$.)

(b) In a manner similar to that of Exercise 9.6, derive the following forward Markovian model:

$$x(k+1) = \Psi(k+1, k)x(k) + y(k) + w_{\mathrm{f}}(k)$$

where

$$\Psi(k+1, k) = \{I - B(k)[Q^{-1}(k) + B^{\mathrm{T}}(k)\Pi_{\mathrm{b}}^{-1}(k+1)B(k)]^{-1} \times B^{\mathrm{T}}(k)\Pi_{\mathrm{b}}^{-1}(k+1)\}\Phi(k+1, k)$$

$$y(k) = B(k)[Q^{-1}(k) + B^{\mathrm{T}}(k)\Pi_{\mathrm{b}}^{-1}(k+1)B(k)]^{-1} \times B^{\mathrm{T}}(k)\Pi_{\mathrm{b}}^{-1}(k+1)\hat{m}_{\mathrm{b}}(k+1)$$

$$E[w_{\mathrm{f}}(k)] = 0$$

$$E[w_{\mathrm{f}}(k)w_{\mathrm{f}}^{\mathrm{T}}(k)] = B(k)[Q^{-1}(k) + B^{\mathrm{T}}(k)\Pi_{\mathrm{b}}^{-1}(k+1)B(k)]^{-1}B^{\mathrm{T}}(k)$$

$$E[x(0)w_{\mathrm{f}}^{\mathrm{T}}(k)] = 0, \qquad k \geqslant 0$$

(c) Show that the following identities hold:

$$\Psi(k+1, k) = \Phi(k+1, k) - B(k)Q(k)B^{\mathrm{T}}(k) \times \Phi^{-\mathrm{T}}(k+1, k)\Pi_{\mathrm{b}}^{-1}(k)$$

$$y(k) = B(k)Q(k)B^{\mathrm{T}}(k)\Phi^{-\mathrm{T}}(k+1, k)\Pi_{\mathrm{b}}^{-1}(k)\hat{m}_{\mathrm{b}}(k)$$

$$E[w_{\mathrm{f}}(k)w_{\mathrm{f}}^{\mathrm{T}}(k)] = B(k)E[\tilde{w}(k)\tilde{w}^{\mathrm{T}}(k)]B^{\mathrm{T}}(k)$$

where

$$E[\tilde{w}(k)\tilde{w}^{\mathrm{T}}(k)] = Q(k) - Q(k)B^{\mathrm{T}}(k) \times \Phi^{-\mathrm{T}}(k+1, k)\Pi_{\mathrm{b}}^{-1}(k) \times \Phi^{-1}(k+1, k)B(k)Q(k)$$

(d) Minimizing the following cost function:

$$J = \|\hat{x}(k+1|N) - \Psi(k+1, k)x(k) - y(k)\|^2_{\tilde{Q}_{\mathrm{f}}^{-1}(k)} + \|x(k) - \hat{x}_{\mathrm{f}}(k|k)\|^2_{P_{\mathrm{f}}^{-1}(k|k)}$$

derive the generalized backward-pass fixed-interval smoother:

$$\hat{x}(k|N) = \hat{x}_{\mathrm{f}}(k|k) + C_{\mathrm{f}}(k)[\hat{x}(k+1|N) - \Psi(k+1, k)\hat{x}_{\mathrm{f}}(k|k) - y(k)], \qquad \hat{x}(N|N) = \hat{x}_{\mathrm{f}}(N|N)$$

$$P(k|N) = P_{\mathrm{f}}(k|k) + C_{\mathrm{f}}(k) \times [P(k+1|N) - P_{\mathrm{f}}(k+1|k)]C_{\mathrm{f}}^{\mathrm{T}}(k), \qquad P(N|N) = P_{\mathrm{f}}(N|N)$$

where

$$\tilde{Q}_{\mathrm{f}}(k) = B(k)[Q^{-1}(k) + B^{\mathrm{T}}(k)\Pi_{\mathrm{b}}^{-1}(k+1)B(k)]^{-1}B^{\mathrm{T}}(k)$$

$$C_{\mathrm{f}}(k) = P_{\mathrm{f}}(k|k)\Psi^{\mathrm{T}}(k+1, k) \times [\Psi(k+1, k)P_{\mathrm{f}}(k|k)\Psi^{\mathrm{T}}(k+1, k) + \tilde{Q}_{\mathrm{f}}(k)]^{-1}$$

and $\{\hat{x}_{\mathrm{f}}(k|k), P_{\mathrm{f}}(k|k)\}$ are given by the following Kalman filter:

$$\hat{x}_{\mathrm{f}}(k|k) = \hat{x}_{\mathrm{f}}(k|k-1) + K_{\mathrm{f}}(k)$$

$$\times [z(k) - H(k)\hat{x}_f(k|k-1)],$$

$$\hat{x}_f(0|-1) = \hat{m}(0)$$

$$P_f(k|k) = [I - K_f(k)H(k)]P_f(k|k-1),$$

$$P_f(0|-1) = \Pi(0)$$

$$K_f(k) = P_f(k|k-1)H^T(k) \times [H(k)P_f(k|k-1)H^T(k) + R(k)]^{-1}$$

$$\hat{x}_f(k+1|k) = \Psi(k+1, k)\hat{x}_f(k|k) + y(k)$$

$$P_f(k+1|k) = \Psi(k+1, k)P_f(k|k)\Psi^T(k+1, k) + \tilde{Q}_f(k)$$

REFERENCES

[1] RAUCH, H. E., TUNG, F. and STRIEBEL, C. T., Maximum Likelihood Estimates of Linear Dynamic Systems, *AIAA J.*, vol. 3, no. 8, 1965, pp. 1445–50.

[2] BRYSON, A. E. and FRAZIER, M., Smoothing for Linear and Nonlinear Dynamic Systems, *Proc. Optimum Syst. Synthesis Conf., U.S. Air Force Tech. Rept.* ASD-TDR-63-119, February 1963.

[3] BRYSON, A. E. and HO, Y. C., *Applied Optimal Control*, Hemisphere, New York, 1975.

[4] MAYNE, D. Q., A Solution of the Smoothing Problem, *Automatica*, vol. 4, 1966, pp. 73–92.

[5] OMATU, S. and SEINFELD, J. H., Filtering and Smoothing for Linear Discrete-Time Distributed Parameter Systems Based on Wiener–Hopf Theory with Applications to Estimation of Air Pollution, *IEEE Trans. Syst., Man, and Cyber.*, vol. SMC-11, no. 12, 1981, pp. 785–801.

[6] MEDITCH, J. S., *Stochastic Optimal Linear Estimation and Control*, McGraw-Hill, New York, 1969.

[7] WATANABE, K., Steady-State Covariance Analysis for a Forward-Pass Fixed-Interval Smoother, *Trans. ASME, J. Dynamic Syst. Meas. Control*, vol. 108, no. 2, 1986, pp. 136–40.

[8] COX, H., On the Estimation of State Variables and Parameters for Noisy Dynamic Systems, *IEEE Trans. Aut. Control*, vol. AC-19, 1964, pp. 5–12.

[9] JAZWINSKI, A. H., *Stochastic Processes and Filtering Theory*, Academic Press, New York, 1970.

[10] WATANABE, K., On the Relationship between the Lagrange Multiplier Method and the Two-Filter Smoother, *Int. J. Control*, vol. 42, no. 2, 1985, pp. 391–410.

[11] DESAI, U. B., WEINERT, H. L. and YUSYPCHUCK, G. J., Discrete-Time Complementary Models and Smoothing Algorithms: The Correlated

Noise Case, *IEEE Trans. Aut. Control*, vol. AC-28, no. 4, 1983, pp. 536–9.

[12] ACKNER, R. and KAILATH, T., Discrete-Time Complementary Models and Smoothing, *Int. J. Control*, vol. 49, no. 5, 1989, pp. 1665–82.

[13] WEINERT, H. L. and DESAI, U. B., On Complementary Models and Fixed-Interval Smoother, *IEEE Trans. Aut. Control*, vol. AC-26, no. 4, 1981, pp. 863–7.

[14] ADAMS, M. B., WILLSKY, A. S. and LEVY, B. C., Linear Estimation of Boundary Value Stochastic Processes—Part I: The Role and Construction of Complementary Models, *IEEE Trans. Aut. Control*, vol. AC-29, no. 9, 1984, pp. 803–11.

[15] ADAMS, M. B., WILLSKY, A. S. and LEVY, B. C., Linear Estimation of Boundary Value Stochastic Processes—Part II: 1-D Smoothing Problems, *IEEE Trans. Aut. Control*, vol. AC-29, no. 9, 1984, pp. 811–21.

[16] WALL, J. E., WILLSKY, A. S. and SANDELL, N. R., On the Fixed-Interval Smoothing Problem, *Stochastics*, vol. 5, 1981, pp. 1–42.

[17] BUCY, R. S. and JOSEPH, P. D., *Filtering for Stochastic Processes with Applications to Guidance*, Wiley, New York, 1969.

[18] THORNTON, C. L. and BIERMAN, G. J., UDU^{T} Covariance Factorization for Kalman Filtering. In C. T. Leondes (ed.), *Control and Dynamic Systems*, Vol. 16, Academic Press, New York, 1977.

[19] BIERMAN, G. J., *Factorization Methods for Discrete Sequential Estimation*, Academic Press, New York, 1977.

[20] BIERMAN, G. J., A New Computationally Efficient Fixed-Interval, Discrete-Time Smoother, *Automatica*, vol. 19, no. 5, 1983, pp. 503–11.

[21] GELB, A. (ed.), *Applied Optimal Estimation*, MIT Press, Cambridge, Massachusetts, 1980.

[22] WATANABE, K., A New Forward-Pass Fixed-Interval Smoother Using the $U-D$ Information Matrix Factorization, *Automatica*, vol. 22, no. 4, 1986, pp. 465–75.

[23] KUGA, H. K., Comment on 'New Computationally Efficient Formula for Backward-Pass Fixed-Interval Smoother and its $U-D$ Factorization Algorithm', *IEE Proc.*, vol. 136, pt. D, no. 6, 1989, pp. 331–2.

[24] KUČERA, V., The Discrete Riccati Equation of Optimal Control, *Kybernetika*, vol. 8, 1972, pp. 430–47.

[25] CHAN, S. W., GOODWIN, G. C. and SIN, K. S., Convergence Properties of the Riccati Difference Equation in Optimal Filtering of Nonstabilizable Systems, *IEEE Trans. Aut. Control*, vol. AC-29, no. 2, 1984, pp. 110–18.

[26] WATANABE, K., Algebraic Solution of a Forward-Pass Fixed-Interval Smoother: Continuous-Time Systems, *Int. J. Syst. Sci.*, vol. 16, no. 10, 1985, pp. 1217–27.

[27] NISHIMURA, T., On the Steady-State Solution of Fixed-Interval Smoother, in *9th SICE Symposium on Control Theory*, 26–28 May 1980, pp. 181–4.

[28] NISHIMURA, T., On the Steady-State Solution of Fixed-Interval and Fixed-Point Smoothers, *Trans. Soc. Instrum. Control Engrs.*, vol. 20, 1984, pp. 1–6 (in Japanese).

[29] FRIEDLAND, B., Optimum Steady-State Position and Velocity Estimation Using Noisy Sampled Position Data, *IEEE Trans. Aero. Electron. Syst.*, vol. AES-9, no. 6, 1973, pp. 906–11.

[30] FATH, A. F., Computational Aspects of the Linear Optimal Regulation Problem, *IEEE Trans. Aut. Control*, vol. AC-14, 1969, pp. 547–50.

[31] WAGDI, M. N., An Adaptive Control Approach to Sensor Failure Detection and Isolation, *J. Guidance and Control*, vol. 5, no. 2, 1982, pp. 118–23.

[32] WATANABE, K. and TZAFESTAS, S. G., Generalized Fixed-Interval Smoothers for Discrete-Time Systems with Predictive Information, *Preprint of 8th IFAC/IFORS Symposium on Identification and Systems Parameter Estimation*, Beijing, 27–31 August 1988, Vol. 2, pp. 872–7.

[33] MONZINGO, R. A., Discrete Optimal Linear Smoothing for Systems with Uncertain Observations, *IEEE Trans. Inf. Theory*, vol. IT-21, no. 3, 1975, pp. 271–5.

[34] MONZINGO, R. A., Discrete Linear Recursive Smoothing for Systems with Uncertain Observations, *IEEE Trans. Aut. Control*, vol. AC-26, no. 3, 1981, pp. 754–7.

[35] BIERMAN, G. J., A Reformulation of The Rauch–Tung–Striebel Discrete Time Fixed Interval Smoother, *Proc. of the 27th Conference on Decision and Control*, Austin, Texas, 1988, pp. 840–4.

[36] WATANABE, K. and TZAFESTAS, S. G., A New Computationally Efficient Formula for Backward-Pass Fixed-Interval Smoother and its U–D Factorization Algorithm. *IEE Proc.*, vol. 136, pt. D, no. 2, 1989, pp. 73–8.

[37] WATANABE, K., Steady-State Error Covariances of Fixed-Point Smoothers, *Int. J. Syst. Sci.*, vol. 18, no. 7, 1987, pp. 1323–37.

[38] BIERMAN, G. J. and THORNTON, C. L., Numerical Comparison of Kalman Filter Algorithms: Orbit Determination Case Study, *Automatica*, vol. 13, 1977, pp. 23–35.

[39] THORNTON, C. L. and BIERMAN, G. J., Gram–Schmidt Algorithms for Covariance Propagation, *Int. J. Control*, vol. 25, no. 2, 1977, pp. 243–60.

[40] THORNTON, C. L. and BIERMAN, G. J., Filtering and Error Analysis via the UDU^{T} Covariance Factorization, *IEEE Trans. Aut. Control*, vol. AC-23, no. 5, 1978, pp. 901–7.

[41] MCREYNOLDS, S. R., Fixed Interval Smoothing: Revisited, *J. Guidance, Control, and Dynamics*, vol. 13, no. 5, 1990, pp. 913–21.

[42] WATANABE, K., A New Fixed-Interval Smoothing Algorithm, *Proc. of the 28th SICE Annual Conference*, Vol. II, 1989, pp. 1103–6.

10

Decentralized Smoothers Based on the Two-Filter Smoother Formula

10.1 INTRODUCTION

The decentralized estimation problem for stochastic systems is an important theme in modern control theory, together with recent progress in microprocessing technology [1]. The need for a solution to this problem arises in many applications, such as power systems [2, 3], freeway ramp metering control systems [4], distributed sensor networks [5], dynamic ship-positioning [6], and map updating and combining [7].

In this chapter, the two-filter smoother, discussed in Chapters 6 and 9, is applied to construct some decentralized smoothing algorithms based on local filtered or smoothed estimates, in which the models used by the local processors are identical to the global model [8]. In Section 10.2, the decentralized smoothing, smoothing update and real-time smoothing problems are solved for a continuous-time linear estimation structure consisting of a central processor and of two local processors [9]. The discrete-time case is also included in Section 10.3.

10.2 DECENTRALIZED SMOOTHING IN CONTINUOUS-TIME SYSTEMS

We first consider the following continuous-time system described by

$$\dot{x}(t) = Ax(t) + Bw(t), \qquad 0 \leqslant t \leqslant T \tag{10.1}$$

where two local stations observe independently the state of equation (10.1), i.e.

$$z_i(t) = H_i x(t) + v_i(t), \qquad i = 1, 2 \tag{10.2}$$

It is assumed that the system noise $w(t)$ and the measurement noises $v_i(t)$, $i = 1, 2$, are zero-mean Gaussian white noise processes with the following statistical properties:

$$E\left\{\begin{bmatrix} w(t) \\ v_1(t) \\ v_2(t) \end{bmatrix} x^{\mathrm{T}}(0)\right\} = 0,$$

$$E\left\{\begin{bmatrix} w(t) \\ v_1(t) \\ v_2(t) \end{bmatrix} [w^{\mathrm{T}}(s) \quad v_1^{\mathrm{T}}(s) \quad v_2^{\mathrm{T}}(s)]\right\}$$

$$= \operatorname{diag}(Q, R_1, R_2)\delta(t - s) \qquad (10.3)$$

The a priori information on the initial condition is assumed to be

$$x(0) \sim N(\hat{x}(0), P(0)) \qquad (10.4a)$$

for the central processor (or central station, or fusion centre) and

$$x(0) \sim N(\hat{x}^{(i)}(0), P^{(i)}(0)), \qquad i = 1, 2 \qquad (10.4b)$$

for the local processors (or local station). It is further assumed that the two local stations are linked to the central processor and not to each other (see Figure 10.1).

Given the measurement data $\{Z^1, Z^2\}$, $Z^i \triangleq \{z_i(t), 0 \leqslant t \leqslant T\}$, $i = 1, 2$, we desire to develop a decentralized method which obtains the global smoothed estimates $\hat{x}(t|T) = E[x(t)|Z^1, Z^2]$ in terms of the local filtered estimates, or local smoothed estimates.

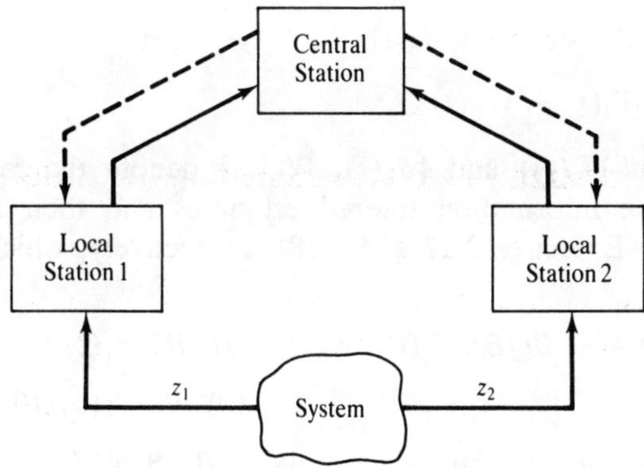

Figure 10.1 A decentralized estimation structure.

Among existing methods, we adopt here the Mayne–Fraser two-filter smoother formula [10–12] as discussed in Chapter 6. If $i = 1$, the smoother for the central processor is described by

$$P_s^{-1}(t)\hat{x}(t|T) = P_f^{-1}(t)\hat{x}_f(t) + P_b^{-1}(t)\hat{x}_b(t) \tag{10.5}$$

$$P_s^{-1}(t) = P_f^{-1}(t) + P_b^{-1}(t) \tag{10.6}$$

where $\{\hat{x}_f(t), P_f(t)\}$ and $\{\hat{x}_b(t), P_b(t)\}$ denote the forward- and backward-time Kalman filtered estimates and error covariances, respectively, which are subject to the equations

$$\dot{\hat{x}}_f(t) = (A - P_f(t)H_1^T R_1^{-1} H_1)\hat{x}_f(t) + P_f(t)H_1^T R_1^{-1} z_1(t), \qquad \hat{x}_f(0) = \hat{x}(0) \tag{10.7}$$

$$\dot{P}_f(t) = AP_f(t) + P_f(t)A^T + BQB^T - P_f(t)H_1^T R_1^{-1} H_1 P_f(t), \qquad P_f(0) = P(0) \tag{10.8}$$

and

$$-\dot{\hat{x}}_b(t) = -(A + P_b(t)H_1^T R_1^{-1} H_1)\hat{x}_b(t) + P_b(t)H_1^T R_1^{-1} z_1(t), \qquad P_b^{-1}(T)\hat{x}_b(T) = 0 \tag{10.9}$$

$$-\dot{P}_b(t) = -AP_b(t) - P_b A^T + BQB^T - P_b(t)H_1^T R_1^{-1} H_1 P_b(t), \qquad P_b^{-1}(T) = 0 \tag{10.10}$$

We can also introduce the information representations for the right-hand sides of equations (10.5) and (10.6). That is, defining $d_f(t) \triangleq P_f^{-1}(t)\hat{x}_f(t)$, $W_f(t) \triangleq P_f^{-1}(t)$, $d_b(t) \triangleq P_b^{-1}(t)\hat{x}_b(t)$ and $W_b(t) \triangleq P_b^{-1}(t)$, we obtain

$$P_s^{-1}(t)\hat{x}(t|T) = d_f(t) + d_b(t) \tag{10.11}$$

$$P_s^{-1}(t) = W_f(t) + W_b(t) \tag{10.12}$$

where $\{d_f(t), W_f(t)\}$ and $\{d_b(t), W_b(t)\}$ denote the forward- and backward-time information filtered estimates and their information matrices (see Exercises 2.17 and 2.18), respectively, which are given by

$$\dot{d}_f(t) = -(A + BQB^T W_f(t))^T d_f(t) + H_1^T R_1^{-1} z_1(t), \qquad d_f(0) = P^{-1}(0)\hat{x}(0) \tag{10.13}$$

$$\dot{W}_f(t) = -W_f(t)A - A^T W_f(t) - W_f(t)BQB^T W_f(t) + H_1^T R_1^{-1} H_1, \qquad W_f(0) = P^{-1}(0) \tag{10.14}$$

and

$$-\dot{d}_b(t) = (A - BQB^T W_b(t))^T d_b(t) + H_1^T R_1^{-1} z_1(t), \qquad d_b(T) = 0 \tag{10.15}$$

$$-\dot{W}_b(t) = W_b(t)A + A^T W_b(t) - W_b(t)BQB^T W_b(t) + H_1^T R_1^{-1} H_1, \qquad W_b(T) = 0 \tag{10.16}$$

10.2.1 Decentralized Smoothing Problems

We are now concerned with decentralized smoothing using two representations of the two-filter smoother indicated above, but based on the local filtered estimates. In order to distinguish the one from the other, we will call the algorithm based on the information filters Algorithm I and that based on the Kalman filters Algorithm II.

Given the measurement data $\{Z^1, Z^2\}$ at the central processor, the global smoothed estimates $\hat{x}(t|T)$ are obtained by equations (10.11) and (10.12), where $\{d_f(t), W_f(t)\}$ reduce to

$$\dot{d}_f(t) = -(A + BQB^T W_f(t))^T d_f(t) + \sum_{i=1}^{2} H_i^T R_i^{-1} z_i(t), \qquad d_f(0) = P^{-1}(0)\hat{x}(0) \tag{10.17}$$

$$\dot{W}_f(t) = -W_f(t)A - A^T W_f(t) - W_f(t)BQB^T W_f(t) + \sum_{i=1}^{2} H_i^T R_i^{-1} H_i, \qquad W_f(0) = P^{-1}(0) \tag{10.18}$$

which are together called the forward-time global information filter, and $\{d_b(t), W_b(t)\}$ reduce to

$$-\dot{d}_b(t) = (A - BQB^T W_b(t))^T d_b(t) + \sum_{i=1}^{2} H_i^T R_i^{-1} z_i(t), \qquad d_b(T) = 0 \tag{10.19}$$

$$-\dot{W}_b(t) = W_b(t)A + A^T W_b(t) - W_b(t)BQB^T W_b(t) + \sum_{i=1}^{2} H_i^T R_i^{-1} H_i, \qquad W_b(T) = 0 \tag{10.20}$$

which constitute the backward-time global information filter.

On the other hand, given the measurement data $\{z_i(s), 0 \leq s \leq t\}$ at the ith local station, the local information filtered values $\{d_f^{(i)}(t), W_f^{(i)}(t)\}$ and $\{d_b^{(i)}(t), W_b^{(i)}(t)\}$ are obtained by

$$\dot{d}_{\text{f}}^{(i)}(t) = -(A + BQB^{\text{T}}W_{\text{f}}^{(i)}(t))^{\text{T}}d_{\text{f}}^{(i)}(t) + H_i^{\text{T}}R_i^{-1}z_i(t),$$
$$d_{\text{f}}^{(i)}(0) = P^{(i)^{-1}}(0)\hat{x}^{(i)}(0) \qquad (10.21)$$

$$\dot{W}_{\text{f}}^{(i)}(t) = -W_{\text{f}}^{(i)}(t)A - A^{\text{T}}W_{\text{f}}^{(i)}(t) - W_{\text{f}}^{(i)}(t)BQB^{\text{T}}W_{\text{f}}^{(i)}(t)$$
$$+ H_i^{\text{T}}R_i^{-1}H_i, \qquad W_{\text{f}}^{(i)}(0) = P^{(i)^{-1}}(0) \qquad (10.22)$$

and

$$-\dot{d}_{\text{b}}^{(i)}(t) = (A - BQB^{\text{T}}W_{\text{b}}^{(i)}(t))^{\text{T}}d_{\text{b}}^{(i)}(t) + H_i^{\text{T}}R_i^{-1}z_i(t),$$
$$d_{\text{b}}^{(i)}(T) = 0 \qquad (10.23)$$

$$-\dot{W}_{\text{b}}^{(i)}(t) = W_{\text{b}}^{(i)}(t)A + A^{\text{T}}W_{\text{b}}^{(i)}(t) - W_{\text{b}}^{(i)}(t)BQB^{\text{T}}W_{\text{b}}^{(i)}(t)$$
$$+ H_i^{\text{T}}R_i^{-1}H_i, \qquad W_{\text{b}}^{(i)}(T) = 0 \qquad (10.24)$$

where $d_{\text{f}}^{(i)}(t) \triangleq P_{\text{f}}^{(i)^{-1}}(t)\hat{x}_{\text{f}}^{(i)}(t)$, $W_{\text{f}}^{(i)}(t) \triangleq P_{\text{f}}^{(i)^{-1}}(t)$, $d_{\text{b}}^{(i)}(t) \triangleq P_{\text{b}}^{(i)^{-1}}(t)\hat{x}_{\text{b}}^{(i)}(t)$ and $W_{\text{b}}^{(i)}(t) \triangleq P_{\text{b}}^{(i)^{-1}}(t)$, in which $\{\hat{x}_{\text{f}}^{(i)}(t), P_{\text{f}}^{(i)}(t)\}$ and $\{\hat{x}_{\text{b}}^{(i)}(t), P_{\text{b}}^{(i)}(t)\}$ represent the ith local forward- and backward-time Kalman filtering estimates and error covariances, respectively.

This leads to the following result based on the information filters.

Theorem 10.1

The decentralized smoothing algorithm based on the local information filters, i.e. Algorithm I, can be implemented in two ways:

(i) *Centralized implementation*

$$\hat{x}(t|T) = P_{\text{s}}(t)\left[q_{\text{f}}(t) + \sum_{i=1}^{2}d_{\text{f}}^{(i)}(t) + q_{\text{b}}(t) + \sum_{i=1}^{2}d_{\text{b}}^{(i)}(t)\right] \qquad (10.25)$$

where $q_{\text{f}}(t)$ and $q_{\text{b}}(t)$ are subject to

$$\dot{q}_{\text{f}}(t) = -(A + BQB^{\text{T}}W_{\text{f}}(t))^{\text{T}}q_{\text{f}}(t)$$
$$+ \sum_{i=1}^{2}(W_{\text{f}}^{(i)}(t) - W_{\text{f}}(t))BQB^{\text{T}}d_{\text{f}}^{(i)}(t),$$
$$q_{\text{f}}(0) = P^{-1}(0)\hat{x}^{(i)}(0) - \sum_{i=1}^{2}P^{(i)^{-1}}(0)\hat{x}^{(i)}(0) \qquad (10.26)$$

$$-\dot{q}_{\text{b}}(t) = (A - BQB^{\text{T}}W_{\text{b}}(t))^{\text{T}}q_{\text{b}}(t)$$
$$+ \sum_{i=1}^{2}(W_{\text{b}}^{(i)}(t) - W_{\text{b}}(t))BQB^{\text{T}}d_{\text{b}}^{(i)}(t), \qquad q_{\text{b}}(T) = 0$$
$$(10.27)$$

(ii) *Decentralized implementation*

$$\hat{x}(t|T) = P_s(t)\left[\sum_{i=1}^{2} q_f^{(i)}(t) + d_f^{(i)} + q_b^{(i)}(t) + d_b^{(i)}(t)\right] \tag{10.28}$$

where $q_f^{(i)}(t)$ and $q_b^{(i)}(t)$ satisfy the following equations

$$\begin{aligned}\dot{q}_f^{(i)}(t) = &-(A + BQB^T W_f(t))^T q_f^{(i)}(t)\\ &+ (W_f^{(i)}(t) - W_f(t))BQB^T d_f^{(i)}(t),\\ &q_f^{(1)}(0) = -P^{(1)^{-1}}(0)\hat{x}^{(1)}(0),\\ &q_f^{(2)}(0) = P^{-1}(0)\hat{x}(0) - P^{(2)^{-1}}\hat{x}^{(2)}(0)\end{aligned} \tag{10.29}$$

$$\begin{aligned}-\dot{q}_b^{(i)}(t) = &(A - BQB^T W_b(t))^T q_b^{(i)}(t)\\ &+ (W_b^{(i)}(t) - W_b(t))BQB^T d_b^{(i)}(t), \qquad q_b^{(i)}(T) = 0\end{aligned} \tag{10.30}$$

Proof

Using the closed-loop transition matrix for the global information filter given in equation (10.17), the local information filter given in equation (10.21) can be rewritten as follows:

$$\begin{aligned}\dot{d}_f^{(i)}(t) = &-(A + BQB^T W_f(t))^T d_f^{(i)}(t)\\ &+ (W_f(t) - W_f^{(i)}(t))BQB^T d_f^{(i)}(t)\\ &+ H_i^T R_i^{-1} z_i(t), \qquad d_f^{(i)}(0) = P^{(i)^{-1}}(0)\hat{x}^{(i)}(0)\end{aligned} \tag{10.31}$$

By superposition, we find that

$$q_f(t) \triangleq d_f(t) - \sum_{i=1}^{2} d_f^{(i)}(t) \tag{10.32}$$

satisfies the differential equation (10.26). This $q_f(t)$ can also be implemented in a decentralized fashion as

$$q_f(t) = \sum_{i=1}^{2} q_f^{(i)}(t) \tag{10.33}$$

where $q_f^{(i)}(t)$ is as given in equation (10.29), and it should be noted that the choice of the initial conditions is not unique.

In the same manner, if

$$q_b(t) \triangleq d_f(t) - \sum_{i=1}^{2} d_b^{(i)}(t) \tag{10.34}$$

is defined for the backward-time information filter, then we understand that $q_b(t)$ is given by equation (10.27). This can also be implemented in a decentralized fashion as

$$q_{\mathrm{b}}(t) = \sum_{i=1}^{2} q_{\mathrm{b}}^{(i)}(t) \tag{10.35}$$

in which $q_{\mathrm{b}}^{(i)}(t)$ are given by equation (10.30). □

In Algorithm II, the global smoothed estimates are obtained by equations (10.5) and (10.6), where $\{\hat{x}_{\mathrm{f}}(t), P_{\mathrm{f}}(t)\}$ satisfy the equations which constitute the forward-time global Kalman filter:

$$\begin{aligned}\dot{\hat{x}}_{\mathrm{f}}(t) &= (A - P_{\mathrm{f}}(t)\sum_{i=1}^{2} H_i^{\mathrm{T}} R_i^{-1} H_i)\hat{x}_{\mathrm{f}}(t) \\ &\quad + P_{\mathrm{f}}(t) \sum_{i=1}^{2} H_i^{\mathrm{T}} R_i^{-1} z_i(t), \qquad \hat{x}_{\mathrm{f}}(0) = \hat{x}(0)\end{aligned} \tag{10.36}$$

$$\begin{aligned}\dot{P}_{\mathrm{f}}(t) &= AP_{\mathrm{f}}(t) + P_{\mathrm{f}}(t)A^{\mathrm{T}} + BQB^{\mathrm{T}} \\ &\quad - P_{\mathrm{f}}(t) \sum_{i=1}^{2} H_i^{\mathrm{T}} R_i^{-1} H_i P_{\mathrm{f}}(t), \qquad P_{\mathrm{f}}(0) = P(0)\end{aligned} \tag{10.37}$$

and $\{\hat{x}_{\mathrm{b}}(t), P_{\mathrm{b}}(t)\}$ satisfy the equations which constitute the backward-time global Kalman filter:

$$\begin{aligned}-\dot{\hat{x}}_{\mathrm{b}}(t) &= -(A + P_{\mathrm{b}}(t)\sum_{i=1}^{2} H_i^{\mathrm{T}} R_i^{-1} H_i)\hat{x}_{\mathrm{b}}(t) \\ &\quad + P_{\mathrm{b}}(t)\sum_{i=1}^{2} H_i^{\mathrm{T}} R_i^{-1} z_i(t), \qquad P_{\mathrm{b}}^{-1}(T)\hat{x}_{\mathrm{b}}(T) = 0\end{aligned} \tag{10.38}$$

$$\begin{aligned}-\dot{P}_{\mathrm{b}}(t) &= -AP_{\mathrm{b}}(t) - P_{\mathrm{b}}(t)A^{\mathrm{T}} + BQB^{\mathrm{T}} \\ &\quad - P_{\mathrm{b}}(t)\sum_{i=1}^{2} H_i^{\mathrm{T}} R_i^{-1} H_i P_{\mathrm{b}}(t), \qquad P_{\mathrm{b}}^{-1}(T) = 0\end{aligned} \tag{10.39}$$

At the ith local station, the corresponding local Kalman filtered estimates $\{\hat{x}_{\mathrm{f}}^{(i)}(t), P_{\mathrm{f}}^{(i)}(t)\}$ and $\{\hat{x}_{\mathrm{b}}^{(i)}(t), P_{\mathrm{b}}^{(i)}(t)\}$ are, respectively, given by

$$\begin{aligned}\dot{\hat{x}}_{\mathrm{f}}^{(i)}(t) &= (A - P_{\mathrm{f}}^{(i)}(t) H_i^{\mathrm{T}} R_i^{-1} H_i)\hat{x}_{\mathrm{f}}^{(i)}(t) \\ &\quad + P_{\mathrm{f}}^{(i)}(t) H_i^{\mathrm{T}} R_i^{-1} z_i(t), \qquad \hat{x}_{\mathrm{f}}^{(i)}(t) = \hat{x}^{(i)}(0)\end{aligned} \tag{10.40}$$

$$\begin{aligned}\dot{P}_{\mathrm{f}}^{(i)}(t) &= AP_{\mathrm{f}}^{(i)}(t) + P_{\mathrm{f}}^{(i)}(t)A^{\mathrm{T}} + BQB^{\mathrm{T}} \\ &\quad - P_{\mathrm{f}}^{(i)}(t) H_i^{\mathrm{T}} R_i^{-1} H_i P_{\mathrm{f}}^{(i)}(t), \qquad P_{\mathrm{f}}^{(i)}(0) = P^{(i)}(0)\end{aligned} \tag{10.41}$$

and

$$-\dot{\hat{x}}_{\mathrm{b}}^{(i)}(t) = -(A + P_{\mathrm{b}}^{(i)}(t) H_i^{\mathrm{T}} R_i^{-1} H_i)\hat{x}_{\mathrm{b}}^{(i)}(t)$$

$$+ P_{\mathrm{b}}^{(i)}(t)H_i^{\mathrm{T}}R_i^{-1}z_i(t), \qquad P_{\mathrm{b}}^{(i)^{-1}}(T)\hat{x}_{\mathrm{b}}^{(i)}(T) = 0 \tag{10.42}$$

$$-\dot{P}_{\mathrm{b}}^{(i)}(t) = -AP_{\mathrm{b}}^{(i)}(t) - P_{\mathrm{b}}^{(i)}(t)A^{\mathrm{T}} + BQB^{\mathrm{T}} - P_{\mathrm{b}}^{(i)}(t)H_i^{\mathrm{T}}R_i^{-1}H_iP_{\mathrm{b}}^{(i)}(t), \qquad P_{\mathrm{b}}^{(i)^{-1}}(T) = 0 \tag{10.43}$$

The following theorem gives the decentralized smoothing algorithms based on the local Kalman filtered estimates (Algorithm II).

Theorem 10.2

The decentralized smoothing algorithms based on the local Kalman filtered estimates are given by

(i) *Centralized implementation*

$$\hat{x}(t|T) = P_{\mathrm{s}}(t)\left[P_{\mathrm{f}}^{-1}(t)\left\{r_{\mathrm{f}}(t) + P_{\mathrm{f}}(t)\sum_{i=1}^{2}P_{\mathrm{f}}^{(i)^{-1}}(t)\hat{x}_{\mathrm{f}}^{(i)}(t)\right\} + P_{\mathrm{b}}^{-1}(t)\left\{r_{\mathrm{b}}(t) + P_{\mathrm{b}}(t)\sum_{i=1}^{2}P_{\mathrm{b}}^{(i)^{-1}}(t)\hat{x}_{\mathrm{b}}^{(i)}(t)\right\}\right] \tag{10.44}$$

where $r_{\mathrm{f}}(t)$ and $r_{\mathrm{b}}(t)$ satisfy

$$\dot{r}_{\mathrm{f}}(t) = \left(A - P_{\mathrm{f}}(t)\sum_{i=1}^{2}H_i^{\mathrm{T}}R_i^{-1}H_i\right)r_{\mathrm{f}}(t) + \sum_{i=1}^{2}(P_{\mathrm{f}}(t)P_{\mathrm{f}}^{(i)^{-1}}(t) - I)BQB^{\mathrm{T}}P_{\mathrm{f}}^{(i)^{-1}}(t)\hat{x}_{\mathrm{f}}^{(i)}(t),$$

$$r_{\mathrm{f}}(0) = \hat{x}(0) - P(0)\sum_{i=1}^{2}P^{(i)^{-1}}(0)\hat{x}^{(i)}(0) \tag{10.45}$$

and

$$-\dot{r}_{\mathrm{b}}(t) = -\left(A + P_{\mathrm{b}}(t)\sum_{i=1}^{2}H_i^{\mathrm{T}}R_i^{-1}H_i\right)r_{\mathrm{b}}(t) + \sum_{i=1}^{2}(P_{\mathrm{b}}(t)P_{\mathrm{b}}^{(i)^{-1}}(t) - I)BQB^{\mathrm{T}}P_{\mathrm{b}}^{(i)^{-1}}(t)\hat{x}_{\mathrm{b}}^{(i)}(t),$$

$$P_{\mathrm{b}}^{-1}(T)r_{\mathrm{b}}(T) = 0 \tag{10.46}$$

(ii) *Decentralized implementation*

$$\hat{x}(t|T) = P_{\mathrm{s}}(t)\left[P_{\mathrm{f}}^{-1}(t)\left\{\sum_{i=1}^{2}r_{\mathrm{f}}^{(i)}(t) + P_{\mathrm{f}}(t)P_{\mathrm{f}}^{(i)^{-1}}(t)\hat{x}_{\mathrm{f}}^{(i)}(t)\right\} + P_{\mathrm{b}}^{-1}(t)\left\{\sum_{i=1}^{2}r_{\mathrm{b}}^{(i)}(t) + P_{\mathrm{b}}(t)P_{\mathrm{b}}^{(i)^{-1}}(t)\hat{x}_{\mathrm{b}}^{(i)}(t)\right\}\right] \tag{10.47}$$

where $r_{\rm f}^{(i)}(t)$ and $r_{\rm b}^{(i)}(t)$ are given by

$$\dot{r}_{\rm f}^{(i)}(t) = \left(A - P_{\rm f}(t)\sum_{i=1}^{2} H_i^{\rm T}R_i^{-1}H_i\right)r_{\rm f}^{(i)}(t) + (P_{\rm f}(t)P_{\rm f}^{(i)^{-1}}(t) - I)\; BQB^{\rm T}P_{\rm f}^{(i)^{-1}}(t)\hat{x}_{\rm f}^{(i)}(t),$$

$$r_{\rm f}^{(1)}(0) = -P(0)P^{(1)^{-1}}(0)\hat{x}^{(1)}(0)$$

$$r_{\rm f}^{(2)}(0) = \hat{x}(0) - P(0)P^{(2)^{-1}}(0)\hat{x}^{(2)}(0) \tag{10.48}$$

and

$$-\dot{r}_{\rm b}^{(i)}(t) = -\left(A + P_{\rm b}(t)\sum_{i=1}^{2} H_i^{\rm T}R_i^{-1}H_i\right)r_{\rm b}^{(i)}(t) + (P_{\rm b}(t)P_{\rm b}^{(i)^{-1}}(t) - I)\; BQB^{\rm T}P_{\rm b}^{(i)^{-1}}(t)\hat{x}_{\rm b}^{(i)}(t),$$

$$P_{\rm b}^{-1}(T)r_{\rm b}^{(i)}(T) = 0 \tag{10.49}$$

Proof

With the aid of the results of Theorem 10.1 and defining

$$r_{\rm f}(t) \triangleq P_{\rm f}(t)q_{\rm f}(t) = \hat{x}_{\rm f}(t) - P_{\rm f}(t)\sum_{i=1}^{2} P_{\rm f}^{(i)^{-1}}(t)\hat{x}_{\rm f}^{(i)}(t) \tag{10.50}$$

we have equation (10.45). This correction value is also obtained by a decentralized method as follows:

$$r_{\rm f}(t) = \sum_{i=1}^{2} r_{\rm f}^{(i)}(t) \tag{10.51}$$

where $r_{\rm f}^{(i)}(t)$ is given by equation (10.48). Similarly, it is found that

$$r_{\rm b}(t) \triangleq P_{\rm b}(t)q_{\rm b}(t) = \hat{x}_{\rm b}(t) - P_{\rm b}(t)\sum_{i=1}^{2} P_{\rm b}^{(i)^{-1}}(t)\hat{x}_{\rm b}^{(i)}(t) \tag{10.52}$$

satisfies the backward-time differential equation (10.46). In a decentralized manner, it can be computed by

$$r_{\rm b}(t) = \sum_{i=1}^{2} r_{\rm b}^{(i)}(t) \tag{10.53}$$

where $r_{\rm b}^{(i)}(t)$ is specified by equation (10.49). □

An advantage of Theorems 10.1 and 10.2 over the algorithm based on the local smoothed estimates $\hat{x}^{(i)}(t|T) = E[x(t)|Z^i]$, which will be discussed later (see also Willsky *et al.* [7] or Levy *et al.* [13]), is that we can see the global filtered estimates in the course of the computations. The local and central processors presented here can proceed in a parallel fashion so that there are no waiting times for the central processor.

Furthermore, comparing with equations (10.5) and (10.44) or (10.47), we readily obtain the solution to the problem of decentralized Kalman filtering.

Corollary 10.1

The centralized filtered estimate $\hat{x}_f(t)$ can be obtained in two ways:

(i) *Centralized implementation*

$$\hat{x}_f(t) = r_f(t) + P_f(t)\sum_{i=1}^{2} P_f^{(i)^{-1}}(t)\hat{x}_f^{(i)}(t) \tag{10.54}$$

where $r_f(t)$ is given by equation (10.45).

(ii) *Decentralized implementation*

$$\hat{x}_f(t) = \sum_{i=1}^{2} r_f^{(i)} + P_f(t)P_f^{(i)^{-1}}(t)\hat{x}_f^{(i)}(t) \tag{10.55}$$

where $r_f^{(i)}(t)$ is subject to equation (10.48).

So far, we have derived the decentralized smoothing algorithms by using the local filtered estimates. In the remainder of this section, we construct alternative decentralized smoothing algorithms based on using the local smoothed estimates. In such a case, $\hat{x}^{(i)}(t|T) = E[x(t)|Z^i]$, $i = 1, 2$ are available, instead of $\{\hat{x}_f^{(i)}(t), P_f^{(i)}(t)\}$ and $\{\hat{x}_b^{(i)}(t), P_b^{(i)}(t)\}$, $i = 1, 2$.

Following this approach, we can develop a decentralized algorithm for the fixed-interval smoother.

Theorem 10.3

The decentralized smoothing algorithm based on the local smoothed estimates is given by

$$\hat{x}(t|T) = \xi(t) + \sum_{i=1}^{2}\hat{x}^{(i)}(t|T) \tag{10.56}$$

where $\xi(t)$ is given by

$$\xi(t) = P_s(t)[P_f^{-1}(t)\xi_f(t) + P_b^{-1}(t)\xi_b(t)] \tag{10.57}$$

$\xi_f(t)$ and $\xi_b(t)$ satisfy the following equations

$$\dot{\xi}_f(t) = \left(A - P_f(t)\sum_{i=1}^{2} H_i^T R_i^{-1} H_i\right)\xi_f(t) - P_f(t)[H_1^T R_1^{-1} H_1 \hat{x}^{(2)}(t|T) + H_2^T R_2^{-1} H_2 \hat{x}^{(1)}(t|T)],$$

$$\xi_f(0) = \hat{x}(0) - P(0)\sum_{i=1}^{2} P^{(i)^{-1}}(0)\hat{x}^{(i)}(0) + \sum_{i=1}^{2}(P(0)P^{(i)^{-1}}(0) - I)\hat{x}^{(i)}(0|T) \qquad (10.58)$$

and

$$-\dot{\xi}_b(t) = -\left(A + P_b(t)\sum_{i=1}^{2} H_i^T R_i^{-1} H_i\right)\xi_b(t) - P_b(t)[H_1^T R_1^{-1} H_1 \hat{x}^{(2)}(t|T) + H_2^T R_2^{-1} H_2 \hat{x}^{(1)}(t|T)],$$

$$P_b^{-1}(T)\xi_b(T) = 0 \qquad (10.59)$$

and where $P_s(t)$, $P_f(t)$ and $P_b(t)$ denote the global smoothing and forward- and backward-time Kalman filtering error covariances, respectively, given by equations (10.6), (10.37) and (10.39).

Proof

See Willsky *et al.* [7] and Levy *et al.* [13]. □

Example 10.1

Consider a third-order system with two output measurements,

$$H_1 = [0 \ \ 1 \ \ 0], \qquad R_1 = 1$$

$$H_2 = [0 \ \ 0 \ \ 1], \qquad R_2 = 1$$

For this example, the compensator $\xi(t)$ in Theorem 10.3 can be calculated from equation (10.57), in which $\xi_f(t)$ and $\xi_b(t)$ are given by

$$\dot{\xi}_f(t) = \left(A - P_f(t)\begin{bmatrix} 0 & 0 & 0 \\ 0 & 1 & 0 \\ 0 & 0 & 1 \end{bmatrix}\right)\xi_f(t) - P_f(t)\begin{bmatrix} 0 \\ \hat{x}_2^{(2)}(t|T) \\ \hat{x}_3^{(1)}(t|T) \end{bmatrix}$$

$$-\dot{\xi}_b(t) = -\left(A + P_b(t)\begin{bmatrix} 0 & 0 & 0 \\ 0 & 1 & 0 \\ 0 & 0 & 1 \end{bmatrix}\right)\xi_b(t)$$

$$- P_b(t)\begin{bmatrix} 0 \\ \hat{x}_2^{(2)}(t|T) \\ \hat{x}_3^{(1)}(t|T) \end{bmatrix}$$

Here, $\hat{x}_2^{(2)}(t|T)$ is the second element of $\hat{x}^{(2)}(t|T)$ and $\hat{x}_3^{(1)}(t|T)$ is the third element of $\hat{x}^{(1)}(t|T)$. □

10.2.2 The Smoothing Update Problem

In this subsection, it is assumed that the first local processor has already computed two local information or Kalman filters. Given measurement data Z^2 at the second local station, instead of reprocessing all the data $\{Z^1, Z^2\}$ to solve $\hat{x}(t|T)$, we desire to update $\{d_f^{(1)}(t), d_b^{(1)}(t)\}$ or $\{\hat{x}_f^{(1)}(t), \hat{x}_b^{(1)}(t)\}$ to incorporate the new measurement data Z^2. These smoothing update problems are essentially due to the results of filtering update problems.

Corollary 10.2

The smoothing update algorithms based on the information filtered estimates result in the following:

(i) *Centralized implementation*

$$\hat{x}(t|T) = P_s(t)[\bar{q}_f(t) + d_f^{(1)}(t) + \bar{q}_b(t) + d_b^{(1)}(t)] \tag{10.60}$$

where $\bar{q}_f(t)$ and $\bar{q}_b(t)$ satisfy

$$\begin{aligned}\dot{\bar{q}}_f(t) &= -(A + BQB^T W_f(t))^T \bar{q}_f(t) \\ &\quad + (W_f^{(1)}(t) - W_f(t))BQB^T d_f^{(1)}(t) \\ &\quad + H_2^T R_2^{-1} z_2(t), \qquad \bar{q}_f(0) = P^{-1}(0)\hat{x}(0) - P^{(1)^{-1}}(0)\hat{x}^{(1)}(0)\end{aligned} \tag{10.61}$$

$$\begin{aligned}-\dot{\bar{q}}_b(t) &= (A - BQB^T W_b(t))^T \bar{q}_b(t) \\ &\quad + (W_b^{(1)}(t) - W_b(t))BQB^T d_b^{(1)}(t) + H_2^T R_2^{-1} z_2(t), \\ &\qquad \bar{q}_b(T) = 0\end{aligned} \tag{10.62}$$

(ii) *Decentralized implementation*

$$\hat{x}(t|T) = P_s(t)[q_f^{(1)}(t) + \bar{q}_f^{(2)}(t) + d_f^{(1)}(t) + q_b^{(1)}(t) + \bar{q}_b^{(2)}(t) + d_b^{(1)}(t)] \tag{10.63}$$

where $\bar{q}_f^{(2)}(t)$ and $\bar{q}_b^{(2)}(t)$ are given by

$$\dot{\bar{q}}_f^{(2)}(t) = -(A + BQB^T W_f(t))^T \bar{q}_f^{(2)}(t) + H_2^T R_2^{-1} z_2(t), \quad \bar{q}_f^{(2)}(0) = P^{-1}(0)\hat{x}(0) \tag{10.64}$$

and

$$-\dot{\bar{q}}_b^{(2)}(t) = (A - BQB^T W_b(t))^T \bar{q}_b^{(2)}(t) + H_2^T R_2^{-1} z_2(t), \quad \bar{q}_b^{(2)}(T) = 0 \tag{10.65}$$

Proof

Using similar ideas in Section 10.2.1, it is found that

$$\bar{q}_f(t) = d_f(t) - d_f^{(1)}(t) \tag{10.66}$$

satisfies equation (10.61). When the correction value $q_f^{(1)}(t)$, which has been calculated from equation (10.29) in the decentralized smoothing problem, is available, it is found that $\bar{q}_f(t)$ can also be decomposed as

$$\bar{q}_f(t) = q_f^{(1)}(t) + \bar{q}_f^{(2)}(t) \tag{10.67}$$

where $\bar{q}_f^{(2)}$ is given by equation (10.64). Similarly, defining

$$\bar{q}_b(t) \triangleq d_b(t) - d_b^{(1)}(t) \tag{10.68}$$

for the backward-time information filters, we get equation (10.62). If $q_b^{(1)}(t)$, obtained from equation (10.30), is used, then $\bar{q}_b(t)$ can also be rewritten as

$$\bar{q}_b(t) = q_b^{(1)}(t) + \bar{q}_b^{(2)}(t) \tag{10.69}$$

where $\bar{q}_b^{(2)}$ satisfies equation (10.65). □

By defining

$$\bar{r}_f(t) \triangleq P_f(t)\bar{q}_f(t) = \hat{x}_f(t) - P_f(t)P_f^{(1)^{-1}}(t)\hat{x}_f^{(1)}(t) \tag{10.70}$$

$$\bar{r}_b(t) \triangleq P_b(t)\bar{q}_b(t) = \hat{x}_b(t) - P_b(t)P_b^{(1)^{-1}}(t)\hat{x}_b^{(1)}(t) \tag{10.71}$$

and by decomposing

$$\bar{r}_f(t) = r_f^{(1)}(t) + \bar{r}_f^{(2)}(t) \tag{10.72}$$

$$\bar{r}_b(t) = r_b^{(1)}(t) + \bar{r}_b^{(2)}(t) \tag{10.73}$$

we obtain the smoothing update algorithm based on the local Kalman filters, which is summarized in the following corollary.

Corollary 10.3

The smoothing update algorithms based on the Kalman filtered estimates are as follows:

(i) *Centralized implementation*

$$\begin{aligned}\hat{x}(t|T) = P_s(t)[&P_f^{-1}\{\bar{r}_f(t) + P_f(t)P_f^{(1)^{-1}}(t)\hat{x}_f^{(1)}(t)\} \\ &+ P_b^{-1}(t)\{\bar{r}_b(t) + P_b(t)P_b^{(1)^{-1}}(t)\hat{x}_b^{(1)}(t)\}]\end{aligned} \tag{10.74}$$

where

$$\begin{aligned}\dot{\bar{r}}_f(t) = &\left(A - P_f(t)\sum_{i=1}^{2} H_i^T R_i^{-1} H_i\right)\bar{r}_f(t) \\ &+ (P_f(t)P_f^{(1)^{-1}}(t) - I)BQB^T P_f^{(1)^{-1}}(t)\hat{x}_f^{(1)}(t) \\ &+ P_f(t)H_2^T R_2^{-1} z_2(t), \\ &\qquad \bar{r}_f(0) = \hat{x}(0) - P(0)P^{(1)^{-1}}(0)\hat{x}^{(1)}(0)\end{aligned} \tag{10.75}$$

$$\begin{aligned}-\dot{\bar{r}}_b(t) = &-\left(A + P_b(t)\sum_{i=1}^{2} H_i^T R_i^{-1} H_i\right)\bar{r}_b(t) \\ &+ (P_b(t)P_b^{(1)^{-1}}(t) - I)BQB^T P_b^{(1)^{-1}}(t)\hat{x}_b^{(1)}(t) \\ &+ P_b(t)H_2^T R_2^{-1} z_2(t), \qquad P_b^{-1}(T)\bar{r}_b(T) = 0\end{aligned} \tag{10.76}$$

(ii) *Decentralized implementation*

$$\begin{aligned}\hat{x}(t|T) = P_s(t)[&P_f^{-1}(t)\{r_f^{(1)}(t) + \bar{r}_f^{(2)}(t) \\ &+ P_f(t)P_f^{(1)^{-1}}(t)\hat{x}_f^{(1)}(t)\} \\ &+ P_b^{-1}(t)\{r_b^{(1)}(t) + \bar{r}_b^{(2)}(t) + P_b(t)P_b^{(1)^{-1}}(t)\hat{x}_b^{(1)}(t)\}]\end{aligned} \tag{10.77}$$

where

$$\begin{aligned}\dot{\bar{r}}_f^{(2)}(t) = &\left(A - P_f(t)\sum_{i=1}^{2} H_i^T R_i^{-1} H_i\right)\bar{r}_f^{(2)}(t) \\ &+ P_f(t)H_2^T R_2^{-1} z_2(t), \qquad \bar{r}_f^{2}(0) = \hat{x}(0)\end{aligned} \tag{10.78}$$

$$-\dot{\bar{r}}_{\mathrm{b}}^{(2)}(t) = -\left(A + P_{\mathrm{b}}(t)\sum_{i=1}^{2} H_i^{\mathrm{T}} R_i^{-1} H_i\right)\bar{r}_{\mathrm{b}}^{(2)}(t)$$
$$+ P_{\mathrm{b}}(t) H_2^{\mathrm{T}} R_2^{-1} z_2(t), \qquad P_{\mathrm{b}}^{-1}(T)\bar{r}_{\mathrm{b}}^{(2)}(T) = 0 \tag{10.79}$$

Note that $\bar{q}_{\mathrm{f}}(t) + d_{\mathrm{f}}^{(1)}(t)$ or $q_{\mathrm{f}}^{(1)}(t) + \bar{q}_{\mathrm{f}}^{(2)}(t) + d_{\mathrm{f}}^{(1)}(t)$ in Corollary 10.2 represents the information filtering update. From this point of view, the global filtered estimates are obtained by $\hat{x}_{\mathrm{f}}(t) = P_{\mathrm{f}}(t)[\bar{q}_{\mathrm{f}}(t) + d_{\mathrm{f}}^{(1)}(t)]$ or $\hat{x}_{\mathrm{f}}(t) = P_{\mathrm{f}}(t)[q_{\mathrm{f}}^{(1)}(t) + \bar{q}_{\mathrm{f}}^{(2)}(t) + d_{\mathrm{f}}^{(1)}(t)]$, if we wish to know them. A similar discussion holds for Corollary 10.3.

If information exchanges between local stations 1 and 2 are allowed, then the centralized implementation in Corollary 10.2 or 10.3 can be replaced by a decentralized one. That is, when local processor 2 receives $d_{\mathrm{f}}^{(1)}$ and $d_{\mathrm{b}}^{(1)}$ or $\hat{x}_{\mathrm{f}}^{(1)}$ and $\hat{x}_{\mathrm{b}}^{(1)}$ from local processor 1 at the same time as it receives its own measurements $z_2(t)$, the correction quantities $\bar{q}_{\mathrm{f}}(t)$ and $\bar{q}_{\mathrm{b}}(t)$ or $\bar{r}_{\mathrm{f}}(t)$ and $\bar{r}_{\mathrm{b}}(t)$ can be computed at processor 2.

For the method using local smoothed estimates, we desire to update $\hat{x}^{(1)}(t|T)$ to incorporate the new measurement data Z^2. A solution to this smoothing update problem is given by the following corollary.

Corollary 10.4

The smoothing update algorithm based on the local smoothed estimates is given by

$$\hat{x}(t|T) = \eta(t) + \hat{x}^{(1)}(t|T) \tag{10.80}$$

where $\eta(t)$ is subject to the two-filter formula:

$$\eta(t) = P_{\mathrm{s}}(t)[P_{\mathrm{f}}^{-1}(t)\eta_{\mathrm{f}}(t) + P_{\mathrm{b}}^{-1}(t)\eta_{\mathrm{b}}(t)] \tag{10.81}$$

where

$$\dot{\eta}_{\mathrm{f}}(t) = \left(A - P_{\mathrm{f}}(t)\sum_{i=1}^{2} H_i^{\mathrm{T}} R_i^{-1} H_i\right)\eta_{\mathrm{f}}(t)$$
$$- P_{\mathrm{f}}(t) H_2^{\mathrm{T}} R_2^{-1}[z_2(t) - H_2\hat{x}^{(1)}(t|T)],$$
$$\eta_{\mathrm{f}}(0) = \hat{x}(0) - P(0)P^{(1)^{-1}}(0)\hat{x}^{(1)}(0)$$
$$+ (P(0)P^{(1)^{-1}}(0) - I)\hat{x}^{(1)}(0|T) \tag{10.82}$$

and

$$-\dot{\eta}_{\mathrm{b}}(t) = -\left(A + P_{\mathrm{b}}(t)\sum_{i=1}^{2} H_i^{\mathrm{T}} R_i^{-1} H_i\right)\eta_{\mathrm{b}}(t)$$

$$- P_b(t)H_2^T R_2^{-1}[z_2(t) - H_2\hat{x}^{(1)}(t|T)],$$

$$P_b^{-1}(T)\eta_b(T) = 0 \qquad (10.83)$$

and where $P_s(t)$, $P_f(t)$ and $P_b(t)$ are given by equations (10.6), (10.37) and (10.39).

Proof

See Willsky *et al.* [7] and Levy *et al.* [13]. □

A variant of the smoothing update problem is the real-time smoothing problem. In this case, given Z^1 and $Z_t^2 \triangleq \{z_2(s), 0 \leq s \leq t\}$, then we want to compute

$$\hat{x}_{rs}(t) \triangleq E[x(t)|Z^1, Z_t^2] \qquad (10.84)$$

in terms of $d_f^{(1)}$, $d_b^{(1)}$ and $z_2(\cdot)$ or of $\hat{x}_f^{(1)}$, $\hat{x}_b^{(1)}$ and $z_2(\cdot)$.

Corollary 10.5

The real-time smoothing algorithms based on the local information filters are as follows:

(i) *Centralized implementation*

$$\hat{x}_{rs}(t) = P_{rs}(t)[\bar{q}_f(t) + d_f^{(1)}(t) + d_b^{(1)}(t)] \qquad (10.85)$$

$$P_{rs}^{-1}(t) = W_f(t) + W_b^{(1)}(t) \qquad (10.86)$$

(ii) *Decentralized implementation*

$$\hat{x}_{rs}(t) = P_{rs}(t)[q_f^{(1)}(t) + \bar{q}_f^{(2)}(t) + d_f^{(1)}(t) + d_b^{(1)}(t)] \qquad (10.87)$$

Proof

In this problem, the measurements $z_2(\cdot)$ are not allowed to be processed by the backward-time information filter of local station 2. Therefore, setting $R_2 = \infty$ in equations (10.20) and (10.24), we have $W_b(t) \equiv W_b^{(1)}(t)$. Moreover, since equation (10.62) with $R_2 = \infty$ gives a trivial solution $\bar{q}_b(t) \equiv 0$, we obtain the desired results from Corollary 10.2. □

From the same reasons as for Corollary 10.5, we have $P_b(t) \equiv P_b^{(1)}(t)$ and $\bar{r}_b(t) \equiv 0$ for the previous smoothing update problem cited in Corollary 10.3. Hence, we have the following corollary without proof.

Corollary 10.6

The real-time smoothing algorithms based on the local Kalman filters are obtained as follows:

(i) *Centralized implementation*

$$\hat{x}_{rs}(t) = P_{rs}(t)[P_f^{-1}(t)\{\bar{r}_f(t) + P_f(t)P_f^{(1)^{-1}}(t)\hat{x}_f^{(1)}(t)\} + P_b^{(1)^{-1}}(t)\hat{x}_b^{(1)}(t)] \tag{10.88}$$

$$P_{rs}^{-1}(t) = P_f^{-1}(t) + P_b^{(1)^{-1}}(t) \tag{10.89}$$

(ii) *Decentralized implementation*

$$\hat{x}_{rs}(t) = P_{rs}(t)[P_f^{-1}(t)\{r_f^{(1)}(t) + \bar{r}_f^{(2)}(t) + P_f(t)P_f^{(1)^{-1}}(t)\hat{x}_f^{(1)}(t)\} + P_b^{(1)^{-1}}(t)\hat{x}_b^{(1)}(t)] \tag{10.90}$$

The real-time smoothing algorithm based on local smoothing is also given by the following corollary.

Corollary 10.7

The real-time smoothing algorithm based on local smoothing is obtained by

$$\hat{x}_{rs}(t) = P_{rs}(t)P_f^{-1}(t)\eta_f(t) + \hat{x}^{(1)}(t|T) \tag{10.91}$$

and

$$P_{rs}^{-1}(t) = P_f^{-1}(t) + P_b^{(1)^{-1}}(t) \tag{10.92}$$

10.3 DECENTRALIZED SMOOTHING IN DISCRETE-TIME SYSTEMS

In this section, we consider the discrete-time linear dynamic system of the form

$$x(k+1) = \Phi x(k) + Bw(k) \tag{10.93}$$

$$z_i(k) = H_i x(k) + v_i(k), \qquad i = 1, 2 \tag{10.94}$$

where $\{w(k),\ v_1(k),\ v_2(k)\}$ are assumed to be zero-mean Gaussian white noises with covariances

$$E\left\{\begin{bmatrix} w(k) \\ v_1(k) \\ v_2(k) \end{bmatrix} [w^{\mathrm{T}}(j) \quad v_1^{\mathrm{T}}(j) \quad v_2^{\mathrm{T}}(j)]\right\} = \mathrm{diag}(Q,\ R_1,\ R_2)\delta_{kj} \tag{10.95}$$

and to be independent of the initial state $x(0)$. For the sake of simplicity, it is assumed that the central processor has the a priori information $x(0) \sim N(\hat{x}(0),\ P(0))$ and the local processors have the same initial conditions. Moreover, the central and local processors are assumed to be constrained by the same conditions as posed in the continuous-time case.

In what follows we will adopt the Mayne–Fraser two-filter smoother [10, 11] as a global smoother, which consists of the forward-time Kalman filter and the backward-time information predictor.

10.3.1 Decentralized Kalman Filtering

Given all measurement data $\{Z_k^1,\ Z_k^2\}$, $Z_k^i \triangleq \{z_i(j), 0 \leqslant j \leqslant k\}$, the global filtering estimate $\hat{x}(k|k) \triangleq E[x(k)|Z_k^1,\ Z_k^2]$ is given by the Kalman filter:

$$\hat{x}(k+1|k) = \Phi \hat{x}(k|k) \tag{10.96}$$

$$\hat{x}(k|k) = \hat{x}(k|k-1) + P_{\mathrm{f}}(k|k) \sum_{i=1}^{2} H_i^{\mathrm{T}} R_i^{-1} \times [z_i(k) - H_i \hat{x}(k|k-1)], \qquad \hat{x}(0|-1) = \hat{x}(0) \tag{10.97}$$

$$P_{\mathrm{f}}(k+1|k) = \Phi P_{\mathrm{f}}(k|k)\Phi^{\mathrm{T}} + BQB^{\mathrm{T}} \tag{10.98}$$

$$P_{\mathrm{f}}^{-1}(k|k) = P_{\mathrm{f}}^{-1}(k|k-1) + \sum_{i=1}^{2} H_i^{\mathrm{T}} R_i^{-1} H_i, \qquad P_{\mathrm{f}}(0|-1) = P(0) \tag{10.99}$$

where the subscript f on $\hat{x}(k|k)$ is omitted for convenience. The filtered estimate can be related to the information-type estimate, i.e. $d_{\mathrm{f}}(k) \triangleq P_{\mathrm{f}}^{-1}(k|k)\hat{x}(k|k)$ is obtained by (see Exercise 2.4 and cf. Anderson and Moore [14])

$$d_f(k) = [I - L(k)B^T]\Phi^{-T} d_f(k-1) + \sum_{i=1}^{2} H_i^T R_i^{-1} z_i(k),$$
$$d_f(0) = P_f^{-1}(0|0)\hat{x}(0|0) \qquad (10.100)$$

where

$$L(k) \triangleq FB(B^T FB + Q^{-1})^{-1}, \; F \triangleq \Phi^{-T} P_f^{-1}(k-1|k-1)\Phi^{-1} \qquad (10.101)$$

Similarly, the local filtered estimate at each station, $\hat{x}^{(i)}(k|k) \triangleq E[x(k)|Z_i^k]$ is then

$$\hat{x}^{(i)}(k+1|k) = \Phi\hat{x}^{(i)}(k|k) \qquad (10.102)$$

$$\hat{x}^{(i)}(k|k) = \hat{x}^{(i)}(k|k-1) + P_f^{(i)}(k|k)H_i^T R_i^{-1} \times [z_i(k) - H_i\hat{x}^{(i)}(k|k-1)], \qquad \hat{x}^{(i)}(0|-1) = \hat{x}(0) \qquad (10.103)$$

$$P_f^{(i)}(k+1|k) = \Phi P_f^{(i)}(k|k)\Phi^T + BQB^T \qquad (10.104)$$

$$P_f^{(i)^{-1}}(k|k) = P_f^{(i)^{-1}}(k|k-1) + H_i^T R_i^{-1} H_i, \qquad P_f^{(i)}(0|-1) = P(0) \qquad (10.105)$$

and the corresponding information expression, $d_f^{(i)}(k) \triangleq P_f^{(i)^{-1}}(k|k)\hat{x}^{(i)}(k|k)$, is reduced to

$$d_f^i(k) = [I - L^{(i)}(k)B^T]\Phi^{-T} d_f^{(i)}(k-1) + H_i^T R_i^{-1} z_i(k), \qquad d_f^{(i)}(0) = P_f^{(i)^{-1}}(0|0)\hat{x}^{(i)}(0|0) \qquad (10.106)$$

where

$$L^{(i)}(k) \triangleq F^{(i)} B(B^T F^{(i)} B + Q^{-1})^{-1}, \quad F^{(i)} \triangleq \Phi^{-T} P_f^{(i)^{-1}}(k-1|k-1)\Phi^{-1} \qquad (10.107)$$

The decentralized filtering algorithm in discrete time is summarized in the following theorem.

Theorem 10.4

The centralized filtering estimate $\hat{x}(k|k)$ in discrete time can be obtained in two ways:

(i) *Centralized implementation*

$$\hat{x}(k|k) = r_f(k) + P_f(k|k)\sum_{i=1}^{2} P_f^{(i)^{-1}}(k|k)\hat{x}^{(i)}(k|k) \tag{10.108}$$

where $r_f(k)$ is given by

$$\begin{aligned} r_f(k) = {} & P_f(k|k)[I - L(k)B^T]\Phi^{-T}P_f^{-1}(k-1|k-1)r_f(k-1) \\ & + P_f(k|k)\sum_{i=1}^{2}[L^{(i)}(k) - L(k)]B^T\Phi^{-T} \\ & \times P_f^{(i)^{-1}}(k-1|k-1)\hat{x}^{(i)}(k-1|k-1) \end{aligned} \tag{10.109}$$

or equivalently

$$\begin{aligned} r_f(k) = {} & P_f(k|k)P_f^{-1}(k|k-1)\Phi r_f(k-1) \\ & + P_f(k|k)\sum_{i=1}^{2} G^{(i)}(k)\hat{x}^{(i)}(k|k-1), \\ & r_f(0) = \hat{x}(0|0) - P_f(0|0)\sum_{i=1}^{2} P_f^{(i)^{-1}}(0|0)\hat{x}^{(i)}(0|0) \end{aligned} \tag{10.110}$$

$$\begin{aligned} G^{(i)}(k) = {} & P_f^{-1}(k|k-1)\Phi P_f(k-1|k-1) \\ & \times P_f^{(i)^{-1}}(k-1|k-1)\Phi^{-1} - P_f^{(i)^{-1}}(k|k-1) \end{aligned} \tag{10.111}$$

(ii) *Decentralized implementation*

$$\hat{x}(k|k) = \sum_{i=1}^{2} r_f^{(i)}(k) + P_f(k|k)P_f^{(i)^{-1}}(k|k)\hat{x}^{(i)}(k|k) \tag{10.112}$$

where

$$\begin{aligned} r_f^{(i)}(k) = {} & P_f(k|k)P_f^{-1}(k|k-1)\Phi r_f^{(i)}(k-1) \\ & + P_f(k|k)G^{(i)}(k)\hat{x}^{(i)}(k|k-1), \\ & r_f^{(1)}(0) = -P_f(0|0)P_f^{(1)^{-1}}(0|0)\hat{x}^{(1)}(0|0), \\ & r_f^{(2)}(0) = \hat{x}(0|0) - P_f(0|0)P_f^{(2)^{-1}}(0|0)\hat{x}^{2}(0|0) \end{aligned} \tag{10.113}$$

Proof

By replacing the closed-loop transition matrix $[I - L^{(i)}(k)B^T]$ in equation (10.106) by $[I - L(k)B^T]$, we have

$$\begin{aligned} d_f^{(i)}(k) = {} & [I - L(k)B^T]\Phi^{-T} d_f^{(i)}(k-1) \\ & + [L(k) - L^{(i)}(k)]B^T\Phi^{-T} d_f^{(i)}(k-1) \\ & + H_i^T R_i^{-1} z_i(k), \qquad d_f^{(i)}(0) = P_f^{(i)^{-1}}(0|0)\hat{x}^{(i)}(0|0) \end{aligned} \tag{10.114}$$

From this equation and equation (10.100), it is found that

$$q(k) \triangleq d_f(k) - \sum_{i=1}^{2} d_f^{(i)}(k) \tag{10.115}$$

satisfies

$$\begin{aligned} q(k) = {} & [I - L(k)B^T]\Phi^{-T} q(k-1) \\ & + \sum_{i=1}^{2} [L^{(i)}(k) - L(k)]B^T\Phi^{-T} d_f^{(i)}(k-1), \\ & q(0) = P_f^{-1}(0|0)\hat{x}(0|0) - \sum_{i=1}^{2} P_f^{(i)^{-1}}(0|0)\hat{x}(0|0) \end{aligned} \tag{10.116}$$

Then, defining

$$r_f(k) \triangleq P_f(k|k) q(k) \tag{10.117a}$$

$$= \hat{x}(k|k) - P_f(k|k) \sum_{i=1}^{2} P_f^{(i)^{-1}}(k|k)\hat{x}^{(i)}(k|k) \tag{10.117b}$$

and substituting equation (10.116) into (10.117a) yields (10.109). The equivalence of equations (10.109) and (10.110) is proved by Watanabe [16]. The decentralized implementation is obvious from equation (10.110). □

Note that Speyer [8] first derived the similar algorithm to the decentralized implementation of Theorem 9.4.

Example 10.2

Consider the system described in Example 9.9, but with the following position and velocity measurements:

$$\Phi = \begin{bmatrix} 1 & \Delta t \\ 0 & 1 \end{bmatrix}, \qquad B = \begin{bmatrix} (\Delta t)^2/2 \\ \Delta t \end{bmatrix}, \qquad Q = \sigma_a^2$$

$$H_1 = [1 \ \ 0], \qquad R_1 = \sigma_x^2$$

$$H_2 = [0 \ \ 1], \qquad R_2 = \sigma_{\dot{x}}^2$$

The local Kalman filter for local station 1 becomes

$$\hat{x}^{(1)}(k|k) = \hat{x}^{(1)}(k|k-1) + P_f^{(1)}(k|k) \times \begin{bmatrix} 1/\sigma_x^2[z_1(k) - \hat{x}_1^{(1)}(k|k-1)] \\ 0 \end{bmatrix}$$

$$P_f^{(1)^{-1}}(k|k) = P_f^{(1)^{-1}}(k|k-1) + \begin{bmatrix} 1/\sigma_x^2 & 0 \\ 0 & 0 \end{bmatrix}$$

$$\hat{x}^{(1)}(k+1|k) = \begin{bmatrix} 1 & \Delta t \\ 0 & 1 \end{bmatrix} \hat{x}^{(1)}(k|k)$$

$$P_f^{(1)}(k+1|k) = \begin{bmatrix} 1 & \Delta t \\ 0 & 1 \end{bmatrix} P_f^{(1)}(k|k) \begin{bmatrix} 1 & 0 \\ \Delta t & 1 \end{bmatrix} + \begin{bmatrix} (\Delta t)^4 \sigma_a^2/4 & (\Delta t)^3 \sigma_a^2/2 \\ (\Delta t)^3 \sigma_a^2/2 & (\Delta t)^2 \sigma_a^2 \end{bmatrix}$$

with initial conditions $\hat{x}^{(1)}(0|-1) = \hat{x}(0)$ and $P_f^{(1)}(0|-1) = P(0)$, where $\hat{x}_1^{(1)}(k|k-1)$ denotes the first element of $\hat{x}^{(1)}(k|k-1)$. The local Kalman filter for local station 2 is

$$\hat{x}^{(2)}(k|k) = \hat{x}^{(2)}(k|k-1) + P_f^{(2)}(k|k) \times \begin{bmatrix} 0 \\ 1/\sigma_x^2[z_2(k) - \hat{x}_2^{(2)}(k|k-1)] \end{bmatrix}$$

$$P_f^{(2)^{-1}}(k|k) = P_f^{(2)^{-1}}(k|k-1) + \begin{bmatrix} 0 & 0 \\ 0 & 1/\sigma_x^2 \end{bmatrix}$$

$$\hat{x}^{(2)}(k+1|k) = \begin{bmatrix} 1 & \Delta t \\ 0 & 1 \end{bmatrix} \hat{x}^{(2)}(k|k)$$

$$P_f^{(2)}(k+1|k) = \begin{bmatrix} 1 & \Delta t \\ 0 & 1 \end{bmatrix} P_f^{(2)}(k|k) \begin{bmatrix} 1 & 0 \\ \Delta t & 1 \end{bmatrix} + \begin{bmatrix} (\Delta t)^4 \sigma_a^2/4 & (\Delta t)^3 \sigma_a^2/2 \\ (\Delta t)^3 \sigma_a^2/2 & (\Delta t)^2 \sigma_a^2 \end{bmatrix}$$

with initial conditions $\hat{x}^{(2)}(0|-1) = \hat{x}(0)$ and $P_f^{(2)}(0|-1) = P(0)$, where $\hat{x}_2^{(2)}(k|k-1)$ denotes the second element of $\hat{x}^{(2)}(k|k-1)$.

The compensator is given by equation (10.109), where $P_f(k|k)$ is given by

$$P_f^{-1}(k|k) = P_f^{-1}(k|k-1) + \begin{bmatrix} 1/\sigma_x^2 & 0 \\ 0 & 1/\sigma_x^2 \end{bmatrix}$$

$$P_f(k+1|k) = \begin{bmatrix} 1 & \Delta t \\ 0 & 1 \end{bmatrix} P_f(k|k) \begin{bmatrix} 1 & 0 \\ \Delta t & 1 \end{bmatrix} + \begin{bmatrix} (\Delta t)^4 \sigma_a^2/4 & (\Delta t)^3 \sigma_a^2/2 \\ (\Delta t)^3 \sigma_a^2/2 & (\Delta t)^2 \sigma_a^2 \end{bmatrix}$$

with initial condition $P_f(0|-1) = P(0)$, $L(k)$ is given by equation (10.101) with

$$F = \begin{bmatrix} 1 & 0 \\ -\Delta t & 1 \end{bmatrix} P_f^{-1}(k-1|k-1) \begin{bmatrix} 1 & -\Delta t \\ 0 & 1 \end{bmatrix}$$

and $L^{(i)}(k)$ is given by equation (10.107) with

$$F^{(i)} = \begin{bmatrix} 1 & 0 \\ -\Delta t & 1 \end{bmatrix} P_f^{(i)^{-1}}(k-1|k-1) \begin{bmatrix} 1 & -\Delta t \\ 0 & 1 \end{bmatrix}$$

10.3.2 Decentralized Smoothing

Given the measurement data $\{Z^1, Z^2\}$, $Z^i \triangleq \{z_i(j), 0 \leq j \leq N\}$, the two-filter form of the fixed-interval smoothed estimate $\hat{x}(k|N) \triangleq E[x(k)|Z^1, Z^2]$ (see also Chapter 9) is given by

$$\hat{x}(k|N) = P_s(k)[P_f^{-1}(k|k-1)\hat{x}(k|k-1) + d_b(k)] \tag{10.118a}$$

or

$$\hat{x}(k|N) = P_s(k)[P_f^{-1}(k|k-1)\Phi\hat{x}(k-1|k-1) + d_b(k)] \tag{10.118b}$$

$$P_s(k) = [P_f^{-1}(k|k-1) + W_b(k)] \tag{10.119}$$

where $d_b(k)$ and $W_b(k)$ are the global backward-time information filtered estimate and the information matrix, which satisfy

$$d_b(k) = \{[I - D\Delta(k)D^T W_b(k+1)]\Phi\}^T d_b(k+1) + \sum_{i=1}^{2} H_i^T R_i^{-1} z_i(k), \qquad d_b(N+1) = 0 \tag{10.120}$$

$$W_b(k) = \Phi^T W_b(k+1)[I - D\Delta(k)D^T W_b(k+1)]\Phi + \sum_{i=1}^{2} H_i^T R_i^{-1} H_i, \qquad W_b(N+1) = 0 \tag{10.121}$$

in which

$$\Delta(k) = [I + D^T W_b(k+1)D]^{-1}, \qquad DD^T \triangleq BQB^T \tag{10.122}$$

The local backward-time information filter at each station is also obtained as

$$d_{\mathrm{b}}^{(i)}(k) = \{[I - D\Delta^{(i)}(k)D^{\mathrm{T}}W_{\mathrm{b}}^{(i)}(k+1)]\Phi\}^{\mathrm{T}}d_{\mathrm{b}}^{(i)}(k)+1)$$
$$+ H_i^{\mathrm{T}}R_i^{-1}z_i(k), \qquad d_{\mathrm{b}}^{(i)}(N+1) = 0 \tag{10.123}$$
$$W_{\mathrm{b}}^{(i)}(k) = \Phi^{\mathrm{T}}W_{\mathrm{b}}^{(i)}(k+1)[I - D\Delta^{(i)}(k)D^{\mathrm{T}}W_{\mathrm{b}}^{(i)}(k+1)]\Phi$$
$$+ H_i^{\mathrm{T}}R_i^{-1}H_i, \qquad W_{\mathrm{b}}^{(i)}(N+1) = 0 \tag{10.124}$$

where

$$\Delta^{(i)}(k) = [I + D^{\mathrm{T}}W_{\mathrm{b}}^{(i)}(k+1)D]^{-1} \tag{10.125}$$

The decentralized smoothing algorithm in discrete time based on the local filtering is given by the following theorem.

Theorem 10.5

The centralized smoothed estimate $\hat{x}(k|N)$ can be implemented as follows:

(i) *Centralized implementation*

$$\hat{x}(k|N) = P_{\mathrm{s}}(k)\{P_{\mathrm{f}}^{-1}(k|k-1)\Phi[r_{\mathrm{f}}(k-1) + P_{\mathrm{f}}(k-1|k-1)$$
$$\times \sum_{i=1}^{2} P_{\mathrm{f}}^{(i)^{-1}}(k-1|k-1)\hat{x}^{(i)}(k-1|k-1)]$$
$$+ q_{\mathrm{b}}(k) + \sum_{i=1}^{2} d_{\mathrm{b}}^{(i)}(k)\} \tag{10.126}$$

where $q_{\mathrm{b}}(k)$ satisfies

$$q_{\mathrm{b}}(k) = \{[I - D\Delta(k)D^{\mathrm{T}}W_{\mathrm{b}}(k+1)]\Phi\}^{\mathrm{T}}q_{\mathrm{b}}(k+1)$$
$$+ \sum_{i=1}^{2}\{D[\Delta^{(i)}(k)D^{\mathrm{T}}W_{\mathrm{b}}^{(i)}(k+1)$$
$$- \Delta(k)D^{\mathrm{T}}W_{\mathrm{b}}(k+1)]\Phi\}^{\mathrm{T}}d_{\mathrm{b}}^{(i)}(k+1),$$
$$q_{\mathrm{b}}(N+1) = 0 \tag{10.127}$$

(ii) *Decentralized implementation*

$$\hat{x}(k|N) = P_{\mathrm{s}}(k)\Bigg\{P_{\mathrm{f}}^{-1}(k|k-1)\Phi$$
$$\times \left[\sum_{i=1}^{2} r_{\mathrm{f}}^{(i)}(k-1) + P_{\mathrm{f}}(k-1|k-1)\right.$$
$$\left.\times P_{\mathrm{f}}^{(i)^{-1}}(k-1|k-1)\hat{x}^{(i)}(k-1|k-1)\right]$$
$$+ \sum_{i=1}^{2} q_{\mathrm{b}}^{(i)}(k) + d_{\mathrm{b}}^{(i)}(k)\Bigg\} \tag{10.128}$$

where

$$\begin{aligned} q_{\mathrm{b}}^{(i)}(k) = {} & \{[I - D\Delta(k)D^{\mathrm{T}}W_{\mathrm{b}}(k+1)]\Phi\}^{\mathrm{T}} q_{\mathrm{b}}^{(i)}(k+1) \\ & + \{D[\Delta^{(i)}(k)D^{\mathrm{T}}W_{\mathrm{b}}^{(i)}(k+1) \\ & - \Delta(k)D^{\mathrm{T}}W_{\mathrm{b}}(k+1)]\Phi\}^{\mathrm{T}} d_{\mathrm{b}}^{(i)}(k+1), \\ & q_{\mathrm{b}}^{(i)}(N+1) = 0 \end{aligned} \tag{10.129}$$

Proof

The quantities having the subscript ‘f’, which are associated with the forward-time filtering, are due to the results of Theorem 10.4. What is left is to derive the quantities having the subscript ‘b’, which are associated with the backward-time filtering. Since $d_{\mathrm{b}}^{(i)}(k)$ in equation (10.123) can be rewritten as

$$\begin{aligned} d_{\mathrm{b}}^{(i)}(k) = {} & \{[I - D\Delta(k)D^{\mathrm{T}}W_{\mathrm{b}}(k+1)]\Phi\}^{\mathrm{T}} d_{\mathrm{b}}^{(i)}(k+1) \\ & + \{D[\Delta(k)D^{\mathrm{T}}W_{\mathrm{b}}(k+1) \\ & - \Delta^{(i)}(k)D^{\mathrm{T}}W_{\mathrm{b}}^{(i)}(k+1)]\Phi\}^{\mathrm{T}} d_{\mathrm{b}}^{(i)}(k+1) \\ & + H_i^{\mathrm{T}} R_i^{-1} z_i(k) \end{aligned} \tag{10.130}$$

it is found, from equations (10.120) and (10.130), that

$$q_{\mathrm{b}}(k) \triangleq d_{\mathrm{b}}(k) - \sum_{i=1}^{2} d_{\mathrm{b}}^{(i)}(k) \tag{10.131}$$

yields the recursive equation (10.127). By decomposing $d_{\mathrm{b}}(k)$ such that

$$d_{\mathrm{b}}(k) = \sum_{i=1}^{2} q_{\mathrm{b}}^{(i)}(k) + d_{\mathrm{b}}^{(i)}(k) \tag{10.132}$$

we have equation (10.128), where $d_{\mathrm{b}}^{(i)}(k)$ is given by equation (10.129). □

For the case when the local smoothed estimates $\hat{x}^{(i)}(k|N) = E[x(k)|Z^i]$ are available, the decentralized smoothing algorithm in discrete time is summarized in the following theorem.

Theorem 10.6

The centralized smoothed estimate $\hat{x}(k|N)$ is given by

$$\hat{x}(k|N) = \xi(k) + \sum_{i=1}^{2} \hat{x}^{(i)}(k|N) \tag{10.133}$$

$$\xi(k) = P_s(k)[P_f^{-1}(k|k-1)\xi_f(k|k-1) + f_b(k)] \tag{10.134}$$

where $P_f(k|k-1)$ is given by equations (10.98) and (10.99), and $P_s(k)$ is given by equation (10.119). $\xi_f(k|k-1)$ and $f_b(k)$ are also subject to

$$\xi_f(k+1|k) = \Phi\xi_f(k|k) \tag{10.135a}$$

$$\begin{aligned}\xi_f(k|k) = &\left[I - P_f(k|k)\sum_{i=1}^{2} H_i^T R_i^{-1} H_i\right]\xi_f(k|k-1)\\ &- P_f(k|k)[H_1^T R_1^{-1} H_1 \hat{x}^{(2)}(k|N)\\ &+ H_2^T R_2^{-1} H_2 \hat{x}^{(1)}(k|N)], \qquad \xi_f(0|-1) = -\hat{x}(0)\end{aligned} \tag{10.135b}$$

and

$$\begin{aligned}f_b(k) = &\{[I - D\Delta(k)D^T W_b(k+1)]\Phi\}^T f_b(k+1)\\ &- [H_1^T R_1^{-1} H_1 \hat{x}^{(2)}(k|N) + H_2^T R_2^{-1} H_2 \hat{x}^{(1)}(k|N)],\\ &\qquad f_b(N+1) = 0\end{aligned} \tag{10.136}$$

Proof

See Watanabe [16]. □

Note that the computation of $\xi_f(k|k-1)$ and $f_b(k)$ can also be achieved in a decentralized manner. That is, they can be readily broken up into two quantities $\{\xi_f^{(1)}(k|k), \xi_f^{(2)}(k|k)\}$ and $\{f_b^{(1)}(k), f_b^{(2)}(k)\}$.

The solution to the smoothing update problem in discrete time is summarized in the following corollary.

Corollary 10.8

The centralized estimate $\hat{x}(k|N)$ in the smoothing update problem can be obtained as follows:

(i) *Centralized implementation*

$$\begin{aligned}\hat{x}(k|N) = &P_s(k)\{P_f^{-1}(k|k-1)\Phi[\bar{r}_f(k-1) + P_f(k-1|k-1)\\ &\times P_f^{(1)^{-1}}(k-1|k-1)\hat{x}^{(1)}(k-1|k-1)]\\ &+ \bar{q}_b(k) + d_b^{(1)}(k)\}\end{aligned} \tag{10.137}$$

where $\bar{r}_f(k)$ is subject to the linear equation:

$$\begin{aligned}\bar{r}_f(k) = {} & P_f(k|k)[I - L(k)B^T]\Phi^{-T}P_f^{-1}(k-1|k-1)\bar{r}_f(k-1) \\ & + P_f(k|k)[L^{(1)}(k) - L(k)]B^T\Phi^{-T} \\ & \times P_f^{(1)^{-1}}(k-1|k-1)\hat{x}^{(1)}(k-1|k-1) \\ & + P_f(k|k)H_2^T R_2^{-1} z_2(k)\end{aligned} \tag{10.138}$$

or

$$\begin{aligned}\bar{r}_f(k) = {} & P_f(k|k)P_f^{-1}(k|k-1)\Phi\bar{r}_f(k-1) \\ & + P_f(k|k)G^{(1)}(k)\hat{x}^{(1)}(k|k-1) + P_f(k|k)H_2^T R_2^{-1} z_2(k), \\ & \bar{r}_f(0) = \hat{x}(0|0) - P_f(0|0)P_f^{(1)^{-1}}(0|0)\hat{x}^{(1)}(0|0)\end{aligned} \tag{10.139}$$

and $\bar{q}_b(k)$ is given by

$$\begin{aligned}\bar{q}_b(k) = {} & \{[I - D\Delta(k)D^T W_b(k+1)]\Phi\}^T \bar{q}_b(k+1) \\ & + \{D[\Delta^{(1)}(k)D^T W_b^{(1)}(k+1) \\ & - \Delta(k)D^T W_b(k+1)]\Phi\}^T d_b^{(1)}(k+1) \\ & + H_2^T R_2^{-1} z_2(k), \qquad \bar{q}_b(N+1) = 0\end{aligned} \tag{10.140}$$

(ii) *Decentralized implementation*

$$\begin{aligned}\hat{x}(k|N) = {} & P_s(k)\{P_f^{-1}(k|k-1)\Phi[r_f^{(1)}(k-1) + \bar{r}_f^{(2)}(k-1) \\ & + P_f(k-1|k-1) \\ & \times P_f^{(1)^{-1}}(k-1|k-1)\hat{x}^{(1)}(k-1|k-1)] \\ & + q_b^{(1)}(k) + \bar{q}_b^{(2)}(k) + d_b^{(1)}(k)\}\end{aligned} \tag{10.141}$$

where $\bar{r}_f^{(2)}(k)$ and $\bar{q}_b^{(2)}$ satisfy

$$\begin{aligned}\bar{r}_f^{(2)}(k) = {} & P_f(k|k)P_f^{-1}(k|k-1)\Phi\bar{r}_f^{(2)}(k-1) \\ & + P_f(k|k)H_2^T R_2^{-1} z_2(k), \qquad \bar{r}_f^{(2)}(0) = \hat{x}(0|0)\end{aligned} \tag{10.142}$$

and

$$\begin{aligned}\bar{q}_b^{(2)}(k) = {} & \{[\Phi - D\Delta(k)D^T W_b(k+1)]\Phi\}^T \bar{q}_b^{(2)}(k+1) \\ & + H_2^T R_2^{-1} z_2(k), \qquad \bar{q}_b^2(N+1) = 0\end{aligned} \tag{10.143}$$

Proof

This derivation is analogous to that carried out previously in the continuous-time case. Hence, we omit the details. □

The smoothing update problem in discrete time for a case when the local smoothed estimate $\hat{x}^{(1)}(k|N)$ is available can be summarized in the following.

Corollary 10.9

The centralized estimate $\hat{x}(k|N)$ in the smoothing update problem is given by

$$\hat{x}(k|N) = \eta(k) + \hat{x}^{(1)}(k|N) \tag{10.144}$$

$$\eta(k) = P_s(k)[P_f^{-1}(k|k-1)\eta_f(k|k-1) + \bar{f}_b(k)] \tag{10.145}$$

where $\eta_f(k|k-1)$ and $\bar{f}_b(k)$ are subject to the following recursions:

$$\eta_f(k+1|k) = \Phi\eta_f(k|k) \tag{10.146a}$$

$$\begin{aligned}\eta_f(k|k) = {} & \left[I - P_f(k|k)\sum_{i=1}^{2} H_i^T R_i^{-1} H_i\right]\eta_f(k|k-1) \\ & + P_f(k|k)H_2^T R_2^{-1}[z_2(k) - H_2\hat{x}^{(1)}(k|N)], \\ & \qquad \eta_f(0|-1) = 0\end{aligned} \tag{10.146}$$

$$\begin{aligned}\bar{f}_b(k) = {} & \{[I - D\Delta(k)D^T W_b(k+1)]\Phi\}^T \bar{f}_b(k+1) \\ & + H_2^T R_2^{-1}[z_2(k) - H_2\hat{x}^{(1)}(k|N)], \qquad \bar{f}_b(N+1) = 0\end{aligned} \tag{10.147}$$

Proof

See Watanabe [16]. □

Now consider the case of $R_2 = \infty$ in equations (10.121) and (10.124). As was mentioned earlier in Section 10.2.2, $W_b(k)$ can be reduced to $W_b^{(1)}(k)$. Consequently, $\Delta(k) \equiv \Delta^{(1)}(k)$, and $\bar{q}_b(k)$ with $R_2 = \infty$ gives $\bar{q}_b(k) \equiv 0$ (or $q_b^{(1)}(k) + \bar{q}_b^{(2)}(k) \equiv 0$). Therefore, the solution to the real-time smoothing problem in discrete time, i.e. $\hat{x}_{rs}(k) = E[x(k)|Z^1, Z_k^2]$ is obtained by the following corollary.

Corollary 10.10

The real-time smoothing algorithms are as follows:

(i) *Centralized implementation*

$$\hat{x}_{rs}(k) = P_{rs}(k)\{P_f^{-1}(k|k-1)\Phi[\bar{r}_f(k-1) + P_f(k-1|k-1) \times P_f^{(1)^{-1}}(k-1|k-1)\hat{x}^{(1)}(k-1|k-1)] + d_b^{(1)}(k)\} \tag{10.148}$$

$$P_{rs}(k) = [P_f^{-1}(k|k-1) + W_b^{(1)}(k)] \tag{10.149}$$

(ii) *Decentralized implementation*

$$\hat{x}_{rs}(k) = P_{rs}(k)\{P_f^{-1}(k|k-1)\Phi[r_f^{(1)}(k-1) + \bar{r}_f^{(2)}(k-1) + P_f(k-1|k-1) \times P_f^{(1)^{-1}}(k-1|k-1)\hat{x}^{(1)}(k-1|k-1)] + d_b^{(1)}(k)\} \tag{10.150}$$

We finally present the real-time smoothing algorithm which computes $\hat{x}_{rs}(k)$ in terms of $\hat{x}^{(1)}(k|N)$. This solution is described in the following corollary.

Corollary 10.11

The real-time smoothing can be implemented in the following way:

$$\hat{x}_{rs}(k) = P_{rs}(k)P_f^{-1}(k|k-1)\eta_f(k|k-1) + \hat{x}^{(1)}(k|N) \tag{10.151}$$

$$P_{rs}^{-1}(k) = [P_f^{-1}(k|k-1) + W_b^{(1)}(k)] \tag{10.152}$$

10.4 SUMMARY

In this chapter, some decentralized smoothing algorithms have been presented for continuous- and discrete-time linear estimation structures consisting of a central processor and of two local processors, in which the local models are assumed to be identical to the global model. For more detailed results based on the local smoothed estimates, we refer the reader to Willsky *et al.* [7], Levy *et al.* [13] and Bello *et al.* [17] for continuous-time systems and to Watanabe [16] for discrete-time systems.

The extension of the results to the case of M local processors is straightforward. We can also treat with a case in which the local model is a subsystem of the global model, i.e. the state-space models available to the local processors differ from the global model [7, 13] (cf. Exercises 10.2 and 10.3). Moreover, decentralized smoothing can be carried out using a variety of smoothing formulas. For example, Watanabe [15] and Watanabe and Tzafestas [18] develop decentralized smoothing algorithms using the forward-pass fixed-interval smoother recursions described in Chapters 6 and 9, respectively. The counterpart results for a backward-pass recursion of discrete-time systems can also be found in Watanabe and Tzafestas [19].

A parallel Kalman filter for a case where the data at various local stations are considered to be time-sequential can be found in Hashemipour *et al.* [20, 21]. Other parallelizing Kalman filters can be found in Desai and Das [22] and Willner *et al.* [23]. Some parallel computing structures of the Kalman filter in systolic array architectures or in a pipelined mechanization are also reported by Andrews [24], Jover and Kailath [25] and Travassos [26]. A parallel smoothing algorithm can be found in Tewfik *et al.* [27], in which the algorithm is based on the even–odd decomposition of the process and of the observations.

EXERCISES

10.1 Suppose that, in the system given by equations (10.1)–(10.4), $Q = 0$ and $\hat{x}(0) = \hat{x}^{(i)}(0) = 0$, $i = 1, 2$. Show how Theorem 10.1 can be adapted.

10.2 Consider a global system described by the following model:

$$\dot{x}(t) = Ax(t) + Bw(t)$$

$$z_i(t) = C_i x(t) + v_i(t), \qquad i = 1, 2$$

where

$$E[w(t)] = 0, \qquad E[w(t)w^{\mathrm{T}}(\tau)] = Q\delta(t - \tau)$$

$$E[v_i(t)] = 0, \qquad E[v_i(t)v_i^{\mathrm{T}}(\tau)] = R_i\delta(t - \tau)$$

and the following local systems:

$$\dot{x}_i(t) = A_i x_i(t) + B_i w_i(t)$$

$$z_i(t) = H_i x_i(t) + v_i(t), \qquad i = 1, 2$$

where

$$E[w_i(t)] = 0, \qquad E[w_i(t)w_i^{\mathrm{T}}(\tau)] = Q_i\delta(t - \tau)$$

Given matrices M_1 and M_2 such that

$$C_i = H_i M_i, \qquad i = 1, 2$$

show how the result of Corollary 10.1 can be modified (see Willsky *et al.* [7]).

10.3 In Exercise 10.2, suppose that $x_1(t) = M_1(t)x(t)$, $M_1 = [I \quad 0]$ and that the global model can be described by

$$x^{\mathrm{T}}(t) = [x_1^{\mathrm{T}}(t)\ s_1^{\mathrm{T}}(t)]$$

$$A = \begin{bmatrix} A_1 & 0 \\ A_{21} & A_{22} \end{bmatrix}, \quad B = \begin{bmatrix} B_1 \\ B_{21} \end{bmatrix}$$

$$C_1 = [H_1 \quad 0], \qquad C_2 = [C_{21} \quad C_{22}]$$

When defining

$$\hat{x}_i^{(j)}(t|t) = E[x_i(t)|Z_t^j]; \qquad \hat{x}_i(t|t) = E[x_i(t)|Z_t^1, Z_t^2]$$

$$\hat{x}^{(j)}(t|t) = E[x(t)|Z_t^j], \quad \hat{x}(t|t) = E[x(t)|Z_t^1, Z_t^2]$$

reconstruct $\hat{x}_1(t|t)$ and $\hat{s}_1(t|t)$ in terms of $\hat{x}_1^{(1)}(t|t)$ and $\hat{x}_2^{(2)}(t|t)$.

10.4 Prove Theorem 10.3 by using the superposition principle of the Hamiltonian system in continuous time.

10.5 Consider again the system in Exercise 10.2 with conditions as shown in Exercise 10.3. Determine the centralized smoothed estimate $\hat{x}(t|T)$ in the decentralized smoothing problem by using $\hat{x}_1^{(1)}(t|T) = E[x_1(t)|Z^1]$ and $z_2(\cdot)$.

10.6 Consider the forward-pass smoother in continuous time given in Section 6.5.2 of Chapter 6 as a global smoother. Develop a decentralized smoothing algorithm using the local smoothed estimates.

10.7 Applying the result of Section 2.6 in Chapter 2, solve the following problems:

(a) Construct a steady-state version of the decentralized smoother given by Theorem 10.2.

(b) Develop a steady-state version of the decentralized Kalman filter given by Corollary 10.1.

(c) Obtain a steady-state version of the decentralized smoother given by Theorem 10.3.

10.8 Prove the equivalence of equations (10.109) and (10.110).

10.9 Consider again the problem of decentralized Kalman filtering posed in Section 10.3.1. Show the following:

(a) At the measurement update,

$$\begin{aligned}\hat{x}(k|k) = P_{\mathrm{f}}(k|k)\{&P_{\mathrm{f}}^{-1}(k|k-1)\hat{x}(k|k-1) \\ &+ \sum_{i=1}^{2}[P_{\mathrm{f}}^{(i)^{-1}}(k|k)\hat{x}^{(i)}(k|k) \\ &- P_{\mathrm{f}}^{(i)^{-1}}(k|k-1)\hat{x}^{(i)}(k|k-1)]\}\end{aligned}$$

$$P_{\mathrm{f}}^{-1}(k|k) = P_{\mathrm{f}}^{-1}(k|k-1) + \sum_{i=1}^{2}[P_{\mathrm{f}}^{(i)^{-1}}(k|k) - P_{\mathrm{f}}^{(i)^{-1}}(k|k-1)]$$

(b) At the time update,

$$\hat{x}(k+1|k) = A\hat{x}(k|k) + \sum_{i=1}^{2}[\hat{x}^{(i)}(k+1|k) - A\hat{x}^{(i)}(k|k)]$$

and the covariance equation is the same as equation (10.98). These algorithms have been studied by Hashemipour *et al.* [20, 21].

10.10 Prove Theorem 10.6 by applying the superposition principle of the Hamiltonian system in discrete time.

10.11 Apply the result of Section 2.3 in Chapter 2 to solve the following problems:

(a) Obtain a steady-state version of the decentralized Kalman filter given by Theorem 10.4.
(b) Develop a steady-state version of the decentralized smoother given by Theorem 10.5.
(c) Construct a steady-state version of the decentralized smoother given by Theorem 10.6.

10.12 Let a global smoother be given by the forward-pass smoother in discrete time shown in Section 9.3 of Chapter 9. Use such a smoother to derive a decentralized smoothing algorithm based on using the local smoothed estimates (see Watanabe and Tzafestas [18]).

10.13 The Bryson–Frazier smoother in discrete time is provided in Exercise 9.5. Use this algorithm to derive a decentralized smoother based on using the local smoothed estimates.

REFERENCES

[1] SANDELL JR, N. R., VARAIYA, P., ATHANS, M. and SAFONOV, M. G., Survey of Decentralized Control Methods for Large Scale Systems, *IEEE Trans. Aut. Control,* vol. AC-23, no. 2, 1978, pp. 108–28.

[2] HASSAN, M. F., SALUT, G., SINGH, M. G. and TITLI, A., A Decentralized Computational Algorithm for the Global Kalman Filter, *IEEE Trans. Aut. Control,* vol. AC-23, no. 2, 1978, pp. 262–8.

[3] SANDERS, C. W., TACKER, E. C., LINTON, T. D. and LING, R. Y.-S., Specific Structures for Large-Scale State Estimation Algorithms Having Information Exchange, *IEEE Trans. Aut. Control,* vol. AC-23, no. 2, 1978, pp. 255–61.

[4] LOOZE, D. P., HOUPT, P. K., SANDELL JR, N. R. and ATHANS, M., On Decentralized Estimation and Control with Application to Freeway Ramp Metering, *IEEE Trans. Aut. Control,* vol. AC-23, no. 2, 1978, pp. 268–75.

[5] VERRIEST, E., FRIEDLANDER, B. and MORF, M., Distributed Processing in Estimation and Detection, *Proc. 1979 IEEE Symp. Dec. Contr.,* Florida, December 1979, pp. 153–8.

[6] GRIMBLE, M. J., Structure of Large Stochastic Optimal and Suboptimal Systems, *IEE Proc.*, vol. 129, no. 5, pt. D, 1982, pp. 167–76.

[7] WILLSKY, A. S., BELLO, M. G., CASTANON, D. A., LEVY, B. C. and VERGHESE, G. C., Combining and Updating of Local Estimates and Regional Maps along Sets of One-Dimensional Tracks, *IEEE Trans. Aut. Control,* vol. AC-27, no. 4, 1982, pp. 799–813.

[8] SPEYER, J. L., Computation and Transmission Requirements for a Decentralized Linear-Quadratic-Gaussian Control Problem, *IEEE Trans. Aut. Control,* vol. AC-24, no. 2, 1979, pp. 266–9.

[9] WATANABE, K., Continuous-Time Decentralized Smoothers Based on Two-Filter Form: Identical Local and Global Models, *Int. J. Syst. Sci.,* vol. 17, no. 7, 1986, pp. 1015–28.

[10] MAYNE, D. Q., A Solution of the Smoothing Problem, *Automatica,* vol. 4, 1966, pp.73–92.

[11] FRASER, D. C. and POTTER, J. E., The Optimum Linear Smoother as a Combination of Two Optimum Linear Filters, *IEEE Trans. Aut. Control,* vol. AC-14, 1969, pp. 387–90.

[12] GELB, A., *Applied Optimal Estimation,* MIT Press, Cambridge, Massachusetts, 1980.

[13] LEVY, B. C., CASTANON, D. A., VERGHESE, G. C. and WILLSKY, A. S., A Scattering Framework for Decentralized Estimation Problems, *Automatica,* vol. 19, no. 4, 1983, pp. 373–84.

[14] ANDERSON, B. D. O. and MOORE, J. B., *Optimal Filtering,* Prentice Hall, Englewood Cliffs, New Jersey, 1979.

[15] WATANABE, K., Decentralized Fixed-Interval Smoothing Algorithms, *Trans. ASME, J. Dynamic Systems Meas. Control,* vol. 108, no. 1, 1986, pp. 86–9.

[16] WATANABE, K., Decentralized Two-Filter Smoother Algorithms for Linear Discrete-Time Systems, *Int. J. Control,* vol. 44, no. 1, 1986, pp. 49–63.

[17] BELLO, M. G., WILLSKY, A. S., LEVY, B. C. and CASTANON, D. A., Smoothing Error Dynamics and Their Use in the Solution of Smoothing and Mapping Problems, *IEEE Trans. Inf. Theory,* vol. IT-32, no. 4, 1986, pp.483–95.

[18] WATANABE, K. and TZAFESTAS, S. G., Discrete-Time Forward-Pass Smoothers in Distributed Sensor Networks, *Int. J. Syst. Sci.,* vol. 19, no. 8, 1988, pp. 1375–85.

[19] WATANABE, K. and TZAFESTAS, S. G., Decentralized Estimation Algorithms for a Backward Pass Fixed-Interval Smoother, *Int. J. Syst. Sci.,* vol. 21, no. 5, 1990, pp.913–31.

[20] HASHEMIPOUR, H. R., ROY, S. and LAUB, A. J., Decentralized Structures for Parallel Kalman Filtering, *IEEE Trans. Aut. Control,* vol AC-33, no. 1, January 1988, pp. 88–94.

[21] HASHEMIPOUR, H. R., ROY, S. and LAUB, A. J., Decentralized Structures for Parallel Kalman Filtering, *Preprints of 10th World Congress on Aut. Control,* vol. 9, Munich, July 1987, pp. 267–72.

[22] DESAI, U. B. and DAS, B., Parallel Algorithms for Kalman Filtering,

Proc. 1985 Amer. Contr. Conf., Boston, Massachusetts, June 1985, pp. 920–1.

[23] WILLNER, D., CHANG, C. B. and DUNN, K. P., Kalman Filter Algorithms for a Multi-Sensor System, *Proc. 15th IEEE Conf. Decision Contr.*, Clearwater, Florida, December 1976, pp. 570–4.

[24] ANDREWS, A., Parallel Processing of the Kalman Filter, *Proc. Int. Conf. Parallel Processing,* Columbus, Ohio, 1981, pp. 216–20.

[25] JOVER, J. M. and KAILATH, T., A Parallel Architecture for Kalman Filter Measurement Update and Parameter Estimation, *Automatica*, vol. 22, no. 1, 1986, pp. 43–57.

[26] TRAVASSOS, R. H., Application of Systolic Array Technology to Recursive Filtering. In S. Y. Kung, H. J. Whitehouse and T. Kailath (eds), *VLSI and Modern Signal Processing*, Prentice Hall, Englewood Cliffs, New Jersey, 1985, pp. 375–88.

[27] TEWFIK, A. H., LEVY, B. C. and WILLSKY, A. S., A New Parallel Smoothing Algorithm, *Proc. 25th Conference on Decision and Control,* Athens, Greece, 1986, pp. 933–7.

11

Multiple Model Adaptive Control in Discrete-Time Systems

11.1 INTRODUCTION

The control of linear stochastic systems with some parameters unknown is a problem of major theoretical and practical importance. However, since the parameter uncertainty renders the optimal control solution expressed as separation theorem [1, 2] unattainable, a number of parameter adaptive suboptimal control strategies have been proposed [3–11].

In this chapter we further apply the partitioned adaptive technique developed in Chapter 3 to the control problem for linear discrete-time stochastic systems with unknown parameters. These adaptive control policies in general are known as *multiple model adaptive control* (MMAC) [12–15], *partitioned adaptive control* [16–19], or *parameter-adaptive self-organizing control* [4, 5], and are known to be effective as an advanced technique for the adaptive control of aircraft [20], for example the F-8C Digital-Fly-By-Wire-Aircraft [21], which has several system models due to different operating conditions.

In Section 11.2 we review the results of the LQG (Linear-Quadratic-Gaussian) control problem for a discrete-time system. We derive the three passive-type MMAC algorithms, the Lee–Sims algorithm [22], the Upadhyay–Lainiotis algorithm [23] and the Deshpande–Upadhyay–Lainiotis [18] or Lainiotis–Upadhyay–Deshpande algorithm [17] in Section 11.3, by applying the dynamic programming (DP) approach [24, 25]. In Section 11.4 we treat with a special case of the adaptive control problem—the joint detection–control problem—using a MMAC algorithm. We further present, in Section 11.5, a fault-tolerant control strategy, due to Montgomery and Caglayan [26] and Montgomery and Price [27] using a multiple model technique.

11.2 LINEAR-QUADRATIC-GAUSSIAN CONTROL

In this section, we present an optimal control strategy, called linear-quadratic-Gaussian (LQG) control, to regulate the stochastic system state with no unknown parameters. The system is described by

$$x(k+1) = \Phi(k+1, k)x(k) + G(k)u(k) + B(k)w(k) \tag{11.1}$$

$$z(k) = H(k)x(k) + v(k) \tag{11.2}$$

for $k = 1, 2, \ldots$, where $x(k) \in \mathbb{R}^n$ and $z(k) \in \mathbb{R}^m$ are the state and observation vectors, respectively, and $u(k) \in \mathbb{R}^r$ is the control vector, and $w(k) \in \mathbb{R}^p$ and $v(k) \in \mathbb{R}^m$ are mutually independent Gaussian system and measurement noise processes independent of the initial state vector $x(0)$. They are assumed to be subject to the following statistical laws:

$$\left.\begin{array}{ll} E[w(k)] = 0, & E[w(k)w^{\mathrm{T}}(j)] = Q(k)\delta_{kj} \\ E[v(k)] = 0, & E[v(k)v^{\mathrm{T}}(j)] = R(k)\delta_{kj} \\ x(0) \sim N(\hat{x}(0), & P(0)) \end{array}\right\} \tag{11.3}$$

where $Q(k) \geqslant 0$, $R(k) > 0$ and $P(0) \geqslant 0$.

Given the measurement data $Z_k \triangleq \{z(1), \ldots, z(k)\}$ and control sequences $U_{k-1} \triangleq \{u(0), \ldots, u(k-1)\}$, we desire to find the control $u(k)$ as a function of Z_k and U_{k-1},

$$u(k) = \phi(k, I_k), \qquad I_k \triangleq \{Z_k, U_{k-1}\} \tag{11.4}$$

so that the following quadratic cost function is minimized:

$$J = E\left\{ \|x(N)\|^2_{Q_c(N)} + \sum_{k=0}^{N-1} \|x(k)\|^2_{Q_c(k)} + \|u(k)\|^2_{R_c(k)} \right\} \tag{11.5}$$

where $Q_c(N)$, $Q_c(k) \geqslant 0$ and $R_c(k) > 0$.

Before presenting the result of the general LQG control problem, it is instructive to examine the special case in which the current value of the state can be measured exactly. The stochastic effects are then due entirely to the system noise.

11.2.1 LQG Control with Exactly Known State

In this case, it is clear that the control policy becomes

$$u(k) = \phi(k, \tilde{I}_k), \qquad \tilde{I}_k \triangleq \{X_k, U_{k-1}\} \tag{11.6}$$

specifying the control $u(k)$ as a function of $X_k \triangleq \{x(0), \ldots, x(k)\}$

and U_{k-1} such that J is minimized subject to the system equations. The result is summarized in the following theorem.

Theorem 11.1

If the state vector can be measured exactly, then the control minimizing equation (11.5) for the system of equations (11.1)–(11.3) is given by

$$u(k) = -F(k)x(k) \tag{11.7}$$

where $F(k)$ is a linear feedback (control) gain matrix obtained from

$$F(k) = [G^{\mathrm{T}}(k)S(k+1)G(k) + R_{\mathrm{c}}(k)]^{-1} \times G^{\mathrm{T}}(k)S(k+1)\Phi(k+1, k) \tag{11.8}$$

and $S(k)$ is obtained from the control Riccati difference equation (RDE) satisfying

$$\begin{aligned} S(k) = Q_{\mathrm{c}}(k) &+ \Phi^{\mathrm{T}}(k+1, k)S(k+1)\Phi(k+1, k) \\ &- F^{\mathrm{T}}(k)[G^{\mathrm{T}}(k)S(k+1)G(k) + R_{\mathrm{c}}(k)]F(k), \\ &S(N) = Q_{\mathrm{c}}(N) \end{aligned} \tag{11.9}$$

Proof

The solution to this problem is obtained via the dynamic programming (DP) method [24, 25].

Optimal cost-to-go
Let $V(\tilde{I}_k, k)$ be the 'optimal cost-to-go' from time k to time N with the information at time k described as

$$V(\tilde{I}_k, k) = \min_{u(k),\ldots,u(N-1)} E\left\{x^{\mathrm{T}}(N)Q_{\mathrm{c}}(N)x(N) + \sum_{j=k}^{N-1} x^{\mathrm{T}}(j)Q_{\mathrm{c}}(j)x(j) + u^{\mathrm{T}}(j)R_{\mathrm{c}}(j)u(j) \mid \tilde{I}_k\right\} \tag{11.10}$$

Then, by the principle of optimality [28, 29], the problem of minimizing J is equivalent to solving the Bellman equation [24]:

$$V(\tilde{I}_k, k) = \min_{u(k)} E\{x^{\mathrm{T}}(k)Q_{\mathrm{c}}(k)x(k) + u^{\mathrm{T}}(k)R_{\mathrm{c}}(k)u(k) + V(\tilde{I}_{k+1}, k+1) \mid \tilde{I}_k\}$$

$$= \min_{u(k)} \{x^{\mathrm{T}}(k)Q_{\mathrm{c}}(k)x(k) + u^{\mathrm{T}}(k)R_{\mathrm{c}}(k)u(k) + E[V(\tilde{I}_{k+1}, k+1)|\tilde{I}_k]\} \quad (11.11)$$

The boundary conditions for equation (11.11) can be found at $k = N$:

$$V(\tilde{I}_N, N) = \min_{u(N)} E[x^{\mathrm{T}}(N)Q_{\mathrm{c}}(N)x(N)|\tilde{I}_N] = x^{\mathrm{T}}(N)Q_{\mathrm{c}}(N)x(N) \quad (11.12)$$

The solution of the Bellman equation

Because $V(\tilde{I}_N, N)$ is quadratic in $x(N)$, and the single state cost functions are quadratic in $x(k)$ and $u(k)$, it is reasonable to expect that $V(\tilde{I}_k, k)$ will be a quadratic function of $x(\cdot)$ for all j, $0 \leqslant j \leqslant N$. We prove by induction that this is true in general; assume that

$$V(\tilde{I}_k, k) = x^{\mathrm{T}}(k)S(k)x(k) + \alpha(k) \quad (11.13)$$

Substituting equation (11.13) into (11.11), we have

$$V(\tilde{I}_k, k) = \min_{u(k)}\{x^{\mathrm{T}}(k)Q_{\mathrm{c}}(k)x(k) + u^{\mathrm{T}}(k)R_{\mathrm{c}}(k)u(k) + E[x^{\mathrm{T}}(k+1)S(k+1)x(k+1) + \alpha(k+1)|\tilde{I}_k]\} \quad (11.14)$$

Note here that

$$E[x(k+1)|\tilde{I}_k] = \Phi(k+1, k)x(k) + G(k)u(k) \quad (11.15)$$

$$E\{[x(k+1) - E[x(k+1)|\tilde{I}_k]][x(k+1) - E[x(k+1)|\tilde{I}_k]]^{\mathrm{T}}|\tilde{I}_k\} = E[B(k)w(k)w^{\mathrm{T}}(k)B^{\mathrm{T}}(k)|\tilde{I}_k] \equiv B(k)Q(k)B^{\mathrm{T}}(k) \quad (11.16)$$

because $u(k)$ is $\tilde{I}_k$-measurable by the fact of equation (11.6). Substitution of equations (11.15) and (11.16) into (11.14) gives

$$V(\tilde{I}_k, k) = \min_{u(k)}\{x^{\mathrm{T}}(k)Q_{\mathrm{c}}(k)x(k) + u^{\mathrm{T}}(k)R_{\mathrm{c}}(k)u(k) + [\Phi(k+1, k)x(k) + G(k)u(k)]^{\mathrm{T}}S(k+1) \times [\Phi(k+1, k)x(k) + G(k)u(k)] + \mathrm{tr}\,[S(k+1)B(k)Q(k)B^{\mathrm{T}}(k)] + \alpha(k+1)\} \quad (11.17)$$

If the square is completed for the terms containing $u(k)$, then

$$
\begin{aligned}
V(\tilde{I}_k, k) = \min_{u(k)} \{ & [u(k) + F(k)x(k)]^{\mathrm{T}} \\
& \times [G^{\mathrm{T}}(k)S(k+1)G(k) + R_{\mathrm{c}}(k)] \\
& \times [u(k) + F(k)x(k)] + x^{\mathrm{T}}(k) \\
& \times [Q_{\mathrm{c}}(k) + \Phi^{\mathrm{T}}(k+1, k)S(k+1) \\
& \times \Phi(k+1, k) - F^{\mathrm{T}}(k) \\
& \times [G^{\mathrm{T}}(k)S(k+1)G(k) \\
& + R_{\mathrm{c}}(k)]F(k)]x(k) \\
& + \mathrm{tr}[S(k+1)B(k)Q(k)B^{\mathrm{T}}(k)] + \alpha(k+1)\}
\end{aligned} \tag{11.18}
$$

which gives

$$u(k) = -F(k)x(k) \tag{11.19}$$

Substituting back into equation (11.18) gives

$$V(\tilde{I}_k, k) = x^{\mathrm{T}}(k)S(k)x(k) + \alpha(k) \tag{11.20}$$

where

$$
\begin{aligned}
S(k) = {} & Q_{\mathrm{c}}(k) + \Phi^{\mathrm{T}}(k+1, k)S(k+1)\Phi(k+1, k) \\
& - F^{\mathrm{T}}(k)[G^{\mathrm{T}}(k)S(k+1)G(k) + R_{\mathrm{c}}(k)]F(k), \\
& \qquad S(N) = Q_{\mathrm{c}}(N)
\end{aligned} \tag{11.21}
$$

$$\alpha(k) = \mathrm{tr}[S(k+1)B(k)Q(k)B^{\mathrm{T}}(k)] + \alpha(k+1), \qquad \alpha(N) = 0 \tag{11.22}$$

The optimal cost then is evaluated at $k = 0$:

$$
\begin{aligned}
J^* = E[V(\tilde{I}_0, 0)] = {} & \hat{x}^{\mathrm{T}}(0)S(0)\hat{x}(0) + \mathrm{tr}[S(0)P(0)] \\
& + \sum_{k=0}^{N-1} \mathrm{tr}[S(k+1)B(k)Q(k)B^{\mathrm{T}}(k)]
\end{aligned} \tag{11.23}
$$

It should be noted that the final term of equation (11.23) represents the degradation of the minimum cost function due to the existence of system noise. Thus, induction now completes the proof. □

It is interesting to note that the linear feedback gain, $F(k)$, and the matrix RDE are exactly the same as in the deterministic (i.e. linear quadratic (LQ)) case [30]. That is, the addition of white Gaussian system noise does not change the optimal control policy and only adds a scalar term $\alpha(k)$ to the minimum expected cost function. The resulting control structure is illustrated in Figure 11.1.

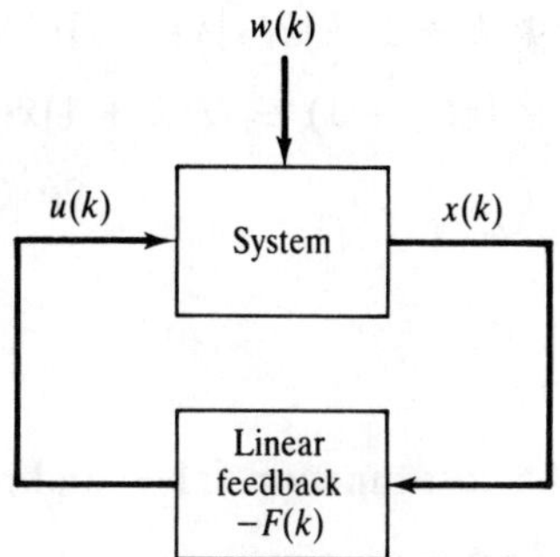

Figure 11.1 Structure of the LQG control problem with exact known state.

11.2.2 LQG Control with Inexactly Known State

In the previous subsection it was assumed that complete knowledge of the system state was available for control. In this subsection, we extend the result to a general case where the state is partly and inexactly measured. The general LQG control problem treated here will be solved by reducing it to the problem considered previously. Theorem 11.1 is generalized in the following theorem.

Theorem 11.2

The control law minimizing equation (11.5) for the system of equations (11.1)–(11.3) is given by

$$u(k) = -F(k)\hat{x}(k|k) \tag{11.24}$$

where

$$F(k) = [G^{\mathrm{T}}(k)S(k+1)G(k) + R_{\mathrm{c}}(k)]^{-1} \times G^{\mathrm{T}}(k)S(k+1)\Phi(k+1, k) \tag{11.25}$$

$$S(k) = Q_{\mathrm{c}}(k) + \Phi^{\mathrm{T}}(k+1, k)S(k+1)\Phi(k+1, k) - F^{\mathrm{T}}(k)[G^{\mathrm{T}}(k)S(k+1)G(k) + R_{\mathrm{c}}(k)]F(k), \quad S(N) = Q_{\mathrm{c}}(N) \tag{11.26}$$

and $\hat{x}(k|k)$ is provided by the Kalman filter:

$$\hat{x}(k+1|k+1) = \Phi(k+1,k)\hat{x}(k|k) + G(k)u(k) + K(k+1) \times [z(k+1) - H(k+1)\hat{x}(k+1|k)],$$
$$\hat{x}(0|0) = \hat{x}(0) \qquad (11.27)$$

Proof

Let us first consider the cost function (11.5) and rewrite it as follows:

$$J = E\{E[x^{\mathrm{T}}(N)Q_c(N)x(N)|I_N]\} + E\left\{\sum_{k=0}^{N-1} E[x^{\mathrm{T}}(k)Q_c(k)x(k)|I_k]\right\} + E\left[\sum_{k=0}^{N-1} u^{\mathrm{T}}(k)R_c(k)u(k)\right] \qquad (11.28)$$

where a fundamental property of conditional expectation was used. Define the conditional mean and covariance of the state by $\hat{x}(k|k) \triangleq E[x(k)|I_k]$ and $P(k|k) \triangleq E\{[x(k) - \hat{x}(k|k)][x(k) - \hat{x}(k|k)]^{\mathrm{T}}|I_k]$, respectively. It is then noted that

$$E[x^{\mathrm{T}}(k)Q_c(k)x(k)|I_k] = \hat{x}^{\mathrm{T}}(k|k)Q_c(k)\hat{x}(k|k) + \mathrm{tr}\,[Q_c(k)P(k|k)] \qquad (11.29)$$

Similarly, we have

$$E[x^{\mathrm{T}}(N)Q_c(N)x(N)|I_N] = \hat{x}^{\mathrm{T}}(N|N)Q_c(N)\hat{x}(N|N) + \mathrm{tr}\,[Q_c(N)P(N|N)] \qquad (11.30)$$

Therefore, J can be rewritten as follows:

$$J = \hat{J} + \mathrm{tr}\,[Q_c(N)P(N|N)] + \sum_{k=0}^{N-1} \mathrm{tr}\,[Q_c(k)P(k|k)] \qquad (11.31)$$

where

$$\hat{J} = E\left\{||\hat{x}(N|N)||^2_{Q_c(N)} + \sum_{k=0}^{N-1} ||\hat{x}(k|k)||^2_{Q_c(k)} + ||u(k)||^2_{R_c(k)}\right\} \qquad (11.32)$$

Since the last two terms of equation (11.31) are independent of I_k, the minimization of J is equivalent to the minimization of $\hat{J}$.

Now the exact equation for $\hat{x}(k|k)$ is given by the Kalman filter equation discussed in Chapter 2:

$$\hat{x}(k+1|k+1) = \Phi(k+1, k)\hat{x}(k|k) + G(k)u(k) \\ + K(k+1)\nu(k+1) \tag{11.33}$$

where $\hat{x}(k|k)$ and $u(k)$ are both I_k-measurable, and $\nu(k+1)$ is a zero-mean white process. Hence the conditional mean and covariance of $\hat{x}(k+1|k+1)$ given I_k are, respectively, given by

$$E[\hat{x}(k+1|k+1)|I_k] = \Phi(k+1, k)\hat{x}(k|k) + G(k)u(k) \tag{11.34}$$

$$\begin{aligned} &E\{[\hat{x}(k+1|k+1) - E[\hat{x}(k+1|k+1)|I_k] \\ &\quad \times [\hat{x}(k+1|k+1) - E[\hat{x}(k+1|k+1)|I_k]]^{\mathrm{T}}|I_k\} \\ &= K(k+1)[H(k+1)P(k+1|k)H^{\mathrm{T}}(k+1) \\ &\quad + R(k+1)]K^{\mathrm{T}}(k+1) \\ &\triangleq \hat{Q}(k) \end{aligned} \tag{11.35}$$

From comparisons between equations (11.33) and (11.1), and equations (11.32) and (11.5), it is obvious that the LQG control problem with inexact information is entirely equivalent to an LQG control problem with perfect information, in which $\hat{x}(k|k)$ replaces $x(k)$, $K(k+1)\nu(k+1)$ replaces $w(k)$, and $\hat{Q}(k)$ replaces $Q(k)$. Therefore, it follows that if

$$\begin{aligned} W(I_k, k) \triangleq \min_{u(k),\ldots,u(N-1)} \; & E\Bigg\{\hat{x}^{\mathrm{T}}(N|N)Q_{\mathrm{c}}(N)\hat{x}(N|N) \\ & + \sum_{j=k}^{N-1} \hat{x}^{\mathrm{T}}(j|j)Q_{\mathrm{c}}(j)\hat{x}(j|j) + u^{\mathrm{T}}(j)R_{\mathrm{c}}(j)u(j)|I_k\Bigg\} \end{aligned} \tag{11.36}$$

then

$$\begin{aligned} W(I_k, k) &= \min_{u(k)} \{\hat{x}^{\mathrm{T}}(k|k)Q_{\mathrm{c}}(k)\hat{x}(k|k) + u^{\mathrm{T}}(k)R_{\mathrm{c}}(k)u(k) \\ &\quad + E[W(I_{k+1}, k+1)|I_k]\} \\ &= \hat{x}^{\mathrm{T}}(k|k)S(k)\hat{x}(k|k) + \alpha(k) \end{aligned} \tag{11.37}$$

Hence, we have

$$u(k) = -F(k)\hat{x}(k|k) \tag{11.38}$$

where $F(k)$ is given by equation (11.25) and $\alpha(k)$ is given by

$$\alpha(k) = \operatorname{tr}[S(k+1)\hat{Q}(k)] + \alpha(k+1), \qquad \alpha(N) = 0 \tag{11.39}$$

where $S(k)$ is subject to equation (11.26). In addition, the optimal cost $\hat{J}^*$ is given by

$$\hat{J}^* = E[W(I_0, 0)]$$
$$= \hat{x}^{\mathrm{T}}(0)S(0)\hat{x}(0) + \sum_{k=0}^{N-1} \mathrm{tr}\,[S(k+1)\hat{Q}(k)] \qquad (11.40)$$

Since J and $\hat{J}$ are related by equation (11.31), the optimal cost function for the controlling equation (11.1) is given by

$$J^* = \hat{x}^{\mathrm{T}}(0)S(0)\hat{x}(0) + \mathrm{tr}\,[Q_c(N)P(N|N)]$$
$$+ \sum_{k=0}^{N-1} \mathrm{tr}\,[Q_c(k)P(k|k)]$$
$$+ \sum_{k=0}^{N-1} \mathrm{tr}\,\{S(k+1)K(k+1)$$
$$\times [H(k+1)P(k+1|k)H^{\mathrm{T}}(k+1)$$
$$+ R(k+1)]K^{\mathrm{T}}(k+1)\} \qquad (11.41)$$

After some algebraic manipulation, we note that equation (11.41) can also be written as follows:

$$J^* = \hat{x}^{\mathrm{T}}(0)S(0)\hat{x}(0) + \mathrm{tr}\,[S(0)P(0)]$$
$$+ \sum_{k=0}^{N-1} \mathrm{tr}\,[S(k+1)B(k)Q(k)B^{\mathrm{T}}(k)]$$
$$+ \sum_{k=0}^{N-1} \mathrm{tr}\,[F^{\mathrm{T}}(k)G^{\mathrm{T}}(k)S(k+1)\Phi(k+1, k)P(k|k)] \qquad (11.42)$$

in which the final term represents the degradation of the minimum cost function due to the estimation of the state. This completes the proof. □

The structure of the general LQG control system is depicted in Figure 11.2. The solution of the LQG control problem can be divided into two parts: first, estimation, which is the computation of the conditional mean estimates of the current state; and second, control, which is the selection of the optimal feedback as if the conditional mean estimate of the current state is the true state of the system. The result is often referred to as the *certainty equivalence principle* or the *separation theorem*. The former term emphasizes the fact that the optimal feedback will treat the conditional mean-state estimate as the true state; it is true that identical control laws apply to both deterministic and stochastic control problems. It should be noted, however, that this is not the case for many problems, but it is often

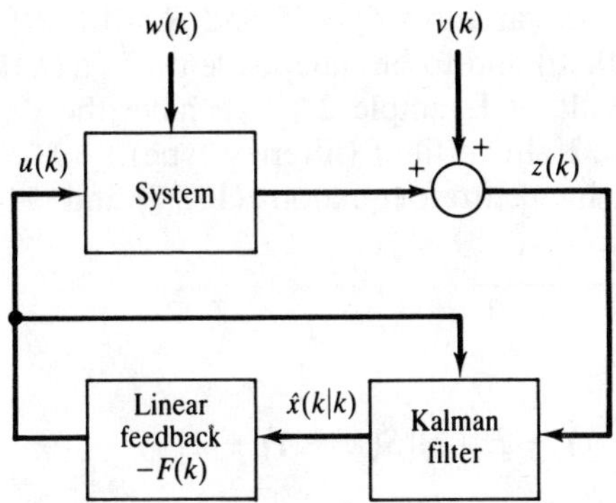

Figure 11.2 Structure of the LQG control problem with inexact measurements.

assumed because this considerably simplifies the computation of a control law as in the (MV or LQG) self-tuning regulator (or controller) [9–11]. On the other hand, the latter term simply indicates that the control problem is solved via two separate problems: estimation and control. The control laws for deterministic and stochastic control problems may be different from each other. Thus, it is concluded that both the certainty equivalence principle and the separation theorem simultaneously hold for the LQG control problem, and that, for a general control problem in which the cost function is not quadratic or the dynamics are nonlinear, if the certainty equivalence principle holds, then the separation theorem is assured, but not vice versa [11].

Example 11.1

Consider the LQG control problem for the following scalar system described by

$$x(k+1) = x(k) + u(k) + w(k)$$

$$z(k) = x(k) + v(k)$$

with the cost function

$$J = E\left[x^2(3) + \sum_{i=1}^{2} u^2(i)\right]$$

It is assumed that $\{w(k)\}$ and $\{v(k)\}$ are zero-mean Gaussian white noise processes with variances $Q = 15$ and $R = 10$. $x(0)$ is assumed to be subject to $N(0,50)$ and to be independent of $\{w(k)\}$ and $\{v(k)\}$.

From the result of Example 2.1, we have the data tabulated in Table 11.1 for the Kalman filter (filtering type).

On the other hand, from equations (11.25) and (11.26), we have

$$F(k) = \frac{S(k+1)}{S(k+1)+1}$$

and

$$\begin{aligned} S(k) &= S(k+1) - F^2(k)[S(k+1)+1] \\ &= S(k+1) - \frac{S(k+1)}{S(k+1)+1} \\ &= \frac{S(k+1)}{S(k+1)+1} \equiv F(k) \end{aligned}$$

Table 11.2 shows the data for $F(k)$ and $S(k)$.

The LQG controls at each stage become

$$u(0) = -0.250\hat{x}(0|0) = 0$$

$$u(1) = -0.333\hat{x}(1|1)$$

and

$$u(2) = -0.5\hat{x}(2|2)$$

Next, examine the optimal cost function for this problem. From equation (11.42) for $N = 3$, we obtain

$$\begin{aligned} J^* = [S(0)P(0)] &+ \sum_{k=0}^{2} [S(k+1)Q] \\ &+ \sum_{k=0}^{2} [F(k)S(k+1)P(k|k)] \end{aligned}$$

Recalling that $P(0) = 50$, we see from Table 11.2 that

$$S(0)P(0) = 0.250 \times 50 = 12.5$$

This term is due to the uncertainty associated with the initial state. Note that for the deterministic three-stage case, we would have $J^* = 0.250x^2(0)$. Similarly, from Table 11.2, the second term of J^* can be computed as

$$Q \sum_{k=0}^{2} S(k+1) = 15[0.333 + 0.5 + 1] = 27.495$$

which is due to the presence of the system noise. Finally, the third

Table 11.1 Data for the Kalman filter

k	$P(k\|k-1)$	$K(k)$	$P(k\|k)$
0	*	*	50
1	65	0.867	8.667
2	23.667	0.703	7.030
3	22.030	0.688	6.878

Table 11.2 Data for $F(k)$ and $S(k)$

k	$F(k)$	$S(k)$
3	*	1
2	0.5	0.5
1	0.333	0.333
0	0.250	0.250

term of J^* can be computed, from Tables 11.1 and 11.2, as follows:

$$\begin{aligned}\sum_{k=0}^{2} F(k)S(k+1)P(k|k) &= 0.250 \times 0.333 \times 50 \\ &\quad + 0.333 \times 0.5 \times 8.667 \\ &\quad + 0.5 \times 1 \times 7.030 \\ &= 9.121\end{aligned}$$

This is due to the estimation of the state, i.e. the filtering error. Consequently, we obtain the minimum cost function as follows:

$$J^* = 12.5 + 27.495 + 9.121 = 49.116$$

11.2.3 Duality of Estimation and Control

It has been shown in Chapter 2 that the estimation error covariance $P(k+1|k)$ and the Kalman filter gain (predictive type) with $S = 0$ are given by

$$\begin{aligned}P(k+1|k) &= \Phi(k+1, k)P(k|k-1)\Phi^{\mathrm{T}}(k+1, k) \\ &\quad + B(k)Q(k)B^{\mathrm{T}}(k) \\ &\quad - K^*(k)[H(k)P(k|k-1)H^{\mathrm{T}}(k) \\ &\quad + R(k)]K^{*\mathrm{T}}(k), \qquad P(0|-1) = P(0) \qquad (11.43)\end{aligned}$$

$$K^*(k) = \Phi(k+1, k)P(k|k-1)H^{\mathrm{T}}(k) \times [H(k)P(k|k-1)H^{\mathrm{T}}(k) + R(k)]^{-1} \tag{11.44}$$

On the other hand, it is observed from previous subsections that the control RDE and control gain for the LQG case are

$$S(k) = \Phi^{\mathrm{T}}(k+1, k)S(k+1)\Phi(k+1, k) + Q_{\mathrm{c}}(k) - F^{\mathrm{T}}(k)[G^{\mathrm{T}}(k)S(k+1)G(k) + R_{\mathrm{c}}(k)]F(k), \quad S(N) = Q_{\mathrm{c}}(N) \tag{11.45}$$

$$F(k) = [G^{\mathrm{T}}(k)S(k+1)G(k) + R_{\mathrm{c}}(k)]^{-1} \times G^{\mathrm{T}}(k)S(k+1)\Phi(k+1, k) \tag{11.46}$$

The similarity between the two solutions is apparent, except that the definition of the matrices and equation (11.43) goes forward, instead of backward as equation (11.45) does. Thus, the linear estimation and control problems are said to be *dual* of each other. One consequence of duality is the fact that the concepts of detectability (or observability) and stabilizability (or controllability), discussed in Chapter 2, can also be defined for control problems. Therefore, all procedures for analyzing the closed-loop system and algorithms for solving the ARE in the estimation problems can be applied to the control problems using the definitions of Table 11.3.

11.3 SOME PASSIVE-TYPE MMAC ALGORITHMS

A system with unknown parameters may be modelled as a set of candidate systems, where each candidate represents one possible model for the system and the parameters are assumed to be constant in the time interval of interest. If θ_i is utilized to index the ith candidate model, the models are described by the following linear difference equation:

$$\theta_i{:}x(k+1) = \Phi_i(k+1, k)x(k) + G_i(k)u(k) + B_i(k)w(k) \tag{11.47}$$

$$z(k) = H_i(k)x(k) + v(k), \qquad i = 1, \ldots, M \tag{11.48}$$

for $k = 1, 2, \ldots$, where M denotes the number of candidate models, and $\{w(k), v(k), x(0)\}$ are assumed to be subject to the following statistical laws:

Table 11.3 Duality of estimation and control

Estimation	*Control*
$\Phi(k+1, k)$	$\Phi^{\mathrm{T}}(k+1, k)$
$B(k)Q(k)B^{\mathrm{T}}(k)$	$Q_{\mathrm{c}}(k)$
$H(k)$	$G^{\mathrm{T}}(k)$
$R(k)$	$R_{\mathrm{c}}(k)$
$P(k+1\|k)$	$S(N-k)$
$P(0)$	$Q_{\mathrm{c}}(N)$
$K^*(k)$	$F^{\mathrm{T}}(k)$
detectability	stabilizability
stabilizability	detectability

$$\left.\begin{aligned} &E[w(k)] = 0, \quad E[w(k)w^{\mathrm{T}}(j)] = Q_i(k)\delta_{kj} \\ &E[v(k)] = 0, \quad E[v(k)v^{\mathrm{T}}(j)] = R_i(k)\delta_{kj} \\ &x(0) \sim N(\hat{x}_i(0), \; P_i(0)) \end{aligned}\right\} \tag{11.49}$$

The matrices $\Phi_i(k+1, k)$, $G_i(k)$ and $H_i(k)$ are, in general, functions of time k, and are subject to uncertainty.

Given the measurement data Z_k, control sequences U_{k-1}, and a set of prior probabilities $p(\theta_i)$ for the candidate models, the problem of this section is to obtain the control sequences $\{u(0), u(1), \ldots, u(N-1)\}$, which approximately minimize the following quadratic cost function:

$$J = E\left\{\|x(N)\|^2_{Q_{\mathrm{c}}(N)} + \sum_{k=0}^{N-1} \|x(k)\|^2_{Q_{\mathrm{c}}(k)} + \|u(k)\|^2_{R_{\mathrm{c}}(k)}\right\} \tag{11.50}$$

where $Q_{\mathrm{c}}(N)$ and $Q_{\mathrm{c}}(k)$ are nonnegative definite matrices, and $R_{\mathrm{c}}(k)$ is a positive definite matrix.

When all the parameters in the system are completely known, we can optimally obtain the solution to the above problem through the so-called separation theorem [1, 2] as discussed in the previous section. However, note that since the present control problem reduces to the nonlinear control problem [18], the optimal solution cannot be obtained and hence many suboptimal (or approximate) solution methods have been proposed [3–5].

In the following, passively adaptive controllers, called multiple model adaptive controllers, which do not account for the fact that future observations will be made, will be presented, where it should be noted that a controller which utilizes such future information is called an actively adaptive controller [31–33].

We demonstrate three passive-type MMAC algorithms by using Bellman's DP method as used in the LQG case.

Theorem 11.3 Lee–Sims method [22]

Given the information data $I_k \triangleq \{Z_k, U_{k-1}\}$, the MMAC algorithm consists of the following:

$$u(k) = -\sum_{i=1}^{M} p(\theta_i|k)F_i(k)\hat{x}_i(k|k) \tag{11.51}$$

$$F_i(k) = N^{-1}(k)G_i^{\mathrm{T}}(k)S_i(k+1)\Phi_i(k+1, k) \tag{11.52}$$

$$N(k) \triangleq \sum_{i=1}^{M} p(\theta_i|k)G_i^{\mathrm{T}}(k)S_i(k+1)G_i(k) + R_c(k) \tag{11.53}$$

where $S_i(k)$ is the control RDE satisfying

$$S_i(k) = Q_c(k) + \Phi_i^{\mathrm{T}}(k+1, k)S_i(k+1)\Phi_i(k+1, k) - F_i^{\mathrm{T}}(k)N(k)F_i(k), \qquad S_i(N) = Q_c(N) \tag{11.54}$$

in which $\hat{x}_i(k|k) \triangleq E[x(k)|I_k, \theta_i]$ is given by

$$\hat{x}_i(k+1|k+1) = \hat{x}_i(k+1|k) + K_i(k+1)v_i(k+1) \tag{11.55}$$

$$\hat{x}_i(k+1|k) = \Phi_i(k+1, k)\hat{x}_i(k|k) + G_i(k)u(k), \qquad \hat{x}_i(0|0) = \hat{x}_i(0) \tag{11.56}$$

$$K_i(k+1) = P_i(k+1|k)H_i^{\mathrm{T}}(k+1)\tilde{P}_i^{-1}(k+1|k) \tag{11.57}$$

$$v_i(k+1) = z(k+1) - H_i(k+1)\hat{x}_i(k+1|k) \tag{11.58}$$

$$\tilde{P}_i(k+1|k) = H_i(k+1)P_i(k+1|k)H_i^{\mathrm{T}}(k+1) + R_i(k+1) \tag{11.59}$$

$$P_i(k+1|k) = \Phi_i(k+1, k)P_i(k|k)\Phi_i^{\mathrm{T}}(k+1, k) + B_i(k)Q_i(k)B_i^{\mathrm{T}}(k), \qquad P_i(0|0) = P_i(0) \tag{11.60}$$

$$P_i(k+1|k+1) = [I - K_i(k+1)H_i(k+1)]P_i(k+1|k) \tag{11.61}$$

and $p(\theta_i|k) \triangleq p(\theta_i|I_k)$ is

$$p(\theta_i|k) = \frac{L(k|\theta_i)p(\theta_i|k-1)}{\sum_{j=1}^{M} L(k|\theta_j)p(\theta_j|k-1)} \tag{11.62}$$

$$L(k|\theta_i) = |\tilde{P}_i(k|k-1)|^{-1/2} \exp[-\tfrac{1}{2}\|v_i(k)\|^2_{\tilde{P}_i^{-1}(k|k-1)}] \tag{11.63}$$

where $|\cdot|$ denotes the determinant of a matrix.

Proof

The cost function (11.50) can be rewritten as equation (11.28). By using the smoothing property of expectations, equation (11.28) can be expressed by

$$J = E\left\{\sum_{i=1}^{M} E[x^{\mathrm{T}}(N)Q_{\mathrm{c}}(N)x(N)|I_N, \theta_i]p(\theta_i|I_N)\right\}$$
$$+ E\left\{\sum_{k=0}^{N-1}\sum_{i=1}^{M} E[x^{\mathrm{T}}(k)Q_{\mathrm{c}}(k)x(k)|I_k, \theta_i]p(\theta_i|I_k)\right\}$$
$$+ E\left[\sum_{k=0}^{N-1} u^{\mathrm{T}}(k)R_{\mathrm{c}}(k)u(k)\right] \tag{11.64}$$

Defining the conditional mean and covariance of the state by $\hat{x}_i(k|k) \triangleq E[x(k)|I_k, \theta_i]$ and $P_i(k|k) \triangleq E\{[x(k) - \hat{x}_i(k|k)]^{\mathrm{T}}|I_k, \theta_i\}$, we note that

$$E[x^{\mathrm{T}}(N)Q_{\mathrm{c}}(N)x(N)|I_N, \theta_i]$$
$$= \hat{x}_i^{\mathrm{T}}(N|N)Q_{\mathrm{c}}(N)\hat{x}_i(N|N) + \mathrm{tr}\,[Q_{\mathrm{c}}(N)P_i(N|N)] \tag{11.65}$$
$$E[x^{\mathrm{T}}(k)Q_{\mathrm{c}}(k)x(k)|I_{\mathrm{k}}, \theta_i]$$
$$= \hat{x}_i^{\mathrm{T}}(k|k)Q_{\mathrm{c}}(k)\hat{x}_i(k|k) + \mathrm{tr}\,[Q_{\mathrm{c}}(k)P_i(k|k)] \tag{11.66}$$

where $\{\hat{x}_i(k|k), P_i(k|k)\}$ are given by equations (11.55)–(11.63) using the results of Chapter 3. From these facts, J can be rewritten as

$$J = \tilde{J} + \sum_{i=1}^{M} \mathrm{tr}\,[Q_{\mathrm{c}}(N)P_i(N|N)]p(\theta_i|I_N)$$
$$+ \sum_{k=0}^{N-1}\sum_{i=1}^{M} \mathrm{tr}\,[Q_{\mathrm{c}}(k)P_i(k|k)]p(\theta_i|I_k) \tag{11.67}$$

where

$$\tilde{J} = E\left\{\sum_{i=1}^{M} [||\hat{x}_i(N|N)||^2_{Q_{\mathrm{c}}(N)}]p(\theta_i|I_N)\right.$$
$$\left.+ \sum_{k=0}^{N-1}\sum_{i=1}^{M} [||\hat{x}_i(k|k)||^2_{Q_{\mathrm{c}}(k)} + ||u(k)||^2_{R_{\mathrm{c}}(k)}]p(\theta_i|I_k)\right\} \tag{11.68}$$

Therefore, the problem of minimizing J is equivalent to solving the following Bellman equation:

$$W(I_k, k) = \min_{u(k)} \left\{ \sum_{i=1}^{M} [\hat{x}_i(k|k) Q_c(k) \hat{x}_i(k|k) + u^T(k) R_c(k) u(k)] p(\theta_i|I_k) + E[W(I_{k+1}, k+1)|I_k \right\} \tag{11.69a}$$

$$= \sum_{i=1}^{M} [\hat{x}_i^T(k|k) S_i(k) \hat{x}_i(k|k) + \alpha_i(k)] p(\theta_i|I_k) \tag{11.69b}$$

$$\triangleq \sum_{i=1}^{M} W_i(I_k, k) p(\theta_i|I_k) \tag{11.69c}$$

Here, note that

$$E[W(I_{k+1}, k+1)|I_k] = \sum_{i=1}^{M} [W(I_{k+1}, k+1)|I_k, \theta_i] p(\theta_i|I_k) \tag{11.70}$$

By introducing the approximation $E[W(I_{k+1}, k+1)|I_k, \theta_i] \simeq E[W_i(I_{k+1}, k+1)|I_k, \theta_i]$, we have

$$W(I_k, k) \simeq \min_{u(k)} \sum_{i=1}^{M} p(\theta_i|I_k) \left\{ \hat{x}_i^T(k|k) Q_c(k) \hat{x}_i(k|k) + u^T(k) R_c(k) u(k) + E[W_i(I_{k+1}, k+1)|I_k, \theta_i] \right\} \tag{11.71}$$

Furthermore, since $\hat{x}_i(k|k)$ and $u(k)$ are both $\{I_k, \theta_i\}$-measurable, it follows that

$$E[\hat{x}_i(k+1|k+1)|I_k, \theta_i] = \Phi_i(k+1, k)\hat{x}_i(k|k) + G_i(k)u(k) \tag{11.72}$$

$$E\{[\hat{x}_i(k+1|k+1) - E[\hat{x}_i(k+1|k+1)|I_k, \theta_i]] \times [\hat{x}_i(k+1|k+1) - E[\hat{x}_i(k+1|k+1)|I_k, \theta_i]]^T|I_k, \theta_i\} = K_i(k+1)\tilde{P}_i(k+1|k)K_i^T(k+1) \triangleq \hat{Q}_i(k) \tag{11.73}$$

Therefore, from equations (11.69)–(11.73), we have

$$W(I_k, k) \simeq \min_{u(k)} \sum_{i=1}^{M} p(\theta_i|I_k) \left\{ \hat{x}_i^T(k|k) \right.$$

$$\times [Q_c(k) + \Phi_i^T(k+1, k)S_i(k+1)\Phi_i(k+1, k)]\hat{x}_i(k|k)$$
$$+ 2\hat{x}_i^T(k|k)\Phi_i^T(k+1, k)S_i(k+1)G_i(k)u(k)$$
$$+ u^T(k)[G_i^T(k)S_i(k+1)G_i(k) + R_c(k)]u(k)$$
$$\left. + \operatorname{tr}[S_i(k+1)\hat{Q}_i(k)] + \alpha_i(k+1) \right\} \tag{11.74}$$

Minimizing the summation of the right-hand side of equation (11.74) with respect to $u(k)$, we obtain

$$u(k) = -N^{-1}(k) \sum_{i=1}^{M} G_i^T(k)S_i(k+1)\Phi_i(k+1, k)\hat{x}_i(k|k)p(\theta_i|I_k) \tag{11.75}$$

where $N(k)$ is given by equation (11.53).

To complete the derivation of the control algorithm, it is necessary to find recursive relationships for $S_i(k)$ and $\alpha_i(k)$. Substituting equations (11.53) and (11.75) into (11.74) yields

$$W(I_k, k) \simeq \sum_{i=1}^{M} p(\theta_i|I_k)\left\{ \hat{x}_i^T(k|k)[Q_c(k) + \Phi_i^T(k+1, k) \right.$$
$$\times S_i(k+1)\Phi_i(k+1, k)$$
$$- F_i^T(k)N(k)F_i(k)]\hat{x}_i(k|k)$$
$$\left. + \operatorname{tr}[S_i(k+1)\hat{Q}_i(k)] + \alpha_i(k+1) \right\} \tag{11.76}$$

Equating similar terms in equations (11.69b) and (11.75) gives the desired equation (11.54) and

$$\alpha_i(k) = \operatorname{tr}[S_i(k+1)\hat{Q}_i(k)] + \alpha_i(k+1), \qquad \alpha_i(N) = 0 \tag{11.77}$$

Finally, the approximate expression for the optimal function $\tilde{J}^*$ is given by

$$\tilde{J}^* = E[W(I_0, 0)]$$
$$= \sum_{i=1}^{M} [\hat{x}_i^T(0|0)S_i(0)\hat{x}_i(0|0) + \alpha_i(0)]p(\theta_i|I_0)$$
$$\simeq \sum_{i=1}^{M} \left\{ \hat{x}_i^T(0)S_i(0)\hat{x}_i(0) + \sum_{k=0}^{N-1} \operatorname{tr}[S_i(k+1)\hat{Q}_i(k)] \right\} p(\theta_i) \tag{11.78}$$

so that the approximate expression for the optimal function J^* reduces to

$$J^* = \sum_{i=1}^{M} \left\{ \hat{x}_i^{\mathrm{T}}(0) S_i(0) \hat{x}_i(0) + \sum_{k=0}^{N-1} \operatorname{tr}[S_i(k+1)\hat{Q}_i(k)] \right\} p(\theta_i)$$
$$+ \sum_{i=1}^{M} \operatorname{tr}[Q_c(N) P_i(N|N)] p(\theta_i|I_N)$$
$$+ \sum_{k=0}^{N-1} \sum_{i=1}^{M} \operatorname{tr}[Q_c(k) P_i(k|k)] p(\theta_i|I_k) \tag{11.79}$$

This completes the proof. □

Thus, in the development presented here, an obvious approximation has been made, since the Bellman equation is not solved completely, but only used to motivate a suboptimal solution. It is true, however, that as the a posteriori probability $p(\theta_i|I_k)$ of the active system approaches 1, the Bellman equation is solved and the control is indeed optimal, i.e. it is the solution to the LQG problem.

An optimal solution in the strict sense would be extremely difficult to obtain, since one should use some control effort in a probing (or learning) mode for system identification and estimation purposes, and some control effort for directly regulating the state vector; regulation and learning are often said to be the dual role of the input and this is automatically included in the optimal strategy. It is well known that such dual control problems [31–34] are not easily solved.

Further, notice that the Riccati difference equation (11.54) for control is explicitly dependent upon the current learning result. Therefore, this equation cannot be solved off-line. Minor modification leads to the following corollary.

Corollary 11.1 Upadhyay–Lainiotis method [23]

The MMAC algorithm presented in Theorem 11.1 can be simplified as follows:

$$u(k) = -\sum_{i=1}^{M} p(\theta_i|k) F_i(k) \hat{x}_i(k|k) \tag{11.80}$$

$$F_i(k) = N^{-1}(k) G_i^{\mathrm{T}}(k) S_i(k+1) \Phi_i(k+1, k) \tag{11.81}$$

$$S_i(k) = Q_c(k) + \Phi_i^{\mathrm{T}}(k+1, k) S_i(k+1) \Phi_i(k+1, k)$$
$$- S_i^*(k+1), \qquad S_i(N) = Q_c(N) \tag{11.82}$$

where

$$S_i^*(k+1) = \Phi_i^{\mathrm{T}}(k+1, k)S_i(k+1)G_i(k)[G_i^{\mathrm{T}}(k)S_i(k+1)G_i(k) + R_c(k)]^{-1}G_i^{\mathrm{T}}(k)S_i(k+1)\Phi_i(k+1, k) \tag{11.83}$$

and the matrix $N(k)$ is obtained from equation (11.53), and the elemental filtered values $\hat{x}_i(k|k)$ and $p(\theta_i|k)$ are provided by equations (11.55)–(11.63).

Proof

Substituting equation (11.81) directly into equation (11.54) gives

$$\begin{aligned} S_i(k) = {} & Q_c(k) + \Phi_i^{\mathrm{T}}(k+1, k)S_i(k+1)\Phi_i(k+1, k) \\ & - \Phi_i^{\mathrm{T}}(k+1, k)S_i(k+1)G_i(k)N^{-1}(k) \\ & \times G_i^{\mathrm{T}}(k)S_i(k+1)\Phi_i(k+1, k) \end{aligned} \tag{11.84}$$

In this equation, the approximation $N(k) \simeq G_i^{\mathrm{T}}(k)S_i(k+1)G_i(k) + R_c(k)$ yields the simplified RDE given by expressions (11.82) and (11.83). □

We note that if the same approximation is utilized in the matrix $N(k)$ of the elemental controller gain $F_i(k)$, then the well-known Deshpande–Upadhyah–Lainiotis (DUL) method [18] can be obtained.

Corollary 11.2 Deshpande–Upadhyay–Lainiotis (DUL) method [18]

The MMAC algorithm given by Corollary 11.1 can be further simplified as follows:

$$u(k) = -\sum_{i=1}^{M} p(\theta_i|k)F_i(k)\hat{x}_i(k|k) \tag{11.85}$$

$$F_i(k) = N_i^{-1}(k)G_i^{\mathrm{T}}(k)S_i(k+1)\Phi_i(k+1, k) \tag{11.86}$$

$$N_i(k) \triangleq G_i^{\mathrm{T}}(k)S_i(k+1)G_i(k) + R_c(k) \tag{11.87}$$

where

$$\begin{aligned} S_i(k) = {} & \Phi_i^{\mathrm{T}}(k+1, k)[S_i^{-1}(k+1) \\ & + G_i(k)R_c^{-1}(k)G_i^{\mathrm{T}}(k)]^{-1}\Phi_i(k+1, k) \\ & + Q_c(k), \qquad S_i(N) = Q_c(N) \end{aligned} \tag{11.88}$$

and the quantities $\hat{x}_i(k|k)$ and $p(\theta_i|k)$ are, of course, given by equations (11.55)–(11.63).

Proof

It is obvious, from the above motivation, that equations (11.52) and (11.53) reduce to equations (11.86) and (11.87), respectively. In addition, noting equation (5A.6) from the discussion of matrix inversion in Appendix 5A, it follows that equations (11.82) and (11.83) can be re-expressed by equation (11.88). □

This algorithm is straightforward and easier to implement than the two previous algorithms. The block diagram for this control strategy is shown in Figure 11.3.

By ignoring the uncertainty in the parameter estimates, we can design the control law as if the estimated parameters were the true system parameters. This approach is commonly called *certainty equiv-*

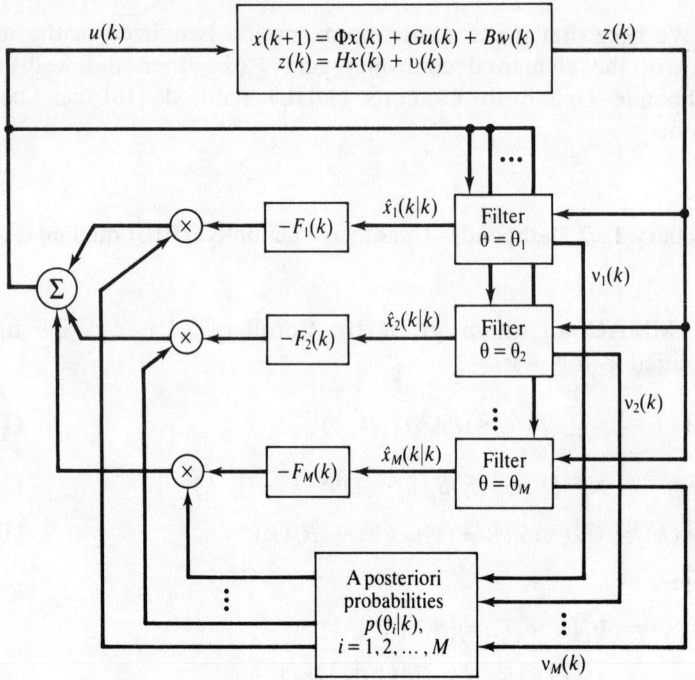

Figure 11.3 MMAC block diagram (DUL method).

alence [32] and involves the separation of the estimation and control problems. The elemental control laws discussed above, i.e. $u_i(k) \triangleq F_i(k)\hat{x}_i(k|k)$, are based on the certainty equivalence principle.

10.4 THE JOINT DETECTION–CONTROL PROBLEM

In this section, we shall deal with the so-called joint detection–control problem by using the DUL algorithm. For the sake of simplicity, we assume that matrices $\Phi_i(k+1, k)$, $G_i(k)$, $Q_i(k)$ and $R_i(k)$ are completely known. We also assume that $\hat{x}_i(0) = \hat{x}(0)$ and $P_i(0) = P(0)$ for all i.

Consider the following two hypothesized observation models:

$$\theta_1 : z(k) = H(k)x(k) + v(k) \tag{11.89a}$$

$$\theta_2 : z(k) = v(k) \tag{11.89b}$$

where it is assumed that a priori probabilities $p(\theta_1)$ and $p(\theta_2) = 1 - p(\theta_1)$ are available.

Now, the DUL algorithm for this joint detection–control problem reduces to

$$u(k) = -F(k)[p(\theta_1|k)\hat{x}_1(k|k) + p(\theta_2|k)\hat{x}_2(k|k)] \tag{11.90}$$

where $F(k)$ is the deterministic control gain given by

$$F(k) = [G^{\mathrm{T}}(k)S(k+1)G(k) + R_{\mathrm{c}}(k)]^{-1} \times G^{\mathrm{T}}(k)S(k+1)\Phi(k+1, k) \tag{11.91}$$

$$S(k) = \Phi^{\mathrm{T}}(k+1, k)[S^{-1}(k+1) + G(k)R_{\mathrm{c}}^{-1}(k)G^{\mathrm{T}}(k)]^{-1} \times \Phi(k+1, k) + Q_{\mathrm{c}}(k), \qquad S(N) = Q_{\mathrm{c}}(N) \tag{11.92}$$

and $\hat{x}_i(k|k)$, $i = 1, 2$, are given respectively by

$$\hat{x}_1(k|k) = \Phi(k, k-1)\hat{x}_1(k-1|k-1) + G(k-1)u(k-1) + K(k)v_1(k), \qquad \hat{x}_1(0|0) = \hat{x}(0) \tag{11.93}$$

$$v_1(k) \triangleq z(k) - H(k)[\Phi(k, k-1)\hat{x}_1(k-1|k-1) + G(k-1)u(k-1)] \tag{11.94}$$

and

$$\hat{x}_2(k|k) = \Phi(k, k-1)\hat{x}_2(k-1|k-1) + G(k-1)u(k-1), \qquad \hat{x}_2(0|0) = \hat{x}(0) \tag{11.95}$$

Here, the filter gain $K(k)$ is provided by the following recursive form:

$$K(k) = P(k|k-1)H^{\mathrm{T}}(k)\widetilde{P}^{-1}(k|k-1) \tag{11.96}$$

$$\widetilde{P}(k|k-1) = H(k)P(k|k-1)H^{\mathrm{T}}(k) + R(k) \tag{11.97}$$

$$P(k|k-1) = \Phi(k, k-1)P(k-1|k-1)\Phi^{\mathrm{T}}(k, k-1) + Q(k-1), \qquad P(0|0) = P(0) \tag{11.98}$$

$$P(k|k) = [I - K(k)H(k)]P(k|k-1) \tag{11.99}$$

Furthermore, the a posteriori probability $p(\theta_1|k)$ is obtained, using the result of equation (3.46), from the following:

$$p(\theta_1|k) = 1 - p(\theta_2|k) = \frac{\rho\Lambda(k)}{1 + \rho\Lambda(k)} \tag{11.100}$$

where $\rho = p(\theta_1)/p(\theta_2)$ and $\Lambda(k)$ is given by the expression

$$\Lambda(k) = \frac{\prod_{j=1}^{k}|\widetilde{P}(j|j-1)|^{-1/2}\exp\left[-\frac{1}{2}\sum_{j=1}^{k}||\nu_1(j)||^2_{\widetilde{P}^{-1}(j|j-1)}\right]}{\prod_{j=1}^{k}|R(j)|^{-1/2}\exp\left[-\frac{1}{2}\sum_{j=1}^{k}||z(j)||^2_{R^{-1}(j)}\right]} \tag{11.101}$$

which is the discrete-time likelihood ratio, $\Lambda(k) \triangleq p(Z_k|\theta_1)/p(Z_k|\theta_2)$, for the binary hypothesis testing (11.89) (cf. Section 3.2).

Example 11.2

In this example, we show a simple simulation using the DUL algorithm in a detection–control problem. The following scalar system is considered

$$x(k+1) = 0.8x(k) + 0.5w(k)$$

$$z(k) = Hx(k) + v(k)$$

and the performance cost is assumed to be given by

$$J = E\left\{x^2(30) + \sum_{k=0}^{29} x^2(k) + u^2(k)\right\}$$

It is also assumed that $w(k) = v(k) \sim N(0,1)$, $x(0) \sim N(2,5)$, and H takes its value from the set $\{1,0\}$ with equal a priori probability 0.5, and that the expectation operator E is calculated as the average of 50

Monte Carlo runs. Note that the true value of H is 1, i.e. $\theta_1\!: H = 1$, $\theta_2\!: H = 0$. Table 11.4 gives the average cost and weighted miss distance squared. Figure 11.4 also shows the evolution of the a posteriori probability of model 1 for one sample path.

Table 11.4 Monte Carlo results for Example 11.2

		Control Policy	
		Optimal control	*MMAC*
Cost	Average	72.8323	72.8635
	Standard deviation	27.0075	27.1555
	Range	25.4167 –139.471	26.1585 –139.95
Weighted miss distance squared	Average	2.11888	2.1298
	Standard deviation	2.60771	2.60337
	Range	4.805×10^{-4} –11.514	1.347×10^{-3} –11.514

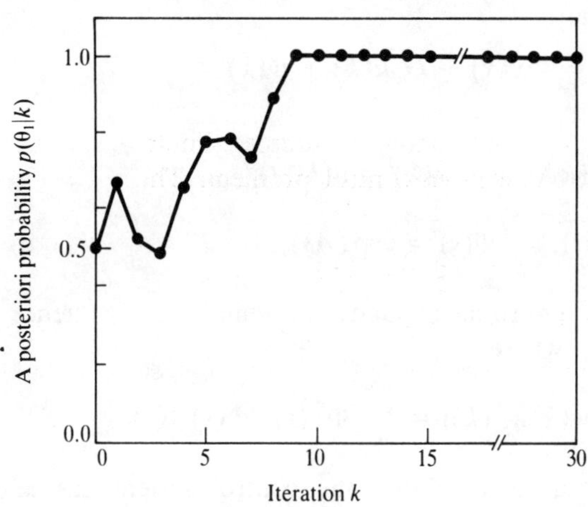

Figure 11.4 Evolution of a posteriori probability for Example 11.2 (one sample run).

In the following section, we shall further discuss a joint detection–control problem in a multiple M-ary hypothesis testing, in which we may generally determine which hypothesis to accept by choosing the hypothesis h_i with the largest likelihood ratio from M hypotheses, and a digital control system which is optimally tolerant of failures in aircraft sensors [26, 27] will be presented.

11.5 APPLICATION TO A FAULT-TOLERANT CONTROL SYSTEM

Consider the equation of motion of an aircraft to be described by

$$\dot{x}(t) = Ax(t) + Gu(t) + w(t) \tag{11.102}$$

where $x(t) \in \mathbb{R}^n$ is the state vector, $u(t) \in \mathbb{R}^r$ is the control vector, and $w(t) \in \mathbb{R}^n$ is a zero-mean Gaussian white noise process with covariance matrix $Q\delta(t-\tau)$. The variable $w(t)$ may represent turbulence, or it may represent uncertainty in the designer's knowledge of the characteristics of the aircraft: basically, it can be thought of as representing the error in calculation of $\dot{x}(t)$, given $x(t)$ and $u(t)$.

We shall deal with the digital control of the plant where the control is considered to be constant with sampling interval Δt, that is $u(t) = u(k\Delta t)$ for $k\Delta t \leqslant t < (k+1)\Delta t$. By integrating (11.102) over each sampling interval, we have the discrete-time equation of motion for the aircraft:

$$x(k+1) = \Phi x(k) + G_{\mathrm{d}}u(k) + w(k) \tag{11.103}$$

where

$$x(k) \triangleq x(k\Delta t), \qquad u(k) \triangleq u(k\Delta t)$$

$$\Phi = \Phi(\Delta t), \qquad \Phi(s) = \exp(As), \qquad G_{\mathrm{d}} = \int_0^{\Delta t} \Phi(s)\,\mathrm{d}sG$$

and $w(k)$ is a zero-mean Gaussian white noise sequence with covariance $Q_{\mathrm{d}}\delta_{kj}$, where

$$Q_{\mathrm{d}} \triangleq E[w(k)w^{\mathrm{T}}(k)] = \int_0^{\Delta t} \Phi^{\mathrm{T}}(s)Q\Phi(s)\,\mathrm{d}s \tag{11.104}$$

Now let us assume that the control system has $M-1$ sensor failure modes for each mode

$$h_i\colon z(k) = H_i x(k) + v_i(k), \qquad i = 1, \ldots, M-1 \tag{11.105}$$

where $v_i(k)$ is a Gaussian white noise sequence where

$$E[v_i(k)] = m_i \tag{11.106}$$

and

$$E\{[v_i(k) - m_i][v_i(j) - m_i]^{\mathrm{T}}\} = R_i\delta_{kj} \tag{11.107}$$

Here, the quantity m_i is an unknown parameter vector. Later we shall solve the problem as if m_i were known and then use the maximum likelihood (ML) estimate of m_i under the ith hypothesis, which is known as the generalized likelihood ratio approach [35].

For the normal unfailed condition, we will assume

$$h_0{:}z(k) = H_0x(k) + v_0(k) \tag{11.108}$$

where $E[v_0(k)] = 0$ and $E[v_0(k)v_0^{\mathrm{T}}(j)] = R_0\delta_{kj}$.

We shall be concerned with selecting the most probable failure state h_i, based on a finite set of measurements Z_k. To do so, we construct a Bayes' risk (or cost function) for the M-ary hypothesis testing [35]:

$$\mathcal{B} = \sum_{i=0}^{M}\sum_{j=0}^{M} P_{h_j}C_{ij}\int_{Y_i} p_{Z|h}(\alpha|h_j)\,\mathrm{d}\alpha \tag{11.109}$$

subject to

$$\sum_{i=1}^{M} P_{h_i} = 1$$

and where the sets Y_i, $0 \leqslant i \leqslant M$, are disjoint and their union represents the entire observation space. P_{h_i} is the a priori probability of hypothesis h_i being true, C_{ij} is an assigned cost of selection of h_i when h_j is true, and $p_{Z|h}(\alpha|h_j)$ is the conditional probability density of the measurement sequence Z_k given that h_j is true. The symbol $\int_{Y_i}$ implies that the integration is carried out over the decision region Y_i in the observation space. Decision regions Y_i are subsets of the observation space such that if $Z_k \in Y_i$ then the hypothesis h_i is to be selected. Note that the integral in equation (11.109) represents nothing more than the probability of making the incorrect decision of selecting h_i when h_j is true for $i \neq j$.

The problem, then, is to choose the boundaries of the decision region Y_i which will result in minimum Bayes' risk. The minimization of Bayes' risk may be performed easily by rewriting $\mathcal{B}$ in the form

$$\mathcal{B} = \sum_{i=0}^{M}\int_{Y_i} \psi_i(\alpha)\,\mathrm{d}\alpha \tag{11.110}$$

where

$$\psi_i(\alpha) = \sum_{j=0}^{M} P_{h_j} C_{ij} p_{Z|h}(\alpha|h_j) \tag{11.111}$$

This function is minimized by selecting h_i at each point α in the observation space such that $\psi_i(\alpha)$ is the smallest of the $M+1$ possible values of $\psi_k(\alpha)$, $0 \leq k \leq M$. Hence, the optimal decision regions are

$$Y_i = \{\alpha | \psi_i(\alpha) = \min_{0 \leq k \leq M} \psi_k(\alpha)\} \tag{11.112}$$

When introducing a dummy hypothesis h_M with a priori probability $P_{h_M} = 0$ such that

$$h_M : z(k) = v_0(k) \tag{11.113}$$

an equivalent decision criterion may be given in terms of likelihood ratios $\Lambda_i(\alpha)$, where

$$\Lambda_i(\alpha) = p_{Z|h}(\alpha|h_i)/p_{Z|h}(\alpha|h_M), \qquad i = 0, \ldots, M-1 \tag{11.114}$$

Dividing each ψ_i by the probability density of Z_k under h_M gives an equivalent decision criterion, $\lambda_i(\alpha)$, in terms of the likelihood ratios

$$\lambda_i(\alpha) = \sum_{j=0}^{M-1} P_{h_j} C_{ij} \Lambda_j(\alpha) \tag{11.115}$$

Then

$$Y_i = \left\{ \alpha | \lambda_i(\alpha) = \min_{0 \leq k \leq M-1} \lambda_k(\alpha) \right\} \tag{11.116}$$

As in the previous section (or in Section 3.3), noting that

$$\begin{aligned} p(Z_k|h_i) &= p(z(k)|Z_{k-1}, h_i) p(Z_{k-1}|h_i) \\ &= p(z(1)|h_i) \prod_{j=2}^{k} p(z(j)|Z_{j-1}, h_j) \end{aligned}$$

we have the likelihood ratio $\Lambda_i(k) \triangleq \Lambda_i(Z_k)$ for the M-ary hypothesis testing, in the form of

$$\Lambda_i(k) = \left[\prod_{j=1}^{k} \frac{|R_0|^{1/2}}{|\widetilde{P}_i(j|j-1)|^{1/2}} \right] \times \exp\left\{ -\frac{1}{2} \sum_{j=1}^{k} [\nu_i^{\mathrm{T}}(j) \widetilde{P}_i^{-1}(j|j-1) \nu_i(j) - z^{\mathrm{T}}(j) R_0^{-1} z(j)] \right\} \tag{11.117}$$

for $i = 0, \ldots, M-1$, where $\nu_i(j)$ is the innovation process under the ith hypothesis given by

$$\nu_i(k) = z(k) - H_i\hat{x}_i(k|k-1) - \bar{m}_i(k) \tag{11.118}$$

$\hat{x}_i(k|k-1)$ and $\widetilde{P}_i(k|k-1)$ are obtained by equations (11.55)–(11.57) and (11.59)–(11.61), but with $\Phi_i(k+1, k) \equiv \Phi$, $G_i(k) \equiv G_d$, $H_i(k) \equiv H_i$, $R_i(k) \equiv R_i$ and $Q_i(k) \equiv Q_d$. Note that, since the true value of $m_i(k)$ is not available, we will use the sample mean of $m_i(j)$, $j = 1, \ldots, k$, which is the ML estimate of m_i at the kth instant under the ith hypothesis.

Considerable simplification occurs if we consider $C_{ij} = 1 (i \neq j)$ and $C_{ii} = 0$. Under these conditions the optimal decision process may be modified without loss of generality to select the maximum of

$$\ln P_{h_i} - \ln \prod_{j=1}^{N_1} |\widetilde{P}_i(j|j-1)|^{1/2} - \frac{1}{2}\sum_{j=1}^{N_1} \nu_i^{\mathrm{T}}(j)\widetilde{P}_i^{-1}(j|j-1)\nu_i(j) \tag{11.119}$$

where N_1 is the total number of measurements used to make the decision. If the steady-state Kalman filter is used and the a priori probabilities of h_i are assumed to be equal, we can further reduce the computational load for the decision process. That is, we may take

$$\tau_i = \frac{N_1}{2}\ln|\widetilde{P}_i| + \frac{1}{2}\sum_{j=1}^{N_1} \nu_i^{\mathrm{T}}(j)\widetilde{P}_i^{-1}\nu_i(j) \tag{11.120}$$

where $\widetilde{P}_i$ denotes the steady-state value of $\widetilde{P}_i(j|j-1)$, and select the hypothesis h_i corresponding to the smallest τ_i, $i = 0, \ldots, M-1$.

The structure of a fault-tolerant control system is schematically indicated in Figure 11.5, where F is the state feedback gain and u_p denotes the pilot input. Note that, unlike the MMAC scheme shown in the previous section, the present state feedback control consists of multiplying the gain F by the estimate of the state corresponding to the hypothesis selected by the detection logic.

Example 11.3 [26]

In this example, we shall apply the method mentioned above to the design of a control system for one space shuttle orbiter configuration at Mach 5 and an altitude of 120000 ft. We consider the lateral dynamics of the vehicle described by equation (11.102), in which

$$A = \begin{bmatrix} -0.0580 & 0 & 0.0170 & -5.791 \\ 1.0 & 0 & 0.5773 & 0 \\ -0.0029 & 0 & -0.0085 & -0.7438 \\ 0.5 & 0.0055 & -0.8660 & -0.0009 \end{bmatrix}$$

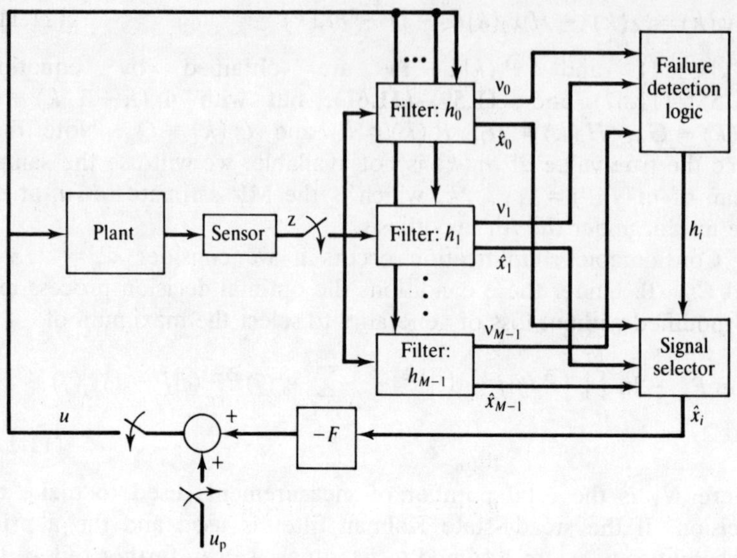

Figure 11.5 Fault-tolerant control system using multiple model technique.

$$G = \begin{bmatrix} 2.256 \\ 0 \\ 0.0553 \\ 0 \end{bmatrix}, \qquad x^{\mathrm{T}}(t) = [p, \phi, r, \beta], \qquad u = \delta_a$$

where p is the roll rate (or velocity) [deg s^{-1}], ϕ is the roll angle [deg], r is the yaw rate [deg s^{-1}], β is the angle of sideslip [deg], and δ_a is the aileron control input [deg].

Table 11.5 shows the level of certainty and (rms) error scale considerations for the variance Q of the system noise $w(t)$, where, a 0 level of certainty implies absolute certainty, 1 implies a high level of certainty, 2 implies only moderate certainty, and 3 implies a low level

Table 11.5 Evaluation of Q

Q component	*Level of certainty*	*Error scale*
Q_1 (p component)	2	0.05 rad s^{-1}
Q_2 (ϕ component)	0	1 rad
Q_3 (r component)	2	0.01 rad s^{-1}
Q_4 (β component)	3	0.001 rad s^{-1}

of certainty. The Q matrix selected is constructed, in terms of the level of certainty and error scale, from the elements of Table 11.5 as follows:

$$Q = \text{diag}\,(2 \times (0.05)^2,\ \ 0 \times (1)^2,\ \ 2 \times (0.01)^2,\ \ 3 \times (0.001)^2)$$

The matrices of the discretized equations of motion using a zero-order-hold with a sampling interval of 0.1 s are given by

$$\Phi = \begin{bmatrix} 0.9798 & -0.0002 & 0.0267 & -0.5752 \\ 0.0992 & 1 & 0.0587 & 0.0310 \\ 0.0021 & 0 & 1.002 & -0.0740 \\ 0.0497 & 0.0006 & -0.0862 & 0.9887 \end{bmatrix}$$

$$G_{\text{d}} = \begin{bmatrix} 0.2240 \\ 0.0139 \\ 0.00536 \\ 0.00538 \end{bmatrix}$$

$$Q_{\text{d}} = \begin{bmatrix} 0.4757 & 0.04757 & -0.006 & 0.0236 \\ 0.04757 & 0.00654 & 0.00100 & 0.00309 \\ -0.006 & 0.00100 & 0.02015 & -0.00185 \\ 0.0236 & 0.00309 & -0.00185 & 0.00211 \end{bmatrix}$$

By way of illustration, consider that the vehicle has three sensors: a roll-rate gyro, a yaw-rate gyro, and a sideslip indicator. There will, therefore, be four hypotheses to consider as follows:

$$h_0{:}H_1 = \begin{bmatrix} 1 & 0 & 0 & 0 \\ 0 & 0 & 1 & 0 \\ 0 & 0 & 0 & 1 \end{bmatrix}, \qquad h_1{:}H_1 = \begin{bmatrix} 0 & 0 & 0 & 0 \\ 0 & 0 & 1 & 0 \\ 0 & 0 & 0 & 1 \end{bmatrix}$$

$$h_2{:}H_2 = \begin{bmatrix} 1 & 0 & 0 & 0 \\ 0 & 0 & 0 & 0 \\ 0 & 0 & 0 & 1 \end{bmatrix}, \qquad h_3{:}H_3 = \begin{bmatrix} 1 & 0 & 0 & 0 \\ 0 & 0 & 1 & 0 \\ 0 & 0 & 0 & 0 \end{bmatrix}$$

The covariances of the measurement error are taken to be

$$R_0 = \text{diag}\,((0.05)^2, \quad (0.01)^2, \quad (0.01)^2)$$

and the failure covariances are assumed to be larger than the unfailed ones, i.e.

$$R_1 = \text{diag}\,(0.025, \quad 0.0001, \quad 0.0001)$$

$$R_2 = \text{diag}\,(0.0025, \quad 0.001, \quad 0.0001)$$

$$R_3 = \text{diag}\,(0.0025, \quad 0.0001, \quad 0.01)$$

Since the state is observable for each hypothesis, we can use the steady-state Kalman filters and obtain their constant gains by using any technique given in Section 2.4. Furthermore, the parameters of signal selector (11.120) become

$$h_0{:}\ln|\tilde{P}_0| = -22.505, \qquad \tilde{P}_0^{-1} = \begin{bmatrix} 300 & 0.6 & 340 \\ 0.6 & 6350 & 550 \\ 340 & 550 & 3552 \end{bmatrix}$$

$$h_1{:}\ln|\tilde{P}_1| = -20.020, \qquad \tilde{P}_1^{-1} = \begin{bmatrix} 40 & 0 & 0 \\ 0 & 6335 & 542 \\ 0 & 542 & 2000 \end{bmatrix}$$

$$h_2{:}\ln|\tilde{P}_2| = -20.439, \qquad \tilde{P}_2^{-1} = \begin{bmatrix} 300 & 0 & 340 \\ 0 & 1000 & 0 \\ 340 & 0 & 2893 \end{bmatrix}$$

$$h_3{:}\ln|\tilde{P}_3| = -18.900, \qquad \tilde{P}_3^{-1} = \begin{bmatrix} 265 & -52.5 & 0 \\ -52.5 & 6106 & 0 \\ 0 & 0 & 100 \end{bmatrix}$$

where the window is set to be $N_1 = 5$. By applying the same technique, under the duality of estimation and control, we can obtain the constant digital feedback gain, $-F = [-4.9 \quad 0.4 \quad 14.5 \quad -6]$, where the gain was determined to be constrained to a control system operating with only roll control.

Figure 11.6 illustrates the fault-tolerant system in operation when failures of increased noise type are introduced. By inspecting each piece of measurement data, it can be seen that the following sensor failure modes have been simulated: $\{h_0, \ h_1, \ h_2, \ h_3\}$. Figure 11.7 shows the detection result of hardover failures, in which a hardover failure in the beta sensor is postulated. Other simulation results can be found in Montgomery and Caglayan [26].

11.6 SUMMARY

After reviewing the so-called LQG control problem, we have presented in this chapter some passive-type MMAC methods for linear discrete-time stochastic systems with unknown constant parameters, using a partitioned adaptive technique described in Chapter 3. Two control examples, joint detection–control and fault-tolerant control, have also been discussed.

Other MMAC algorithms can be found, for example, in Stein and

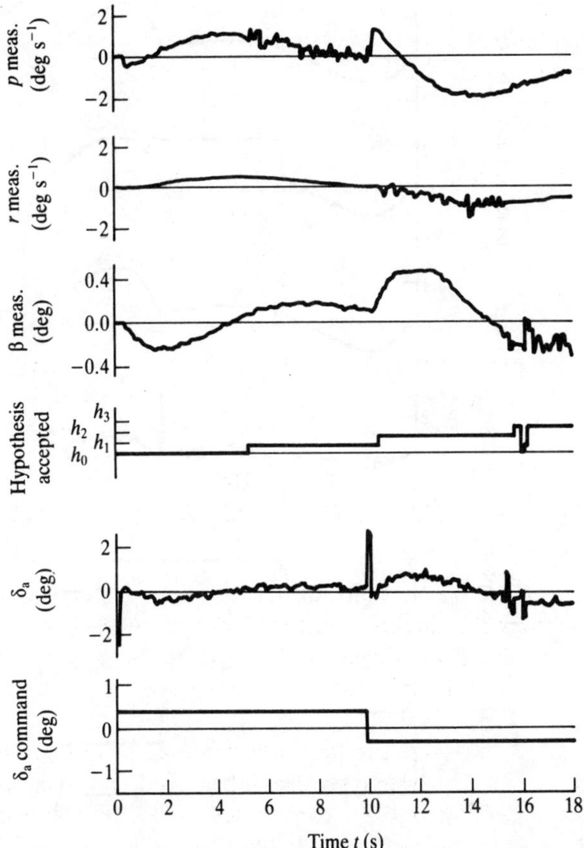

Figure 11.6 Operation of fault-tolerant system during failures resulting in increased noise.

Saridis [36], Saridis and Rao [37], and Wenk and Bar-Shalom [38, 39].

This chapter has been basically devoted to the presentation of results for the *non-switching parameter* case. The MMAC problems for the *switching parameter* case are very difficult to solve. One solution is given in Lee and Sims [22], and further information can be found in Sworder [40], VanLandingham and Moose [41], Moose *et al.* [42], Tugnait [43], Mariton [44, 45], and Watanabe and Tzafestas [46] (see also next chapter). A brief review of the state-of-the-art in linear stochastic control systems with multiplicative and additive noise can also be found in Panossian [47].

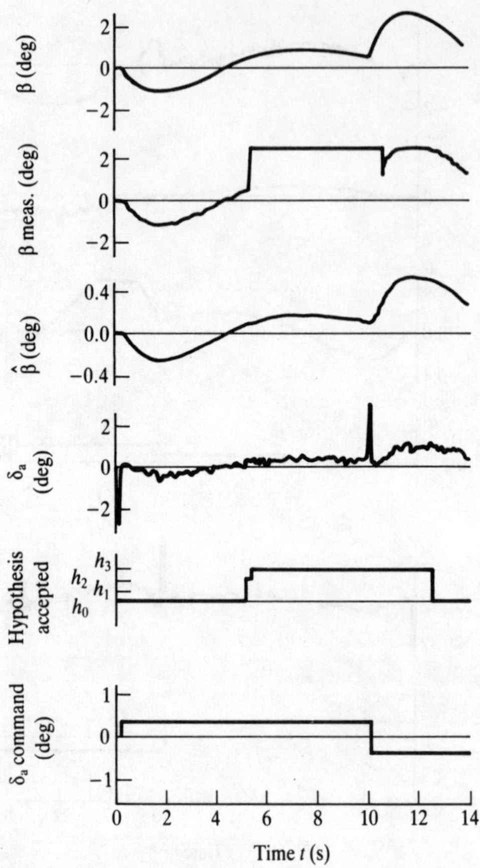

Figure 11.7 Operation under saturation hardover failure.

The problem of multiple model adaptive control in a distributed-sensor network is solved in Watanabe and Tzafestas [48]. Some extensions of the results to distributed-parameter systems can also be found in Tzafestas [49, 50], Tzafestas and Stavroulakis [51] and Watanabe *et al*. [52]. We refer the reader to Stavroulakis [53, 54] for many theoretical results on distributed-parameter systems.

EXERCISES

11.1 Consider the following function

$$f(u) = u^T S u + x^T u + u^T x$$

where u, $x \in \mathbb{R}^n$ and $S > 0$. Show that $f(u)$ can be rewritten by

$$f(u) = (u + S^{-1}x)^{\mathrm{T}} S(u + S^{-1}x) - x^{\mathrm{T}} S^{-1} x$$

and that if

$$u = -S^{-1}x$$

then the minimum of $f(u)$ is

$$f_{\min}(u) = -x^{\mathrm{T}} S^{-1} x$$

11.2 Consider the following stochastic system in continuous time:

$$\mathrm{d}x(t)/\mathrm{d}t = A(t)x(t) + G(t)u(t) + B(t)w(t)$$

where $w(t) \sim N(0, Q(t))$, $x(0) \sim N(\hat{x}(0), P(0))$, and these are independent of each other. The following cost function is to be minimized using the principle of optimality:

$$J = E\left\{ \|x(t_{\mathrm{f}})\|^2_{Q_{\mathrm{c}}(t_{\mathrm{f}})} + \int_{t_0}^{t_{\mathrm{f}}} [\|x(t)\|^2_{Q_{\mathrm{c}}(t)} + \|u(t)\|^2_{R_{\mathrm{c}}(t)}]\,\mathrm{d}t \right\}$$

where $Q_{\mathrm{c}}(t_{\mathrm{f}})$, $Q_{\mathrm{c}}(t) \geqslant 0$ and $R_{\mathrm{c}}(t) > 0$.

(a) Applying the limiting procedure shown in Chapter 2 and using the results of Theorem 11.1, show that the LQG optimal control policy is given by

$$u(t) = -F(t)x(t)$$
$$F(t) = R_{\mathrm{c}}^{-1}(t)G^{\mathrm{T}}(t)S(t)$$

where $S(t)$ satisfies the following RDE:

$$\mathrm{d}S(t)/\mathrm{d}t = -A^{\mathrm{T}}(t)S(t) - S(t)A(t) - Q_{\mathrm{c}}(t) + S(t)G(t)R_{\mathrm{c}}^{-1}(t)G^{\mathrm{T}}(t)S(t), \qquad S(t_{\mathrm{f}}) = Q_{\mathrm{c}}(t_{\mathrm{f}})$$

and the optimal cost J^* is given by

$$J^* = \hat{x}^{\mathrm{T}}(0)S(0)\hat{x}(0) + \mathrm{tr}\,[S(0)P(0)] + \int_{t_0}^{t_{\mathrm{f}}} \mathrm{tr}\,[S(t)B(t)Q(t)B^{\mathrm{T}}(t)]\,\mathrm{d}t$$

(b) Further, consider the following observation system:

$$z(t) = H(t)x(t) + v(t)$$

where $v(t) \sim N(0, R(t))$ is assumed to be independent of $\{w(t)\}$ and $x(0)$. By using the limiting procedure and the results of Theorem 11.2, show that

$$u(t) = -F(t)\hat{x}(t|t)$$

where $F(t)$ is given in part (a), $\hat{x}(t|t)$ is given by Theorem 2.11 and the optimal cost J^* is

$$J^* = \hat{x}^{\mathrm{T}}(0)S(0)\hat{x}(0) + \mathrm{tr}[S(0)P(0)] + \int_{t_0}^{t_f} \mathrm{tr}[S(t)B(t)Q(t)B^{\mathrm{T}}(t) + F^{\mathrm{T}}(t)G^{\mathrm{T}}(t)S(t)P(t|t)]\,\mathrm{d}t$$

or

$$J^* = \hat{x}^{\mathrm{T}}(0)S(0)\hat{x}(0) + \mathrm{tr}[Q_{\mathrm{c}}(N)P(N|N)] + \int_{t_0}^{t_f} \mathrm{tr}[Q_{\mathrm{c}}(t)P(t|t) + S(t)K(t)R(t)K^{\mathrm{T}}(t)]\,\mathrm{d}t$$

where $K(t) \triangleq P(t|t)H^{\mathrm{T}}(t)R^{-1}(t)$.

11.3 Verify expression (11.42).

11.4 Solve the following problems.

(a) Consider a cost function which contains coupled (or cross-product) weighting described by

$$J = E\left\{\|x(N)\|^2_{Q_{\mathrm{c}}(N)} + \sum_{k=0}^{N-1} [\|x(k)\|^2_{Q_{\mathrm{c}}(k)} + 2x^{\mathrm{T}}(k)M(k)u(k) + \|u(k)\|^2_{R_{\mathrm{c}}(k)}]\right\}$$

Show that, in the LQG control problem with exactly or inexactly known state, the controller gain is given by

$$F^*(k) = [G^{\mathrm{T}}(k)S(k+1)G(k) + R_{\mathrm{c}}(k)]^{-1} \times [G^{\mathrm{T}}(k)S(k+1)\Phi(k+1, k) + M^{\mathrm{T}}(k)]$$

where $S(k)$ is subject to the following equation

$$S(k) = Q_{\mathrm{c}}(k) + \Phi^{\mathrm{T}}(k+1, k)S(k+1)\Phi(k+1, k) - F^{*\mathrm{T}}(k)[G^{\mathrm{T}}(k)S(k+1)G(k) + R_{\mathrm{c}}(k)]F^*(k), \qquad S(N) = Q_{\mathrm{c}}(N)$$

Hint: Use the following facts:

$$x^{\mathrm{T}}Q_{\mathrm{c}}x + 2x^{\mathrm{T}}Mu + u^{\mathrm{T}}R_{\mathrm{c}}u = x^{\mathrm{T}}Q_{\mathrm{c}}^*x + u^{*\mathrm{T}}R_{\mathrm{c}}u^*$$

where

$$Q_{\mathrm{c}}^* = Q_{\mathrm{c}} - MR_{\mathrm{c}}^{-1}M^{\mathrm{T}}$$

$$u^* = u + R_{\mathrm{c}}^{-1}M^{\mathrm{T}}x$$

and

$$x(k+1) = \Phi^*(k+1, k)x(k) + G(k)u^*(k) + B(k)w(k)$$

where

$$\Phi^*(k+1, k) = \Phi(k+1, k) - G(k)R_c^{-1}M^T(k)$$

(b) Consider again a similar problem for the following continuous-time cost function:

$$J = E\left\{\|x(t_f)\|^2_{Q_c(t_f)} + \int_{t_0}^{t_f} [\|x(t)\|^2_{Q_c(t)} + 2x^T(t)M(t)u(t) + \|u(t)\|^2_{R_c(t)}]\,dt\right\}$$

Show that the control gain is given by

$$F^*(t) = R_c^{-1}(t)[G^T(t)S(t) + M^T(t)]$$
$$\dot{S}(t) = -A^T(t)S(t) - S(t)A(t) - Q_c(t) + F^{*T}(t)R_c(t)F^*(t), \qquad S(t_f) = Q_c(t_f)$$

11.5 Consider the time-invariant version of the discrete-time system of equations (11.1)–(11.3), but with the following modified cost function:

$$J_s = \lim_{N\to\infty} J/N$$

It is assumed that (Φ, D) is stabilizable, (C, Φ) is detectable and $P(0) \geq 0$ (see Chapter 2). Defining

$$GR_c^{-1}G^T = C_cC_c^T$$
$$Q_c = D_c^TD_c$$

it can be deduced, from the duality of estimation and control, that if (D_c, Φ) is detectable, (Φ, C_c) is stabilizable and $S(N) \geq 0$, then there exists a unique stabilizing solution S_s of the discrete-time ARE:

$$S - \Phi^TS\Phi + \Phi^TSG[G^TSG + R_c]^{-1}G^TS\Phi - Q_c = 0$$

(a) Show that, for the LQG control problem with exactly known state, the optimal cost J_s^* is reduced to

$$J_s^* = \text{tr}\,[SBQB^T]$$

(b) In the LQG control problem with inexactly known state, show that the optimal cost J_s^* is given by

$$J_s^* = \text{tr}\,[Q_cP_f + SK[HPH^T + R]K^T]$$

where

$$P_f = [I - KH]P$$
$$K = PH^T[HPH^T + R]^{-1}$$

in which P is subject to equation (2.55), or

$$J_s^* = \mathrm{tr}\,[SBQB^T + F^T G^T S \Phi P_f]$$

where

$$F = [G^T S G + R_c]^{-1} G^T S \Phi$$

11.6 Consider the discrete-time linear control problem subject to the following matrix coefficients:

$$\Phi = \begin{bmatrix} 0.9 & 0.1 \\ 0 & 0.4 \end{bmatrix}, \quad G = \begin{bmatrix} 1 \\ 1 \end{bmatrix}$$

$$Q_c = \begin{bmatrix} 1 & 0 \\ 0 & 1 \end{bmatrix}, \qquad R_c = 20$$

(a) Use the eigenvalue–eigenvector method to solve the ARE.
(b) Use Kleinman's iterative method to solve the ARE.
(c) Use the doubling algorithm to solve the ARE.
(d) Compute the control gain by applying the Chandrasekhar-type algorithm with $S(N) = 0$. In this case, are there any merits in using such an algorithm?

11.7 Derive the steady-state version of Exercise 11.2 by setting

$$J_s = \lim_{t_f \to \infty} J/t_f$$

11.8 Consider the LQG control problem of a robot manipulator given in Exercise 2.19 with the following cost function:

$$J_s = \lim_{t_f \to \infty} \frac{1}{t_f} E\left\{\int_{t_0}^{t_f} [x_1^2(t) + x_2^2(t) + u^2(t)]\,dt\right\}$$

Obtain the optimal cost for this problem, applying any algorithm presented in Chapter 2 to solve the AREs.

11.9 Consider the continuous-time linear control problem with the following matrix coefficients.

$$A = \begin{bmatrix} 0 & 1 \\ 0 & 0 \end{bmatrix}, \quad G = \begin{bmatrix} 1 \\ 1 \end{bmatrix}, \quad Q_c = \begin{bmatrix} 1 & 0 \\ 0 & 1 \end{bmatrix}, \quad R_c = 1$$

(a) Use the eigenvalue–eigenvector method to solve the ARE.
(b) Apply Kleinman's iterative method to obtain the solution of the ARE.
(c) Use the matrix sign function method to solve the ARE.
(d) Obtain the solution of the ARE by applying the doubling algorithm. Repeat the problem with the time steps $\Delta = 1 \times 10^{-3}$, $\Delta = 1 \times 10^{-5}$ and $\Delta = 1 \times 10^{-7}$.

11.10 Suppose that $\Phi_i(k+1, k) = \Phi(k+1, k)$, $G_i(k) = G(k)$, $H_i(k) = H(k)$, $Q_i(k) = Q(k)$, $R_i(k) = R(k)$ and $x(0) \sim N(\hat{x}_i(0), P_i(0))$, $i = 1, \ldots, M$. Write down the MMAC algorithm based on the Deshpande–Upadhyay–Lainiotis method for this case.

11.11 Suppose that $\Phi_i(k+1, k) = \Phi(k+1, k)$, $G_i(k) = G(k)$, $H_i(k) = H(k)$, $Q_i(k) = Q(k)$, $R_i(k) = R(k)$, $\hat{x}_i(0) = \hat{x}(0)$ and $P_i(0) = P(0)$. Show how the MMAC algorithm given by Theorem 11.1 can be modified when the unknown parameter is included only in the control weighting matrices, i.e. $\{Q_c^i(N), Q_c^i(k), R_c^i(k)\}$, $i = 1, 2, \ldots, M$. What happens in the evaluation of the a posteriori probabilities $p(\theta_i|Z_k)$ for such a case?

11.12 Consider an aircraft longitudinal autopilot problem [37] with known lifting time constant and measurement noise variance. The discretized version of the equation of motion in sampling period $\Delta t = 0.1$ s is

$$x(k+1) = \begin{bmatrix} 1 & \Delta t \\ \omega_0^2 \Delta t & (1 - 2\zeta_i \omega_0) \end{bmatrix} x(k)$$

$$+ \begin{bmatrix} 0 \\ \omega_0^2 \Delta t \end{bmatrix} u(k) + \begin{bmatrix} 0 \\ \sqrt{10\Delta t} \end{bmatrix} w(k)$$

$$z(k) = [1 \quad 0] x(k) + v(k)$$

where $x(k) = [x_1(k) \quad x_2(k)]$, $x_1(k)$ is the perturbation from the angle of attack with $x(0) \sim N(\hat{x}(0), P(0))$ and $x_2(k)$ is its velocity. Furthermore, $\omega_0 = -1$ is undamped pitch natural frequency, ζ_i is a damping ratio, $w(k) \sim N(0, \Delta t \sigma_i^2)$ and $v(k) \sim N(0,0.1)$. Using the parameter set

$$\sigma_1 = 1.0, \qquad \sigma_2 = 2.0$$

$$\zeta_1 = 0.1, \qquad \zeta_2 = 0.0$$

develop an MMAC algorithm based on the DUL method to minimize the following cost function:

$$J = E\left\{ \sum_{k=0}^{N-1} [x_1^2(k) + x_2^2(k) + u^2(k)] \right\}$$

11.13 Give a counterexample to show that the control law in Section 11.4 is not optimal (see Wernersson [55, 56]).

11.14 Consider the following continuous-time system described by the n-dimensional state-space model

$$\mathrm{d}x(t) = A(t, \theta)x(t)\,\mathrm{d}t + G(t, \theta)u(t) + B(t, \theta)\,\mathrm{d}w(t), \qquad t \geqslant t_0$$

$$\mathrm{d}z(t) = H(t, \theta)x(t)\,\mathrm{d}t + \mathrm{d}v(t)$$

with cost function:

$$J = E\left\{ \|x(t_f)\|^2_{Q_c(t_f)} + \int_{t_0}^{t_f} [\|x(t)\|^2_{Q_c(t)} + \|u(t)\|^2_{R_c(t)}]\,\mathrm{d}t \right\}$$

where $Q_c(t_f)$, $Q_c(t) \geqslant 0$ and $R_c(t) > 0$. Verify that if the unknown parameter vector θ belongs to the discrete set $\Theta = [\theta_1, \theta_2, \ldots, \theta_M]$, then the continuous-time version of the MMAC shown in Corollary 11.2 is given by

$$u(t) = -\sum_{i=1}^{M} F_i(t)\hat{x}(t|t, \theta_i)p(\theta_i|t)$$

$$F_i(t) = R_c^{-1}(t)G^T(t, \theta_i)S(t, \theta_i)$$

$$\frac{dS(t, \theta_i)}{dt} = -A^T(t, \theta_i)S(t, \theta_i) - S(t, \theta_i)A(t, \theta_i) + S(t, \theta_i)G(t, \theta_i)R_c^{-1}(t)G^T(t, \theta_i)S(t, \theta_i) - Q_c(t), \quad S(t_f, \theta_i) = Q_c(t_f)$$

where $\hat{x}(t|t, \theta_i)$ is obtained by the continuous-time Kalman filter matched to θ_i and $p(\theta_i|t)$ is given by Corollary 3.1 of Chapter 3 (see also Lainiotis *et al*. [16, 17]).

REFERENCES

[1] MEIER III, L., LARSON, R. E. and TETHER, A. J., Dynamic Programming for Stochastic Control of Discrete Systems, *IEEE Trans. Aut. Control*, vol. AC-16, 1971, pp. 767–75.

[2] WITSENHAUSEN, H. S., Separation of Estimation and Control for Discrete Time Systems, *Proc. IEEE*, vol. 59, no. 11, November 1971, pp. 1557–66.

[3] ASHER, R. B., ANDRISANI II, D. and DORATO, P., Bibliography on Adapative Control Systems, *Proc. IEEE*, vol. 64, no. 8, August 1976, pp. 1226–40.

[4] SARIDIS, G. N., *Self-Organizing Control of Stochastic Systems*, Marcel Dekker, New York, 1977.

SARIDIS, G. N., Toward the Realization of Intelligent Controls, *Proc. IEEE*, vol. 67, no. 8, 1979, pp. 1115–33.

[5] SARIDIS, G. N., Application of Pattern Recognition Methods to Control Systems, *IEEE Trans. Aut. Control*, vol. AC-26, no. 3, June 1981, pp. 638–45.

[6] EGARDT, B., *Stability of Adaptive Controllers*, Springer-Verlag, Berlin, 1979.

[7] LANDAU, I. D., *Adaptive Control, The Model Reference Approach*, Marcel Dekker, New York, 1979.

[8] LANDAU, I. D. and TOMIZUKA, M., *Theory and Practice of Adaptive Control Systems*, Ohmusha, Tokyo, 1981 (in Japanese).

[9] ÅSTRÖM, K. J. and WITTENMARK, B., *Computer Controlled Systems*, Prentice Hall, Englewood Cliffs, New Jersey, 1984.

[10] GOODWIN, G. C. and SIN, K. S., *Adaptive Filtering Prediction and Control*, Prentice Hall, Englewood Cliffs, New Jersey, 1984.

[11] GRIMBLE, M. J. and JOHNSON, M. A., *Optimal Control and Stochastic Estimation*, Vol. 2, Wiley, Chichester, 1988.

[12] GREENE, C. S., An Analysis of the Multiple Model Adaptive Control Algorithm, Ph.D. dissertation, MIT, Cambridge, Massachusetts, August 1978.

[13] BINDER, Z., FONTAINE, H., MAGALHAES, M. F. and BAUDOIS, D., About a Multimodel Control Methodology, Algorithms, Multiprocessors, Implementation and Application, *Preprints of IFAC 8th Triennial World Congress*, vol. VII, August 1981, pp. 89–94.

[14] ASHER, R. B., SIMS, C. S. and SEBESTA, H. R., Optimal Open-Loop Feedback Control for Linear Systems with Unknown Parameters, *Inf. Sci.*, vol. 11, 1976, pp. 265–77.

[15] WATANABE, K., A Passive-Type Multiple Model Adaptive Control (MMAC) of Linear Discrete-Time Stochastic Systems with Uncertain Observation Subsystem, *Int. J. Syst. Sci.*, vol. 15, no. 6, 1984, pp. 647–59.

[16] LAINIOTIS, D. G., DESHPANDE, J. G. and UPADHYAY, T. N., Optimal Adaptive Control: A Nonlinear Separation Theorem, *Int. J. Control*, vol. 15, no. 5, May 1972, pp. 877–88.

[17] LAINIOTIS, D. G., UPADHYAY, T. N. and DESHPANDE, J. G., Optimal Adaptive Control of Linear Systems with Unknown Measurement Subsystems, *Inf. Sci.*, July 1973, pp. 217–33.

[18] DESHPANDE, J., UPADHYAY, T. N. and LAINIOTIS, D. G., Adaptive Control of Linear Stochastic Systems, *Automatica*, vol. 9, 1973, pp. 107–15.

[19] LAINIOTIS, D. G., Partitioning: A Unifying Framework for Adaptive Systems, II: Control, *Proc. IEEE*, vol. 64, no. 8, August 1976, pp. 1182–98.

[20] ATHANS, M. and WILLNER, D., A Practical Scheme for Adaptive Aircraft Flight Control Systems, *NASA Technical Note*, NASA TN D-7647, 1973, pp. 315–36.

[21] ATHANS, M., CASTANON, D., DUNN, K.-P., GREENE, C. S., LEE, W. H., SANDELL JR, N. R. and WILLSKY, A. S., The Stochastic Control of the F-8C Aircraft Using a Multiple Model Adaptive Control (MMAC) Method—Part I: Equilibrium Flight, *IEEE Trans. Aut. Control*, vol. AC-22, no. 5, October 1977, pp. 768–80.

[22] LEE, A. Y. and SIMS, C. S., Adaptive Estimation and Stochastic Control for Uncertain Models, *Int. J. Control*, vol. 19, no. 3, 1974, pp. 625–39.

[23] UPADHYAY, T. N. and LAINIOTIS, D. G., Joint Adaptive Plant and Measurement Control of Linear Stochastic Systems, *IEEE Trans. Aut. Control*, vol. AC-19, no. 5, October 1974, pp. 567–71.

[24] BELLMAN, R., *Dynamic Programming*, Princeton University Press, Princeton, New Jersey, 1957.

[25] BERTSEKAS, D. P., *Dynamic Programming and Stochastic Control*, Academic Press, New York, 1976.

[26] MONTGOMERY, R. C. and CAGLAYAN, A. K., Failure Accommodation in Digital Flight Control Systems by Bayesian Decision Theory, *J. Aircraft*, vol. 13, no. 2, February 1976, pp. 69–75.

[27] MONTGOMERY, R. C. and PRICE, D. B., Failure Accommodation in Digital Flight Control Systems Accounting for Nonlinear Aircraft Dynamics, *J.*

Aircraft, vol. 13, no. 2, February 1976, pp. 76–82.

[28] ÅSTRÖM, K. J., *Introduction to Stochastic Control Theory*, Academic Press, New York, 1970.

[29] MEDITCH, J. S., *Stochastic Optimal Linear Estimation and Control*, McGraw-Hill, New York, 1969.

[30] OWENS, D. H., *Multivariable and Optimal Systems*, Academic Press, London, 1981.

[31] TSE, E. and BAR-SHALOM, Y., Actively Adaptive Control for Nonlinear Stochastic Systems, *Proc. IEEE*, vol. 64, no. 8, August 1976, pp. 1172–81.

[32] BAR-SHALOM, Y. and TSE, E., Concepts and Methods in Stochastic Control. In C. T. Leondes (ed.), *Control and Dynamic Systems: Advances in Theory and Applications*, Vol. 12, Academic Press, New York, 1976.

BAR-SHALOM, Y. and TSE, E., Dual Effect, Certainty Equivalence and Separation in Stochastic Control, *IEEE Trans. Aut. Control*, vol. AC-19, 1974, pp. 494–500.

[33] BAR-SHALOM, Y., Stochastic Dynamic Programming: Caution and Probing, *IEEE Trans. Aut. Control*, vol. AC-26, no. 5, October 1981, pp. 1184–95.

[34] FEL'DBAUM, A. A., *Optimal Control Systems*, Academic Press, New York, 1965.

[35] SAGE, A. P. and MELSA, J. L., *Estimation Theory and Applications to Communications and Control*, Robert E. Krieger, New York, 1979.

[36] STEIN, G. and SARIDIS, G. N., A Parameter Adaptive Control Technique, *Automatica*, vol. 5, no. 6, 1969, pp. 731–40.

[37] SARIDIS, G. N. and RAO, T. K., A Learning Approach to the Parameter Adaptive Self-Organizing Control Problem, *Automatica*, vol. 9, no. 5, 1972, pp. 589–98.

[38] WENK, C. J. and BAR-SHALOM, Y., A Multiple Model Adaptive Dual Control Algorithm for Stochastic Systems with Unknown Parameters, *Proc. 18th IEEE Conference on Decision and Control*, vol. 2, 1979, pp. 723–30.

[39] WENK, C. J. and BAR-SHALOM, Y., A Multiple Model Adaptive Dual Control Algorithm for Stochastic Systems with Unknown Parameters, *IEEE Trans. Aut. Control*, vol. AC-25, no. 4, 1980, pp. 703–10.

[40] SWORDER, D. D., Control of System Subject to Sudden Change in Character, *Proc. IEEE*, vol. 64, no. 8, August 1976, pp. 1219–25.

[41] VANLANDINGHAM, H. F. and MOOSE, R. L., Digital Control of High Performance Aircraft Using Adaptive Estimation Techniques, *IEEE Trans. Aero. and Electron. Syst.*, vol. AES-13, no. 2, March 1977, pp. 112–19.

[42] MOOSE, R. L., VANLANDINGHAM, H. F. and ZWICKE, P.E., Digital Set Point Control of Nonlinear Stochastic Systems, *IEEE Trans. Industrial Electron. and Control Instrum.*, vol. IECI-25, no. 1, February 1978, pp. 39–45.

[43] TUGNAIT, J. K., Control of Stochastic Systems with Markov Interrupted

Observations, *IEEE Trans. Aero. and Electron. Syst.*, vol. AES-19, no. 2, 1983, pp. 232–9.

[44] MARITON, M., Joint Estimation and Control of Jump Linear Systems with Multiplicative Noises, *Trans. ASME, J. Dynamic Syst. Meas. Control*, vol. 109, March 1987, pp. 24–8.

[45] MARITON, M., On the Influence of Noise on Jump Linear Systems, *IEEE Trans. Aut. Control*, vol. AC-32, no. 12, December 1987, pp. 1094–7.

[46] WATANABE, K. and TZAFESTAS, S. G., Stochastic Control for Systems with Faulty Sensors, *Trans. ASME, J. Dynamic Syst. Meas. Control*, vol. 112, March 1990, pp. 143–7.

[47] PANOSSIAN, H. V., Review of Linear Stochastic Optimal Control Systems and Applications, *Trans. ASME, J. Vibration, Acoustics, Stress, and Reliability in Design*, vol. 111, 1989, pp. 399–403.

[48] WATANABE, K. and TZAFESTAS, S. G., A Hierarchical Multiple Model Adaptive Control of Discrete-time Stochastic Systems for Sensor and Actuator Uncertainties, *Automatica*, vol. 26, no. 5, 1990, pp. 875–86.

[49] TZAFESTAS, S. G., Parameter Adaptive Control of Stochastic Distributed Systems, *Int. J. Syst. Sci.*, vol. 11, no. 12, 1980, pp. 1397–1433.

[50] TZAFESTAS, S. G., Adaptive Control of Stochastic Distributed Parameter Systems: A Survey, *Large Scale Syst.*, vol. 6, 1984, pp. 221–48.

[51] TZAFESTAS, S. G. and STAVROULAKIS, P., Partitioned Adaptive Filtering and Control of Distributed Systems with Space-Dependent Unknown Parameters, *Mathematics and Computers in Simulation*, vol. XXIII, 1981, pp. 206–12.

[52] WATANABE, K., YOSHIMURA, T. and SOEDA, T., Optimal Adaptive Estimation and Stochastic Control for Distributed-Parameter Systems, *IEEE Trans. Aut. Control*, vol. AC-27, no. 1, 1982, pp. 216–19.

[53] STAVROULAKIS, P., *Distributed Parameter Systems Theory, Part I: Control*, Hutchinson Ross, Stroudsburg, Pennsylvania, 1983.

[54] STAVROULAKIS, P., *Distributed Parameter Systems Theory, Part II: Estimation,* Hutchinson Ross, Stroudsburg, Pennsylvania, 1983.

[55] WERNERSSON, Å., Comments on: Optimal Stochastic Control for Discrete-Time Linear Systems with Interrupted Observations, *Automatica*, vol. 10, 1974, pp. 113–15.

[56] WERNERSSON, Å., Comments on: Optimal Adaptive Control of Linear Systems with Unknown Measurement Subsystems, *Inf. Sci.*, vol. 8, 1975, pp. 89–93.

12

Multiple Model Adaptive Filtering and Control for Markovian Jump Systems

12.1 INTRODUCTION

So far, we have been concerned with multiple model adaptive filtering and control for systems with unknown constant parameters. In this final chapter, we further extend the partition theorem introduced in Chapter 3 to systems having abruptly changing (or jump) parameters whose evolutions are not observed. Most of the ideas discussed in Chapter 3 carry over to such a case in a natural and simple way.

The consideration of system models with jumps is motivated by the potential applicability of such models to a large class of realistic problems:

1. Fault detection for a dynamic system with *failures* in components or subsystems (see, for example, Willsky [1]; Gustafson *et al*. [2]; Willsky [3]; Watanabe [4, 5]).
2. Target tracking for a moving vehicle with sudden manoeuvres (see, for example, Ricker and Williams [6]; Moose *et al* [7]; Chang and Tabaczynski [8]).
3. Approximation of a *nonlinear* system by a set of linearized models to cover the entire dynamic range (see, for example, VanLandigham and Moose [9]; Moose *et al*. [10]).

There are two main approaches to the estimation of systems having jumps. One approach is to use the multiple model adaptive filter (MMAF) developed for systems having unknown constant parameters, setting an upper (or lower) bound on the a posteriori probabilities of parameters [2–5]. Another approach is to extend the MMAF approach to systems with possibly unknown, time-varying parameters, which are modelled as a finite state Markov (or semi-Markov) chain with known transition probabilities. The latter approach seems to be more natural than the former, if the transition probabilities are completely known.

However, a complication in the Markovian parameter case is that the number of 'multiple models' grows exponentially with time due to the Markovian nature of unknown parameters. As a result, the optimal estimator requires a bank of elemental estimators growing in number exponentially with time, which renders the optimal solution impractical. To circumvent this problem, a number of successful suboptimal algorithms have appeared, among them the random sampling algorithm (RSA) [11], the detection–estimation algorithm (DEA) [12–14], the generalized pseudo-Bayes algorithm (GPBA) [15–19], and the interacting multiple model algorithm (IMMA) [20–22].

In this chapter, we are concerned with the GPBA approach to state estimation and system structure detection for linear discrete-time Markovian jump systems, in which the unknown time-varying parameters are modelled by a Markov chain with known transition probabilities. In Section 12.2, a formal statement of the state estimation problem is presented. In Section 12.3, the optimal solution to the filtering problem is described. Three GPBAs are introduced in Section 12.4. The problems of system structure detection and control are discussed in Sections 12.5 and 12.6 respectively.

12.2 STATEMENT OF THE ESTIMATION PROBLEM

Let $\theta(k) \in \mathfrak{S} = \{1, 2, \ldots, s\}$, $k = \{1, 2, \ldots\}$, denote a finite-state discrete Markov chain with completely known time-invariant transition probabilities

$$p_{ij} \triangleq \Pr\{\theta(k) = j | \theta(k-1) = i\}, \qquad i, j \in \mathfrak{S} \tag{12.1}$$

and initial probability distribution $p_i = \Pr\{\theta(0) = i\}$, $i \in \mathfrak{S}$. Let $\pi = [p_{ij}]$ denote the $s \times s$ transition probability matrix. The system state equation is given by

$$x(k+1) = \Phi(\theta(k+1))x(k) + B(\theta(k+1))w(k) \tag{12.2}$$

where $x(k) \in \mathbb{R}^{n_x}$ is the system state, and $w(k) \in \mathbb{R}^p$ is a zero-mean Gaussian white noise sequence with covariance Q. The observation equation associated with equation (12.2) is modelled by

$$z(k) = H(\theta(k))x(k) + D(\theta(k))v(k) \tag{12.3}$$

where $z(k) \in \mathbb{R}^m$ is the observation vector, and $v(k) \in \mathbb{R}^m$ is a zero-mean Gaussian white measurement noise with covariance R such that $D_i R D_i^{\mathrm{T}} > 0$ $(i \in \mathfrak{S})$ where $D(\theta(k)) \in \{D_i,\ i = 1, 2, \ldots, s\}$. The initial state is assumed to be subject to

$$x(0) \sim N(\hat{x}(0), P(0)) \tag{12.4}$$

Finally, $x(0)$, $\{w(k)\}$, $\{(v(k)\}$ and $\{\theta(k)\}$ are assumed to be mutually independent.

The objective here is to find the minimum mean-squared error (MMSE) state estimate $\hat{x}(k|k)$ of $x(k)$ given the observations $Z_k \triangleq \{z(i), 1 \leq i \leq k\}$.

12.3 OPTIMAL SOLUTION OF THE ESTIMATION PROBLEM

Define a Markov chain state sequence $I(k)$ as follows:

$$I(k) \triangleq \{\theta(1), \ldots, \theta(k)\} \tag{12.5}$$

and let $I_j(k)$ denote a specific sequence from the space of all possible sequences from $I(k)$ which contains s^k elements. If the state estimate conditioned on a specific sequence is written

$$\hat{x}_j(k|k) \triangleq E[x(k)|I_j(k), Z_k] \tag{12.6}$$

and the associated estimation error covariance is written

$$P_j(k|k) \triangleq E\{[x(k) - \hat{x}_j(k|k)][x(k) - \hat{x}_j(k|k)]^{\mathrm{T}}|I_j(k), Z_k\} \tag{12.7}$$

then we have the following theorem.

Theorem 12.1

The MMSE filtered estimate $\hat{x}(k|k) = E[x(k)|Z_k]$ and the associated estimation error covariance $P(k|k) = E\{[x(k) - \hat{x}(k|k)]$ $[x(k) - \hat{x}(k|k)]^{\mathrm{T}}|Z_k\}$ are calculated from

$$\hat{x}(k|k) = \sum_{j=1}^{s^k} \hat{x}_j(k|k) p(I_j(k)|Z_k) \tag{12.8}$$

and

$$P(k|k) = \sum_{j=1}^{s^k} \{P_j(k|k) + [\hat{x}_j(k|k) - \hat{x}(k|k)] \times [\hat{x}_j(k|k) - \hat{x}(k|k)]^{\mathrm{T}}\} p(I_j(k)|Z_k) \tag{12.9}$$

where $p(I_j(k)|Z_k)$ is the a posteriori probability of $I_j(k)$ given Z_k, which is subject to

$$p(I_j(k)|Z_k) = \frac{p(z(k)|I_j(k), Z_{k-1})p(I_j(k)|Z_{k-1})}{\sum_{l=1}^{s^k} p(z(k)|I_l(k), Z_{k-1})p(I_l(k)|Z_{k-1})} \tag{12.10}$$

in which $p(z(k)|\cdot)$ denotes the conditional probability density function of the observation $z(k)$ given the past observations Z_{k-1} and the particular state mode sequence $I_j(k)$.

Proof

The proof is apparent from the result of Corollary 3.2 of Chapter 3. □

It should be noted that

$$p(I_j(k)|Z_{k-1}) = p(\theta(k)|Z_{k-1}, I_l(k-1))p(I_l(k-1)|Z_{k-1}) \tag{12.11}$$

where $I_j(k) = \{\theta(k), I_l(k-1)\}$ with $\theta(k) \in \mathfrak{S}$, i.e. $I_l(k-1)$ is a subsequence of $I_j(k)$ and $\theta(k)$ is the last or most recent element of $I_j(k)$. Since $\theta(k)$ is modelled as a Markov chain, we have

$$\begin{aligned} p(\theta(k)|Z_{k-1}, I_l(k-1)) &= p(\theta(k)|I_l(k-1)) \\ &= p(\theta(k)|\theta(k-1) = i), \qquad i \in \mathfrak{S} \end{aligned} \tag{12.12}$$

where the first step in equation (12.12) follows from conditional independence of $\{\theta(k)\}$ and $\{z(k)\}$, and the second step follows from the Markovian nature of $\theta(k)$. Note that equation (12.12) is just an element of the transition probability matrix π.

Now, with the initial information given in equation (12.4) and sequence $I_j(k)$, we can obtain $\{\hat{x}_j(k|k), P_j(k|k)\}$ recursively by applying Kalman filters matched to sequences $I_j(k)$, $j = 1, \ldots, s^k$. Moreover, the weighting probability $p(I_j(k)|Z_k)$ can be computed from the information supplied by the same Kalman filters, because $p(z(k)|I_j(k), Z_{k-1})$ is Gaussian. Thus, the optimal estimator given in equation (12.8) can be implemented with a weighted sum of s^k estimates $\hat{x}_j(k|k)$.

12.4 GENERALIZED PSEUDO-BAYES ALGORITHMS

It is important to note that the optimal estimator (12.8) requires an exponentially increasing memory and computational capacity with time. One has to resort, therefore, to suboptimal schemes to circumvent this difficulty. In the effort to reduce the large amounts of memory required by the optimal estimator, various suboptimal estimators have been described in a number of papers as indicated in Section 12.1. In this section we shall focus on the GPBAs only.

12.4.1 Fundamental Approach

The fundamental approach was first proposed by Ackerson and Fu [15] in the context of the switching environment problem. This pseudo-Bayes estimator was given in such a way that it would have to calculate and store only one state estimate at each instant. It is assumed that $p(x(k)|Z_k)$ is normally distributed with mean $\hat{x}(k|k)$ and covariance $P(k|k)$; in truth, $p(x(k)|Z_k)$ is the sum of s^k separate Gaussian distributions, as we have seen in the previous section. The resulting algorithm is summarized in the following corollary.

Corollary 12.1 Ackerson–Fu (AF) method [15]

The estimator $\hat{x}(k|k)$ can be calculated from

$$\hat{x}(k|k) = \sum_{i=1}^{s} \hat{x}_i(k|k)p(\theta(k) = i|Z_k) \tag{12.13}$$

and the estimation error covariance matrix $P(k|k)$ is given by

$$P(k|k) = \sum_{i=1}^{s} P_i(k|k)p(\theta(k) = i|Z_k) \tag{12.14}$$

where the a posteriori probability of $\theta(k)$ is

$$p(\theta(k) = j|Z_k) = \frac{p(z(k)|\theta(k) = j, Z_{k-1})\sum_{i=1}^{s} p_{ij}p(\theta(k-1) = i|Z_{k-1})}{\sum_{l=1}^{s} p(z(k)|\theta(k) = l, Z_{k-1})\sum_{i=1}^{s} p_{il}p(\theta(k-1) = i|Z_{k-1})} \tag{12.15}$$

Here $\hat{x}_i(k|k) = E[x(k)|\theta(k) = i, Z_k]$ and $P_i(k|k) = E\{[x(k) - \hat{x}_i(k|k)][x(k) - \hat{x}_i(k|k)]^{\mathrm{T}}|\theta(k) = i, Z_k\}$ can be calculated by using $\hat{x}(k-1|k-1)$ and $P(k-1|k-1)$ as follows:

$$\hat{x}_i(k|k-1) = \Phi(\theta(k) = i)\hat{x}(k-1|k-1), \qquad \hat{x}(0|0) = \hat{x}(0) \tag{12.16}$$

$$P_i(k|k-1) = \Phi(\theta(k)=i)P(k-1|k-1)\Phi^{\mathrm{T}}(\theta(k)=i) + B(\theta(k)=i)QB^{\mathrm{T}}(\theta(k)=i), \qquad P(0|0) = P(0) \tag{12.17}$$

$$\hat{x}_i(k|k) = \hat{x}_i(k|k-1) + K_i(k)v_i(k) \tag{12.18}$$

$$P_i(k|k) = [I - K_i(k)H(\theta(k)=i)]P_i(k|k-1) \tag{12.19}$$

where

$$K_i(k) \triangleq P_i(k|k-1)H^{\mathrm{T}}(\theta(k)=i)V_i^{-1}(k|k-1) \tag{12.20}$$

$$v_i(k) \triangleq z(k) - H(\theta(k)=i)\hat{x}_i(k|k-1) \tag{12.21}$$

$$V_i(k|k-1) \triangleq H(\theta(k)=i)P_i(k|k-1)H^{\mathrm{T}}(\theta(k)=i) + D(\theta(k)=i)RD^{\mathrm{T}}(\theta(k)=i) \tag{12.22}$$

Furthermore, $p(z(k)|\theta(k)=i,\ Z_{k-1})$ can be obtained by

$$p(z(k)|\theta(k)=i,\ Z_{k-1}) \sim N\{H(\theta(k)=i)\hat{x}_i(k|k-1),\ V_i(k|k-1)\} \tag{12.23}$$

Proof

See Ackerson and Fu [15]. □

This algorithm is very simple to implement (see also Figure 12.1). However, it should be noted that, as suggested by Tugnait and Haddad [12], Tugnait [13] and Akashi and Kumamoto [11], and as will be seen in later simulations, the performance of this algorithm is inferior to that of other GPBAs.

12.4.2 The One-Step Measurement Update Approach

The generalization of the above algorithm was proposed first by Jaffer and Gupta [16] for systems with interrupted observations.

The essential assumption of the GPBA is that the probability density of the system state at time k conditioned on Z_k and the Markov chain state sequence $I_j(k, k-1) \triangleq \{\theta(i), k-n \leqslant i \leqslant k\}$, $j = 1, \ldots, s^{n+1}$, $n \geqslant 1$, is Gaussian. That is, it is assumed that

$$p(x(k)|Z_k,\ I_j(k,\ k-1)) \sim N(\hat{x}_j(k|k),\ P_j(k|k)) \tag{12.24}$$

Under this assumption, the state estimate $\hat{x}(k|k)$ and the associated

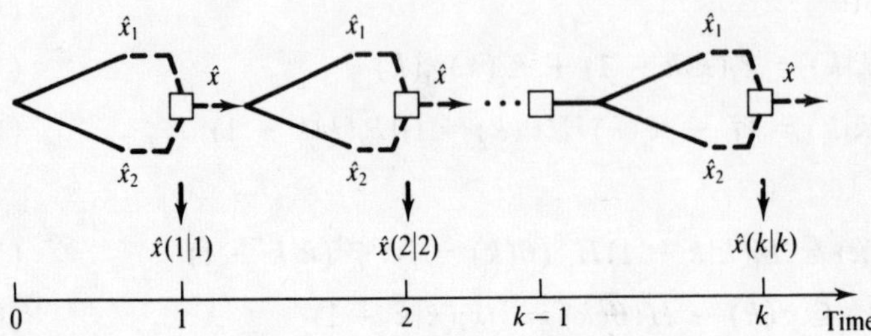

Figure 12.1 Time evolution of elemental Kalman filters using the pseudo-Bayes method, where $s = 2$.

state estimation error covariance matrix $P(k|k)$ are approximated by

$$\hat{x}(k|k) \simeq \sum_{j=1}^{s^{n+1}} \hat{x}_j(k|k)p(I_j(k, k-n)|Z_k) \tag{12.25}$$

$$P(k|k) \simeq \sum_{j=1}^{s^{n+1}} \{P_j(k|k) + [\hat{x}_j(k|k) - \hat{x}(k|k)] \times [\hat{x}_j(k|k) - \hat{x}(k|k)]^{\mathrm{T}}\}p(I_j(k, k-n)|Z_k) \tag{12.26}$$

Note, however, that for $k \leqslant n+1$ this suboptimal method coincides exactly with the optimal one given by Theorem 12.1.

Here it is important to note that, given an observation $z(k)$, we can consider two approaches to updating the conditional estimates which have been obtained up to time $k-1$. One is that the s^{n+1} conditional estimates $\hat{x}_j(k|k)$ can be obtained by updating the s^n one-step past conditional estimates at time $k-1$. This method is called here the *one-step measurement update approach*. Another is based on computing the s^{n+1} conditional estimates by updating the s n-step past conditional estimates at time $k-n$, but with the condition that $z(k-1), \ldots, z(k-n+1)$ have already been stored. This method, which will be described in the next subsection, is called here the *multistep* (or *n-step*) *measurement update approach*.

For the sake of convenience, we introduce the hypothesis:

$$h_{i_k}: \theta(k) = i_k, \qquad i_k \in \mathfrak{S} \tag{12.27}$$

Then, expressions (12.25) and (12.26) can be realized in the one-step measurement update approach as in the following corollary.

Corollary 12.2 Modified Jaffa–Gupta (JG) method

It is assumed that, for $k > n + 1$, the following quantities at time $k - 1$ are available:

$$\hat{x}_{i_{k-n}, \ldots, i_{k-1}}(k-1|k-1) \triangleq E[x(k-1)|h_{i_{k-n}}, \ldots, h_{i_{k-1}}, Z_{k-1}]$$

$$P_{i_{k-n}, \ldots, i_{k-1}}(k-1|k-1)$$
$$\triangleq E\{[x(k-1) - \hat{x}_{i_{k-n}, \ldots, i_{k-1}}(k-1|k-1)]$$
$$\times [x(k-1) - \hat{x}_{i_{k-n}, \ldots, i_{k-1}}(k-1|k-1)]^{\mathrm{T}}|h_{i_{k-n}}, \ldots, h_{i_{k-1}}, Z_{k-1}\}$$

and $p(h_{i_{k-n}}, \ldots, h_{i_{k-1}}|Z_{k-1})$. Then, given h_{i_k} and $z(k)$, expressions (12.25) and (12.26) can be rewritten as follows:

$$\hat{x}(k|k) = \sum_{i_{k-n}=1}^{s} \ldots \sum_{i_k=1}^{s} \hat{x}_{i_{k-n}, \ldots, i_k}(k|k)p(h_{i_{k-n}}, \ldots, h_{i_k}|Z_k) \tag{12.28}$$

$$\begin{aligned} P(k|k) = \sum_{i_{k-n}=1}^{s} \ldots \sum_{i_k=1}^{s} \{ & P_{i_{k-n}, \ldots, i_k}(k|k) \\ & + [\hat{x}_{i_{k-n}, \ldots, i_k}(k|k) - \hat{x}(k|k)] \\ & \times [\hat{x}_{i_{k-n}, \ldots, i_k}(k|k) - \hat{x}(k|k)]^{\mathrm{T}}\}p(h_{i_{k-n}}, \ldots, h_{i_k}|Z_k) \end{aligned} \tag{12.29}$$

where $\hat{x}_{i_{k-n}, \ldots, i_k}(k|k)$ and $P_{i_{k-n}, \ldots, i_k}(k|k)$ are obtained recursively by the parallel Kalman filters:

$$\hat{x}_{i_{k-n}, \ldots, i_k}(k|k-1) = \Phi(\theta(k) = i_k)\hat{x}_{i_{k-n}, \ldots, i_{k-1}}(k-1|k-1) \tag{12.30}$$

$$\begin{aligned} P_{i_{k-n}, \ldots, i_k}(k|k-1) = {} & \Phi(\theta(k) = i_k) \\ & \times P_{i_{k-n}, \ldots, i_{k-1}}(k-1|k-1)\Phi^{\mathrm{T}}(\theta(k) = i_k) \\ & + B(\theta(k) = i_k)QB^{\mathrm{T}}(\theta(k) = i_k) \end{aligned} \tag{12.31}$$

$$\begin{aligned} & K_{i_{k-n}, \ldots, i_k}(k) = \\ & \qquad P_{i_{k-n}, \ldots, i_k}(k|k-1)H^{\mathrm{T}}(\theta(k) = i_k)V^{-1}_{i_{k-n}, \ldots, i_k}(k|k-1) \end{aligned} \tag{12.32}$$

$$\begin{aligned} \hat{x}_{i_{k-n}, \ldots, i_k}(k|k) = {} & \hat{x}_{i_{k-n}, \ldots, i_k}(k|k-1) + K_{i_{k-n}, \ldots, i_k}(k) \\ & \times [z(k) - H(\theta(k) = i_k)\hat{x}_{i_{k-n}, \ldots, i_k}(k|k-1)] \end{aligned} \tag{12.33}$$

$$\begin{aligned} P_{i_{k-n}, \ldots, i_k}(k|k) = {} & [I - K_{i_{k-n}, \ldots, i_k}(k)H(\theta(k) = i_k)] \\ & \times P_{i_{k-n}, \ldots, i_k}(k|k-1) \end{aligned} \tag{12.34}$$

in which $V_{i_{k-n},\ldots,i_k}(k|k-1)$ is defined by

$$\begin{aligned} &V_{i_{k-n},\ldots,i_k}(k|k-1) \\ &= H(\theta(k)=i_k)P_{i_{k-n},\ldots,i_k}(k|k-1)H^{\mathrm{T}}(\theta(k)=i_k) \\ &\quad + D(\theta(k)=i_k)RD^{\mathrm{T}}(\theta(k)=i_k) \end{aligned} \tag{12.35}$$

The initial conditions for the above equations are

$$\hat{x}_{i_0}(0|0) = \hat{x}(0), \qquad P_{i_0}(0|0) = P(0) \tag{12.36}$$

for all $i_0 \in \mathfrak{S}$. The a posteriori probability $p(h_{i_{k-n}}, \ldots, h_{i_k}|Z_k)$ can be computed recursively as follows:

$$p(h_{i_{k-n}}, \ldots, h_{i_k}|Z_k) = \frac{p(z(k)|h_{i_{k-n}}, \ldots, h_{i_k}, Z_{k-1})p(h_{i_{k-n}}, \ldots, h_{i_k}|Z_{k-1})}{\sum_{i_{k-n}=1}^{s} \cdots \sum_{i_k=1}^{s} p(z(k)|h_{i_{k-n}}, \ldots, h_{i_k}, Z_{k-1})p(h_{i_{k-n}}, \ldots, h_{i_k}|Z_{k-1})} \tag{12.37}$$

where

$$\begin{aligned} &p(z(k)|h_{i_{k-n}}, \ldots, h_{i_k}, Z_{k-1}) \\ &\sim N[H(\theta(k)=i_k)\hat{x}_{i_{k-n},\ldots,i_k}(k|k-1), V_{i_{k-n},\ldots,i_k}(k|k-1)] \end{aligned} \tag{12.38}$$

and

$$p(h_{i_{k-n}}, \ldots, h_{i_k}|Z_{k-1}) = p_{i_{k-1},i_k}p(h_{i_{k-n}}, \ldots, h_{i_{k-1}}|Z_{k-1}) \tag{12.39}$$

Before going on to the next stage, we have to store the following quantities:

$$p(h_{i_{k-n+1}}, \ldots, h_{i_k}|Z_k) = \sum_{i_{k-n}=1}^{s} p(h_{i_{k-n}}, \ldots, h_{i_k}|Z_k) \tag{12.40}$$

$$\hat{x}_{i_{k-n+1},\ldots,i_k}(k|k) = \sum_{i_{k-n}=1}^{s} \frac{\hat{x}_{i_{k-n},\ldots,i_k}(k|k)p(h_{i_{k-n}}, \ldots, h_{i_k}|Z_k)}{p(h_{i_{k-n+1}}, \ldots, h_{i_k}|Z_k)} \tag{12.41}$$

$$P_{i_{k-n+1},\ldots,i_k}(k|k) = \sum_{i_{k-n}=1}^{s} \frac{P^*_{i_{k-n},\ldots,i_k}(k|k)p(h_{i_{k-n}}, \ldots, h_{i_k}|Z_k)}{p(h_{i_{k-n+1}}, \ldots, h_{i_k}|Z_k)} \tag{12.42}$$

where

$$\begin{aligned} P^*_{i_{k-n},\ldots,i_k}(k|k) = P_{i_{k-n},\ldots,i_k}(k|k) &+ [\hat{x}_{i_{k-n},\ldots,i_k}(k|k) - \hat{x}_{i_{k-n+1},\ldots,i_k}(k|k)] \\ &\times [\hat{x}_{i_{k-n},\ldots,i_k}(k|k) - \hat{x}_{i_{k-n+1},\ldots,i_k}(k|k)]^{\mathrm{T}} \end{aligned} \tag{12.43}$$

Proof

See Jaffer and Gupta [16]. □

It should be noted that Chang and Athans [17] and Tugnait [13] use an equation similar to (12.29) to evaluate the covariance matrix $P(k|k)$, whereas in Corollary 12.1 (see also Ackerson and Fu [15]) an approximate representation is used, and in the original paper of Jaffer and Gupta [16] this covariance matrix is not computed. In addition, equation (12.43) in the original paper is also approximated by

$$P^*_{i_{k-n},\ldots,i_k}(k|k) \simeq P_{i_{k-n},\ldots,i_k}(k|k) \tag{12.44}$$

Because of this, Corollary 12.2 is often referred to as the *modified* Jaffer–Gupta method [19].

As can be seen from the above results, the one-step measurement update method requires the storage of s^n estimates (each of n_x dimensions), s^n covariances ($n_x(n_x+1)/2$ dimensions) and s^n a posteriori probabilities (scalar), regardless of the number of storages which the optimal estimator described in Section 12.3 needs. As $n \to k-1$, of course, the method approaches the optimal algorithm. Figure 12.2 shows the time evolution of elemental Kalman filters using this method with $s = n = 2$.

12.4.3 Multistep Measurement Update Approach

As discussed in the previous subsection, this approach is first based on reprocessing the past $n-1$ observations, i.e. $z(k-1), \ldots, z(k-n+1)$, with s parallel filtered estimates as starting conditions at time $k-n$. The idea of this method is very similar to that of an algorithm for fixed-lag smoothing for a lag of $n-1$ units of time. The resulting algorithm is summarized in the following corollary.

Corollary 12.3 ***n*-step measurement update method**

Suppose that, for $k > n+1$, the following quantities at time $k-n$ are provided:

$$\hat{x}_{i_{k-n}}(k-n|k-n) \triangleq E[x(k-n)|h_{i_{k-n}}, Z_{k-n}]$$

$$\begin{aligned} P_{i_{k-n}}(k-n|k-n) \triangleq\ & E\{[x(k-n) - \hat{x}_{i_{k-n}}(k-n|k-n)] \\ & \times [x(k-n) - \hat{x}_{i_{k-n}}(k-n|k-n)]^{\mathrm{T}} \\ & \times |h_{i_{k-n}}, Z_{j-n}\} \end{aligned}$$

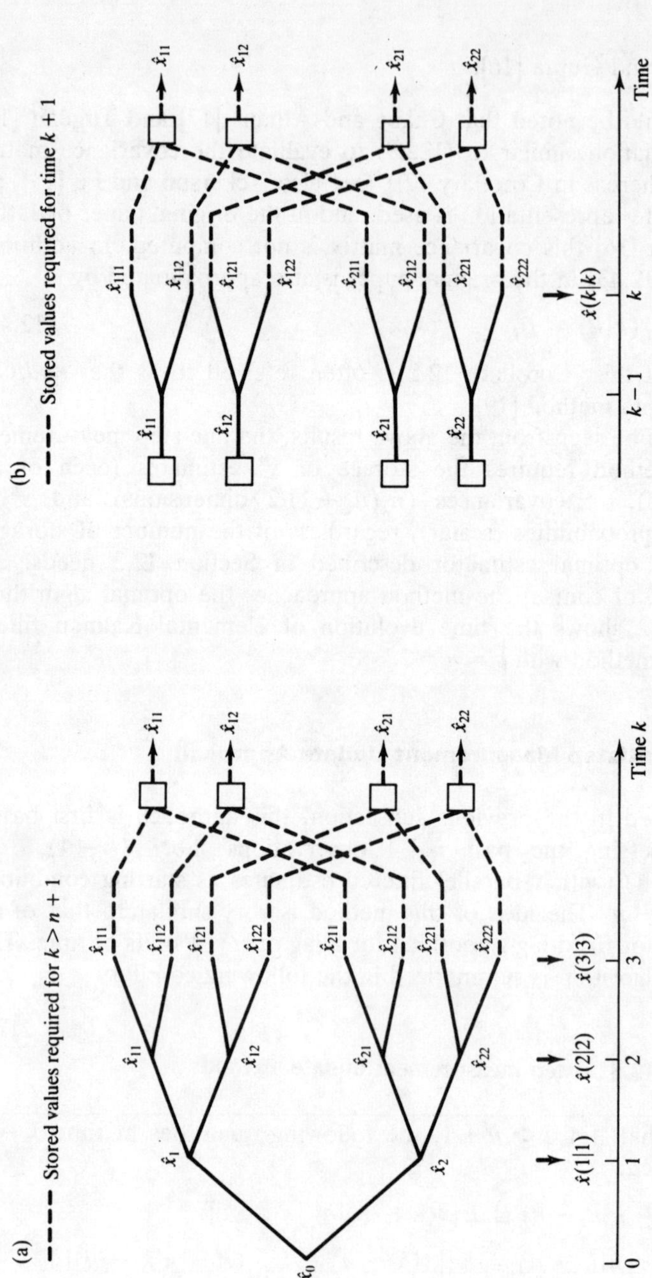

Figure 12.2 Time evolution of elemental Kalman filters using the one-step measurement update method, where $s = n = 2$, (a) for $k \leq n+1$; (b) for $k > n+1$.

and $p(h_{i_{k-n}}|Z_{k-n})$. Then, given $h_{i_{k-n+1}}$ and Z_{k-n+1} at time $k-n+1$, we compute

$$\hat{x}_{i_{k-n}, i_{k-n+1}}(k-n+1|k-n+1) \triangleq E[x(k-n+1)|h_{i_{k-n}}, h_{i_{k-n+1}}, Z_{k-n+1}]$$

and

$$\begin{aligned} P_{i_{k-n}, i_{k-n+1}}&(k-n+1|k-n+1) \\ &\triangleq E\{[x(k-n+1) - \hat{x}_{i_{k-n}, i_{k-n+1}}(k-n+1|k-n+1)] \\ &\quad \times [x(k-n+1) - \hat{x}_{i_{k-n}, i_{k-n+1}}(k-n+1|k-n+1)]^{\mathrm{T}} \\ &\quad \times |h_{i_{k-n}}, h_{i_{k-n+1}}, Z_{k-n+1}\} \end{aligned}$$

by using the s^2 parallel Kalman filters

$$\begin{aligned} \hat{x}_{i_{k-n}, i_{k-n+1}}&(k-n+1|k-n) \\ &= \Phi(\theta(k-n+1) = i_{k-n+1})\hat{x}_{i_{k-n}}(k-n|k-n) \end{aligned} \tag{12.45}$$

$$\begin{aligned} &P_{i_{k-n}, i_{k-n+1}}(k-n+1|k-n) \\ &= \Phi(\theta(k-n+1) = i_{k-n+1})P_{i_{k-n}}(k-n|k-n) \\ &\quad \times \Phi^{\mathrm{T}}(\theta(k-n+1) = i_{k-n+1}) \\ &\quad + B(\theta(k-n+1) = i_{k-n+1})QB^{\mathrm{T}}(\theta(k-n+1) = i_{k-n+1}) \end{aligned} \tag{12.46}$$

$$\begin{aligned} &K_{i_{k-n}, i_{k-n+1}}(k-n+1) \\ &= P_{i_{k-n}, i_{k-n+1}}(k-n+1|k-n)H^{\mathrm{T}}(\theta(k-n+1) = i_{k-n+1}) \\ &\quad \times V^{-1}_{i_{k-n}, i_{k-n+1}}(k-n+1|k-n) \end{aligned} \tag{12.47}$$

$$\begin{aligned} &\hat{x}_{i_{k-n}, i_{k-n+1}}(k-n+1|k-n+1) \\ &= \hat{x}_{i_{k-n}, i_{k-n+1}}(k-n+1|k-n) \\ &\quad + K_{i_{k-n}, i_{k-n+1}}(k-n+1)[z(k-n+1) - H(\theta(k-n+1) \\ &\quad = i_{k-n+1})\hat{x}_{i_{k-n}, i_{k-n+1}}(k-n+1|k-n)] \end{aligned} \tag{12.48}$$

$$\begin{aligned} &P_{i_{k-n}, i_{k-n+1}}(k-n+1|k-n+1) \\ &= [I - K_{i_{k-n}, i_{k-n+1}}(k-n+1)H(\theta(k-n+1) = i_{k-n+1})] \\ &\quad \times P_{i_{k-n}, i_{k-n+1}}(k-n+1|k-n) \end{aligned} \tag{12.49}$$

where $V_{i_{k-n},i_{k-n+1}}(k-n+1|k-n)$ is defined as

$$\begin{aligned} &V_{i_{k-n},i_{k-n+1}}(k-n+1|k-n) \\ &= H(\theta(k-n+1)=i_{k-n+1})P_{i_{k-n},i_{k-n+1}}(k-n+1|k-n) \\ &\quad \times H^{\mathrm{T}}(\theta(k-n+1)=i_{k-n+1}) \\ &\quad + D(\theta(k-n+1)=i_{k-n+1}) \\ &\quad \times RD^{\mathrm{T}}(\theta(k-n+1)=i_{k-n+1}) \end{aligned} \tag{12.50}$$

and the initial conditions for these quantities are provided by equations (12.36). In addition, we compute the probability density functions

$$\begin{aligned} &p(z(k-n+1)|h_{i_{k-n}}, h_{i_{k-n+1}}, Z_{k-n}) \\ &\sim N[H(\theta(k-n+1)=i_{k-n+1})\hat{x}_{i_{k-n},i_{k-n+1}}(k-n+1|k-n), \\ &\quad V_{i_{k-n},i_{k-n+1}}(k-n+1|k-n)] \end{aligned} \tag{12.51}$$

Continuing in the same manner for time stages $k-n+2, \ldots, k$, we compute the parallel Kalman filters $s^3, \ldots, s^{n+1}$. Then, expressions (12.25) and (12.26) can be rewritten as follows:

$$\hat{x}(k|k) = \sum_{i_k=1}^{s} \hat{x}_{i_k}(k|k)p(h_{i_k}|Z_k) \tag{12.52}$$

$$\begin{aligned} P(k|k) = \sum_{i_k=1}^{s} \{&P_{i_k}(k|k) + [\hat{x}_{i_k}(k|k) - \hat{x}(k|k)] \\ &\times [\hat{x}_{i_k}(k|k) - \hat{x}(k|k)]^{\mathrm{T}}\} \\ &\times p(h_{i_k}|Z_k) \end{aligned} \tag{12.53}$$

where $\hat{x}_{i_k}(k|k) = E[x(k)|h_{i_k}, Z_k]$, $P_{i_k}(k|k) = E\{[x(k) - \hat{x}_{i_k}(k|k)]$ $[x(k) - \hat{x}_{i_k}(k|k)]^{\mathrm{T}}|h_{i_k}, Z_k\}$ and $p(h_{i_k}|Z_k)$ are as follows:

$$\hat{x}_{i_k}(k|k) = \sum_{i_{k-n}=1}^{s} \cdots \sum_{i_{k-1}=1}^{s} \hat{x}_{i_{k-n},\ldots,i_k}(k|k)p(h_{i_{k-n}}, \ldots, h_{i_{k-1}}|h_{i_k}, Z_k) \tag{12.54}$$

$$\begin{aligned} P_{i_k}(k|k) = \sum_{i_{k-n}=1}^{s} \cdots \sum_{i_{k-1}=1}^{s} \{&P_{i_{k-n},\ldots,i_k}(k|k) \\ &+ [\hat{x}_{i_{k-n},\ldots,i_k}(k|k) - \hat{x}_{i_k}(k|k)] \\ &\times [\hat{x}_{i_{k-n},\ldots,i_k}(k|k) - \hat{x}_{i_k}(k|k)]^{\mathrm{T}}\} \\ &\times p(h_{i_{k-n}}, \ldots, h_{i_{k-1}}|h_{i_k}, Z_k) \end{aligned} \tag{12.55}$$

$$p(h_{i_k}|Z_k) = \frac{\sum_{i_{k-n}=1}^{s} \cdots \sum_{i_{k-1}=1}^{s}[\prod_{j=0}^{n-1} F(i, j, k, n)]p(h_{i_{k-n}}|Z_{k-n})}{\sum_{i_{k-n}=1}^{s} \cdots \sum_{i_{k-1}=1}^{s}\sum_{i_k=1}^{s}[\prod_{j=0}^{n-1} F(i, j, k, n)]p(h_{i_{k-n}}|Z_{k-n})} \quad (12.56)$$

where

$$F(i, j, k, n) \triangleq p(z(k-j)|h_{i_{k-n}}, \ldots, h_{i_{k-j}}, Z_{k-1-j})p(h_{i_{k-j}}|h_{i_{k-1-j}}) \quad (12.57)$$

Here, the probability $p(h_{i_{k-n}}, \ldots, h_{i_{k-1}}|h_{i_k}, Z_k)$ can be given by

$$p(h_{i_{k-n}}, \ldots, h_{i_{k-1}}|h_{i_k}, Z_k) = \frac{[\prod_{j=0}^{n-1} F(i, j, k, n)]p(h_{i_{k-n}}|Z_{k-n})}{\sum_{i_{k-n}=1}^{s} \cdots \sum_{i_{k-1}=1}^{s}[\prod_{j=0}^{n-1} F(i, j, k, n)]p(h_{i_{k-n}}|Z_{k-n})} \quad (12.58)$$

where $\hat{x}_{i_k}(k|k)$, $P_{i_k}(k|k)$ and $p(h_{i_k}|Z_k)$ must be stored as the starting conditions for the $(k+n)$th stage.

Proof

The result is very similar to Corollary 12.2. Hence, the derivations of equations (12.56) and (12.58) are merely verified here. Using Bayes' rule we first obtain

$$p(h_{i_{k-n}}, \ldots, h_{i_{k-1}}|h_{i_k}, Z_k) = \frac{p(h_{i_{k-n}}, \ldots, h_{i_{k-1}}, h_{i_k}|Z_k)}{p(h_{i_k}|Z_k)} \quad (12.59)$$

In this equation, the numerator is written as

$$\begin{aligned}
&p(h_{i_{k-n}}, \ldots, h_{i_{k-1}}, h_{i_k}|Z_k) \\
&= \frac{p(h_{i_{k-n}}, \ldots, h_{i_k}, z(k), \ldots, z(k-n+1), Z_{k-n})}{p(z(k), \ldots, z(k-n+1), Z_{k-n})} \\
&= \{p(z(k)|h_{i_{k-n}}, \ldots, h_{i_k}, z(k-1), \ldots, z(k-n+1), Z_{k-n}) \\
&\quad \times p(h_{i_k}|h_{i_{k-n}}, \ldots, h_{i_{k-1}}, z(k-1), \ldots, z(k-n+1), Z_{k-n}) \\
&\quad \times p(z(k-1)|h_{i_{k-n}}, \ldots, h_{i_{k-1}}, z(k-2), \ldots, \\
&\quad z(k-n+1), Z_{k-n}) \\
&\quad \times p(h_{i_{k-1}}|h_{i_{k-n}}, \ldots, h_{i_{k-2}}, z(k-2), \ldots, z(k-n+1), Z_{k-n}) \\
&\quad \vdots \\
&\quad \times p(z(k-n+1)|h_{i_{k-n}}, h_{i_{k-n+1}}, Z_{k-n})p(h_{i_{k-n+1}}|h_{i_{k-n}}, Z_{k-n}) \\
&\quad \times p(h_{i_{k-n}}|Z_{k-n})\}/p(z(k), \ldots, z(k-n+1), Z_{k-n})
\end{aligned} \quad (12.60)$$

The denominator in equation (12.60) is modified as follows:

$$
\begin{aligned}
&p(z(k), \ldots, z(k-n+1) | Z_{k-n}) \\
&= \sum_{i_{k-n}=1}^{s} \cdots \sum_{i_k=1}^{s} p(z(k), \ldots, z(k-n+1) | h_{i_{k-n}}, \ldots, h_{i_k}, Z_{k-n}) \\
&\quad \times p(h_{i_{k-n}}, \ldots, h_{i_k} | Z_{k-n})
\end{aligned} \tag{12.61}
$$

Noting that

$$
\begin{aligned}
&p(z(k), \ldots, z(k-n+1) | h_{i_{k-n}}, \ldots, h_{i_k} | Z_{k-n}) \\
&= \{p(z(k) | h_{i_{k-n}}, \ldots, h_{i_k}, z(k-1), \ldots, z(k-n+1), Z_{k-n}) \\
&\quad \times p(h_{i_k} | h_{i_{k-n}}, \ldots, h_{i_{k-1}}, z(k-1), \ldots, z(k-n+1), Z_{k-n}) \\
&\quad \times p(z(k-1) | h_{i_{k-n}}, \ldots, h_{i_{k-1}}, z(k-2), \ldots, \\
&\quad z(k-n+1), Z_{k-n}) \\
&\quad \times p(h_{i_{k-1}} | h_{i_{k-n}}, \ldots, h_{i_{k-2}}, z(k-2), \ldots, \\
&\quad z(k-n+1), Z_{k-n}) \\
&\quad \vdots \\
&\quad \times p(z(k-n+1) | h_{i_{k-n}}, h_{i_{k-n+1}}, Z_{k-n}) p(h_{i_{k-n+1}} | h_{i_{k-n}}, Z_{k-n}) \\
&\quad \times p(h_{i_{k-n}} | Z_{k-n}) p(Z_{k-n})\} / p(h_{i_{k-n}}, \ldots, h_{i_k} | Z_{k-n}) p(Z_{k-n})\}
\end{aligned} \tag{12.62}
$$

yields

$$
\begin{aligned}
&p(z(k), \ldots, z(k-n+1) | Z_{k-n}) \\
&= \sum_{i_{k-n}=1}^{s} \cdots \sum_{i_k=1}^{s} p(z(k) | h_{i_{k-n}}, \ldots, h_{i_k}, z(k-1), \ldots, \\
&\quad z(k-n+1), Z_{k-n}) \\
&\quad \times p(h_{i_k} | h_{i_{k-1}}) p(z(k-1) | h_{i_{k-n}}, \ldots, h_{i_k}, z(k-1), \ldots, \\
&\quad z(k-n+1), Z_{k-n}) \\
&\quad \times p(h_{i_{k-1}} | h_{i_{k-2}}), \ldots, p(z(k-n+1) | h_{i_{k-n}}, h_{i_{k-n+1}}, Z_{k-n}) \\
&\quad \times p(h_{i_{k-n+1}} | h_{i_{k-n}}) p(h_{i_{k-n}} | Z_{k-n})
\end{aligned} \tag{12.63}
$$

on invoking equation (12.12). On the other hand, it follows that

$$
\begin{aligned}
p(h_{i_k} | Z_k) = \sum_{i_{k-n}=1}^{s} \cdots \sum_{i_{k-1}=1}^{s} & p(h_{i_k} | h_{i_{k-n}}, \ldots, h_{i_{k-1}}, Z_k) \\
& \times p(h_{i_{k-n}}, \ldots, h_{i_{k-1}} | Z_k)
\end{aligned} \tag{12.64}
$$

Now we note that

$$\begin{aligned}
&p(h_{i_k}|h_{i_{k-n}}, \ldots, h_{i_{k-1}}, Z_k) \\
&= p(h_{i_k}|h_{i_{k-n}}, \ldots, h_{i_{k-1}}, z(k), \ldots, z(k-n+1), Z_{k-n}) \\
&= \frac{p(h_{i_{k-n}}, \ldots, h_{i_k}, z(k), \ldots, z(k-n+1), Z_{k-n})}{p(h_{i_{k-n}}, \ldots, h_{i_{k-1}}, z(k), \ldots, z(k-n+1), Z_{k-n})}
\end{aligned} \tag{12.65}$$

and

$$\begin{aligned}
&p(h_{i_{k-n}}, \ldots, h_{i_{k-1}}|Z_k) \\
&= p(h_{i_{k-n}}, \ldots, h_{i_{k-1}}|z(k), \ldots, z(k-n+1), Z_{k-n}) \\
&= \frac{p(h_{i_{k-n}}, \ldots, h_{i_{k-1}}, z(k), \ldots, z(k-n+1), Z_{k-n})}{p(z(k), \ldots, z(k-n+1), Z_{k-n})}
\end{aligned} \tag{12.66}$$

From equations (12.65) and (12.66),

$$\begin{aligned}
&p(h_{i_k}|h_{i_{k-n}}, \ldots, h_{i_{k-1}}, Z_k)p(h_{i_{k-n}}, \ldots, h_{i_{k-1}}|Z_k) \\
&= \frac{p(h_{i_{k-n}}, \ldots, h_{i_k}, z(k), \ldots, z(k-n+1), Z_{k-n})}{p(z(k), \ldots, z(k-n+1)|Z_{k-n})p(Z_{k-n})}
\end{aligned} \tag{12.67}$$

Noting that this equation is completely identical with equation (11.60), we find that equation (11.65) can be rewritten as follows:

$$\begin{aligned}
&p(h_{i_k}|Z_k) = \sum_{i_{k-n}=1}^{s} \cdots \sum_{i_{k-1}=1}^{s} p(z(k)|h_{i_{k-n}}, \ldots, h_{i_k}, z(k-1), \ldots, \\
&z(k-n+1), Z_{k-n})p(h_{i_k}|h_{i_{k-1}})p(z(k-1)|h_{i_{k-n}}, \ldots, h_{i_{k-1}}, \\
&z(k-2), \ldots, z(k-n+1), Z_{k-n})p(h_{i_{k-1}}|h_{i_{k-2}}, \ldots, \\
&p(z(k-n+1)|h_{i_{k-n}}, h_{i_{k-n+1}}, Z_{k-n})p(h_{i_{k-n+1}}|h_{i_{k-n}}) \\
&\times p(h_{i_{k-n}}|Z_{k-n})/p(z(k), \ldots, z(k-n+1)|Z_{k-n})
\end{aligned} \tag{12.68}$$

Combining equations (12.63) and (12.68) gives equation (12.56). Combining equations (12.59), (12.60) and (12.68) yields (12.58). This completes the proof. □

Figure 12.3 shows the time evolution of elemental Kalman filters for this approach with $s = n = 2$. Unlike the one-step measurement update approach described in the previous subsection, the present approach requires the storage of $s \times n$ estimates (each of n_x dimensions), $s \times n$ covariances ($n_x(n_x + 1)/2$ dimensions), $s \times n$ a posteriori probabilities (scalar) and $n - 1$ observations (m dimensions). It is then interesting to note that the method requires geometrically growing storage with s and n, whereas the one-step measurement update method requires exponentially growing storage with s and n.

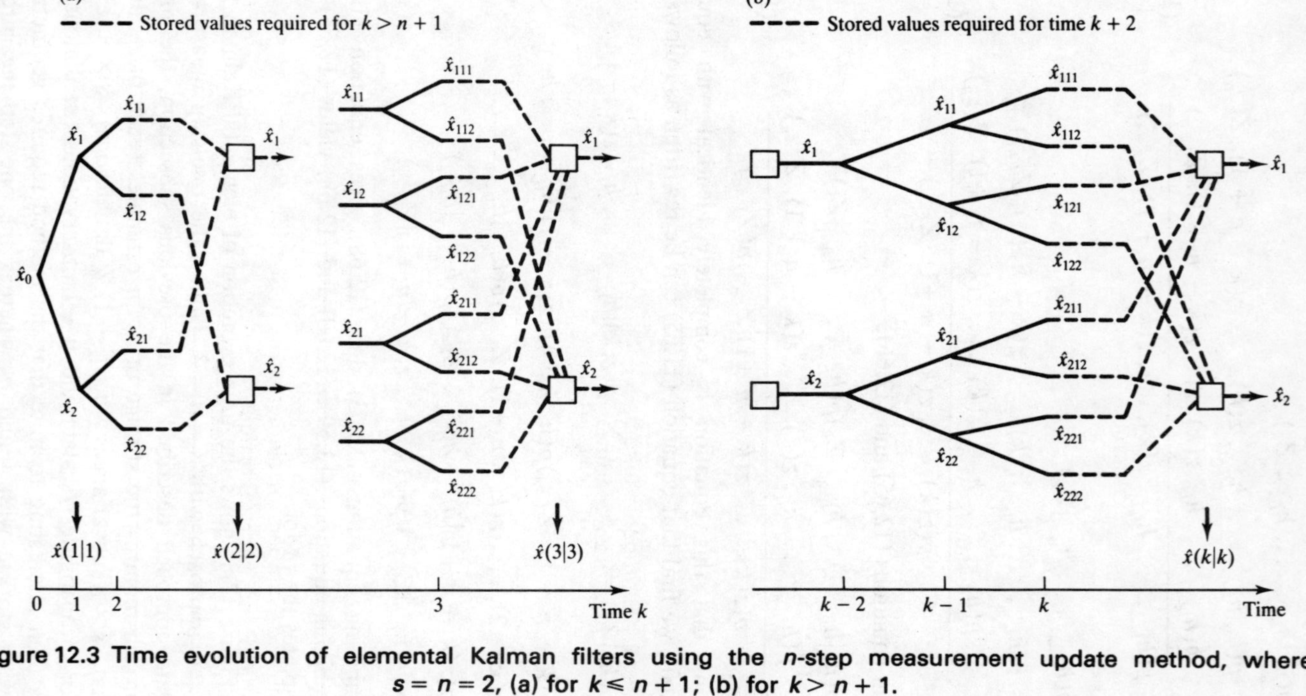

Figure 12.3 Time evolution of elemental Kalman filters using the n-step measurement update method, where $s = n = 2$, (a) for $k \leq n + 1$; (b) for $k > n + 1$.

Table 12.1 gives their storage requirements:

$$T_1 = s^n[n_x + n_x(n_x + 1)/2 + 1] \tag{12.69}$$

for the one-step measurement update method and

$$T_n = sn[n_x + n_x(n_x + 1)/2 + 1] + (n - 1)m \tag{12.70}$$

for the n-step measurement update method, where $n_x = 10$ and $m = 5$ are assumed. It is found from this table that for storage requirements a remarkable difference between the two methods appears as s and n become large. Note that for $n = 1$ both methods are completely identical in the computational form; this is in fact the algorithm of Chang and Athans [17].

However, it should be noted that the computational speed of the n-step measurement update algorithm is slightly slower than that of the one-step measurement update algorithm. This fact can be confirmed by comparing the numbers of elemental Kalman filters required for both methods, i.e.

$$E_1 = s^{n+1} \text{ and } E_n = \sum_{i=1}^{n} s^{i+1} \tag{12.71}$$

It is seen from Table 12.2 that a difference between the two methods

Table 12.1 Storage requirements for GPBAs, where the left-hand and right-hand figures denote the one- and n-step measurement update methods, respectively

	$s = 2$		$s = 3$		$s = 4$		$s = 5$	
$n = 1$	132	132	198	198	264	264	330	330
$n = 2$	264	269	594	401	1056	533	1650	665
$n = 3$	528	406	1782	604	4224	802	8250	1000
$n = 4$	1056	543	5346	807	16896	1071	41250	1335
$n = 5$	2112	680	16038	1010	67584	1340	206250	1670

Table 12.2 Numbers of elemental Kalman filters for GPBAs, where the left-hand and right-hand figures denote the one- and n-step measurement update methods, respectively

	$s = 2$		$s = 3$		$s = 4$		$s = 5$	
$n = 1$	4	4	9	9	16	16	25	25
$n = 2$	8	12	27	36	64	80	125	150
$n = 3$	16	28	81	117	256	336	625	775
$n = 4$	32	60	243	360	1024	1360	3125	3900
$n = 5$	64	124	729	1089	4096	5456	15625	19525

disappears as s becomes large for any n. This would seem to indicate that the n-step measurement update method is effective for reducing the storage requirements of systems with a relatively large number of Markov chain states.

Example 12.1

In this example, we consider a simple system described by the following equations:

$$x(k+1) = 1.04x(k) + w(k)$$

$$z(k) = x(k) + D(\theta(k))v(k), \qquad k = 1, 2, \ldots$$

$$s = 2, \text{ i.e. } \theta(k) \in \{1, 2\}$$

It is assumed that the initial state is subject to $x(0) \sim N(30, 400)$, $\{w(k)\}$ and $\{v(k)\}$ are mutually independent zero-mean Gaussian white noise sequences with covariances $Q = 0.1$ and $R = 1.0$, respectively. The parameter process $\theta(k)$ is modelled by a Markov chain with transition probability matrix:

$$\pi = \begin{bmatrix} 0.3 & 0.7 \\ 0.7 & 0.3 \end{bmatrix}$$

We here take $p_1 = p_2 = 0.5$. Furthermore, we have $D(1) = 10$ and $D(2) = 1$. A similar example has been reported by several authors [11–14]. The system means that the model $\theta(k) = 1$ represents a failure mode due to the abrupt change of measurement noise. Figure 12.4 depicts the transition probability diagram for this problem.

Four suboptimal approaches—the Ackerson–Fu (AF) method

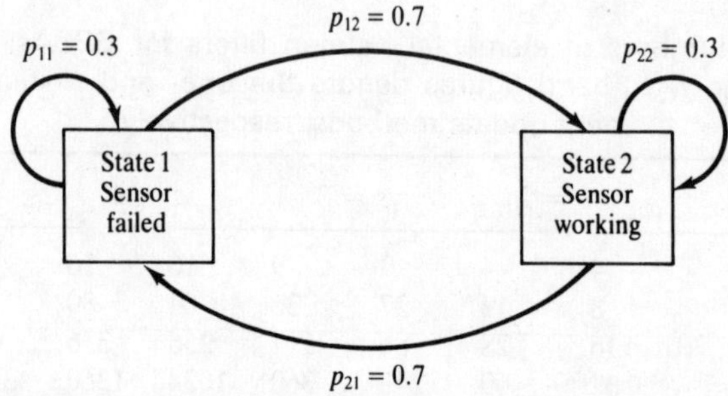

Figure 12.4 Transition probability diagram for Example 12.1.

given by Corollary 12.1; the Jaffer–Gupta (JG) method given by Corollary 12.2, but with approximation (12.44), the modified JG method given by Corollary 12.2, and the n-step measurement update method given by Corollary 12.3—have been implemented to compare their performances for the estimation problem. Figure 12.5 shows the time evolutions of rms estimation errors. The performances were evaluated by averaging over 50 Monte Carlo runs. A lower bound is obtained by running a Kalman filter which knows the true values of the switching parameters. The information in Figure 12.5 is also summarized in Table 12.3 after averaging 30 time stages.

From Figure 12.5 and Table 12.3, it is observed that the performances of the modified JG and the n-step measurement update methods are superior to those of the other two methods; of the latter, the AF method outperforms the JG method. In particular, the n-step measurement update method has the best estimation performance.

Example 12.2

Here the system of Example 12.1 is again considered, except that now

$$\pi = \begin{bmatrix} 0.85 & 0.15 \\ 0.15 & 0.85 \end{bmatrix}$$

and in the truth model the structural state sequence $\{\theta(k),\ k \geqslant 1\}$ is fixed. But the four suboptimal methods are designed by using switching models with the transition statistics postulated here. The objective is to test the ability of the algorithm to identify the system structure if it is held constant for a sufficiently long interval of time, even though the design transition statistics and the true transitions statistics are mismatched.

The fixed sequence used for the truth model was selected as follows:

$$\theta(k) = \begin{cases} 1, & 1 \leqslant k \leqslant 9,\ 20 \leqslant k \leqslant 30 \\ 2, & 10 \leqslant k \leqslant 19 \end{cases}$$

In Figure 12.6, the a posteriori probabilities have been plotted after averaging over 50 Monte Carlo runs. It is seen from Figure 12.6 and Table 12.4 that increasing n does not necessarily lead to improved estimation performance. In other words, we note that all suboptimal methods with $n = 1$, except for the AF method, have indicated a good performance in the estimation problem under consideration.

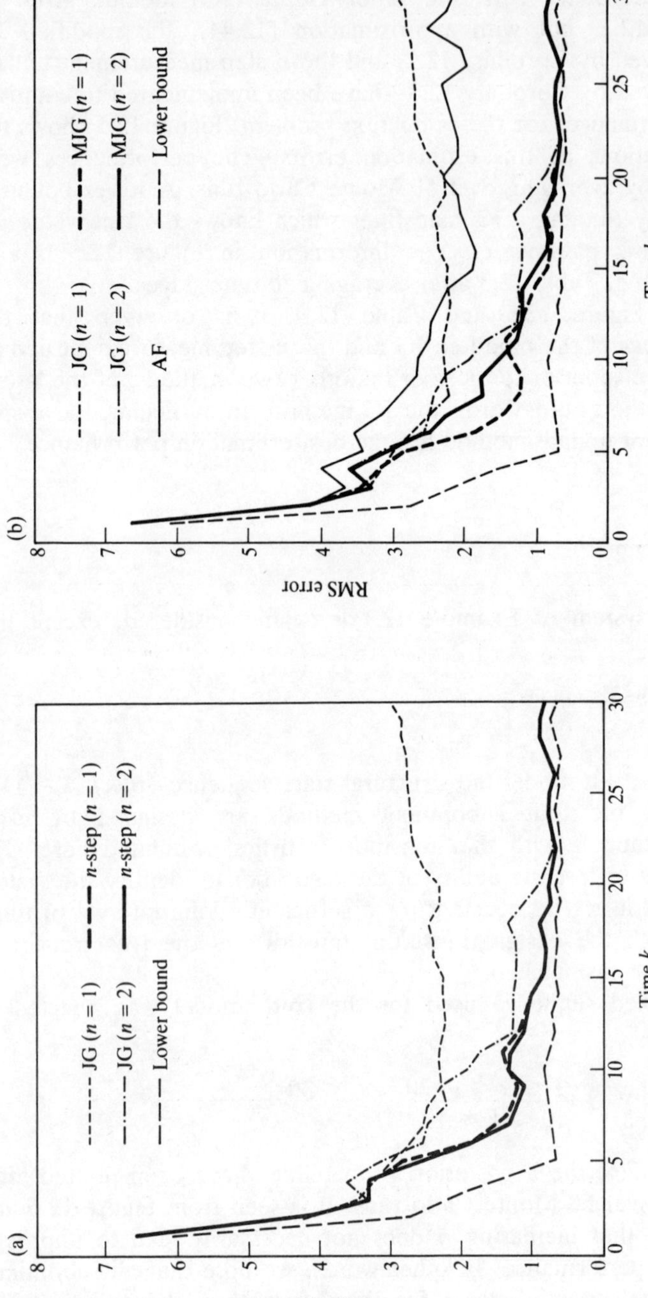

Figure 12.5 Comparison of rms errors in state estimation in Example 12.1 due to suboptimal algorithms: (a) JG and n-step measurement methods; (b) AF, JG and modified JG methods.

Table 12.3 RMS state estimation error for Example 12.1

Algorithm	*Average rms error in state estimation*
AF method	2.677
JG method with	
$n = 1$	2.763
$n = 2$	1.743
$n = 3$	1.749
Modified JG method with	
$n = 1$	1.482
$n = 2$	1.596
$n = 3$	1.695
n-step measurement update method with	
$n = 1$	1.482
$n = 2$	1.449
$n = 3$	1.439
Lower bound	1.009

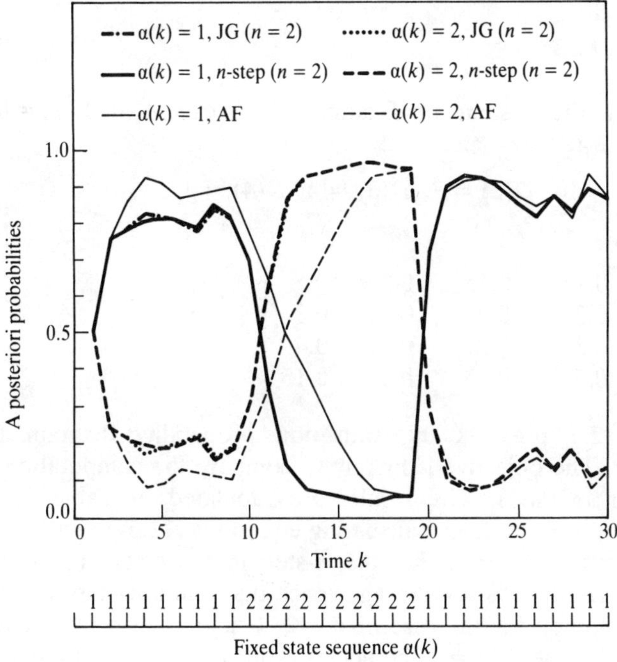

Figure 12.6 Comparison of a posteriori probabilities of system structural states due to suboptimal algorithms in Example 12.2: AF, JG and n-step measurement methods.

Table 12.4 RMS state estimation error for Example 12.2

Algorithm	*Average rms error in state estimation*
AF method	3.034
JG method with	
$n = 1$	2.263
$n = 2$	2.258
Modified JG method with	
$n = 1$	2.237
$n = 2$	2.239
n-step measurement update method with	
$n = 1$	2.237
$n = 2$	2.239
Lower bound	2.046

Example 12.3

Consider the system of Example 12.1 except that $p_i = 0.25$ for $i = 1, \ldots, 4$,

$$D(1) = 10,\ D(2) = 7,\ D(3) = 4,\ D(4) = 1$$

and

$$\pi = \begin{bmatrix} 0.15 & 0.15 & 0.2 & 0.5 \\ 0.2 & 0.2 & 0.3 & 0.3 \\ 0.3 & 0.2 & 0.1 & 0.4 \\ 0.5 & 0.15 & 0.2 & 0.15 \end{bmatrix}$$

Figure 12.7 presents the transition probability diagram for this problem. The objective is mainly to compare the computation time of the four methods. The results are tabulated in Table 12.5. It is important to note, from calculating equations (12.69) and (12.70) with $n_x = m = 1$ and $s = 4$, that the n-step measurement update method requires about 50% less storage space than the one-step measurement update method for the case of $n = 2$. It is also seen from Table 12.5 that in the case of $n = 2$ the actual ratio of two computational speeds for the JG (or modified JG) method and the n-step measurement update method is close to an ideal ratio (i.e. 1.25) shown in Table 12.2, as would be expected.

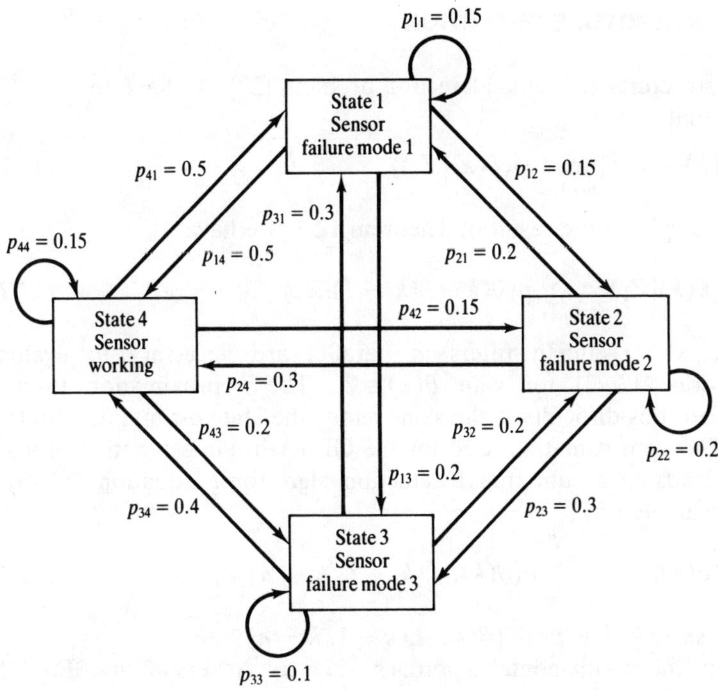

Figure 12.7 Transition probability diagram for Example 12.3.

Table 12.5 RMS state estimation error for Example 12.3

Algorithm	*Average rms error in state estimation*	*Average computation time* (s)
AF method	2.379	1.664
JG method with		
$n = 1$	2.258	3.322
$n = 2$	1.739	10.442
$n = 3$	1.727	39.480
Modified JG method with		
$n = 1$	1.704	3.650
$n = 2$	1.633	11.722
$n = 3$	1.676	44.422
n-step measurement update method with		
$n = 1$	1.704	3.770
$n = 2$	1.693	12.886
$n = 3$	1.688	49.100
Lower bound	1.127	0.646

12.5 STRUCTURE DETECTION

In a structure- (or fault-)detection problem [23], we have to find $\hat{\theta}(k)$ such that

$$\hat{\theta}(k) = \arg\{\max_{\theta(k)\in\mathfrak{S}} p(\theta(k)|Z_k)\} \tag{12.72}$$

Now, applying the result of Theorem 12.1, we have

$$p(\theta(k)|Z_k) = \sum_{j=1}^{s^{k-1}} p(\theta(k), I_j(k-1)|Z_k) \tag{12.73}$$

Thus, s^{k-1} Kalman filters in parallel are necessary to evaluate equation (12.73) for gain $\theta(k) \in \mathfrak{S}$. The approximation used to alleviate this difficulty is the same as for the state estimation problem.

The approximation used for the GPBA in the estimation problem also leads to a suboptimal detection algorithm. Equation (12.73) is approximated by

$$p(\theta(k)|Z_k) \simeq \sum_{j=1}^{s^n} p(\theta(k), I_j(k-1, k-n)|Z_k) \tag{12.74}$$

because $I_j(k, k-n) = \{\theta(k), I_j(k-1, k-n)\}$.

In the fundamental approach due to Ackerson and Fu [15], equation (12.15) can be directly used, instead of applying expression (12.74). In the one-step measurement update approach, using equation (12.37) or (12.40), the suboptimal detection algorithm (12.74) can be written as

$$p(h_{i_k}|Z_k) = \sum_{i_{k-n}=1}^{s} \cdots \sum_{i_{k-1}=1}^{s} p(h_{i_{k-n}}, \ldots, h_{i_k}|Z_k) \tag{12.75}$$

or

$$p(h_{i_k}|Z_k) = \sum_{i_{k-n+1}=1}^{s} \cdots \sum_{i_{k-1}=1}^{s} p(h_{i_{k-n+1}}, \ldots, h_{i_{k-1}}, h_{i_k}|Z_k) \tag{12.76}$$

On the other hand, in the n-step measurement update approach, expression (12.74) can now be directly provided by equation (12.56). Therefore, we need no additional computations to choose the normal or fault state corresponding to the largest value of expression (12.74).

Example 12.4

Reconsider Example 12.1 to examine the detection performances of the four methods. The results of averaging over 50 Monte Carlo runs and 30 time stages are tabulated at Table 12.6. From this table, it is

Table 12.6 Probability of error in structure detection for Example 12.4

Algorithm	*Average probability of error in structure detection*
AF method	0.3327
JG method with	
$n = 1$	0.1973
$n = 2$	0.1907
$n = 3$	0.1833
Modified JG method with	
$n = 1$	0.1767
$n = 2$	0.1827
$n = 3$	0.1840
n-step measurement update method with	
$n = 1$	0.1767
$n = 2$	0.1733
$n = 3$	0.1753
Lower bound	*

observed that the modified JG and the n-step measurement update methods outperform the other two methods. Note that an increase in n leads to an improvement in estimation performance, but not necessarily to an improvement in detection performance.

12.6 THE PROBLEM OF CONTROL OF SYSTEMS WITH FAULTY SENSORS

In this section, we further consider the following discrete-time system:

$$x(k+1) = \Phi x(k) + Gu(k) + Bw(k) \tag{12.77}$$

$$z(k) = H(\theta(k))x(k) + D(\theta(k))v(k) \tag{12.78}$$

where $u(k) \in \mathbb{R}^r$ is a control input, and other terms are as described in Section 12.2.

The objective is to find control sequences $\{u(k), 0 \leqslant k \leqslant N-1\}$ which minimize a quadratic cost function:

$$J = E\left[\|x(N)\|^2_{Q_c(N)} + \sum_{k=0}^{N-1} \|x(k)\|^2_{Q_c} + \|u(k)\|^2_{R_c}\right] \tag{12.79}$$

where $Q_c(N)$, $Q_c \geqslant 0$ and $R_c > 0$. When defining the control inputs $U_{k-1} \triangleq \{u(i), 0 \leqslant i \leqslant k-1\}$, we expect that the controller is of the form

$$u(k) = \phi(k, Z_k, U_{k-1}) \tag{12.80}$$

where ϕ denotes a measurable function of Z_k and U_{k-1}.

The above formulation is for the problem of control of linear discrete-time systems with faulty sensors, sometimes called the *jump-linear-quadratic-Gaussian* (JLQG) control problem. The anomaly sensors are assumed to be modelled by a finite-state Markov chain whose transition probabilities are completely known. As we have seen in the previous chapter, however, the problem under consideration does not provide an optimal solution.

In the following, we shall extend the 'passive' MMAC technique due to Deshpande, Upadhyay and Lainiotis (DUL) [24], provided in Corollary 11.2, to the system of equations (12.77) and (12.78) with cost function (12.79). The result is summarized in Theorem 12.2.

Theorem 12.2

The DUL MMAC algorithm for equations (12.77) and (12.78) becomes

$$u(k) = -F(k)\sum_{j=1}^{s^k} \hat{x}_j(k|k)p(I_j(k)|Z_k, U_{k-1}), \qquad 0 \leqslant k \leqslant N-1 \tag{12.81}$$

where the deterministic control gain matrix $F(k)$ is given by

$$F(k) = [G^{\mathrm{T}}S(k+1)G + R_c]^{-1}G^{\mathrm{T}}S(k+1)\Phi \tag{12.82}$$

in which the matrices $S(k)$ are given by the matrix RDE:

$$\begin{aligned} S(k) = \Phi^{\mathrm{T}}[&S(k+1) - S(k+1)G \\ &\times \{G^{\mathrm{T}}S(k+1)G + R_c\}^{-1}G^{\mathrm{T}}S(k+1)]\Phi \\ &+ Q_c, \qquad S(N) = Q_c(N) \end{aligned} \tag{12.83}$$

The a posteriori probability of $I_j(k)$, $p((I_j(k)|\cdot)$, is provided by the equation (12.10).

Proof

The derivation is apparent from the results of Sections 11.3 and 11.4. □

As discussed in Section 11.4, the controller gain $F(k)$ in this problem does not depend upon the future evolution of the structure state sequences $I(k)$. $\hat{x}_j(k|k)$ and $p(I_j(k)|Z_k, U_{k-1})$ can, of course, be computed by applying Kalman filters matched to sequences $I_j(k)$, $j = 1, \ldots, s^k$. For large k, however, we must resort to a suboptimal estimator as shown in Section 12.4 to reduce the computational burden.

The following examples due to Tzafcstas and Watanabe [25] and Watanabe and Tzafestas [26] illustrate MMACs using a suboptimal estimator.

Example 12.5

Consider the following system:

$$x(k+1) = 1.04x(k) + 1.2u(k) + w(k)$$

$$z(k) = H(\theta(k))x(k) + D(\theta(k))v(k)$$

with cost function

$$J = E\left\{x^2(30) + \sum_{k=0}^{29}[x^2(k) + u^2(k)]\right\}$$

where $\theta(k) \in \{1, 2\}$, $H(1) = 1$, $D(1) = 10$, $H(2) = 1$ and $D(2) = 1$. Other conditions are the same as in Example 12.1.

The Monte Carlo runs were performed using MMACs with the four suboptimal estimators applied to Example 12.1. In addition, a lower bound was provided by realizing an LQG controller which knows the true values of the switching parameters. The cost function was averaged over 50 Monte Carlo runs.

In Table 12.7, the average cost functions for various MMACs are tabulated together with rms estimation errors averaged over 30 time stages and 50 Monte Carlo runs.

The certainty equivalence principle holds for this case [27, 28]. Therefore, the dual effects of control are known to be absent. The dual effect refers to the fact that in a stochastic problem the control has, in general, two effects: it affects the state (control action) and it affects the uncertainty of the state (augmented by the possible unknown parameters) [29]. Therefore, the accuracy of state estimation will be control-dependent if the control has a dual effect. In others words, the MMAC given by equation (12.81) is optimal and the only approximation involved is that due to the use of a suboptimal filter. It is observed from Table 12.7 that the n-step measurement update method outperforms other methods. Note also that

Table 12.7 Performance measures for Example 12.5

Algorithm	*Average cost*	*RMS state estimation error*
AF method	3085	2.677
JG method with		
$n = 1$	3085	2.763
$n = 2$	2966	1.583
$n = 3$	3010	1.851
Modified JG method with		
$n = 1$	2950	1.482
$n = 2$	2967	1.577
$n = 3$	3000	1.799
n-step measurement update method with		
$n = 1$	2950	1.482
$n = 2$	2946	1.449
$n = 3$	2945	1.439
Lower bound	2891	1.009

increasing the number n in the JG (or modified JG) method does not necessarily lead to improved control performance.

Example 12.6

Consider again the system of Example 12.5 except that

$$x(k+1) = -0.9x(k) + 1.2u(k) + w(k)$$

and $H(1) = 0$. This observation model represents the physical situation where an observation may at any time not contain the desired signal, and is thus called an *interrupted observation*. The certainty equivalence principle does not hold for this case and the dual effects of control are known to be present [30, 31]. The average control cost and rms state estimation error are displayed in Table 12.8. Observe that the estimation performance of the JG and modified JG methods is better than that of the n-step measurement update method. Note, however, that the improved state estimation does not necessarily translate into lower control cost; this seems to be attributed to the presence of the dual effect. As discussed in Chapter 11, the MMAC is passively adaptive in that it does not seek actively to exploit this dual effect. Clearly, a dual control law should do better in this

Table 12.8 Performance measures for Example 12.6

Algorithm	*Average cost*	*RMS state estimation error*
AF method	2888	3.522
JG method with		
$n = 1$	2735	2.657
$n = 2$	2722	2.068
$n = 3$	2692	2.042
Modified JG method with		
$n = 1$	2749	2.768
$n = 2$	2724	2.093
$n = 3$	2692	2.036
n-step measurement update method with		
$n = 1$	2749	2.768
$n = 2$	2709	2.369
$n = 3$	2706	2.348
Lower bound	2590	1.353

situation. The MMAC laws discussed here can form a basis for designing dual control laws following the approach of Wenk and Bar-Shalom [32, 33]. Note, however, that dual control laws are in general computationally demanding [34].

12.7 SUMMARY

In this chapter, the problem of state estimation, system structure detection and control has been discussed for linear discrete-time systems with abruptly changing parameters. The changes were modelled by a Markov chain with known transition probabilities. Since the optimal solution was impractical, some suboptimal multiple model adaptive filter (MMAF) and multiple model adaptive control (MMAC) algorithms based on the pseudo-Bayes or generalized pseudo-Bayes algorithm (GPBA) technique were introduced to reduce the computational burdens of the 'optimal' MMAF and MMAC.

No discussion was provided in this chapter to establish: the feasibility of knowing the transition probabilities for practical problems, the robustness of the Markov approach to errors in transition probability assignments, and the feasibility of estimating the transition probabilities.

We are very interested in enlightenment concerning these issues. On the feasibility of estimating the transition probabilities, see, for example, Sawaragi *et al.*, [35], Tugnait and Haddad [36, 37], and Tugnait [38, 39].

Although we have considered the MMAC for a system with faulty sensors in Section 12.6, we will be able to treat a case when the state equation includes some faulty actuators. However, the control does not simplify to Theorem 12.2; that is, we need s^{N+1} sequences of control gains in general. An approximate solution for such a JLQG control problem can be found in Watanabe and Tzafestas [40].

Some control-related work on the same class of models as decribed in this chapter may be further found in Sworder [41, 42], Wonham [43], and Mariton [44] for continuous-time systems, in which the parameters are assumed to be observed. For some discrete-time versions in deterministic environments, i.e. for jump-linear-quadratic (JLQ) control problems, see Blair and Sworder [45], Chizeck *et al.* [46] and Birdwell *et al.* [47]. Further work on the JLQ or JLQG control problem can be found in Mariton [48].

The IMMA technique extended to distributed-sensor networks is reported by Chang and Bar-Shalom [49], and the algorithm is applied to tracking a manoeuvring target in a cluttered and low-detection environment.

An important class of system models, called *composite sources* [50], can be cast into the framework of systems with interrupted observations. A composite source is a random process consisting of a collection of subsources and a switch process which randomly selects among these subsources [51]. This type of random process structure can be found in the areas of bioelectric signal processing [52], speech modelling [53], image modelling [54], and so on.

EXERCISES

12.1 Consider a scalar system of the form

$$x(k+1) = a(\theta(k+1))x(k) + w(k)$$
$$z(k) = x(k) + v(k)$$

where $\theta(k) = \{1, 2\}$. Using the sampling data $\{z(k)\}$, $k = 1, 2, 3$, develop the optimal MMAF (Theorem 12.1) and the associated elemental Kalman filters at every sampling instant.

12.2 Consider a system described by the following state-space model:

$$x(k+1) = \Phi x(k) + Gu(k) + Bw(k)$$

$$z(k) = Hx(k) + v(k)$$

where the sequences $\{w(k)\}$ and $\{v(k)\}$ are governed by one of s multivariate Gaussian distributions. Each allowable Gaussian distribution is such that

$$E[w^{(i)}(k)] = \mu_w^{(i)}$$

$$E\{[w^{(i)}(k) - \mu_w^{(i)}][w^{(i)}(j) - \mu_w^{(i)}]^{\mathrm{T}}\} = Q^{(i)}\delta_{kj}$$

$$E\{[w^{(i)}(k) - \mu_w^{(i)}][w^{(r)}(j) - \mu_w^{(r)}]^{\mathrm{T}}\} = 0, \; i \neq r$$

$$E[v^{(i)}(k)] = \mu_v^{(i)}$$

$$E\{[v^{(i)}(k) - \mu_v^{(i)}][v^{(i)}(j) - \mu_v^{(i)}]^{\mathrm{T}}\} = R^{(i)}\delta_{kj}$$

$$E\{[v^{(i)}(k) - \mu_v^{(i)}][v^{(r)}(j) - \mu_v^{(r)}]^{\mathrm{T}}\} = 0, \; i \neq r$$

$$E\{[w^{(i)}(k) - \mu_w^{(i)}][v^{(r)}(j) - \mu_v^{(r)}]^{\mathrm{T}}\} = 0, \text{ for all } i, j, k, r$$

$$i, r = 1, 2, \ldots, s$$

Define a Markov chain state sequence $I(k)$:

$$I(k) \triangleq \{i_1, i_2, \ldots, i_k\}$$

where i_j is the index of the distribution from which $w(j)$ and $v(j)$ are sampled, i.e. i_j represents the Markov chain state at time j. Let $I_j(k)$ denote a specific sequence from the space of the sequences $I(k)$ such that

$$I_j(k) = \{I_m(k-1), i_k = l\}$$

(a) Derive the mean and covariance of the Gaussian distributions $p(z(k)|I_j(k), Z_{k-1})$ and the elemental Kalman filter conditioned on $\{I_j(k), Z_k\}$.

(b) Show that if $R^{(i)} = R$ and $Q^{(i)} = Q$, $i = 1, 2, \ldots, s$, then the pseudo-Bayes estimator given by Corollary 12.1 reduces to

$$\hat{x}(k|k) = \Phi\hat{x}(k-1|k-1) + \bar{\mu}_w + K(k) \times [z(k) - H\Phi\hat{x}(k-1|k-1) - H\bar{\mu}_w - \bar{\mu}_v]$$

where

$$\bar{\mu}_w = \sum_{j=1}^{s} \mu_w^{(j)} p(i_k = j|Z_k)$$

$$\bar{\mu}_v = \sum_{j=1}^{s} \mu_v^{(j)} p(i_k = j|Z_k)$$

12.3 Consider a system described by the following state-space model:

$$x(k+1) = \Phi x(k) + Bw(k)$$

$$z(k) = \theta(k)x(k) + v(k)$$

where $\{\theta(k)\}$ is a Markov sequence of binary random variables taking on values 0 or 1 at each stage. Defining $I_j(k)$ as a particular sequence of $I(k) \triangleq \{\theta(1), \theta(2), \ldots, (\theta)(k)\}$ which contains 2^k elements, develop the optimal elemental Kalman filtered estimates $\hat{x}_j(k|k) \triangleq E[x(k)|I_j(k), Z_k]$, the associated estimation error covariances $P_j(k|k) \triangleq E\{[x(k) - \hat{x}_j(k|k)][x(k) - \hat{x}_j(k|k)]^{\mathrm{T}}|I_j k, Z_k\}$, and the a posteriori probabilities $p(I_j(k)|Z_k)$ in terms of their one-step past values.

12.4 Reconsider the system of Exercise 12.1.

(a) Develop a generalized pseudo-Bayes estimator using the one-step measurement update method with $s = n = 2$ given in Corollary 12.2.

(b) Draw the block diagram.

12.5 Consider again Exercise 12.4, but using the n-step measurement update method with $s = n = 2$ presented by Corollary 12.3.

12.6 Show that the estimators developed in Exercises 12.4 and 12.5 have the same structure for $n = 1$.

12.7 Develop the elemental Kalman filters for the system described in Exercise 12.2 by applying the one- (or n)-step measurement method with $s = 2$ and $n = 1$.

12.8 Consider an n-dimensional system

$$x(k+1) = \Phi(\theta(k+1))x(k) + G(\theta(k+1))u(k) + B(\theta(k))w(k)$$

$$z(k) = H(\theta(k))x(k) + D(\theta(k))v(k)$$

with quadratic cost given by equation (12.79). Develop an MMAC algorithm as in Theorem 12.2.

12.9 Consider the system described in Exercise 12.8. Suppose that there is no system noise and $x(k)$ is observed exactly. Show that the control algorithm can be expressed in the following form:

$$u(k) = F_{i_k}(k)x(k), \qquad \text{if } \theta(k) = i_k$$

where

$$F_{i_k}(k) = -N_{i_k}^{-1}(k) \sum_{i_{k+1}=1}^{s} p_{i_k, i_{k+1}} G^{\mathrm{T}}(\theta(k+1) = i_{k+1}) \times S_{i_{k+1}}(k+1)\Phi(\theta(k+1) = i_{k+1})$$

$$N_{i_k}(k) = R_c + \sum_{i_{k+1}=1}^{s} p_{i_k, i_{k+1}} G^{\mathrm{T}}(\theta(k+1) = i_{k+1}) \times S_{i_{k+1}}(k+1)G(\theta(k+1) = i_{k+1})$$

and

$$S_{i_k}(k) = Q_c + \sum_{i_{k+1}=1}^{s} p_{i_k, i_{k+1}} \Phi^{\mathrm{T}}(\theta(k+1) = i_{k+1})S_{i_{k+1}}(k+1) \times \Phi(\theta(k+1) = i_{k+1}) - F_{i_k}^{\mathrm{T}}(k)N_{i_k}(k)F_{i_k}(k), \qquad S_{i_N}(N) = Q_c(N)$$

REFERENCES

[1] WILLSKY, A. S., A Survey of Design for Failure Detection in Dynamic Systems, *Automatica*, vol. 12, 1976, pp. 601–11.

[2] GUSTAFSON, D. E., WILLSKY, A. S., WANG, J. Y., LANCASTER, M. C. and TRIEBWASSER, J. H., ECG/VCG Rhythm Diagnosis using Statistical Signal Analysis, I. Identification of Persistent Rhythms, *IEEE Trans. Biomedical Engng.*, vol. BME-25, no. 4, 1978, pp. 344–53.

[3] WILLSKY, A. S., Failure Detection in Dynamic Systems, *AGARD graph* no. 109, 1980.

[4] WATANABE, K., A Multiple Model Adaptive Filtering Approach to Fault Diagnostic Systems. In R. J. Patton, P. M. Frank and R. N. Clark (eds), *Fault Diagnosis in Dynamic Systems,* Prentice Hall, London, 1989.

[5] WATANABE, K., A Decentralized Multiple Model Adaptive Filtering for Discrete-Time Stochastic Systems, *Trans. ASME, J. Dynamic Syst. Meas. Control*, vol. 111, September 1989, pp. 371–7.

[6] RICKER, G. G. and WILLIAMS, J. R., Adaptive Tracking Filter for Maneuvering Targets, *IEEE Trans. Aero. Electron. Syst.,* vol. AES-14, no. 1, 1978, pp. 185–93.

[7] MOOSE, R. L., VANLANDINGHAM, H. F. and MCCABE, D. H., Modeling and Estimation for Tracking Maneuvering Targets, *IEEE Trans. Aero. Electron. Syst.*, vol. AES-15, no. 3, 1979, pp. 448–56.

[8] CHANG, C. B. and TABACZYNSKI, J. A., Application of State Estimation to Target Tracking, *IEEE Trans. Aut. Control*, vol. AC-29, no. 2, 1984, pp. 98–109.

[9] VANLANDINGHAM, H. F. and MOOSE, R. L., Digital Control of High Performance Aircraft Using Adaptive Estimation Techniques, *IEEE Trans. Aero. Electron. Syst.*, vol. AES-13, no. 2, 1977, pp. 112–19.

[10] MOOSE, R. L., VANLANDINGHAM, H. F. and ZWICKE, P. E., Digital Set Point Control of Nonlinear Stochastic Systems, *IEEE Trans. Industrial Electron. and Control Instrum.*, vol. IECI-25, no. 1, 1978, pp. 39–45.

[11] AKASHI, H. and KUMAMOTO, H., Random Sampling Approach to State Estimation in Switching Environments, *Automatica*, vol. 13, 1977, pp. 429–34.

[12] TUGNAIT, J. K. and HADDAD, A. H., A Detection Estimation Scheme for State Estimation in Switching Environments, *Automatica*, vol. 15, 1979, pp. 477–81.

[13] TUGNAIT, J. K., Detection and Estimation for Abruptly Changing Systems, *Automatica*, vol. 18, no. 5, 1982, pp. 607–15.

[14] MATHEWS, V. J. and TUGNAIT, J. K., Detection and Estimation with Fixed Lag for Abruptly Changing Systems, *IEEE Trans. Aero. Electron. Syst.*, vol. AES-19, no. 5, 1983, pp. 730–9.

[15] ACKERSON, G. A. and FU, K. S., On State Estimation in Switching Environments, *IEEE Trans. Aut. Control*, vol. AC-15, no. 1, 1970, pp. 10–17.

[16] JAFFER, A. G. and GUPTA, S. C., On Estimation of Discrete Process under

Multiplicative and Additive Noise Conditions, *Inf. Sci.*, vol. 3, 1971, pp. 267–76.

[17] CHANG, C. B. and ATHANS, M., State Estimation for Discrete Systems with Switching Parameters, *IEEE Trans. Aero. Electron. Syst.*, vol. AES-14, no. 3, 1978, pp. 418–25.

[18] SUGIMOTO, S. and ISHIZUKA, I., Identification and Estimation Algorithms for a Markov Chain plus AR Process, *Proc. IEEE Int. Conf. on Acoustics, Speech and Signal Processing*, ICASSP 83, 1983, pp. 247–50.

[19] WATANABE, K. and TZAFESTAS, S. G., A New Generalized Pseudo-Bayes Estimation for Discrete-Time Systems with Jump Failure Parameters, *Preprints of 12th IMACS World Congress on Scientific Computation*, vol. 2, 1988, pp. 427–30.

[20] BLOM, H. A. P., An Efficient Filter for Abruptly Changing Systems, *Proc. 23rd IEEE Conf. on Decision and Control*, 1984, pp. 656–8.

[21] BLOM, H. A. P., An Efficient Decision-Making-Free Filter for Processes with Abrupt Changes, *Proc. IFAC Identification and Systems Parameter Estimation 1985*, 1985, pp. 631–6.

[22] BLOM, H. A. P. and BAR-SHALOM, Y., The Interacting Multiple Model Algorithm for Systems with Markovian Switching Coefficients, *IEEE Trans. Aut. Control*, vol. AC-33, no. 8, 1988, pp. 780–3.

[23] VAN TREES, H. L., *Detection, Estimation, and Modulation Theory*, Part I, Wiley, New York, 1968.

[24] DESHPANDE, J. G., UPADHYAY, T. N. and LAINIOTIS, D. G., Adaptive Control of Linear Stochastic Systems, *Automatica*, vol. 9, 1973, pp. 107–15.

[25] TZAFESTAS, S. G. and WATANABE, K., Stochastic Control Algorithms for Systems with Markovian Faulty Sensors, *Preprint of 12th IMACS World Congress on Scientific Computation*, vol. 2, 1988, pp. 681–4.

[26] WATANABE, K. and TZAFESTAS, S. G., Stochastic Control for Systems with Faulty Sensors, *Trans. ASME, J. Dynamic Syst. Meas. Control*, vol. 112, March 1990, pp. 143–6.

[27] AKASHI, H., KUMAMOTO, H. and NOSE, K., Application of Monte Carlo Method to Optimal Control for Linear Systems under Measurement Noise with Markov Dependent Statistical Property, *Int. J. Control*, vol. 22, no. 6, 1975, pp. 821–36.

[28] TUGNAIT, J. K., Control of Stochastic Systems with Markov Interrupted Observations, *IEEE Trans. Aero. Electron. Syst.*, vol. AES-19, no. 2, 1983, pp. 232–9.

[29] BAR-SHALOM, Y., Stochastic Dynamic Programming: Caution and Probing, *IEEE Trans. Aut. Control*, vol. AC-26, no. 5, 1981, pp. 1184–95.

[30] FUJITA, S. and FUKAO, T., Optimal Stochastic Control for Discrete-Time Systems with Interrupted Observations, *Automatica*, vol. 8, 1972, pp. 425–32.

[31] WERNERSSON, Å., Comments on 'Optimal Stochastic Control for Discrete-Time Systems with Interrupted Observations', *Automatica*, vol. 10, 1974, pp. 113–15.

[32] WENK, C. J. and BAR-SHALOM, Y., A Multiple Model Adaptive Dual

Control Algorithm for Stochastic Systems with Unknown Parameters, *IEEE Trans. Aut. Control*, vol. AC-25, no. 4, 1980, pp. 703–10.

[33] WENK, C. J. and BAR-SHALOM, Y., Model Adaptive Dual Control of MIMO Stochastic Systems, *Proc. 20th IEEE Conf. on Decision and Control*, San Diego, California, December 1981, pp. 821–4.

[34] GRIFFITHS, B. E. and LOPARO, K. A., Optimal Control of Jump-Linear Gaussian Systems, *Int. J. Control*, vol. 42, no. 4, 1985, pp. 791–819.

[35] SAWARAGI, Y., KATAYAMA, T. and FUJISHIGE, S., Adaptive Estimation for a Linear System with Interrupted Observation, *IEEE Trans. Aut. Control*, vol. AC-18, no. 2, 1973, pp. 152–4.

[36] TUGNAIT, J. K. and HADDAD, A. H., State Estimation under Uncertain Observations with Unknown Statistics, *IEEE Trans. Aut. Control*, vol. AC-24, no. 2, 1979, pp. 201–10.

[37] TUGNAIT, J. K. and HADDAD, A. H., Adaptive Estimation in Linear Systems with Unknown Markovian Noise Statistics, *IEEE Trans. Inf. Theory*, vol. IT-26, no. 1, 1980, pp. 66–78.

[38] TUGNAIT, J. K., Adaptive Estimation and Identification for Discrete Systems with Markov Jump Parameters, *IEEE Trans. Aut. Control*, vol. AC-27, no. 5, 1982, pp. 1054–65.

[39] TUGNAIT, J. K., On Identification and Adaptive Estimation for Systems with Interrupted Observations, *Automatica*, vol. 19, no. 1, 1983, pp. 61–73.

[40] WATANABE, K. and TZAFESTAS, S. G., A Multiple Model Adaptive Control for Jump Linear Stochastic Systems, *Int. J. Control*, vol. 50, no. 5, 1989, pp. 1603–17.

[41] SWORDER, D. D., Control of Systems Subject to Sudden Change in Character, *Proc. IEEE*, vol. 64, no. 8, August 1976, pp. 1219–25.

[42] SWORDER, D. D., Control of Systems Subject to Small Measurement Disturbances, *Trans. ASME, J. Dynamic Syst. Meas. Control*, vol. 106, 1984, pp. 182–8.

[43] WONHAM, W. M., Random Differential Equations in Control Theory. In A. T. Bharuda-Reid (ed.), *Probabilistic Methods in Applied Mathematics*, Vol. II, Academic Press, New York, 1970, pp. 131–212.

[44] MARITON, M., Joint Estimation and Control of Jump Linear Systems with Multiplicative Noises, *Trans. ASME, J. Dynamic Syst. Meas. Control*, vol. 109, March 1987, pp. 24–8.

[45] BLAIR, W. P. and SWORDER, D. D., Feedback Control of a Class of Linear Discrete Systems with Jump Parameters and Quadratic Criteria, *Int. J. Control*, vol. 21, no. 5, 1975, pp. 833–41.

[46] CHIZECK, H. J., WILLSKY, A. S. and CASTANON, D., Discrete-Time Markovian-Jump Linear Quadratic Optimal Control, *Int. J. Control*, vol. 43, no. 1, 1986, pp. 213–31.

[47] BIRDWELL, J. D., CASTANON, D. A. and ATHANS, M., On Reliable Control System Design, *IEEE Trans. Syst., Man, and Cyber.*, vol. SMC-16, no. 5, 1986, pp. 703–11.

[48] MARITON, M., *Jump Linear Systems in Automatic Control*, Marcel Dekker, New York, 1990.

[49] CHANG, K. C. and BAR-SHALOM, Y., Distributed Adaptive Estimation with Probabilistic Data Association, *Automatica*, vol. 25, no. 3, 1989, pp. 359–69.

[50] BERGER, T., *Rate Distortion Theory*, Prentice Hall, Englewood Cliffs, New Jersey, 1971.

[51] TUGNAIT, J. K., An Addendum and Correction to: 'On Identification and Adaptive Estimation for Systems with Interrupted Observations', *Automatica*, vol. 20, no. 1, 1984, pp. 135–6.

[52] SANDERSON, A. C., SEGEN, J. and RICHEY, E., Hierarchical Modeling of EEG Signals, *IEEE Trans. Pattern Analysis and Machine Intelligence*, vol. PAMI-2, no. 5, 1980, pp. 405–15.

[53] SCHAFER, R. W. and RABINER, L. R., Digital Representations of Speech Signals, *Proc. IEEE*, vol. 63, no. 4, 1975, pp. 662–77.

[54] RAJALA, S. A. and DEFIGUIREDO, R. J. P., Adaptive Nonlinear Image Restoration by a Modified Kalman Filtering Approach, *IEEE Trans. Acoustics, Speech, and Signal Processing*, vol. ASSP-29, no. 5, 1981, pp. 1033–42.

APPENDIX A

Brief Review of Linear Algebra

This appendix summarizes some results from matrix theory and linear algebra [1–6] used in this book.

A.1 MATRICES AND VECTORS

An $n \times m$ (*rectangular*) matrix A consists of a collection of nm elements

a_{ij} $(i = 1, \ldots, n;\ j = 1, \ldots, m)$

written in an array of n rows and m columns:

$$A = \begin{bmatrix} a_{11} & a_{12} & \cdots & a_{1m} \\ a_{21} & a_{22} & \cdots & a_{2m} \\ \vdots & \vdots & & \vdots \\ a_{n1} & a_{n2} & \cdots & a_{nm} \end{bmatrix}$$

If all of the elements a_{ij} are zero, A is called a *zero matrix* or *null matrix*. If all of the elements of an $n \times n$ (*square*) matrix are zero except for those along the principal diagonal, as follows:

$$A = \begin{bmatrix} a_{11} & 0 & \cdots & 0 \\ 0 & a_{22} & \cdots & 0 \\ \vdots & \vdots & \ddots & \vdots \\ 0 & 0 & \cdots & a_{nn} \end{bmatrix}$$

then A is called *diagonal*. Moreover, if $a_{ii} = 1$ for all i, then the matrix is called the *identity matrix* and is denoted by I.

A square matrix is *symmetric* if $a_{ij} = a_{ji}$ for all values of i and j from 1 to n. It follows that a diagonal matrix is always symmetric.

An *upper triangular* matrix is a square matrix all of whose elements below the principal diagonal are zero, as follows:

$$A = \begin{bmatrix} a_{11} & a_{12} & \cdots & a_{1n} \\ 0 & a_{22} & \cdots & a_{2n} \\ \vdots & \vdots & \ddots & \vdots \\ 0 & 0 & \cdots & a_{nn} \end{bmatrix}$$

Similarly, a *lower triangular* matrix is a square matrix all of whose elements above the principal diagonal are zero.

A matrix composed of a single column, i.e. an $n \times 1$ matrix, is called an n-vector or a *column n-vector* and is denoted by

$$x = \begin{bmatrix} x_1 \\ x_2 \\ \vdots \\ x_n \end{bmatrix}$$

Thus, x_i is the ith scalar element. A *row n-vector* is a matrix with one row and n columns.

A matrix can be decomposed not only into its scalar elements, but also into arrays of elements called matrix *block-decompositions* (or partitions)

$$A = \begin{bmatrix} A_{11} & A_{12} & \cdots & A_{1i} \\ \vdots & \vdots & & \vdots \\ A_{k1} & A_{k2} & \cdots & A_{ki} \end{bmatrix}, \qquad x = \begin{bmatrix} x_1 \\ x_2 \\ \vdots \\ x_k \end{bmatrix}$$

where A_{ki} denotes the kith block (matrix) element and x_k denotes the kth block (vector) element. If a square matrix A can be block-decomposed in such a way that $A_{ij} = 0$ (the zero matrix) for all blocks for which $i \neq j$ and such that all blocks A_{ii} are square, then the matrix A is called *block diagonal*.

A.2 EQUALITY, ADDITION AND MULTIPLICATION

Two $n \times m$ matrices A and B are equal if and only if $a_{ij} = b_{ij}$ for all i and j. The *sum* of two $n \times m$ matrices A and B is defined by the following expression

$$C = A + B = [a_{ij} + b_{ij}]$$

The *difference* is defined similarly. In addition, the following properties hold:

1. $A + B = B + A$
2. $A + (B + C) = (A + B) + C$
3. $A + 0 = 0 + A = A$

The *product* of an $n \times m$ matrix A by a scalar b is the $n \times m$ matrix

$$C = bA = Ab = [ba_{ij}]$$

The product of an $n \times m$ matrix A and an $m \times p$ matrix B is the $n \times p$ matrix C defined by

$$C = AB = [a_{ij}][b_{ij}] = \left[\sum_{k=1}^{m} a_{ik}b_{kj}\right] = [c_{ij}]$$

Two matrices which can be multiplied in this way, i.e. where the number of columns in A equals the number of rows in B, are said to be *conformable*.

The order of the matrices in the product is significant, and AB can be described as premultiplying B by A or postmultiplying A by B. For general conformable matrices, we obtain the following:

4. $A(BC) = (AB)C$
5. $IA = AI = A$
6. $0A = A0 = 0$
7. $A(B + C) = AB + AC$
8. In general, $AB \neq BA$, even for A and B both square
9. $AB = 0$ in general does not imply that A or B is 0.

Given that matrix block-decompositions are conformable, matrix operations upon block-decomposed matrices obey the same rules of equality, addition, and multiplication. For example,

10. $$\begin{bmatrix} A_{11} & A_{12} \\ A_{21} & A_{22} \end{bmatrix} + \begin{bmatrix} B_{11} & B_{12} \\ B_{21} & B_{22} \end{bmatrix} = \begin{bmatrix} A_{11} + B_{11} & A_{12} + B_{12} \\ A_{21} + B_{21} & A_{22} + B_{22} \end{bmatrix}$$

11. $$\begin{bmatrix} A_{11} & A_{12} \\ A_{21} & A_{22} \end{bmatrix} \begin{bmatrix} B_{11} & B_{12} \\ B_{21} & B_{22} \end{bmatrix} = \begin{bmatrix} A_{11}B_{11} + A_{12}B_{21} & A_{11}B_{12} + A_{12}B_{22} \\ A_{21}B_{11} + A_{22}B_{21} & A_{21}B_{12} + A_{22}B_{22} \end{bmatrix}$$

A.3 TRANSPOSITION

Suppose A is an $n \times m$ matrix. The *transpose* of A, denoted A^{T}, is an $m \times n$ matrix that satisfies $[a_{ij}]^{\mathrm{T}} = [a_{ji}]$ for all i and j. Thus, transposition can be interpreted as interchanging the roles of the rows and columns of a matrix. It is easy to establish the following:

1. If x is an n-vector, x^{T} is a row vector.
2. $(A^{\mathrm{T}})^{\mathrm{T}} = A$
3. $(A + B)^{\mathrm{T}} = A^{\mathrm{T}} + B^{\mathrm{T}}$
4. $(AB)^{\mathrm{T}} = B^{\mathrm{T}}A^{\mathrm{T}}$
5. If A is a *symmetric* matrix, $A^{\mathrm{T}} = A$.
6. If x and y are n-vectors, $x^{\mathrm{T}}y$ is a scalar and xy^{T} is a square $n \times n$ matrix; xx^{T} is symmetric as well.
7. If A is a symmetric $n \times n$ matrix and B is a general $m \times n$ matrix, then $C = BAB^{\mathrm{T}}$ is a symmetric $m \times m$ matrix.
8. If A and B are both symmetric $n \times n$ matrices, $(A + B)$ is also symmetric but (AB) generally is not.
9. $\begin{bmatrix} A & B \\ C & D \end{bmatrix}^{\mathrm{T}} = \begin{bmatrix} A^{\mathrm{T}} & C^{\mathrm{T}} \\ B^{\mathrm{T}} & D^{\mathrm{T}} \end{bmatrix}$

A.4 DETERMINANTS

The *determinant* of a square $n \times n$ matrix A is a scalar value that arises naturally in the solution of sets of linear equations; it is denoted by $|A|$ or $\det A$. The evaluation of the determinant can be found in terms of lower-order determinants through use of *Laplace's expansion*. That is, it can be performed recursively through

$$|A| = \sum_{j=1}^{n} a_{ij}C_{ij}$$

for any fixed $i = 1, 2, \ldots, n$. Here, C_{ij} is the *cofactor* of a_{ij}, defined as $C_{ij} = (-1)^{i+j}M_{ij}$, and M_{ij} is the *minor* of a_{ij}, defined as the determinant of the $(n-1) \times (n-1)$ matrix formed by deleting the ith row and jth column of the $n \times n$ matrix A. For $n = 1, 2, 3$ we can also determine the determinant of the matrix A through *Sarrus' method*:

$$\det[a_{11}] = a_{11} \text{ for } n = 1$$

$$\det\begin{bmatrix} a_{11} & a_{12} \\ a_{21} & a_{22} \end{bmatrix} = a_{11}a_{22} - a_{12}a_{22} \text{ for } n = 2$$

$$\det\begin{bmatrix} a_{11} & a_{12} & a_{13} \\ a_{21} & a_{22} & a_{23} \\ a_{31} & a_{32} & a_{33} \end{bmatrix} = a_{11}a_{22}a_{33} + a_{12}a_{23}a_{31} + a_{13}a_{21}a_{32}$$

$$- a_{11}a_{32}a_{23} - a_{12}a_{21}a_{33}$$

$$- a_{13}a_{22}a_{31} \text{ for } n = 3$$

Moreover, we obtain the following results:

1. $|A^{\mathrm{T}}| = |A|$
2. If all the elements of any row or column of A are zero, $|A| = 0$.
3. If any row (column) of A is a scalar multiple of any other row (column), then $|A| = 0$.
4. If a scalar multiple of any row (column) is added to any other row (column) of A, the value of the determinant is unchanged.
5. If A and B are $n \times n$ matrices, $|AB| = |A||B|$.
6. If A is diagonal or triangular, then $|A|$ equals the product of its diagonal elements: $|A| = \prod_{i=1}^{n} a_{ii}$.
7. $\left|\begin{bmatrix} A & B \\ 0 & C \end{bmatrix}\right| = |A||C|$

A.5 MATRIX INVERSION AND SINGULARITY

Given a square matrix, if there exists a matrix such that both premultiplying and postmultiplying it by A yields the identity, then this matrix is called the *inverse* of A, and is written A^{-1}; thus we have $AA^{-1} = A^{-1}A = I$. If a square matrix A does not possess such an inverse, then $|A| = 0$, and A is said to be *singular*.† If $|A| \neq 0$, the inverse is unique, and A is called *nonsingular*. If the matrix A is nonsingular, then A^{-1} can be evaluated as follows:

$$A^{-1} = \frac{[\mathrm{adj}\, A]}{|A|} = \frac{[C_{ij}]^{\mathrm{T}}}{|A|}$$

where [adj$A$] is the adjoint of A, defined as the $n \times n$ matrix whose

† Instead, a type of approximate inverse called the *pseudo-* or *generalized* inverse of A can be defined [7].

ij element is the cofactor C_{ij}. In addition, the following properties hold:

1. If A is nonsingular, then so is A^{-1}, and $(A^{-1})^{-1} = A$.
2. $(AB)^{-1} = B^{-1}A^{-1}$ if all indicated inverses exist.
3. $(A^{-1})^{\mathrm{T}} = (A^{\mathrm{T}})^{-1}$ if A is nonsingular.
4. $|A^{-1}| = |A|^{-1} = 1/|A|$ if $|A| \neq 0$.
5. If A^{-1} and B^{-1} exist, then

$$\begin{bmatrix} A & 0 \\ C & B \end{bmatrix}^{-1} = \begin{bmatrix} A^{-1} & 0 \\ -B^{-1}CA^{-1} & B^{-1} \end{bmatrix}$$

and

$$\begin{bmatrix} A & D \\ 0 & B \end{bmatrix}^{-1} = \begin{bmatrix} A^{-1} & -A^{-1}DB^{-1} \\ 0 & B^{-1} \end{bmatrix}$$

6. If A^{-1} exists, then

$$\begin{bmatrix} A & D \\ C & B \end{bmatrix}^{-1} = \begin{bmatrix} A^{-1} + E\Delta^{-1}F & -E\Delta^{-1} \\ -\Delta^{-1}F & \Delta^{-1} \end{bmatrix}$$

where $\Delta = B - CA^{-1}D$, $E = A^{-1}D$, and $F = CA^{-1}$.

If A is such that its inverse equals its transpose, $A^{-1} = A^{\mathrm{T}}$, then A is called *orthogonal*. If A is orthogonal, $AA^{\mathrm{T}} = A^{\mathrm{T}}A = I$, and $|A| = \pm 1$.

A.6 LINEAR INDEPENDENCE AND RANK

A set of k n-vectors $x_1, x_2, \ldots, x_k$ is said to be *linearly independent* if the only real or complex numbers $\{\alpha_1, \alpha_2, \ldots, \alpha_k\}$ for which

$$\sum_{i=1}^{k} \alpha_i x_i = 0$$

are $\alpha_1 = \alpha_2 = \ldots = \alpha_k = 0$. Otherwise, the $\{x_i\}$ will be said to be *linearly dependent*.

The *rank* of an $m \times n$ matrix A is the positive integer q such that some $q \times q$ submatrix of A, formed by deleting $(m - q)$ rows and $(n - q)$ columns, is nonsingular, while no $(q + 1) \times (q + 1)$ submatrix is nonsingular. The rank of A is also the maximum number of linearly independent rows of A and the maximum number of linearly independent columns of A.

A useful inequality on rank is *Sylvester's inequality*, which states that for the $m \times n$ matrix A and the $n \times p$ matrix B,

$$\text{rank } A + \text{rank } B - n \leqslant \text{rank } (AB) \leqslant \min \{\text{rank } A, \text{rank } B\}$$

If rank A is equal to the number of columns or the number of rows of A, A is often said to have *full rank*. If A is an $n \times n$ matrix, the statement 'rank $A = n$' is equivalent to the statement 'A is nonsingular'.

A.7 THE RANGE SPACE AND NULL SPACE OF A MATRIX

Let A be an $m \times n$ matrix. The *range space* of A, denoted $\mathcal{R}(A)$, is the set of all vectors Ax, where x ranges over the set of all n-vectors. The range space is often called the *image* of A and written Im$\{A\}$. The range space has dimension equal to the rank of A, that is the maximal number of linearly independent vectors in $\mathcal{R}(A)$ is rank A. The *null space* of A, denoted $\mathcal{N}(A)$, is the space of all solutions y of the equation $Ay = 0$. The null space is often called the *kernel* of A and written Ker$\{A\}$. The dimension of $\mathcal{N}(A)$ is called the *nullity* of A. It is useful to note that

$$\text{rank } A + \text{nullity } A = n$$

A.8 EIGENVALUES AND EIGENVECTORS

The equations $Ax = \lambda x$ for some $n \times n$ matrix, which can also be written

$$(A - \lambda I)x = 0$$

possesses a nontrivial solution if and only if

$$|A - \lambda I| = 0$$

The nth order polynomial $f(\lambda) = |A - \lambda I|$ is called the *characteristic polynomial* of A, and the equation $f(\lambda) = 0$ is called its *characteristic equation*. The n *eigenvalues* of A are the (not necessarily distinct) roots of this equation, and nonzero solutions to

$$Ax_i = \lambda_i x_i$$

corresponding to the roots λ_i, are called *eigenvectors*. It is also true that for a general matrix A,

$$|A| = \prod_{i=1}^{n} \lambda_i$$

If A is singular, A possesses at least one zero eigenvalue.

Let the eigenvalues of the $n \times n$ matrix A be the distinct values $\lambda_1, \lambda_2, \ldots, \lambda_n$, and let the associated eigenvectors be $e_1, e_2, \ldots, e_n$. Then, if $E = [e_1, e_2, \ldots, e_n]$, E is nonsingular, and $E^{-1}AE$ is a diagonal matrix whose ith diagonal element is λ_i, $i = 1, 2, \ldots, n$. Moreover, if A is also symmetric, then the eigenvalues are all real and E is orthogonal.

As defined up to this point, eigenvectors are *right eigenvectors* in the sense that they appear as columns on the right-hand side of A in the equation

$$Ae_i = \lambda_i e_i$$

It is also possible to consider *left eigenvectors* that are multiplied as rows on the left-hand side of A in the form

$$f_i^{\mathrm{T}} A = \lambda_i f_i^{\mathrm{T}}$$

where f_i is an n-dimensional column vector. It is easy to see that the right and left eigenvalues (not eigenvectors) are identical, and that

$$f_j^{\mathrm{T}} e_i = 0 \quad \text{for } i \neq j$$

which is referred to as an *orthogonality* relation.

A.9 QUADRATIC FORMS AND POSITIVE (SEMI)DEFINITENESS

If A is a symmetric $n \times n$ matrix and x is an n-vector, then the scalar quantity $x^{\mathrm{T}}Ax$ is called a *quadratic form*, because when written as

$$x^{\mathrm{T}}Ax = \sum_{i=1}^{n} \sum_{j=1}^{n} a_{ij} x_i x_j$$

it is quadratic in the entries x_i of x.

If $x^{\mathrm{T}}Ax > 0$ for all $x \neq 0$, the quadratic form is said to be *positive definite*, as is the matrix A itself, often written notationally as $A > 0$. If $x^{\mathrm{T}}Ax \geqslant 0$ for all $x \neq 0$, the quadratic form and A are termed *positive semidefinite* (or *nonnegative definite*), denoted as $A \geqslant 0$. Moreover, by the notation $A > B$ $(A \geqslant B)$ is meant that $(A - B)$ is positive definite (semidefinite). Similarly, if $x^{\mathrm{T}}Ax < 0$ for all nontrivial x, we employ the term *negative definite*, written as $A < 0$.

There are simple tests, known as *Sylvester's theorem*, for positive definiteness and positive semidefiniteness. For A to be positive definite, all *leading principal minors* must be positive, i.e.

$$a_{11} > 0, \begin{vmatrix} a_{11} & a_{12} \\ a_{12} & a_{22} \end{vmatrix} > 0, \begin{vmatrix} a_{11} & a_{12} & a_{13} \\ a_{12} & a_{22} & a_{23} \\ a_{13} & a_{23} & a_{33} \end{vmatrix} > 0, \ldots, |A| > 0$$

For A to be positive semidefinite, all minors whose diagonal entries are diagonal entries of A must be nonnegative. For example, for a 3×3 matrix A,

$$a_{11}, a_{22}, a_{33} \geqslant 0, \begin{vmatrix} a_{11} & a_{12} \\ a_{12} & a_{22} \end{vmatrix}, \begin{vmatrix} a_{11} & a_{13} \\ a_{13} & a_{33} \end{vmatrix}, \begin{vmatrix} a_{22} & a_{23} \\ a_{23} & a_{33} \end{vmatrix} \geqslant 0$$

$$\begin{vmatrix} a_{11} & a_{12} & a_{13} \\ a_{12} & a_{22} & a_{23} \\ a_{13} & a_{23} & a_{33} \end{vmatrix} \geqslant 0$$

The following properties hold:

1. If A is positive definite, it is nonsingular, and its inverse A^{-1} is also positive definite.
2. The symmetric matrix A is positive definite (positive semidefinite) if and only if all its eigenvalues are positive (nonnegative).
3. If A and B are positive semidefinite, so is $A + B$; and if one is positive definite, so is $A + B$.

A.10 EXPONENTIAL OF A SQUARE MATRIX

Let A be a square matrix. Then it can be shown that the series

$$I + A + \frac{1}{2!} A^2 + \frac{1}{3!} A^3 + \ldots$$

converges, in the sense that the (i,j) entry of the partial sums of the series converges for all i and j. The sum is defined as $\exp(A)$. It follows that

$$\exp(At) = \sum_{i=0}^{\infty} \frac{A^i t^i}{i!}$$

Other properties are: $p(A)\exp(At) = \exp(At)p(A)$ for any polynomial $p(A)$, e.g. $p(A) = \sum_{i=0}^{r} a_i A^i$, where the a_i are scalars, and $\exp(-At) = [\exp(At)]^{-1}$.

A.11 TRACE

Let A be an $n \times n$ matrix. Then the trace of A, denoted $\text{tr}[A]$, is defined as

$$\text{tr}[A] = \sum_{i=1}^{n} a_{ii}$$

An important property is that

$$\text{tr}[A] = \sum_{i=1}^{n} \lambda_i$$

where the λ_i are eigenvalues of A. Other properties are as follows:

1. $\text{tr}[A] = \text{tr}[A^{\mathrm{T}}]$
2. $\text{tr}[A_1 + A_2] = \text{tr}[A_1] + \text{tr}[A_2]$
3. If B is an $n \times m$ matrix and C is an $m \times n$ matrix, so that BC is $n \times n$ and CB is $m \times m$, then

   $$\text{tr}[BC] = \text{tr}[CB] = \text{tr}[B^{\mathrm{T}}C^{\mathrm{T}}] = \text{tr}[C^{\mathrm{T}}B^{\mathrm{T}}]$$
4. $\text{tr}[A^{\mathrm{T}}A] = \sum_{i=1}^{n}\sum_{j=1}^{n} a_{ij}^2$
5. If x and y are n-vectors and A is an $n \times n$ matrix, then

   $$\text{tr}[xy^{\mathrm{T}}] = \text{tr}[x^{\mathrm{T}}y] = x^{\mathrm{T}}y$$
   $$\text{tr}[Axy^{\mathrm{T}}] = \text{tr}[y^{\mathrm{T}}Ax] = y^{\mathrm{T}}Ax = x^{\mathrm{T}}A^{\mathrm{T}}y$$

A.12 SIMILARITY

If A and B are $n \times n$ matrices and T is a nonsingular $n \times n$ matrix, and $A = T^{-1}BT$, then A and B are said to be related by a *similar transformation*, or are simply termed *similar*. Similarity is an equivalence relation. Thus:

1. A is similar to A.
2. If $A = T^{-1}BT$, then $B = TAT^{-1}$, i.e. if A is similar to B, then B is similar to A.
3. If A is similar to B and B is similar to C, then A is similar to C.

Finally, if A and B are similar, their determinants, eigenvalues, eigenvectors, characteristic polynomials and trace are equal; also if A is positive definite, then so is B, and vice versa.

A.13 JORDAN CANONICAL FORM

Although not all square matrices are diagonalizable, it is always possible to obtain very close to a diagonal matrix via a similar transformation. In fact, it can be shown that an $n \times n$ matrix A can be reduced to the *Jordan canonical form* by using an $n \times n$ matrix T as follows:

$$T^{-1}AT = \begin{bmatrix} J_1 & & & 0 \\ & J_2 & & \\ 0 & & \ddots & \\ & & & J_s \end{bmatrix},$$

$$J_i = \begin{bmatrix} \lambda_i & * & & \\ & \lambda_i & * & 0 \\ 0 & & \ddots & * \\ & & & \lambda_i \end{bmatrix} (n_i \times n_i)$$

where $\{\lambda_1, \ldots, \lambda_s\}$ are the eigenvalues of A, n_i being the number of times λ_i occurs, i.e. $n_1 + n_2 + \ldots + n_s = n$, J_i are the *Jordan blocks* of A, and the * on the superdiagonal are either 1 or 0 depending on further characteristics of the matrix A. In general, T and the λ_i are complex. By allowing diagonal blocks

$$\begin{bmatrix} \alpha & -\beta \\ \beta & \alpha \end{bmatrix}$$

to replace diagonal elements $\alpha \pm j\beta$, we obtain a 'real Jordan form' for a real A with T real. If A is *skew symmetric*, i.e. $A = -A^{\mathrm{T}}$, symmetric or orthogonal, T may be chosen orthogonal.

A.14 NORMS AND INNER PRODUCTS

A.14.1 Norms

The *norm* of a vector x, denoted by $\|x\|$, is a measure of the size or length of x. There is no unique definition, but the following postulates must be satisfied:

1. $\|x\| \geqslant 0$ for all x, with equality if and only if $x = 0$.
2. $\|ax\| = |a|\|x\|$ for any scalar constant a and for all x.
3. $\|x + y\| \leqslant \|x\| + \|y\|$ for all x and y.

Let $x_i (i = 1, \ldots, n)$ denote the components of the n-dimensional vector x. Three possible definitions of $\|x\|$ are:

$$\|x\| = \left[\sum_{i=1}^{n} x_i^2\right]^{1/2}$$

$$\|x\| = \max_i |x_i|$$

$$\|x\| = \sum_{i=1}^{n} |x_i|$$

The first expression is the so-called *Euclidean norm*.

The norm of an $n \times n$ matrix A, written $\|A\|$, can similarly be defined in an $(n \times n)$-dimensional vector space. It must satisfy similar postulates as follows:

1. $\|A\| \geqslant 0$ for all A, with equality if and only if all a_{ij} are zero.
2. $\|aA\| = |a|\|A\|$ for any constant a and for all A.
3. $\|A + B\| \leqslant \|A\| + \|B\|$ for all A and B.

A convenient definition of $\|A\|$ is in terms of an associated vector norm:

$$\|A\| = \max_{\|x\|=1} \|Ax\|$$

The particular vector norm used must be settled in order to fix the matrix norm. Corresponding to three vector norm definitions listed above, the matrix norms become, respectively

$$\|A\| = [\lambda_{\max}(A^{\mathrm{T}}A)]^{1/2}$$

$$\|A\| = \max_i \left(\sum_{j=1}^{n} |a_{ij}|\right)$$

$$\|A\| = \max_j \left(\sum_{i=1}^{n} |a_{ij}|\right)$$

The first expression is an explicit formula for the Euclidean norm. Other possible choices for the definition of $\|A\|$ are:

$$\|A\| = \max_{i,j} |a_{ij}|$$

$$\|A\| = \sum_{i,j=1}^{n} |a_{ij}|$$

$$\|A\| = \left(\sum_{i,j=1}^{n} a_{ij}^2\right)^{1/2} = \{\mathrm{tr}\,[A^{\mathrm{T}}A]\}^{1/2}$$

Important properties of matrix norms are

$$\|Ax\| \leqslant \|A\|\|x\|$$

$$\|A + B\| \leqslant \|A\| + \|B\|$$

and

$$\|AB\| \leqslant \|A\|\|B\|$$

A.14.2 Inner Products

The inner product (or scalar product) of two column n-dimensional vectors x and y denoted (x, y) or $\langle x, y \rangle$ is defined by

$$\langle x, y \rangle \triangleq x^{\mathrm{T}} y = \sum_{i=1}^{n} x_i y_i$$

The inner product of x and x is given by

$$\langle x, x \rangle = \left[\sum_{i=1}^{n} x_i^2\right] = \|x\|^2$$

Analogously to the inner product of two vectors, we may define an inner product of two $n \times n$ matrices A and B:

$$\langle A, B \rangle = \mathrm{tr}[A^{\mathrm{T}} B] = \sum_{i,j=1}^{n} a_{ij} b_{ij}$$

The *Schwartz inequality* states that

$$|x^{\mathrm{T}} y| = \left(\sum_{i=1}^{n} x_i y_i\right) \leqslant \|x\|\|y\|$$

where the equality signs hold if and only if x and y are linearly dependent (i.e. $x = \lambda y$ for some scalar λ).

A.15 DIFFERENTIATION AND INTEGRATION

Let A be an $n \times n$ matrix function of a scalar variable t, such as time. Then

$$\mathrm{d}A(t)/\mathrm{d}t = \dot{A}(t) = (\mathrm{d}a_{ij}/\mathrm{d}t)$$

It follows that

1. $\mathrm{d}[A^{\mathrm{T}}(t)]/\mathrm{d}t = [\mathrm{d}A(t)/\mathrm{d}t]^{\mathrm{T}}$
2. $\mathrm{d}[A(t)B(t)]\,\mathrm{d}t = \dot{A}(t)B(t) + A(t)\dot{B}(t)$
3. $\mathrm{d}[A^{-1}(t)]/\mathrm{d}t = -A^{-1}[\mathrm{d}A(t)/\mathrm{d}t]A^{-1}(t)$

4. $\mathrm{d}[\exp(At)]/\mathrm{d}t = A\exp(At) = \exp(At)A$, where A is time-invariant.

Similarly, the integral of a matrix is defined in a straightforward way as

$$\int A(t)\,\mathrm{d}t = \left(\int a_{ij}\,\mathrm{d}t\right)$$

Let s be a scalar function of a vector x. Then

$$\frac{\mathrm{d}s}{\mathrm{d}x} = \text{a row vector whose } i\text{th entry is } \frac{\partial s}{\partial x_i}$$

Suppose z is a vector function of a vector x. Then

$$\frac{\mathrm{d}z}{\mathrm{d}x} = \text{a matrix whose } ij \text{ entry is } \frac{\partial z_i}{\partial x_j}$$

Suppose s is a scalar function of a matrix A. Then

$$\frac{\mathrm{d}s}{\mathrm{d}A} = \text{a matrix whose } ij \text{ entry is } \frac{\partial s}{a_{ij}}$$

Moreover, for vectors x and y as functions of a vector c, for constant vector e and for constant matrices A and B, we obtain:

5 $\partial c/\partial c = I$

6 $\partial(Ax)/\partial c = A\partial x/\partial c$, and thus, $\partial(Ac)/\partial c = A$

7a $\partial(x^{\mathrm{T}}Ay)/\partial c = x^{\mathrm{T}}A\partial y/\partial c + y^{\mathrm{T}}A^{\mathrm{T}}\partial x/\partial c$

7b $\partial(e^{\mathrm{T}}Ac)/\partial c = e^{\mathrm{T}}A$

7c $\partial(c^{\mathrm{T}}Ae)/\partial c = e^{\mathrm{T}}A^{\mathrm{T}}$

7d $\partial(c^{\mathrm{T}}Ac)/\partial c = c^{\mathrm{T}}A + c^{\mathrm{T}}A^{\mathrm{T}} = 2c^{\mathrm{T}}A$, if $A = A^{\mathrm{T}}$

7e $\partial[(e - Bc)^{\mathrm{T}}A(e - Bc)]/\partial c = -2(e - Bc)^{\mathrm{T}}AB$, if $A = A^{\mathrm{T}}$

REFERENCES

[1] BELLMAN, R., *Introduction to Matrix Analysis*, TATA McGraw-Hill, New Delhi, 1974.

[2] SCHULTZ, D. G. and MELSA, J. L., *State Functions and Linear Control Systems*, McGraw-Hill, New York, 1967.

[3] BIERMAN, J. B., *Factorization Methods for Discrete Sequential Estimation*, Academic Press, New York, 1977.

[4] LUENBERGER, D. G., *Introduction to Dynamic Systems*, Wiley, New York, 1979.
[5] KAILATH, T., *Linear Systems*, Prentice Hall, Englewood Cliffs, New Jersey, 1980.
[6] CHEN, C.-T., *Linear System Theory and Design*, Holt, Rinehart and Winston, New York, 1984.
[7] BARNETT, S., *Matrices in Control Theory*, Robert E. Krieger, Malabar, Florida, 1984.

APPENDIX B

Brief Review of Probability Theory

This appendix introduces some of the elementary ideas of probability [1–4] used in this text. For a more advanced treatment, see, for example, Doob [5], Papoulis [6], Feller [7], and Wong and Hajek [8]. We also briefly review results concerning constant-parameter estimators [9–11].

B.1 PROBABILITY THEORY

B.1.1 Probability Axioms

An *event* is defined as some specific outcome of an experiment in which chance or uncertainty is involved. In this connection, an event A is said to occur if and only if the observed outcome of the experiment is an element of A. The set of all possible outcomes of such an experiment is called the *sample space* of the experiment. Denote this set by Ω and its elements by ω.

Suppose that the experiment is performed N times and that, in these N trials, an event A occurs $N(A)$ times. Then we say that the *probability* of the event A is given by the expression

$$P(A) = \lim_{N\to\infty} \frac{N(\mathrm{A})}{N}, \qquad 0 \leqslant N(A) \leqslant N$$

assuming that the indicated limit exists. Note that $P(A)$ is sometimes denoted as $\Pr(A)$.

Now define a *probability space* $(\Omega, \mathcal{F}, P)$ consisting of the following:

(1) a sample space Ω of elements ω;
(2) a σ-algebra $\mathcal{F}$, which is a class of sets A_i, each of which is a subset of Ω (i.e. $A_i \subset \Omega$), such that if A_i is an element of $\mathcal{F}$ (i.e. if $A_i \in \mathcal{F}$), then:

(a) $\bar{A}_i \in \mathcal{F}$, where $\bar{A}_i$ is the complement of A_i $(\bar{A}_i = \Omega - A_i)$,

(b) $\Omega \in \mathcal{F}$ and then the empty set $\varnothing \in \mathcal{F}$,

(c) if $A_1, A_2, \ldots, \in \mathcal{F}$, then their union and intersection are also in $\mathcal{F}$ $(\cup_i A_i \in \mathcal{F}$ and $\cap_i A_i \in \mathcal{F})$, where all possible finite and countably infinite unions and intersections are included;

(3) the *probability function* (or *probability measure*) $P(\cdot)$, which is a real scalar-valued function defined on $\mathcal{F}$ that assigns a value, $P(A)$, to each $A \in \mathcal{F}$ such that:

(a) $P(A) \geqslant 0$ for all $A \in \mathcal{F}$,

(b) $P(\Omega) = 1$,

(c) if $A_1, A_2, \ldots$ are elements of $\mathcal{F}$ and are disjoint, or mutually exclusive—i.e. if $A_i \cap A_j = \varnothing$ for all $i \neq j$—then $P(\cup_{i=1}^{N} A_i) = \sum_{i=1}^{N} P(A_i)$ for all finite and countably finite N.

For our purposes, it suffices to consider Ω as the set of points in n-dimensional Euclidean space, $\mathbb{R}^n$, and to let $\mathcal{F}$ be the class of subsets in Ω of the form $\{\omega : \omega \leqslant a, \omega \in \Omega\}$, where ω is an n-vector and a is an n-vector of specified value. This particular σ-algebra is called a *Borel field*.

B.1.2 Conditional Probability

We consider two events A and B in a sample space Ω. The probability of the *joint* occurrence of A and B is $P(A \cap B)$, written sometimes $P(AB)$, and that of B alone is $P(B)$. We denote the *conditional probability* of A occurring given that B has already occurred by $P(A|B)$ and define it by the relation

$$P(A|B) = \frac{P(A \cap B)}{P(B)}$$

under the assumption $P(B) \neq 0$.

B.1.3 Independence

Events $A_1, A_2, \ldots, A_n$ in Ω are *mutually independent* if and only if

$$P(A_{i_1} \cap A_{i_2} \cap \ldots \cap A_{i_k}) = P(A_{i_1})P(A_{i_2}) \ldots P(A_{i_k})$$

for all integers $i_1, \ldots, i_k$ selected from 1, 2, ..., n with no two the same. Given three events A, B, $C \in \Omega$, it is possible for each pair to be mutually independent, that is $P(AB) = P(A)P(B)$, $P(BC) = P(B)P(C)$ and $P(CA) = P(C)P(A)$ but that $P(ABC) \neq P(A)P(B)P(C)$.

Two events A and B are *conditionally independent* given an event C if

$$P(AB|C) = P(A|C)P(B|C)$$

If A_i, $i = 1, 2, \ldots, n$ are mutually disjoint and $\cup A_i = \Omega$, then

$$P(B) = \sum_{i=1}^{n} P(B|A_i)P(A_i)$$

for arbitrary B.

B.1.4 Bayes' Rule

If $P(B) \neq 0$,

$$P(A|B) = \frac{P(B|A)P(A)}{P(B)}$$

If A_i, $i = 1, 2, \ldots, n$ are mutually disjoint and $\cup A_i = \Omega$,

$$P(A_j|B) = \frac{P(B|A_j)P(A_j)}{\sum_{i=1}^{n} P(B|A_i)P(A_i)}$$

B.1.5 Random Variables

It is desirable to introduce a mapping from the sample space to the real numbers in order to facilitate qualitative analysis. This is done via the *random variable*.

Definition B.1

A random variable is a real-valued function $X(\omega)$ on Ω such that every set $\{\omega : X(\omega) \leq x\}$, x real, is an element of $\mathcal{F}$, where $\omega \in \Omega$.

B.1.6 Probability Distribution and Density Functions

In view of the definition of a random variable $X(\omega)$, the function

$$F_X(x) \triangleq P(X(\omega) \leq x)$$

is defined for all real x and is called the *distribution function* of the random variable X. The subscript X on F identifies the random variable. The distribution function is monotonic nondecreasing, so that $\lim_{x\to\infty} F_X(x) = 1$, $\lim_{x\to-\infty} F_X(x) = 0$, and it is continuous from the right. Hence, $0 \leq F_X(x) \leq 1$.

A random variable X is called *continuous* if there exists a *probability density function* $p_X(x)$ such that

$$F_X(x) = \int_{-\infty}^{x} p_X(\xi)\, d\xi, \qquad -\infty < x < \infty$$

Note here that a more common notation of the density function is $p(x)$. The density function exists if the distribution function is absolutely continuous, i.e. the number of points at which $F_X(\cdot)$ is not differentiable is countable. It is immediately evident that $F_X(x)$ is then continuous at all x, and that

$$p_X(x) = \frac{dF_X(x)}{dx}$$

at all x at which the derivative exists.

B.1.7 Pairs of Random Variables

Suppose that X and Y are random variables. Then $F_{X,Y}(x, y) = P\{(X \leq x) \cap (Y \leq y)\}$ is the *joint probability distribution function* of random variables X and Y. If the derivative exists, the joint probability density function is

$$p_{X,Y}(x, y) = \frac{\partial^2}{\partial x \partial y} F_{X,Y}(x, y)$$

Given $F_{X,Y}(x, y)$, it follows that $F_X(x) = F_{X,Y}(x, \infty)$ and

$$p_X(x) = \int_{-\infty}^{\infty} p_{X,Y}(x, y)\, dy$$

B.1.8 Conditional Distribution and Densities

Suppose $A = \{X(\omega) \leq x\}$ and B is any event. From the conditional probability,

$$P(X(\omega) \leq x | B) = \frac{P\{(X(\omega) \leq x) \cap B\}}{P(B)}$$

which is just the distribution function

$$F_{X|B}(x|B) = \frac{P\{(X(\omega) \leq x) \cap B\}}{P(B)}$$

Now suppose $B = \{y < Y(\omega) \leq y + \Delta y\}$. Letting $\Delta y \to 0$, one obtains

$$p_{X|Y}(x|y) = \frac{p_{X,Y}(x, y)}{p_Y(y)} = \frac{p_{X,Y}(x, y)}{\int p_{X,Y}(x, y)\,\mathrm{d}x}$$

if $p_{X,Y}(x, y)$ and $p_Y(y) \neq 0$ exist. $p_{X|Y}(x|y)$ is called the *conditional probability density function* of X given Y. We also have the important formulas:

$$p_{X|Y}(x|y) \geq 0$$

$$\int_{-\infty}^{+\infty} p_{X|Y}(x|y)\,\mathrm{d}x = 1$$

$$p_X(x) = \int_{-\infty}^{+\infty} p_{X|Y}(x|y)p_Y(y)\,\mathrm{d}y$$

B.1.9 Random Vectors, Marginal and Conditional Densities

Suppose n random variables $X_1, X_2, \ldots, X_n$ define a random n-vector X. We have

$$F_X(x) = P\{(X_1 \leq x_1) \cap (X_2 \leq x_2) \cap \ldots \cap (X_n \leq x_n)\}$$

and

$$p_X(x) = \frac{\partial^n}{\partial x_1 \partial x_2 \ldots \partial x_n} F_X(x)$$

Marginal densities can be obtained by integration:

$$p_{X_1,X_2,X_3}(x_1, x_2, x_3) = \int_{-\infty}^{+\infty}\int_{-\infty}^{+\infty} \cdots \int_{-\infty}^{+\infty} p_X(x)\,\mathrm{d}x_4\,\mathrm{d}x_5 \ldots \mathrm{d}x_n$$

Conditional densities can also be found as

$$p_{X_1,X_2,\ldots,X_l|X_{l+1},X_{l+2},\ldots,X_n}(x_1, x_2, \ldots, x_l|x_{l+1}, x_{l+2}, \ldots, x_n)$$

$$= \frac{p_{X_1,X_2,\ldots,X_n}(x_1, x_2, \ldots, x_n)}{p_{X_{l+1},X_{l+2},\ldots,X_n}(x_{l+1}, x_{l+2}, \ldots, x_n)}$$

B.1.10 Independent Random Variables

X and Y are independent random variables if the events $\{X(\omega) \leqslant x\}$ and $\{Y(\omega) \leqslant y\}$ are independent for all x and y:

$$F_{X,Y}(x, y) = F_X(x)F_Y(y)$$

or

$$p_{X,Y}(x, y) = p_X(x)p_Y(y)$$

or

$$p_{X|Y}(x|y) = p_X(x)$$

B.1.11 Expectation and Variance

The *expectation*—also called the *mean, average* or *first moment* (see Section B.1.12)—of a random variable X is defined by

$$E[X] \triangleq \int_{-\infty}^{+\infty} x p_X(x)\,\mathrm{d}x$$

where the integral is assumed absolutely convergent. The *variance*—or second central moment (see Section B.1.12)—σ^2 is given by the expression

$$\sigma^2\{X\} \triangleq E[(X - E[X])^2] = \int_{-\infty}^{+\infty} (x - E[X])^2 p_X(x)\,\mathrm{d}x$$
$$= E[X^2] - (E[X])^2$$

Chebyshev's inequality states that

$$P\{|X - E[X]| > \varepsilon\} \leqslant \frac{\sigma^2}{\varepsilon^2}$$

The *standard deviation* of X, $\sigma(X)$, is given by the square root of the variance.

For vector X, the variance is replaced by the *covariance matrix*

$$E\{(X - E[X])(X - E[X])^{\mathrm{T}}\}$$

The variance is always nonnegative, and the covariance matrix nonnegative definite symmetric. When $g(X)$ is a function of a random variable X, its expected value is defined as

$$E[g(X)] = \int_{-\infty}^{+\infty} g(x)p_X(x)\,\mathrm{d}x$$

Similarly, we write

$$E[g(X, Y)] = \int_{-\infty}^{+\infty}\int_{-\infty}^{+\infty} g(x,y)p_{X,Y}(x, y)\,dx\,dy$$

for the expected value of the function of the two random variables X and Y.

The expectation operator is linear. Also, if X_i represents mutually independent random variables, then

$$E[X_1, X_2, \ldots, X_n] = E[X_1]E[X_2]\ldots E[X_n]$$

If they are also of zero mean,

$$E\left[\left(\sum_{i=1}^{n} X_i\right)^2\right] = \sum_{i=1}^{n} E[X_i^2]$$

B.1.12 Moments

The kth *moment* of a random variable X is $m_k = E[X^k]$. The kth *central moment* (or kth *moment of X about the mean*) is $\mu_k = E[(X - E[X])^k]$. Note that the second central moment is just the variance of X. The *joint moments* of two random variables X and Y are given by the set of numbers $E[X^kY^l]$. $E[XY]$ is the *correlation* of X and Y. The *joint central moments* are defined in an obvious manner and $E[(X - E[X])(Y - E[Y])]$ is the covariance of X and Y. If $E[XY] = E[X]E[Y]$, X and Y are called *uncorrelated*, and if $E[XY] = 0$, they are called *orthogonal*. Independent random variables are always uncorrelated.

B.1.13 Characteristic Function

A random variable X may also be specified in terms of its *characteristic function*, defined by

$$\phi_X(s) \triangleq E[\exp(jsX)], \qquad j^2 = -1$$

The variable s can take on complex values. Evidently, we see that ϕ_X is just the Fourier transform of the density function

$$\phi_X(s) = \int_{-\infty}^{+\infty} \exp(jsx)p_X(x)\,dx$$

Therefore, if the characteristic function is absolutely integrable, the inverse Fourier transform

$$p_X(x) = \frac{1}{2\pi}\int_{-\infty}^{+\infty} \exp(-jsx)\phi_X(s)\,\mathrm{d}s$$

gives the density function. If $X_1, X_2, \ldots, X_n$ are n random variables, the *joint characteristic function* is

$$\phi_X(s_1, s_2, \ldots, s_n) = E\left\{\exp\left(j\sum_{i=1}^{n} s_i X_i\right)\right\}$$

We have $\phi_X(0) = 1$ and $|\phi_X(s_1, s_2, \ldots, s_n)| \leq 1$ for all real s_i. The moments $E[X^k]$ of X are related to ϕ_X by

$$E[X^k] = (1/j^k)\left.\frac{\mathrm{d}^k}{\mathrm{d}s^k}\phi_X(s)\right|_{s=0}$$

If X and Y are jointly distributed, $\phi_X(s) = \phi_{X,Y}(s_1, 0)$ or $\phi_Y(s) = \phi_{X,Y}(0, s_2)$. If they are independent,

$$\phi_{X,Y}(s_1, s_2) = \phi_X(s_1)\phi_Y(s_2)$$

and conversely. If $\{X_i\}$ is a set of independent random variables and $Z = X_1 + X_2 + \ldots + X_n$, then

$$\phi_Z(s) = \phi_{X_1}(s)\phi_{X_2}(s)\ldots\phi_{X_n}(s)$$

B.1.14 Conditional Expectation

The *conditional expected value* of a scalar-valued function g of a random n-vector X with respect to another random vector Y is defined as

$$E[g(X)|Y = y] = \int_{-\infty}^{+\infty}\ldots\int_{-\infty}^{+\infty} g(x)p_{X|Y}(x|y)\,\mathrm{d}x_1\ldots\mathrm{d}x_n$$

The *conditional mean* of X given $Y = y$ is defined as

$$E[X|Y = y] = \int_{-\infty}^{+\infty}\ldots\int_{-\infty}^{+\infty} xp_{X|Y}(x|y)\,\mathrm{d}x_1\ldots\mathrm{d}x_n$$

and the corresponding *conditional covariance matrix* as

$$E\{[X - E(X|Y = y)][X - E(X|Y = y)]^{\mathrm{T}}\}$$

$$= \int_{-\infty}^{+\infty}\ldots\int_{-\infty}^{+\infty}[x - E(X|y)][x - E(X|y)]^{\mathrm{T}}p_{X|Y}(x|y)\,\mathrm{d}x_1\ldots\mathrm{d}x_n$$

Let X, Y, Z be jointly distributed random variables (or vectors); c, d fixed constants; $g(\cdot)$ a scalar-valued function. Then the following results hold:

1. $E[X|Y] = E[X]$, if X and Y are independent.
2. $E[X] = E\{E[X|Y]\}$
3. $[g(Y)X|Y] = g(Y)E[X|Y]$
4. $E[g(Y)X] = E\{g(Y)E[X|Y]\}$
5. $E[c|Y] = c$
6. $E[g(Y)|Y] = g(Y)$
7. $E[cX + dZ|Y] = cE[X|Y] + dE[Z|Y]$
8. $E[XY|Z] = E[X|Z]E[Y|Z]$, if X and Y are conditionally independent of Z.

B.1.15 Central Limit Theorem

If the random variables $\{X_i\}$ are mutually independent, the distribution of

$$Y_n = \frac{1}{n} \sum_{i=1}^{n} X_i$$

is approximately Gaussian (normal), with mean $(1/n) \sum_{i=1}^{n} \mu_i$ and variance $(1/n) \sum_{i=1}^{n} \sigma_i^2$, where μ_i and σ_i^2 are the mean and variance of X_i. As $n \to \infty$, the approximation becomes more accurate.

B.1.16 Convergence of Random Sequences

Let $\{X_n, n = 1, 2, \ldots\}$ be a sequence of random variables. There are a number of ways in which the sequence might converge (as $n \to \infty$). We define three modes of convergence here.

C1 $X_n \to X$ almost surely (a.s.), or with probability 1 (w.p.1), if $\lim_{n\to\infty} X_n(\omega) = X(\omega)$ for almost all ω (almost all realizations). That is, it holds except perhaps on an event A such that $P(A) = 0$.

C2 $X_n \to X$ in mean square if $\lim_{n\to\infty} E[||X_n - X||^2] = 0$. We sometimes write $\text{l.i.m.} X_n = X$ and call X the limit in the mean of $\{X_n\}$.

C3 $X_n \to X$ in probability if for all $\varepsilon > 0$, $\lim_{n\to\infty} P[||X_n - X|| \geqslant \varepsilon] = 0$.

Now it is known that:

(a) C1 implies C3.
(b) C2 implies C3.
(c) C3 implies that a subsequence of $\{X_n\}$ satisfies C1.
(d) C3 and $|X_n| < c$, for some c, all $n \geq$ some n_0, and almost all ω, implies C2.

The following notion is often helpful:

C4 $X_n \to X$ in pth mean if $\lim_{n\to\infty} E[||X_n - X||^p] = 0$.

B.1.17 Hölder's Inequality

Let $p > 1$, $q > 1$ and $(1/p) + (1/q) = 1$. For two random variables $X(\omega)$ and $Y(\omega)$, if $E[|X|^p] < \infty$ and $E[|Y|^q] < \infty$, then

$$E[|XY|] \leq (E[|X|^p])^{1/p}(E[|Y|^q])^{1/q}$$

This is Hölder's inequality. *Schwartz's inequality* is the name given to the special case where $p = q = 2$.

B.1.18 Minkowski's Inequality

Let p be a real number such that $p \geq 1$. For two random variables $X(\omega)$ and $Y(\omega)$, if $E[|X|^p] < \infty$ and $E[|Y|^p] < \infty$, then

$$(E[|X + Y|^p])^{1/p} \leq (E[|X|^p])^{1/p} + (E[|Y|^p])^{1/p}$$

B.1.19 Jensen's Inequality

Let $f(x)$ be a convex function with the lower bound. If $f(X(\omega))$ is differentiable, then

$$f(E[X]) \leq E[f(X)]$$

B.1.20 Dominated Convergence Theorem

If $X_n \to X$ (w.p.1) and there exists an integrable random variable $Y(\omega)$ such that $|X| \leq Y$, then

$$\lim_{n\to\infty} E|X_n - X| = 0$$

B.2 STOCHASTIC PROCESSES

B.2.1 Stochastic Processes

Definition B.2

A stochastic process is a family of random vectors (or scalars) $\{x(t), t \in T\}$ indexed by a parameter t all of whose values lie in some appropriate index set T.

The parameter t will refer to time in our applications. In particular, we shall need only the following two time index sets. The first is the set of discrete time instants $T = \{t_k; k = 0, 1, \ldots\}$, where $t_k < t_{k+1}$. To simplify the notation, we replace t_k by k so that our index set becomes $T = \{k : k = 0, 1, \ldots\}$. In this case, we have a discrete-time stochastic process.

The second time index set is one which involves intervals on the time axis such as $T = \{t : 0 \leqslant t \leqslant T\}$ or $T = \{t : t \geqslant t_0\}$. Here, the process is termed a *continuous-time stochastic process*.

The phrase 'family of random vectors' means that a stochastic process is composed of a collection or *ensemble* of random vectors over the index set. The ensemble may contain either a countable or nondenumerable number of elements. The combination of the notions of time and ensemble implies that a stochastic process is actually a function of two variables. That is readily indicated by the notation $\{x(t, \omega), t \in T \text{ and } \omega \in \Omega\}$. For a fixed value of t, x is a vector-valued function on Ω; that is, it is a random vector. If, on the other hand, ω is fixed, then x is a vector-valued function of time which is one possible *realization* or *sample function* of the process.

If the sample space Ω is discrete, the term *chain* is commonly used instead of the word *process*. Also, note that we write a discrete-time stochastic process as $x(k)$.

B.2.2 Probability Law of a Stochastic Process

Let $\{X(k), k \in T\}$ be a stochastic process. For any finite set $\{k_1, \ldots, k_n\} \in T$, the set of all probability densities $p(x(k_1), x(k_2), \ldots, x(k_n))$ (or the corresponding distribution functions $F(x(k_1), x(k_2), \ldots, x(k_n))$) serves to define the probability structure of the random process. Similarly, we can define conditional densities in the usual way.

The mean $m(k)$ of a process is the time function $E[x(k)]$. The *(auto)correlation* is the set of quantities $E[x(k_1)x^{\mathrm{T}}(k_2)]$. The covariance is the set of quantities $E\{[x(k_1) - m(k_1)][x(k_2) - m(k_2)]\}$ for all k_1 and k_2. When $k_1 = k_2$, the covariance is nonnegative definite symmetric. The first-order densities of a process are the set of densities $p(x(k))$ for all k, and the second-order densities the set $p(x(k_1), x(k_2))$ for all k_1 and k_2. The mean and covariance can be obtained entirely from the first- and second-order densities.

A process has uncorrelated, orthogonal, or independent increments if, for all finite sets $\{k_i : k_i < k_{i+1}\} \in T$, $x(k_i) - x(k_{i+1})$ is a sequence of uncorrelated, orthogonal, or independent random variables. $\{x(k)\}$ and $\{y(k)\}$ are (i) uncorrelated, (ii) orthogonal, or (iii) independent processes according as (i) $E[x(k_1)y^T(k_2)] = E[x(k_1)]E[y^T(k_2)]$ for all k_1 and k_2, (ii) $E[x(k_1)y^T(k_2)] = 0$ for all k_1 and k_2, and (iii) for any sets $\{k_i\}$ and $\{l_i\}$, the vector random variables $[x^T(k_1), \ldots, x^T(k_n)]^T$ are independent of the vector random variable $[y^T(l_1), \ldots, y^T(l_n)]^T$.

B.2.3 Markov Processes

A stochastic process $\{x(k), k \in T\}$ is called a *Markov process*, if, for any finite parameter set $\{k_i : k_i < k_{i+1}\} \in T$, and for every λ,

$$P(x(k_n, \omega) \leqslant \lambda | x(k_1), \ldots, x(k_{n-1})) = P(x(k_n, \omega) \leqslant \lambda | x(k_{n-1}))$$

This property can also be written in terms of density functions

$$p(x(k_n)|x(k_1), \ldots, x(k_{n-1})) = p(x(k_n)|x(k_{n-1}))$$

where $k_1 < k_2 < \ldots < k_n$. A Markov process is sometimes called a *first-order Markov process*. A *second-order Markov process* is one where at most two pieces of information are all that affect the future, i.e.

$$p(x(k_n)|x(k_1), \ldots, x(k_{n-1})) = p(x(k_n)|x(k_{n-1}), x(k_{n-2}))$$

for $k_1 < k_2 < \ldots < k_n$. Third- and higher-order Markov processes can be defined similarly.

B.2.4 Martingale

Let $X(t)$ be adapted to a nondecreasing family of σ-algebras $\{\mathcal{F}_t\}$ with $E[|X(t)|] < \infty$. $(X(t), \mathcal{F}_t)$ is called a *martingale* (*supermartingale, submartingale*) if

$$E[X(t)|\mathcal{F}_s] = X(s) \ (\leqslant X(s), \geqslant X(s))$$

for any $t \geq s$, $s, t \in T$. This definition also holds for a discrete-time process.

B.2.5 Stationary Processes

A stochastic process $\{x(k)\}$ is *strict-sense stationary*, or simply *stationary*, if its associated probability densities are unaffected by time transition; i.e. for an arbitrary integer n and times $k_1, \ldots, k_n$ and $\tau \in T$,

$$p(x(k_1), \ldots, x(k_n)) = p(x(k_1 + \tau), \ldots, x(k_n + \tau))$$

It is *asymptotically stationary* if

$$\lim_{\tau \to \infty} p(x(k_1 + \tau), \ldots, x(k_n + \tau))$$

exists. The stochastic processes $\{x(k)\}$ and $\{y(k)\}$ are jointly stationary if $\{[x^{\mathrm{T}}(k) \quad y^{\mathrm{T}}(k)]^{\mathrm{T}}\}$ is stationary. If $\{x(k)\}$ is stationary, then $E[x(k)] = m$, independent of k, and

$$R(k_1, k_2) \triangleq E\{[x(k_1) - m][x(k_2) - m]^{\mathrm{T}}\} = R(k_1 - k_2)$$

A process is *wide-sense stationary* (or weakly stationary, or covariance stationary) if its first- and second-order densities are invariant under time transition. Then, its mean is constant, and its covariance, $R(k_1, k_2)$, is of the form $R(k_1 - k_2)$. Stationary processes are wide-sense stationary. Both the covariance and autocorrelation $C(k_1, k_2) = C(k_1 - k_2)$ are even where the arguments are scalar; i.e. $R(-k) = R(k)$, $C(-k) = C(k)$, and $C(0) \geq |C(k)|$ for all k. In the vector process case,

$$R(-k) = R^{\mathrm{T}}(k)$$

$$C^{\mathrm{T}}(-k) = C(k)$$

B.2.6 Ergodic Processes

Certain stationary processes are *ergodic*. The basic idea behind ergodicity is that time averages can be replaced by an expectation over the set of experiment outcomes. There are two approaches. One says that a process $\{x(k)\}$ is ergodic if for any suitable function $f(\cdot)$, the following limit exists almost surely:

$$E[f\{x(k)\}] = \lim_{N \to \infty} \frac{1}{2N + 1} \sum_{i=-N}^{N} f(\{x(i)\})$$

If $\{x(k)\}$ is Gaussian with covariance sequence, R_k, the following condition is sufficient for ergodicity:

$$\sum_{k=-\infty}^{+\infty} |R_k| < \infty$$

Alternatively, one seeks for a given $f(\cdot)$ conditions for the limit to exist as a mean square limit. A sufficient condition is then that

$$\sum_{k=-\infty}^{+\infty} |R_k^f| < \infty$$

where R^f denotes the covariance of $f(\cdot)$. Taking

$$f(\{x(k)\}) = x(k) \quad \text{and} \quad f(\{x(k)\}) = x(k)x(k+l)$$

leads to the concepts of ergodicity in the mean and in the covariance function; these last two concepts are valid for processes which are wide-sense stationary.

B.2.7 Power Spectrum

If $\{x(k)\}$ is a discrete-time random process that is wide-sense stationary, the *power spectrum* is, assuming it exists for some z,

$$\Phi(z) = \sum_{k=-\infty}^{+\infty} z^{-k} R_k, \qquad z = \exp(j\lambda) \qquad (-\pi < \lambda < \pi)$$

or

$$\Phi(\lambda) = \sum_{k=-\infty}^{+\infty} \exp(-j\lambda k) R_k$$

$\Phi(z)$ is nonnegative if a scalar, or nonnegative definite if a matrix, for all z such that $|z| = 1$. Also, $\Phi(z) = \Phi^{\mathrm{T}}(z^{-1})$. Finally, we have

$$\begin{aligned} E[x(k)x^{\mathrm{T}}(k)] &= \frac{1}{2\pi j} \int_{|z|=1} \frac{\Phi(z)\,\mathrm{d}z}{z} \\ &= \frac{1}{2\pi} \int_{-\pi}^{\pi} \exp(j\lambda k)\Phi(\lambda)\,\mathrm{d}\lambda \end{aligned}$$

B.2.8 White Noise

White noise processes usually have zero mean, and, when stationary, are processes whose power spectrum is constant.

Constancy of the power spectrum is equivalent to

$$E[x(k)x^{\mathrm{T}}(l)] = C\delta_{kl}$$

for some constant matrix C, where δ_{kl} denotes the Kronecker delta function. Similarly, for continuous-time processes, we have

$$E[x(t)x^{\mathrm{T}}(\tau)] = C\delta(t-\tau)$$

where $\delta(t-\tau)$ denotes the Dirac delta function.

B.3 GAUSSIAN RANDOM VARIABLES, VECTORS AND PROCESSES

B.3.1 Gaussian Random Variable

X is a *Gaussian* or, equivalently, *normal* random variable if its probability density is given by

$$p_X(x) = (1/2\pi\sigma^2)^{1/2}\exp\left[-\frac{1}{2}\left(\frac{x-\mu}{\sigma}\right)^2\right]$$

where one can evaluate $E[X]=\mu$ and $E\{(X-E[X])^2\}=\sigma^2$, respectively. In this evaluation, we can make use of the integral

$$\int_{-\infty}^{+\infty}\exp\left(-\tfrac{1}{2}x^2\right)\mathrm{d}x = (2\pi)^{1/2}$$

If the random variable is Gaussian with mean μ and variance σ^2, we sometimes write $X \sim N(\mu,\sigma^2)$. The mode of p_X, i.e. the value of x maximizing p_X, is μ.

If $X \sim N(\mu_X,\sigma_X^2)$ and $Y \sim N(\mu_Y,\sigma_Y^2)$, with X and Y independent, then $(X+Y) \sim N(\mu_X+\mu_Y,\sigma_X^2+\sigma_Y^2)$. Moreover, if $X \sim N(\mu,\sigma^2)$, then the characteristic function of X is $\phi_X(s) = \exp[j\mu s-(\sigma^2 s^2/2)]$, and conversely.

B.3.2 Gaussian Random Vector

Let X be a random n-vector. If X has a nonsingular covariance matrix, we say that X is Gaussian or normal if and only if its probability density and characteristic functions are given by

$$p_X(x) = [(2\pi)^{n/2}|\Sigma|^{1/2}]^{-1}\exp[-\tfrac{1}{2}(x-m)^{\mathrm{T}}\Sigma^{-1}(x-m)]$$

$$\phi_X(s) = \exp[js^{\mathrm{T}}m - \tfrac{1}{2}s^{\mathrm{T}}\Sigma s]$$

where

$$m = E[X]$$
$$\Sigma = E\{[X - m][X - m]^{\mathrm{T}}\}$$

If the covariance of X is singular, then X lies in a proper subspace of n-space (w.p.1); in fact, for any vector α in the null space of the covariance matrix Σ, we have $\alpha^{\mathrm{T}}(X - m) = 0$ for all X (w.p.1). We cannot define the Gaussian property via the probability density, but we can still define it via the characteristic function: X is Gaussian if and only if for some m and Σ, $\phi_X(s) = \exp[js^{\mathrm{T}}m - \frac{1}{2}s^{\mathrm{T}}\Sigma s]$.

B.3.3 Joint Densities, Marginal Densities and Conditional Densities

To say that X_1 and X_2 are joint Gaussian random variables is the same as saying that the random vector $X = [X_1 \quad X_2]^{\mathrm{T}}$ is a Gaussian random vector. All marginal densities derived from a Gaussian random vector are themselves Gaussian; for example, if $X = [X_1 \quad X_2 \ldots X_n]^{\mathrm{T}}$ is Gaussian, then $\hat{X} = [X_1 \quad X_2 \ldots X_m]^{\mathrm{T}}$ is Gaussian. All conditional densities formed by conditioning some entries in a Gaussian random vector on other entries are Gaussian. If $X = [X_1 \quad X_2]^{\mathrm{T}} \sim N(m, \Sigma)$, then X_1 conditioned on X_2 is Gaussian with mean $m_1 + \Sigma_{12}\Sigma_{22}^{-1}(x_2 - m_2)$ and covariance $\Sigma_{11} - \Sigma_{12}\Sigma_{22}^{-1}\Sigma_{12}^{\mathrm{T}}$. Here, the m_i and Σ_{ij} are the obvious submatrices of m and Σ.

B.3.4 Linear Transformations

Let $X \sim N(m, \Sigma)$. Let $Y = Ax + b$, where A is a constant matrix, b a constant vector and Y a random vector. Then $Y \sim N(Am + b, A\Sigma A^{\mathrm{T}})$. This follows from

$$\begin{aligned}
\phi_Y(s) &= E\{\exp(js^{\mathrm{T}}Y)\} = E\{\exp(js^{\mathrm{T}}Ax + js^{\mathrm{T}}b)\} \\
&= E\{\exp(js^{\mathrm{T}}b)\exp(js^{\mathrm{T}}Ax)\} = \exp(js^{\mathrm{T}}b)\phi_X(A^{\mathrm{T}}s) \\
&= \exp[js^{\mathrm{T}}(Am + b) - \tfrac{1}{2}s^{\mathrm{T}}A\Sigma A^{\mathrm{T}}s]
\end{aligned}$$

In particular, if X and Y are jointly Gaussian, $X + Y$ is Gaussian.

B.3.5 Uncorrelated Gaussian Variables

Let X and Y be uncorrelated and Gaussian. Then they are independent. This follows from the equality $\phi_{X,Y}(s_1, s_2) = \phi_X(s_1)\phi_Y(s_2)$.

B.3.6 Conditional Expectation

Suppose X and Y are jointly Gaussian. Then $E[X|Y]$, which is a function of Y, is of the form $AY + b$, for a constant matrix A and vector b, and is therefore Gaussian. In fact, if

$$\begin{bmatrix} X \\ Y \end{bmatrix} \sim N\left(\begin{bmatrix} m_x \\ m_y \end{bmatrix}, \begin{bmatrix} \Sigma_{xx} & \Sigma_{xy} \\ \Sigma_{yx} & \Sigma_{yy} \end{bmatrix}\right)$$

then $E[X|Y] \triangleq m_{x|y}$ is given by

$$m_x + \Sigma_{xy}\Sigma_{yy}^{-1}(Y - m_y)$$

and $E\{[X - m_{x|y}][X - m_{x|y}]^{\mathrm{T}}|Y\} \triangleq \Sigma_{x|y}$ is also given by

$$\Sigma_{xx} - \Sigma_{xy}\Sigma_{yy}^{-1}\Sigma_{xy}^{\mathrm{T}}$$

B.3.7 Gaussian Random Processes

A random process $\{x(k)\}$ is a Gaussian or normal random process if, for any set of points $\{k_i : i = 1, \ldots, n\} \in T$, the random variables $x(k_1), \ldots, x(k_n)$ are jointly Gaussian, i.e.

$$p(x(k_1), \ldots, x(k_n)) = [(2\pi)^{n/2}|\Sigma|^{1/2}]^{-1} \exp[-\tfrac{1}{2}(x - m)^{\mathrm{T}}\Sigma^{-1}(x - m)]$$

where

$$x \triangleq [x(k_1), \ldots, x(k_n)]^{\mathrm{T}}, \qquad m \triangleq E[x]$$

and

$$\Sigma^{(ij)} \triangleq E\{[x(k_i) - m^{(i)}][x(k_j) - m^{(j)}]\}, \qquad m^{(i)} \triangleq E[x(k_i)]$$

A complete probabilistic description of the process is provided by $E[x(k_i)]$ and $\mathrm{Cov}[x(k_i), x(k_j)]$ for all k_i and k_j. A (vector) random process that is both Gaussian and a first-order Markov process is called a *Gauss–Markov process*.

B.3.8 Wiener Processes

If $\{w(k)\}$ is a white noise, discrete-time, Gaussian process and $w(k) \sim N(0, 1)$ for each k, the process $\{x(k)\}$ defined by $x(k+1) = x(k) + w(k)$, $k \geq 0$, where $x(0) \sim N(m_0, \Sigma_0)$, is a *Wiener* (or *Brownian motion*) *process*.

B.4 ESTIMATES AND ESTIMATORS

Consider the problem of estimating a vector-valued constant parameter x. Let the parameter be observed through the observations

$$z(i) = h(i, x, w(i)), \qquad i = 1, 2, \ldots$$

where i denotes a discrete-time index, $w(i)$ is noise, and h is a vector-valued function. For k observations, find a function

$$\hat{x}(k) = \hat{x}[k, Z_k]$$

where $Z_k \triangleq \{z(1), \ldots, z(k)\}$. This function is usually called an *estimator*, and the value of the function is the *estimate*.

B.4.1 Some Properties of Estimates

Since the estimator is a function of the observations, the estimate is itself a random vector and one can describe its properties in a statistical sense. The following properties [9, 10] are useful in characterizing estimates.

1. An estimate $\hat{x}(k)$ is said to be *unbiased* if

   $$E[\hat{x}(k) - x] = 0$$

   which means that the estimation error has mean zero. Otherwise, it is said to be *biased*. An estimate is also said to be *asymptotically unbiased* if

   $$\lim_{k\to\infty} E[\hat{x}(k) - x] = 0$$

2. An estimate $\hat{x}(k)$ is said to be *consistent* if

   $$\lim_{k\to\infty} P[\|\hat{x}(k) - x\| \geqslant \varepsilon] = 0 \text{ for all } \varepsilon > 0$$

   The sequence of estimates $\{\hat{x}(k)\}$ is said to converge *in probability* to the parameter x (cf. Section B.1.16). We can also define an estimate to be consistent if

   $$\lim_{k\to\infty} E[\|\hat{x}(k) - x\|^2] = 0$$

   This is called convergence in mean square.
3. An unbiased estimate $\hat{x}_1(k)$ is said to be *efficient* with respect to another unbiased estimate $\hat{x}_2(k)$ if

   $$E\{[x - \hat{x}_1(k)][x - \hat{x}_1(k)]^{\mathrm{T}}\} \leqslant E\{[x - \hat{x}_2(k)][x - \hat{x}_2(k)]^{\mathrm{T}}\}$$

Efficient estimators are usually discussed in terms of the *Cramer–Rao inequality*. This inequality gives a lower bound for the error covariance of an estimator.

4. In the estimation of a nonrandom parameter x with an unbiased estimator $\hat{x}(k)$, the covariance of the error in the estimator is bounded below by the *Fisher information matrix* $\mathcal{J}$ (see also Exercise 2.7):

$$E\{[x - \hat{x}(k)][x - \hat{x}(k)]^{\mathrm{T}}|x\} \geqslant \mathcal{J}^{-1}$$

where

$$\mathcal{J} \triangleq E\left\{\left[\frac{\partial}{\partial x}\ln p(Z_k|x)\right]^{\mathrm{T}}\left[\frac{\partial}{\partial x}\ln p(Z_k|x)\right]\middle| x\right\}$$

$$= -E\left\{\frac{\partial^2}{\partial x^2}\ln p(Z_k|x)\middle| x\right\}$$

where $p(Z_k|x)$ is the likelihood function. Equality holds in the above equation if and only if

$$\left[\frac{\partial}{\partial x}\ln p(Z_k|x)\right]^{\mathrm{T}} = C(x)[x - \hat{x}(k)]$$

where $C(x)$ is a matrix whose elements can be a function of x.

5. For a random parameter x estimated by an unbiased estimator $\hat{x}(k)$, the covariance is bounded below by a similar expression. Let a known joint probability density function of x and Z_k be denoted by $p(x, Z_k)$. Then

$$E\{[x - \hat{x}(k)][x - \hat{x}(k)]^{\mathrm{T}}|x\} \geqslant \mathcal{J}^{-1}$$

where

$$\mathcal{J} \triangleq E\left\{\left[\frac{\partial}{\partial x}\ln p(x, Z_k)\right]^{\mathrm{T}}\left[\frac{\partial}{\partial x}\ln p(x, Z_k)\right]\right\}$$

$$= -E\left\{\frac{\partial^2}{\partial x^2}\ln p(x, Z_k)\right\}$$

Equality holds if and only if

$$\left[\frac{\partial}{\partial x}\ln p(x, Z_k)\right]^{\mathrm{T}} = C[x - \hat{x}(k)]$$

where C is independent of x and Z_k. The lower bound is achieved with equality if and only if $p(x|Z_k)$ is Gaussian.

B.4.2 Maximum Likelihood Estimation

In this approach, there is no a priori probability density function $p(x)$ and the a posteriori probability density function $p(x|Z_k)$ cannot be defined. Instead, we can define the *likelihood function* $p(Z_k|x)$. It is reasonable to choose as an estimate of x that value $\hat{x}_{\mathrm{ML}}(k)$ which maximizes $p(Z_k|x)$. A necessary condition for $\hat{x}_{\mathrm{ML}}(k)$ to maximize the likelihood function is

$$\frac{\partial}{\partial x}\, p(Z_k|x) = 0$$

or equivalently

$$\frac{\partial}{\partial x} \ln p(Z_k|x) = 0$$

The above two equations are sometimes called the *likelihood equation* and *log likelihood equation*.

The following facts are known for the maximum likelihood estimator [9]:

1. A maximum likelihood estimator $\hat{x}_{\mathrm{ML}}(k)$ is a consistent estimator.
2. A maximum likelihood estimator $\hat{x}_{\mathrm{ML}}(k)$ is asymptotically Gaussian with mean x and covariance $(k\mathcal{J})^{-1}$.
3. A maximum likelihood estimator $\hat{x}_{\mathrm{ML}}(k)$ is asymptotically efficient in the sense that the Cramer–Rao lower bound is achieved asymptotically.

B.4.3 Bayesian Estimation

For a random parameter, we can have an a priori probability density function $p(x)$, and apply a procedure known as *Bayesian estimation*. An estimate $\hat{x}(k)$ is then chosen to minimize the Bayes' risk $\mathcal{B}$ in estimation [11] (see Chapter 11 for the Bayes' risk in M-ary hypothesis testing) which is defined as the expected value of the cost function $C(\tilde{x})$, where $\tilde{x} \triangleq x - \hat{x}(k)$. That is, the Bayes' risk can be expressed as

$$\begin{aligned}
\mathcal{B} &= E[C(x - \hat{x}(k))] \\
&= \int_{-\infty}^{\infty}\int_{-\infty}^{\infty} C(\tilde{x})p(x,\, Z_k)\,\mathrm{d}x\,\mathrm{d}Z_k \\
&= \int_{-\infty}^{\infty} I(\hat{x}(k))p(Z_k)\,\mathrm{d}Z_k
\end{aligned}$$

where

$$I(\hat{x}(k)) \triangleq \int_{-\infty}^{\infty} C(\tilde{x})p(x|Z_k)\,\mathrm{d}x$$

If $C(\tilde{x})$ is assumed to be nonnegative, then $I(\hat{x}(k))$ is also nonnegative, and minimizing $I(\hat{x}(k))$ is equivalent to minimizing the risk.

The optimal estimate $\hat{x}(k)$ which minimizes the risk will clearly depend on the cost function assignment. The two cases are given in the following subsections.

B.4.4 Minimum Mean-Square Error Estimate

The first cost assignment is the quadratic- (or squared-)error cost function shown in Figure B.1. For this case, $I(\hat{x}(k))$ becomes

$$I(\hat{x}(k)) = \int_{-\infty}^{\infty} \|x - \hat{x}(k)\|^2 p(x|Z_k)\,\mathrm{d}x$$

which can also be interpreted as the *mean-square error* (MSE), i.e. $I(\hat{x}(k)) = E[\|x - \hat{x}(k)\|^2 | Z_k]$. Minimizing $I(\hat{x}(k))$ above gives the desired estimate as

$$\hat{x}_{\text{MMSE}}(k) = \int_{-\infty}^{\infty} xp(x|Z_k)\,\mathrm{d}x$$

which is just the mean of the a posteriori probability density function $p(x|Z_k)$. This estimate is called the *minimum mean-square error* (MMSE) or *minimum variance* (MV) estimate.

When ϕ is defined as any function of the data Z_k, the error in the estimate $\hat{x}_{\text{MMSE}}(k)$ is orthogonal to ϕ in the sense that

$$E\{[x - \hat{x}_{\text{MMSE}}(k)]\phi^{\mathrm{T}}\} = 0$$

which is called the *orthogonal projection lemma*.

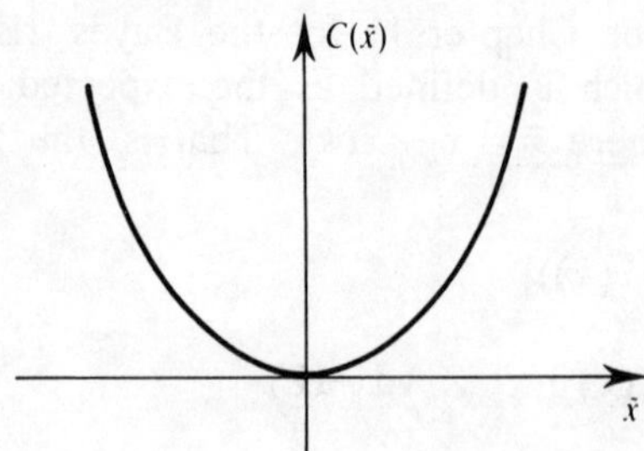

Figure B.1 A quadratic-error cost function for a scalar parameter.

B.4.5 Maximum A Posteriori Estimate

Another cost assignment is the uniform cost function depicted in Figure B.2. This cost function assigns zero cost to all estimates that are within small errors and assigns $1/\varepsilon$ cost to all others. For the uniform cost function, $I(\hat{x}(k))$ becomes

$$I(\hat{x}(k)) = 1 - \frac{1}{\varepsilon} \int_{\hat{x}-\varepsilon/2}^{\hat{x}+\varepsilon/2} p(x|Z_k)\,dx$$

The most important case occurs when we let ε approach zero ($\varepsilon \geqslant 0$) so that $C(\tilde{x}) = \delta(\|x - \hat{x}\|)$, which is the Dirac delta function. Thus, to minimize $I(\hat{x}(k))$, one should choose the value $\hat{x}(k)$ that maximizes the a posteriori probability density function $p(x|Z_k)$. We call this estimate the *maximum a posteriori* estimate, $\hat{x}_{\mathrm{MAP}}(k)$, which is defined by

$$\frac{\partial}{\partial x} p(x|Z_k)\bigg|_{x=\hat{x}_{\mathrm{MAP}}(k)} = 0$$

We can also obtain $\hat{x}_{\mathrm{MAP}}(k)$ by maximizing $\ln p(x|Z_k)$ with respect to x. Using Bayes' rule and taking the logarithm of the a posteriori probability density, we have

$$\ln p(x|Z_k) = \ln p(Z_k|x) + \ln p(x) - \ln p(Z_k)$$

Since $p(Z_k)$ does not contain terms in x, maximizing $p(x|Z_k)$ is equivalent to maximizing $p(x, Z_k)$. Thus, the MAP estimate can be obtained from

$$\frac{\partial}{\partial x} \ln p(Z_k|x) + \frac{\partial}{\partial x} \ln p(x) = 0$$

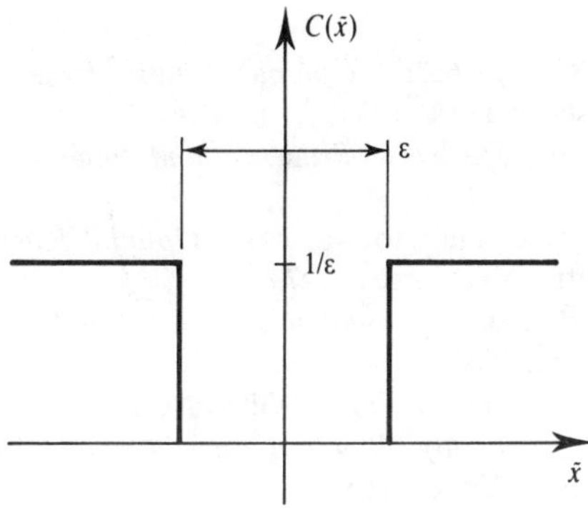

Figure B.2 A uniform-error cost function for a scalar parameter.

or

$$\frac{\partial}{\partial x} p(x, Z_k) = \frac{\partial}{\partial x} [p(Z_k|x)p(x)] = 0$$

Note here that the a posteriori probability density function $p(x|Z_k)$ is sometimes called an *unconditional* likelihood function, because of the random nature of x. On the other hand, $p(Z_k|x)$ is sometimes called a *conditional* likelihood function.

B.4.6. Least-Squares Estimation

Another common estimation approach for nonrandom parameters is the *least-squares* (LS) method. For observations

$$z(j) = h(j, x) + w(j), \qquad j = 1, 2, \ldots$$

the least-squares estimate of x at time k is that which minimizes the sum of the squared errors

$$\frac{\partial}{\partial x} \left\{ \frac{1}{2} \sum_{i=1}^{k} [z(i) - h(i, x)]^{\mathrm{T}} [z(i) - h(i, x)] \right\} \bigg|_{x=\hat{x}_{\mathrm{LS}}(k)} = 0$$

The minimizing criterion above makes no assumptions about the noise $w(j)$. With an additional assumption on the noise $w(j)$ that is subject to $w(j) \sim N(0, R)$, we obtain several properties for linear $\hat{x}_{\mathrm{LS}}(k)$ (see Sorenson [9] and Exercise 2.9).

REFERENCES

[1] MEDITCH, J. S., *Stochastic Optimal Linear Estimation and Control*, McGraw-Hill, New York, 1969.

[2] JAZWINSKI, A. H., *Stochastic Processes and Filtering Theory*, Academic Press, New York, 1970.

[3] ANDERSON, B. D. O. and MOORE, J. B., *Optimal Filtering*, Prentice Hall, Englewood Cliffs, New Jersey, 1979.

[4] CHEN, H. F., *Recursive Estimation and Control for Stochastic Systems*, Wiley, New York, 1985.

[5] DOOB, J. L., *Stochastic Processes*, Wiley, New York, 1952.

[6] PAPOULIS, A., *Probability, Random Variables, and Stochastic Processes*, McGraw-Hill, New York, 1965.

[7] FELLER, W., *An Introduction to Probability Theory and Its Applications*, Wiley, New York, 1968.

[8] WONG, E. and HAJEK, B., *Stochastic Processes in Engineering Systems*, Springer-Verlag, New York, 1985.

[9] SORENSON, H. W., *Parameter Estimation*, Marcel Dekker, New York, 1980.

[10] THERRIEN, C. W., *Decision, Estimation and Classification*, Wiley, New York, 1989.

[11] SAGE, A. P. and MELSA, J. L., *Estimation Theory with Applications to Communications and Control*, Robert E. Krieger, New York, 1979.

APPENDIX C

Brief Review of Linear System Theory

The purpose of this appendix is to provide several results of linear system theory [1–4] that are used frequently in this text. For more extensive treatments, standard textbooks by Casti [5] and Kailath [6] should be consulted.

C.1 CONTINUOUS-TIME LINEAR SYSTEM MODELS

Consider the following state-space model

$$\dot{x}(t) = A(t)x(t) + B(t)w(t), \qquad x(t_0) = x_0 \tag{C.1}$$

where $x \in \mathbb{R}^n$ and $w(t) \in \mathbb{R}^p$. The solution of equation (C.1) is

$$x(t) = \Phi(t, t_0)x_0 + \int_{t_0}^{t} \Phi(t, \tau)B(\tau)w(\tau)\,\mathrm{d}\tau$$

where the (state) transition matrix $\Phi(t, t_0)$ is the solution of the matrix differential equation

$$\mathrm{d}\Phi(t, t_0)/\mathrm{d}t = A(t)\Phi(t, t_0), \qquad \Phi(t_0, t_0) = I,\ t \geqslant t_0$$

which has the properties

$$\Phi(t_2, t_1) = \Phi(t_2, \tau)\Phi(\tau, t_1), \text{ for all } t_1, t_2, \tau \geqslant t_0$$

$$\Phi(t_2, t_1) = \Phi^{-1}(t_1, t_2), \text{ for all } t_1, t_2 \geqslant t_0$$

If A is constant, the solution of $x(t)$ can be rewritten as

$$x(t) = \exp[A(t - \tau)]x_0 + \int_{t_0}^{t} \exp[A(t - \tau)]B(\tau)w(\tau)\,\mathrm{d}\tau$$

The matrix differential equation

$$\mathrm{d}X(t)/\mathrm{d}t = AX(t) + X(t)B + C(t), \qquad X(t_0) = X_0$$

also occurs commonly. The solution of this equation may be represented by

$$X(t) = \exp[A(t - t_0)]X_0 \exp[B(t - t_0)] + \int_{t_0}^{t} \exp[A(t - \tau)]C(\tau)\exp[B(t - \tau)]\,d\tau$$

C.2 SAMPLING OF A CONTINUOUS-TIME LINEAR SYSTEM

Consider again the continuous-time system given by equation (C.1). Then

$$x(kT + T) = \Phi(kT + T, kT)x(kT) + \int_{kT}^{kT+T} \Phi(kT + T, \tau)B(\tau)w(\tau)\,d\tau$$

or

$$x(k + 1) = \Phi(k + 1, k)x(k) + w(k) \tag{C.2}$$

with obvious definitions. If w is a white noise process, and $E[w(t)w^{\mathrm{T}}(\tau)] = Q(t)\delta(t - \tau)$, then

$$E[w(k)w^{\mathrm{T}}(j)] = \delta_{kj} \int_{kT}^{kT+T} \Phi(kT + T, \tau)B(\tau)Q(\tau)B^{\mathrm{T}}(\tau) \times \Phi^{\mathrm{T}}(kT + T, \tau)\,d\tau$$

Consider the following observation system

$$z(t) = H(t)x(t) + v(t) \tag{C.3}$$

where $z(t), v(t) \in \mathbb{R}^m$ and $E[v(t)v^{\mathrm{T}}(s)] = R(t)\delta(t - s)$. A sampling process is usually defined by

$$z(k) = \frac{1}{\Delta}\int_{kT}^{kT+\Delta} z(t)\,dt$$

The result is

$$z(k) = H(k)x(k) + v(k)$$

where

$$H(k) \triangleq \frac{1}{\Delta}\int_{kT}^{kT+\Delta} H(t)\,dt$$

$$v(k) \triangleq \frac{1}{\Delta}\int_{kT}^{kT+\Delta} \left[H(t)\int_{T}^{t}\Phi(t, \tau)B(\tau)w(\tau)\,d\tau + v(t)\right]dt$$

If $w(t)$ and $v(t)$ are independent white noise processes, then we find as $\Delta \to 0$ that $\{w(k)\}$ and $\{v(k)\}$ also approach independence, while

$$E[v(k)v^{\mathrm{T}}(k)] \to \frac{1}{\Delta^2} \int_{kT}^{kT+\Delta} R(\tau)\,\mathrm{d}\tau.$$

C.3 DISCRETE-TIME LINEAR SYSTEM MODELS

Consider a discrete-time state-space model

$$x(k+1) = \Phi(k+1, k)x(k) + B(k)w(k),$$
$$x(0) = x_0,\ k = 0, 1, 2, \ldots \qquad \text{(C.4)}$$

The solution of (C.4) is

$$x(k) = \Phi(k, 0)x_0 + \sum_{i=0}^{k-1} \Phi(k, i+1)B(i)w(i)$$

where

$$\Phi(k, i) = \Phi(k, k-1)\Phi(k-1, k-2)\ldots$$
$$\Phi(i+1, i) \qquad (k > i),\ \Phi(k, k) = I$$

which constitutes the transition matrix for

$$x(k+1) = \Phi(k+1, k)x(k)$$

Here, the following properties are held

$$\Phi(k, i) = \Phi(k, l)\Phi(l, i)$$
$$\Phi(k, i) = \Phi^{-1}(i, k) \text{ for all } k,\ l \text{ and } i \text{ with } k \geqslant l \geqslant i$$

When Φ is time-invariant,

$$x(k) = \Phi^k x_0 + \sum_{i=0}^{k-1} \Phi^{k-i-1} B(i)w(i)$$

The matrix difference equation in discrete time, which is called the *Lyapunov equation*,

$$X(k+1) = \Phi X(k)\Phi^{\mathrm{T}} + C(k), \qquad X(0) = X_0$$

can also be written as follows:

$$X(k) = \Phi^k X_0 \Phi^{\mathrm{T}^k} + \sum_{i=0}^{k-1} \Phi^{k-i-1} C(i) \Phi^{\mathrm{T}^{k-i-1}}$$

It is easy to prove that the solution of the following algebraic Lyapunov equation (ALE):

$$X = \Phi X \Phi^{\mathrm{T}} + C$$

can be obtained by

$$X = \sum_{l=0}^{\infty} \Phi^{l} C \Phi^{\mathrm{T}^{l}}$$

if Φ is asymptotically stable.

C.4 REACHABILITY AND OBSERVABILITY FOR CONTINUOUS-TIME SYSTEMS

Consider the following continuous-time system

$$\dot{x}(t) = A(t)x(t) + G(t)u(t) \tag{C.5}$$

$$z(t) = H(t)x(t) \tag{C.6}$$

where $x(t) \in \mathbb{R}^n$, $u(t) \in \mathbb{R}^r$ and $z(t) \in \mathbb{R}^m$.

C.4.1 Reachability, Controllability and Stabilizability

Definition C.1

The state of a continuous-time system is said to be *reachable* (from the zero state) at time t if there exists a $\tau \leqslant t$ and an input $u(t)$ which transfers the zero state at time τ to the state x at time t.

Definition C.2

The continuous-time system given by equation (C.5) is said to be *controllable* (to the zero state) if, given any initial state $x(\tau)$, there exists a $t \geqslant \tau$ and a $u(t)$ such that $x(t) = 0$.

These concepts of reachability and controllability are depicted in Figures C.1 and C.2, respectively.

Theorem C.1

A necessary and sufficient condition for system reachability (controllability) at time τ is that $W(\tau, t)$ $(W(s, \tau))$ is positive definite for

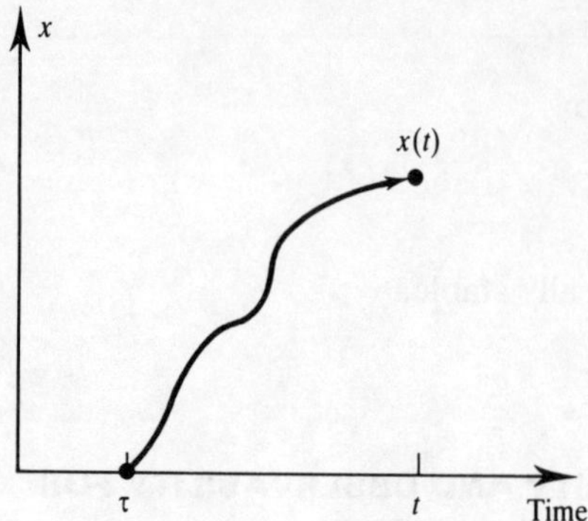

Figure C.1 Reachability.

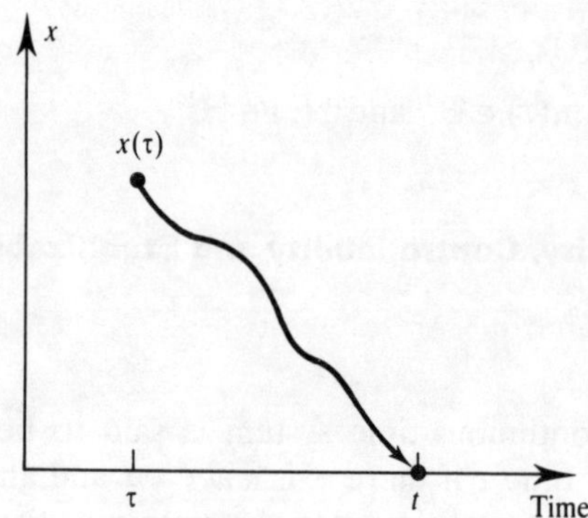

Figure C.2 Controllability.

some $t \geqslant \tau$ $(s \leqslant \tau)$, where W is an $n \times n$ Gramian matrix given by

$$W(\tau, t) = \int_{\tau}^{t} \Phi(\tau, \lambda) G(\lambda) G^{\mathrm{T}}(\lambda) \Phi^{\mathrm{T}}(\tau, \lambda) \, \mathrm{d}\lambda$$

A stronger type of reachability (controllability) known as *uniform* reachability (uniform controllability) holds if and only if for some $\sigma > 0$ there exists a positive constant α_1 such that $W(\tau, \tau + \sigma) \geqslant \alpha_1$ for all τ.

Controllability and reachability are entirely different concepts. However, they coincide in special cases, one of which is when the system is linear and time-invariant.

Theorem C.2

A linear time-invariant system, i.e. $A(t) = A$ and $G(t) = G$, is completely reachable if and only if it is completely controllable.

Theorem C.3

A linear time-invariant system or the pair (A, G) is completely reachable (completely controllable) if and only if

$$\operatorname{rank} W = n$$

where $W \triangleq [G, AG, \ldots, A^{n-1}G]$ and the matrix W is called the reachability (controllability) matrix.

That the controllability matrix W has rank n implies that there are n linearly independent columns in W. It may be that we can find n such columns in the *partial* controllability matrix

$$W_j = [G, AG, \ldots, A^{j-1}G], \qquad 1 \leqslant j \leqslant n$$

and the smallest integer j such that W_j has rank n is called the *controllability index* of (A, G).

The condition for controllability can be weakened to that of *stabilizability*.

Definition C.3

A system (A, G) is *stabilizable* if there exists an $r \times n$ matrix K such that the matrix $A + GK$ is asymptotically stable, i.e. that for all eigenvalues, λ_i, of the matrix $A + GK$ the inequality $\mathrm{Re}(\lambda_i) < 0$ holds for $i = 1, \ldots, n$.

C.4.2 Observability, Reconstructibility and Detectability

Definition C.4

The state $x(\tau)$ of the system given by equations (C.5) and (C.6) is *observable* if it can be determined by means of the future values $z(t)$, $t > \tau$, of the output variable and if the interval $t - \tau$ is finite.

Definition C.5

The state $x(\tau)$ of the system given by equations (C.5) and (C.6) is *reconstructible* if it can be determined by means of the past values $z(s)$, $s < \tau$, of the output variable and if the interval $\tau - s$ is finite.

Theorem C.4

The system given by equations (C.5) and (C.6) or the pair $(H(t), A(t))$ is completely observable (completely state-reconstructible) if and only if $M(\tau, t)$ $(M(s, \tau))$ is positive definite for $t \geqslant \tau$ $(s \leqslant \tau)$ where the Gramian matrix $M(\tau, t)$ is defined by

$$M(\tau, t) = \int_{\tau}^{t} \Phi^{\mathrm{T}}(\sigma, \tau) H^{\mathrm{T}}(\sigma) H(\sigma) \Phi(\sigma, \tau)\, \mathrm{d}\sigma$$

A stronger type of observability (state reconstructibility) is obtained by the following theorem.

Theorem C.5

The linear system given by equations (C.5) and (C.6) or the pair $(H(t), A(t))$ is uniformly completely observable (uniformly completely state-reconstructible) if and only if for some $\sigma > 0$ there exist positive constants α_i, $i = 1, 2$ $(\alpha_i, i = 3, 4)$, such that

$$0 < \alpha_1 I \leqslant M(\tau, \tau + \sigma) \leqslant \alpha_2 I$$

$$(0 < \alpha_3 I \leqslant M(\tau - \sigma, \tau) \leqslant \alpha_4 I)$$

hold for all τ.

In the time-invariant case, state reconstructibility is equivalent to observability.

Theorem C.6

The linear time-invariant system or the pair (H, A) is completely observable if and only if it is completely state-reconstructible.

Theorem C.7

The linear time-invariant system or the pair (H, A) is completely observable (completely state-reconstructible) if and only if

$$\text{rank } M = n$$

where $M \triangleq [H^{\mathrm{T}} \, A^{\mathrm{T}} H^{\mathrm{T}} \ldots (A^{\mathrm{T}})^{n-1} H^{\mathrm{T}}]$ and the matrix M is called the observability (reconstructibility) matrix.

The *observability index* of (H, A) can be defined in a similar way to the controllability index, as the smallest integer j such that

$$M_j = [H^{\mathrm{T}} \; A^{\mathrm{T}} H^{\mathrm{T}} \ldots (A^{\mathrm{T}})^{j-1} H^{\mathrm{T}}], \qquad 1 \leqslant j \leqslant n$$

has full rank n.

The condition for observability can be weakened to that of detectability.

Definition C.6

The pair (H, A) is *detectable* if there exists an $n \times m$ matrix R such that $A + RH$ is asymptotically stable, i.e. for all eigenvalues λ_i of the matrix $A + RH$ the inequality $\mathrm{Re}(\lambda_i) < 0$ holds for $i = 1, \ldots, n$.

C.5 REACHABILITY AND OBSERVABILITY FOR DISCRETE-TIME SYSTEMS

Suppose that

$$x(k+1) = \Phi x(k) + G u(k), \qquad x(0) = x_0 \tag{C.7}$$

$$z(k) = H x(k) \tag{C.8}$$

where $x(k) \in \mathbb{R}^n$, $u(k) \in \mathbb{R}^r$ and $z(k) \in \mathbb{R}^m$. The definitions of reachability, reconstructibility and stabilizability are the same as those defined for continuous-time systems.

C.5.1 Complete Reachability

The pair (Φ, G) is completely reachable if any of the following equivalent conditions holds.

1. rank $W = n$, where $W \triangleq [G \quad \Phi G \ldots \Phi^{n-1} G]$.
2. $y^{\mathrm{T}} \Phi^i G = 0$ for $i = 0, 1, \ldots, n-1$ implies $y = 0$.
3. $y^{\mathrm{T}} G = 0$ and $y^{\mathrm{T}} \Phi = \lambda y^{\mathrm{T}}$ for some constant λ implies $y = 0$.
4. There exists an $r \times n$ matrix K such that the eigenvalues of $\Phi + GK$ can take arbitrary value.
5. There exists $\{u(k)\}$ for $k \in [0, n-1]$ such that $x(n)$ takes an arbitrary value, provided that $x_0 = 0$ in equation (C.7).

If the pair (Φ, G) is not completely reachable and rank $W = n_{\mathrm{r}}$, there exists a nonsingular matrix T such that

$$T\Phi T^{-1} = \begin{matrix} n_{\mathrm{r}} \\ n - n_{\mathrm{r}} \end{matrix} \overset{\begin{matrix} n_{\mathrm{r}} & \quad n - n_{\mathrm{r}} \end{matrix}}{\begin{bmatrix} \Phi_{11} & \Phi_{12} \\ 0 & \Phi_{22} \end{bmatrix}}, \qquad TG = \begin{matrix} n_{\mathrm{r}} \\ n - n_{\mathrm{r}} \end{matrix} \begin{pmatrix} G_1 \\ 0 \end{pmatrix}$$

with (Φ_{11}, G_1) completely reachable.

C.5.2 Complete Controllability

The pair (Φ, G) is completely controllable if any of the following equivalent conditions holds.

1. Range $\Phi^n \subset$ Range $[G \quad \Phi G \ldots \Phi^{n-1} G]$.
2. $y^{\mathrm{T}} \Phi^i G = 0$ for $i = 0, 1, \ldots, n-1$ implies $y^{\mathrm{T}} \Phi^n = 0$.
3. $y^{\mathrm{T}} G = 0$ and $y^{\mathrm{T}} \Phi = 0 = \lambda y^{\mathrm{T}}$ implies $\lambda = 0$ or $y = 0$.
4. There exists $\{u(k)\}$ for $k \in [0, n-1]$ such that $x(n) = 0$, given an arbitrary x_0 in equation (C.7).

C.5.3 Complete Stabilizability

The pair (Φ, G) is completely stabilizable if any of the following equivalent conditions holds.

1. $y^{\mathrm{T}} G = 0$ and $y^{\mathrm{T}} \Phi = \lambda y^{\mathrm{T}}$ for some constant λ implies $|\lambda| < 1$ or $y = 0$.
2. There exists an $r \times n$ matrix K such that $|\lambda_i(\Phi + GK)| < 1$ for all i.
3. If

$$T\Phi T^{-1} = \begin{pmatrix} \Phi_{11} & \Phi_{12} \\ 0 & \Phi_{22} \end{pmatrix}, \qquad G = \begin{pmatrix} G_1 \\ 0 \end{pmatrix}$$

with (Φ_{11}, G_1) completely controllable, then $|\lambda_i(\Phi_{22})| < 1$.

If there exists a vector $y \neq 0$ such that $y^{\mathrm{T}}G = 0$ and $y^{\mathrm{T}}\Phi = \lambda y^{\mathrm{T}}$, then λ is said to be an *uncontrollable eigenvalue* (or *mode*) of the pair (Φ, G). On the other hand, λ is said to be an *unobservable eigenvalue* (or *mode*) of the pair (H, Φ), if there exists a vector $x \neq 0$ such that $Hx = 0$ and $\Phi x = \lambda x$. Clearly, the pair $(\Phi, G)[(H, \Phi)]$ is stabilizable [detectable] if and only if the uncontrollable [unobservable] modes are stable.

C.5.4 Complete Observability, Reconstructibility, and Detectability

From the principle of *duality* [7], we can say that the pair (H, Φ) is completely observable, reconstructible, or detectable according as $(\Phi^{\mathrm{T}}, H^{\mathrm{T}})$ is completely reachable, controllable, or stabilizable, respectively.

C.6 CANONICAL DECOMPOSITION OF LINEAR SYSTEMS

Time-invariant linear systems having a finite number of independent output variables and a finite order of individual subsystems may be decomposed into four partial systems [8, 9]: reachable and observable partial systems; reachable and unobservable partial systems; unreachable and observable partial systems; and unreachable and unobservable partial systems.

Theorem C.8

The realization of a dynamic system is minimum if and only if it is reachable and observable.

REFERENCES

[1] JAZWINSKI, A. H., *Stochastic Processes and Filtering Theory*, Academic Press, New York, 1970.

[2] ANDERSON, B. D. O. and MOORE, J. B., *Optimal Filtering*, Prentice Hall, Englewood Cliffs, New Jersey, 1979.

[3] STREJC, V., *State Space Theory of Discrete Linear Control*, Wiley, New York, 1981.

[4] O'REILLY, J., *Observers for Linear Systems*, Academic Press, New York, 1983.
[5] CASTI, J. L., *Dynamical Systems and Their Applications: Linear Theory*, Academic Press, New York, 1977.
[6] KAILATH, T., *Linear Systems*, Prentice Hall, Englewood Cliffs, New Jersey, 1980.
[7] KALMAN, R. E., FALB, P. L. and ARBIB, M. A., *Topics in Mathematical Systems Theory*, McGraw-Hill, New York, 1969.
[8] GILBERT, E. G., Controllability and Observability in Mathematical Control Systems, *SIAM J. Control*, vol. 1, 1963, pp. 128–51.
[9] KALMAN, R. E., Mathematical Description of Linear Dynamical Systems, *SIAM J. Control*, vol. 1, 1963, pp. 152–92.

APPENDIX D

Brief Review of Stability Theory

D.1 LOCAL EQUILIBRIUM STABILITY CONDITIONS

The various definitions of stability can be broadly classified as those which deal with the equilibrium of the null solution of free or unforced systems and those which consider the dynamic response of systems subject to various classes of forcing functions or inputs [1–7].

Consider the general vector set of homogeneous differential equations,

$$\dot{x}(t) = f(t, x(t)),\ x(t_0) = x_0,\ f(t, 0) = 0 \text{ for all } t \tag{D.1}$$

where $x(t) \in \mathbb{R}^n$, $t \in [0, \infty)$, the function f is sufficiently smooth, and it is assumed that the equilibrium state x_e of equations (D.1) can always be set to zero, so that the equilibrium state x_e and the null solution to equations (D.1) are considered throughout as equivalent.

Definition D.1

The zero solution $x(t; x_0, t_0) = 0$ for all $t \geqslant t_0$ of the differential equation $\dot{x}(t) = f(t, x(t))$ is said to be *Lyapunov stable* if and only if for each $t_0 > 0$ and each $\xi > 0$ there is a $\delta(\xi, t_0) > 0$ such that $\|x_0\| < \delta(\xi, t_0)$ implies that $\|x(t; x_0, t_0)\| < \xi$ for all $t \geqslant t_0$. Otherwise, it is called *unstable*.

A simplified schematic illustration is given in Figure D.1. If the equations (D.1) are linear, i.e.

$$\dot{x}(t) = A(t)x(t), \qquad x(t_0) = x_0 \tag{D.2}$$

then the conditions of Definition D.1 can be simplified.

Definition D.2

The zero solution $x(t; x_0, t_0) = 0$ for all $t \geqslant t_0$ of the linear system

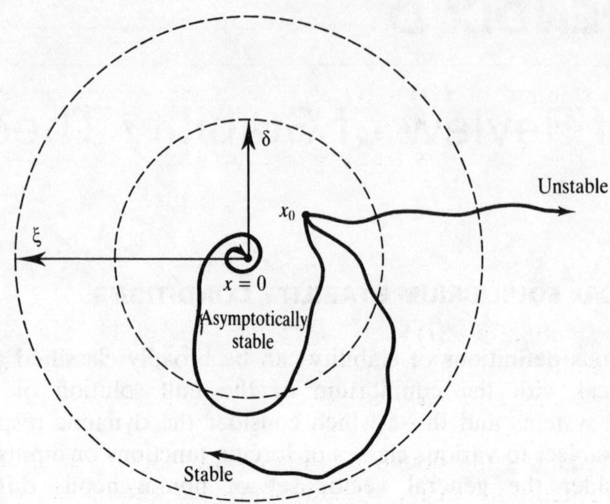

Figure D.1 Stability.

(D.2) is called stable *sensu* Lyapunov if and only if for each $t_0 \geqslant 0$ there is a finite constant $N(t_0)$ such that $\|x(t; x_0, t_0)\| \leqslant N(t_0)\|x_0\|$ for all $t \geqslant t_0$.

Theorem D.1

The zero solution of $\dot{x}(t) = A(t)x(t)$ is stable *sensu* Lyapunov if there is a constant N (which may depend on t_0) such that $\|\Phi(t, t_0)\| \leqslant N(t_0)$ for all $t \geqslant t_0$, where $\Phi(t, t_0)$ is the state transition matrix of system (D.2).

For nonstationary systems it is crucial to distinguish between *uniform* and *nonuniform* stability properties. The zero solution to equations (D.1) is *uniformly stable* if the δ in Definition D.1 is independent of t_0. Therefore, if the N of Theorem D.1 is independent of t_0, and $\|\Phi(t, t_0)\| \leqslant N$ for all $t \geqslant t_0$, then the stability of the linear system (D.2) is uniform.

If the upper bound $N(t_0)\|x_0\|$ in Definition D.2 is replaced by a bound $B(x_0, t_0)$ which may depend on each solution, we then have bounded or Lagrange stability. In addition, if this bound B is independent of t_0 the null solution is uniformly bounded, which is also equivalent to uniform stability for linear systems.

Definition D.3

The zero solution of $\dot{x}(t) = f(t, x(t))$ is *asymptotically stable* if it is stable and if there exists a $\delta(t_0) > 0$ such that $\|x_0\| < \delta(t_0)$ implies that $\|x(t; x_0, t_0)\| \to 0$ as $t \to \infty$ (see also Figure D.1).

Theorem D.2

The zero solution of $\dot{x}(t) = A(t)x(t)$ is asymptotically stable if there exists a finite number α (which may depend on t_0) such that $\|\Phi(t, t_0)\| \leqslant \alpha$ for all $t \geqslant t_0$ and $\|\Phi(t, t_0)\| \to 0$ as $t \to \infty$.

If in addition the bound α is independent of t_0 and the above limit holds uniformly in t_0, or equivalently for some positive scalar ξ there is a $T(\xi)$ which is independent of t_0 such that $\|\Phi(t, t_0)\| < \xi$ for $t_0 + T(\xi) \leqslant t$, then the linear system (D.2) is *uniformly asymptotically stable*.

The stability properties for the linear homogeneous system (D.2) can be summarized completely in terms of its state transition matrix $\Phi(t, t_0)$.

Theorem D.3

The zero solution to system (D.2), i.e. the equilibrium state x_e of (D.2), is:

(i) stable if and only if there exists a constant $N > 0$ such that

$$\|\Phi(t, t_0)\| \leqslant N \text{ for all } t \geqslant t_0$$

(ii) uniformly stable if and only if there exists a constant $N > 0$ such that $\|\Phi(t, t_0)\Phi^{-1}(s, t_0)\| \leqslant N$ for $t_0 \leqslant s < t < \infty$

(iii) asymptotically stable if and only if $\|\Phi(t, t_0)\| \to 0$ as $t \to \infty$

(iv) uniformly asymptotically stable if there exist positive constants N, α such that $\|\Phi(t, t_0)\Phi^{-1}(s, t_0)\| \leqslant N \exp[-\alpha(t - s)]$ for $t_0 \leqslant s \leqslant t < \infty$.

An equivalent condition to condition (iv) for uniformly asymptotic stability of system (D.2) is if there exist positive constants N_1, α_1 such that $\|\Phi(t, t_0)\| \leqslant N_1 \exp[-\alpha_1(t - s)]$ for $t_0 \leqslant s \leqslant t < \infty$. This inequality is a form of *exponential stability* and is clearly a sufficient condition for uniform asymptotic stability.

Theorem D.4

The null solution to the linear time-invariant system

$$\dot{x}(t) = Ax(t) \tag{D.3}$$

is

(i) stable if and only if every characteristic value of A has real part no greater than zero, that is $\mathrm{Re}(\lambda_i) \leq 0$ for all i, and those with $\mathrm{Re}(\lambda_i) = 0$ are distinct
(ii) asymptotically stable if and only if every characteristic value of A has negative real part, i.e. $\mathrm{Re}(\lambda_i) < 0$ for all i.

D.2 STABILITY IN THE LARGE

The properties of stability and asymptotic stability are local; that is, it is only known that there exists some region in the state space surrounding the equilibrium state such that all solutions initiating in that region are stable or asymptotically stable. In the latter case, the region of validity of convergence is called the *domain of attraction*. If the domain of attraction is the whole state space, then any motion returns to the equilibrium whatever the initial perturbation may be.

Definition D.4

The zero solution of equations (D.1), i.e. the equilibrium state x_e, is called *asymptotically stable in the large*, or *globally asymptotically stable*, if it is stable and *every* solution of equations (D.1) tends to zero as $t \to \infty$.

Definition D.5

The zero solution of equations (D.1) is *exponentially asymptotically stable in the large* if there exists an $\alpha > 0$ and for any $\beta > 0$ there exists an $N(\beta) > 0$ such that if $\|x_0\| \leqslant \beta$ then $\|x(t; x_0, t_0)\| \leqslant N(\beta) \exp[-\alpha(t - t_0)]\|x_0\|$ for all $t \geqslant t_0$.

Theorem D.5

If the zero solution of the linear system (D.2) is asymptotically stable, it is asymptotically stable in the large. Morcover, if the zero solution of system (D.2) is uniformly asymptotically stable, it is exponentially asymptotically stable in the large. In this case, we can find an $N > 0$ independent of β.

D.3 SOME RESULTS DUE TO THE SECOND METHOD OF LYAPUNOV

Theorem D.6

Consider the nonlinear system (D.1). Suppose there exists a scalar function $V(x, t)$ with continuous first partial derivatives with respect to x and t such that $V(0, t) = 0$ and

(i) $V(x, t)$ is positive definite, i.e. there exists a continuous, nondecreasing scalar function α such that $\alpha(0) = 0$ and, for all t and all $x \neq 0$

$$0 < \alpha(\|x\|) \leqslant V(x, t)$$

(ii) there exists a continuous scalar function γ such that $\gamma(0) = 0$ and the derivative $\dot{V}$ of V along the motion starting at t, x satisfies, for all t and all $x \neq 0$,

$$\begin{aligned}\dot{V}(x, t) &= \left.\frac{\mathrm{d}V(x(\tau; x_0, t))}{\mathrm{d}\tau}\right|_{\tau=t} \\ &= \lim_{h\to 0} [V(x + hf(x, t), t + h) - V(x, t)]/h \\ &= \partial V/\partial t + (\operatorname{grad} V)^{\mathrm{T}} f(x, t) \\ &\leqslant -\gamma(\|x\|) < 0;\end{aligned}$$

(iii) there exists a continuous, nondecreasing scalar function such that $\beta(0) = 0$ and, for all t,

$$V(x, t) \leqslant \beta(\|x\|)$$

(iv) $\alpha(\|x\|) \to \infty$ with $\|x\| \to \infty$

then the equilibrium state $x_e = 0$ is uniformly asymptotically stable in the large; $V(x, t)$ is called a *Lyapunov function* of system (D.1).

Theorem D.7

Consider a continuous-time, linear dynamic system

$$\dot{x}(t) = A(t)x(t) + G(t)u(t)$$

subject to the restrictions

(i) $\|A(t)\| \leqslant c_1 < \infty$ for all t

(ii) $0 < c_2 \leqslant \|G(t)x\| \leqslant c_3 < \infty$ for all $\|x\| = 1$ and all t.

Then the following propositions are equivalent:

(a) Any uniformly bounded input

$$\|u(t)\| \leqslant c_4 < \infty, \; t \geqslant t_0$$

gives rise to a uniformly bounded response for all $t \geqslant t_0$

$$\|x(t)\| = \|\Phi(t, t_0)x(t_0) + \int_{t_0}^{t} \Phi(t, \tau)G(\tau)u(\tau)\,\mathrm{d}\tau\|$$

$$\leqslant c_5(c_4, \|x_0\|) < \infty$$

(b) For all $t \geqslant t_0$, $\int_{t_0}^{t} \|\Phi(t, \tau)\|\,\mathrm{d}\tau \leqslant c_6 < \infty$;

(c) The equilibrium state $x_e(t) = 0$ of the input-free system is uniformly asymptotically stable;

(d) There exist positive constants c_7, c_8 such that

$$\|\Phi(t, t_0)\| \leqslant c_7 \exp[-c_8(t - t_0)] \text{ for all } t \geqslant t_0,$$

(e) Given any positive definite matrix $Q(t)$ continuous in t and satisfying for all $t \geqslant t_0$

$$0 < c_9 I \leqslant Q(t) \leqslant c_{10} I < \infty$$

the scalar function defined by

$$V(x, t) = \int_{t}^{\infty} \|\Phi(\tau, t)x\|^2_{Q(\tau)}\,\mathrm{d}\tau$$

$$= \|x\|^2_{P(t)}$$

exists and is a Lyapunov function for the input-free system satisfying the requirements of Theorem D.6, with its derivative along the free motion starting at (x, t) being

$$\dot{V}(x, t) = -\|x\|^2_{Q(t)}$$

Corollary D.1

The equilibrium state $x_e(t) = 0$ of the continuous-time, free, linear, stationary dynamic system (D.3) is asymptotically stable if and only if given any symmetric, positive definite matrix Q there exists a symmetric, positive definite matrix P which is the unique solution of the set of $n(n+1)/2$ linear equations

$$A^T P + PA = -Q$$

which are called algebraic Lyapunov equations (ALEs) in continuous time, and $\|x\|^2_P$ is a Lyapunov function for (D.3).

Corollary D.2

The real parts of the eigenvalues of a constant matrix A are less than σ if and only if given any symmetric, positive definite matrix Q there exists a symmetric, positive definite matrix P which is the unique solution of the set of $n(n+1)/2$ linear equations

$$-2\sigma P + A^T P + PA = -Q$$

D.4 SOME STABILITY RESULTS ON DISCRETE-TIME SYSTEMS

Stability concepts in discrete-time systems are analogous to their continuous-time counterparts, except that trajectories take discrete jumps from one sample to the next. The following theorem is important.

Theorem D.8

Consider the discrete-time, free dynamic system:

$$x(k+1) = \Phi x(k) \tag{D.5}$$

where Φ is an $n \times n$ matrix. Then

(i) the equilibrium state $x_e = 0$ of system (D.5) is stable if and only if Φ has no eigenvalues outside the unit circle and if the eigenvalues on the unit circle correspond to Jordan blocks of order 1
(ii) the equilibrium state $x_e = 0$ of system (D.5) is globally asymptotically stable if and only if $|\lambda_i(\Phi)| < 1$, $i = 1, \ldots, n$.

We also have the following corollaries.

Corollary D.3

Given any $Q \triangleq GG^{\mathrm{T}} \geqslant 0$ such that (Φ, G) is completely reachable, the equilibrium state $x_e = 0$ of system (D.5) is globally asymptotically stable if and only if there exists a unique $P > 0$ such that

$$\Phi^{\mathrm{T}} P \Phi - P = -Q$$

which is called an ALE in discrete time (cf. Exercise 5.14).

Corollary D.4

The eigenvalues of a constant matrix Φ are less than ρ in absolute value if and only if, given any symmetric matrix $Q > 0$, the linear equation

$$\rho^{-2} \Phi^{\mathrm{T}} P \Phi - P = -Q$$

has a unique, symmetric, matrix solution $P > 0$.

REFERENCES

[1] KALMAN, R. E. and BERTRAM, J. E., Control Systems Analysis and Design via the 'Second Method' of Lyapunov, *Trans. ASME, J. Basic Engng.*, vol. 82, 1960, pp. 371–93.
[2] LASALLE, J. P. and LEFESCHETZ, S., *Stability by Liapunov's Direct Method with Applications*, Academic Press, New York, 1961.
[3] YOSHIZAWA, T., *Stability Theory by Liapunov's Second Method*, Gakujutsutosho Printing Co., Tokyo, 1966.
[4] WILLEMS, J. C., *Stability Theory of Dynamic Systems*, Thomas Nelson and Sons, London, 1970.
[5] CASTI, J. L., *Dynamical Systems and Their Applications: Linear Theory*, Academic Press, New York, 1977.

[6] HARRIS, C. J. and MILES, J. F., *Stability of Linear Systems*, Academic Press, New York, 1980.

[7] CASTI, J. L., *Nonlinear Systems Theory*, Academic Press, New York, 1985.

[8] KALMAN, R. E. and BERTRAM, J. E., Control Systems Analysis and Design via the 'Second Method' of Lyapunov: II Discrete-Time Systems, *Trans. ASME, J. Basic Engng.*, vol. 82, 1960, pp. 394–400.

[9] ANDERSON, B. D. O. and MOORE, J. B., *Optimal Filtering*, Prentice Hall, Englewood Cliffs, New Jersey, 1979.

Index

Absolutely continuous, 76, 112
Ackerson–Fu (AF) method, 482, 496
Adaptive controller
  actively, 449
  passively, 449
Adaptive estimator, 1
Adaptive flight control, 7
Adjoint vector, *see* Lagrange multiplier (vector)
Algebraic Lyapunov equation (ALE)
  continuous-time, 135, 571
  discrete-time, 198, 368, 557, 572
Algebraic Riccati equation (ARE)
  continuous-time, 46
  discrete-time, 27, 366
A posteriori probability, 74
  density function, 149
A priori probability, 74
Asymptotic and convergence properties of partitioned adaptive filters, 125–143
Asymptotic stationary process, *see* Stationary process, asymptotic
Automated rhythm analysis, 3
Average, 535

Backward Hamiltonian (system) matrix, *see* Hamiltonian (system) matrix, backward
Backward symplectic matrix, *see* Symplectic matrix, backward
Backward-time information filter
  in continuous time, 70, 404
  in discrete time, 64, 339, 424
Backward-time Kalman filter
  formal, 64, 69, 233, 392
  strict, 233, 340
Backward-time model, *see* Formal backward-time mode, in continuous time or in discrete time
Backward partitioning filter, *see* Partitioning filter, backward
Backward-pass fixed-interval smoother
  generalized, 398
  in continuous time, 233
  in discrete time, 381
Backward transmission operator, *see* Transmission operator, backward
Backward source (contribution) vector(s), *see* Source (contribution) vector(s), backward
Backwards Markovian (state) model, *see* Markovian (state) model, backwards
Batch processing, 257, 260
Bayes' risk, 76
  for M-ary hypothesis testing, 461
  in estimation, 549
Bayes' rule, 86, 532
Bayesian estimation, *see* Estimation, Bayesian
Bayesian learning, *see* Learning, Bayesian
Bearing (measurement), 119, 259
Bearing-only TMA (problem), 257, 258
  equation of motion for, 273, 274
Bellman equation, 438
Bhattacharyya
  coefficient, 91, 121
  distance, 91
Bias correction filter
  and predictor, 284–93, 309–25
  stability of, 302–9
Bias correction smoother, 293–301
Bias-free
  filtered estimate, 287, 312
  fixed-interval smoothed estimate, 298
  predicted estimate, 292
Bias state, 285, 310
Bierman rank two $U-D$ smoother, *see* Rank two $U-D$ smoother, Bierman
Binary hypothesis testing, *see* Hypothesis testing, binary
Blending matrix, 264
  at filtering instant, 312
  at predicting instant, 313
  for smoothing, 299
  generalized, 287
Block-diagonalized matrix, 60
Block-triangularized matrix, 60
Borel field, 74, 109, 531
Brownian motion, *see* Wiener process
Bryson–Frazier smoother
  in discrete time, 392, 395
  in continuous time, 248

Candidate model, 100, 448
Canonical decomposition of linear systems, 563
Canonical structural model, 284
Central limit theorem, 538
Central moment, *see* Moment(s), central
Central station, *see* Processor(s), central
Certainty equivalence, 456
  principle, 444, 505, 506
Chain, 540
Chandrasekhar(-type) algorithm(s)
  dual, 244, 299
  in continuous time, 56, 242, 301
  in discrete time, 38, 253
Characteristic
  equation, 521
  function, 536
Cholesky decomposition
  upper square root free, 377, 378
  lower square root free, 378
Closed-loop
  system matrix, 46
  transition matrix, 153, 212
Cofactor, 518
Coloured noise, 98
Complementary models, 246, 339
Composite sources, 508
Computational approach
  to the continuous-time Riccati equation, 51–61
  to the discrete-time Riccati equation, 34–9
Conditional covariance matrix, 537
Conditional expected value, 537
Conditional mean, 537
Conditional mode estimator, 45
Conditional moment estimator, 45
Conditional partition theorem, *see* Partition theorem, conditional
Conditional probability
  density function, 534
  distribution function, 534
Conditionally independent events, 532
Control
  elemental, 457
  of systems with faulty sensors, 503–7
  gain, 438
  optimal, 437
  sequences, 437, 503
Controllability
  complete, 562
  Gramian matrix, 558
  index, 559
  matrix, 559
  uniform, 558
Controllable, 30, 49, 557
Controller
  adaptive, *see* Adaptive controller
  gain, *see* Control, gain
Controlling the number of elemental Kalman filters, 90–6
Convergence in performance, 128
Convergence of random sequences
  almost surely (a.s.), 538
  with probability 1 (w.p.1), 538
  in mean square, 538
  limit in the mean (l.i.m.), 538
  in probability, 538
Correlation, 536, 540
  time, 194
Cost function, 437, 469
  quadratic, 437, 449, 550
Covariance(s), 15, 39
  matrix, 535
Cramer–Rao inequality, 548

Data compression, 265
Decentralized estimation, 402
Decentralized smoothing
  in continuous time, 402–18
  in discrete time, 418–30
Decentralized Kalman filtering
  in continuous time, 411
  in discrete time, 419–24, 432
Delta algorithm, 180
Density function, *see* Probability, density function
Dependent multipartitioning filter
  in continuous time, 156, 157
  in discrete time, 170, 171
Dependent partitioning filter
  in continuous time, 152, 153
  in discrete time, 164, 165
Detectability, 15
Detectable, 30, 48, 561
Detection and classification of cardiac arrhythmias, 3, 4
Detection–estimation algorithm (DEA), 479
Discrete parameter space, *see* Parameter space, discrete
Distributed parameter systems, 108, 184, 468
Distributed-sensor network, 246, 468, 508
Distribution function, 533
Divergence, 120
  of filter, 305
Doubling algorithm, 181
  in continuous time, 61
  in discrete time, 39, 182, 197
Dominated convergence theorem, 126, 539
Doppler radar (receiver), 97, 235
Dot sum, 202

Dual Chandrasekhar algorithm(s), *see* Chandrasekhar(-type) algorithm(s), dual
Dual control, 454, 507
Dual effect (or role), 454, 505, 506
Duality, 564
  of estimation and control, 447, 448
Deshpande–Upadhyay–Lainiotis (DUL) method, 455, 504
Dynamic programming (DP), 345, 438

Eigenvalue(s), 521
Eigenvector(s), 521
  left, 522
  right, 522
  orthogonality relation of, 522
Eigenvalue–eigenvector method
  in continuous time, 56, 57
  in discrete time, 34
Ensemble of random vector(s), 540
Equilibrium state, 565
Ergodic process, 542
Error covariance
  filter(ed), 17, 40, 152
  predicted, 17, 218
  smoothed, 81, 152, 239
Estimation
  Bayesian, 549
  least-squares (LS), 65, 552
  maximum likelihood (ML), 549
  weighted least-squares (WLS), 65
Estimate(s), 547–52
  asymptotically unbiased, 547
  biased, 547
  consistent, 547
  efficient, 547
  filtered, 16, 152
  maximum a posteriori probability (MAP), 134, 551
  maximum likelihood (ML), 134, 138
  minimum mean-square error (MMSE), 16, 550
  minimum variance (MV), 16, 331, 550
  predicted, 16, 218
  smoothed, 81, 152, 239
  unbiased, 21, 547
Estimator, 547–52
  optimal, 2, 481
  self-organizing, 1
  suboptimal, 482
  unbiased, 21
Euclidean norm
  of vector, 526
  of matrix, 526
Event, 530
Expectation, 535
  conditional, 546
Exponential of a square matrix, 523
Exponential stability, 568
Extended Kalman filter
  in continuous time, 44, 45
  in discrete time, 25, 26

Failure(s), 496
  of increased noise, 466
  (saturation) hardover, 466
Failure (or Fault) detection, 325, 478
  and identification (FDI), 4
Fault tolerant control system, 460–6
Feedback gain, *see* Control, gain
Filter, *see* Kalman filter
Filtered (or filtering) estimate
  global, 410, 419
  local, 403, 420
Filter error covariance, *see* Error covariance, filter(ed)
Filter gain, *see* Kalman filter, gain
Filtering, *see* Filter
First moment, 535, *see also* Moment(s)
First-order filter, 259
Fisher information matrix, 548
  in continuous time, 68
  in discrete time, 62
Fixed-interval smoother
  bias correction, 297
  partitioning, 294
Fixed-interval smoothing problem, 233, 236
Fixed-point smoothed estimate, 23
Fixed-point smoother
  in continuous time, 249
  in discrete time, 66, 67
Fixed-point smoother due to Kailath–Frost, 81
Fixed-lag smoother
  in continuous time, 249
  in discrete time, 67, 68
Formal backward-time information filter
  in continuous time, 70
  in discrete time, 64
Formal backward-time Kalman filter
  in continuous time, 69
  in discrete time, 64
Formal backward-time model
  in continuous time, 69
  in discrete time, 63, 120
Forward Hamiltonian (system) matrix, *see* Hamiltonian (system) matrix, forward
Forwards Markovian (state) model, *see* Markovian (state) model, forwards
Forward partitioning filter, *see* Partitioning filter, forward

Forward partitioning smoother, *see* Partitioning smoother, forward
Forward-pass smoothing algorithm, *see* Smoothing algorithm(s), forward-pass
Forward-pass fixed-interval smoother
  generalized, 394
  in continuous time, 238
  in discrete time, 334–48
  $U-D$ factorized, 357
Forward source contribution vector, *see* Source (contribution) vector(s), forward
Forward transmission operator, *see* Transmission operator, forward
Friedland's two-stage bias correction filter, 325
Fubini's theorem, 110, 137
Fusion centre, *see* Processor(s), central

Gauss–Markov process, 546
Gaussian
  random processes, 546
  random variable, 544
  random vector, 544
  sum (approximation), 80, 92
  white noise process (or sequence), 15, 85, 331, 403
Generalized bias correction filter
  in continuous time, 287, 288
  in discrete time, 312, 313
Generalized bias correction predictor
  in continuous time, 292, 293
  in discrete time, 314
Generalized Chandrasekhar(-type) algorithm(s), 240–5
  dual version of, 241
Generalized likelihood ratio approach, 461
Generalized pseudo-Bayes algorithm(s) (GPBA), 479, 481–500
Generalized two-filter smoother (formula)
  in continuous time, 236
  in discrete time, 341
Generalized $X-Y$ functions, 240–5
Global filtered estimate, *see* Filtered estimate, global
Global information filter, *see also* Information filter, global
  backward-time, 405
  forward-time, 405
Global Kalman filter, *see* Kalman filter, global
Global smoothed estimate, *see* Smoothed estimate, global

Hamiltonian, 47
Hamiltonian matrix method, 52
Hamiltonian system (or equation)
  continuous-time, 47, 205
  discrete-time, 28, 331
Hamiltonian (system) matrix
  backward, 47
  forward, 28, 47
Hölder's inequality, 126, 131, 539
Hypothesis testing (or detection)
  binary, 75
  M-ary, 460

Identifiability condition, 140
Independent multipartitioning filter
  in continuous time, 157, 158
  in discrete time, 173
Independent partitioning filter
  in continuous time, 154, 155
  in discrete time, 166, 167
Independent random variables, 535
Indicator variable, 114
Inertial navigation system (INS), 96
Infinitesimal generator, 203
Information filter
  gain, 351
  global, 405, 424
  in continuous time, 69
  in discrete time, 62, 63
  local, 405, 424
Information matrix, 352
  Fisher, 63, 68
Initial layer, 207
Initial source vector, 208
Inner product
  in the Euclidean space, 77
  of two matrices, 527
  of two vectors, 527
Innovation
  process (or sequence), 22, 108
  model, 22
  properties of, 108–13
Intelligent control system, 1
Interacting multiple model algorithm (IMMA), 479, 508
Interrupted observation, 506
Ito stochastic calculus, 42

Jaffer and Gupta (JG) method, 497
  modified, 485–7
Jensen's inequality, 110, 539
Joint central moments, 536
Joint characteristic function, 537
Joint detection, estimation and system indentification, 114
Joint detection–control problem, 458–60
Joint detection–estimation problem, 81–4
Joint moments, 536

Joint probability
  density function, 533
  distribution function, 533
Jordan block(s), 525
Jordan canonical form, 525
Joseph form, 62, 63, 346, 396
Jump-linear-quadratic (JLQ) control, 508
Jump-linear-quadratic-Gaussian (JLQG) control, 504, 508
Jumps, 478

Kalman filter
  continuous-time, 39–45
  discrete-time, 15–26
  elemental, 80, 90, 493
  extended, 25, 26, 44, 45 *see also* Extended Kalman filter
  gain, 16, 40
  global, 408
  in a pipelined mechanization, 431
  in systolic array architecture, 431
  linearized, 45
  local, 408
  nominal, 152, 164
  parallel (or parallelizing), 431
  steady-state, 27, 46
Kalman-Bucy filter, *see* Kalman filter, continuous-time
Kleinman's iterative method
  in continuous time, 57, 58
  in discrete time, 35–7
Kullback(–Leibler) information measure(s), 120, 144

Lagrange multiplier method, 330
Lagrange multiplier (vector), 154, 177, 205, 331
Lainiotis filter, 147
Lainiotis' partition theorem, *see* Partition theorem, Lainiotis'
Lambda algorithm, *see* Partitioned numerical algorithms
Laplace's expansion, 518
Leading principal minor, 523
Learning, 2
  Bayesian, 126
Learning control system, 1
Least-squares (LS) estimation, 552
  linear, 65
  weighted, 65
Lee–Sims method, 450
Left reflection operator, *see* Reflection operator, left
Likelihood equation, 549
  log, 549
Likelihood function(s), 549
  conditional, 117, 552
  in discrete time, 88
  unconditional, 552
Likelihood ratio, 120
  conditional, 117
  for M-ary hypothesis testing, 462
  function, 77
  log, 120
  theta (θ)-conditional, 77
  unconditional, 78, 144
Limiting procedure, 40
Linearized Kalman filter, *see* Kalman filter, linearized
Linearly
  dependent, 520
  independent, 520
Linear minimum variance (MV) estimate, 16
  estimator, 21
Linear-quadratic-Gaussian (LQG) control, 437–48, 469
Linear system models
  continuous-time, 554, 555
  discrete-time, 556, 557
Local filtered estimate, *see* Filtered estimate, local
Local information filter, *see* Information filter, local
Local Kalman filter, *see* Kalman filter, local
Local smoothed estimate, *see* Smoothed estimate, local
Local station, *see* Processor(s), local
Lyapunov equation, *see also* Algebraic Lyapunov equation (ALE)
  discrete-time, 556
Lyapunov function, 305, 570
Lyapunov stable, 565

Marginal density, 534
M-ary hypothesis testing, *see* Hypothesis testing, M-ary
Manoeuvre predictor, 264
Markov chain
  finite-state, 479
  semi-, 478
Markov jump systems, 479
Markov process, 541
  first-order, 541
  second-order, 541
Markovian (state) model
  backwards, 234, 393
  forwards, 397
Martingale, 109, 541
  sub-, 541
  super-, 541

Matrix
  block-decomposition(s) (or partition (s)), 516
  block-diagonal, 516
  characteristic polynomial of, 521
  conformable, 517
  determinant of, 518
  diagonal, 515
  difference of, 516
  differentiation of, 527
  full rank, 521
  identity, 515
  image of, 521
  integration of, 528
  inverse of, 519
  kernel of, 521
  lower triangular, 516
  nonsingular, 519
  norm of, 526
  null space of, 521
  nullity of, 521
  orthogonal, 520
  product of, 516
  range space of, 521
  rank of, 520
  rectangular, 515
  similar, 524
  singular, 519
  skew symmetric, 525
  square, 515
  sum of, 516
  symmetric, 515, 518
  trace of, 524
  transpose of, 517
  upper triangular, 515
  zero (or null), 515
Matrix exponential series expansion, 99
Matrix inversion lemma, 184, 185
Matrix sign function method, 58–61
Maximum a posteriori probability (MAP) estimate, *see* Estimate(s), maximum a posteriori probability (MAP)
Maximum likelihood (ML) estimate, *see* Estimate(s), maximum likelihood (ML)
Maximum likelihood (ML) identification, 169
  model approximation via, 145
Mayne–Fraser two-filter smoother (or formula)
  in continuous time, 231, 404
  in discrete time, 332
Mean, 535
Mean-square error (MSE), 550
Measure, *see* Probability measure
Measurement update, 17, 352, 382
Minimum mean-square error (MMSE) estimate, *see* Estimate(s), minimum mean-square error (MMSE)
Minimum variance (MV) estimate, *see* Estimate(s), minimum variance (MV)
Minkowski's inequality, 130–539
Minor, 518
Mixture density function, 92
Modal matrix, 59
Modified weighted Gram–Schmidt (MWGS) time update algorithm, 379, 380
Moment(s), 536
  central, 536
  joint, 536
  joint central, 536
Monte Carlo simulation (or run), 101, 459, 497
Motion of an antenna, 182
Multipartitioning filter
  dependent, 156, 170
  in continuous time, 155–60
  independent, 157, 173
  in discrete time, 169–75
Multiple model, 8, 436
Multiple model adaptive control (MMAC), 8, 436
  in continuous time, 473
  in discrete time, 448–57
Multiple model adaptive filter (MMAF)
  in continuous time, 80
  in discrete time, 87
Multiple model adaptive filtering, 8
Multi-step measurement update approach, 487–96
Mutually independent events, 531

Negative definite, 522
Negative exponential method, 54–6
Noise-free measurement model, 22
Noise parameter adaptation, 84
Nominal filter gain, 153
Nominal Kalman filter, *see also* Kalman filter, nominal zero-, 230
Nominal scattering matrix
  backward(-time), 223
  forward(-time), 209
Non-Gaussian density, 80
Non-Gaussian initial state, 80, 96
Nonlinear
  filter, 2
  model, 25, 45
  tracking system, 118
Nonnegative definite, 522
Non-switching parameter, 467

Norm in the Euclidean space, 77
Normal random processes, *see* Gaussian random processes
n-step measurement update approach, *see* Multistep measurement update approach

Observability
  Gramian matrix, 560
  index, 167, 561
  matrix, 561
Observable, 559
Omega receiver, 97
One-step measurement update approach, 483–7
Optimal control, *see* Control, optimal
Optimal cost, 440, 444
Optimal cost-to-go, 438
Optimal estimator, *see* Estimator, optimal
Optimal quadratic performance (or risk), 127, 133
Orthogonal projection lemma(s), 21, 550
Orthogonal random variables, 536

Parallel processing, 257
Parameter-adaptive self-organizing control, 436
Parameter
  abruptly changing (or jump, or switching), 467, 478
  constant (or non-switching), 448, 467, 478
  identification, 168
  time-varying, 478
  true, 85
  unknown, 73
Parameter space
  continuous, 79, 129
  discrete, 79, 87, 125, 133
Partition theorem
  conditional, 185
  discrete-time, 85
  Lainiotis', 74
Partitioned adaptive control, 436
Partitioned adaptive filtering, 74
Partitioned numerical algorithms, 175–82
  for time-invariant systems, 179, 180
  doubling, 180, 181
Partitioning filter
  backward, 220–45
  dependent, 152, 159, 164, 174
  forward, 205–12
  in continuous time, 147–55
  independent, 154, 166
  in discrete time, 160–9
Partitioning predictor, 218
Partitioning smoother
  forward, 212–19
Passive bearing-only tracking, 257
Perturbed Kalman filter, 218
Positive definite, 522
Positive semidefinite, 522
Power spectrum, 543
Predicting, *see* Predictor
Predictive information, 6, 120, 397
Predictor, 236, 264, 339
Principle of optimality, 438
Probability, 530
  conditional, 531
  density function, 533
  distribution function, 533
  function (or measure), 531
  space, 530
Probing, 454
Process, 540
Processor(s)
  central, 403
  local, 403
Pseudo-Bayes estimator, 482
Pseudolinear partitioned tracking filter, 257, 266
Pseudolinear partitioning filter, 259–64
  in bath processing, 260
  in recursive processing, 261
Pseudolinear partitioning fixed-point smoother, 262
Pseudolinear tracking filter, 257

Quadratic form, 522
Quadratic mean convergence of partitioned adaptive filter, 127

Radon–Nikodym derivative, 77
Random sampling algorithm (RSA), 479
Random variable, 532
Rank-one (factorization) update algorithm, 380, 381
Rank two $U - D$ smoother
  arithmetic operation counts for, 385–91
  Bierman, 381–5
  Watanabe, 361
Rauch–Tung–Striebel (RTS) smoother, 232, 332, 395
Reachability
  complete, 561
  Gramian matrix, 558
  index, 559
  matrix, 559
  uniform, 558
Reachable, 557
Real-time smoothing algorithms, 417, 418, 429
Reconstructible, 560

Reconstructibility
  Gramian matrix, 560
  matrix, 561
Redheffer star product, *see* Star product
Redheffer's scattering theory, 201–5, *see also* Scattering theory
Reflection operator
  left, 206
  right, 206
Residual, 22
Riccati difference equation (RDE), 17, 438
Riccati differential equation (RDE), 40, 469
Right reflection operator, *see* Reflection operator, right
Robot manipulator, 70
Root mean square (rms) error, 101

Sample function (or realization) of the process, 540
Sample space, 530
Sarrus' method, 518
Satellite tracking system, 347
Scattering matrix, 201
  nominal, 206, 213
Scattering theory
  continuous-time, 201–18
  discrete-time, 250–2
Second method of Lyapunov, 569–71
Schwartz inequality, 95, 527, 539
Second-order filter, 259
Second-order process, 109
Self-organization, 1
Self-organizing estimator, *see* Estimator, self-organizing
Self-tuning regulator
  LQG, 445
  MV, 445
Separation theorem, 444
Shaping filter, 98
Sigma (σ)-field (or algebra), 74, 530
Similar transformation, 524
Smoothed estimate
  global, 403
  local, 410
Smoother
  fixed-interval, 23, 233, 238, 297, 330
  fixed-lag, 23, 67, 68
  fixed-point, 23, 66, 67, 176
Smoother gain, 361, 388
Smoothing, *see* Smoother
Smoothing algorithm(s)
  forward-pass, 237
  generalized two-filter, 231–6
  parallel, 431
  Weinert–Desai, 237–40
Smoothing property of conditional expectation, 75, 110
Smoothing update algorithms (or problem), 413, 427, 429
Source (contribution) vector(s), 202
  backward, 206
  forward, 206
  nominal, 208, 214
Spacecraft, 235, 290
Space shuttle orbitor, 463
Stabilizability, 15, 559
  complete, 562
Stabilizable, 30, 48, 559
Stabilized form, *see* Joseph form
Stabilizing solution, 28, 46, 367
Stability
  of continuous-time Kalman filter, 45–51
  of discrete-time Kalman filter, 26–34
  theory, 565–72
Stable
  asymptotically, 305, 567
  exponentially asymptotically, 568
  globally asymptotically, 568
  sensu Lyapunov, 566
  uniformly, 566
  uniformly asymptotically, 305, 567
Standard deviation, 535
Star product, 202
(State) transition matrix, 554, 556
Stationary process, 542
  strict-sense, 542
  asymptotically, 542
  wide-sense, 542
Statistically linearized filter, 45
Steady-state
  bias correction filter, 306–9
  fixed-interval smoother, 369
  Kalman filter, *see* Kalman filter, steady-state
  smoothed (error) covariance, 368
Stochastic differential equation(s), 42, 116
Stochastic process(es), 540–4
  continuous-time, 540
  discrete-time, 540
Stokes' identity, 242
Strong solution, 28, 46
Structurally partitioning filter, 278–84
Structure and parameter adaptive estimation
  in continuous time, 73–84
  in discrete time, 85–90
Structure detection, 502
Sudden manoeuvre, 478
Switching parameter, *see* Parameter, abruptly changing
Sylvester's inequality, 521
Sylvester's theorem, 523

Symplectic, 28
Symplectic matrix
  backward, 29, 366
  forward, 28
System parameter adaptation, 84
(System) structure adaptation, 2, 84

Target tracking, 5, 478
Time update, 17, 352, 382
Tracking motion analysis (TMA) problem, 257–9
  two-dimensional, 258
  three-dimensional, 273, 275
Tracking of a target trajectory, 81
Tracking of ship position, 5, 6
Tracking problem
  exponentially correlated acceleration, 194
Transition matrix method, *see* Hamiltonian matrix method
Transition probability matrix, 479
Transmission operator
  backward, 206
  forward, 206
Triangular UDU^{T} factorization, 377, 378
Two-filter smoother (or formula)
  generalized, 236, 341
  Mayne–Fraser, 231, 332
  Mehra, 231
Two-point boundary-value problem (TPBVP), 205, 331
Two-stage bias correction
  estimators, 278
  filter, 278, 290, 305, 312
  fixed-interval smoothers, 278
  predictor, 278, 314

$U - D$ factorization, 353
$U - D$ factorized
  backward(-time) information filter, 353–7
  forward-pass fixed-interval smoother, 357–65
$U - D$ filter algorithms, 378–80
$U - D$ information matrix factorization, 348–65
$U - D$ Kalman filter for the rank-two smoother, 381–3
$U - D$ measurement update (algorithm), 379
Unconditional likelihood ratio function, *see* Likelihood ratio, unconditional
Uncontrollable, 30, 49
  eigenvalue (or mode), 563
Uncorrelated, 536
Unique stabilizing solution, 31, 50
Unobservable, 265
  eigenvalue (or mode), 563
Unstable, 565
Upadhyay-Lainiotis method, 454

Variance, 3, 535, *see also* Moment(s), central
Vector
  column, 516
  differentiation of, 528
  norm of, 525
  row, 516
  transpose of, 518
VSTOL aircraft, 373

Watanabe rank-two $U - D$ smoother, *see* Rank-two $U - D$ smoother, Watanabe
Weighted sum, 2, 87
Weighting matrix, 65
Weinert–Desai smoothing algorithm(s), *see* Smoothing algorithm(s), Weinert–Desai
Weinert–Desai fixed-interval smoother, 238
Whitening filter, 22
White noise, 543
Wiener process, 42, 546

X–Y functions, *see* Chandraskhar(-type) algorithm(s)

Zero-mean, 3, 4, 15, 39